AF673948

Flora of Eastern Ghats

Volume 7

The Authors

Prof. T. Pullaiah obtained his M. Sc. (1973) and Ph. D. (1976) degrees in Botany from Andhra University. He was a Post-doctoral Fellow at Moscow State University, Russia, during 1976-1978. He travelled widely in Europe and visited Universities and Botanic Gardens in about 17 countries. He joined Sri Krishnadevaraya University as Lecturer (1979) and became Professor (1993). He held several positions in the University as Dean, Faculty of Life Sciences; Head, Department of Botany; Chairman, Board of Studies in Botany; Head, Department of Sericulture; Co-ordinator and Chairman, Board of Studies in Biotechnology; Principal, Sri Krishnadevaraya University College. He retired from active Service in 2011 and was selected by UGC as UGC-BSR Faculty Fellow in the Department of Botany, Sri Krishnadevaraya University. He has published 75 books, 335 research papers and 35 popular articles. His books include Flora of Andhra Pradesh (5 volumes), Biodiversity in India (8 volumes). Flora of Telangana (3 volumes), Encyclopaedia of World Medicinal Plants (5 volumes), Ethnobotany of India (5 volumes), Global Biodiversity (4 volumes), Orchids of Eastern Ghats and Monograph on *Brachystelma* and *Ceropegia* in India. He was the Principal-Investigator of 20 major Research Projects totalling more than a Crore of Rupees funded by DBT, DST, CSIR, UGC, BSI, WWF, GCC etc. Under his guidance, 54 students obtained their Ph. D. degrees and 34 students their M. Phil. Degrees. He is the recipient of Best Teacher Award from the Government of Andhra Pradesh, Prof. P. Maheshwari Gold Medal and Dr. G.Panigrahi Memorial Award of Indian Botanical Society and Prof. Y. D. Tiagi Gold Medal of the Indian Association for Angiosperm Taxonomy. He was President of Indian Association for Angiosperm Taxonomy (2013) and President of the Indian Botanical Society (2014). He was a member of Species Survival Commission of International Union for Conservation of Nature and Natural Resources (IUCN).

Dr. S. Karuppusamy is Assistant Professor in the Department of Botany, The Madura College, Madurai, Tamil Nadu. He obtained his M.Sc. degree from Madurai Kamaraj University, Madurai and Ph.D. degree from Gandhigram Rural Institute, Deemed Univerrsity, Tamil Nadu. Later on he worked as Post Doctoral Fellow in the Department of Biotechnology, and Research Associate in the Department of Botany, Sri Krishnadevaraya University, Anantapur. He has published more than 120 research papers in reputed journals and co-authored five books – Flora of Eastern Ghats volume 3 and 4, Flora of Andhra Pradesh volume 5, Revised Flora of Andhra Pradesh Volume 3 and *Caralluma (Sensu lato)*, Orchids of Eastern Ghats and Monograph on *Brachystelma* and *Ceropegia* in India.. He is the recipient of Prof. Y.S.Murthy Gold Medal from Indian Botanical Society and Rolla S. Rao award from Indian Association for Angiosperm Taxonomy. He is a Fellow of Indian Association for Angiosperm Taxonomy and Indian Botanical Society. He is the Principal-Investigator of 3 major Research Projects funded by UGC, SERB and NIF. Under his guidance, 4 students obtained their Ph. D. degrees.

Flora of Eastern Ghats

Volume 7

Grasses, Gymnosperms, Additions, Key to the Families and Floristic Analysis

T. Pullaiah

Department of Botany,
Sri Krishnadevaraya University
Anantapuramu – 515 003
Andhra Pradesh

S. Karuppusamy

Department of Botany
The Madura College,
Madurai – 625 011
Tamil Nadu

2020

Regency Publications

A Division of

Astral International Pvt. Ltd.

New Delhi – 110 002

ISBN: 9789390371396 (Int. Edition)

Published by : **Daya Publishing House®**
A Division of
Astral International Pvt. Ltd.
– ISO 9001:2015 Certified Company –
4736/23, Ansari Road, Darya Ganj
New Delhi-110 002
Ph. 011-43549197, 23278134
E-mail: info@astralint.com
Website: www.astralint.com

Abbreviations Used in General Text and Enumeration

BLAT: Blatter Herbarium, St. Xavier College, Mumbai

CAL: Central National Herbarium, Howrah, Calcutta (now called as Kolkata)

cm: centimeter

DD: Herbarium, BSI, DehraDun

et al.: *et alia* (and others, of people)

FBI: Hooker's Flora of British India

Fig.: Figure

Fl.: Flowering season

Fr.: Fruiting season

Gamble: Gamble's Flora of Presidency of Madras

m: meter

MH: Madras Herbarium, BSI, Southern Circle, TNAU campus, Coimbatore

mm: millimeter(s)

Ori.: Oriya

p.p.: *pro-parte*
sine. coll.: Without collector's name
s.n.: Without number
SGH: Sri Ganesan Herbarium, The Madura College, Madurai
SKU: Sri Krishnadevaraya University, Anantapur
Sq. km: Square kilo-meter(s)
SVU: Sri Venkateswara University, Tirupati
Tam.: Tamil
Tel.: Telugu
var.: varities (variety)
Vern.: Vernacular name (s)

Abbreviations – Districts

AGL: Angul
ATP: Anantapur (now called Anantapuramu)
BLR: Balasore
BG: Bargarh
BGR: Bolangir
CDP: Cuddapah (Now called as Kadapa)
CGP: Chengalput
CMT: Coimbatore
CTK: Cuttack
CTR: Chittoor
DGL: Dindigul
DGR: Deogarh
DKL: Dhenkanal
DMP: Dharmapuri
EG: East Godavari
GJM: Ganjam
GJP: Gajapathi
GNT: Guntur
KDM: Kandhmal
KGR: Krishnagiri
KHD: Kalahandi
KJR: Kendujhar (Keonjhar)
KRN: Krishna

KRT: Koraput
KMM: Khammam
KNL: Kurnool
MBJ: Mayurbhanj
MDS: Madras
MDU: Madurai
MBNR: Mahabubnagar
NA: North Arcot
NLR: Nellore
NMK: Namakkal
NPD: Naupada
PLN: Pulbhani
PRK: Prakasam
RGD: Rayagadh
SA: South Arcot
SBP: Sambalpur
SKLM: Srikakulam
SM: Salem
SPR: Subamapur (Sonepur)
TRP: Tiruchirapalli
TRV: Tiruvannamalai
VJN: Vijayanagaram
VLR: Vellore
VSKP: Visakhapatnam
WG: West Godavari

Abbreviations – Plant Explorers

AC: Alok Chorghe
AL: A. Laxmaiah
AM: A.Mohan
AMK: A.Mukherji
AMR: A.MadhusudanaReddy
ANH: A.N.Henry
ANS: A. Narayana Swamy
APD & SP: A.P.Das & Sauris Panda
AU: A.Ugraiah

AVNR: A.V.N.Rao
AWL: A.W.Lushington
BO: Bourne
BR: B.Ravi Prasad Rao
BSS: B.Sadasivaiah
BSN: B.Suryanarayana
CAB: C.A.Barber
CECF: C.E.C. Fischer
CJP: C. Jayaprada
CPSM: C.P. Sreemadhavan
CS: C. Subbarayudu
DAM: D. AliMoulali
DBD: D.B. Deb
DCSR: D.C.S. Raju
DN: D. Narasimhan
DRC: D. Rangachari
EKK: E.K. Krishnan
EV: E.Vajravelu
GO: G. Obulesu
GP: G. Panigrahi
GVS: G.V. Subba Rao
HFM: H.F. Mooney
HHH: H.H. Haines
HOS: H.O. Saxena
HS: H. Santapau
JLE: J.L.Ellis
JS: J. Joseph
JSG: J.S. Gamble
KCJ: K.C. Jacob
KMM: K.M. Matthew
KMS: K.M. Sebastine
KNS: K.N. Subramanian
KRM: K. Ramamurthy
KS: K. Subramanyam
KSM: K. Sri RamaMurthy
MB: M. Brahmam

MBV: M.B. Viswanatham
MCB: M. Chandrabose
MCK: M. Chennakesavulu
MM: M. Mohanan
MRR: M. Rama Rao
MSR: M.S. Ramaswami
NPBK: N.P. Balakrishnan
NRR: N. Rama Rao
NVG: N.Venugopal
NY: N.Yesoda
PMR: P.M. Rajeswaramma
PP: P. Perumal
PVK: P.Venkanna
PVP: P.V. Prasanna
RCS: R. Chandrasekaran
RHB: R.H. Beddome
RKM: R. Krishnamohan
RR: R. Rajan
RSR: Rolla S.Rao
RVK: R.Vijaya Kumar
RVR: R.R.Venkata Raju
R.V.Reddy: R.Venkateswara Reddy
SKK: S. Karthikeyan
SKS: S. Karuppusamy
SKW: S.K.Wagh
SP & *AP*: Sauris Panda & A.P. Das
SRS: S.R. Srinivasan
SS: S. Sunitha
SSR: S. Sandhya Rani
SX: Saxena
TDCK: T. Dharma Chandra Kumar
TP: T. Pullaiah
TRS: T. Ravishakar
VNS: V. Narayanaswami
VRK: V. Ramakrishnaiah
VS: V. Srinivasulu

POACEAE (nom. alter. GRAMINEAE)

KEY TO THE SUB-FAMILIES

1. Mature spikelets falling entire from their pedicels; all alike or differing in sex and structure, perfect spikelets with 2 heteromorphous florets, the upper bisexual, the lower male or barren .... **Panicoideae**
1. Mature spikelets breaking up, leaving more or less persistent glumes on the pedicel or if falling entire then not consisting of heteromorphous florets .... **Pooideae**

KEY TO THE TRIBES OF PANICOIDEAE

1. Male and female spikelets in separate inflorescence on the same plant or the female spikelets at the base of the inflorescence and male above them, lemma hyaline or membranous, thinner than the glumes .... **Maydeae**
1. Male and female spikelets in same inflorescence, all bisexual or with male or barren; male or barren spikelets near to the bisexual spikelet, lemma of the fertile floret indurated:
    2. Spikelets usually in pairs or in threes, one sessile and other pedicelled, those of each pair homogamous or heterogamous, glumes more or less rigid and firmer than lemma, upper lemma usually awned or reduced to an awn .... **Andropogoneae**
    2. Spikelets solitary, more or less similar, glumes membranous, lower lemma similar to upper glume in texture, upper lemma rigid, often chartaceous or crustaceous, awned or unawned .... **Paniceae**

KEY TO THE GENERA OF THE TRIBE MAYDEAE

1. Axes of female spikelets not fused; Male and female spikelets in the same inflorescence:
    2. Female spikelets completely enclosed in a metamorphosed leaf-sheath through which the rachis grows out and bears the male spikelets above .... **Coix**
    2. Female spikelets not enclosed in a false involucre as above:
        3. Racemes with male spikelets at the apex, female spikelets imbedded in each joint of the rachis, so that the smooth cartilaginous axis and the lower glume form a nearly cylindric mass .... **Tripsacum**

3. Racemes entirely male or with 1 or more female spikelets at the base, lower glume of female spikelets thickly coriaceous embracing the rachis and enclosing the upper glume and floret..**Chionachne**

1. Axes of female spikelets fused, male and female spikelets in different inflorescence ....**Zea**

KEY TO THE GENERA OF THE TRIBE ANDROPOGONEAE

1. Spikelets solitary .... **Dimeria**

1. Spikelets in pairs or in threes:

2. Spikelets in each pair usually dissimilar, the sessile bisexual and pedicelled male or neuter:

3. Inflorescence of solitary raceme:

4. Spikelets awned:

5. All spikelets heterogamous:

6. Racemes supported by spathes:

7. Sessile spikelets pitted; perennial; nodes densely bearded ....**Dichanthium (foveolatum)**

7. Sessile spikelets not pitted; annual; nodes glabrous .... **Schizachyrium**

6. Racemes not supported by spathes:

8. Ligule short membranous; inflorescence a terminal raceme ....**Sehima**

8. Ligule elongate, acute; inflorescence a panicle .... **Andropogon (fastigiatus)**

5. Lowest pair of spikelets homogamous:

9. Joints and pedicels glabrous, upper lemma not 2-lobed ....**Heteropogon**

9. Joints and pedicels hairy, upper lemma 2-lobed, awned from the sinus... **Parahyparrhenia**

4. Spikelets unawned:

10. Sessile spikelet globose; lower glume pitted all over; pedicelled spikelets reduced..........................**Hackelochloa**

10. Sessile spikelet not globose; lower glume not pitted; pedicelled spikelets well developed:

11. Sessile spikelets with pectinate margins .... **Eremochloa**

11. Sessile spikelets without pectinate margins:

12. Racemes laterally compressed, joints and pedicels fused to receive sessile spikelet:

13. Ligule membranous, ciliate, lower glume of the sessile spikelet not winged........................ **Hemarthria**

13. Ligule a fringe of hairs, lower glume of the sessile spikelet broadly winged.................... **Manisuris**

12. Racemes cylindric, joints and pedicels free:

14. Ligule a short ciliate rim; sessile spikelet ovoid, keeled; pedicelled spikelet well developed:

15. Sessile spikelet solitary at each node of the axis; joints and spikelets reduced towards the apex of the raceme and appendaged .... **Rottboellia**

15. Sessile spikelets in pairs at each node of the axis; all joints and spikelets uniformly developed, apex not appendaged:

16. Lower glume of sessile spikelets not winged on keels....................................... **Mnesithea**

16. Lower glume of sessile spikelets with a broad wing on each keel..................... **Coelorachis**

14. Ligule a frimbriate membrane rim; sessile spikelet oblong, not keeled; the pedicelled spikelet suppressed or rudimentary .... **Ophiuros**

3. Inflorescence of racemes more than one paired, digitate, fascicled, corymbose or paniculate:

17. Spikelets in threes.. **Chrysopogon *p.p.***

17. Spikelets in pairs:

18. Racemes supported by the spathes:

19. Involucral spikelets present:

20. Involucral spikelets with strong bristles, sessile, persistent.. **Themeda**

20. Involucral spikelets glabrous, pedicelled, deciduous .... **Iseilema**

19. Involucral spikelets absent:

21. Basal joints of the racemes bulbous .... **Apluda**

21. Basal joints of the racemes not bulbous:

22. Joints thickened upwards, toothed at apex... **Andropogon (pumilus)**

22. Joints not thickened upwards, not toothed at apex:

23. Perennials, erect, upper lemma of the sessile spikelets 2-lobed, awned in the sinus .... **Cymbopogon**

23. Annuals, prostrate, rooting at the nodes, upper lemma of the sessile spikelets not cleft, passing into the awn .... **Pseudanthistiria**

18. Racemes not supported by spathes:

24. Inflorescence a panicle:

25. Spikelets with many pairs of spikelets arranged in whorled racemes, lodicules glabrous:

26. Perennial, robust grass, nodes glabrous, primary branches of the panicle whorled, spikelets laterally compressed .... **Chrysopogon** ***p.p.***

26. Annual, decumbent grass, nodes villous, primary branches of the panicle solitary or alternate, spikelets dorsally compressed .... **Pseudosorghum**

25. Spikelets with a pair of spikelets arranged in the panicle, lodicules hairy:

27. Ciliate on pedicel, internodes of spikelets and lodicules.. **Sorghum**

27. Glarbrous on pedicel, internodes of spikeletes and below half of lodicules .... **Hemisorghum**

24. Inflorescence not a panicle, digitate racemes:

28. Lower glume of sessile spikelet 2-toothed.......... **Thelepogon**

28. Lower glume of sessile spikelet not toothed:

29. Upper lemma awned from the back, below the middle .... **Arthraxon**

29. Upper lemma awned from the apex or from the sinus:

30. Upper lemma well developed, awn arise from the sinus:.................................... **Ischaemum**

30. Upper lemma reduced to a translucent base of an awn, awn arises from the apex of the lemma

31. Pedicels and rachis internodes with a translucent base of an awn, awn arises from the apex of the lemma:

32. Racemes 1-5-spikeletted .... **Capillipedium**

32. Racemes more than 8-spikeleted .... **Bothriochloa**

31. Pedicels and rhachis internodes without a translucent median line .... **Dichanthium**

2. Spikelets in each pair usually similar:

33. Spikelets arranged in panicle, and enclosed in long, white, silky hairs:

34. Spikelet awned .... **Spodiopogon**

34. Spikelet unawned:

35. Panicle narrow, contracted, rachis quite tough, all spikelets pedicelled .... **Imperata**

35. Panicle effuse, rachis fragile, spikelets 2-nate, one sessile and other pedicelled .... **Saccharum**

33. Spikelets arranged in several pairs of racemes, and not enclosed in silky hairs:

36. Racemes supported by spathes:

37. Basal sheaths woolly .... **Eulaliopsis**

37. Basal sheaths glabrous:

38. Racemes few to many:

39. Spikelets binate, both pedicelled, leaf-blades linear-lanceolate, flat .... **Eulalia (contorta)**

39. Spikelets binate, one sessile and other pedicelled, leaf-blades filiform-linear .... **Lophopogon**

38. Racemes solitary .... **Pogonatherum**

36. Racemes not supported by spathes:

40. Racemes 2-20, pedicelled spikelets well developed:

41. Culms tufted, lower glume of sessile spikelet channeled .... **Microstegium**

41. Culms straggling, lower glume of sessile spikelet not channeled:

42. Spikelets many; stamens 3 .... **Eulalia *p.p.***

42. Spikelets up to 4; stamens 2 .... **Phacelurus**

40. Racemes 1-2, pedicelled spikelets reduced .... **Apocopis**

KEY TO THE GENERA OF THE TRIBE PANICEAE

1. Spikelets unawned:

2. Inflorescence of spiciform racemes:

3. Ligule absent .... **Echinochloa** *p.p.*

3. Ligule present:

4. Spikelets not in clusters, solitary, binate or ternate:

5. Lower glume well developed:
    6. Lower glume turned away from the axis:
        7. Spikelets flattened or depressed, upper lemma mucronate .... **Urochloa**
        7. Spikelets usually turgid, upper lemma acute.. **Paspalidium**
    6. Lower glume turned towards the axis .... **Brachiaria**
5. Lower glume usually absent, if present minute, scale like:
    8. Spikelets elliptic or elliptic-oblong:
        9. Leaf blades rounded at apex, sheaths compressed and keeled on the back; spikelets sessile, upper lemma involute at margin .... **Axonopus**
        9. Leaf blades acute or acuminate, sheaths not as above; spikelets pedicelled, upper lemma flat at margin .... **Digitaria**
    8. Spikelets orbicular, oblong or ovate .... **Paspalum**

4. Spikelets 1-6, clustered:
    10. Spikelets solitary or 2-3, enclosed in an involucre of bristles .... **Cenchrus**
    10. Spikelets 1-6, intermixed with small scale like empty glumes .... **Trachys**

2. Inflorescence of panicle:
    11. Spikelets covered by involucral bristles:
        12. Bristles solitary, born by the prolongation of the branch beyond the terminal spikelet........................... **Psuedoraphis**
        12. Bristles numerous, born of highly modified bristle or spikelike branchlets:
            13. Spikelets ellipsoid, 2-3 mm long, bristles rigid, barbed, involucre persistent, upper lemma rugose .... **Setaria**
            13. Spikelets lanceolate, 6-7 mm long, bristles plumose, involucre deciduous, upper lemma smooth .... **Pennisetum**
    11. Spikelets not covered by involucral bristles:
        14. Spikelets without an annular callus at the base:
            15. Spikelets gibbose:
                16. Spikelets dorsally compressed, erect grass................ **Sacciolepis**
                16. Spikelets laterally compressed, prostrate grass:
                    17. Upper lemma with a crest like structure at apex .... **Cyrtococcum**

17. Upper lemma without a crest like structure .... **Pseudechinolaena**

15. Spikelets not gibbose:

18. Upper lemma with 2 lateral, basal appendages; caryopsis oblong .... **Ichnanthus**

18. Upper lemma without basal appendages; caryopsis usually ellipsoid-ovoid:

19. Upper glume more or less equaling the spikelet .... **Panicum**

19. Upper glume much shorter than the spikelet .... **Ottochloa**

14. Spikelets with an annular callus at the base .... **Eriochloa**

1. Spikelets awned:

20. Upper lemma glabrous along the margin, muticous, lower lemma awned or muticous, upper glume awned or aristate:

21. Lower lemma awned:

22. Upper glume entire, awned:

23. Culms spongy below, inflorescence a thyrosid panicle, branches erect, appressed to the rachis......... **Hymenachne**

23. Culms tough below, inflorescence a loose panicle of spiciform racemes.................................... **Echinochloa** *p.p.*

22. Upper glume emarginate or 2-lobed, aristate .... **Melinis**

21. Lower and upper glume and lower lemma awned.............. **Oplismenus**

20. Upper lemma densely ciliate along the margins, shortly aristulate, lower lemma muticous, upper glume unawned, acuminate .... **Alloteropsis**

KEY TO THE TRIBES OF THE SUB-FAMILY POOIDEAE

1. Tall, woody, arborescent or shrubs; leaf blades usually with a petiole like base, articulate with the sheaths; lodicules 3 ....**Bambuseae**

1. Annual or perennials, herbs; leaf blades not articulate on the sheaths, lodicules 2:

2. Inflorescence of solitary, digitate or scattered spikes borne on opposite sides of the rachis, spikelets sessile:

3. Spikelets on opposite sides of the rachis of solitary spikes:

4. Ovary hairy at the apex, grains often longitudinally grooved, lodicules hairy, lemmas 3-5-nerved ....**Triticeae**

4. Ovary glabrous, grains not grooved, lodicules glabrous, lemmas 1-3-nerved .... **Leptureae**

3. Spikelets secund on a tough rachis, inflorescence of digitate or solitary spikes .... **Chlorideae**

2. Inflorescence of panicles or spikelike, open or contracted, spikelets sessile or pedicelled:

5. Spikelets with one fertile floret:

6. Glumes minute or suppressed, fertile lemma and palea similar in texture .... **Oryzeae**

6. Glumes well developed, fertile lemma and palea not similar:

7. Spikelets with 2 or 3 florets, the lower male or barren, the upper bisexual:

8. Spikelets with 3 florets .... **Phalarideae**

8. Spikelets with 1 fertile floret:

11. Spikelets awned:

12. Spikelets 1-flowered, awn single:

13. Spikes solitary, cylindrical glumes persistent .... **Perotideae**

13. Panicle many-branched, spikelets falling entire .... **Garnotieae**

12. Spikelets many-flowered, awn trifid .... **Aristideae**

11. Spikelets unawned:

14. Panicle; spikelets ellipsoid, terete, grows usually in dry localities .... **Sporoboleae**

14. Spike or spiciform racemes; spikelets laterally compressed, grows in sea-shores and dry localities .... **Zoysieae**

7. Spikelets with 2 florets:

9. Lower floret empty and without a palea, glumes up to half the length of the spikelet, spikelets falling along with pedicel .... **Thysanolaeneae**

9. Lower floret male or barren, palea present, glume as long as the spikelet, spikelets falling leaving the glumes on the rachis:

10. Glumes equal in size obtuse or mucronate, upper lemma unawned .... **Isachneae**

10. Glumes unequal, the lower acute, upper acuminate, upper lemma awned .... **Arundinelleae**

5. Spikelets with two or more fertile florets:

15. Leaf-blades usually broad, ovate-lanceolate, lemmas bear tuberculate-based hairs on the margins .... **Centotheceae**

15. Leaf blades linear-lanceolate, lemmas without tuberculate-based hairs on the margins:

16. Spikelets-awned, 2-3-flowered:

17. Inflorescence of panicle, effuse, lemmas 2, sterile lemmas absent .... **Arundinelleae**

17. Inflorescence of contracted panicle or racemes, lemmas more than 2, sterile lemmas present:

18. Tufted grass; lemmas 9-nerved, 9-awned; caryopsis ovoid .... **Pappophoreae**

18. Stoloniferous grass; lemmas 7-9-nerved, 1-awned; caryopsis obovate- elliptic............ **Pommereulleae**

16. Spikelets unawned, 4-many-flowered .... **Eragrosteae**

KEY TO THE GENERA OF TRIBE ARUNDINEAE

1. Plants reed like:

2. Spikelets 3-flowered; ligule membranous; callus not bearded; margin of lemma pilose .... **Arundo**

2. Spikelets 3-10-flowered; ligule a ciliate rim; callus bearded; margin of lemma glabrous.. **Phragmites**

1. Plant not reed like... **Zenkeria**

KEY TO THE GENERA OF TRIBE CHLORIDEAE

1. Spikelets usually awned:

2. Inflorescence usually a solitary spike:

3. Glumes awned .... **Melanocenchris**

3. Glumes aristate or mucronate:

4. Annual; perfect lemma broadly ovate, not keeled, truncate .... **Tetrapogon**

4. Perennial; perfect lemma lanceolate, keeled, 2-fid:

5. Awn of spikelets not more than 8 mm long................ **Enteropogon**

5. Awn of spikelets 12-15 mm long ....**Schoenefeldia**

2. Spikes with 2-20 racemes:

6. Spike digitate; lemma laterally compressed; grains trigonous to subterete .... **Chloris**

6. Spike racemiform; lemma not laterally compressed; grains ovoid.. **Gymnopogon**

1. Spikelets unawned:
    7. Spikes solitary, terminal:
        8. Spikelets more or less sunk in the rachis, 1-2-flowered........... **Oropetium**
        8. Spikelets not sunk in the rachis, only one-flowered ....**Microchloa**
    7. Spikes of 2-8 racemes, terminal, umbellate ....**Cynodon**

KEY TO THE GENERA OF TRIBE ERAGROSTEAE

1. Spikelets awned:
    2. Inflorescence of digitate racemes or spike:
        3. Spike solitary, cylindric, often interrupted at base ....**Elytrophorus**
        3. Digitate racemes 1-many:
            4. Racemes whorled or scattered, spikelets 8-20-flowered; caryopsis oblong .... **Acrachne**
            4. Racemes umbelled, spikelets 3-5-flowered; caryopsis subglobose .... **Dactyloctenium**
    2. Inflorescence panicle or spiciform racemes:
        5. Lemmas entire .... **Dinebra**
        5. Lemmas emarginate. 2-4-fid at apex:
            6. Inflorescence of solitary spikes; upper glume shortly 2-toothed at the apex, lemmas and paleas glabrous on the keels and nerves.. **Tripogon**
            6. Inflorescence a panicle; upper glume not toothed at the apex, lemmas and paleas pilose on keels and nerves .... **Leptochloa** *p.p.*
1. Spikelets unawned:
    7. Inflorescence of panicle, lax or contracted:
        8. Upper glume 3-nerved .... **Eragrostis**
        8. Upper glume 1-nerved:
            9. Panicle short, branches straight, more or less spreading, spikelets closely packed................................. **Desmostachya**
            9. Panicle lax, branches slender or triquetrous, flexuous, spikelets distant... **Leptochloa** *p.p.*
    7. Inflorescence of digitate racemes or spikes:
        10. Spikes aggregate of dense globose heads .... **Coelachyrum**
        10. Spikes not in dense globose heads:
            11. Spikes solitary... **Eragrostiella**
            11. Spikes 2-10...**Eleusine**

KEY TO THE GENERA OF TRIBE ISACHINEAE

1. Panicle loose, spikelets many, glumes deciduous, upper lemma becoming indurated .... **Isachne**
1. Panicle contracted or spiciform, spikelets few, glumes persistent, upper lemma membranous........ **Coelachne**

KEY TO THE GENERA OF TRIBE ORYZEAE

1. Spikelets awned:
    2. Leaf-blades ovate-oblong; glumes absent........ **Hygroryza**
    2. Leaf-blades linear; glumes present, minute, oblong or subulate........ **Oryza** *p.p.*
1. Spikelets unawned:
    3. Glumes absent or reduced, a hyaline rim........ **Leersia**
    3. Glumes present, minute, oblong-lanceolate........ **Oryza** *p.p.*

KEY TO THE GENERA OF TRIBE POEAE

1. Lemma awned; stamen 1........ **Vulpia**
1. Lemma not awned; stamen 3:
    2. Lemma very broadly ovate-elliptic, inflated, cordate at base........ **Briza**
    2. Lemma not as above .... **Poa**

KEY TO THE GENERA OF TRIBE BAMBUSEAE

1. Ovary appendage broadly conical and fleshy:
    2. Internodes solid; spikelets in globose congested heads, rhachilla internodes short, always unarmed, lodicules 0........ **Dendrocalamus**
    2. Internodes hollow; spikelets in heads on spicate branchlets, rachilla internode elongate, sometimes armed, lodicules 3 .... **Bambusa**
1. Ovary appendage elongate, stiff and steeple-like:
    3. Flowering glumes abxially glabrous; spikelets deciduous along with glumes........ **Pseudoxytenanthera**
    3. Flowering glumes abaxilly densely villous; spikelets disarticulating above persistent glumes .... **Pseudostachyum**

KEY TO THE GENERA OF ARUNDINELLEAE

1. Lemma of upper floret densey bearded on dorsal surface about the middle........ **Jansenella**
1. Lemma of upper floret glabrous or scaberulous, not as above........ **Arundinella**

The following tribes contain only one genus in this flora.

Agrostideae..**Polypogon**

Aristideae...**Aristida**

Centotheceae ....**Centotheca**

Garnotieae..**Garnotia**

Lepturea ....**Lepturus**

Pappophoreae ....**Enneapogon**

Perotideae ....**Perotis**

Phalarideae ....**Phalaris**

Pommereulleae ....**Pommereulla**

Sporoboleae ....**Sporobolus**

Thysanolaeneae...**Thysanolaena**

Triticeae...**Triticum**

Zoysieae ....**Tragus**

ACRACHNE Wight & Arnott ex Chiovenda

Acrachne racemosa (Heyne ex Roth) Ohwi, Bull. Tokyo Sci. Mus. n. 18:1. 1947; Bor, Grass. India 487. 1960; Moulik, Grass. Bamb. India 2: 584. 1997. *Eleusine racemosa* Heyne ex Roth, Syst. Veg. 2: 583. 1817. *E. verticillata* Roxb., Fl. Ind. 1:346. 1820; FBI 7: 295. 1897; Fischer. 3:1840. 1934.

Annuals, culms erect, tufted, to 70 cm; nodes glabrous. Leaf sheaths glabrous; ligules hairy; leaf blades linear-lanceolate, ovate-acuminate, base rounded, acuminate, margins scaberulous. Racemes spiciform; spikelets ovate, 6-18-flowered; lower glume ovate, keeled, 3-nerved, acute and mucronate; upper glume ovate-lanceolate, keeled, acuminate, awned; lower lemma broadly ovate, 3-nerved, awned; upper lemma ovate, 2-nerved, awned; palea minute; stamens 3. Caryopsis oblong, transversely rugose.

Common near streams in forests throughout Eastern Ghats. Fl. & Fr.: July - December.

Upper Ahobilam (KNL), *BR* & *AMR* 20074; Rollapenta (PKM), *BSS* & *SKB* 29393; Ethipothala water falls (GNT), *VRK* 5792; Bhajaranga kona (KDP), *KVS* 29716; Pathalaganga, near Srisailam (KNL), *JLE* 16833 (MH); Guvvalacheruvu (KDP), *KS* 6423 (CAL & MH); Diguvametta (PKM), *JLE* 42189 (MH); Bollapalle RF (GNT), *S.India Flora* 17427 (MH); Kondapalli (KSN), *CAB* 4563 (MH); Boragharo (GJM), *VNS* 5974 (MH); Pachidya (GJM), *VNS* 5835 (MH).

INDIA: Plains of India.

WORLD: Africa, SW Asia, Afghanistan, Pakistan, Sri Lanka, Myanmar, N Australia.

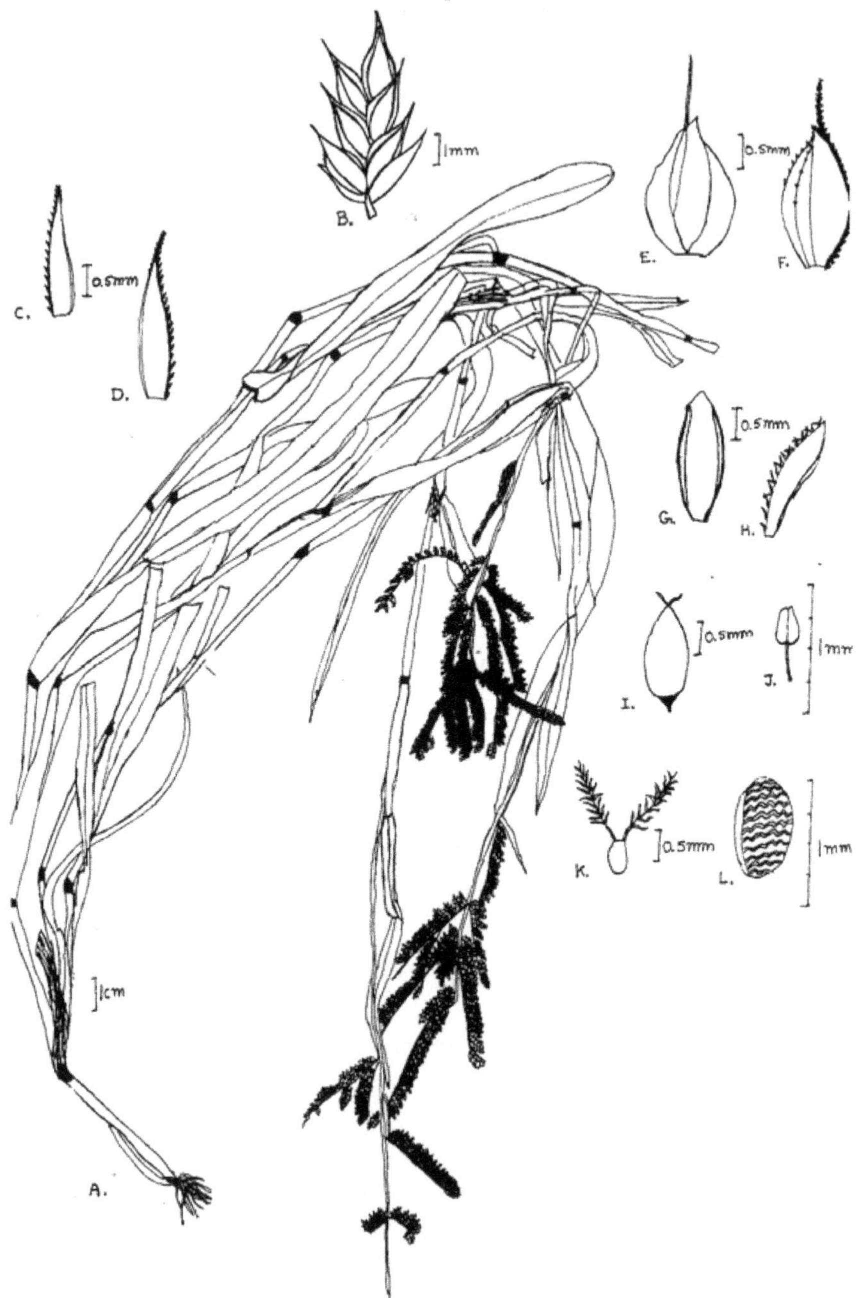

Figure 1. **Acrachne racemosa** Ohwi

A. Twig, B. Spikelet, C. Lower glume, D. Upper glume, E & F. Lower lemma front and backside views, G & H. Upper lemma front and side views, I. Palea, J. Stamen, K. Pistil, L. Caryopsis.

ALLOTEROPSIS J. Presl ex C. Presl

1. Annual; leaf blades lanceolate, base cordate, margins pectinate; raceme rachis glabrous .... **A. cimicina**
1. Perennial; leaf blades linear, base narrow, margins smooth; raceme rachis pilose .... **A. semialata**

Alloteropsis cimicina (L.) Stapf, in Prain, Fl. Trop. Afr. 9: 487.1919; Fischer 3: 1766. 1934; Bor, Grass. India 276. 1960; Matthew, Fl. Tamilnadu Carnatic 3: 1799. 1983. *Milium cimicinum* L., Mant. Alt. 184.1771. *Axonopus cimicinus* (L.) P. Beauv., Ess. Agrost. 12. 1812; FBI 7: 64. 1896.

Annuals, culms erect, tufted, 80-100 cm high. Leaf sheaths covered with tubercle based hairs; ligules fimbriate; leaf blades ovate-lanceolate, 2-10 x 1-1.5 cm, margins tubercle based hairy, apex acute. Racemes 4-10, digitate, terminal, 5-15 cm long; spikelets broadly elliptic, 2-4 x 2 mm, 2-flowered, lower staminate or barren, upper bisexual; lower glume ovate-lanceolate, 2 x 1 mm, acuminate, 3-nerved; upper glume broadly ovate, 2-4 x 2 mm, ciliate, acuminate, 5-nerved; lower lemma ovate, 1.5-3 x 1 mm, acuminate, 5-nerved; lower palea elliptic, 1.5-2 x 0.5 mm, hyaline, bilobed at apex; stamens 3; lodicules truncate, 0.5 mm; upper lemma elliptic, 2-3 x 1mm, aristate, margin ciliate; upper palea ovate, 2-3 x 1 mm, 2-keeled, chartaceous, papillose; stamens 3; stigmas plumose. Caryopsis ovate.

Common as undergrowth in forests and in open dry habitats throughout Eastern Ghats. Fl. & Fr.: June–March.

Kekati RF (ATP), *TP* & *NY* 773; Chennampalli (ATP), *PVP* 3311; Kalasamudram (ATP), *BSS* & *KRKS* 37905; Upper Ahobilam (KNL), *TP* & *RVR* 2713; Dhone (KNL), *PVP* 2696; Balapalli RF (KDP), *PVP* 3453, 3456, 3457, *A.M.Rao* 454 (BSID), *JLE* 14273 (MH); Palakonda hills (KDP), *CS* & *DAM* 7697; Tirumala hills (CTR), *RVR* & *GO* 2419, *KS* 6866 (MH, CAL); Mahanandi (KNL), *JLE* 42337 (MH); Baninikonda (KDP), *JSG* 20869 (MH); Panappakkam (CTR), *S. India Flora* 15759 (MH); Nakirekal (GNT), *S. India Flora* 17476 (MH); Kondapalli RF (KSN), *CAB* 4592 (MH); Polavaram Agency (WG), *DCSR* 322 (CAL); Rampa hill (EG), *MSR* 1606 (CAL); Dharakonda (VSKP), *S. India Flora* 17083 (MH); Manuguru (KMM), *RCS* 104257 (MH & BSID); Purunakote section, Satkosia Tiger Reserve (AGL), *KCM* 5174 (BSID); Dunaei dam (GJM), *AC* 10210 (BSID); Kattupalli (CPT), *DN* 914 (MH); Tirukkalukundrum hill (CPT), *DN* 636 (MH); Tippukadu RF (NA), *KRM* 17689 (MH).

INDIA: Throughout India.

WORLD: Old World Tropics.

Alloteropsis semialata (R.Br.) Hitchc., Contrib. US Nat. Herb. 12, 210. 1909; Bor, Grass. India 276. 1960; Saxena & Brahmam, Fl. Orissa 4: 2238. 1996; Moulik, Grass. Bamb. India 1: 70 1997. *Panicum semialatum* R.Br., Prodr. 192. 1810. *Axonopus semialatus* (R.Br.) Hook.f., Fl. Brit. India 7: 64. 1896 var. *semialata*.

Perennials; culms erect, tufted, to 130 cm high; basal sheaths covered with silky hairs; nodes bearded. Leaf sheaths 4-20 cm, mouth hairy, basal ones densely silky hairy; ligules truncate, 0.3 mm, membranous or fimbriate; leaf blades 10-20 x 0.4-1 cm, linear-lanceolate, margin scabrid, green or glaucous. Racemes 3-5, sub-digitate, rhachis and pedicels hairy; spikelets ovate, 4-6 x 3 mm, awned, lower glume ovate, 3.5 x 2 mm, aristate, chartaceous, 3-nerved; upper glume broadly ovate, 5-5.5 x 2 mm, chartaceous, 5-nerved, winged, ciliate; lower lemma ovate-lanceolate, 4.5 x 2.5 mm, coriaceous, 3-nerved; palea minute; upper lemma oblong-lanceolate, 4.5 x 2 mm, coriaceous, keeled; lodicules 2; stamens 3, anther 1 mm; ovary oblong.

Rare in forests, usually in dried pond beds and sandy soils in Similipahar in Northern Eastern Ghats, Telangana Nallamalais and Talakona in Southern Eastern Ghats. Fl. & Fr.: July–November.

Talakona (CTR), *BR* & *BSS* 30621; Joranda, Similipahar (MBJ), *SX* 348, *SX* & *MB* 4369 (RRL-B); Appapur (MBNR), *BR* & *AMR* 26802.

INDIA: Subtropical Himalaya from Kumaon eastwards to Sikkim, Naga and Khasi hills, Manipur, Bihar, Odisha, Andhra Pradesh.

WORLD: Myanmar, Mauritius, China, Philippines, Australia and Tropical Africa.

ANDROPOGON Linnaeus

1. Inflorescence of solitary raceme .... **A. fastigiatus**
1. Inflorescence of two to five racemes:
    2. Inflorescence of 3-5 racemes, some times only two; culms 60-200 cm; lower leaves 10-40 cm long.............................. **A. chinensis**
    2. Inflorescence of two racemes; culms 20-90 cm; leaves up to 20 cm long:
        3. Lower glume of sessile spikelet concave, strongly compressed so that margins almost meet in the middle line........................... **A. lividus**
        3. Lower glume of the sessile spikelet slightly concave .... **A. pumilus**

Andropogon chinensis (Nees) Merr., Philipp. J. Sci., C. 12.: 101. 1917. *Homoeatherum chinensis* Nees, Intr. Nat. Syst. Bot. Ed. 2: 448. 1836. *Andropogon ascinodis* C.B.Clarke, J. Linn. Soc. Bot. 25: 87. t. 36. 1889; Fischer 3: 1753. 1934; Saxena & Brahmam, Fl. Orissa 4: 2240. 1996. *A. apricus* sensu Hook.f., FBI 7: 169. 1896.

Perennial caespitose grass, culms 60-200 cm long, Leaf sheaths rounded, narrow; leaf blade linear, 10-40 cm x 1-8 mm, setaceous, glaucous, pubescent or glabrescent beneath; ligule scarious, oblong. Inflorescence of 3-5 racemes, sheathed with a spathiform leaf-sheath; sessile spikelets 4.3-5 mm long, callus shortly bearded; lower glume linear, margins strongly inflexed and the centre deeply depressed (grooved) between the keels, upper glume cymbiform, hyaline, 1-nerved, with a fine awn, 3 mm long; lower lemma of both spikelets narrow, hyaline, ciliate; upper lemma short, with awn 2-2.5 cm or more. Pedicelled spikelets: lower glume 2-cuspidate from the shortly excurrent keels and often with a fine scabrid awn 6.2-7.5 mm long, keels ciliate; upper glume and lower lemma as in sessile spikelet; upper lemma shorter, finely awned with 3 stamens and a 3-fid quadrate scale.

Sporadic in hillls forests in Norther Eastern Ghats in Odisha. Fl. & Fr.: October–November.

Simlipahar, *V. Naik* Acc. Nos. 3896, 3897(RRL-B).

INDIA: Manipur, Bihar, Odisha.

WORLD: Myanmar.

Andropogon fastigiatus Sw., Prodr. Veg. Ind. Occ. 26. 1788; FBI 7: 167. 1896. *Diectomis fastigiata* (Sw.) Kunth in Humb. & Bonpl., Nov. Gen. & Sp. 1: 193; t. 64. 1816; Bor, Grass. India 135. 1960; Moulik, Grass. Bamb. India 1: 275. 1997.

A perennial, tufted grass; culms up to 50 cm high, decumbent below, fastigiately branched upwards, leafy. Leaf sheaths terete; ligule elongate, acute; blade 7-30 x 0.1-0.2 cm, narrowly linear, acuminate, scabrid. Raceme single, enclosed by spathe, 2-3.5 cm long, fragile; joints much shorter than the spikelets, villous with white hairs, apex strongly toothed; spathes 7.5 cm, linear-lanceolate. Sessile spikelet: 5 mm, oblong; callus with long hairs; lower glume linear-lanceolate, 6-nerved, bicuspidate, keeled, margins thick, apex ciliate; upper glume shorter, cymbiform, 3-nerved, bidentate, lobes obtuse, awned, apex ciliate; lower lemma linear-oblong, 2-keeled, 2-nerved, margins ciliate; upper lemma cymbiform, 2-toothed, awned; palea minute, hyaline; lodicules 2; stamens 3. Pedicelled spikelets: 7 mm, neuter; lower glume ovate-lanceolate, 2-cuspidate, awned, ciliate, many-nerved; upper glume much smaller, membranous, awned; lower lemma very small, oblong, hyaline.

Occasional in open places in southern (Lakshmaiah *et al.*, 2012) and Northern Eastern Ghats. Fl. & Fr.: August–December.

On the way to Cherukonda (VSKP), *GVS* 42751 (BSID); Jacum, Karlapat Wildlife Sanctuary (KHD), *AC* 2448 (BSID); Kuldhia WLS, Devgiri (BLR), *AC* 2362 (BSID).

INDIA: West Bengal, Bihar, Odisha, Andhra Pradesh.

WORLD: Tropics of both hemispheres.

Andropogon lividus Thwaites, Enum. Pl. Zeyl. 367. 1864; FBI 7:209. 1896; Bor, Grass. India 91.1960; Moulik, Grass. Bamb. India 1:208. 1997; B.R.P.Rao *et al.*, J. Econ. Taxon. Bot. 35: 606. 2011.

Perennials; culms erect, tufted, to 90 cm high. Leaf sheaths keeled; ligules membranous, 1-3 mm; leaf blades 5-20 x 0.5 cm, linear-lanceolate. Racemes 2, 2-3 cm long, purplish, sparsely villous. Spikelets binate. Sessile spikelets 4.5 x 1 mm, elliptic–lanceolate, awned, awn 1.5 cm; lower glume 4.2 x 0.9 mm, slightly concave, oblong- lanceolate, 2-nerved, keeled, scabrid; upper glume 4.5 x 0.8 mm, oblong-lanceolate, boat-shaped; lower lemma 4 x 0.7 mm, oblong-lanceolate, tubercle-based hairs in the upper half; lower palea absent; upper lemma 3 x 0.6 mm, oblong, 2-cleft, awned, awn 1.5 cm; upper palea 0.6 x 0.3 mm, ovate; lodicules 2, truncate, 0.5 mm; ovary 0.9 x 0.3 mm, oblong; style 1 mm; stigma 2.5 mm. Pedicelled spikelet 4.5 x 1 mm, elliptic-lanceolate; pedicel 3 x 0.6 mm; lower glume 4.5 x 1.2 mm, oblong-lanceolate, acute, 9-nerved, 2-keeled, scabrid; upper glume 4.3 x 0.8 mm, oblong-lanceolate, 3-nerved, acute, 2-keeled, scabrid; lower lemma absent; upper lemma

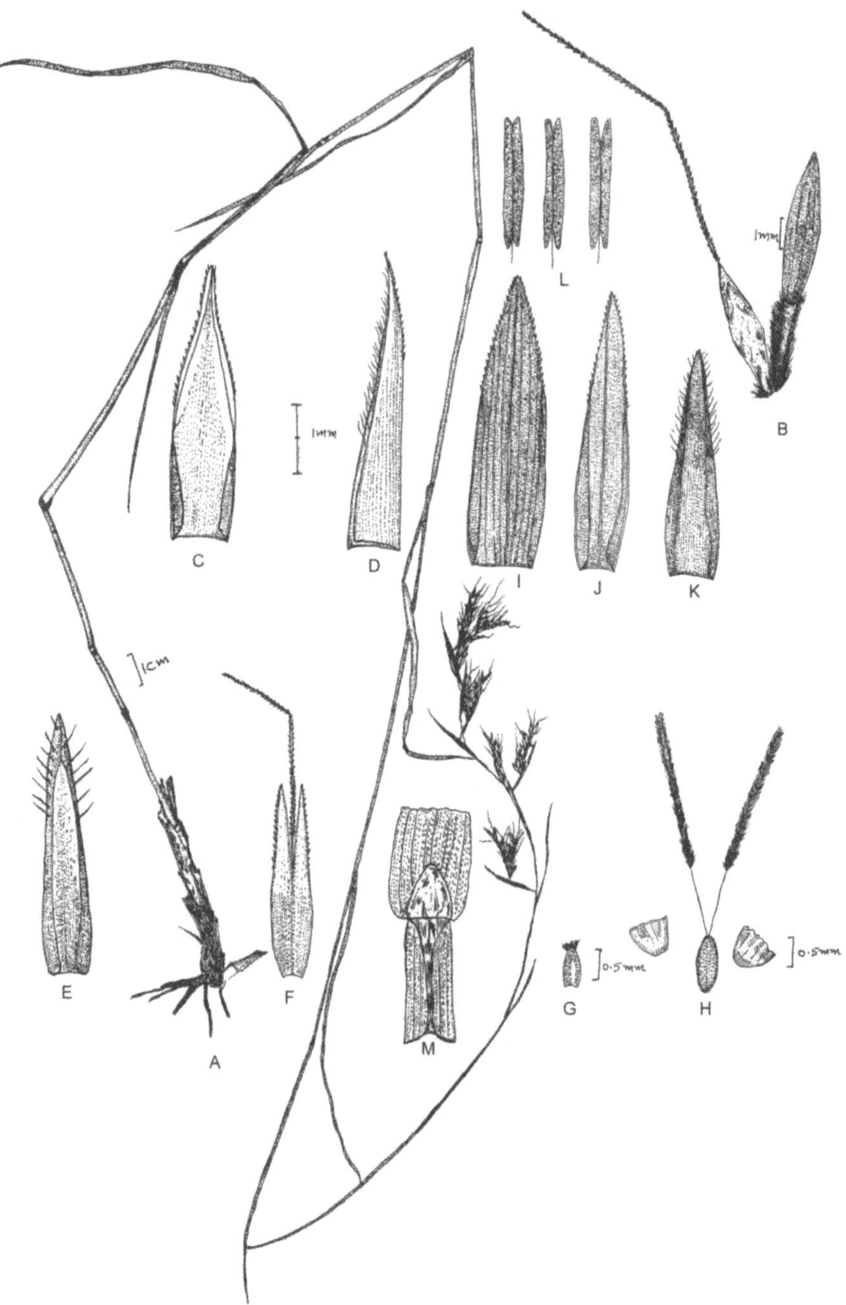

Figure 2. **Andropogon lividus** Thw.

A. Habit; B. Sessile & pedicelled Spikelet; C. Sessile spikelet lower glume; D. Upper glume; E. Lower lemma; F. Upper lemma; G. Upper palea; H. Lodicules & Pistil; I. Pedicelled spikelet lower glume; J. Upper glume; K. Upper palea; L. Stamens; M. Ligule (After Rao ***et al.***, 2011).

3.5 x 0.8 mm, 2-nerved, 2-keeled, ciliate, hyaline; lower and upper paleas absent; stamens 3; anthers 2 x 0.3 mm.

Occasional on hill tops in southern Eastern Ghats. Fl. & Fr.: October–January.

Samrajyam Konda (ATP), *BR & KVS* 29194.

INDIA: Kerala, Andhra Pradesh.

WORLD: Tropical Asia.

Andropogon pumilus Roxb., Fl. Ind. 1: 277. 1820; FBI 7: 170, 1896; Fischer 3: 1753. 1934; Bor, Grass. India 93. 1960; Matthew, Fl. Tamilnadu Carnatic 3: 1799. 1983.

An annual, stoloniferous grass; culms up to 30 cm, tufted, slender, nodes glabrous. Leaf sheaths 1-3 cm, loose, glabrous; ligule membranous; blades 3-7 cm, linear, glabrous, base cordate, apex acuminate. Racemes 2, subtended by a narrow glabrous spathe; racemes 2 cm, digitate pedunculate; spikelets in pairs, sessile bisexual and pedicelled staminate or neuter. *Sessile spikelet*: 4 mm; lower glume linear, concave, 5-nerved, bidentate; upper glume boat-shaped, 3-nerved; lemma linear-oblong, keeled, upper awned; palea minute or absent; stames 3, stigma plumose. *Pedicelled spikelet:* 3 mm, muticous, similar to sessile spikelet.

Common grass on drier areas of Northern and Southern Eastern Ghats. Fl. & Fr.: Sept.–December. Tam.: *Kavattam pillu, Ponvandukaal pillu.*

Muchukota RF (ATP), *PVP* 2569; Guttur kona (ATP), *NY* 981; Gutturu (ATP), *BR* & *P.Priyadarsini* 37143 (BSID); Dharmapuri RF (ATP), *NY* 887; Batrepalli (ATP), *KRKS* 39601; Kammavaripalli (KNL), *TP* & *PVP* 3378; Phirangipuram (GNT), *CAB* 4618 (MH); Vandalur (CPT), *S.India Flora* 11568 (MH); Betkote, Karlapat Wildlife Sanctuary (KHD), *AC* 2469 (BSID).

INDIA: Rajasthan, Madhya Pradesh, Odisha, Andhra Pradesh.

WORLD: Nepal, Indo-China.

APLUDA Linnaeus

Apluda mutica L., Sp. Pl. 92. 1753; Fischer 3: 1750. 1934; Bor, Grass. India 93. 1960; Matthew, Fl. Tamilnadu Carnatic 3: 1800. 1983. *A. varia* Hack. subsp. *aristata* (L.) Hack., Monogr. Phan. 6: 196. 1889; FBI 7: 150. 1896. *A. aristata* L., Amoen. Acad. 4. 303. 1759; Fischer 3: 1750. 1934.

A leafy, perennial, tufted grass. Culms erect, up to 3 m high, glabrous, nodes glabrous. Leaf sheaths 3.5 cm long, glabrous; ligule membranous; blades 6-20 x 0.2-0.5 cm, linear-lanceolate, base narrowed, apex cuspidate, glabrous. Panicle 3-90 cm long, terminal; spikelets in threes, one sessile, bisexual, the rest pedicelled, one staminate and the other rudimentary. *Sessile spikelet*: 2-flowered, lower male, upper bisexual, 6 mm; lower glume oblong, coriaceous, many-nerved, margins scabrid; upper glume oblong, acuminate, coriaceous, 5-nerved, apex hairy; lower lemma broadly ovate, chartaceous, 5-nerved, apex scabrid; palea oblong, 2-nerved, apex scabrid, male; upper lemma hyaline, deeply bidentate, awned; palea small, ovate, bisexual; stamens 3, stigmas plumose. Caryopsis ellipsoid. *Pedicelled spikelet*: 5

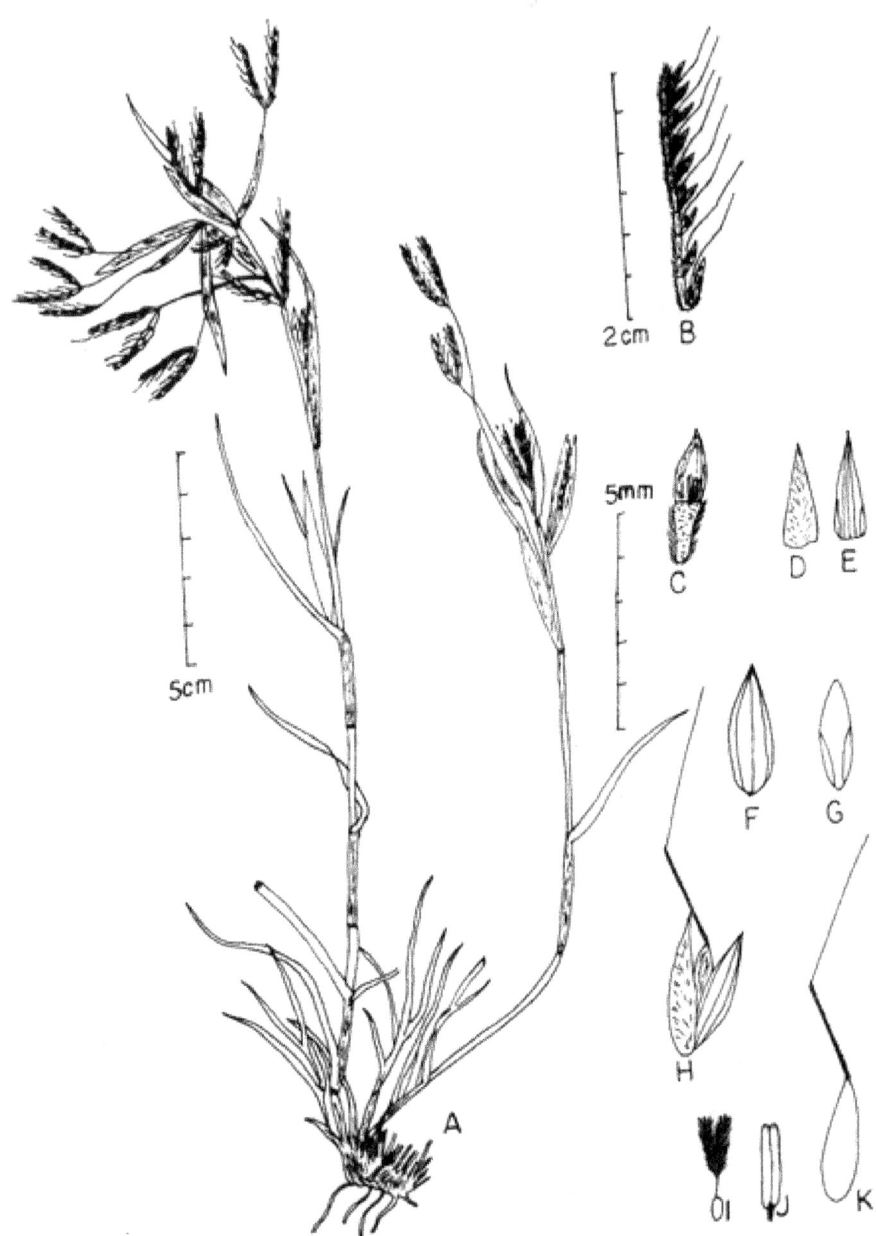

Figure 3. **Andropogon pumilus** Roxb.

A. Twig, B. Spike, C. Pedicelled spikelet, D. Lower glume (dorsal view), E. Lower, glume (ventral view), F. Upper glume, G. Lower lemma, H. Sessile spikelet, I. Pistil, J. Anther, K. Upper lemma.

mm, lanceolate, bisexual, awnless; glumes unequal, many-nerved, lower glume ovate, coriaceous, hairy; upper glume lanceolate, coriaceous, apex scabrid; lower lemma lanceolate, many-nerved, chartaceous, apex scabrid; palea oblong, shortly bidentate apex, scabrid; upper lemma oblong, many-nerved, chartaceous; palea oblong; stamens 3, stigmas plumose.

Common along the hill slopes through out Eastern Ghats. Fl. & Fr.: September–December. Tam.: *Moongilpul, Mandapul.*

Muchukota RF (ATP), *TP* & *NY* 971; Upper Ahobilam RF (KNL), *TP* 2727 & 2718; South Dhone RF (KNL), *RVR* 1602; Rollapenta (KNL), *PVP* & *RVR* 2337; Tirumala hills (CTR), *MSG* 12158, 12185, 12151, 12154; Peddamantanala (PKM), *BSS* & *SKB* 29397 (BSID); Macherla RF (GNT), *VRK* 3274; Loyapalle (GNT), *VRK* 3844; Kadapa (KDP), *JSG* 20854 & *EKK* 78528 (DD); Balapalle (KDP), *JSG* 11164 (CAL), *JLE* 15704 (CAL); Balaparallagutta (CTR), *GVS* 46792 (CAL); Diguvametta (PKM), *JLE* 32557 (CAL); Kondapalli (KSN), *CAB* 8120 (MH); Godavari Gorge (WG), *BO* 3474 (CAL); Ravigudem bungalow (WG), *DCSR* 514 (CAL); Chinturu, Maredumilli (EG), *MM* 102608 (BSID); Sunkarimetta (VSKP), *NPBK* 719 (CAL); Salur forest (VJN), *NPBK* 1121 (CAL); Thumukunta (MBNR), *BR* & *MV* 40726; Umamaheswaram (MBNR), *SRS* 111373; Nilagiri view poit (MBNR), *SRS* 109716 (BSID); Ramavaram (KMM), *RR* 112581 (BSID); Nellippakka RF (KMM), RCS 98903 (MH); Gingee to Chettupattu (SA), *KRM* 60307 (MH); Tippukadu RF (NA), *KRM* 17928 (MH); way to Komatiyur (NA), *KS* 7472 (MH); Vandalur (CPT), *S.India Flora* 11546 (MH); Hogainakkal RF (SLM), *EV* 21912 (MH); Ghat Road, Salem to Yercaud (SLM), *KS* 7588 (MH); Tikarpada, Satkosia Tiger Reserve, *KCM* 7058 (BSID); Majhipda section, Satkosia Tiger Reserve, *KCM* 6160 (BSID); Hatigarh, Similipal Biospher Reserve (MBJ), *AC* 2353 (BSID); Baniabasa, Similipal B.R. (MBJ), *Bahali* & *Agrawala* 673 (BSID); Daringabadi (GJM), *CAB* 1354 (MH).

INDIA: Throughout India.

WORLD: Arabia and Socotra, throughout tropical SE Asia, Australia and New Caledonia.

Note: A very much variable plant, occurring in awned and unawned forms. Dr. N. L. Bor upholds Hackel's opinion that the varieties pass into one another by numerous intermediates and for that matter several other equally distinct varieties could be set up.

APOCOPIS Nees

1. A perennial, densely tufted grass, nerves of lower glume reaching hyaline upper margin:
   2. Upper lemma very narrowly oblong or linear, bidentate, awn 10-20 mm long .... **A. courtallumensis**
   2. Upper lemma oblong, bifid, awn 4.5 mm long or 0 .... **A. paleacea**
1. An annual, decumbent grass, nerves of lower glume not reaching hyaline upper margin:
   3. Raceme solitary, up to 2 cm long .... **A. vaginata**
   3. Racemes 2, above 2 cm long........ **A. mangalorensis**

Apocopis courtallumensis (Steud.) Henr., Blumea 4: 524. 1941; Bor, Grass. India 95. 1960. *Andropogon courtallumensis* Steud., Syn. Pl. Glumac. 1: 377. 1854. *Apocopis wightii* Nees ex Thw., Enum. Pl. Zeyl. 365. 1864; FBI 7: 142. 1896; Fischer 3: 1825. 1934.

A perennial, densely tufted grass; culms up to 60 cm high. Leaf sheaths glabrous, softly hairy; blades 1.5-13 x 0.1-0.4 cm, narrowly lanceolate, flat, more or less hairy from tuberculate based hairs. Racemes usually 2, 2-4 cm long; rachis and pedicels rufous hairy; spikelets 2 mm; lower glume obcordate-obovate, truncate, yellow with a red-brown apex or dark brown with a yellow tip, glabrous-densely rufous hairy; upper glume narrowly lanceolate, 2-keeled, inflected, scabrid; lower lemma oblong, apex ciliate; upper lemma linear, 2-dentate, awned, geniculate; palea very short, convolute.

Rare in drier localities of Northern Eastern Ghats. Fl. & Fr.: September–December.

Chinalldasakaru (EG), *VNS* 51 (CAL); Dhanei (GJM), *SX* & *MB* 4075 (RRL-B); Paniganda (GJM), *SX* & *MB* 1959 (RRL-B), *JSG* 21722 (CAL).

INDIA: Bihar, Odisha, Andhra Pradesh, Central India, Deccan Peninsula.

WORLD: Sri Lanka, Myanmar.

Apocopis mangalorensis (Hochst. ex Steud.) Henrard, Blumea 4: 523. 1941; Bor, Grass. India 96. 1960; Kumar *et al.*, Indian J. Forestry 40: 79. 2017. *Amblyachyrum mangalorense* Hochst. ex Steud., Syn. Pl. Glumac. 1: 413. 1854.

Annual tufted grass; culms geniculate, 8.5-33 cm; nodes with purple dots below. Leaf sheath 1.8-3 cm, more or less ribbed, sparsely tuberculate; ligule 0.3-0.5 mm, membranous, hayaline with hairy apex; leaf blades lanceolate, 1-2.5 x 0.3 cm, acuminate at apex, rounded to truncate at base, tuberculate hairy on both surfaces. Racemes 2, digitate, exerted from spathe; peduncle 4-13 cm long, more or less ribbed; spikelets paired, alternate, closely imbricate, pedicelled spikelets reduced to pedicels; joints terete, 1-1.2 mm. Sessile spikelet obcordate to oblong with broader apex, brownish to yellowish. Lower glume broadly obovate to obcordate, truncate at ends, broadly retuse at apex, chartaceous, 5-7-nerved; upper glume oblong, truncate at ends, apex ciliate to ciliolate, membranous to chartaceous, 3-nerved. Florets 2; lower male, upper bisexual and awned. Lower lemma oblong to narrowly elliptic, membranaceous; upper lemma oblong, acute at 2-lobed, awned from sinus, flat, membranaceous to chartaceous,1-nerved; upper palea oblong to elliptic, truncate at ends, ciliolate at apex, membranaceous, faintly 2-nerved, 2-keeled, margins infolded; stamens 2. Caryopsis oblong to elliptic, acute, brownish with minute plumose hairs towards style apex.

Rare in Kolli hills and Shervarayan hills of Tamil Nadu and in Seshachalam hills in Andhra Pradesh. Fl. & Fr.: August–February.

Way to Rudragal from Talakona (KDP), *BSS* & *P. Priyadarsini* 32932 (SKU); Kolli hills (NMK), *SKS* 302 (SGH); Perungudimedu, near Chennai (CPT), *sine coll. s.n.* (MH); Vandalur (CPT), *sine coll. s.n.* (MH).

INDIA: Peninsula.

WORLD: Myanmar and Sri Lanka.

Apocopis paleacea (Trin.) Hochr., Bull. N.Y.Bot. Garden 6: 262. 1910; Bor, Grass. India 96. 1960. *Ischaemum paleaceum* Trin., Mem. Acad. Imp. Sci. St. Petersbourg, Ser. 6. 2: 293. 1832. *Apocopis royleana* Nees, Ann. Mag. Nat. Hist. 7: 142. 1841; FBI 7: 142. 1896.

Perennial erect grass, 30-60 cm high, rootstock creeping. Leaves linear-lanceolate, 1.2-5 x 0.5-1 cm, sheaths and blades with spreading hairs, margin scaberulous, apex acute or acuminate. Racemes 2-4, paired or digitate, 1.5-4 cm long, spikelets usually sessile only, distichously second on the rusty or whitish-villous rachis, spikelets 3-5 mm long with a short brownish or whitish bearded callus; lower glume adnate to the callus at base, rather shorter than upper glume, broad, dorsally convex embracing the upper glume; lower lemma very broadly lanceolate-oblong, ciliolate, convolute and its palea similar, upper lemma as long, oblong, hyaline.

Rare in Northern Eastern Ghats of Odisha.

Nawna, Simlipahar (MBJ), *SX* 360 (RRL-B).

INDIA: Meghalaya, Manipur, West Bengal, Bihar, Odisha.

WORLD: China, India to Indo-China

Apocopis vaginata Hack., Oesterr. Bot. Z. 41: 8. 1891; Bor, Grass. India 96. 1960. *A. wightii* Nees ex Thw. var. *vaginata* (Hack.) Hook. f., Fl. Brit. India7: 143. 1896.

An annual, tufted, decumbent grass; culms up to 20 cm high, much branched from the base; nodes glabrous. Leaf sheaths lax, longer than the internodes, hairy near the mouth; ligule a small membrane; leaf blade 2-5 x 0.2-0.3 cm, linear, base narrow, apex acute tuberculate based hairs. Racemes 19-22 mm long, closely appressed together; peduncle slightly hairy with brown hairs; joints ciliate; spikelets 5 mm, oblong; lower glume obovate, membranous, truncate, apex ciliate, hyaline, densely hairy in the lower portions, 7-9-nerved; upper glume oblong, truncate margins, hyaline, 3-nerved, lower lemma linear-oblong, hyaline, obtuse apex, ciliate, nerveless; upper lemma linear, membranous, truncate, shortly bidentate, awned; palea small, oblong, truncate with ciliate apex; stigmas plumose. Caryopsis small, narrowly oblong.

Rare in hills in northern Eastern Ghats. Fl. & Fr.: October–January

Bison hills (EG), *CAB* 5086 (MH); Sonabera (KRT), *HFM* 3692 (DD); Dimili, Kotgarh Wildlife Sanctuary (KDM), *AC* 2428 (BSID); Talkudar, Rajgangapur, Sundargarh, *AC* 10262 (BSID).

INDIA: Uttar Pradesh, Madhya Pradesh, Odisha, Andhra Pradesh, South India.

WORLD: India to Indo-China.

ARISTIDA Linnaeus

1. Small annuals; lower glume 1.2-1.8 mm, upper glume 2-2.5 mm long; lemma 1-1.2 mm long; branches of awn unequal, central 5 mm long .... **A. cumingiana**
1. Annuals or perennials, much larger in all their parts:
    2. Glumes muticous .... **A. adscensionis**
    2. Glumes aristate:
        3. Panicle effuse, pyramidal with spreading branch .... **A. hystrix**
        3. Panicle contracted:
            4. Awn with a column:
                5. The articulation is between the top of the column *i.e.*, just below the awns .... **A. mutabilis**
                5. The articulation is between the top of the lemma and the base of the column:
                    6. Glumes of variable length; lemma with lengthy awns .... **A. stocksii**
                    6. Glumes and lemma not as above .... **A. funiculata**
            4. Awn sessile:
                7. Lateral awns very much reduced .... **A. redacta**
                7. Lateral awns well developed .... **A. setacea**

Aristida adscensionis L., Sp. Pl. 82. 1753; FBI 7: 221. 1896; var. **adscensionis;** Bor, Grass. India 407. 1960; Matthew, Fl. Tamilnadu Carnatic 3: 1801. 1983; Moulik, Grass. Bamb. India 2: 584. 1997. *A. depressa* Retz., Observ. Bot. 4. 22. 1786; Fischer 3: 1809. 1934.

A perennial, tufted grass; culms up to 40 cm high, erect or geniculately ascending, glabrous; nodes glabrous. Leaf sheaths 25–40 cm long, glabrous; ligule a ridge of fine soft hairs; leaf blades 2-4 x 0.1-0.2 cm, narrowly linear, convolute, base cordate, apex sub-acute. Panicle contracted, 9-16 cm long; spikelet 1-flowered, linear, 0.6-1 cm; lower glume membranous, narrowly ovate, 1-nerved; upper glumes membranous linear-lanceolate, acute, minutely aristate; lemma elliptic, awned, trifid, middle awn longer than lateral awns; palea narrowly linear; lodicules 2; stamens 3, stigmas plumose. Caryopsis as long the lemma, linear.

Common in plains and slopes of open forests of most of the hill ranges of Eastern Ghats. Fl. & Fr.: August-September.

Ramallakota RF (KNL), *RVR* 1594; Palakonda hills (KDP), *CS* 8242; Kalyani dam (CTR), *RVR & GO* 2438; Tirumala hills (CTR), *MSG* 12155; Macherla RF (GNT), *VRK* 3552, *PCN* 135 (CAL & DD) & *ACA* 134 (DD); Chelama (KNL), *JLE* 22072 (MH); Bughanka (KDP), *KS* 6444 (MH); Kadapa, *EKK* 78153 (DD); Kondveedu fort (GNT), *CAB* 4710 (MH); Rampachodavaram (EG), *MM* 102659 (BSID); Lakshmipuram

(VSKP), *CAB* 1830 (MH); Nattavaram (VSKP), *CAB* 1873 (MH); Tippukadu RF (NA), KRM 17694 (MH); Hogainakkal (SLM), *EV* 20643 (MH); Motijharan (SBP), *HFM* 600 (DD); Sirumalai hills (DGL), *SKS* 403 (SGH); Punakote, Satkosia Wildlife Sanctuary (AGL), *AC* 2388 (BSID); Mandargiri hill (AGL), AC 2369 (BSID).

INDIA: Throughout India.

WORLD: Pantropical.

Aristida cumingiana Trin. & Rupr., Sp. Gram. Stip. 141. 1842 & in Mem. Acad. Sci. Petersb. 6, 7: 141. 1843; FBI 7: 224. 1896; Bor, Grass. India 409. 1960; Saxena & Brahmam, Fl. Orissa 4: 2248. 1996.

An annual, tufted, very slender grass, 7.5-30 cm high. Leaves basal or subbasal, very narrow or filiform, convolute, 2.5-7.5 cm long. Panicle decompound, forming greater part of the plant, with suberect or effuse branches; spikelets small, often purple, 2-2.5 mm long at the ends of the capillary branches and pedicels; lower glume 1.2-1.8 mm; upper glume 2-2.5 mm; lemma 1-1.2 mm; awn inarticulate, very fine.

Occasional in open grasslands of Eastern Ghats of Odisha. Fl.: October.

Sirkagutu and Pudaipada (KJR), *HFM* 2830 (DD); Gursung, Bamra (SBP), *HFM* 2830 (DD).

INDIA: West Bengal, Bihar, Odisha.

WORLD: From Senegal and Ethiopia through India and China to the Philippines.

Aristida funiculata Trin. & Rupr., Sp. Gram. Stip. 159. 1842; FBI 7: 226. 1896; Fischer 3: 1809. 1934; var. **funiculata**; Bor, Grass. India 410. 1960; Matthew, Fl. Tamilnadu Carnatic 3: 1802. 1983.

An annual, tufted grass; culms up to 50 cm high, geniculately ascending; nodes glabrous. Leaf sheaths 4-8 cm long, glabrous; ligule a small ciliate membrane; leaf blade 6-16 x 0.1-0.2 cm, linear, convolute, base cordate, apex acuminate. Panicle lax, narrow, 22 cm long; spikelets 0.8-1 cm, pedicelled, 1-flowered; glumes linear, hyaline, arista elongate; lemma truncate, awn trifid, subequal; stamens 3, stigmas plumose. Caryopsis linear, spiny.

Common grass along the hill slopes and open forests. Fl. & Fr.: August–September. Tel.: *Cheepuru gaddi.*

Batrepalli (ATP), *KRKS* & *BSS* 39608; Kalasamudram (ATP), *KRKS* 40449; Mahanandi (KNL), *VBH* 86682 (BSID); Kondaveedu fort (GNT), *CAB* 4699 (MH); Kondapalli (KSN), *CAB* 4568 (MH); Silhapatni, Khariar (NPD), *HFM* 3138 A (DD); Bafga Khol, Rajgangpur, Sundargarh, *AC* 10258 (BSID).

INDIA: Rajasthan, Concan southwards, Odisha, Andhra Pradesh.

WORLD: Tropical E. Africa, through Arabia to Middle East, Pakistan.

Aristida hystrix L.f., Suppl. Pl. 113. 1781; FBI 7: 225. 1896; Fischer 3: 1809. 1934; Bor, Grass. India 410. 1960; Matthew, Fl. Tamilnadu Carnatic 3: 1802. 1983.

A perennial, tufted grass; culms up to 60 cm high, diffuse with creeping rootstock, ascending; nodes glabrous. Leaf sheaths 2-4 cm long, densely pilose at the mouth; ligule of soft hairs; blade 3-9 x 0.1-0.2 cm, linear, base cordate, sparsely pilose on the margins, apex acuminate. Panicle lax, pyramidal, 10-20 cm, terminal; spikelet 1.5-2.5 cm, linear; glumes linear, aristate; lemma linear, 3-nerved, awn trifid, sub-equal; palea hyaline; stamens 3, stigmas plumose. Caryopsis linear, to 7 mm, prickly.

Common grass in plains and lower slopes in open forests. Fl. & Fr.: September–December. Tel.: *Dollubagaddi*; Tam.: *Oosipillu*.

Kalasamudram RF (ATP), *PVP* 2602; Gooty (APT), *JSG* 20822 (MH); Gani RF (KNL), *RVR* 1512; Palakoda (KDP), *CS* 8254; Bakarapet RF (CTR), *RVR* & *GO* 2443; Gooty (ATP), *JSG* 20822 (MH & DD); Balapalle (KDP), *JLE* 15781 (MH & CAL); Narasimhakonda (NLR), *J.Swamy* & *S.Nagaraju* 14631 (BSID); Macherla RF (GNT), *SCA* 134 (CAL); Karaka (VSKP), *CAB* 1645 (H); Ramanagutta (KMM), *RR* 107993 (BSID); Sirumalai hills (DGL), *SKS* 405 (SGH); Vandalur (CPT), *S.India Flora* 11545 (MH); Majhpada section, Satkosia Tiger Reserve (AGL), *KCM* 6147 (BSID).

INDIA: Peninsular India, Madhya Pradesh.

WORLD: Sri Lanka.

Aristida mutabilis Trin. & Rupr., Sp. Gram. Stip. 150. 1842; FBI 7: 226. 1896; Fischer 3: 1809. 1934; Bor, Grass. India 411. 1960.

An annual, tufted grass; culms up to 40 cm high; nodes glabrous. Leaf sheaths 3-7 cm long, glabrous; ligule membranous; leaf blade 5-12 x 0.2-0.5 cm, linear, convolute, base cordate, apex acuminate, glabrous. Panicle lax, sub cylindric, 15 cm long, spikelets 1-1.5 cm, linear, 1-flowered, glumes linear, keeled, 3-nerved, shortly awned; lemma cylindric, scaberulous, callus shortly bearded, awn trifid, subequal; stamens 3, stigmas plumose. Caryopsis narrowly cylindric.

Occasional grass in plains of Southern Eastern Ghats. Fl. & Fr.: July–September

Pennahobilam (ATP), *PVP* 2653; Ongole hills (PKM), *sine coll., s.n.* (MH).

INDIA: S. & N.W. India

WORLD: Mauritania to India, perhaps also in Tropical East Africa.

Aristida redacta Stapf, Kew Bull. 1892: 85. 1892; FBI 7: 227. 1896; Fischer 3: 1809. 1934; Bor, Grass. India 412. 1960.

A perennial, tufted grass; culms up to 50 cm high, simple or proliferously branched; nodes glabrous. Leaf sheaths 3-8 cm long, glabrous; ligule shortly and densely ciliate; blade 8-15 x 0.2 cm, linear, convolute, base cordate, rough margins, apex acuminate, hairy above, smooth below. Panicle effuse, 25 cm; spikelets 10 mm, linear, pedicelled; glumes linear, subequal, shortly aristate, 1-nerved, 1-keeled, keel scabrous; lemma chartaceous, callus hairy, awn trifid, subequal; stamens 3, stigmas plumose.

Rare grass in dry habitat of hillocks in southern Eastern Ghats. Fl. & Fr.: September–April.

Talupula (ATP), *KRKS* 41994; Bakkannagaripalli (KDP), *KRKS* & *BSS* 41991; Obulapuram thanda (PKM), *BSS* & *SKB* 32940; Kurnool, *MH* Acc No. 57329 (MH).

INDIA: Peninsular, C. & E. India.

WORLD: Bangladesh.

Aristida setacea Retz., Observ. Bot. 4: 22. 1786; FBI 7: 225; 1896; Fischer 3: 1809. 1934; Bor, Grass. India 412. 1960; Matthew, Fl. Tamilnadu Carnatic 3: 1803. 1983.

A perennial, tufted, grass; culms up to 2 m high, erect or geniculately ascending from a woody base; nodes glabrous. Leaf sheaths 2-10 cm long, glabrous; ligule of short hairs; leaf blade 10-25 x 0.1-0.3 cm, linear, convolute, base cordate, apex acuminate, glabrous. Panicle contracted, 30 cm, branches erect; spikelets 1-1.5 cm, linear, 1-flowered, pedicelled; glumes linear, convolute, awned; lemma linear, awn trifid, subequal; palea hyaline, linear to narrowly ovate, 1-2-nerved; lodicules 2; stamens 3; stigma plumose. Caryopsis linear, to 2 mm, prickly.

Common in open lands. Fl. & Fr.: September–April. Tel.: *Gaddiparakalu, Cheepurugaddi.*

Kalasamudram RF (ATP), *PVP* 2604 & 2608; North Dhone RF (KNL), *RVR* 1612; Owk RF (KNL), *RVR* 1476, 3041; Kammavaripalli (KNL), *TP* & *PVP* 3392; Mobbuchintalapalle (KDP), *SSR* 33666; Way to Akasaganga (CTR), *MSG* 12144; Koraprolu (PKM), *RVK* 18328; Chalama (KNL), *MPG* 106 (DD); Mogilipenta (KDP), *JSG* 21332 (DD); Madaraju Gudur (NLR), *PMR* 16437; Macherla RF (GNT), *SCA* 132 (CAL & DD); Gobanapalam (WG), *KS* 5137 (CAL); Near Tiger Camp, Maredumilli (EG), *MM* 102578 (BSID); Araku valley (VSKP), *NPBK* 596; Salur (VJN), *NPBK* 962 (CAL); on way to Umamaheswaram (MBNR), *SRS* 111309 (BSID); Vatavarlapalli (MBNR), *VBH* 84968 (BSID); Lankapally hills (KMM), *RR* 106015 (BSID); Sirumalai hills (DGL), *SKS* 410 (SGH); Tippukadu RF (NA), *KRM* 17700 (MH); Gomukhanandi MHEP, Kalrayans (SA), *KRM* 77580, 83871 (MH); Gingee to Chettupattu (SA), *KRM* 60311 (MH); Guindy park RF (MDS), *ANH* 47195 (H); Betta Mogilalam (SLM), *S.India Flora* 9829 (MH); Alagar hills (DGL), *K.Altaf Ahamed Kabeer* 113279 (MH); Kuanaria, Satkosia Tiger Reserve, *KCM* 8342 (BSID); Lasari, Kotgarh Wildlife Sanctuary, *AC* 2433 (BSID); Balipadar (GJM), *AC* 10221 (BSID); Mandargiri hill (AGL), *AC* 2368 (BSID); Purunakote, Satkosia Tiger Reserve, *AC* 2378 (BSID); Rukeekonda (GJM), *VNS* 82981(MH).

INDIA: Bihar, Odisha, Madhya Pradesh, Andhra Pradesh, Karnataka, Tamil Nadu.

WORLD: Sri Lanka, Myanmar to Mascarene Islands.

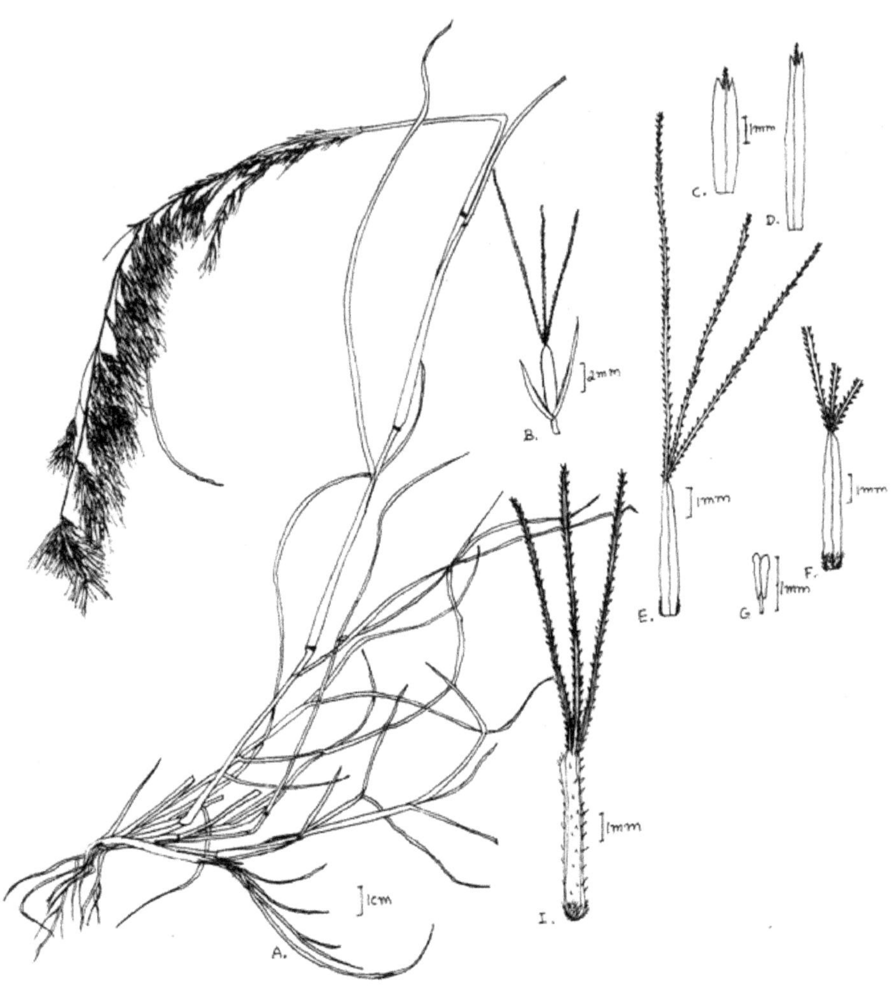

Figure 4. **Aristida setacea** Retz.

A. Twig, B. Spikelet, C. Lower glume, D. Upper glume, E & F. Lemma front and back views, G. Stamen, H. Caryopsis.

Aristida stocksii (Hook.f.) Domin, Biblioth. Bot. 20: 338. 1915; Kumar *et al.*, Rheedea 26(2): 161. 2016. *A. funiculata* var. *stocksii* Hook.f., Fl. Brit. India 7: 227. 1896; Bor, Grass. India 412. 1960.

Annual or perennial, tufted. Culms 25–35 cm high, woody at base, glabrous, rounded. Leaf leaf sheath *c.* 2.3 x 0.2 cm, ribbed, glabrous; ligules membranous, 0.2–0.3 mm, fimbriate at apex; leaf blades linear, to 6.3 x 0.2 cm, convolute, serrulate at margins, acuminate at apex, veins thickened, scabrid on ventral side. Inflorescence 15–20 cm long, contracted or slightly lax panicles, twisted when dry. Spikelets with single floret, fertile spikelets pedicelled; glumes extraordinarily unequal, aristate, membranous, glabrous, 1-veined, scabrid on vein, persistent; lower glume elliptic-lanceolate, 2.9–3.1 x 0.2–0.3 cm, aristate (arista *c.* 2 mm long), upper glume similar to lower glume, to 0.7 x 0.1 mm, aristate (arista *c.* 1.5 mm long), fertile lemma elliptic-oblong, subterete, *c.* 5 mm long, margins convolute, covering most of palea, coriaceous, without keel, 3-veined, articulation between lemma apex and column base; column very long, twisted, to 4.6 cm long, having white long hairs; awns of lemma 6–8 cm long, scabrid, callus barbed with white hairs, tip of the callus curved, glossy, glabrous, palea without keels, anthers 3; stigma plumose. Caryopses not seen.

Gregarious in alkaline sandy soils in dry deciduous forests in Anantapuram district. Fl. & Fr.: September–December.

Kalasamudram Reserve Forest (ATP), *S. Salamma* & *M. Anil Kumar* 49201.

INDIA: Maharashtra, Karnataka, Andhra Pradesh. Endemic.

ARTHRAXON P.Beauvois

1. Lower glume of sessile spikelets rounded on the back, not keeled, spikelets laterally compressed:
    2. Pedicelled spikelets, at least in the upper part of the raceme developed, spikelets up to 3 mm:
        3. Stamens 2 .... **A. lancifolius**
        3. Stamens 3 .... **A. depressus**
    2. Pedicelled spikelets not developed, pedicels more or less reduced, spikelets up to 5 mm:
        4. Spikelets 3-5 mm long; anthers 0.5 mm long:
            5. Stamens 2 .... **A. hispidus**
            5. Stamens 3 .... **A. nudus**
        4. Spikelets 6-8 mm long; anthers over 1.5 mm long .... **A. castratus**
1. Lower glume of sessile spikelets flat on the back, 2-keeled, spikelets dorsally compressed:
    6. Lower glume of sessile spikelet pectinately spinulose; conspicuous on keels from base to apex .... **A. lanceolatus**
    6. Lower glume of sessile spikelet faintly pectinately spinulose on keels from base to apex .... **A. prionodes**

Arthraxon castratus (Griff.) Naray. ex Bor in Kanjilal, Fl. Assam 5: 376. 1940; Bor, Grass. India 99. 1960; Matthew, Fl. Tamilnadu Carnatic 2: 1804. 1983; Saxena & Brahmam, Fl. Orissa 4: 2252. 1996 *Andropogon castratus* Griff., Not. Pl. Asiat. 3: 89. 1851. *Arthraxon rudis* (Steud.) Hochst., Flora 39: 188. 1856; FBI 7: 144. 1896; Fischer, 3: 1728. 1934; Mooney, Suppl. Bot. Bihar & Orissa 188. 1950.

Annual grass, culms 25-50 cm from a prostrate, widely rambling and rooting base. Leaf sheath terete, smooth, striate or with outer margin ciliate, ligule 2.5 mm long, membranous; blade 2-7.5 x 0.5-1.5 cm, margins in lower part with bulbous-based hairs, upwards ciliolate, upper surface usually sparsely appressed-pilose, especially near the base and the margin, lower surface glabrous. Inflorescence with 1-5 branches, longest lowest branch 1.7-7 cm long, joints slightly ciliate; only sessile spikelet present, linear-lanceolate, 6-8 mm long; callus naked; glumes unequal, lower glume rounded on the back, lanceolate, 3.3-6.8 mm long; upper glume 4-7 mm long; lemmas with long-pilose margins, lower lemma lanceolate, 2-5 mm long, upper lemma lanceolate, 2.5 mm long; palea 1-2 mm long; anther 3, over 1.5 mm long. Caryopsis 2.2-3 mm long. Pedicelled spikelet reduced to 1.8 mm long stipe, pilose.

Near Kholgam, Sambalpur, in moist open glade near stream in mixed deciduous forest (Matthew, 1983; Mooney, *loc. cit.)*. Fl. & Fr.: October - December.

INDIA: Assam, Manipur, Odisha, Madhya Pradessh, Tamil Nadu.

WORLD: Bangladesh, Myanmar, Sri Lanka, Malesia, Vietnam.

Arthraxon depressus Stapf ex Fischer, Bull. Misc. Inform. 1933: 349. 1933; Fischer 3: 1728. 1934; Bor, Grass. India 99. 1960; Matthew, Fl. Tamilnadu Carnatic 2: 1804. 1983.

Annual tufted grass; culm up to 18 cm, geniculate, nodes glabrous. Leaf sheaths to 2.5 cm, terete; ligule membranous, to 2 mm; blade ovate to oblong-lanceolate, 2.5-3.5 x 0.8-1.5 cm, tuberculate pilose hairy, acute to acuminate at apex. Racemes 2-4, to 5 cm; joints 3-4 cm, pubescent. Sessile spikelets oblong, to 6.5 mm, base cuneate. Glumes unequal; lower glume linear-lanceolate, to 6 mm, 7-nerved, margin not reflexed, smooth; upper glume to 6.3 mm, obscurely 3-nerved, apex acuminate, erose; lemma pilose on margin; lower lemma to 5.5 mm; upper lemma to 4.8 mm; awn to 8 mm; palea to 2.5 mm, truncate at apex; stamens 3. Pedicelled spikelet a stipe, pilose.

Hill slopes of Shervarayan and Kolli hills above 1300 m. Fl. & Fr.: August–December.

Shervarayan hills (SM), *SKS* 529 (SGH).

INDIA: Tamil Nadu (Endemic).

Arthraxon hispidus (Thunb.) Makino, Bot. Mag. (Tokyo) 26: 214. 1912; Fischer 3: 1728. 1934; Bor, Grass. India 99. 1960; Welken, Blumea 27: 266. 1981. *Phalaris hispida* Thunb., Fl. Jap. 44. 1784. *Arthraxon ciliaris* P. Beauv. Ess. Agrost. 111, 152. 11. f. 6. 812; FBI 7: 147. 1896. *A. micans* (Nees) Hochst., Flora 39: 188. 1856; Bor, Grass. India 101. 1960. *Bathratherum micans* Nees, Edinb. New Phil. J. 18:

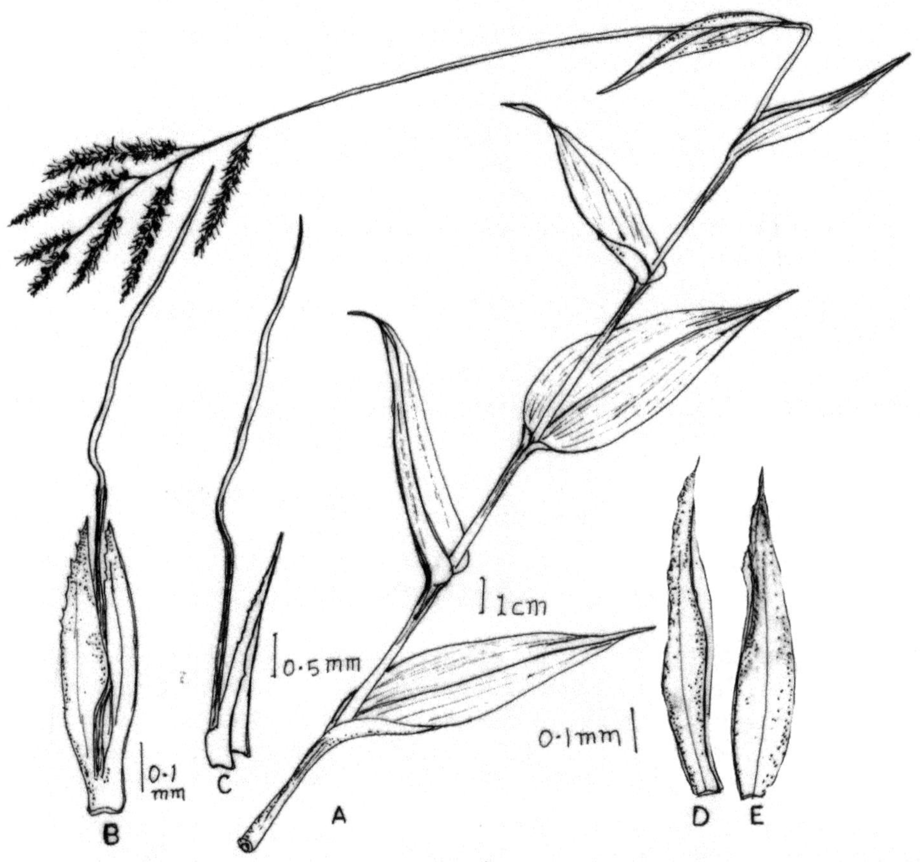

Figure 5. **Arthraxon depressus** Stapf. ex C.E.C. Fischer.

A. Culm, B. Spikelet, C. Upper glume, D. Upper lemma, E. Lower lemma.

182. 1835. *Arthraxon submutius* (Nees ex Steud.) Hochst. *loc.cit.* FBI 7:145. 1896; 144; Bor, *loc.cit.*102. *A. breviaristatus* Hack. in DC. Monogr. Phan. 6: 350. 1889; FBI.; Bor, *loc.cit.* 99. *A. intermis* Hook. f., Fl. Brit. India 7:145. 1896; Bor, *loc. cit.*100. *A. quartinianus* (A. Rich.) Nash in N. Amer. Flora 17: 99. 1812; Bor, *loc. cit.*102; Fischer, *loc.cit.*3: 1728. 1934. *Alectoridia quartiniana* A. Rich., Tent. Fl. Abyss. 2: 488.t.99.1850.

An annual, tufted grass; culms up to 35 cm high, prostrate, rooting base; nodes shortly bearded. Leaf sheaths 1.5-2.5 cm, hirsute in the upper part with tuberculate based hairs; ligule membranous with ciliate apex; blades 2.5-5 x 0.5-2 cm, ovate-lanceolate, broad, base amplexicaul, apex acuminate, tuberculate based hairs on both surfaces, ciliate towards the base. Racemes 1-9, 25 cm long, terminal; rachis fragile; spikelets 3-5 mm, solitary, oblong-lanceolate, laterally compressed; glumes subequal; lower lanceolate, sub chartaceous, scaberulous towards apex, 7-9-nerved; upper obliquely lanceolate, mucronate, 3- or 5-nerved; lower lemma oblong, hyaline, 2-nerved; upper lemma narrowly linear-lanceolate, awned; stamens 2, stigmas plumose. Caryopsis oblong-elliptic to oblanceolate, acute, greenish brown.

Rare in marshy localities in Northern Eastern Ghats. Fl. & Fr.: September–February.

Near Sileru (VSKP), *GVS* 42717 (MH); Chintapalli (VSKP), *JLE* 37149 (MH); Bison hills (EG), *CAB* 5086 (MH); Mahendragiri (GJP), *SX* & *MB* 3674 (RRL-B); Paniganda (GJM), *SX* & *MB* 1976 (RRL-B); Chunabatta, Satkosia Wildlife Sanctuary (AGL), *AC* 2384 (BSID).

INDIA: Odisha, Andhra Pradesh, Kerala

WORLD: Africa, Central Asia, Pakistan, Nepal, Sri Lanka, Myanmar, Philippines, Indonesia, Malaysia, Japan, Australia.

Arthraxon lanceolatus (Roxb.) Hochst., Flora 39: 188. 1856; FBI 7: 143. 1896. var. **echinatus** (Nees) Hack. Mon. Androp. 348. 1889. *Bathratherum echinatum* Nees, Edinb. New Phil. J. 18: 181. 1835. *Arthraxon echinatus* (Nees) Hochst., Flora 39: 188. 1856; Fischer 3: 1728. 1934; Bor, Grass. India 99. 1960. *A. spathaceus* Hook. f., Fl. Brit. India 7: 145. 1896.

An annual, tufuted grass; culms up to 40 cm high, geniculately ascending from the base; nodes glabrous. Leaf sheaths 2-3 cm long, tuberculate ciliate-glabrous; ligule fimbriate; leaf blades 3-4.5 x 0.5-1 cm; ovate-lanceolate, base amplexicaul, margin tuberculate ciliate, apex acuminate, glabrous. Racemes 3.5-5 cm, terminal; spikelets 0.5-0.6 cm. alternate; glumes unequal speculate; lower glume linear, 6-nerved, echinulate on the back; upper 3-4-nerved; lemmas linear, ciliate, upper awned, geniculate; palea with notched apex; stamens 3; stigmas plumose. Caryopsis narrowly linear.

Occasional along the streams and moist places of northern and southern Eastern Ghats. Fl. & Fr.: September–March.

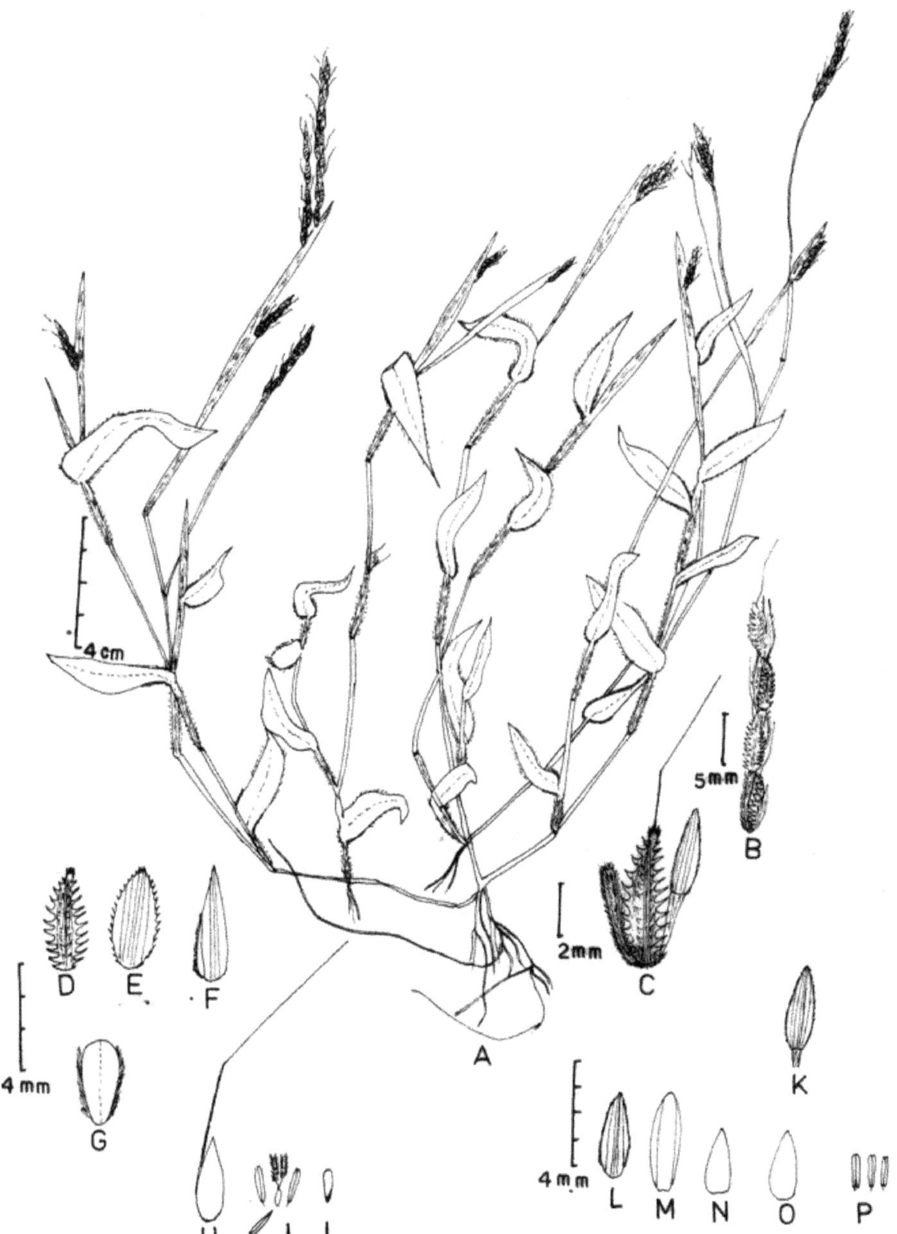

Figure 6. **Arthraxon lanceolatus** (Roxb.) Hochst. var. **echinatus** (Nees) Hack.

A. Twig, B. Raceme C. Sessile and pedicelled spikelets, D. Sessile spikelet lower glume front view, E. Back view of lower glume, F. Upper glume, G. Lower lemma, H. Upper lemma, I. Pistil and stamens, J. Grain,, K. Pedicelled spikelet, L. Lower glume, M. Upper glume, N. Lower lemma, O. Upper lemma, P. Stamens.

Alurkona, near stream, (ATP), *PVP* 3491; Isakgundam-Ghat road valley (KDP), *R.V. Reddy* 7944; Maddileti stream (KNL), *RVR* 3072; Rangapuram RF (KNL), *RVR* 1702; Papavinasanam (CTR), *MSG* 12167; Tangelakunta (KDP), *JSG* 21230 (MH); Borpuram (MBNR), *BSS* & *SKB* 32287 (SKU & BSID); Yenkur (KMM), *RR* 108572 (BSID).

INDIA: Assam, Uttar Pradesh, Odisha, Central and South India.

WORLD: South Africa, Indo-China.

Arthraxon lancifolius (Trin.) Hochst., Flora 39. 188. 1856; Bor, Grass. India 100. 1960; Fischer 3: 1729. 1934. *Andropogon lancifolius* Trin., Mem. Acad. Sci. Petersb. Ser. 6, 2: 271. 1832. *Arthraxon microphyllus* (Trin.) Hochst., Flora 39: 188. 1856; FBI 7: 147. 1896. *Andropogon microphyllus* Trin., Mem. Acad. Sci. Petersb. Ser. 6, 2: 275. 1832.

An annual, tufted grass; culms up to 60 cm high, decumbent at base; nodes slightly bearded. Leaf sheaths 0.1-0.15 cm long, glabrous; ligule membranous; blades 1.5-3 x 0.4-0.8 cm, ovate-lanceolate, base amplexicaul, apex acute. Panicle densely flowered, spicate; rachis capillary; spikelets 3 mm, laterally compressed, paired ciliate, narrowly lanceolate, sessile bisexual, pedicelled empty; lower glume lanceolate, minutely puberulous, 2-cuspidate, 4-nerved; upper glume narrowly lanceolate, tapering into a very fine awn-like point; lower lemma shorter than upper glume, lanceolate, epaleate; upper lemma ovate-lanceolate, hyaline with an awn; stamens 2.

Common grass of damp habitats growing in colonies in rock crevices of most of the hill ranges. Fl. & Fr.: August–December.

Gundlabrahmeswaram (KNL), *PVP* & *RVR* 2372; Maredumilli (EG), *TP* & *MSG* 12266; Tirumala hills (CTR), *MSG* 12182; Palakonda hills (KDP), *CS* 8233; Way to Diguvametta (PKM), *JLE* 22165 (CAL); Chinturu, Maredumilli (EG), *MM* 102607 (BSID); Muzhumamidivalasa, Maredumilli (EG), *MM* 102522 (BSID); Yenkur (KMM), *RR* 108571, 108572 (BSID); Labangi, Satkosia Wildlife Sanctuary (AGL), *AC* 2395 (BSID); Tulka section, Satkosia Tiger Reserve (AGL), *KCM* 5620 (BSID); Betkote, Karlapat Wildlife Sanctuary (KHD), *AC* 2461, 2462, 2467 (BSID); Jacum, Karlapat Wildlife Sanctuary (KHD), *AC* 2438 (BSID); Dimili, Kotgarh Wildlife Sanctuary (KDM), *AC* 2425 (BSID).

INDIA: Throughout India.

WORLD: Asia, Africa, Sri Lanka, Myanmar and China.

Arthraxon nudus (Steud.) Hochst., Flora 39: 188. 1856; Bor, Grass. India101. 1960. Moulik, Grass. Bamb. India 1: 261. 1997; A.M. Reddy *et al.*, J. Econ Taxon. Bot. 38: 21. 2014. *Andropogon nudus* Steud. Syn. Pl. Glum., 1: 383. 1854.

Annuals, culms creeping, to 90 cm long, nodes bearded, rooting from lower nodes. Leaf sheaths 3-6 cm, villous and ciliate along the margins; ligule 2 mm, a ciliolate membrane; leaf blades 4-5.5 x 0.8-1.8 cm, ovate-lanceolate, cordate at base, ciliate along the margins. Racemes subdigitate, 3-10 in number, 2-5 cm, joint linear,

3 mm, angular, rhachis fragile at base; spikelets paired, fertile spikelets sessile, sterile spikelets pedicelled. Sessile spikelets 3-3.5 x 1.3 mm, dorsally or laterally compressed, lanceolate, callus sparsely bearded; lower floret empty, upper florest bisexual; lower glumes 3.5 x 12 mm, lanceolate, membranous to coriaceous, 9-nerved, scabrid along the nerves, upper glume 3.5 x 1 mm, lanceolate, margin hyaline, retrorsely ciliate, 1-keeled, 3-nerved, lower lemma 1.8 x 0.7 mm, linear, 2-nerved, delicate, hyaline, glabrous, upper lemma1.8 x 1.5 mm, 1-nerved, deeply notched, awned, hyaline, awn 9-13 mm, slender, palea absent; stamens 3, anthers 0.3 mm, ovary 0.5 mm, oblong, styles 0.5 mm, stigmas 0.5 mm, pinkish. Caryopsis terete. Pedicelled spikelet reduced to a pedicel or subulate point or completely suppressed.

Rare in moist shade localities of the forests. Fl. & Fr.: August–February.

Chintapalli (VSKP), *AMR* 25616 (SKU); Devarapalli, Maredumilli (EG), *MM* 100793 (MH); Near Sileru approach (VSKP), *GVS* 42717 (MH).

Note: This taxon is closely rlated to *Arthraxon hispidus*, but can be distinguished by having 3 erect stamens, minute anthers and long slender pedicels.

Arthraxon prionodes (Steud.) Dandy in Andr., Fl. Pl. Sudan 3: 399. 1956; Bor, Grass. India 101. 1960. *Andropogon prionodes* Steud., Syn. Pl. Glumac. 1: 383 1854. *Arthraxon lanceolatus* sensu Hook.f., FBI 7: 143. 1896 *p.p.* non (Roxb.) Hochst. 1856.

A perennial, tufted grass; culms up to 60 cm high, hairy, trailing; nodes glabrous. Leaf sheaths slightly pubescent; leaf blade 4-20 x 2-7 cm, lanceolate-ovate, base cordate, apex acuminate, tuberculate ciliate on margin. Racemes 2-8, 7 cm long; rachis slightly hairy; spikelets 5 mm long, linear-lanceolate. Lower glume hyaline, 5-7-nerved, acute, pectinately spinulose on keels, faint; upper glume lanceolate, 3-nerved; lower lemma shorter, linear-oblong, acute; upper lemma hyaline, awned; stamens 3, stigmas plumose.

Occasional in moist and shady localities in Nallamalais. Fl. & Fr.: September–December.

Bairanifalls, Diguvametta (PKM), *JLE* 32469 (CAL); Bhajaranga kona (KDP), *BR* & *KVS* 29117; Ramandurg (Ballari), *S. India Flora* 14430 (MH).

INDIA: Andhra Pradesh.

WORLD: Afghanistan, Pakistan, Bhutan, China, Myanmar, Thailand, Vietnam, E.Africa, SW Asia.

ARUNDINELLA Raddi

1. Upper lemma with two setae, one on each side of the awn at the tip
    2. Perennials .... **A. setosa**
    2. Annuals:
        3. Whole plant covered with tubercle-based hairs; spikelets 7.5-9.5 mm long, acuminate.................................. **A. tuberculata**

3. Plants entirely covered with tubercle-based hairs; spikelets at most 5-6 mm long .... **A. nervosa**

1. Upper lemma devoid of setae, with a single awn:
   4. Spikelets not more than 2 mm long; blades elliptic and lanceolate:
      5. Blades elliptic; annual; panicle effuse .... **A. pumila**
      5. Blades lanceolate; perennial; panicle cylindric ....**A. bengalensis**
   4. Spikelets more than 2 mm long, not as above:
      6. Perennial, glabrous-sparsely scabrid grasses:
         7. Panicle lax, axis glabrous...**A. nepalensis**
         7. Panicle contracted, axis scabrid..**A. purpurea**
      6. Annual, hispid grass .... **A. ciliata**

Arundinella bengalensis (Spreng.) Druce, Rep. Bot. Exchang. Cl. Brit. Isles 605. 1916; Bor, Grass. India 421. 1960. *Panicum bengalensis* Spreng., Syst. Veg. 1: 311. 1825. *Arundinella wallichii* Nees ex Steud., Syn. Pl. Glum. 1: 114. 1854; FBI 7: 75. 1896.

A perennial, tufted grass; culms up to 60 cm high, woody, creeping; nodes slightly bearded. Leaf sheath smooth, scabrid villous; ligule a ridge of hairs; blades 15-35 x 0.8 -1.6 cm, radical, lanceolate, base narrowed, apex acuminate, villous. Panicle 30 cm long, cylindrical; branches up to 5 cm long, lower sometimes whorled; rachis angular, scabrid; spikelets ovoid, 3-4 seriate, secund, subsessile; glumes ovate, acute, 3-5-nerved, hispid; lower lemma oblong, obtuse, 5-nerved; upper lemma ovoid, subacute, nearly glabrous, occasionally awn present.

Occasional in Northern Eastern Ghats. Fl. & Fr.: September-December

Araku valley (VSKP), *NPBK* 10810 (CAL & MH), Araku (VSKP), *NPBK* 10810 (MH), *GVS* 21568 (CAL, MH); Taprang forest (KHD), *HFM* 1186 (DD); Sonabeda (KRT), *HFM* 3684 (DD), Jenabil Range, Similipal B.R. (MBJ), *AC* 2322 (BSID); Mahendragiri (GJM), *VNS* 5714 (MH).

INDIA: Assam, Nagaland, Sikkim, West Bengal, Bihar, Odisha, Andhra Pradesh.

WORLD: SE Asia.

Arundinella ciliata (Roxb.) Nees ex Miq., Nieuwe Verh. Konink. Nederl. Inst. 3. Pt. 4. 30. 1851; Bor, Grass. India 421. 1960; Matthew, Fl. Tamilnadu Carnatic 3: 1806. 1983. *Holcus ciliatus* Roxb., Fl. Ind.1. 321. 1820. *Arundinella agrostoides* Hook. f., Fl. Brit. India 7: 71.1896. *A. holcoides* auct. non (Kunth) Trin.: Fischer 3: 1801. 1934; Saxena & Brahmam, Fl. Orissa 4: 2259. 1996.

An annual, hispid grass; culms up to 50 cm high, tufted; nodes bearded. Leaf sheaths 4-7.5 cm, pilose; ligule a ridge of hairs; blade 9-10 x 0.6-1 cm, linear-lanceolate, base cordate, apex acute, hispid. Panicle contracted, up to 25 cm long; spikelets 0.2-0.25 cm, pedicelled; glumes tuberculate-pilose, 5-nerved, lower awned; lower lemma ovate-lanceolate, 3-nerved, paleated; palea ovate-lanceolate; upper lemma elliptic, usually female; palea oblong; awn to 4 mm; stamens 3. Caryopsis ellipsoid.

Occasional on hill slopes, rocky places and dry habitats in Niyamgiri hill in Odisha, Visakhapatnam hills, Horsley hills in Andhra Pradesh and southern Eastern Ghats of Tamil Nadu. Fl. & Fr.: September - December. Tam.: *Masapul*.

Horsley hills (CTR), *PVP* 3497, *CECF* 55422 (MH & CAL), *S.India Flora* 15485 (MH); Cherukonda (VSKP), *GVS* 42745 (MH); Balmadies estate–Yercaud (SLM), *KS* 7519 (MH); Deomali hill (KPT), *AC* 4228 (BSID); Niyamagiri hills (KHD), *AC* 2473 (BSID).

INDIA: West Bengal, Bihar, Odisha, Andhra Pradesh.

WORLD: Myanmar, Philippines.

Arundinella nepalensis Trin., Gram, Panic. 61 1826; Bor, Grass. India 423. 1960; Fischer 3: 1801. 1934. *A. brasilensis* Hook. f.; non Raddi; FBI 7: 73. 1896.

A perennial grass; culms up to 80 cm high, creeping, tufted and intricately branched; nodes glabrous. Leaf sheaths 3-5 cm, glabrous; ligule a minute ridge, fringed with hairs; blades 5-15 **x** 0.2 cm, ciliate at the base. Panicle lax, pyramidal, 40 cm long, rachis glabrous; spikelets 0.5-0.8 cm, ovoid-lanceolate, 2-flowered, pedicelled; glumes ovate-lanceolate, 3-nerved, acuminate; lower lemma elliptic, bifid, 5-nerved; upper lemma lanceolate, sparsely bearded at the base, awned; palea narrowly lanceolate; stamens 3; ovary ellipsoid.

Very rare grass in Nallamalais and Similipal Biosphere Reserve. Fl. & Fr.: July–December.

Chalama (KNL), *MPG* 106 (CAL); Chahala, Similipal RF (MBJ), *A.R.K.Sastry* & *G.P.Singh* 12282 (BSID).

INDIA: South & NE India, E to C. Himalaya.

WORLD: Africa, Australia, Pakistan, Bhutan, Nepal, China, Thailand, Vietnam.

Arundinella nervosa (Roxb.) Nees ex Hook. & Arn., Bot. Beechy Voy. 237. 1836; Bor, Grass. India 424. 1960; Moulik, Grass. Bamb. India 59. 1997; Fischer 3: 1801. 1934. *Holcus nervosus* Roxb., Fl. Ind. 1: 320. 1820.

Annuals; culms erect, slender, to 50 cm high; nodes glabrous. Leaf sheaths 2-4 cm, covered with tubercle-based hairs, margins villous; ligules 1 mm, fimbriate; leaf blades 3-8 x 0.5-1 cm, linear-lanceolate, acute-acuminate, rounded at base, covered with tubercle-based hairs. Panicles 5-10 cm; spikelets elliptic-lanceolate, 4-4.5 x 1 cm; lower glume oblong-lanceolate, 4 x1.3 mm, 3-nerved, chartaceous, acuminate; upper glume ovate-lanceolate, 4.5 x 1.3 mm, acuminate, chartaceous, 5-nerved, both the glumes covered with tubercle based hairs. Florets 2, lower floret male, upper floret bisexual; lower lemma elliptic or oblong, 2.8 x 1 mm, 5-nerved, chartaceous, acute; lower palea lanceolate, 2.5 x 0.8 mm, hyaline; stamens 3, anthers 1.2 mm; lodicules 2, 0.3 mm; upper lemma oblong-elliptic, 2 x 1.4 mm, notched, awned with 2 setae, 1.5 mm on either side, subcoriaceous, awn 5 mm, column 2-3 mm; upper palea lanceolate, 1.7 x 0.6 mm, 2-nerved; stamens 3, anthers 1.5 mm; ovary ovate, 0.5 mm, styles 0.4 mm; stigmas 0.5 mm.

Common in grasslands and in open forests in Southern Eastern Ghats. Fl. & Fr.: October–January.

Chinnapalle (ATP), *KRKS* 40372; Kalasamudram (ATP), *KRKS* 40454; Gounivaripalli (ATP), *BR* & *P.Priyadarsini* 37199, 37309 (SKU & BSID); Rachakuntaalli (KDP), *KRKS* 37906; Tirumala hills (CTR), *AMR* 25674; Yerragondapalem (PKM), *BSS* & *SKB* 29252.

INDIA: Peninsular India.

Note: The spikelets of this species are exactly like those of *A. setosa* Trin., except with densely covered tubercle-based hairs.

Arundinella pumila (Hochst.) Steud., Syn. Pl. Glum. 1, 114. 1854; Fischer 3: 1801. 1934; Bor, Grass. India 423. 1960; Matthew, Fl. Tamilnadu Carnatic 3: 1806. 1983. *Acratherum pumilum* Hochst. ex A. Rich., Tent. Fl. Abyss. 2. 414, t.100. 1851. *Arundinella tenella* Nees ex Steud. *loc.cit.* 115; FBI 7:71. 1896.

An annual, slender grass, culms erect, up to 40 cm high, sparsely tuberculate-pilose; nodes sparsely bearded. Leaf sheaths 2-8 cm long, glabrous with ciliate margins; ligule a fringe of hairs; blade 4-12 x 0.5-1.3 cm, ovate-lanceolate, membranous, base narrowed, apex acuminate, veins prominent, with long soft hairs. Panicle effuse, capillary, branched; spikelets 2 mm, pedicelled; glumes ovate-lanceolate, thinly membranous, lower 3-nerved, acute; upper 5-nerved, acuminate; lower lemma ovate, obtuse, paleate; upper lemma lanceolate, awn to 3 mm; stamens 3. Caryopsis narrow, ellipsoid.

Occasional grass in most of the hills. Fl. & Fr.: July-November.

Kalasamudram (ATP), *BR* & *P.Priyadarsini* 37081; Way to Ramanapenta (KNL), *JLE* 22202 (MH & CAL); Bison hills (EG), *CAB* 5037 & 5056 (MH); Near Tiger camp, Maredumilli (EG), *MM* 102580 (BSID); Gudem valley (VSKP), *VNS* 497 (CAL); Sapparlagedda (VSKP), *GVS* 42772 (MH); Labangi, Satkosia Wildlife Sanctuary (AGL), *AC* 2394 (BSID); Dimli, Kotgah Wildlife sanctuary (KHD), *AC* 2421 (BSID); Jacum, Karlapat Wildlife sanctuary (KHD), *AC* 2446, 2447 (BSID).

INDIA: Throughout India.

WORLD: Tropical Africa, Sri Lanka, Nepal, Myanmar, SE Asia.

Arundinella purpurea Hochst. ex Steud., Syn. Pl. Glum. 1, 115. 1854. var. **purpurea;** Bor, Grass. India 42. 1960. *A. fuscata* Hook.f., Fl. Brit. India7: 74. 1896; non Nees ex Buese 1854; Fischer 3: 1802. 1934.

A perennial grass, culms up to 50 cm high, erect, tomentose. Leaf sheaths 4-7 cm long, scabrid panicle contracted, 5 cm long; axis scabrid. Spikelets lanceolate, 4-8 mm long, pedicelled; glumes chartaceous, 3-4-nerved, bristly along the margins; lemma chartaceous; stamens 3. Caryopsis ellipsoid, free within the lemma.

Rare grass is Nallamalais. Fl. & Fr.: August–December.

Nallamalais (KNL), *V.S.Raman* 94235 (MH & CAL).

INDIA: South India.

WORLD: Indo-China.

Arundinella setosa Trin., Gram. Panic. 63. 1826; FBI 7: 70. 1896; Fischer, 3: 1801. 1934; Bor, Grass. India 424. 1960; Matthew, Fl. Tamilnadu Carnatic 3: 1807. 1983.

A perennial, stoloniferous grass; culms up to 90 cm high, glabrous at the base, slender; nodes glabrous, Leaf sheaths 4-9 cm long, loose, tuberculate-pilose; ligule membranous, ciliate at apex; blade 5-12 x 0.4-0.6 cm, base narrow, apex acuminate. Panicle 20 cm, effuse, terminal; spikelets 3.5 mm. pedicelled, 1-2-flowered; glumes oblong, 5-nerved, acuminate, tuberculate-pilose, upper awned; lemma oblong-lanceolate, 3-nerved, both paleated, upper sparsely scaberulous, awn to 1 cm; paleas small, narrowly lanceolate, keeled, obtuse, stamens 3, stigmas plumose. Caryopsis oblong.

Occasional in upper hills of most of the hill ranges. Fl. & Fr.: July-December. Tel.: *Pathi oopagaddi.*

Kekati RF (ATP), *TP& NY* 774; Dharmapuri RF (ATP), *NY* 893; Kalasamudram RF (ATP), *EC* 2911; Gani RF (KNL), *TP* & *RVR* 1774; Gundam - Uruchintala (KNL), *TP& RVR* 3022; Guvvalacheruvu valley near bungalow (KDP), *R.V. Reddy* 8037; Tirumala hills (CTR), *RVR* & *GO* 2430; Chelama (KNL), *JLE* 18057, 22031 (CAL); Mogilikuppa (KDP), *JSG* 21307 (MH); Balapalle (KDP), *S.India Flora* 15412 (MH); Palamaner RF (CTR), *KCJ* 440 (MH); Way to Sileru (VSKP), *JLE* 37171 (MH); Way to Cherukonda peak (VSKP), *GVS* 42739 (MH); Karlapat (KHD), *HFM* 3474 (DD); Mahendragiri (GJP), *SX* & *MB* 3653, 3981 (RRL-B); Similipahar (MBJ), *SX* 3814, *SX* & *MB* 4251, 4804 (RRL-B); Tulka section, Satkoia Tiger Reserve, *KCM* 5279 (BSID); Sirumalai hills (DGL), *SK* 415 (SGH); Banigocha, Satkosia Tiger Reserve (AGL), *KCM* 8355 (BSID); Jenabil Range, Similipal B.R. (MBJ), *AC* 2338 (BSID); Mahendragiri hills (GJP), *AC* 4208, 4211, 4218 (BSID); Kambakkam hills (CPT), *sine coll. s.n.* (MH).

INDIA: Bihar, Odisha, Andhra Pradesh, Western Himalaya, Khasia hills, Central India, Nilgiri.

WORLD: Nepal, SE Asia, China, Malesia.

Note: This grass can be distinguished from other species of *Arundinella* by its spikelets being glabrous and length 4 mm.

Arundinella tuberculata Munro ex Lisboa, J. Bombay Nat. Hist. Soc. 5: 344. 1891; FBI 7: 69. 1896; Bor, Grass. India 425. 1960; Moulik, Grass. Bamb. India 1: 61. 1997.

Annuals, culms erect or decumbent at the base, to 120 cm; nodes glabrous. Leaf sheaths 4-12, covered with tubercle based hairs, margins villous. Ligules a ridge of hairs. Panicles 15-35 cm; spikelets ovoid, acute, 10-12 x 1-1.5 mm; lower glume oblong-lanceolate, 4-4.5 x 1-1.5 mm, 3-nerved, chartaceous, acuminate, covered with tubercle based hairs; upper glume ovate, acute, 3-3.5 x 1-1.5 mm, awn 3 mm,

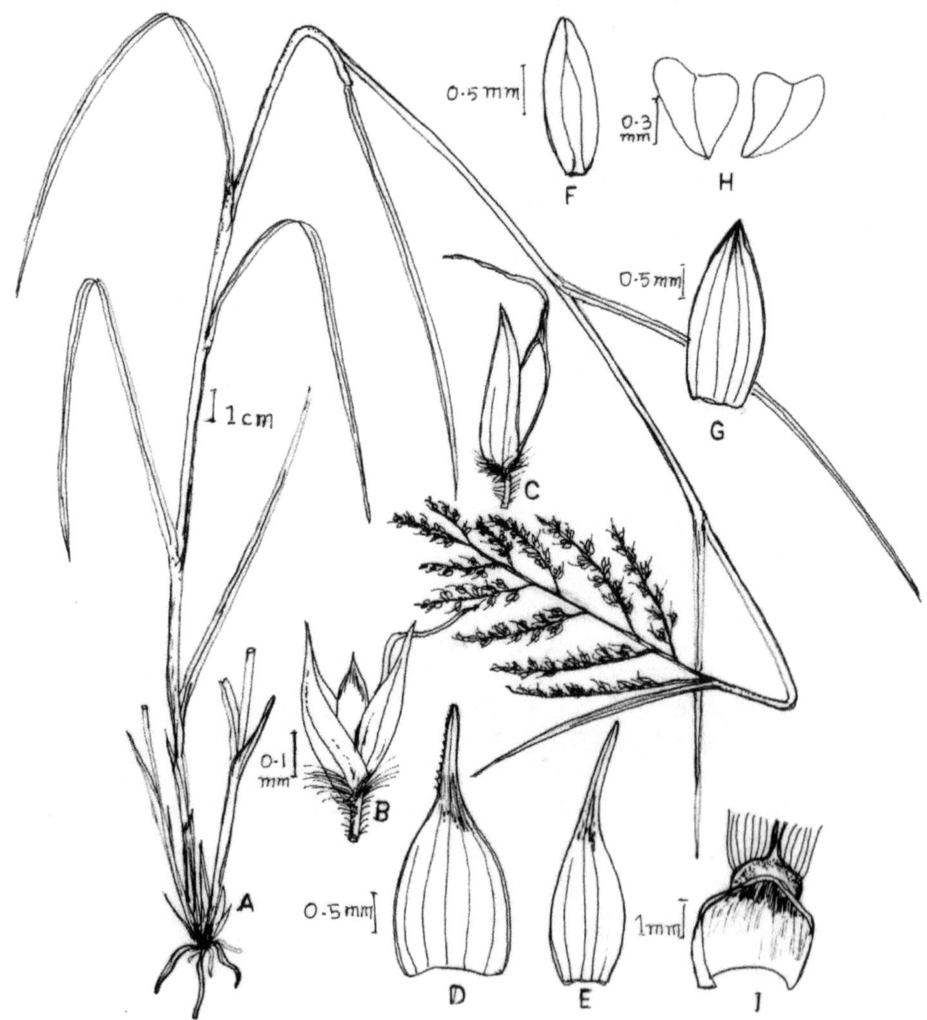

Figure 7. **Arundinella setosa** Trin.

A. Plant, B. Spikelet, C. Empty spikelet, D. Upper glume, E. Lower glume, F. Upper lemma, G. Lower lemma, H. lodicules, I. Collar.

5-nerved, covered with tubercle based hairs. Florets 2; lower floret male or neuter, upper bisexual; lower lemma, 3-3.5 x 1-1.5 mm, elliptic-oblong, 5-nerved, 2-keeled, glabrous, subcoriaceous; lower palea oblong-lanceolate, 2.5 x 0.7 mm, hyaline, 2-keeled; upper lemma lanceolate, acute, 2.2 x 0.7 mm, notched, awned with 2 setae, 1.7 mm on either side, awn 10.5 mm, subcrustaceous when ripe, brown, 2-keeled, keels scabrid; upper palea lanceolate, 2.5 x 0.7 mm, 2-nerved, 2-keeled, hyaline; stamens 3, filament 0.3 mm, anthers 0.9 x 0.2 mm; ovary 0.2 x 0.1 mm, styles 0.5 mm, stigmas 0.5 mm; lodicules 2, 0.1 x 0.1 mm, ovate, hyaline.

Occasional in drier localities of Southern Eastern Ghats (Sadasivaiah *et al.*, 2008). Fl. & Fr.: October–February.

Batrepalle RF (ATP), *BR* & *KVS* 29193; Batrepalli (ATP), *BR* & *P.Priyadarsini* 37062 (SKU & BSID); Talakona (CTR), *BSS & SKB* 30562; Samrajya konda (ATP), *KVS & MSB* 29192; Vatavarlapalli (MBNR), *BSS* & *SKB* 30573 (SKU).

INDIA: Peninsular & C. India. Endemic to India.

ARUNDO Linnaeus

Arundo donax L., Sp. Pl. 81. 1753; Bor, Grass. India 413. 1960; FBI 7: 302. 1806; Fischer 3: 1805. 1934; Matthew, Fl. Tamilnadu Carnatic 3: 1807. 1983.

A tall, perennial, grass; culms stout, up to 5 m high, creeping below, erect, yellow; many-noded, glabrous. Leaf sheath 2-7 cm long, tightly covered the internodes, glabrous; ligule ciliate; blades 17-26 x 1-2.5 cm, distichous, linear-lanceolate, base cordate, apex long tapering to a very fine point, broad, glabrous. Panicle up to 30 cm long, erect, large decompound, branches fascicled, with silky hairs; spikelets laterally compressed, oblong-lanceolate, 0.7-1.2 cm, 3-flowered, rachilla jointed below the floret, glabrous; glumes linear-lanceolate, obtuse, 1-nerved; lemma lanceolate, pilose, nerveless; stamens 3, stigmas plumose. Caryopsis oblong.

Common in plains usually near the water courses. Fl. & Fr.: July–December. Tam.: *Mudampul, Naanal; Naanathattai.*

Balapalle RF (KDP), *PVP* 3449; Gandi temple (KDP), *KRKS* & *BSS* 39069; Kailasakona (CTR), *PVP* 2143; Nellore (NLR), *MSG* 12203; Near Amudalakona (CTR), *GVS* 4587 (MH); Kailasakona (CTR), *DRC* 1721 (MH); Palanka (PKM), *J.Swamy, LR* & *SNR* 7533 (BSID); Tiger camp, Maredumilli (EG), *MM* 105034 (MH); Paderu - Narsipatnam (VSKP), *GVS* 47360 (CAL, MH); Sirumalai hills (DGL), *SKS* 508 (SGH); Bhuvanagiri to Parangipettai (SA), *KRM* 53568 (MH).

INDIA: Punjab, Odisha, Andhra Pradesh.

WORLD: Mediterranean region eastwards to Myanmar, N. Africa, Pakistan.

AXONOPUS P. Beauvois

Axonopus compressus (Sw.) P. Beauv., Ess. Agrost. 12(154): 167. 1812; Bor, Grass. India 278. 1960; Matthew, Fl. Tamilnadu Carnatic 3: 1808. 1983. *Milium compresseum* Sw., Nov. Gen. Pl. 24. 1788.

An annual grass, culms up to 25 cm high, decumbent, spread on the floor like mat; nodes densely bearded with white hairs. Leaf sheaths compressed, 4-7 cm,

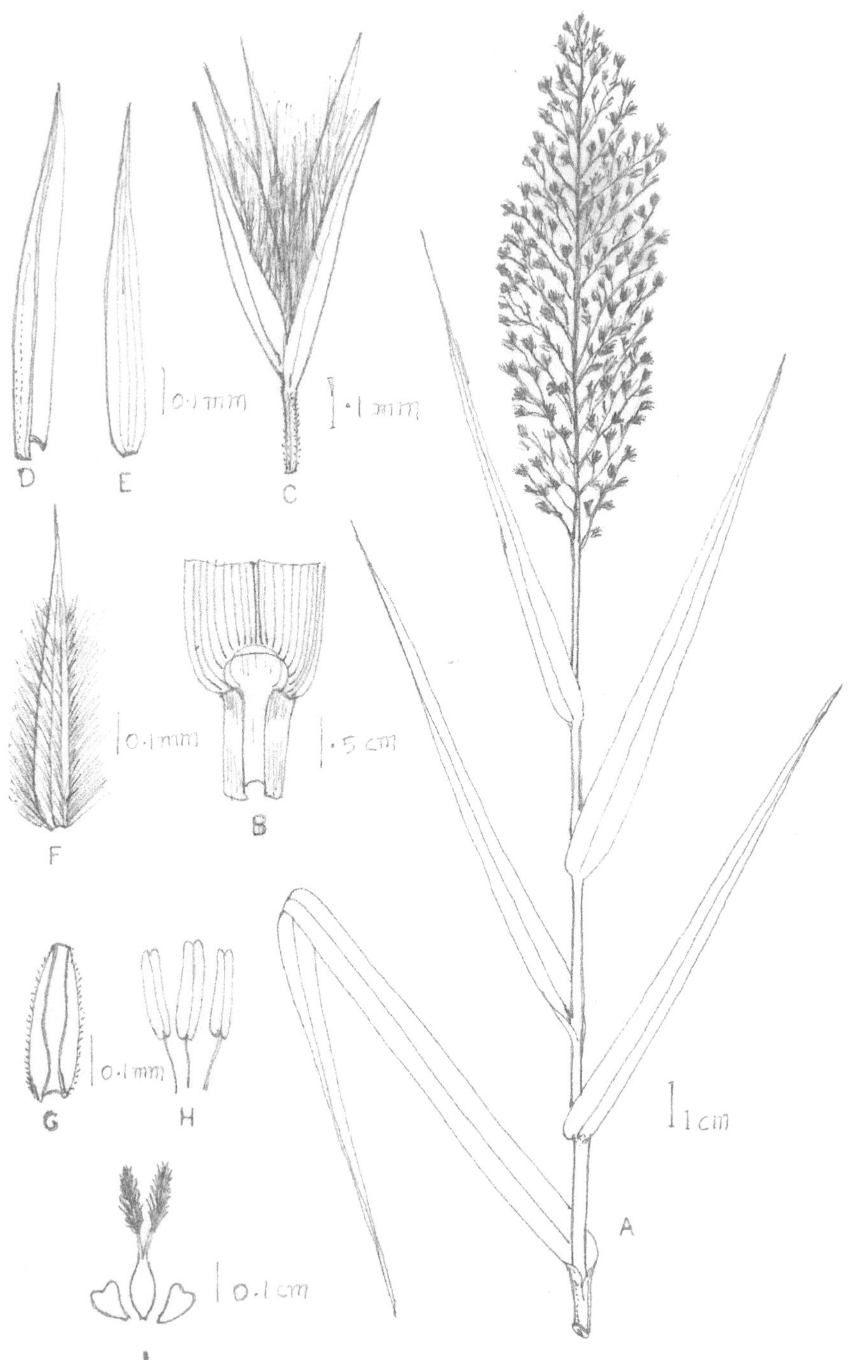

Figure 8. **Arundo donax** L.

A. Culm, B. Collar, C. Spikelet, D. Upper glume, E. Lower glume, F. Lemma, G. Palea, H. Stamens, I. Ovary with lodicules.

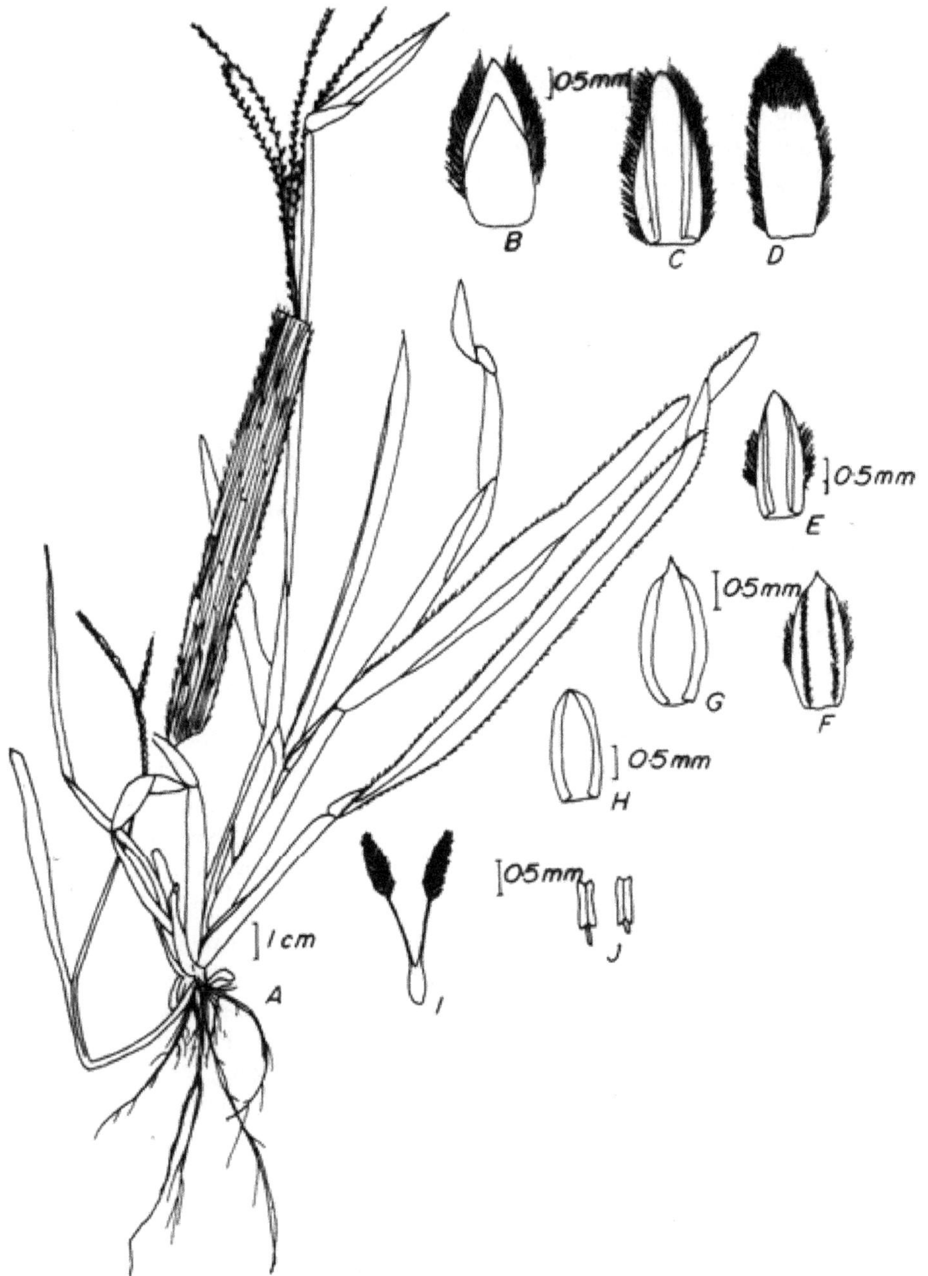

Figure 9. **Axonopus compressus** P. Beauv.

A. Twig, B. Spikelet, C & D. Upper glume front and back views, E & F. Lower lemma front and back views, G. Upper lemma, H. Palea, I. Pistil, J. Stamens.

keeled, glabrous; ligule membranous; blades 5-20 x 0.8-1.5 cm, oblong, flat, rarely folded, base rounded, margin with tuberculate based hairs, apex obtuse-acute, prominently nerved, glabrous. Spikes up to 12 cm long, sub digitate; rachis flexuous; spikelets 2.5 mm, oblong, secund, alternate, deciduous. Lower glume absent; upper glume oblong, obtuse, hairy on the outer surface, nerves obscure; lower lemma oblong, acute, hairy on the outer surface, nerves obscure, membranous; upper lemma oblong, scaberulous, apiculate at apex, coriaceous; palea oblong, glabrous; stamens 3, stigmas plumose. Caryopsis obovoid-ellipsoid.

In moist areas, along the road sides and water courses. Fl. & Fr: October-March

Tirumala hills (CTR), *MSG* 12168; Talakona (CTR), *BR* & *P.Priyadarsini* 30533; Maredumilli (EG), *TP* & *MSG* 12264, *MM* 102514 (MH); Ramandrug (Ballari), *S.India Flora* 15935 (MH).

INDIA: Introduced and naturalized.

WORLD: Native of Tropical America, widely introduced elsewhere.

BAMBUSA Schreber *nom. cons.*

1. A large, thorny shrub; ovary glabrous .... **B. bambos**
1. A large, thorn-less shrub; ovary hairy:
    2. Caespitose; palea not ciliate; anthers obtuse .... **B. tulda**
    2. Culms solitary; palea long-ciliate; anther apiculate .... **B. nutans**

Bambusa bambos (L.) Voss, Vilm. Blumengartn. ed. 3 1: 1189. 1895. *Arundo bambos* L., Sp. Pl. 1: 81. 1753. *Bambusa arundinacea* (Retz.) Willd., Sp. Pl. ed. 4. 2: 245. 1799; FBI 7: 395. 1896; Fischer 3: 1859. 1934; Matthew, Fl. Tamilnadu Carnatic 2: 1808. 1983; Saxena & Brahmam, Fl. Orissa 4: 2265. 1996; Moulik, Grass. Bamb. India 1: 22. 1997. *Bambos arundinacea* Retz., Observ. Bot. 5: 24. 1788.

A large thorny, tufted shrub; culms up to 30 m high, stout, erect, internodes hollow; nodes prominent, the lower producing horizontal, armed with 2-3 stout, recurved spines; leaf sheaths 1- 4 cm, glabrous; ligule narrow, entire or fringed with pale hairs; blade 10-15 x 1-15 cm, lanceolate, appressed hairy, base rounded, margin ciliate-scabrous, apex acuminate, midrib narrow, nerves 4-6. Panicles compound, sub-secund, recemosely arranged spikes; spikes 1-3 cm long, in loose clusters, lanceolate, acute, glabrous; glumes 2 or 0, ovate-lanceolate. acute, many-nerved; lemmas 3-7, the upper most 1-3 male or neuter, 3-nerved, obtuse; palea truncate, with 2 ciliate keels; stamens 3; style short. Caryopsis elliptic, dark.

Common in all hill ranges. Fl. & Fr.: Very infrequent. Tel.: *Mulla veduru*; Tam.: *Mungil*.

Lankamalai RF (KDP), *SRSR* 13176; Peddamantanala (PKM), *RVK* 15973; Mellavagu (GNT), *VRK* 5738; Bhupatipalem (EG), *TP & MSG* 12249; Siglabylu (KDP), *JSG* 20860 (CAL & DD); Mamandur (CTR), *DRC* 1291; Talakona (CTR), *DRC* 1303; Tirumala hills (CTR), *DRC* 1356; Kanupartha (NLR), *JSG* 21813 (DD).

INDIA: Throughout India, except Himalayas and Indo-Gangetic plains.

WORLD: Myanmar, Sri Lanka.

Bambusa nutans Wall. ex Munro, Trans. Linn. Soc.London 26: 92. 1868; Saxena & Brahmam, Fl. Orissa 4: 2266. 1996.

Culms solitary from a creeping rhizome, 3.7-10 cm diam.; internodes 37-45 cm; culm-sheaths 15-30 cm, densely covered with black deciduous bristles at back; blades triangular-acuminate, with rather small, fimbriate auricles. Leaves linear-lanceolate, covered with pellucid glands, softly downy beneath, margins scaberulous, ending in a twisted scabrid point, base generally obtuse or subacute, not cordate, rarely rounded but attenuate into the distinct 3 mm long petiole, nerves 8-9 each side of the midrib; ligule long, obtuse, hairy. Spikelets in bracteated fascicles, many empty or gemmiparous, not flattened, fertile few, 1.8-2.5 cm long; glumes 2-3, lemmas 3-7 fertile, all ovate, acute, mucronate, many-nerved, puberulous within; palea shorter, ovate, flattened, keels overlapping, long-ciliate; anther tips long apiculate or slightly penicillate; ovary subovoid.

Odisha (Jain *et al.*, J. Bombay Nat. Hist. Soc. 72: 767. 1975).

INDIA: Lower Himalaya, Uttarakhand, Sikkim, Assam, Odisha.

WORLD: Bangladesh.

Bambusa tulda Roxb., Fl. Ind. 2: 193. 1832; Fischer 3: 1859.1934: FBI7: 387. 1896.

A thornless shrub, culms up to 7 m high, 50 cm in diameter; nodes not swollen, internodes 25 cm. white-fringed below the nodes. Culms sheaths 15-16.5 x 15-17 cm, broad below, narrowed upwards, apex rounded or sub-triangular, glabrous-hairy, blades of culms sheath sparingly hairy inside. Leaf sheaths glabrous ending in a smooth callus and oblong, long-fringed hairs; ligule narrow, obscure; blades 15-17 x 0.7-4 cm, base rounded, apex subulate, twisted, puberulous below, glabrescent upside, nerves 6-10 pairs; petioles often hairy. Panicle effuse, leafless, compound; spikelets 2.5 cm long, on spiciform branches; rachilla clavate, fragile, glumes 1-3; lemmas ovate-lanceolate, usually mucronate; palea 2-keeled, apex penicillate; stamens 6; ovary hairy.

Rare in Northern Eastern Ghats. Fl. & Fr.: Very infrequent. Gregarious flowering seen in the year 2012 (Naithani *et al.*, 2012b).

Lambasingi ghat (VSKP), *JSG* 21797 (MH & DD).

INDIA: Assam, West Bengal, Bihar, Odisha, Andhra Pradesh.

WORLD: Bhutan, Bangladesh, Nepal, China, Thailand, Vietnam.

BOTHRIOCHLOA Kuntze

1. Lower racemes shorter than the rachis...**B. bladhii**
1. Lower racemes longer than the rachis:
    2. Lower glume of the sessile spikelets pitted:

3. Stoloniferous grass, nodes sparsely bearded, racemes digitate .... **B. pertusa**

3. Tufted grass, nodes densely bearded, spikes umbellate ...... **B. kuntzeana**

2. Lower glume of the sessile spikelets not pitted .... **B. pseudoischaemum**

Bothriochloa bladhii (Retz.) S.T. Blake, Proc. Roy. Soc. Queensland 80: 62. 1969. *Andropogon bladhii* Retz., Observ. Bot. 2: 27. 1781. *Dichanthium glabrum* (Roxb.) Jain & Deshp., Bull. Bot. Surv. India 20: 134. 1978 (1979) var. *glabrum*; Deshp. in Fasc. Fl. India 15: 13. 1984. *Andropogon glaber* Roxb. Fl. Ind. 1: 267, 1820. *A. intermedius* auct. non R. Br. 1810; FBI 7: 175. 1896. *Amphilophis glabra* (Roxb.) Stapf in Fl. Trop. Afr. 9: 172 1917; Fischer 3: 1732. 1934. *Bothriochloa glabra* (Roxb.) Camus, Ann. Soc. Lyon, 1930; Bor, Grass. India 107. 1960.

A perennial, tufted, coarse grass; culms up to 1 m high; nodes bearded. Leaf sheaths 10-20 cm, glabrous; ligule truncate; blades 15-30 x 0.4-0.6 cm, linear, base truncate, apex acuminate, glabrous, midrib prominent. Panicle 10-15 cm long, narrowly oblong; lower branches shorter than upper. Spikelets 6 mm, 2-nate. *Sessile spikelet:* 3 mm, oblong, bisexual; glumes equal, keeled, hairy margins; lower truncate; upper lanceolate; lower lemma oblong, nerveless, hyaline; upper narrow, awned; stamens 3, stigmas plumose. Caryopsis oblong, slightly compressed. *Pedicelled spikelet:* 3 mm, oblong, male or barren, similar to sessile but awnless.

Occasional in the forests. Fl. & Fr.: July–November.

Peddacheruvu RF (KNL), *PVR* & *RVR* 2180; Pamuleru (EG), *TP* & *MSG* 12270; Rollapenta to Peccheruvu (KNL), *JLE* 42267 (MH); Pulusumamella (KNL), *NRR* & *VBH* 83728 (BSID); Diguvametta (PKM), *JLE* 42179 (MH); Thimmapuram (EG), *CAB* 5092 (MH); Bison hills (EG), *CAB* 5092 (MH); Chinturu, Maredumilli (EG), *MM* 102606 (BSID); Valamuru, Maredumilli (EG), *MM* 100867A (BSID); Antarla (VSKP), *JSG* 21798 (DD); Vankachintha (VSKP), *GVS* 42603 (MH); Balagudem (VSKP), *VNS* 4559 (MH); Kuthadya hill (GJM), *VNS* 5884 (MH); Jenabil Range, Similipal B.R. (MBJ), *AC* 2306, 2319, 2332 (BSID); Chahala Tarban Road, on way to Daldali, Similipal Biosphere Reserve (MBJ), *ARKS* & *GPS* 12235 (BSID); Jacum, Karlapat Wildlife Sanctuary (KHD), *AC* 2441 (BSID).

INDIA: Throughout India.

WORLD: Old world Tropics.

Bothriochloa kuntzeana (Hack.) Henr., Blumea 3: 456. 1940; Bor, Grass. India 108. 1960; Moulik, Grass. Bamb. India 1: 266. 1997. *Dichanthium kuntzeana* (Hack.) Jain & Deshp., Bull. Bot. Surv. India 20: 134. 1978 (1979). *Andropogon kuntzeana* Hack. in DC., Monogr. Phan. 6: 478. 1889; FBI 7: 175. 1896. *A. pertusus* (L.) Willd. var. *longifolia* Hack. in DC., Monogr. Phan. 6: 482. 1889; FBI 7: 174. 1896. *Amphilophis kuntzeana* (Hack.) Haines, Bot. Bihar Orissa 1031. 1924; Fischer 3: 1732. 1934.

A perennial, tufted grass; culms up to 30 cm high, terete, branches from base; nodes densely bearded. Leaf sheaths terete, glabrous; ligule membranous; blade 10-

40 x 0.3-1 cm, linear, cauline, margins scaberulous, apex acuminate, glabrous. Spikes corymbose, 30 cm long, densely fascicled; joints and pedicels plumosely ciliate with long hairs. Sessile spikelet: linear-lanceolate, 5 mm, bisexual; callus with long hairs; lower glume membranous, 7-9-nerved, pitted, glabrous; upper glume lanceolate, apex acumimate, margins ciliate; lower lemma lanceolate, glabrous; upper lemma with a geniculate awn; lodicules 2; stamens 3. Pedicelled spikelet: linear-lanceolate, 5 mm, male; lower glume linear, 9-11-nerved, keels ciliate, scabrid at apex; upper glume lanceolate, 5-nerved; lower lemma ovate-oblong, ciliate, nerveless; upper lemma linear, small or some times absent.

Occasional in southern Eastern Ghats. Fl. & Fr.: August–December.

Bhupathayyakona (CTR), *BR* & *P.Priyadarsini* 37323 (SKU & BSID); Nungambakkam (CPT), *sine coll. s.n.* (MH).

INDIA: Peninsular, C. & E. India

WORLD: Nepal.

Bothriochloa pertusa (L.) A. Camus, Ann. Soc. Lyon. 1930, n.s. 76: 164. 1931; Bor, Grass. India 109. 1960; Matthew, Fl. Tamilnadu Carnatic 3: 1809. 1983; Moulik, Grass. Bamb. India 1: 267. 1997. *Dichanthium pertusum* (L.) Clayton, Kew Bull. 32: 4. 1977; Deshp. in Fasc. Fl. India 15: 22. 1984. *Holcus pertusus* L., Mant. Alt. 301. 1771. *Andropogon pertusus* (L.) Willd., Sp. Pl. 4: 922. 1806; FBI 7: 173. 1896. *Amphilophis pertusus* (L.) Nash ex Stapf, Agric. New W. Ind. 15: 179. 1916; Fischer 3: 1731. 1934.

A perennial, stoloniferous grass; culms up to 50 cm high, ascending; nodes sparsely hairy. Leaf sheaths 2-6 cm long, glabrous; ligule membranous; blades 3-10 x 0.2-0.4 cm, linear, base truncate, apex acuminate, glabrous. Racemes 2-6.5 cm long, slender, digitate, silky hairy; spikelets 6 mm, 2-nate, pitted. *Sessile spikelet*: 4 mm, oblong-lanceolate, pitted; callus bearded; lower glume elliptic-oblong, bearded below, dorsally pitted, subchartaceous; upper glume ovate-lanceolate, finely pointed at the tip; lower lemma lanceolate, nerveless, hyaline; upper lemma reduced to a slender awn; stamens 3; stigmas plumose. Caryopsis oblong. *Pedicelled spikelet*: 4 mm, male or barren, similar to sessile spikelet, but awnless.

Common in plains and hills of southern Eastern Ghats. Fl. & Fr.: July–January. Tam.: *Chinnakoraipul*.

Guttur (ATP), *BR* & *P.Priyadarsini* 37151 (BSID); Horsley hills (CTR), *PVP* & *DAM* 4670; Peapuly (KNL), *PVP* 2657 & 3162; Chelama (KNL), *MPG* 103 (DD); Rollapenta (KNL), *JLE* 42266 (MH); Kondaveedu fort (GNT), *CAB* 4700 (MH); Kondapalli (KSN), *CAB* 8174 (MH); Anantagiri (VSKP), *GVS* 19561 (CAL); Yellandu-Karepally (KMM), *RR* 105952 (BSID); Mysamma temple (MBNR), *SRS* 108622 (BSID); Guddaegunda, Koilkonda (MBNR), *SRS* 110769 (BSID); Tippukadu RF (NA), *KRM* 17925 (MH); Mandargiri hill (AGL), *AC* 2364, 2367 (BSID); Tikarpada, Satkosia Wildlife Sanctuary (AGL), *AC* 2402 (BSID); Kapilas hill (Dhenkanal), *Sine coll.* 11226 (BSID); Berhapur (GJM), *VNS* 4720 (MH).

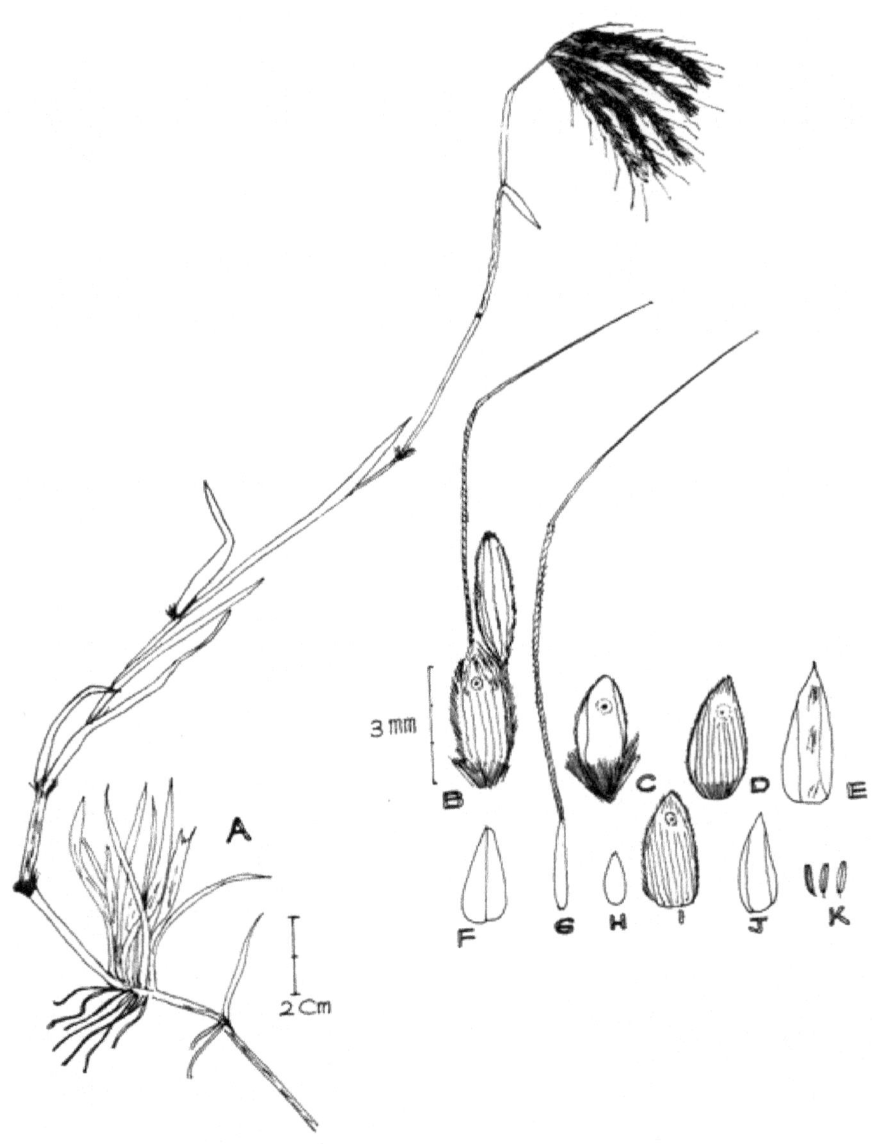

Figure 10. **Bothriochloa pertusa** (L.) A. Camus
[Syn.: *Dichanthium pertusum* (L.) Clayton]

A. Twig, B. Spikelet, C. Lower glume of sessile spikelet front view, D. Back view, E. Upper glume, F. Lower lemma, G. Upper lemma, H. Grain, I. Lower glume or pedicelled spikelet, J. Lemma, K. Stamens.

INDIA: Throughout India.

WORLD: Arabia eastwards to SE Asia, Pakistan, Tropical Africa.

Bothriochloa pseudoischaemum (Nees ex Stued.) Henr., Blumea 3: 457. 1940; Bor, Grass. India 109. 1960; Moulik, Grass. Bamb. India 1: 267. 1997; Matthew, Fl. Tamilnadu Carnatic 3: 1810. 1983. *Dichanthium pseudoischaemum* (Nees ex Steud.) Jain & Deshp., Bull. Bot. Surv. India. 20: 134. 1978 (1979); Deshp. in Fasc. Fl. India 15: 23. 1984. *Andropogon pseudoischaemum* Nees ex Steud. Syn. Pl. Glum. 1: 380. 1854; FBI 7: 172. 1896. *A. ischaemum* sensu Roxb., Fl. Ind. 1: 264. 1820. (non L. 1753). *Amphilophis pseudoischaemum* (Nees ex Steud.) Fischer, Fl. Madras 3: 1731. 1934.

A perennial, tufted grass; culms up to 2 m high, branching from the base; nodes bearded. Leaf sheaths 5-15 cm long, hispid at mouth; ligule membranous, ciliate; blades 15-30 x 0.5-1 cm, linear, base rounded, apex acuminate, scaberulous. Panicle of 5-10 racemes, 8 cm long, joints and pedicels pilose; spikelets 1 cm, 2-nate. *Sessile spikelet:* 5 mm, elliptic-lanceolate, bsexual; lower glume elliptic-lanceolate, silky, 2-keeled, subchartaceous, bidentate, 5-7-nerved; upper glume lanceolate, apex aristate, 3-nerved; lower lemma lanceolate, nerveless, hyaline; upper lemma reduced to an awn; stamens 3; stigmas plumose. Caryopsis oblong, compressed. *Pedicelled spikelet:* 5 mm, elliptic-lanceolate, male or barren, similar to sessile spikelet, but awnless.

Common on hills of southern Eastern Ghats. Fl. & Fr.: July–September.

Pennahobilam (ATP), *TP* & *NY* 367; Kalasamudram RF (ATP), *EC* 2871; Alurkona (ATP), *PVP* 3492; Dhone (KNL), *PVP* 2699; Peapuly (KNL), *PVP* & *RVR* 2664; Guvvalacheruvu (KDP), *R.V.Reddy* 7992, *KS* 6399 (CAL); Palakonda hills (KDP), *CS* & *DAM* 7700 & *PVP* 3434; Talakona RF (CTR), *PVR* 2321; Papavinasanam (CTR), *PVP* 3695; Tirumala hills (CTR), *MSG* 12165; Loyapalle (GNT), *VRK* 3863; Kothakota (ATP), *JSG* 21157 (DD); Bairani falls, Diguvametta (PKM), *JLE* 32467 (MH); Chinthuru, Maredumilli (EG), *MM* 102606 (MH); Hogainakkal (SLM), *KCJ* 17980 (MH), *EV* 20655 (MH).

INDIA: South India.

WORLD: Sri Lanka.

BRACHIARIA (Trinius) Grisebach

1. Spikelets solitary, scattered on effused panicle .... **B. deflexa**
1. Spikelets paired, grouped on short racemes:
    2. Spikelets imbricate or approximate:
        3. Rachis flat, margins scaberulous:
            4. Tufted grass:
                5. Annual grass .... **B. subquadripara**
                5. Perennial grass .... **B. mutica**

4. Stoloniferous grass .... **B. distachya**

3. Rachis trigonous or terete:

6. Panicle linear, racemes closely appressed to the rachis .... **B. eruciformis**

6. Panicle oblong-lanceolate, branches erect:

7. Spikelets up to 2.5 cm long .... **B. reptans**

7. Spikelets 3-5 mm long:

8. Annual .... **B. ramosa**

8. Perennial:

9. Culms not robust; spikelets turgid, apiculate at apex .... **B. brizantha**

9. Culms robust; spikelets not turgid, acute at apex .... **B. semiverticillata**

2. Spikelets distant:

10. Spikelets up to 2.5 mm; leaf margin semi-undulate .... **B. semiundulata**

10. Spikelets above 2.5 mm; leaf margin not undulate:

11. Blades narrowly linear, base narrow, sheaths glabrous; spikelets numerous .... **B. remota**

11. Blades linear–lanceolate, base rounded and subcordate, sheaths ciliate; spikelets few .... **B. kurzii**

Brachiaria brizantha (Hochst. ex A. Rich) Stapf in Prain, Fl. Trop. Afr. 9: 531. 1919; Bor, Grass. India 281. 1960. *Panicum brizanthum* Hochst. ex A. Rich., Tent. Fl. Abyss. 2: 363. 1851.

Perennial grass, rooting at nodes, culms 50-130 cm high, surface smooth; nodes glabrous. Leaf sheaths 5-8 x 1 cm, densely hairy, margin ciliate, slightly compressed, keeled. Ligule a tuft of hairs or fimbriate, 0.5 mm. Leaf blade 10-25 x 1-1.5 cm, ovate-lanceolate, surface puberulous, base cordate, margin serrate, apex acuminate. Inflorescence pseudopanicle, to 15 cm; racemes 25, 5-8 cm; rachis hairy, triquetrous; spikelets solitary, secund, alternately arranged on the rachis, subsessile, heterogamous, 4-5.5 x 1.5–2 mm, turgid, ovate-elliptic, hairy, lower lemma ovate, palea obovate, 2-keeled, chartaceous; stamens 3; upper floret bisexual, upper lemma ovate, membranous, palea ovate-elliptic, 2-keeled, membranous; stamens 3; stigma bifid, ovary elliptic; lodicules 2.

Naturalised in Northern Eastern Ghats (A.M. Reddy *et al.*, 2002). Fl. & Fr.: August–December.

Chintapalli (VSKP), *AMR* & *BR* 24586.

INDIA: Introduced for fodder

WORLD: Native to Africa, widely naturalized throughout the humid and subhumid tropics.

Brachiaria deflexa (Schumach.) C. E. Hubb. ex Robyns, Bull. Jard. Bot. Etat. 9: 181. 1932; Bor, Grass. India 281. 1960. *Panicum deflexum* Schumach., Beskr. Guin. Pl. 63. 1827. *Pseudobrachiaria deflexa* (Schumach.) Launert, Mitt. Bot. Staatssamml. Muenchen 8: 158. 1970.

An annual grass; culms up to 50 cm high, geniculateiy ascending at base, rooting at nodes; nodes glabrous. Leaf sheath compressed, glabrous; ligule membranous; blades 7-15 x 0.3-0.6 cm, linear-lanceolate, base rounded, apex acuminate, glabrous. Racemes up to 8 cm long; spikelets 3 mm, oblong-ellipsoid, distant, rarely in pairs; glumes subequal, half the length of the spikelet.

Occasional in open forests. Fl. & Fr.: July-December.

Kadapa, *KS* 6855 (CAL); Panapakkum (CTR), *MH* 14154 (MH); Ellur (KSN), *CAB* 7905 (MH); Polavaram agency (WG), *DCSR* 330 (CAL).

INDIA: Uttar Pradesh, Maharashtra, Odisha, Andhra Pradesh.

WORLD: Pakistan, Senegal to Yemen and Southwards to South Africa.

Brachiaria distachya (L.) Stapf in Prain, Fl. Trop. Afr. 9: 565. 1919; Fischer 3: 1769. 1934; Bor, Grass. India 285. 1960; Matthew, Fl. Tamilnadu Carnatic 3: 1811. 1983. *Panicum distachyon* L., Mant. Pl. 183. 1771; FBI 7: 37. 1896.

An annual, stoloniferous grass; culms up to 80 cm high, rooting at nodes; nodes glabrous. Leaf sheaths 1.5-3 cm, margins ciliate; ligule a fimbriate membrane; leaf blades 2-4 x 0.3-0.5 cm, linear-narrowly lanceoalate, base cordate, margins cartilaginous, apex acute, pubescent-glabrous. Racemes with 2 or 3 branches, 4 cm long; rachis flat, ribbon like; peduncle hirsute. Spikelets 3 mm, ellipsoid, secund, glabrous; glumes unequal, 5-nerved, lower glume small, ovate; upper oblong, acute, lemma 3-nerved, paleated; lower lemma oblong-ovate, margin incurved; palea oblong, obtuse; upper lemma ovate, coriaceous; palea similar to upper lemma; stamens 3; stigmas plumose. Caryopsis ovoid, compressed.

Common in open forests and cultivated fields. Fl. & Fr.: August- October. Vern.: Tel.: *Koranna gaddi.*

Vemula (KDP), *KRKS* & *BSS* 41951; Papavinasanam (CTR), *DAM* & *PVP* 3681; Rampachodavariun (EG), *TP* & *MSG* 12277; Macherla RF (GNT), *MPG* 128 (CAL); Jaggayyapeta (KSN), *CAB* 8211 (MH); Peddavegi (WG), *VNS* 82390 (MH); Godavari Deepakunka (EG), *GVS* 24550 (MH); Sitanur (CPT), *S.India Flora* 11197 (MH); Tippukadu RF (NA), *KRM* 17690 (MH); Near Melkondai RF (SA), *KRM* 60381 (MH); Mahendragiri way (GJP), *SX* & *MB* Acc. No. 2776 (RRL-B); Similipahar (MBJ), *SX* & *MB* 5198, 5383 (RRL-B).

INDIA: Plains of India.

WORLD: Pakistan, Sri Lanka, Myanmar, Thailand, introduced in Africa.

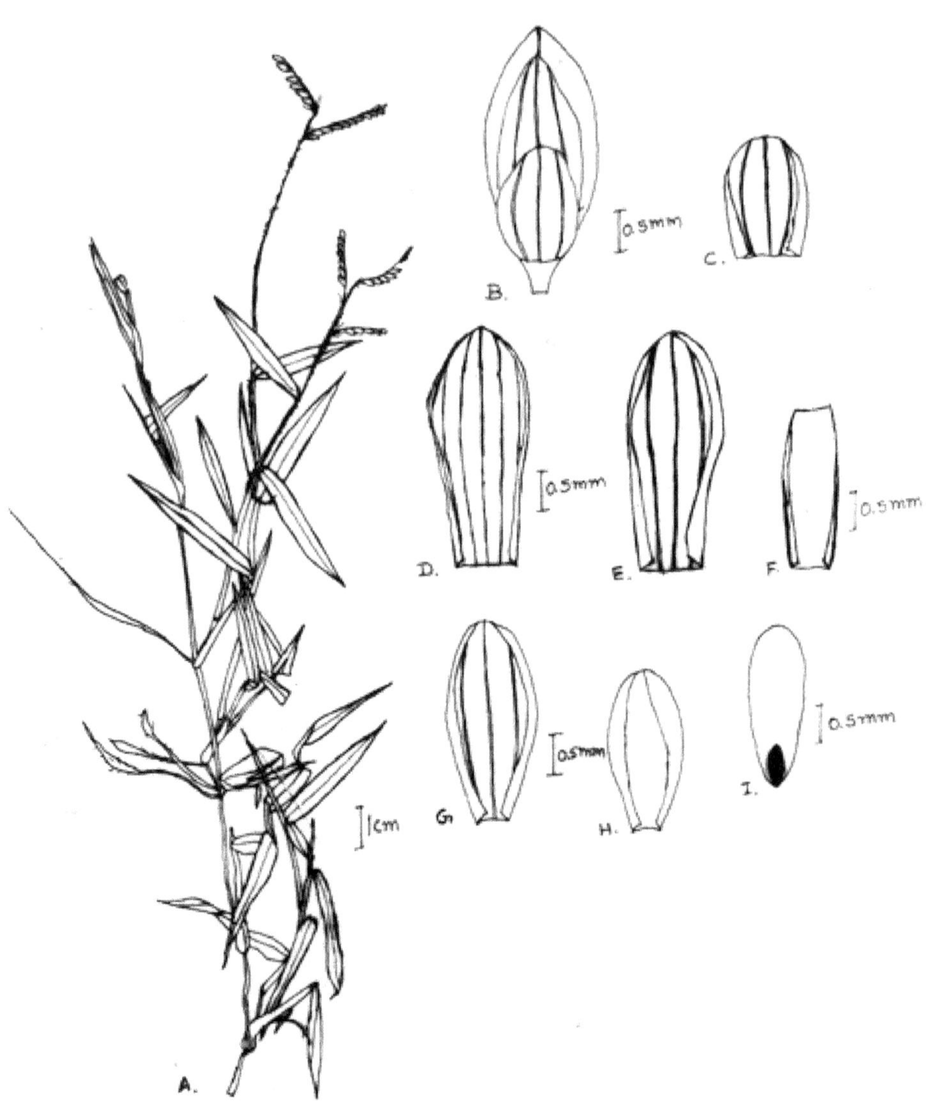

Figure 11. **Brachiaria distachya** Stapf

A. Twig, B. Spikelet, C. Lower glume, D. Upper glume, E. Lower lemma, F. Palea, G. Upper lemma, H. Palea, I. Caryopsis.

Brachiaria eruciformis (Sm.) Griseb. in Ledeb., Fl. Ross. 4. 469. 1856; Bor, Grass. India 283. 1960; Fischer 3: 1769.1934; Matthew, Fl. Tamilnadu Carnatic 3: 1812. 1983. *Panicum eruciformis* Sm. in Sibth & Sm., F1. Graeca 1.44 1806. *P. isachne* Roth ex Roem. & Schult., Syst. Veg. 2.458. 1817; FBI 7: 28. 1896.

An annual grass, culms up to 40 cm high, geniculately ascending; nodes densely pubescent. Leaf sheaths 2-4.5 cm long, slightly pubescent; ligule hairy; blades 1.5-4.5 x 0.2-0.4 cm, lanceolate, base rounded, apex acuminate, glabrous. Panicle 5.5 cm long, subsecund, racemosely arranged spikes; rachis triquetrous, winged; spikelets ellipsoid, 2 mm, densely pubescent, subacute to obtuse; glumes dissimilar, lower a minute scale; upper broadly ovate, densely pubescent, 5-nerved; lower lemma lanceolate, slightly pubescent, 3-nerved; palea truncate; upper lemma crustaceous, smooth, 3-nerved; palea similar to upper lemma, nerveless; stamens 3; stigma plumose. Caryopsis ovoid, 1.5 mm, dark.

Common in fields and moist localities of all hill ranges. Fl. & Fr: August-December. Tel.: *Domakalu gaddi.*

Hottabetta (ATP), *KVS* & *AL* 28070; Penukonda (ATP), *BR* & *P.Priyadarsini* 37172; Ramallakota (KNL), *RVR* 1844; Velpula (KDP), *KRKS* & *KRS* 37679; Sangam (NLR), *PMR* 23050; Tadipatri (ATP), *JSG* 21305 (MH & DD); Yerrabalem (KDP), *PCN* 164 (CAL & DD); Kadapa, *MH* Acc. No. 88501 (MH); Narsaraopet (GNT), *CAB* 4543 (MH); Macherla RF (GNT), *MRS* 307 (DD); Kondapalli (KSN), *CAB* 7973 (MH); Bhimavaram (WG), *CAB* 4347; Anantagiri (VSKP), *GVS* 19612 (CAL); near Gade Gedda (VSKP), *GVS* 42642 (MH); Gajapathinagaram (VJN), *VNS* 4595 (MH); Ramandurg (Ballari), *S.India Flora* 14385 (MH); Kambukudi (NA), *KS* 6544 (MH); Hogainakkal (KGR), *EV* 23546 (MH).

INDIA: Throughout India.

WORLD: India westwards to Spain and N. Africa.

Brachiaria kurzii (Hook. f.) A. Camus in Lecomte, Fl. Indo-Chine 7: 438. 1922. Fischer 3: 1770. 1934; Bor, Grass. India 283. 1960. *Panicum kurzii* Hook. f., Fl. Brit. India 7: 38. 1896.

An annual, tufted grass, culms up to 30 cm long, weak, decumbent. Leaf sheaths ciliate; blades 5-15 x 0.2-0 4 cm long, linear-lanceolate, base rounded or sub-cordate, margin tuberculate based hairs, especially near the base. Racemes spiciform; spikes usually few; spikelets elliptic, 3 mm, distant; lower glume elliptic; upper glume elliptic-oblong, glabrous; lemmas dissimilar; lower elliptic-oblong; palea similar to its lemma, with well developed, incurved flaps below the middle, male; upper lemma crustaceous, margin incurved; palea similar to its lemma, keeled, bisexual; stamens 3; stigmas plumose. Caryopsis elliptic, compressed.

Occasional in Rampa hills. Fl. & Fr.: September–January.

Bhupathayyakona (CTR), *PP* & *BR* 37327; Rampa (EG), *DCSR* 362 (CAL); Ethakonda (EG), *VNS* 109 (CAL).

INDIA: NE India, Uttar Pradesh, Bihar, West Bengal, Odisha, Andhra Pradesh, Tamil Nadu.

WORLD: Bangladesh.

Brachiaria mutica (Forssk.) Stapf in Prain, Fl. Trop. Afr. 9. 526. 1919; Bor, Grass. India 284. 1960; Fischer 3: 1769. 1934; FBI 7: 34. 1896. *Panicum muticum* Forssk., Fl. Aegypt.-Arab. 20. 1775.

A perennial, tufted grass, culms up to 80 cm high, prostrate; nodes bearded. Leaf sheaths 15 cm long, pubescent; leaf blade 12-20 x 0.5-1.5 cm, linear, hirsute, base rounded, margin scabrid, apex acuminate, midrib prominent. Spiciform branched racemes to 20 cm, rachis broadly winged, scaberulous, nodes densely pubescent; spikelets 4 mm, ovate-lanceolate, pedicelled with long hairs; glumes dissimilar; lower ovate, minute; upper ovate, mucronate, 5-nerved; lemma broadly ovate, obtuse, hyaline, 7-nerved; palea elliptic, obtuse, 2-nerved; stamens 3; stigma plumose. Caryopsis elliptic.

Cultivated and run wild. Fl. & Fr.: September–February.

Anantapuram (ATP), *PVP* 2493; Way to Kappaladoruvu (NLR), *MSG* 10896; Gandavaaram (NLR), *PMR* 22397; Atchampeta (GNT), *VRK* 5875; Tikarpada section, Satkosia Tiger Reserve (AGL), *KCM* 6166 (BSID).

INDIA: Assam, West Bengal, Odisha, Andhra Pradesh, Tamil Nadu.

WORLD: Tropical regions of Africa and America, introduced into many warm countries.

Brachiaria ramosa (L.) Stapf in Oliver, Fl. Trop. Afr. 9: 542. 1919; Fischer 3: 1770. 1934; Bor, Grass. India 284. 1960; Matthew, Fl. Tamilnadu Carnatic 3: 1812. 1983. *Panicum ramosum* L., Mant. Pl. 29. 1767; FBI 7: 36. 1896.

An annual, tufted grass; culms up to 50 cm high; nodes pubescent. Leaf sheaths 0.4-0.8 cm long, glabrous, margin hairy; ligule membranous with a row of hairs; blades 6-12 x 0.4-1 cm, linear-lanceolate; base rounded, margin scaberulous, apex acuminate, slightly pubescent, densely hairy near the base. Panicle 9-12 cm long; rachis triquetrous, scaberulous; spikelets 0.3 cm, broadly elliptic-oblong, acuminate, pubescent or glabrescent, paired, one sessile, other pedicelled; glumes membranous, acuminate, puberulous; lower glume broadly ovate, 5-nerved, upper ovate-lanceolate, 7-nerved; lower lemma similar to upper glume; upper lemma ovate, sub crustaceous, smooth, 3-nerved; palea similar to upper lemma, nerveless; stamens 3; stigma plumose. Caryopsis ovoid.

Common in cultivated fields and waste lands of all hill ranges. Fl. & Fr.: August–September. Tam.: *Sanampul*.

Pennahobilam (ATP), *TP* & *NY* 369; Kalasamudram (ATP), *BR* & *P.Priyadarsini* 37100 (BSID); Kammavaripalle (KNL), *TP* & *PVP* 3383; Dhone RF (KNL), *PVP* 2693; Guvvalacheruvu (KDP), *PVP* & *GO* 3433, *R.V.Reddy* 8484; Anupu (GNT), *VRK* 3514; Gundlabrahmeswaram (KNL), *JLE* 16909 (CAL); Kadapa, *EKK* 78535 (DD); Duggamakonda (KDP), *KS* 6355 (MH); Panapakkam (CTR), *S.India Flora* 15783

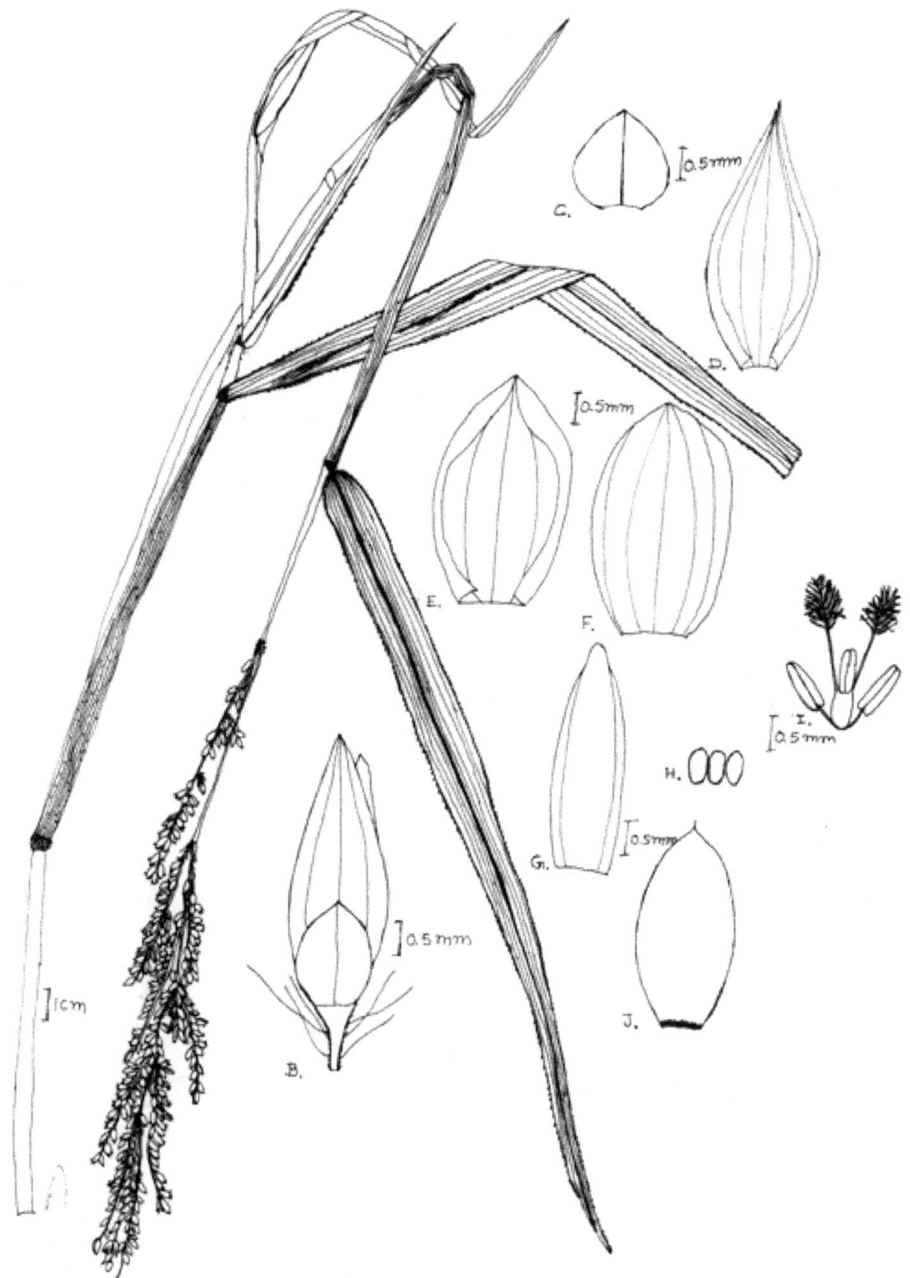

Figure 12. **Brachiaria mutica** (Forssk.) Stapf

A. Twig, B. Spikelet, C. Lower glume, D. Upper glume, E. Lower lemma, F. Upper lemma, G. Palea, H. Lodicules.

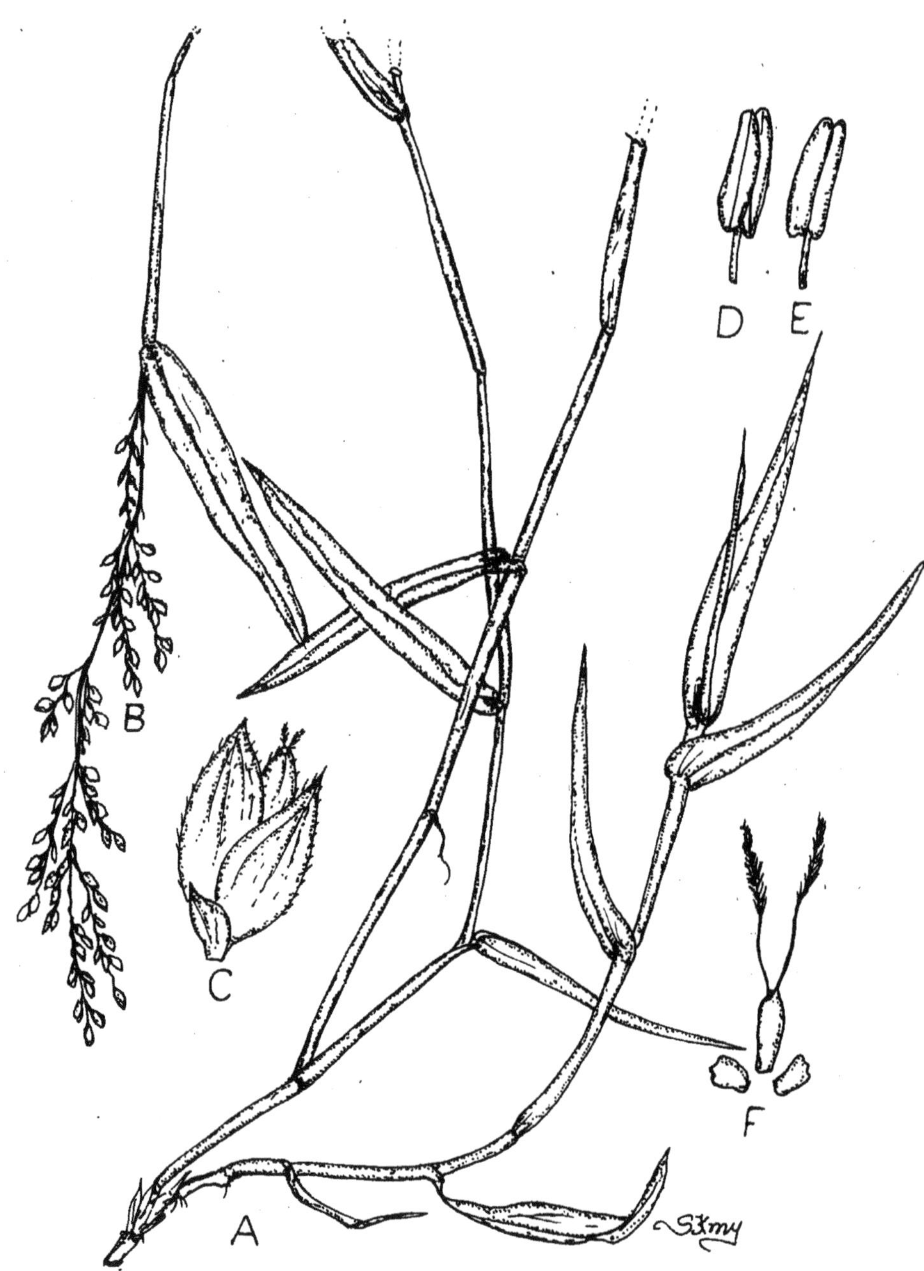

Figure 13. **Brachiaria ramosa** (L.) Stapf.

A. Habit, B. Inflorescence, C. Flower, D & E. Stamnes, F. Pistil & lodicules.

(MH); Kondapalli hills (KSN), *CAB* 7995 (MH); Gokavaram (EG), *CAB* 8258 (MH); Galikonda (VSKP), *GVS* 19586 (CAL); Gajapathinagaram (VJN), *VNS* 4588 (MH); Krishnapuram (VSKP), *CAB* 1991 (MH); Baduga near Parvathipuram (VJN), *VNS* 4532 (MH); Ramanagutta (KMM), *RR* 107979 (BSID); Ramandurg (Ballari), *S.India Flora* 15894 (MH); Chotkei, Purunakote section, Satkosia Tiger Reserve (AGL), *KCM* 5244 (BSID); Tulka, Satkosia Tiger Reserve (AGL), *KCM* 8322 (BSID); Jacum, Karlapat Wildlife Sanctuary (KHD), *AC* 2440 (BSID); Sithanur (CPT), *S.India Flora* 11214 (MH); Kottaimedu (SA), *KRM* 60203 (MH); Bargur (SLM), *S.India Flora* 15511 (MH); Santhamarandahalli (SLM), *S.India Flora* 9754 (MH); Curangi (GJM), *VNS* 5800 (MH).

INDIA: Throughout India.

WORLD: Senegal to Yemen and south-wards to Malawi and S. Africa, Pakistan, Tropical Asia.

Brachiaria remota (Retz.) Haines, Bot. Bihar Orissa 1005. 1924; Fischer 3: 1770.1934; Bor, Grass. India 285. 1960; Matthew, Fl. Tamilnadu Carnatic 3: 1813. 1983. *Panicum remotum* Retz., Observ. Bot. 4: 17. 1786; FBI 7: 38. 1896.

An annual, tufted grass; culms up to 50 cm high, weak, decumbent at base; nodes glabrous. Leaf sheaths 9 cm long, glabrous, margins ciliate towards upwards; ligule membranous; blades 5-10 x 0.3-0.7 cm, linear, base rounded, apex acuminate, glabrous. Panicle open, 10 cm long; branches alternate, spreading, filiform; spikelets 5 mm, ellipsoid, distant, sparsely hairy, secund; glumes membranous, acuminate, lower broadly ovate; upper ovate-lanceolate; lower lemma similar to upper glume; upper lemma subcrustaceous, ovate; palea similar to its lemma; stamens 3; stigmas plumose. Caryopsis ellipsoid.

Occasional along wastelands usually under shade. Fl. & Fr: August-October.

Balapalle RF (KDP), *PVP* 3473; Guvvalacheruvu (KDP), *R.V. Reddy* 8009; Tirumala hills (CTR), *RVR & GO* 2401, *MSG* 12183; Mangapuram (CTR), *RVR & GO* 2447; Penchalakona (NLR), *MSG* 12193; Anupu (GNT), *VRK* 3514; Thummalabailu (KDP), MH 15376 (MH); G. Konduru (KSN), *PV* 5865 (MH); Kondapalli RF (KSN), *P.Venkanna* 5313 (MH); Nathavaram (KSN), *CAB* 1880 (MH); Chatrapur (GJM), *CECF* 1126 (CAL); Santha Marandahalli (SLM), *S.India Flora* 9744 (MH); Pachamalai (TP), *KMS* 6161 (MH).

INDIA: Odisha, Andhra Pradesh, Tamil Nadu.

WORLD: Sri Lanka.

Brachiaria reptans (L.) C. A. Gardner & C.E Hubb. in Hook. Icon. Pl. Sub. t 3363. 1938; Bor, Grass. India 285. 1960; Matthew, Fl. Tamilnadu Carnatic 3: 1814. 1983. *Panicum reptans* L., Syst. Nat. (ed. 10) 870. 1759. *P. prostratum* Lam., Tabl. Encycl. 1: 171. 1791; FBI 7: 33. 1896. *Urochloa reptans* (L.) Stapf in Oliver, Fl. Trop. Afr. 9: 601. 1920; Fischer 3: 1775. 1934.

An annual grass; culms up to 40 cm, prostrate, rooting at the nodes; nodes bearded. Leaf sheaths 2-3 cm long, glabrous; ligule membranous; blades 4-6 x 0.7-1 cm, ovate-lanceolate, base amplexicaul, basally margins ciliate, apically scaberulous, apex acuminate, glabrous, midrib prominent. Panicle of 5-9 racemes, up to 7 cm long, densely crowded spikelets on trigonous rachis; spikelets 0.2 cm, ovoid, paired, alternate, shortly pedicelled; lower glume broadly ovate-sub orbicular, 1-nerved; upper glume apiculate, 7-nerved; lower lemma 5-nerved, paleated, male; upper lemma indurated, apiculate, 3-nerved; palea similar to lemma, bisexual; stamens 3; stigmas plumose. Caryopsis broadly oblong, flattened

Common weed in cultivated and open forests except in the wettest of all hill ranges. Fl.& Fr.: June-September.

Kalasamudram RF (ATP), *PVP* 2615; Gulamallebad RF (KNL), *RVR* 1407; Gani (KNL), *TP & AA* 1939; Mantralayam (KNL), *MSG* 12207; Balarachapalli (KDP), *PVP* 3446; Talakona (CTR), *MSG & AA* 12235; N.S. Right canal (GNT), *VRK* 3551; Bukkapatnam (ATP), *JSG* 20927 (DD); Nandavaram (NLR), *MSR* 1177 (CAL); Palagudem (WG), *VNS* 44960 (MH); Polavaram agency (WG), *DCSR* 160 (CAL); Chinturu, Maredumilli (EG), *MM* 101377 (MH); Anantagiri (VSKP), *MRS* 106 (DD); Gajapathinagaram (VJN), *VNS* 4575 (MH); Poosaripatty (DMP), *P.Venu* 111748 (BSID); Vannipakka Kammarpalayam (CPT), *DN* 134 (MH); Dimli, Kotgarh Wildlife Sanctuary (KDM), *AC* 2424 (BSID); Palakuppam (SA), *KNS* 4038 (MH).

INDIA: Plains of India from Punjab to Assam, Odisha, Tamil Nadu, Andhra Pradesh.

WORLD: Polynesia to northern Australia.

Brachiaria semiundulata (Hochst.) Stapf in Oliver, Fl. Trop. Africa 9: 556. 1919; Fischer 3: 1770. 1934; Bor, Grass. India 285. 1960; Matthew, Fl. Tamilnadu Carnatic 2: 1814. 1983. *Panicum semiundulatum* Hochst. Flora 24: 715. 1841. *P. villosum* Hook.f., Fl. Brit. Ind. 7: 34. 1896. *Urochloa semiundulata* (Hoschst. ex A. Rich.) Ashalatha & V.J. Nair, Bull. Bot. Surv. India 35: 30. 1993.

Tufted annual grass, culms to 30 cm high; nodes glabrous. Leaf sheath 1.5-4.5 cm; ligule to 1 mm; blade ovate, 1.5-3.5 x 0.7-1 cm, villous, cartilaginous at margins, round at base, apex obtuse. Racemes to 4 cm; rachis triquetrous, pilose. Spikelets ovoid, 2 x 0.7 mm, paired, densely villous, shortly stipitate. Glumes hyaline, acute; lower glume to 1 mm, 1-nerved; upper glume to 2 mm, 5-nerved, mucronate; lower lemma similar to upper glume, mucronate; palea to 1.8 mm; upper lemma rugose, to 1.8 mm, glabrous; stamens 3. Caryopsis obovoid.

Slopes of Eastern Ghats of Tamil Nadu. Fl. & Fr.: August–November.

Sirumalai hills (DGL), *SKS* 510 (SGH); *J. Pallithanam* 9660 (RHT, K).

INDIA: Tamil Nadu.

WORLD: Sri Lanka, East Africa.

Brachiaria semiverticillata (Rottler) Alston in Trimen, Handb. Fl. Ceylon 6 (Suppl.): 318. 1931; Fischer 3: 1770. 1934; Bor, Grass. India 286. 1960; Matthew, Fl. Tamilnadu Carnatic 2: 1815. 1983.

Culms erect, 1.5-2 m, stout, to 5 mm wide, glabrous. Leaf sheaths to 5 cm, glabrous or pubescent; ligule to 1 mm; leaf blades oblong, 5-15 x 0.5-1.5 cm, flat or

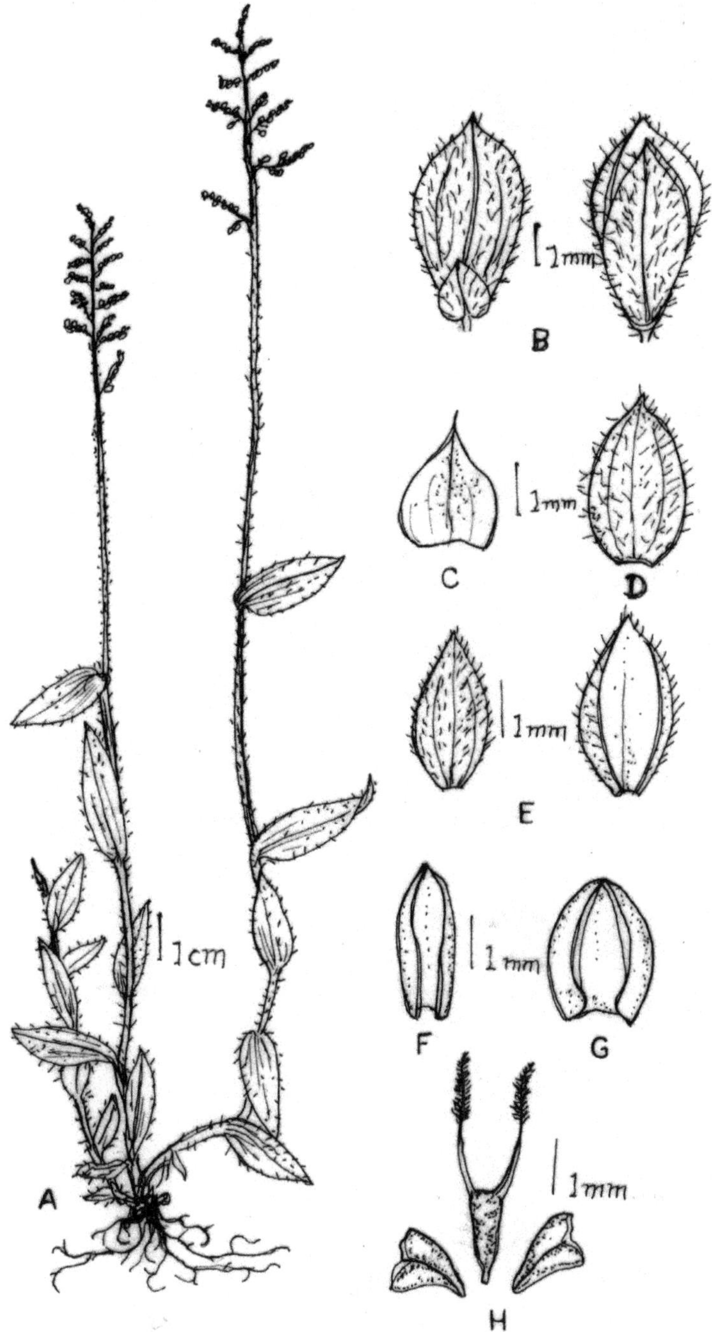

Figure 14. **Brachiaria semisundulata** (Hochst.) Stapf.,

A. Plant, B. Spikelets, C. Upper glume, D. Lower glume, E. Lemma, F. Outer palea, G. Inner palea, H. Ovary with lodicules.

inrolled, pubescent or glabrous. Racemes open, 5-20 cm long; branches spreading, 3-8 cm, alternte, axils at base woolly-hirsute; rachis triquetrous, narrowly winged or not, densely pubescent; pedicel 0.5-3.5 cm, scabrid, scattered-pilose. Spikelets oblong, 3.5 x 1.5 mm, usually paired, solitary at ends, somewhat contiguous, obtuse, appressed to rachis or not, stipe to 0.5 mm. Glumes acute, lower glume to 1.8 mm, 5-nerved; upper glume to 3 mm, 7-nerved; lower lemma similar and equal to upper glume, 5-nerved; upper lemma rugose, to 3 mm; palea to 2.5 mm. Caryopsis obovoid, to 1.8 mm.

Common in foothills on moist slopes in Tamilnadu Carnatic (Matthew, 1983).

Shervarayan hills (SLM), *KMM & NV* 15850, 17429, 17670, 19399, 24417, 24919, 26068 (RHT, K).

INDIA:Tamil Nadu.

WORLD: Sri Lanka.

Brachiaria subquadripara (Trin.) Hitch., Lingnan Sci. J. 7: 214. 1931. *Panicum subquadriparum* Trin., Gram. Panic. 145. 1826. *Brachiaria miliiformis* (J. Presl.) Chase in Contrib. U.S. Natl. Herb. 22: 35. 1920; Fischer 3: 1226. 1934; Bor, Grass. India 283. 1960. *Panicum miliiforme* J. Presl in C. Presl, Reliq. Haenk 1: 300. 1830.

Tufted annual grass, culms 20-50 cm high; nodes glabrescent. Leaf sheath keeled above, ciliate on one margin; ligule membranous with long hairs from apex and back; leaf blade linear-lanceolate, 7-15 cm long, margin serrulate, apex acute to acuminate. Panicles of 2-6 alternating racemes; raceme 3-7 cm long, puberulous below; spikelets obovate to oblanceolate, 3.2–3.5 x ca 1.2 mm; glumes unequal; lower ovate, cupulate, acute to obtuse; upper glume boat-shaped; florets 2, lower barren, upper bisexual; lower lemma ovate to elliptic, chartaceous, faintly 5-nerved, its palea smaller; upper lemma chartaceous-subcoriaceous, acute, 3-nerved, rugulose dorsally, its palea elliptic, rugulose; stamens 3, stigma plumose. Caryopsis obovate to elliptic, greenish.

Rare in moist places in Tirumala hills. Fl. & Fr.: December.

Papavinasanam (CTR), *DAM & PVP* 3681 (SKU).

INDIA: North India, Andhra Pradesh.

WORLD: Mexico, Panama, Tropical Africa, Madgascar, Sri Lanka, Nepal, China, Malaysia, Thailand, Australia, Pacific Islands.

BRIZA Linnaeus

Briza maxima L., Sp. Pl. 70: 1753; FBI 7: 336. 1896; Fischer 3: 1849. 1934; Bor, Grass. India 527. 1960.

Annual tufted grass; culms geniculate, 26-55 cm high, more or less ribbed; nodes brownish. Leaves linear-lanceolate, 11-15 x 0.5 cm, acuminate, midnerve prominent, margins serrulate; ligule truncate to acute or obtuse, 2.6-3.5 mm long, membranous, hyaline. Panicle 5-9 cm, with 3-6 spikelets, pendulous when mature; peduncle more or less ribbed, smooth; rachis sparsely serrulate to barbed. Spikelets broadly ovate-lanceolate, 1-2 x 0.8-1.4 cm, acute at apex, rounded at base, greenish to golden yellowish, margins more or less serrate; pedicels 1.5-2 cm, slender and more or less capillary, barbellate below and smooth near apex. Lower glume broadly ovate to elliptic, boat-shaped, imbricate and shorter than lowest floret, 6 x 3 mm, chartaceous, faintly 7-9-nerved, firmly keeled, greenish to purple brown, purple tinged near apex, margins membranous, hyaline; upper glume 6.3 x 3.5 mm, similar to lower glume. Florets 8-16, closely imbricate; rachilla zigzag, enclosed by basal part of lemma. Lowest lemma broadly ovate-elliptic, boat shaped, 7 x 4.5 mm, 1-keeled, dorsally ciliate on exposed area, other portions glabrous, greenish yellow, at time purple tinged at apex. Lowest palea obovate, 3.5 x 2.5 mm, obtuse to truncate at apex, chartaceous, 2-nerved, 2-keeled, keels winged, ciliate or ciliolate from the base to just below the apex; stamens 3; anthers 1.5 mm long, purplish; filaments 0.4-.1.3 mm long, hyaline; ovary ovate to elliptic, 0.8 -1 x 0.5 mm, acute; style 0.8 mm long, hyaline; stigma plumose, 3.5 mm long, white. Caryopsis obovate to broadly elliptic, acute to subobtuse, 2.2 x 1.3 mm, shallowly concave dorsally, brownish.

Roadsides in hill ranges of high altitudes in Northern Eastern Ghats. Fl. & Fr.: July–March.

Aska (GJM), *sine. coll.* 53824 (MH).

INDIA: Meghalaya, Tamil Nadu, Kerala, Odisha.

WORLD: Native of Mediterranean, widely introduced and becoming naturalized.

Note: We included this taxon based on single old collection from Ganjam probably collected by C.A. Barber. No recent collection anywhere in Eastern Ghats ranges.

CAPILLIPEDIUM Stapf

1. Culms weak, decumbent and trailing, much geniculately branched .... **C. filiculme**
1. Culms erect, robust, branched:
    2. Nodes densely bearded, callus villously bearded; panicles contracteed .... **C. huegelii**
    2. Nodes glabrous or shortly hairy; callus shortly bearded; panicle often somewhat open:
        3. Lower glume of sessile spikelet flat on back; leaves conspicuously narrowed at base .... **C. assimile**
        3. Lower glume of sessile spikelet channeled; leaves slightly narrowed or round at base .... **C. parviflorum**

Capillipedium assimile (Steud.) A. Camus in Lecomte, Fl. Gen. Indoch. 7: 314. 1922; Bor, Grass. India 110. 1960. *Andropogon assimilis* Steud. in Zoll. Syst. Verz. 58. 1854; FBI 7: 18 1896. *Dichanthium assimile* (Steud.) Deshp., Fasc. Fl. India 15: 6. 1984. *Capillipedium glaucopsis* (Steud.) Stapf in Hook., Ic. Pl. t. 3085.

1922; Fischer 3: 1730. 1934. *Andropogon glaucopsis* Steud., Syn. Pl. Glum. 1: 397. 1854. *A. schmidii* Hook.f., Fl. Brit. India 7: 180. 1896.

Erect or decumbent, perennial grass, culms striate, glabrous; nodes bearded. Leaf sheaths striate, glabrous except at apex, ciliate on margins at apex; ligules short, ciliate; leaf blades up to 18.5 x 1 cm, linear, base narrowed, margins scaberulous, apex acuminate, glabrous or hispid on both sides. Panicles up to 14 cm long, lax, decompound; racemes up to 1.4 cm long; sessile spikelets up to 2.5 mm long, narrowly elliptic-oblong; spikelets 2-nate; sessile spikelets up to 2.5 mm long, narrowly elliptic-oblong; callus bearded; lower glume similar in size and shape to that of the spikelets, flat on the back, truncate at apex, margins incurved and ciliate in the upper half, keels ciliate, 2-nerved; upper glume as long as the spikelet, narrowly elliptic-oblong, acuminate, 3-nerved, margins ciliate in the upper half; lower lemma empty, hyaline, lobed, long ciliate at apex, nerveless; palea wanting, upper lemma invariably female and reduced to a hyaline linear stipe passing into a puberulous awn up to 0.9 cm long; palea wanting; lodicules 2, linear-oblong, 2-horned; ovary linear, styles 2, stigmas plumose. Grains linear-oblong, reddish brown, apiculate. Pedicelled spikelets: up to 0.5 cm long, linear-lanceolate, awnless, acute, male; pedicels grooved, long ciliate; lower glume as long as the spikelet, linear-lanceolate, acute, reddish-brown, 7-nerved, keels ciliate; upper glume up to 0.4 cm long, linear-lanceolate, acute, strongly 3-nerved, margins ciliate except in the lower part; lower lemma up to 0.35 cm long, hyaline, linear-oblong, obtuse, ciliate, nerveless, epaleate; lodicules 2; stamens 3.

Occasional in hill forests of northern Eastern Ghats. Fl. & Fr.: August–November.

Bison hills (EG), *CAB* 5090 (MH); Anantagiri (VSKP), *GVS* 19576 (CAL); Ebul RF (VSKP), *VNS* 5741 (CAL); Meghasani, Smilipahar (MBJ), *SX & MB* 4258, 4259 (RRL-B); Mahendragiri (GJP), *SX & MB* 3696 (RRL-B); Gandhamardhan (BG), *SX & MB* 5881 (RRL-B); Malayagiri (DK), *MB & Dhal* 8361 (RRL-B); Karlapat (KHD), *MD & Dhal* 7602 (RRL-B); Barakamara, Jeabil, Similipal Biosphere Reserve (MBJ), *ARKS & GPS* 12402 (BSID); Hatigarh, Similipal Bioshpere Reserve (MBJ), *AC* 2351 (BSID); Labangi, Satkosia Wildlife Sanctuary (AGL), *AC* 2397 (BSID); Niyamgiri hill (KHD), *AC* 2470 (BSID).

INDIA: Throughout.

WORLD: Pakistan, Nepal, Bhutan, Bangladesh, Myanmar, Indonesia, China, Japan.

Capillipedium filiculme (Hook. f.) Stapf in Hook. f., Ic. Pl. Sub. tab. 3085. 1922; Fischer 3: 1730. 1934; Bor, Grass. India 111. 1960; Moulik, Grass. Bamb. India 1: 269. 1997. *Andropogon filiculme* Hook. f., Fl. Brit. India 7: 18 1896. *Dichanthium filiculme* (Hook. f.) Jain & Deshp., Bull. Bot. Surv. India 20: 134. 1978 (1979).

A perennial tufted grass; culms up to 1 m high, decumbent or geniculately ascending below, filiform, glabrous; nodes bearded with white hairs. Leaf sheaths 3-10 cm long; ligule membranous; blades 2-10 x 0.4-0.6 cm, lanceolate, base narrow, apex acuminate. Panicle 10 cm long; branches few, filiform, scented; spikelets 3 mm, 2-nate; pedicels 2 mm, filiform. *Sessile spikelet:* 3 mm, oblong, lanceolate, acute,

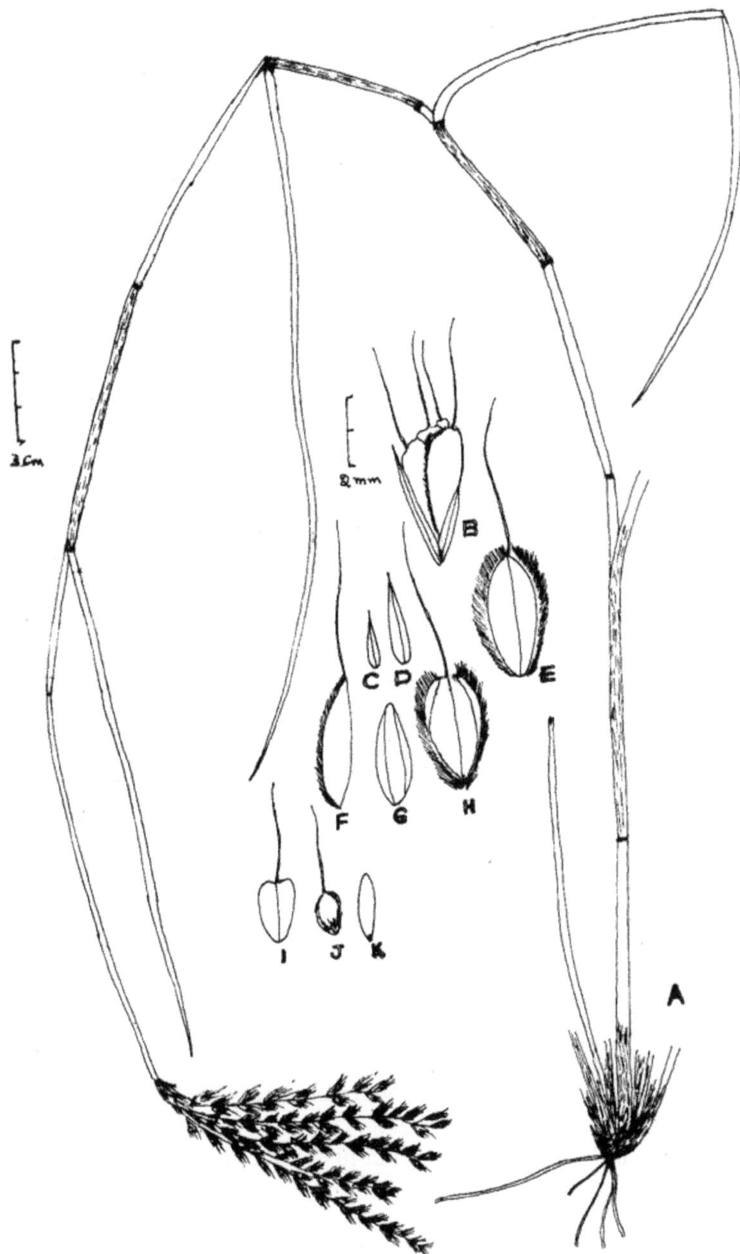

Figure 15. **Capillipedium filiculme** (Hook. f.) Stapf
[Syn.: *Dichanthium filiculme* (Hook. f.) Jain & Deshp.]

A. Twig, B. Spikelets, C. Pedicelled spikelet, D. Lower glume, E. Upper glume, F. Lemma, G. Stamen, H. Sessile spikelet, I & J. Lower glume front and back views, K. Upper glume, L. Lower lemma, M. Upper lemma, N. Pistil, O. Stamens.

sparsely hairy; lower lemma orbicular-ovate, obtuse, hyaline, small; upper lemma reduced to a geniculate awn, hairy; stamens 3, stigmas plumose. *Pedicelled spikelet:* 4 mm, oblong-lanceolate, unawned, male; lower glume oblong-lanceolate, sparsely hairy; upper glume broadly oblong; hairy in the upper portions; lower lemma small, ovate, hyaline; stamens 3.

Occasional in Nallamalais. Fl. & Fr.: August–January.

Ahobilam (KNL), *TP& GNVF* 7059 (SKU).

INDIA: Peninsular India. Endemic.

Capillipedium huegelii (Hack.) Stapf in Hook. Ic. Pl. Sub. tab. 3685 1922; Fischer 3: 1730. 1934; Bor, Grass. India 111. 1960. *Andropogon huegelii* Hack. in DC. Monogr. Phan. 6: 492. 1889; FBI 7: 180. 1896. *Dichanthium huegelii* (Hack.) Jain & Deshp., Bull. Bot. Surv. India 20: 135. 1978 (1979); Deshp. in Fasc. Fl. India 15: 15. 1984.

A perennial, tufted grass; culms up to 1 m high, branching from the base, red; nodes bearded. Leaf sheaths 5-15 cm long, ciliate at mouth, glabrous; ligule membranous; leaf blades 10-25 x 0.6-1 cm, linear, flat, base narrow, apex acuminate, glabrous. Panicle 9 cm long; branches capillary with bearded axis; spikelets 6 mm, 2-nate. *Sessile spikelet:* 3 mm, oblong-ovate, bisexual; lower glume ovate, villous below the middle, a dorsal pit, 11-nerved, chartaceous; upper glume boat-shaped; lower lemma nerveless, hyaline; upper lemma stipiform, awned; stamens 3; stigmas plumose. Caryopsis ovate, slightly compressed. *Pedicelled spikelet:* 3 mm, male or barren, similar to sessile spikelet, but awnless.

Occasional near water courses and streams in Nallamalais and Shevaroys. Fl. & Fr.: August- December.

Peddacheruvu RF (KNL), *PVP* & *RVR* 2177; Pachagundala (KNL), *BSS* & *SKB* 29278; Chelema (KNL), *JLE* 17949 (CAL & MH); Diguvametta (PKM), *JLE* 32545 (MH, CAL & BSID); Manjakuttai, Yercaud (SLM), *DBD* 31410 (MH).

INDIA: Central India southwards.

WORLD: Indian Subcontinent, Myanmar.

Note: Bor (*l.c.*) states that *C. huegelii* (Hack.) A. Camus and *C. assimile* (Steud.) A. Camus are conspecific. The characters are quite overlapping and distinguished characters are feeble. In Flora of Orissa, Saxena and Brahmam (1996) and in Grasses of Tamil Nadu Althaf and V.J. Nair (2009) treated *C. huegelii* as a synonym under *C. assimile.*

Capillipedium parviflorum (R.Br.) Stapf in Prain, Fl. Trop. Africa 9: 169. 1917; Fischer 3: 1893. 1935; Bor, Grass. India 112. 1960; Saxena & Brahmam, Fl. Orissa 4: 2283. 1996. *Holcus parviflorus* R.Br., Prodr. 199. 1810. *Andropogon micranthus* Kunth, Revis. Gramin. 1: 165. 1829; FBI 7: 178. 1896.

Perennial, erect, tufted grass, 0.6-1.35 m high, simple or sparingly branched, nodes bearded. Leaf sheath hirsute or hairy, ligule short, truncate, leaf blade linear, 15-30 cm x 4.3-6.2 mm, flat, apex setaceous acuminate, glabrous, pubescent or hirsute particularly just above the ligule, base slightly narrowed or rounded, margins rough. Panicle decompound, effuse, purple, oblong or narrowly pyramidal, 5-20 cm long; racemes reduced to one sessile and 2 pedicelled spikelets; sessile spikelet linear-oblong to elliptic-oblong, 2.8-4 mm long, usually dark purple; lower glume channeled along the middle on the back, minutey truncate; upper glume sub-3-keeled, also with inflexed margins; lower lemma short, obtuse, nerveless; upper lemma with an awn of 1.2-1.8 cm long. Pedicelled spikelet usually male; pedicel hairy, channeled; lower glume 5-7-nerved, scarcely keeled, incurved; upper glume with incurved villous margins, 3-5-nerved; lower lemma broadly oblong, upper lemma linear, hyaline.

Frequent in hill forest in Northern Eastern Ghats in Odisha. Fl. & Fr.: September–January.

Gandhamardhan, *SX & MB* 6150 (RRL-B); Mahendragiri (GJP), *SX & MB* 3640, 3652, 3886 (RRL-B); Niyamgiri hill (KHD), *AC* 2472 (BSID).

CENCHRUS Linnaeus

1. Involucral bristles retrorsely barbellate, rigid, spinose; lower glume absent .... **C. biflorus**
1. Involucral bristles antrorsely barebellate, flexuous, not spinose; lower glume present:
    2. Bristles connate at base only .... **C. ciliaris**
    2. Bristles connate into a cup .... **C. setigerus**

Cenchrus biflorus Roxb., Fl. Ind. 1: 238. 1820; FBI 7: 89. 1896; Bor, Grass. India 287. 1960; Matthew, Fl. Tamilnadu Carnatic 3: 1816. 1983. *C. barbatus* Schumach., Beskr. Guin. Pl. 43. 1827; Fischer 3: 1749. 1934. *C. catharticus* Del., Cat. Hort. Monsp. 1838; FBI 7: 90. 1896.

An annual, erect or ascending grass; culms up to 30 cm high; nodes glabrous. Leaf sheaths 1.5-5.5 cm long, glabrous, keeled, ciliate near the mouth; ligule pubescent ring; blade 4-9 x 0.2- 0.5 cm, linear-lanceolate, base rounded, apex acuminate, glabrous. Racemes 5-10 cm long, solitary, cylindric; upper branches reduced to subesessible clusters, involucral bristles rigid, apex retrorsely barbellate; outer bristles acicular, erect, spreading; inner bristles ciliate at base, grooved at base; spikelets 5 mm, subglobose; lower glume 0; upper glume lanceolate, acute, 3-nerved; lower lemma lanceolate, 5-nerved; palea containing a male floret or empty; upper lemma scaberulous on nerves, palea similar to its lemma, bisexual; stamens 3. Caryopsis oblong, hard.

Common in foot hills of southern Eastern Ghats. Fl. & Fr.: July–Nov.

Muchukota RF (ATP), *PVP* 2560; Gani RF (KNL), *RVR* 1763; Vemula (KDP), *KRKS* & *BSS* 41950; Ingalur (KDP), *BR* & *KVS* 29111; V.P. South (GNT), *VRK* 5810; Gazulapalle (KNL), *KCJ* 83138 (MH); Vinukonda (GNT), *CAB* 4767 (MH).

INDIA: Hotter parts of India.

WORLD: Tropical Africa, Western Asia, Arabia, Australia.

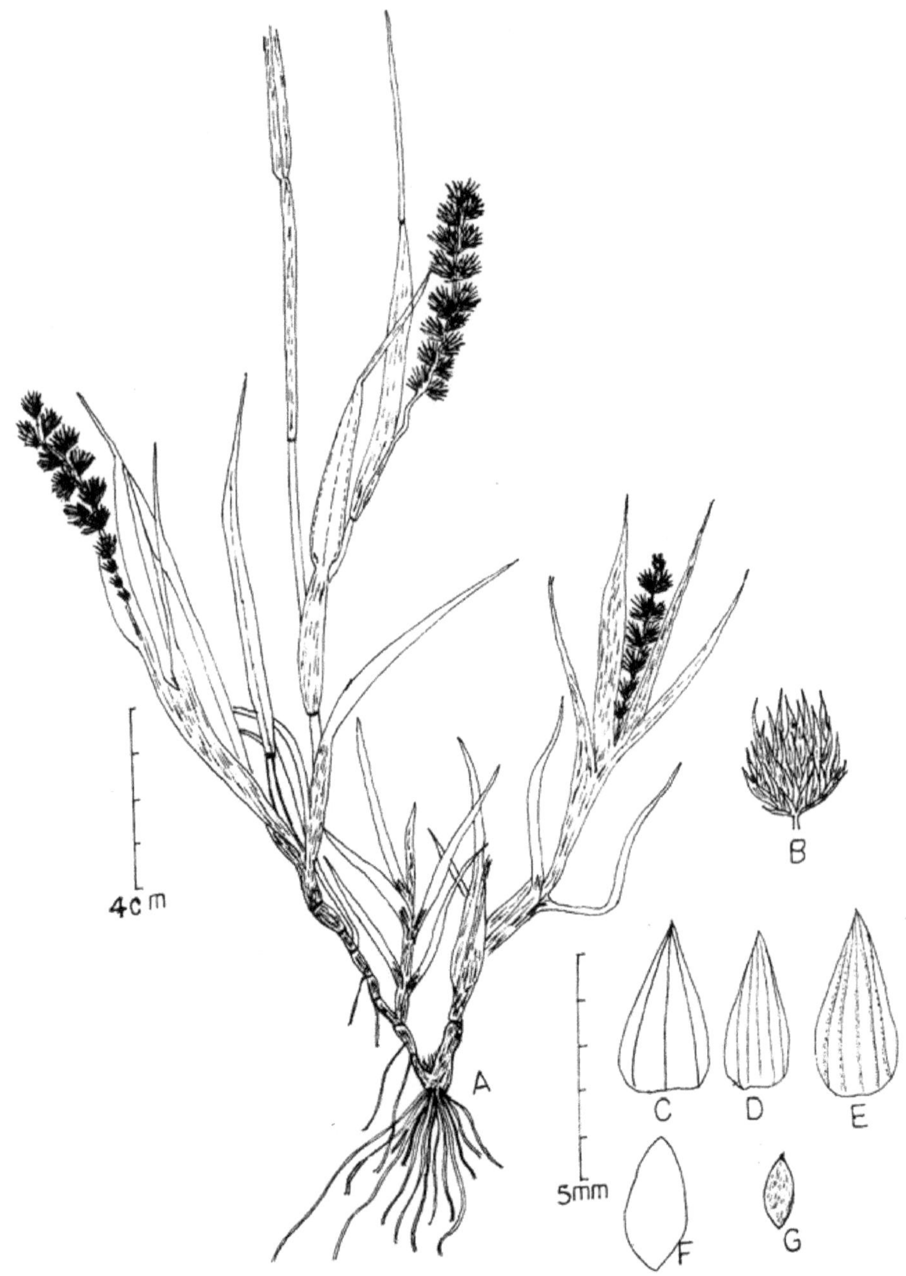

Figure 16. **Cenchrus biflorus** Roxb.

A. Twig, B. Spikelet with still supporting bristles, C. Upper glume, D. Lower lemma, E. Upper lemma, F. Palea, G. Caryopsis.

Cenchrus ciliaris L., Mant. Pl. 302. 1771; Fischer 3: 1749. 1934; Bor, Grass. India 287. 1960; Matthew, Fl. Tamilnadu Carnatic 3: 1816. 1983. *Pennisetum cenchroides* Rich. in Pers., Syn. Pl. 1: 72. 1805, *nom. superfl;* FBI 7: 88. 1896. *P. ciliare* Link., Hort. Berol. 1: 213. 1827.

A perennial, tufted grass; culms up to 50 cm high, erect or decumbent, much branched. Leaf sheaths 4-6 cm long, keeled, glabrous, ciliate near the mouth; ligule a pubescent ring; blade 7-15 x 02-0.5 cm, linear-lanceolate, base rounded, apex acuminate, glabrous-sparsely tuberculate pilose. Racemes 6-10 cm long, solitary, cylindric; involucral bristles antrorsely barbellate; outer filiform, not ciliate, inner flattened, thickened base, densely ciliate, connate below into a ring; spikelets 6 mm, oblong-lanceolate, 1-3 in each involucral; glumes ovate, hyaline, 1-nerved; lemmas ovate-lanceolate, acuminate, 5-nerved; paleas lanceolate, lower usually male, upper bisexual; stamens 3. Caryopsis oblong, greyish brown, apex beaked.

Common in the lower hill ranges. Fl. & Fr.: July–January. Tam.: *Kolukkattaipullu*.

Akuthotapalle (ATP), *PVP* 2634; Gooty fort (ATP), *PVP* 3670; Gooty (ATP), *JSG* 800 (MH) & 20824 (MH & DD); Kailasakona (CTR), *DRC* 1738 (MH); Kotappakonda Sacred Grove (GNT), *J.Swamy* & *S.Nagaraju* 14124 (BSID); Sirumalai hills (DGL), *SKS* 521 (SGH).

INDIA: Hotter parts of India.

WORLD: Africa, SW Asia, Pakistan, China; introduced in America and Australia.

Cenchrus setigerus Vahl, Enum. Pl. 2:395. 1806; Fischer 3: 1794. 1934; Bor, Grasses India 290. 1960. *C. biflorus* Hook.f., Fl. Brit. India 7: 89. 1896, non Roxb. 1820.

Perennial, forming clumps from a bulbous base; culms 5-80 cm high, geniculately ascending. Leaf-blades 2-20 cm long, 2-7 mm wide. Panicle 2-12 cm long; involucres cup-shaped, 3-7 mm long; inner spines short, flattened, connate for a quarter to half their length to form a cup, glabrous or obscurely puberulous and grooved on the face, occasionally sparsely ciliate on the margins, green to dark purple, antrorsely scaberulous, narrowly triangular at the tip; outer spines few, short, often suppressed, rarely almost as long as the inner. Spikelets 1-3 per burr, 3-5 mm long.

Occasional in grasslands in southern Eastern Ghats (B.R.P. Rao, 2008; Lakshmaiah *et al.*, 2012). Fl. & Fr.: August-January.

BT Dam (ATP), *BR* & *KVS* 29179 (SKU).

INDIA: Native to NW India.

WORLD: E and NE Africa, SW Asia, Pakistan and China.

CENTOTHECA Desvaux

Centotheca lappacea (L.) Desv., Nouv. Bull. Soc. Philom. 2: 189. 1810; FBI 332. 1896; Fischer 3: 1848. 1934; Bor, Grass. India 457. 1960; Matthew, Fl. Tamilnadu Carnatic 3: 1817. 1983. *Cenchrus lappaceus* L. Sp. Pl. 1488. 1763.

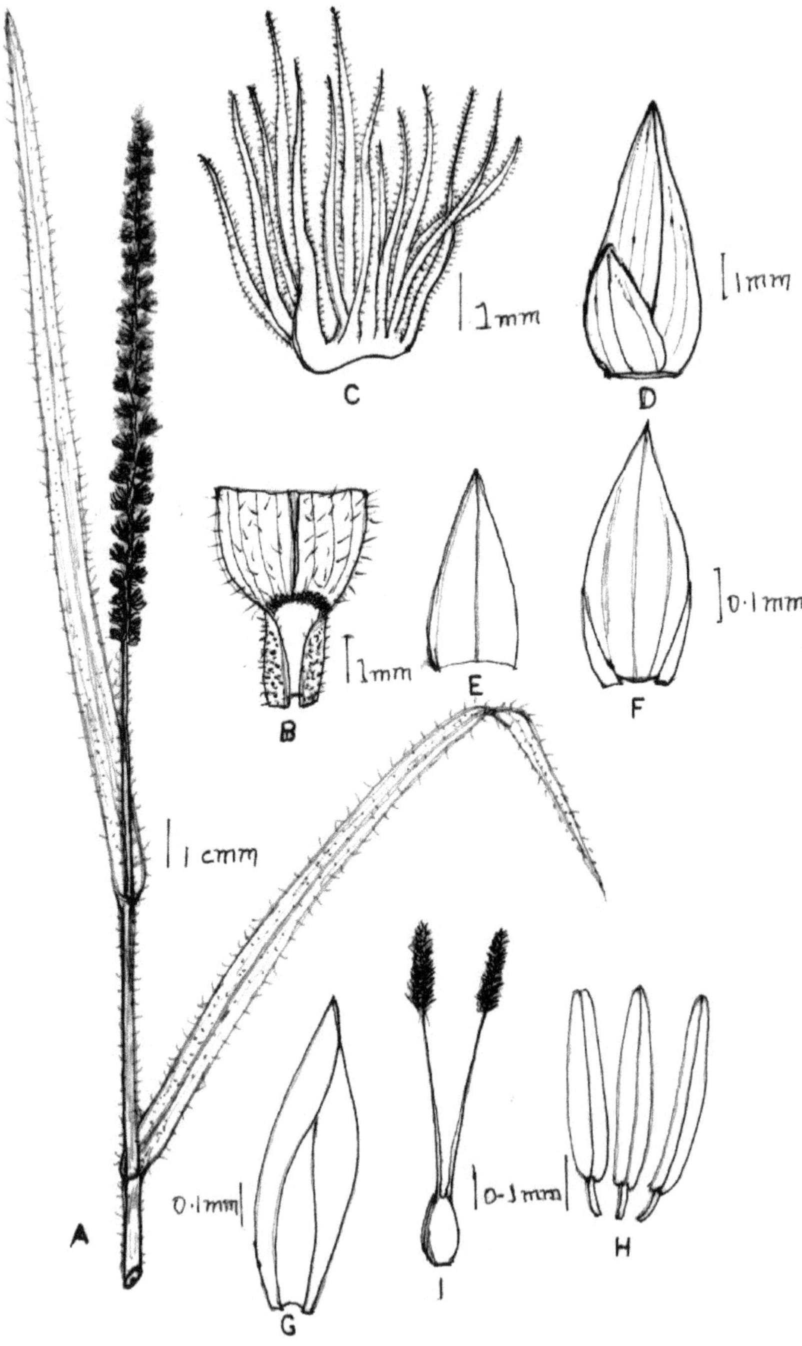

Figure 17. **Cenchrus ciliaris** L.

A. Culm, B. Collar, C. Bract, D. Spikelet, E. Upper glume, F. Lower glume, G. Palea, H. Stamens, I. Ovary.

Figure 18. **Centotheca lappacea** (L.) Desr.

A. Twig, B. Spikelet, C. Lower glume, D. Upper glume, E. Unarmed lower glume, F. Armed upper lemma, G. Palea, H. Caryopsis.

A perennial, tufted grass; culms up to 50 cm high, decumbent; nodes glabrous. Leaf sheaths 3.5-6.5 cm long, more than nodes, loose, glabrous; ligule membranous; blades flat, 5-8.5 x 1.5-2 cm, obovate-elliptic, prominently veined, base narrowed, apex acute. Panicle lax, 10 cm long; spikelets up to 5 cm, secund, laterally compressed, with 2 bisexual florets and one sterile floret; glumes unequal, ovate-lanceolate, persistent; lower lemma oblong, 9-nerved, mucronate; upper lemma oblong, with bulbous based, reflexed bristles on the margins towards apex; palea oblong, 2-keeled; stamens 3. Caryopsis oblong-obvoid.

Occasional in forests, often found in shady places and hill tops. Fl. & Fr.: October - January

Gundlabrahmeswaram (KNL), *PVP & RVP* 2683, *JLE* 23799 (CAL); Tirumala hills (CTR), *PVP* 3687; Maredumilli (EG), *TP & MSG* 12254; Midway of Kailasakona (CTR), *GVS* 46993 (MH & CAL); Isukagundalu, G.B.M. WLS (PKM), *J.Swamy* & *LR* 14418 (BSID); Bison hill (EG), *CAB* 5083 (MH); Near Tiger Camp, Mardumilli (EG), *MM* 102593 (BSID); Gurguria, Similipahar (MBJ), *SX* & *MB* 5205 (RRL-B); Kuludi (DK), *MB* & *Dhal* 8318 (RRL-B); Koinpur (GJM), *SX* 1253 (RRL-B); Mahendragiri (GJP), *SX* & *MB* 3897 (RRL-B); Jenabil Range, Similipal B.R. (MBJ), *AC* 2307 (BSID).

INDIA: Throughout Eastern, Central and Southern India.

WORLD: Bhutan, Nepal, Sri Lanka, Indonesia, Malaysia, Myanmar, Philippines, Thailand, Vietnam,China, Polynesia and Tropical Africa.

CEPHALOSTACHYUM Munro

Cephalostachyum pergracile Munro, Trans. Linn. Soc. London. 26: 141. 1868; Hook.f., Fl. Brit. India 7: 413. 1897; Naithani *et al.*, Indian Forester 137: 528. 2011. *Schizostachyum pergracile* (Munro) R. B. Majumdar in Karthikeyan *et al.*, Fl. Ind. Enum. Monocot. 282. 1989.

A deciduous, arboreous, caespitose bamboo; culms erect, glaucous-green, somewhat white-puberulous below the nodes. 10-13 m high, 5-7.5 cm thick; internodes 30-45 cm long; walls very thin; culm-sheaths much shorter than internodes, 10-15 cm long, 15-20 cm broad, densely covered with black, stiff, deciduous hairs, afterward polished chestnut-brown; blade 5 cm long, ovate, cordate, cuspidate, auricles rounded, edges with long stiff curved white bristles; ligule very narrow, entire. Leaves linear-lanceolate, 15-35 x 2.5-3.25 cm, base rounded or cuneate, petiole 5 mm long; sheaths glabrous. Inflorescence a large panicle with verticels of long drooping, filiform spikes, bearing distant broad heads of spikelets supported by small chaffy sheathing bracts; rhachis very slender, wiry, 3.25-5 cm between the clusters. Spikelets in bracteate clusters, 1.25-1.75 cm long, without glumes, but 1-2 sterile florets, followed by a fertile floret, ending in a terminal sterile floret or filiform, lemma 1.25-1.75 cm long, ovate-lanceolate, many-nerved, densely pale-hairy, mucronate; palea as long as the lemma, 2-keeled, the keels close together, ciliate, apex deeply bifid, mucronate; lodicules 2, narrow, 5 mm long, lanceolate, somewhat obtuse and ciliate at tip, 3-5-nerved; stamens with narrow filaments; anthers purple; ovary smooth, sub-globular at base; style 3- cornered, stigmas 2-3, stout, recurved. Caryopsis obovate-oblong, shining, 1.25 cm long, grooved on one side, ending in a 1.25 cm long beak.

Common in moist and dry mixed deciduous forests of Sukumamidi North and South beats in Khammam district (Naithani *et al.*, 2011). Frequently found near streams associated with *Dendrocalamus strictus.*

INDIA: West and south west India.

WORLD: China, Tropical Asia, Indo-China, Nepal, Myanmar.

CHIONACHNE R. Brown

Chionachne gigantea (J.Koenig) Veldkamp, Blumea 47: 559. 2002. *Coix gigantea* J. Koenig, Naturforscher (Halle) 23: 211. 1788; Fischer 3: 1705. 1934; Bor, Grass. India 264. 1960. *Chionachne koenigii* (Spreng.) Thwaites, Enum. Pl. Zeyl. 357. 1864; Fischer 3:1706. 1934; Bor, Grass. India 263. 1960; Matthew, Fl. Tamilnadu Carnatic 2: 1818. 1983; Saxena & Brahmam, Fl. 4: 2286. 1996. *Coix koenigii* Spreng. in L., Syst. Veg. 1: 239. 1824. *Polytoca barbata* (Roxb.) Stapf ex Hook. f., Fl. Brit. India 7: 102. 1896.

A perennial, tufted grass; culms up to 2 m; nodes bearded. Leaf sheaths 2-6 cm long, glabrous, margin ciliate; ligule membranous; blades oblong-lanceolate, base rounded, margin scaberulous, apex acuminate, glabrous, midrib prominent. Racemes terminal, 5 cm long, umbellate. Female spikelets at base. Male spikelets at the top of the racemes; spathes 3 cm, oblong, apex setose, spatheoles 1.5 cm, open, with small setaceous apex. Male spikelet: 2-flowered, one pedicelled, other sessile, rachis not articulate. Spikelets 8 mm, oblong; lower glume shallowly concave, narrowly winged, 7-8-nerved; upper glume chartaceous, narrowed, 6-nerved; lemmas similar, paleas shortly aristate; stamens 3. Female spikelet: solitary, oblong, 4 mm, dorsiventrally compressed; glumes dissimilar, lower glume obovate, coriaceous, margins incurved; upper ovate, membranous; lemmas hyaline; stigmas plumose. Caryopsis orbicular, compressed.

Common on hill slopes in the forests. Fl. & Fr.: July-December. Vern.: Tel.: *Goluva gaddi.*

Gundlabrahmeswaram (KNL), *PVP & RVR* 2373, *JLE* 16918 (MH); Penukonda (ATP), *S.India Flora.* 14323 (MH); Balapalle (KDP), *JLE* 14949 (CAL), 15739 (MH); Kadapa, *JSG* 11192 (CAL); Diguvametta (PKM), *JLE* 32472 (CAL); Isukagundalu, GBWEL (PKM), *J.Swamy* & *L.Rasingam* 14421 (BSID); Chinnagora (VSKP), *KCJ* 17202 (MH); Chintapalli (VSKP), *J.Venkateswrlu s.n.* (MH); way to Birmisala (VSKP), *GVS* 42660 (MH); Way to Tekupalli (KMM), *PVS* 84052 (BSID); Ragada forest (GJM), *SX* & *MB* 3285 (RRL-B); Similipahar (MBJ), *SX* & *MB* 5127, 5190 (RRL-B); Jenabil Range, Similipal B.R. (MBJ), *AC* 2326 (BSID); Chunabatti, Satkosia Wildlife Sanctuary (AGL), *AC* 2381 (BSID); Katrang, Satkosia Tiger Reserve (AGL), *KCM* 6799 (BSID).

INDIA: Warmer parts of India.

WORLD: Pakistan, Nepal, Sri Lanka, Myanmar, Malesia.

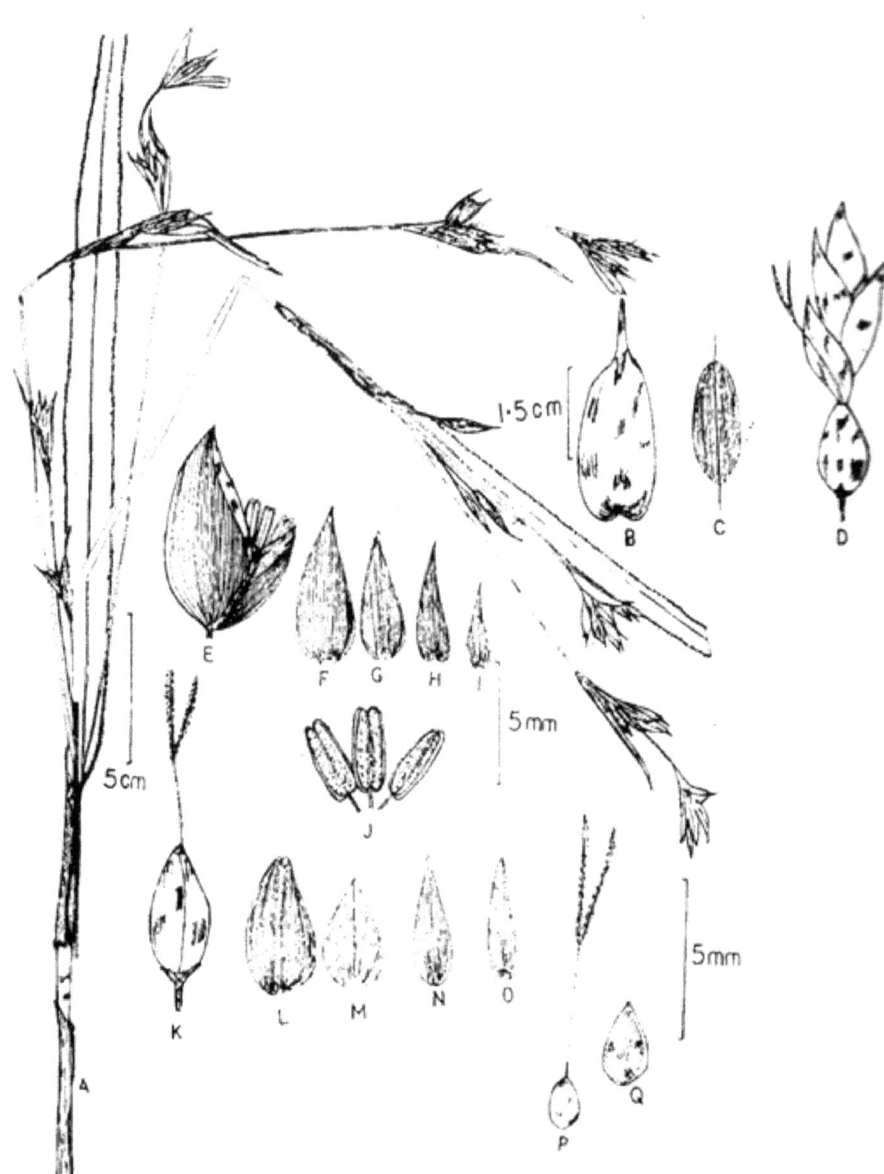

Figure 19. **Chionachne gigantea** (J.Koenig) Veldkamp [Syn.: C. koenigii (Spreng.) Thw.]

A. Twig, B. Spathe, C. Spatheole, D. Male and female spikelet, E. Male spikelets, F. Lower glume, G. Upper glume.

CHLORIS Swartz

1. Spikes racemosely arranged.. **C. roxburghiana**
1. Spikes umbelled:
    2. Perfect florets 2; culms puberulous below the inflorescence........ **C. bournei**
    2. Perfect florets 1; culms glabrous below the inflorescence:
        3. Empty lemma above the fertile floret rudimentary to well developed:
            4. Spikes spreading; lower lemma glabrous......................... **C. pycnothrix**
            4. Spikes closely crowded; lower lemma with spreading hairs:
                5. Lowest lemma with a crown of spreading hairs at the apex, margins shortly ciliate..................................... **C. virgata**
                5. Lowest lemma without a crown of spreading hairs at the apex, margin with long hairs .... **C. gayana**
        3. Empty lemma above the fertile lemma well developed:
            6. Culms puberulous below the inflorescence; spikelets 5-awned .... **C. quinquesetica**
            6. Culms glabrous below the inflorescence; spikelets 3-4-awned:
                7. Sterile lemmas 2, 3-awned, keels of the fertile lemma densely ciliate on the back..................................... **C. barbata**
                7. Sterile lemmas 3, 4-awned, keels of the fertile lemma glabrous on the back .... **C. montana**

Chloris barbata Sw., Fl. Ind. Occid. 1: 200. 1797; FBI 7: 292. 1896; Fischer 3: 1838. 1934; Bor, Grass. India 465. 1960; Matthew, Fl. Tamilnadu Carnatic 3: 1820. 1983.

An annual, tufted grass; culms up to 60 cm high, creeping at base; nodes glabrous. Leaf sheaths 4-9 cm long, glabrous; ligule membranous; blades linear-lanceolate, base rounded, apex acuminate, glabrous. Terminal, digitate spikes 7 cm long; spikelets 3 mm, 3-flowered; glumes linear, densely ciliate on the margin; fertile lemma obovate, densely stiff ciliate on margins, awned; both sterile lemmas boat-shaped, awned; stamens 3, stigmas plumose. Caryopsis oblong, compressed.

Common in plains and forests of all hill ranges. Fl. & Fr.: July–December.

Amagondapalem RF (ATP), *NY* 1052; Kalasamudram (ATP), *PVP* 2613; Peapuly (KNL), *PVP* 2659 & 2688; Gulemallebad RF (KNL), *RVR* 1702; Palakonda hills (KDP), *CS* 8268; Tirumala hills (CTR), *MSG* 12142; BN Kandriga (CTR), *BR & P.Priyadarsini* 37347; Gundlabrahmeswaram (KNL), *JLE* 22066 (CAL); Motakotla tank (KDP), *KS* 7819 (CAL); Macherla RF (GNT), *MPG* 166 (CAL); Kondapalli (KSN), *sine coll. s.n.* (MH); Simhachalam hills (VSKP), *GVS* 19377 (CAL); Avadi (CPT), *sine coll. s.n.* (MH); Veerambakkam (NA), *MBV* 712 (MH); Madanancheri (NA), *MBV* 782 (MH);

Palakuppam (SA), *VNS* 4056 (MH), *CAB* 6073 (MH), Edaikal RF (SA), *KRM* 60391 (MH); Vedambium (SLM), *R. Ansari* 50027 (MH).

INDIA: Plains of India.

WORLD: Pantropical.

Chloris bournei Rang. & Tadul., J. Indian Bot. 2. 189. 1921; Fischer 3: 1838. 1934; Bor, Grass. India 466. 1960.

An annual, tufted grass; culms up to 90 cm high; nodes glabrous. Leaf sheaths 5-10 cm long, glabrous; ligule membranous; blades 3-15 x 0.2-0.4 cm, linear-lanceolate, glabrous. Spikes 3-8, 9 cm long, digitate; peduncle puberulous below the spikes; spikelets 7-8 mm, elliptic, perfect florets 2, 4-awned, empty lemmas 2. Glumes unequal, lower glume elliptic, membranous; upper glume elliptic; fertile lemma obovate, back scaberulous, both margins and keels ciliate, awned; palea narrow; fertile lemma II broader than the first, awned; empty lemmas 2, sub-globose, awned; stamens 3; stigmas plumose.

Rare along agricultural field margins, especially in black cotton soils. Fr. & Fl.: June-November.

Way to Gollalaguduru (KDP), *KRKS* & *BSS* 41962; Yerrabalem (KDP), *PCN* 165 (CAL DD); Macherla RF (GNT), *SCA* 130 (CAL & DD); Nandyal farm (KNL), *KCJ* 83140 (MH): Guntur, *GVN* 16804 (MH); Gidikuppa (EG), *CAB* 5348 (MH); Ammapettai (SA), *VNS* 4171 (MH).

INDIA: South India. Endemic.

Chloris gayana Kunth ex Stapf in Dyer, Fl. Cap. 7: 642. 1900; Bor, Grass. India 466. 1960.

A perennial, stoloniferous, introduced grass; culms up to 1 m high, simple or branched below; nodes glabrous. Leaf blades 15-40 x 0.2-0.8 cm, linear, flat, apex acuminate. Spikes 6-15, 15 cm long, umbellate, sessile, greenish or brownish; spikelets 3-4- flowered, perfect floret one, 3 mm, 2-awned; glumes very unequal; fertile 1st lemma lanceolate, pallid, bidentate ciliate along the marginal nerves and keel, shortly bearded below the tip, awned; upper florets sterile, 2nd lemma projecting from the side of the lowest lemma, lanceolate, scaly-ciliate on the margins, awned; 3rd lemma reduced to a scabrid, oblong or clavate, awnless scale; 4th lemma reduced to a clavate awnless scale or sometimes absent.

Introduced and frequent along the irrigation channels. Fl. & Fr.: August-December.

Amagondapalem RF (ATP), *NY* 1052 (SKU).

INDIA: Introduced.

WORLD: From Senegal eastwards to the Sudan and south to South Africa, Pakistan.

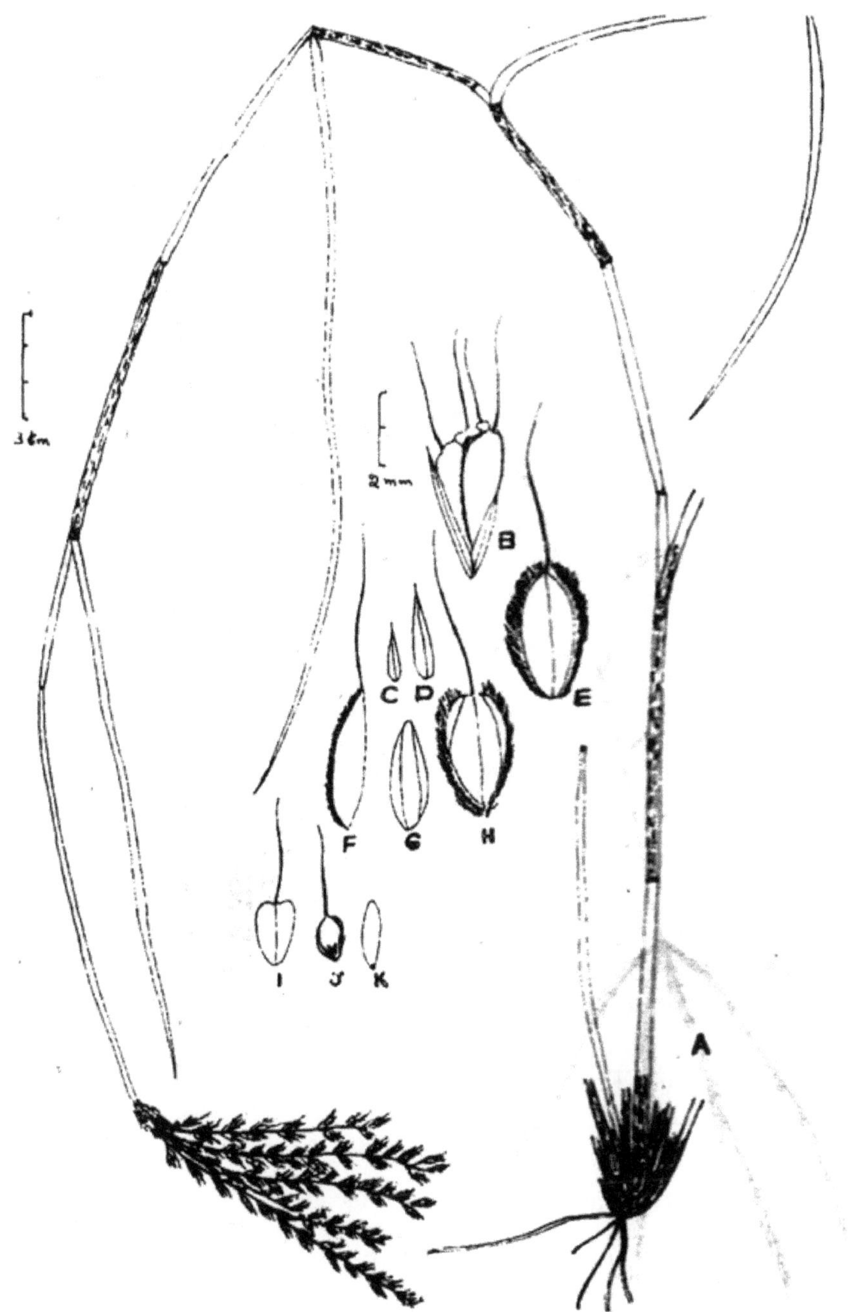

Figure 20. **Chloris bournei** Rang. & Tadul.

A. Twig, B. Spikelet, C. Lower glume, D. Upper glume, E. I Fertile lemma back view, F. Lateral view, G. Palea, H. II Ferile lemma, I. III Sterile lemma, J. IV Sterile lemma, K. Grain.

Chloris montana Roxb., Fl. Ind. 1: 331.1820; FBI 7: 292. 1896; Fischer 3: 1838. 1934; Bor, Grass. India 466. 1960; Matthew, Fl. Tamilnadu Carnatic 3: 1821. 1983.

A perennial, tufted grass; culms up to 80 cm high; nodes glabrous. Leaf sheaths 2-5 cm; ligule membranous, short; blades 3-12 x 0.2-0.4 cm linear, involute. Spikes 4-8, 5-6 cm long. erect, digitate, terminal; spikelets 4 mm, unilateral, 4-awned; lower glume mucronulate; upper glume keeled, aristate, scabrid on keels; fertile lemma I ciliate on margin, glabrous at back, awned: palea narrow; sterile lemma 3, awned. Caryopsis fusiform.

Occasional in Eastern Ghats. Fl. & Fr.: August- January.

Gounivaripalli (ATP), *BR* & *P.Priyadarsini* 37308 (SKU); Motakatla tank (KDP), *KS* 7821 (CAL & MH); Yerrabalem (KDP), *PCN* 167 (CAL); Panappakam (CTR), *S.India Flora* 15788 (MH); Gajapathinagaram (VJN), *VNS* 4579 (MH); Balugam, near Parvathipuram (SKLM), *VNS* 4558 (MH); Soraiyur (NA), *S.V. Parthasarathi* 87074 (MH); Eachankadu–Palakuppam (SA), *VNS* 4087 (MH); Kottaimedu (SA), *KRM* 60199 (MH); Avadi (CPT), *sine coll. s.n.* (MH); Vandalur (CPT), *S.India Flora* 11610 (MH); Kottur near Krishnagiri (SLM), *KCJ* 20104 (MH).

INDIA: Throughout India.

WORLD: Indo-China.

Chloris pycnothrix Trin., Gram. Unifl. 234. 1824; W. Clayton in Hutch. & Dalziel, Fl. W. Trop. Afr. 3(2): 400. 1972; Matthew, Fl. Tamilnadu Carnatic 3: 1821. 1983.

An annual, stoloniferous grass; culms 15-30 cm, creeping, rooting at nodes; nodes glabrous. Leaf sheaths to 6.5 cm; ligule to 1 mm; blades 3.5-7 x 0.2-0.3 cm, apically obtuse, scabrid along margins. Spikes flexuous, 5-6 cm. Spikelets 2.5-3 mm, 2-flowered, 2-awned. Lower glume to 2 mm, mucronate; upper glume to 3 mm, shortly aristate; fertile lemma lanceolate, to 3 mm, scabrid at apical margin, awn 1.3-1.5 cm; palea oblong, 2.5 mm, scabrid along keels; stamens 3; sterile lemma to 0.7 mm, awn to 4.5 mm. Caryopsis linear, to 2 mm.

Common in mosit places of Shervarayan hills of Tamil Nadu. Fl. & Fr.: March–July.

Shervarayan hills, Kaka shola (SM), *SKS* 1328 (SGH); Shervarayan temple (SM), *KMM & JB* 224000 (RHT, K).

INDIA: Tamil Nadu.

WORLD: Africa, Arabia and Brazil.

Chloris quinquesetica Bhide, J. Proc. Asiat. Soc. Beng. n.s. 8. 311. 1912; Bor, Grass. India 466. 1960.

A perennial, stoloniferous grass; culms up to 30 cm high, rooting at lower nodes; nodes glabrous. Leaf sheaths 1.5-3 cm long, keeled, glabrous, ciliate at the mouth; ligule a fimbriate membrane; leaf blades 2-5 x 0.3-0.5 cm, lanceolate, base truncate, apex acute. Spikes 2-4 cm long, crowded in a short peduncled racemose

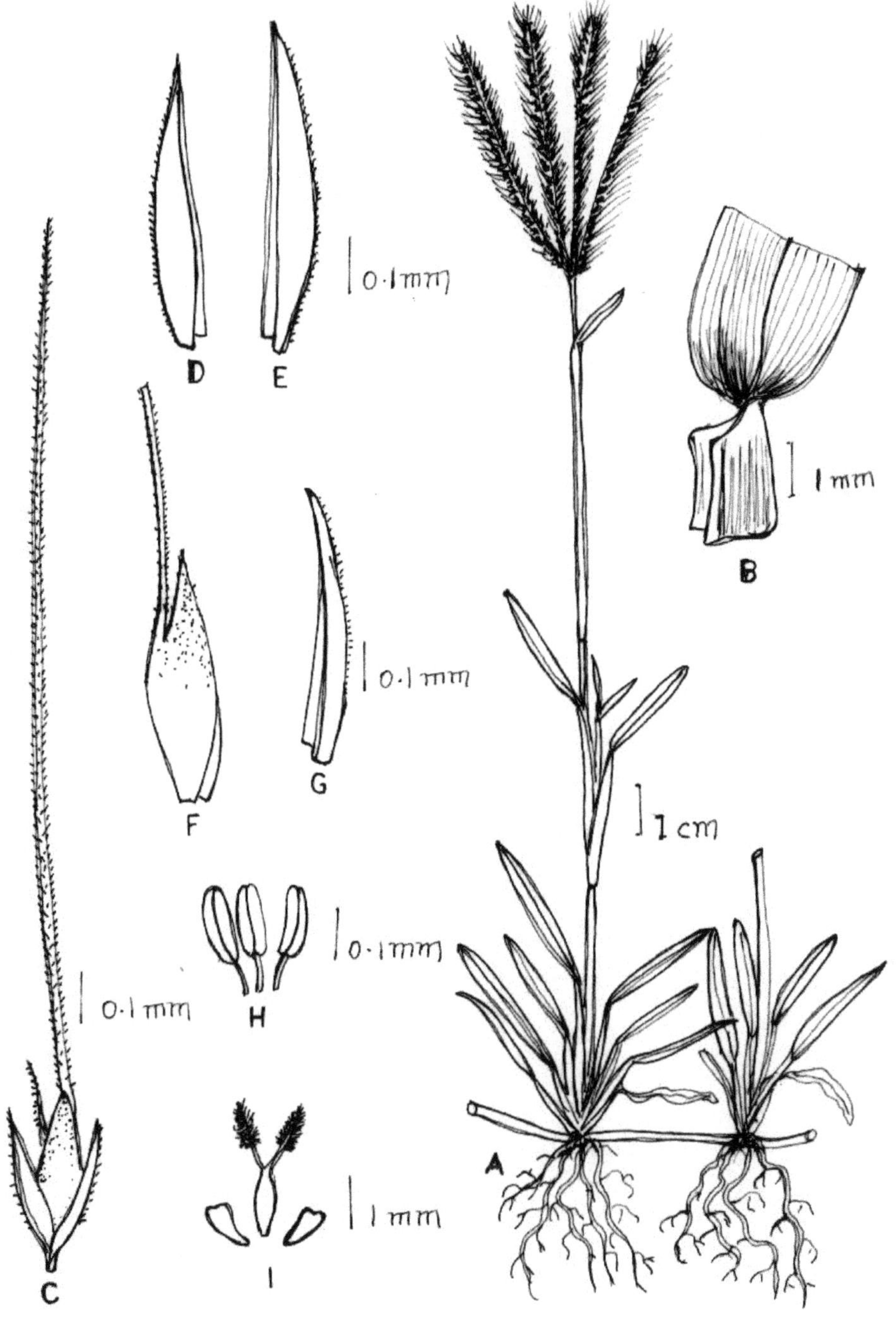

Figure 21. **Chloris pycnothrix** Trin.

A. Plant, B. Collar, C. Spikelet, D. Upper glume, E. Lower glume, F. Upper lemma, G. Lower lemma, H. Stamens, I. Ovary with lodicules.

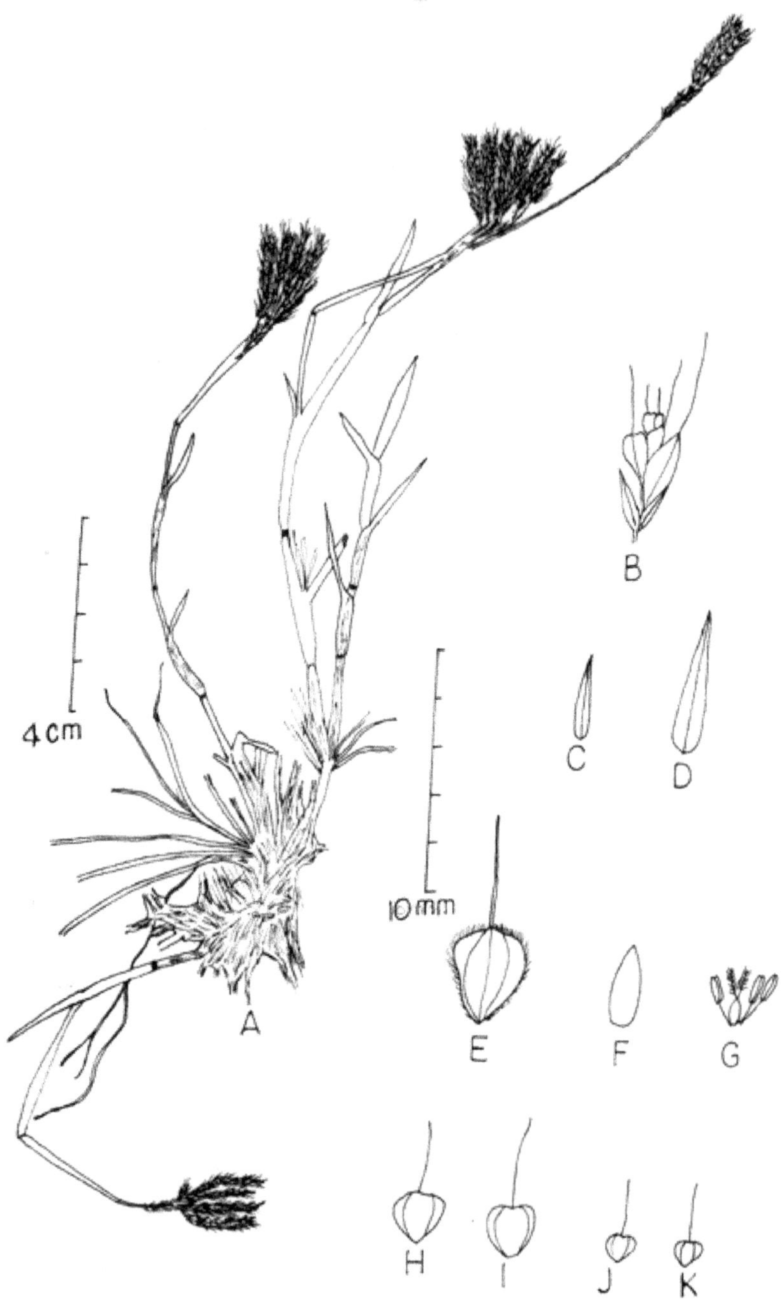

Figure 22. **Chloris quinquesetica** Bhide

A. Twig, B. Spikelet, L. Lower glume, D. Upper glume, E. Fertile lemma, F. Palea, G. Pistil and stamens, H. I Sterile lemma, I. II Sterile lemma, J. III Sterile lemma, K. IV Sterile lemma.

fascicle; peduncles puberulous below the inflorescence; spikelets 0.2 cm, I, II, IV & VII lemmas empty, III lemma perfect, all awned; glumes elliptic-oblong, 1-nerved, membranous; perfect lemma elliptic-obovate, 3-nerved, lateral nerves densely bearded with long white hairs, awned; palea as long as the lemma, elliptic; sterile lemmas broadly ovate, apex truncate, awned, epaleate; stamens 3; stigmas plumose. Caryopsis plano-convex or trigonous.

Occasional grass in southern Eastern Ghats. Fl. & Fr.: July-November.

Sanevaripalli (ATP), *KRKS* 40156; Dorigallu (ATP), *KRKS* 40469; Kammavaripalle (KNL), *TP & PVP* 2776; Peddarangapuram (KDP), *KRKS* & *BSS* 40487; Narasaraopet (GNT), *KCJ* 4718, 5784 (MH); Vijayaroi (KSN), *KCJ* 8243 (MH).

INDIA: West to NW India.

WORLD: Arabia, Oman, Pakistan.

Chloris roxburghiana Schult., Syst. Veg. 2. Mant. 329. 1827; Bor, Grass. India 468. 1960. *C. polystachya* Roxb., Fl. Ind. 1, 332, 1820; FBI 7: 292. 1896, non Lagasca (1816); Fischer 3: 1838.1934.

A perennial, tufted grass; culms up to 1 m high, erect or occasionally geniculately ascending; nodes glabrous. Leaf sheaths 3-8 cm long, glabrous, ciliate at mouth; ligule membranous; blades 9-22 x 0.3-0.6 cm, linear, folded, base rounded, apex acuminate. Spikes 9 cm long, racemosely arranged, terminal, compact, feathery heads of numerous (12-30) spikes; spikelets 0.25 cm long, 3-flowered, 3-awned; glumes elliptic-obovate, strongly 1-nerved, scaberulous on nerves (back side), upper with mucronate apex; perfect lemma oblong-obovate, 1-nerved, ciliate on keels and margins, awned; palea as long as the lemma, elliptic; sterile lemmas 2, broadly ovate, truncate apex, awned; stamens 3; stigmas plumose. Caryopsis linear-oblong.

Occasional in plains and foot hills of south Eastern Ghats. Fl. & Fr.: July–December.

Tirumala hills (CTR), *RVR* & *GO* 2428; Talakona RF (CTR), *PVP* 2322; Eeduru (NLR), *PMR* 22327; Sirumalai hills (DGL), *SKS* 304 (SGH).

INDIA: Peninsular to Northwest India.

WORLD: Tropicl Africa, Arabia, Madagascar.

Chloris virgata Sw., Fl. Ind. Occid 1: 203. 1797; FBI 7: 291. 1896; Fischer 3: 1838. 1934: Bor, Grass. India 468. 1960.

An annual, leafy, stoloniferous grass; culms up to 60 cm high, geniculately ascending or creeping, rooting at the nodes; nodes glabrous. Leaf sheaths 2-10 cm long, glabrous; ligule membranous; blades 3-10 x 0.1-0.4 cm, narrowly linear, base rounded, margin sparsely tuberculate base hairs, scaberulous, apex acuminate, glabrous. Spikes 4-8, 6 cm long, digitate, crowded at the apex of peduncle; spikelets 3 cm long, ovate-lanceolate, perfect only one; rachis hairy; glumes unequal, membranous, awned; fertile lemma ciliate along the margins, keels with crown of hairs, awned; callus pilose; sterile lemma I emerging from the side of fertile lemma,

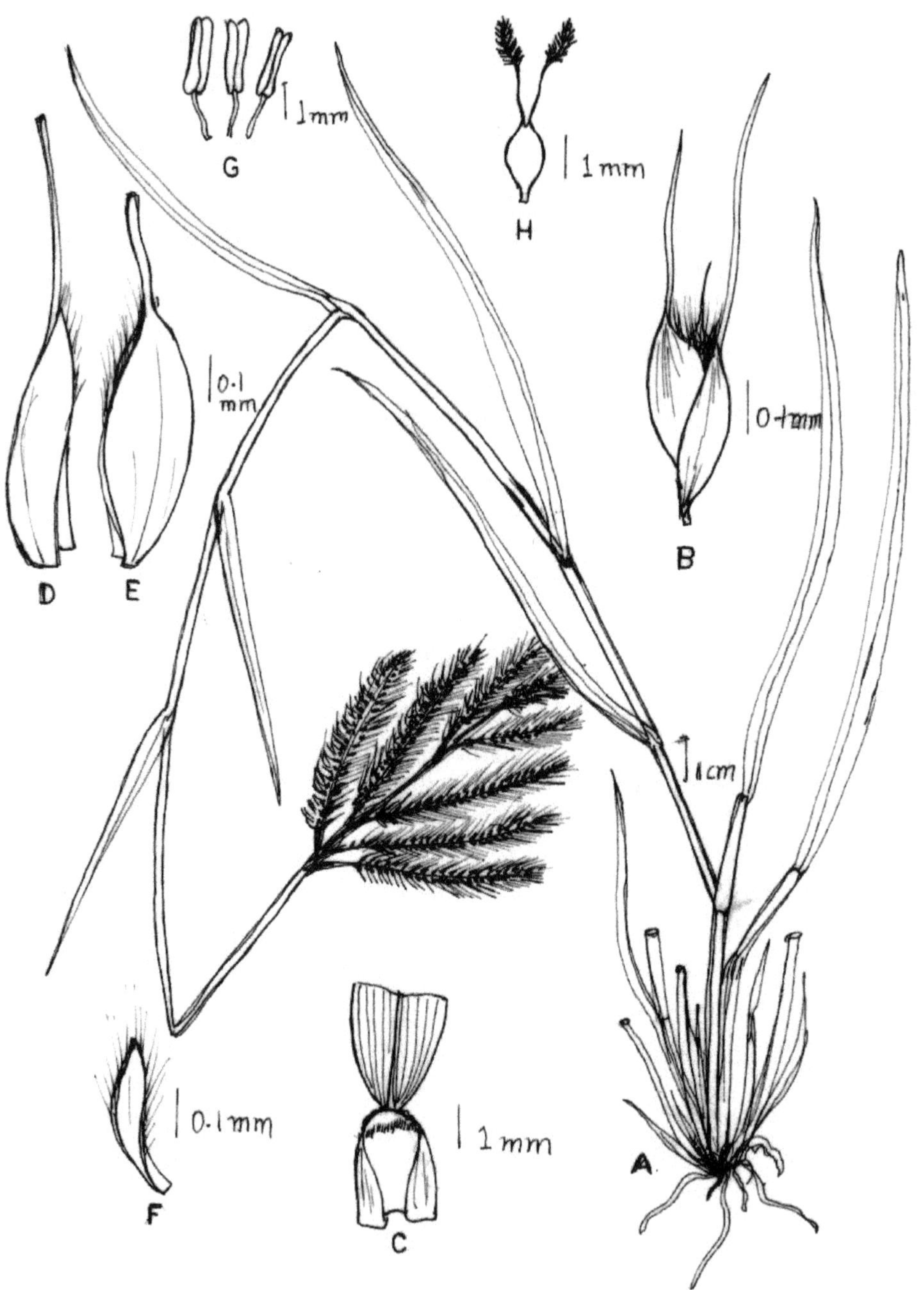

Figure 23. **Chloris roxburghiana** Schult.

A. Plant, B. Collar, C. Spikelet, D. Upper glume, E. Lower glume, F. Lemma, G. Stamens, H. Ovary.

oblong, awned; sterile lemma II oblanceolate, clavate, awnless; stamens 3, stigmas plumose. Caryopsis ellipsoid, compressed.

Occasional in waste lands. Fl. & Fr.: January-November.

Gooty fort (ATP), *MSG* 12239; Peapuly (KNL), *PVP* 2684; Chelama (KNL), *JLE* 22066 (MH); Kondapalli hills (KSN), *CAB* 7999 (MH); Mandapeta (EG), *CAB* 4287 (MH); Gaddangi (EG), *CAB* 12636 (MH); Bison hill (EG), *CAB* 5095 (MH); Borapuram (MBNR), *BR* & *BSS* 32286 (BSID); Metigiri (SLM), *sine coll. s.n.* (MH); Kattupalli (CPT), *DN* 912 (MH); Barapar (SBP), *HFM* 3591 (DD); Sirumalai hills (DGL), *KS* 5370 (MH).

INDIA: Plains of India.

WORLD: Tropics.

CHRYSOPOGON Trinius *nom. cons.*

1. Racemes composed of several to many spikelet pairs; lower glume tuberculate:
   2. Joints of pedicels glabrous; callus of sessile spikelets glabrous .... **C. zizanioides**
   2. Joints of pedicels with a basal tuft of hairs; callus of sessile spikelets bearded .... **C. lawsonii**
1. Racemes reduced to ternate spikelets; lower glume smooth, not tuberculate:
   3. Pedicels half the length of the sessile spikelets or longer:
      4. Pedicels glabrous:
         5. Lemmas in spikelets awned .... **C. aciculatus**
         5. Lemmas in spikelets awnless .... **C. nodulibarbis**
      4. Pedicels villous with rusty or pale hairs:
         6. Lower glume of the pedicelled spikelet with a long awn, upper glume not or very short awn:
            7. Sessile spikelet up to 6 mm long; lower glume of sessile spikelet aristate .... **C. asper**
            7. Sessile spikelet above 6 mm long; lower glume of sessile spikelet not aristate:
               8. Leaf blades hairy with long white hairs; spikelets in whorls of trinate or groups:
                  9. Leaves linear, up to 45 cm long, 20 mm wide, hairy with long (caducous) white hairs and much smaller hairs .... **C. verticillatus**
                  9. Leaves linear-lanceolate, up to 30 cm long, 1 cm wide, quite glabrous .... **C. hamiltonii**

8. Leaf blades glabrous; spikelets binate, not in whorls........ **C. lancearius**

6. Both glumes of the pedicelled spikelets awned........ **C. orientalis**

3. Pedicels shorter than half the length of the sessile spikelet:

10. Blades glabrous, occasionally with a few tubercle-based hairs, not puberulous:

11. Blades conspicuously distichous; panicle narrow, interrupted........ **C. polyphyllus**

11. Blades not conspicuously distichous; panicle oblong, continuous:

12. Pedicelled spikelets shorter than the sessile, hairs chocolate-brown; lower glume of pedicelled spikelet not awned........ **C. hackelii**

12. Pedicelled spikelets equal or longer than the sessile, hairs golden or golden- brown; lower glume of pedicelled spikelet awned:

13. Upper glume of the sessile spikelet pectinate-ciliate in the lower 2/3 or 3/4........ **C. fulvus**

13. Upper glume of the sessile spikelet glabrous on the back or shortly ciliate in the upper 1/4 only........ **C. serrulatus**

10. Blades puberulous........ **C. velutinus**

Chrysopogon aciculatus (Retz.) Trin., Fund. Agrost. 188. 1820; Fischer 3: 1738. 1934; Bor, Grass. India 115. 1960; Matthew, Fl. Tamilnadu Carnatic 3: 1822. 1983. *Andropogon aciculatus* Retz., Observ. Bot. 5: 22. 1789; FBI 7: 188. 1896.

A perennial, tufted grass; culms up to 60 cm high, geniculately ascending below, nodes glabrous. Leaf sheaths 3-5 cm, glabrous; ligule a minute fimbriate rim; blades 2-7 x 0.4-0.6 cm, oblong, flat, base rounded, margins scabrid, apex acute. Panicle narrow, up to 10 cm long; spikelets 2-nate. *Sessile spikelet*: Lower glume flat, stiff hispid on the back, bidentate; upper glume lanceolate, folded, scabrid on the back; lower lemma linear-lanceolate, sparsely pilose; upper lemma narrow, hyaline awned, paleated; stamens 3. Caryopsis laterally compressed. *Pedicelled spikelet*: lanceolate, 5 mm; pedicel glabrous; lower glume thin, lanceolate: upper glume short, acuminate, ciliate; lower lemma oblong, hyaline; upper lemma smaller, broader, ciliate; stamens 3. Caryopsis to 1.5 mm, hairy.

Occasional in damp places. Fl. & Fr.: July–October.

Gooty (ATP), *KVS & AL* 29717; Kailasanatha kona (CTR), *MSG* 12236; Bhupatipalem (EG), *TP & MSG* 12241; Ethakonda (EG), *VNS* 543 (CAL); Anantagiri (VSKP), *GVS* 19559 (MH); Gokavaram (EG), *CAB* 8259 (MH); Maredumilli (EG), *MM* 105149 (BSID); Anantagiri (VSKP), *GVS* 19559 (MH); Koriguttalu (KMM), *RR*

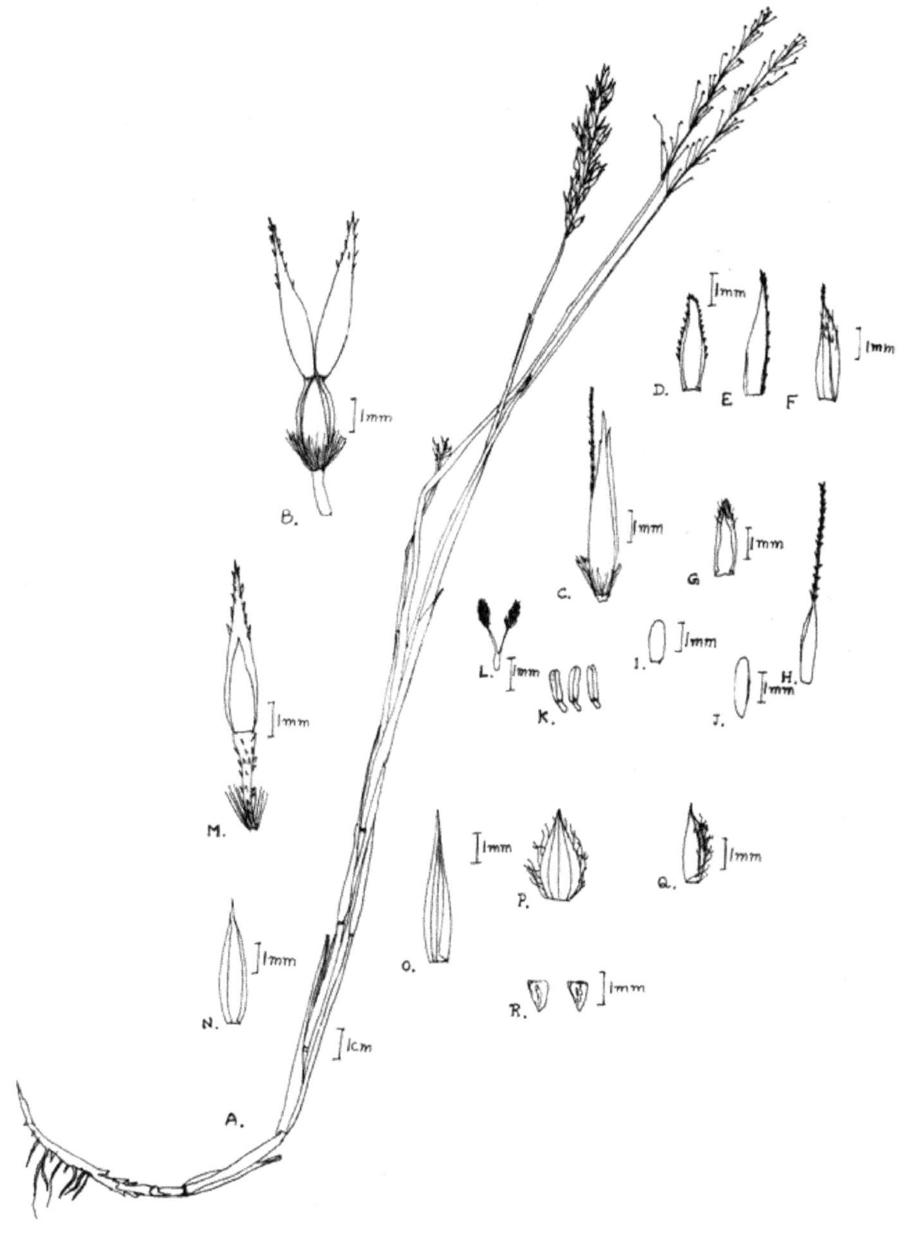

Figure 24. **Chrysopogon aciculatus** Trin.

A. Twig, B. Spikelets, C. Sessile spikelets, D. Lower glume, E & F. Upper glume and front views, G. Lower lemma, H. Upper lemma. I. Palea, J. Caryopsis, K. Stamens, L. Pistil, M. Pedicelled spikelet, N. Lower glume, O. Upper glume, P. Lower lemma, Q. Upper lemma, R. Lodicules.

106040 (BSID); Dunaei dam (GJM), *AC* 10208 (BSID); Balipadar (GJM), *AC* 10215 (BSID); Kuthadya hill (GJM), *VNS* 5927 (MH).

INDIA: Tropical parts of India.

WORLD: Nepal, Sri Lanka, Myanmar, Indo-China, China, Malesia, Polynesia, Australia.

Chrysopogon asper (B. Heyne ex Hook. f.) Blatt. & McCann, Bombay Grasses 68. 1935; Fischer 3: 1738. 1934; Bor, Grass. India 116. 1960. *Andropogon asper* B. Heyne ex Hook. f., Fl. Brit. India 7: 189. 1896.

A perennial, stoloniferous grass; culms stout, high, up to 2 m; nodes glabrous. Leaf sheaths 3-15 cm long, broad, hirsute below, upper glabrous, ciliate near mouth; ligule a fimbriate rim; blades distichous, 12-40 x 1-2 cm, broadly linear, base cordate, spinulose beneath the midrib, margins spinulosely serrulate, apex acute. Panicle 15 cm long, terminal, narrow; spikelets arranged in whorls, 3, 2 pedicelled, centre one sessile. *Sessile spikelet*: 6 mm, callus long villous; lower glume oblong, hispid beneath the tip, 3-nerved; upper glume elliptic, margin scaberulous, hispid below the awn, 7-nerved; lower lemma oblong- elliptic, 2-nerved, ciliate; upper lemma oblong, 3-nerved, awned; paleas minute, linear; stamens 3, stigmas plumose. *Pedicelled spikelet*: 8 mm, villous; glumes thin, lanceolate, lower apex bidentate, awned, 7-nerved, margin scaberulous; upper 3-nerved; lower lemma oblong, 2-nerved, ciliate, obtuse; upper lemma narrowly lanceolate, 1-nerved; palea narrowly linear, ciliate; stamens 3.

Occasional in the forests of southern Eastern Ghats. Fl. & Fr.: August-November.

Kalasamudram (ATP), *KRKS* & *BSS* 40321; Guvvalacheruvu (KDP), *R.V. Reddy* 8041; Rachakuntapalli (KDP), *KRKS* & *BSS* 40176; Tirumala hills (CTR), *MSG* 12146 & 12148; Talakona RF (CTR), *PVP* 2325, *JHFB* 112950 (BSID); Papavinasanam (CTR), *PVP* 3674; Balapalle (KDP), *JSG* 11193 (CAL); Kadapa (KDP), *EKK* 79518 (DD); Kannapenta (NLR), *JSG* 21737 (DD).

INDIA: Peninsular India (endemic).

Chrysopogon fulvus (Spreng.) Chiov., F1. Somala 327. 1929; Bor, Grass. India 166. 1960; Matthew, Fl. Tamilnadu Carnatic 3: 1823. 1983. *Pollinia fulva* Spreng., Pl. Minus Cognit. Pug. 2: 10. 1815. *Chrysopogon montanus* Trin. ex Spreng., Entadeck. 2: 93. 1821; Fischer 3: 1739. 1934. *Andropogon monticola* Roem. & Schult., Syst. Veg. 2. 665. 1827; FBI 7: 192. 1896.

A perennial, tufted grass; culms up to 40 cm high, erect or geniculately ascending; nodes glabrous. Leaf sheaths 1-6, glabrous, ciliate near mouth; ligule membranous; blade 6-15 x 0.2-0.5, linear, base truncate, margin sparsely scaberulous, apex acuminate. Panicle 7 cm long, terminal; spikelets 7.5 cm, in whorls, 3-flowered, 2-pedicelled, 1 sessile. *Sessile spikelet*: 6 mm, narrowly elliptic, hermaphrodite; callus villose, lower glume narrow-oblong, keeled, margins scaberulous, apically hispid; upper glume boat-shaped, 3-nerved, densely hispid on back, awned; lower lemma elliptic, ciliate; upper lemma minute, awned; palea narrowly linear; stamens 3, stigmas plumose. *Pedicelled spikelet*: 6 mm, narrowly elliptic, male or barren;

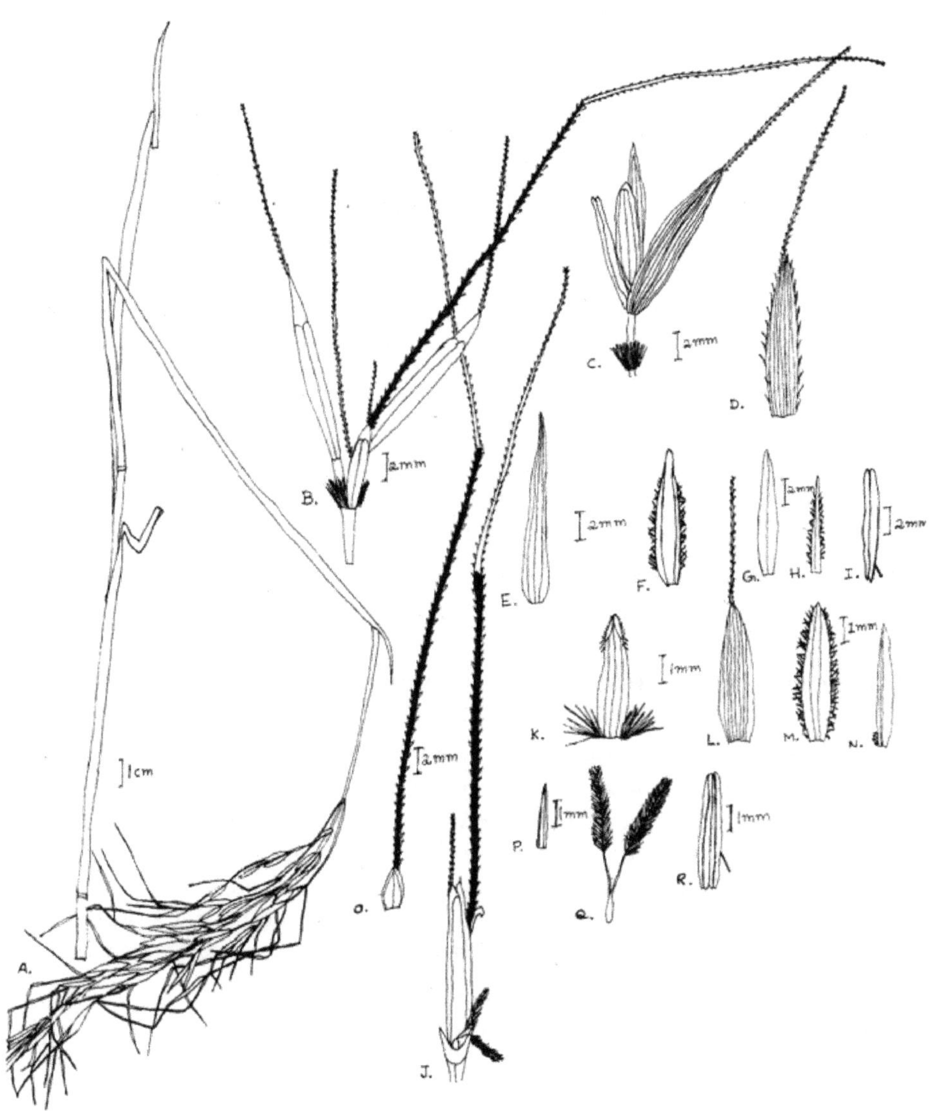

Figure 25. **Chrysopogon asper** (Heyne ex Hook. f.) Blatt. & McCann

A. Twig, B. Spikelets, C. Pedicelled spikelet, D. Lower glume, E. Upper glume, F. Lower lemma, G. Upper lemma, H. Palea, I. Stamen, J. Sessile spikelet, K. Lower glume, L. Upper glume, M. Lower lemma, N. Palea, O. Upper lemma, P. Palea, Q. Pistil, R. Stamen.

pedicel villose; lower glume oblong, margins scaberulous, 7-nerved, chartaceous below; upper glume elliptic, 3-nerved; lemma ovate, 2-nerved, ciliate; palea minute, narrowly lanceolate; stamens 3.

Common in plains and hilly tracts of all hill ranges. Fl. & Fr: July-November.

Yadiki RF (ATP), *TP* 467; Peapuly (KNL), *PVP* 2679; Madhavaram RF (KNL), *RVR* 1640; Palakonda (KDP), *CS & DAM* 7636; Bhupathaiah kona (CTR), *BR* & *P.Priyadarsini* 37317; Chintala thanda (GNT), *VRK* 3535; Chelama (KNL), *JLE* 22083 (MH), *MPG* 108, 110 & 111 (MH); Seshachalam hills (KDP), *S.India Flora* 15355 (MH); Nellore, *CECF* 4195 (MH); Kamalapuram (KDP), *JSG* 11206 (MH); Kadapa, *EKK* 795779 & 78541 (DD); Chintakunta (KDP), *JSG* 21310 (DD); Diguvametta (PKM), *MPG* 114 (MH); Kondapalli (KSN), *CAB* 4397 (MH); Kondapudi (EG), *CAB* 5231 (MH); Umamaheswaram (MBNR), *SRS* 111339 (BSID); Koriguttalu (KMM), *RR* 106040 (BSID); Yenkur (KMM), *RR* 108573 (BSID); Muttur hills (SLM), *CAB* 6045 (MH); Hoginakkal RF (SLM), *EV* 21919 (MH); Guindy park (MDS), *ANH* 47158 (MH); Tirukklukundram hill (CPT), *DN* 639 (MH); Takarai RF (SA), *KRM* 52807 (MH); Tippukadu RF (NA), *KRM* 17693 (MH); Beldungri (SBP), *SX* & *MB* 2344 (RRL-B); Jenabil Range, Similipal B.R. (MBJ), *AC* 2350 (BSID); Mandragiri hill (AGL), *AC* 2371 (BSID); Barga Khol, Rajgangapur, Sundargarh, *AC* 10255 (BSID); Jacum, Karlapat Wildlife Sanctuary (KHD), *AC* 2451 (BSID).

INDIA: Warmer parts of India.

WORLD: Tropics of Asia and East Africa.

Chrysopogon hackelii (Hook. f.) C.E.C.Fischer, Fl. Madras 3: 1739. 1934; Bor, Grass. India 117. 1960. *Andropogon hackelii* Hook. f., Fl. Brit. India 7: 194. 1896.

A perennial, tufted grass; culms up to 2 m; nodes glabrous. Leaf sheaths 5-6 cm, scabrid, keeled; ligule a fringe of hairs; blades 10-22 ´ 0.5-0.8 cm, linear, apex acuminate. Panicle 15 cm long; branches whorled; spikelets 2-nate. *Sessile spikelets*: 7 mm; lower glume 4-nerved; upper glume cymbiform, 3-nerved, bidentate, glabrous, awned; lower lemma very narrow, 3-nerved, hyaline, awned; stamens 3. *Pedicelled spikelet*: 3- 4 mm; pedicel villous along the margins, shorter than the sessile spikelet; lower lemma 2-nerved; upper lemma narrowly lanceolate, 1-nerved; stamens 3.

Rare on hills, along grassy slopes. Fl. & Fr: July-November.

Pulikallu (ATP), *BR* & *ANS* 38191; Guvvalacheruvu (KDP), *KS* 6428 (MH & CAL); Godavari gorge (WG), *BO* 3467 (CAL & MH).

INDIA: Peninsular India. Endemic.

Chrysopogon hamiltonii (Hook.f.) Haines, Bot. Bihar Orissa 1036. 1924; Bor, Grass. India 117. 1960; Moulik, Grass & Bamb. India 1: 241. 1997. *Andropogon hamiltonii* Hook.f., Fl. Brit. India 7: 190. 1897.

Perennials, erect, tufted, to 1.8 m high. Leaf sheaths 2-12 cm, glabrous; ligules 0.3 mm, membranous, ciliate; leaf blades 10-30 x 0.6 cm, linear, lanceolate, acute, glabrous, scaberulous below. Panicle 10-20 cm long; spikelets in groups of threes, one sessile and two pedicelled. *Sessile spikelet: c.* 6 x 1.2 mm, lanceolate, awned, awn *c*. 7.6 cm long; lower glume *c*. 5 x 1.3 mm, lanceolate, rigid; upper glume *c*. 6 x 1.6 mm, lanceolate-oblong, awned, awn *c*. 5.5 mm long, 3-nerved; lower lemma *c*. 5.8 mm, lanceolate, hyaline, 2-nerved, ciliate; upper lemma *c*. 6 x 0.7 mm, lanceolate, awned, awn *c*. 7.6 cm, 3-nerved, hyaline; palea absent; lodicules 2, *c*. 0.6 mm; stamens 3, anthers *c*. 4.5 x 0.8 mm; ovary *c*. 0.7 x 0.4 mm, oblong, styles *c*. 0.8 mm, stigma 2.8 mm long. *Pedicelled spikelet*: 11 x 1.5 mm, lanceolate, awned, awn 11 mm long, pedicel 5 mm long; lower glume *c*. 9 x 1.6 mm, lanceolate, awned, awn *c*. 11 m long, 7-nerved, keeled, scabrid; upper glume *c*. 7.3 x 1.5 mm, lanceolate, 3-nerved, keeled, ciliate; lower lemma *c*. 8.5x 1.5 mm, 2-nerved, keeled, ciliate, hyaline; upper lemma *c*. 6 x 1 mm, keeled, ciliate, hyaline; palea absent; lodicule 2, *c*. 0.5 mm; stamens 3, 0.6 x 1 mm.

Common on hill slopes, usually in dry habitats in Anantapuram district (B.R.P.Rao *et al*., 2012). Fl. & Fr.: October–December.

Gounivaripalli-Samrajyamkonda (ATP), *BR* & *KVS* 28188 (SKU).

INDIA: Bihar, Andhra Pradesh. Endemic.

Chrysopogon lancearius (Hook. f.) Haines, Bot. Bihar Orissa 1036. 1924; Bor, Grass. India 118. 1960; Saxena & Brahmam, Fl. Orissa 4: 2294. 1996. *Andropogon lancearius* Hook. f., Fl. Brit. India 7: 190. 1896.

A perennial, tufted grass; culms up to 90 cm long, leafy; nodes glabrous. Leaf sheaths keeled, glabrous; ligule a ciliate line; blades 15-30 x 5-15 cm, lanceolate, base narrowed, margins spinulose, apex acuminate, midrib prominant. Panicle up to 15 cm, narrow, branches whorled; spikelets 2-nate. *Sessile spikelet:* 5 mm, oblong; lower glume oblong, chartaceous, truncate, upper glume oblong, chartaceous, shortly ciliate on the keels; lower lemma linear-oblong, hyaline; upper lemma linear, hyaline, awned; stamens 3. *Pedicelled spikelet*: 7 cm, lanceolate; pedicels white, flat, ciliate with long brownish hairs; glumes lanceolate, membranous, glabrous; lower lemma shorter, linear-oblong, hyaline, ciliate; upper lemma hyaline.

Occasional in hills. Fl. & Fr.: July-November.

Kondagorulapenta (GNT), *MPG* 119 (MH); Bison hills (EG), *CAB* 5112 (MH).

INDIA: E. Himalaya, E. India.

WORLD: Bhutan, Bangladesh.

Chrysopogon lawsonii (Hook.f.) Veldkamp, Austrobaileya 5: 515. 1999. *Vetiveria lawsonii* (Hook, f.) Blatt. & McCann, J. Bombay Nat. Hist. Soc. 32: 409, 1928; Fischer 3:1733. 1934; Bor, Grass. India 258. 1960; Moulik, Grass. Bamb. India 1: 256. 1997. *Andropogon lawsonii* Hook. f., Fl. Brit. India 7:187. 1896.

A perennial, tufted, robust grass; culms up to 2 m high; nodes glabrous. Leaf sheaths 5-20 cm long, glabrous; ligule a small ring of fimbriate membrane, apex with long hairs; leaf blade 10-20 x 0.4-0.6 cm, linear, base truncate, apex acuminate, glabrous. Panicle up to 20 cm long; branches capillary, whorled; racemes up to 6 cm long; spikelets lanceolate, 2-nate. *Sessile spikelets*: 5 mm, linear, laterally compressed, bisexual; glumes equal, coriaceous; lower linear, spinulose; upper boat-shaped, keels ciliate, spinulose at apex, awned; lemma hyaline; lower lemma linear, margins ciliate; upper lemma linear, 2-lobed, awned; palea linear, ciliate; lodicules 2; stamens 3, stigmas plumose. Caryopsis linear-oblong. *Pedicelled spikelet*: 6 mm, lanceolate, male; lower glume lanceolate, chartaceous; upper glume lanceolate, awned, chartaceous; lemma lanceolate hyaline; stamens 3.

Occasional grass in marshy areas. Fl. & Fr: August- March.

Salur (VJN), *MCK* 25235; Peccheruvu (KNL), *TRS* & *PVS* 79525 (BSID).

INDIA: Peninsular India.

WORLD: Nepal.

Chrysopogon nodulibarbis (Hochst. ex Steud.) Henrard, Blumea 4: 534 1941. *Andropogon zeyalnicus* Steud., Syn. Pl. Glumac. 1: 397 1854; Hook. f., FBI 7: 192. 1896. *Chrysopogon zeylanicus* (Nees) Steud., Enum. Pl. Zeyl. 366 1864; Fischer, 3: 1739. 1934. *Sorghum nodulibarbis* (Steud.) Kuntze, Revis. Gen. Pl. 2: 792 1891.

Perennial grass; culms 80-150 cm high; nodes glabrous. Leaves usually collected at base, 20-30 x 0.4-1 cm, linear-lanceolate; sheath sharply keeled; ligule a fimbriate membrane. Panicle 10-20 cm long, lax with few spikelets. *Sessile spikelets* 6-8 mm long; lower glume 6 x 1.5 mm with two lines of bristles on dorsal side near the margins; upper glume similar but not with bristles; lower lemma delicate, 1-nerved; awn 40 mm long; anthers 3 mm long. *Pedicelled spikelets* to 14 mm long; glumes unequal, awnless; upper floret male; lemmas awnless.

Fl. & Fr.: December–February.

Mottur Hill (SLM), *CAB* 6043 (MH).

INDIA: Kerala, Tamil Nadu.

WORLD: Sri Lanka.

Note: This grass species is not accounted for Tamil Nadu grasses (Althaf Ahamed Kabeer and Nair, 2009). We included this taxin based on a sheet collected by C.A.Barber and housed in MH.

Chrysopogon orientalis (Desv.) Camus in Lecomte, Fl. Gen. de 1′ Indo-Chine 7: 332. 1922; Fischer 3: 1739. 1934; Bor, Grass. India 118. 1960; Matthew, Fl. Tamilnadu Carnatic 3: 1823. 1983. *Rhaphis orientalis* Desv., Opuse. 69. 1831. *Andropogon wightianus* Nees ex Steud., Syn. Pl. Glum. 1: 395. 1854; FBI 7: 191. 1896. *Chrysopogon wightianus* (Nees ex Steud.) Thwaites, Enum. Pl. Zeyl. 366. 1864.

A perennial, tufted grass; culms up to 8 cm high, ascending at base, forms mat; nodes glabrous. Leaf sheaths 5-9 cm, shortly ciliate-glabrous at mouth; ligule very short, villous; leaf blades 5-10 x 0.2-0.5 cm, oblong, basal cauline, flat, base cordate, apex acuminate, densely pubescent-glabrous on both surfaces. Panicle 15 cm long, contracted; branches with reddish-brown hair at tip; spikelets 2-nate. *Sessile spikelet*: 8 mm long, cylindric; callus long, densely bearded with rusty hairs; lower glume laterally compressed, glabrous below, hispid above, bidentate; upper

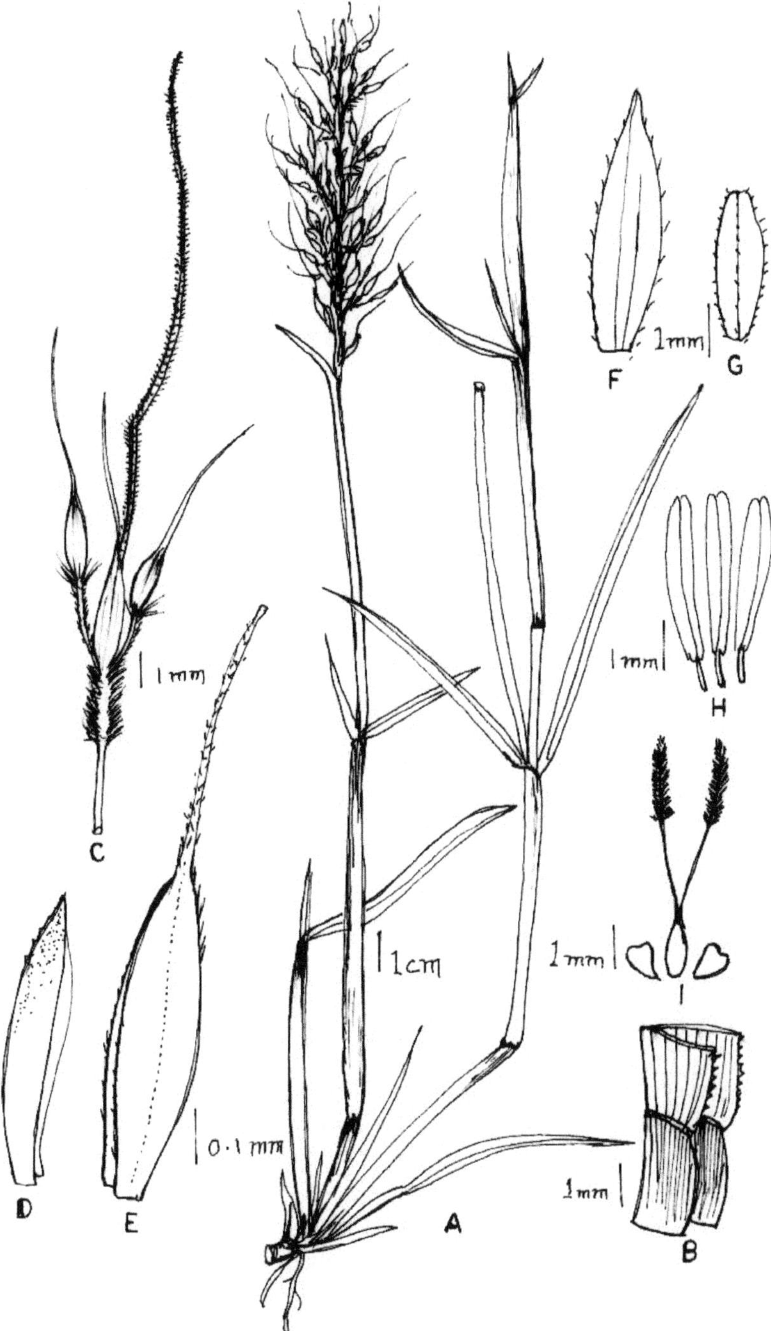

Figure 26. **Chrysopogon orientalis** (Desv.) Camus

A. Camus, A. Plant, B. Collar, C. Spikelets, D. Upper glume, E. Lower glume, F. Lemma, G. Palea, H. Stamens, I. Ovary with lodicules.

glume chartaceous, hispid on keels, bidentate, awned; lower lemma linear-oblong, ciliated; upper lemma awned, bidentate. *Pedicelled spikelet*: 12 cm long, lanceolate, pubescent; pedicel truncate, margins shortly ciliate; lower glume ciliate, awned; upper glume lanceolate, awned; lower lemma oblong, ciliate; upper lemma very narrow, ciliate; stamens 3.

Occasional in open forests. Fl. & Fr.: October-December.

Penchalakona hills (NLR), *MSG* 12187, *S.India Flora* 12791 (MH); Kanupatri (NLR), *JSG* 21737 (CAL); Penchalakonda (NLR), *CECF* 12791 (MH); Balapalle (KDP), *JSG* 11193 (MH); Arumanipenta (KDP), *S.India Flora* 15411 (MH); Kambakkam hills (CTR), *sine coll. s.n.* (MH); Mahendragiri (GJP), *CECF* 95 (CAL); Sirumalai hills (DGL), *SKS* 413 (SGH); Kolli hills (NMK), *SKS* 572 (SGH).

INDIA: Peninsular India.

WORLD: Myanmar.

Chrysopogon polyphyllus (Hack. ex Hook. f.) Blatt. & McCann, J. Bombay Nat. Hist. Soc. 32: 416. 1928; Fischer 3: 1739. 1934; Bor, Grass. India 118. 1960. *Andropogon polyphyllus* Hack. ex Hook. f., Fl. Brit. India 7: 194. 1896.

A perennial, tufted grass; culms up to 90 cm long, simple or fastigiately branched, glabrous. Leaf sheaths terete, appressed; ligule a ciliate membrane; blades 15-20 x 0.3-0.5 cm, narrow, flat, base cordate, margin minutely scabrid, apex acuminate, glabrous. Panicle 15 cm long, densely flowered, whorled close; spikelets 2-nate. *Sessile spikelet*: 4 mm, white or pale purplish; callus bearded at base with long hairs; lower glume obtuse, glabrous, keel ciliate at tip; upper glume keeled, ciliate, awned; lower lemma linear, hyaline, ciliate; upper awned. *Pedicelled spikelet*: narrowly lanceolate, acuminate, glabrous longer than the sessile spikelet; pedicel villous at the tip only.

Rare in middle Eastern Ghats

Devipatnam (EG), *JSG* 15781 (MH).

INDIA: Central India southwards.

WORLD: Myanmar.

Chrysopogon serrulatus Trin., Mem. Acad. Sci. Petersb. 6, 2: 318. 1832; Bor, Grass. India 118. 1960. *Andropogon monticola* Schult. var. *trinii* Hook. f., Fl. Brit. India 7: 193. 1896.

A perennial, coarse, tufted grass; culms up to 50 cm high, slender. Leaf sheaths upper terete, lower more or less compressed; blades 14-45 x 0.3-0.6 cm, base rounded, margins scaberulous, apex acuminate, glabrous. Panicle 10 cm, of several or few whorls of 3-13; branches capillary, unequal, lower long, upper short, bearing solitary spikes, glabrous; spikes variable in size; spikelets 2-nate. *Sessile spikelet*: 6 cm; lanceolate, bisexual; callus with short or long rusty red hairs; lower glume narrow, compressed, keeled, apex rounded; upper glume broader, bidentate, keels scabrid; lower lemma hyaline, oblong, obtuse, ciliate; upper lemma linear, hyaline, awned. *Pedicelled spikelet*: 8 mm, male or neuter, linear-lanceolate; pedicels with long rusty red hairs; lower glume lanceolate, membranous, keels scabrid, glabrous; upper glume lanceolate, acuminate, ciliate; lower lemma oblong, hyaline, obtuse; upper lemma narrower, hyaline; stamens 3.

Rare. Fl. & Fr.: July-November.

Penchalakona (NLR), *PMR* 19504; Balabuthrapuram (EG), *BO* 3226(CAL).

INDIA: NW India.

WORLD: Afghanistan, Pakistan, Nepal, Myanmar, Madagascar, Tropical Africa.

Chrysopogon velutinus (Hook. f.) Bor, Grass. India 119. 1960; Fischer 3: 1739. 1934. *Andropogon velutinus* Hook. f., F1. Brit. India 7: 194. 1896,

A perennial, tufted grass; culms up to 40 cm long, compressed, leafy at base. Leaf sheaths strongly compressed, keeled, puberulous, lower ones spinulosely ciliate, upper pubescent; ligule a ciliate rim; blades 10-20 x 0.6-0.8 cm, linear, flat, distichous, apex acuminate, pubescent. Panicle 8 cm long; branches spreading with sparsely shortly white ciliate rachis; spikelets 2-nate. *Sessile spikelet* 7 mm, pale; callus very short rufous, bearded; lower glume keeled, keel ciliate above; upper glume rufous, hairy, aristate; lemmas hyaline, oblong, awned. *Pedicelled spikelet*: 5 mm, oblong unawned; pedicel cuneate, sparsely villous; lower glume rufous, hispid in lower portion; upper glume lanceolate, acuminate, ciliate; lemmas hyaline, oblong; palea hyaline.

Extremely rare and endemic to Kadapa district.

Kadapa district, *Wight* 2314 (MH); Katam, Chillavaripalli (ATP), *J. Swamy & S. Nagaraju* 7977 (BSID).

INDIA: Andhra Pradesh, Endemic.

Note: This grass is extremely rare and reported to be endemic to Kadapa district in Andhra Pradesh.

Chrysopogon verticillatus (Roth) Trin. ex Steud., Nom, Bot. ed. 2. 1: 360 1840; Fischer 3: 1739. 1934; Bor, Grass. India 119. 1960. *Andropogon verticillatus* Roxb., Fl. Ind. 1: 267. 1820; FBI 7: 189. 1896.

A perennial, tufted grass; culms up to 40 cm high, terete, glabrous; lower nodes often bearded. Leaf sheaths glabrous; leaf blades 35-50 x 1-1.5 cm, linear, base rounded, margins spinulose, apex acuminate, sparsely hairy towards base. Panicle 30 cm, pyramidal, lax; branches horizontal; spikes narrow, pale, spikelets few. *Sessile spikelet* 4 mm; linear, bisexual; lower glume hyaline, 4-nerved, acute; upper glume linear, bidentate, apex ciliate, awned; lower lemma oblong, obtuse, keeled, keels retrorsely ciliate; upper lemma very narrow, hyaline, 3-nerved, obscurely bidentate, awned. *Pedicelled spikelet*: 1 cm, lanceolate; pedicel margins ciliate; lower glume 2-nerved, awned; upper glume lanceolate, acuminate, ciliate, 3-nerved; lower lemma very narrow, ciliate, 2-nerved; upper lemma linear, 1-nerved, bidentate; stamens 3.

Rare grass. Fl. & Fr.: July–November.

Gopalapuram (NLR), *MSG* 10886; Sirivaka (EG), *BO* 3420 (MH); Bison hills (EG), *CAB* 5112 (MH); Chahaala Rest House, Similipal Biosphere Reserve (MBJ), *ARKS* & *GPS* 12285 (BSID); Jenabil Range, Similipal B.R., *AC* 2337 (BSID); Mahendragiri hills (GJP), *AC* 4219 (BSID).

INDIA: Odisha, Andhra Pradesh, Karnataka, Kerala, Tamil Nadu. Endemic.

Chrysopogon zizanioides (L.) Roberty, Bull. Inst. Franc. Afrique Noire, A 22: 106. 1960. *Vetiveria zizaniodes* (L.) Nash in Small, Fl. South Souht-East U.S. 67. 1903; Fischer 3: 1733. 1934; Bor, Grass. India 258. 1960; Matthew, Fl. Tamilnadu Carnatic 3: 1913. 1983. *Phalaris zizanioides* L., Mant. Pl. 2:183. 1771. *Andropogon squarrosus* auct. non. L. f.; FBI 7: 186. 1896; Bor, *loc.cit.*

A perennial, tufted, robust, aquatic grass; culms up to 2 m high; nodes glabrous. Leaf sheaths 4-10 cm long, glabrous; ligule a scarious rim; blades 5-25 x 0.5-1.2 cm long, linear, flat, base truncate, apex acuminate, glabrous. Panicle up to 25 cm long; branches whorled, capillary; racemes up to 10 cm long; spikelets lanceolate, 2-nate. *Sessile spikelet:* 4 mm, lanceolate, bisexual; glumes equal, coriaceous; lower glume lanceolate, spinulose; upper boat-shaped, keeled, keel ciliate, scabrid on back; lower lemma linear, hyaline, keel winged, wings ciliate on margin, epaleated; upper lemma oblong, hyaline, 2-lobed, lobes mucronate, awned, paleated; palea linear, hyaline, nerveless; stamens 3; stigmas plumose. Caryopsis oblong, slightly oblique at apex. *Pedicelled spikelet:* 4 mm, lanceolate, male; glumes equal, coriaceous; lower glume lanceolate, spinulose; upper glume boat-shaped, keeled, keels ciliate, scabrid at apex; lower lemma lanceolate, hyaline, margins ciliate; upper lemma oblong, 2-lobed, lobes mucronate; stamens 3.

Common in marshy plains and near ponds through out Eastern Ghats. Fl. & Fr.: August-October. Tel.: *Vattiveru.*

Lower Ahobilam RF (KNL), *TP & RVR* 2753; Palakonda hills (KDP), *PVP* 3436; B.N.Kandriga (CTR), *BR* & *P.Priyadarsini* 37341; Maredumilli (EG), *TP & MSG* 12262; Jonnavalasa tank (VJN), *MCK* 18842; Pecheruvu (KNL), *NRR* & *VBH* 83735, 84000 (BSID); Rajampet (KDP), *JSG* 11233 (CAL & DD); Kadapa, *EKK* 78257 (DD); Kambakkam hills (CTR), *S.India Flora* 16441 (MH); Barmakonda RF (VSKP), *KCJ* 17147 (MH); Vatavarlapally (MBNR), *KI* 18729; Kurumapuram RF (SA), *KRM* 60274 (MH); Gingee (SA), *S.India Flora* 17955 (MH); Tippukadu RF (NA), *KRM* 17926 (MH); Tikarpada, Satkosia Wildlife Sanctuary (AGL), *AC* 2399 (BSID); Badrama Wildlife Sanctuary (SBP), *AC* 10243 (BSID); Purunakote, Satkosia Tiger Reserve (AGL), *KCM* 5291 (BSID).

INDIA: Throughout India.

WORLD: Pakistan, Nepal, Sri Lanka, Myanmar, SE Asia.

COELACHNE R. Brown

Coelachne simpliciuscula (Wight. & Arn. ex Steud.) Munro ex Benth., J. Linn. Soc. Bot. 19: 93. 1881; Bor, Grass. India 576. 1960; Matthew, Fl. Tamilnadu Carnatic 3: 1824. 1983. *Panicum simpliciusculum* Wight & Arn. ex Steud., Syn.

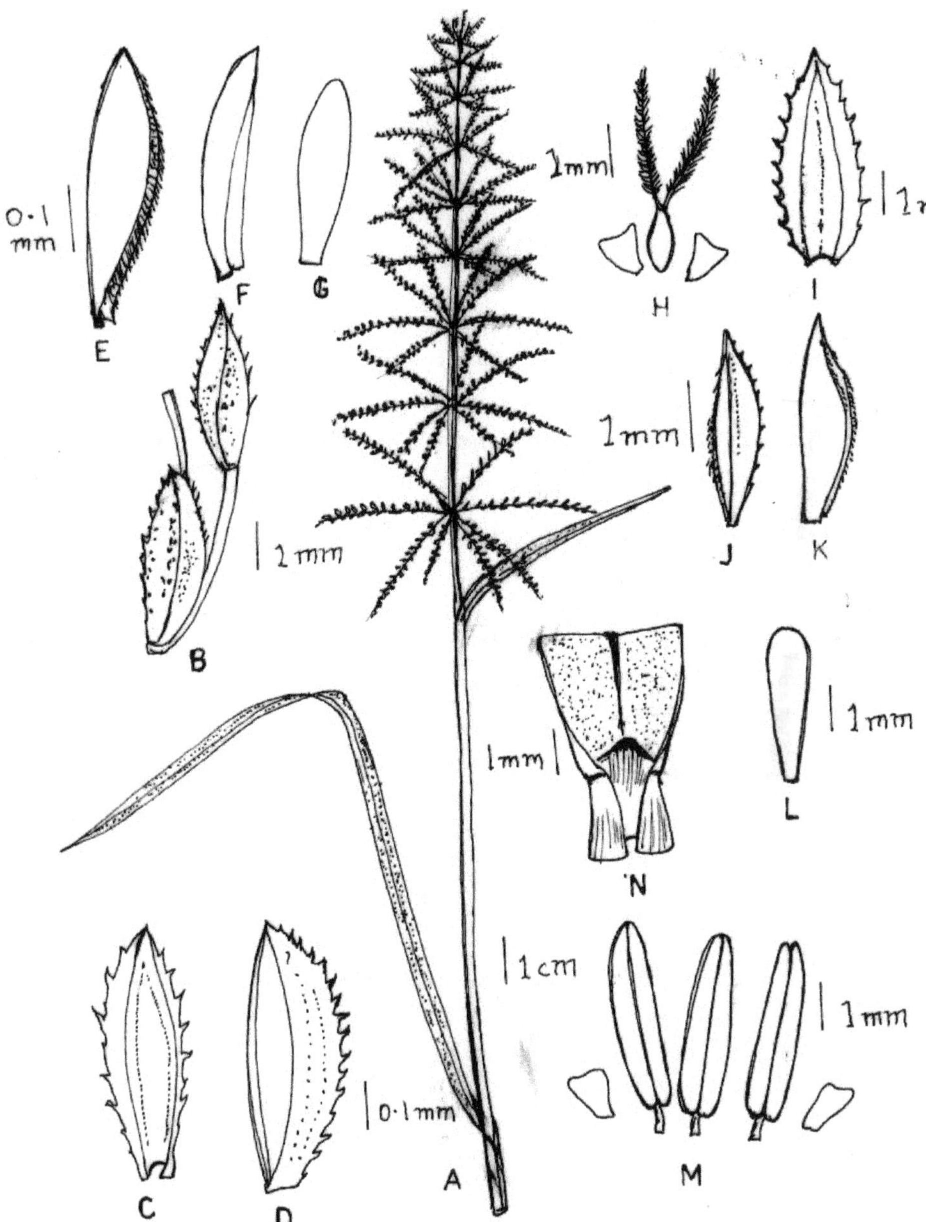

Figure 27. **Chrysopogon zizanioides** (L.) Roberty.

A. Culm, B. Spikelets, Sessile spikelet: C. Upper glume, D. Lower glume, E. Upper lemma, F. Lower lemma, G. Palea, H. Ovary with lodicules, Pedicelled spikelet: I. Upper glume, J. Lower glume, K. Lemma, L. Palea, M. Stamens and lodicules, N. Collar.

Pl. Glumac 1: 96. 1854. *Coelachne pulchella* R. Br. var. *simpliciuscula* (Wight & Arn. ex Steud.) Hook. f., Fl. Brit. India 7: 271. 1896; Fischer 3: 1804. 1934.

An annual grass; culms up to 20 cm long, decumbent; nodes bearded. Leaf sheaths 1-3 cm; ligule a fimbriate rim; blades 2-5 x 0.3-0.6 cm, oblong flat, pubescent above on nerves. Panicle up to 10 cm long, oblong, interrupted; spikelets 2-flowered, pedicelled bisexual, glumes persistent, ovate of suborbicular, glabrous, sparsely ciliate at base; palea equal to lemma; upper lemma cilate at base; palea equal to lemma; upper lemma cilate at base; palea narrow; stamens 3. Caryopsis ellipsoid.

Rare along streams and marshy places in Northern Eastern Ghats. Fl. & Fr.: May–October

Minumuluru (VSKP), *GVS* 30061 (CAL & MH); Jalaput (VSKP), *GVS* 19722 (MH); Mahendragiri (GJP), *SX* & *MB* 3689 (RRL-B); Similipahar (MBJ), *SX* 464 (RRL-B).

INDIA: More or less throughout India.

WORLD: SE Asia, China, Madagascar.

COELACHYRUM Hochst & Nees

Coelachyrum lagopoides (Burm.f.) Senaratna, Grass. Sri Lanka 79. 1956; Bor, Grass. India 488. 1960; Matthew, Fl. Tamilnadu Carnatic 3: 1825. 1983. *Cynosurus lagopoides* Burm.f., Fl. Ind. 29. 1768. *Coelochyropsis lagopoides* (Burm.f.) Bor, Ann. Natur. Hist. Mus. Wien.75: 25. 1972; Gayathri & Pullaiah in Pullaiah, Fl. Andhra Pradesh 3: 1157. 1997; Moulik, Grass. Bamb. India 2: 585.1997. *Eleusine lagopoides* (Burm. f.) Merr., Philipp. J. Sci. 19:339.1921; Fischer 3: 1840.1934. *E. brevifolia* (Willd.) R. Br. ex Hook. f., Fl. Brit. India 7: 294.1896.

Perennials, culms creeping or erect, to 50 cm, rooting at nodes; nodes glabrous. Leaf sheaths glabrous; ligule membranous; leaf blades oblong-lanceolate, margins ciliate. Spikes numerous, crowded into a globose head; spikelets elliptic-ovate, many–flowered; lower glume lanceolate, 1-nerved, keeled, chartaceous; upper glume ovate-lanceolate, chartaceous, acute, 1-nerved; upper lemma elliptic-lanceolate, 3-nerved, ciliate, hyaline, keels winged; palea oblong, coriaceous, base bearded; stamens 3. Caryopsis trigonous or ovate, transversely rugose.

Occasional in plains and forest peripheries in central and southern Eastern Ghats. Fl. & Fr.: July-December.

Kalasamudram (ATP), *AMR* 24521, *KVS* & *BSS* 28992; Dorigallu (ATP), *KRKS* & *BSS* 40471; Gounivaripalli (ATP), *BR* & *P.Priyadarsini* 37301; Kammavaripalli (KNL), *TP* 2775; Peapuly (KNL), *PVP* 3189; Ahobilam (KNL), *BR* & *BSS* 32265; Varikuntapadu (NLR), *PMR* 16472; Panapakkam (CTR), *S.India Flora* 15786 (MH); Vinukonda (GNT), *CAB* 4764 (MH); Avadi (CPT), *sine coll. s.n.* (MH); Vandalur (CPT), *S.India Flora* 11110 (MH); Hogainakkal RF (SLM), *EV* 20662 (MH).

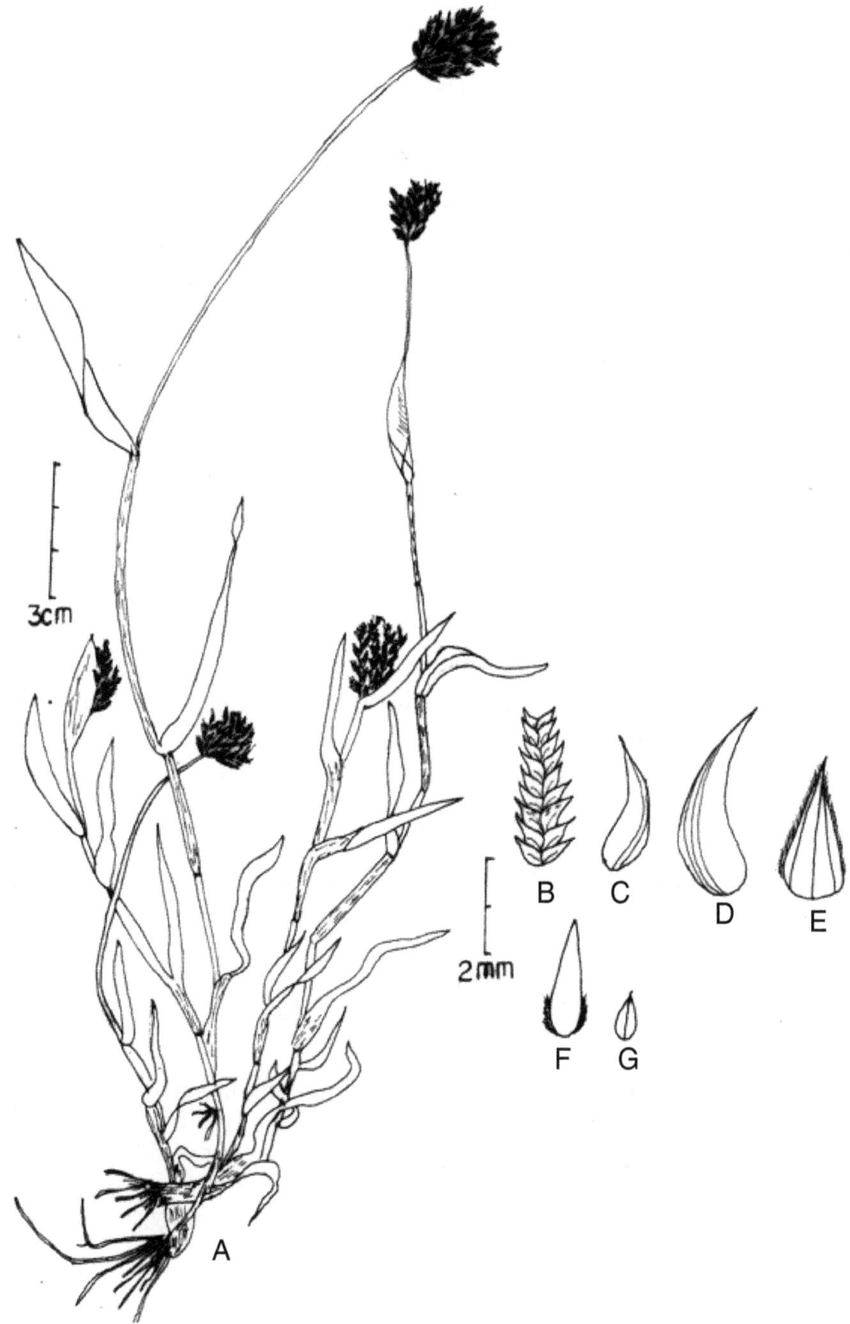

Figure 28. **Coelachyrum lagopoides** (Burm.f.) Senartna.
(Syn.: *Coelachyropsis lagopoides* Ber.)

A. Twig, B. Spikelets, C. Lower glume, D. Upper glume, E. Lemma, F. Palea, G. Caryopsis.

INDIA: Coasts of South India.

WORLD: Bangladesh and Sri Lanka.

Note: The genus *Coelochyropsis* Bor (1960) was treated as synonym to *Coelachyrum* by Clayton & Renovoize (1986), same concept is followed in the present study.

COELORACHIS Brongniart

Coelorachis clarkei (Hack.) Blatt. & McCann, J. Bombay Nat. Hist. Soc. 32: 33. 1927. *Rottboellia clarkei* Hack., Oesterr. Bot. Z. 41: 8. 1891; FBI 7: 156. 1896. *Manisuris clarkei* (Hack.) Bor in Santapu, Fl. Khandala 357. 1953 & Grass. India 191. 1960. *Mnesithea clarkei* (Hack.) de Koning & Sosef, Blumea *31: 290 1986.*

Annuals, culms erect, 20-40 cm high. Leaves linear or lanceolate, 10-15 x 0.8-1.3 cm. Inflorescence compound, paniculate or fastigiated racemes. Spikelets appressed, in pairs; sessile spikelets obovate, 2.5-3 mm long; glumes dissimilar, exceeding apex of florets; lower glume obovate, gibbous, upper glume oblong, membranous; florets 2, lower barren, upper bisexual; lower lemma oblong, hyaline; fertile lemma oblong, hyaline.

Occasional along roadsides in Northern Eastern Ghats. Fl. & Fr.: September–October.

Way to Arhikal, Lakshmipuram (VSKP), *NRR* & *DN* 84226 (BSID); Harishankar (BL), *SX* & *MB* 6249 (RRL-B); Jacum, Karlapat Wildlife Sanctuary (KHD), *AC* 2444 (BSID).

INDIA: Bihar, Madhya Pradesh, Odisha, Gujarat, Peninsular India.

COIX Linnaeus

1. Lower glume of the male spikelets narrowly or hardly winged .... **C. lacryma-jobi**
1. Lower glume of the male spikelets broadly winged .... **C. aquatica**

Coix aquatica Roxb., Fl. Ind. 3: 571. 1832; FBI 7: 100. 186; Bor, Grass. India 264. 1960.

A floating and creeping grass; culms up to 3 m high, succulent, floating roots at the nodes. Leaf blades linear to linear-elliptic, 20-60 x 1.5-2.5 cm, margins hispid, apex acute, glandular hairs on the upper surface. Spikes terminal and from exterior axils, peduncled, solitary or in pairs. Male spikes drooping, many-lobed; spikelets 3-seriate, centre one pedicelled, female; ripe involucre abruptly constricted at the neck into a beak.

Floating and creeping in tanks and lakes in Northern Eastern Ghats. Fl. & Fr: October-January

Biccavole (EG), *BO* 3221 (CAL); Madupadu (EG), *VJVS* 96705 (MH).

INDIA: West Bengl, Bihar, Madhya Pradesh, Odisha, Andhra Pradesh, Concan.

WORLD: SE Asia.

Coix lacryma-jobi L., Sp. Pl. 972. 1753; FBI 7: 100. 1896, Fischer 3: 1705. 1934; Bor, Grass. India 264. 1960; Matthew, Fl. Tamilnadu Carnatic 3: 1826. 1983.

A tall, perennial leafy, monoecious grass; culms up to 1 m high, rooting at lower nodes; nodes thick, glabrous. Leaf sheaths 4-6 cm long, glabrous; ligule a very narrow membrane; leaf blades 15-30 x 1-2 cm, convolute, base cordate, margins spinulosely serrate, apex acuminate, midrib thick, glabrous. Racemes 5 mm long, axillary and terminal, lower spikelets solitary, female, male spikelets protrude through the apex of female spikelets; peduncle very long, up to 9 mm. *Female spikelets:* enclosed in a hardened, polished involucre, 9 mm; glumes broadly ovate, acuminate, margin incurved, coriaceous, lemmas ovate, acuminate, sub-coriaceous; palea ovate-oblong, sub-coriaceous; styles basally united, stigmas long, plumose. *Male spikelet*: Racemes up to 40 cm long, spikelet 9 mm, oblong, 2-flowered; lower glume oblong, obtuse, coriaceous, winged, wings scabrid; upper glume oblong, sub-coriaceous; lower lemma oblong, chartaceous; palea oblong; upper lemma lanceolate; palea hyaline, lanceolate; stamens 3. Caryopsis broadly ovate, obtuse, furrowed in the middle.

Occasional along water courses and moist localities. Fl. & Fr.: October–February. Tam.: *Kaatukundumani.*

Nulakamaddi dam Sesharayi (EG), *VNS* 287 (CAL); Madhiveedu (EG), *PVS* & *NRR* 76859 (MH); Rampachodavaram (EG), *MM* 101350 (BSID); Dummakonda RF (EG), *GVS* 68630 (MH); Simlipahar (MBJ), *V.Naik* Acc. No. 3836 (RRL-B); Sirumalai hills (DGL), *SKS* 316 (SGH); Kolli hills (NMK), *SKS* 571 (SGH); Brooklyn–Yercaud (SLM), *KS* 6576 (MH); Yercaud (SLM), *AVNR* 27475 (MH), *DBD* 31411 (MH).

INDIA: Throughout.

WORLD: Native of Tropical Asia but now widely distributed in the Tropics.

CYMBOPOGON Sprengel

1. Sessile spikelet without an awn:
    2. Sessile spikelets linear–lanceolate; lower glume usually shallowly concave in the lower half .... **C. citratus**
    2. Sessile spikelets oblong–oblanceolate; lower glume flat on the back in the lower half............ **C. nardus** var. **nardus**
1. Sessile spikelets awned:
    3. Panicle narrow, of short, dense fascicles of raceme pairs; joints and pedicels villous all over; hairs long, concealing the sessile spikelets............ **C. jwarancusa**

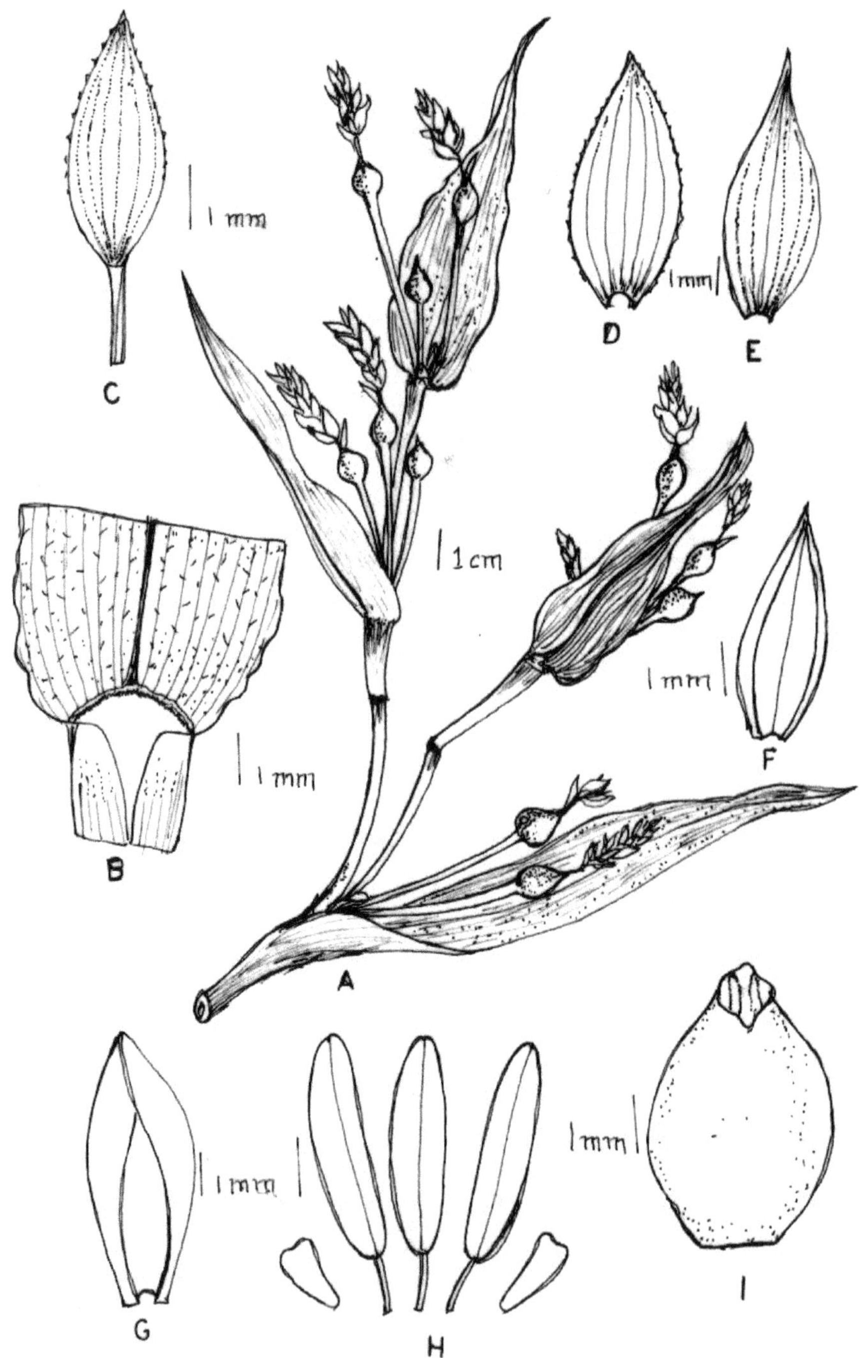

Figure 29. **Coix lacryma-jobi** L.

A. Plant, B. Collar, C. Spikelet, D. Upper glume, E. Lower glume, F. Lemma, G. Palea, H. Stamens and lodicules, I. Bract.

3. Panicle large, decompound or compound, sometimes narrow, but the hairs not covering the joints and pedicels:
    4. Lower glume of the sessile spikelet with a deep slot-like depression occupying most of its length............................**C. gidarba**
    4. Lower glume of sessile spikelet without a slot, either flat, convex, shallowly concave or with a deep narrow slit in the lower half:
        5. Lower glume of the sessile spikelet with a narrow slit-like groove in the lower half which appears as a rib on the inner surface:
            6. Leaf blades cordate and amplexicaul at the base, linear-lanceolate, acuminate...**C. martinii**
            6. Leaf blades narrow and rounded at the base ....**C. caesius**
        5. Lower glume of the sessile spikelet flat in the lower half:
            7. Pedicel of the lowest pedicelled spikelet in the sessile raceme not swollen:
                8. Inflorescence very congested .... **C. nardus** var. **confertiflorus**
                8. Inflorescence lax and loose:
                    9. Lower glume of the sessile spikelet with a concave goove from the base to near apex .... **C. pendulus**
                    9. Lower glume of the sessile spikelet flat, puckered or wrinkled, without a concave groove...........................**C. flexuosus** var. **flexuosus**
            7. Pedicel of the lowest pedicelled spikelet in the sessile raceme swollen ....**C. coloratus**

Cymbopogon caesius (Hook. & Arn.) Stapf, Bull. Misc. Inform. Kew 1906: 341 & 366. 1906; Fischer 3: 1757. 1934; Bor, Grass. India 125. 1960; Matthew, Fl. Tamilnadu Carnatic 3: 1827. 1983. *Andropogon caesius* Nees ex Hook. & Arn. in Hook. Bot. Beechey. Voy. 244. 1838. *A. schoenanthus* L. subsp. *gencinus* var. *caesius* Hack. in DC. Monogr. Phan. 6. 610. 1889; FBI 7: 205. 1896.

A perennial, tufted grass; culms up to 2 m high, erect or ascending, more or less wiry; nodes hairy. Leaf sheaths shorter than the internodes, 5-10 cm long, glabrous; ligule rounded-truncate, scarious; leaf blades 5-25 x 0.5-0.9 long, linear-lanceolate, base rounded, apex tapering with a long setaceous point, glabrous. Spatheate panicle up to 30 cm long; spathes 5 cm long, narrow-lanceolate; spatheoles 2 cm long; narrow-lanceolate; racemes up to 6 mm long, 2-nate, rachis white-villous along the margins. *Sessile spikelet*: oblong, 4 mm, bisexual; lower glume flat in the upper half and with a deep v-shaped groove in the lower; upper glume narrow, sparsely ciliate; lower lemma oblong, truncate; upper lemma stiptiform, ciliate, deeply bifid, awned, epaleated; stamens 3; stigmas plumose. Caryopsis oblong. *Pedicelled spikelet*: linear-lanceolate, 4 mm, male or barren; lower glume 7-10-nerved; upper glume 3-nerved, hyaline; lemmas oblong, 2-nerved, truncate; epaleated; stamens 3.

Abundant in open forests. Fl. & Fr.: August-March.

Muchukota (ATP), *PVP* 2563; Peapuly (KNL), *PVP* 2669; Kammavaripalli (KNL), *TP* & *PVP* 3400; Guvvalacheruvu (KDP), *PVP* 3419; *KS* 6821 (MH); Chinnakudala (KDP), *KRKS* & *BSS* 40466; Near Bapanakanda falls (KDP), *KS* 6821 (CAL); Horsley hills (CTR), *MSG & AA* 12217 & 12219; Nallapadu (GNT), *VRK* 6763; Jammalakonda (CTR), *GVS* 45807 (MH); Near Tiger camp, Maredumilli (EG), *MM* 102579 (BSID); Guddaegudem RF (MBNR), *SRS* 110760 (BSID); Erloo to Mahabubnagar (MBNR), *SRS* 104515 (BSID); Pennagaram RF (SLM), *CAB* 6041 (MH); Manchavadi ghat (SLM), *KS* 7578 (MH); Palakuppam (SA), *VNS* 80170 (MH); Gingee RF (SA), *KMS* 12380 (MH); Tippukadu RF (NA), *KRM* 17699 (MH); Kottiyur RF, Yelagiri hills (NA), *MBV* 613 (MH); Vandalur (CPT), *S.India Flora* 11612 (MH).

INDIA: Hotter regions of India.

WORLD: Pakistan, Sri Lanka, Arabia, East Africa from Sudan to South Africa.

Cymbopogon citratus (DC.) Stapf, Bull. Misc. Inform. Kew 1906: 322. 1906; Fischer 3: 1756. 1934; Bor, Grass. India 126. 1960. *Andropogon citratus* DC., Cat. Hort. Monsp. 78. 1813; FBI 7: 210. 1896.

A perennial, densely tufted grass; culms up to 70 cm high; internodes usually coated with wax. Leaf sheaths loose at base, cinnamon-coloured inside; leaf blades in dense tuft from short annular rhizome, 10 cm long, broad, rough on both sides. Panicle pyramidal; spikelets awnless or reduced to a bristle not visible outside the spikelet, 2-nate. *Sessile spikelet:* linear-lanceolate, 6 mm; lower glume linear-lanceolate, acuminate, concave in lower half on back side.

Cultivated, grown for its aromatic oil. Fl. & Fr.: July–December. Tel.: *Vasana gaddi.*

INDIA: Cultivated.

WORLD: Cultivated in Tropical Asia and South America.

Cymbopogon coloratus (Hook. f.) Stapf, Bull. Misc. Inform. Kew 1906: 321. 1906; Fischer 3: 1756. 1934; Bor, Grass. India 126. 1960; Matthew, Fl. Tamilnadu Carnatic 3: 1828. 1983. *Andropogon nardus* L. var. *coloratus* Hook. f., Fl. Brit. India 7: 206. 1896. *Andropogon coloratus* Nees in Wight, Cat. Ind. Pl. N. 1703. 1883.

A tufted perennial grass; culms up to 2 m high; nodes bearded. Leaf sheaths 5-15 cm, basal tomentose, upper glabrous; ligule fimbriate membrane; leaf blade 10-40 x 0.4-0.8 cm, flat, linear–lanceolate, base rounded, apex acuminate, scabrid. Panicle dense, up to 8 cm, spathes 2 cm long; spatheoles up to 1.5 cm; racemes 4 cm, 2-nate, lowest pedicel of sessile racemes swollen, densely hairy along the margins. *Sessile spikelet:* oblong, 5 mm; lower glume coriaceous, oblong; upper glume hyaline, sparsely hairy, minutely aristate; lower lemma ciliate, keeled, hyaline; upper lemma deeply bifid, awned; stamens 3; stigmas plumose. *Pedicelled spikelet:* oblong, male or barren, 5 mm; glumes keeled: lemmas hyaline, epaleated; stamens 3.

Common in the forest of southern Eastern Ghats. Fl. & Fr.: August–December. Tel.: *Boda gaddi.*

Muchukota RF (ATP), *TP* & *NY* 231; Tadipatri–Bugga (ATP), *TP* 442; Rangapuram RF (KNL), *RVR* 1500; Rachakuntapalli (KDP), *KRKS* & *SSR* 33622; Talakona RF (CTR), *PVP* 2131; Peddamantanala (PKM), *RVK* 15834; Chelama (KNL), *JLE* 17962 (CAL); Chelama (KNL), *JLE* 16788 (MH); Balapalle (KDP), *JSG* 11195 (CAL), *JLE* 14931 (CAL & MH); Tirumala hills (CTR), *KS* 6876 (CAL); Somasila (NLR), *PV* 110212; Veligonda (NLR), *S.India Flora* 12760 (MH); Diguvametta (PKM), *MPG* 117 (DD); Kondaveedu Sacred Grove (GNT), *J.Swamy* & *S.Nagaraju* 14108 (BSID); Kondapalli (KSN), *CAB* 4396 (MH); Umamaheswaram top (MBNR), *SRS* 111333 (BSID); Tippukadu RF (NA), *KRM* 16626 (MH); Usur RF, near Vellore (NA), *KCJ* 6361 (MH).

INDIA: Peninsular to central India.

WORLD: Indo-China.

Cymbopogon flexuosus (Nees ex Steud.) Wats. in Atkins., Gaz. N. W. Prov. Ind. 392. 1882; var **flexuosus**; Fischer 3: 1756. 1934; Bor, Grass. India 127. 1960. *Andropogon flexuosus* Nees ex Steud., Syn. Pl. Glum 1. 388. 1854. *A. nardus* L. var. *flexuosus* Hack. in DC. Monogr. Phan. 6, 603. 1889; FBI 7: 207. 1896.

A perennial, tufted grass; culms up to 3 m high; nodes glabrous. Leaf sheaths 6-15 cm long, glabrous; ligule membranous; blade 15-50 x 1-1.5 cm, lanceolate, base attenuate, margins scaberulous, apex acuminate, glabrous. Panicle decompound, very lax, up to 30 cm long; spathes 3 cm; spatheole to 1.5 cm; racemes 2 cm; spikelets 2-nate. *Sessile spikelet*: 3 mm, oblong, bisexual; lower glume oblong, chartaceous, back with 1-2 slot like depressions, nerveless; upper glume boat-shaped; lower lemma hyaline; upper lemma narrow, awned, epaleated; stamens 3, stigmas plumose. *Pedicelled spikelet:* male or barren, oblong, acumiate; pedicel hairy; lower glume 10-nerved; upper glume 3-nerved; lemma hyaline, ciliate, epaleated; stamens 3.

Common along the hill tops and in open forests through out Eastern Ghats. Fl. & Fr.: July–December.

Kekati RF (ATP), *TP* & *NY* 781 & 782; Amagondapalem RF (ATP), *NY* 1025; Singavaram (KNL), *AA* 2057; Tirumala hills (CTR), *MSG* 12152, *KCJ* 470 (MH);

Nallamalai (KNL), *V.S.Raman s.n.* (MH); Guvvalacheruvu (KDP), *KS* 6821 (MH); Kambakkam hills (CTR), *MCB* 45085 (MH); Sunkarimetta (VSKP), *GVS* 21632 (CAL); Araku valley (VSKP), *GVS* 32884 (CAL); Sirumalai hills (DGL), *SKS* 406 (SGH); Devarabetta (SLM), *KCJ* 18053 (MH); Kolli hills (NMK), *SKS* 578 (SGH); Tulkaa section, Satkosia Tiger Reserve (AGL), *KCM* 5636 (BSID); Jenabil range, Similipal B.R. (MBJ), *AC* 2347 (BSID); Labangi, Satkosia Wildlife Sanctuary (AGL), *AC* 2393, 2396 (BSID).

INDIA: Warmer regions of India.

WORLD: Myanmar, Thailand.

Cymbopogon gidarba (Ham. ex Steud.) A. Camus, Bull. Mus. Natl. Hist. Nat. 26: 562. 1920; Haines, Bot. Bihar Orissa 1048. 1924; Fischer 3: 1757. 1934; Bor, Grass. India 128. 1960. *Andropogon gidarba* Ham. ex Steud., Syn. Pl. Gluma. 1: 387. 1854; FBI 7: 208. 1896.

An aromatic, tufted, perennial; culms up to 75 cm high. Leaf sheaths terete, glabrous; ligule short, truncate; leaf blades 4-40 x 0.4-0.6 cm, linear, margin ciliate at base, apex acuminate, glabrous. Panicles narrow, slender; branches simple; spikelets 6 mm long, 2-nate, joints and pedicels shortly villous. *Sessile spikelet*: pale or dark green, bisexual; lower glume puberulous or quite glabrous in the slot: upper glume cymbiform, 1-nerved, acute; lower lemma lanceolate, 2-nerved; upper lemma linear, bidentate, ciliate, awned; stamens 3, stigmas plumose. *Pedicelled spikelet:* longer than sessile spikelet, lanceolate, acute.

Common in open forests. Fl. & Fr.: July–December.

Palamaner RF (CTR), *KCJ* 457 (MH); Kambakkam hills (CTR), *sine coll. s.n.* (MH); Kakulakonda (NLR), *KCJ* 385; (MH); Kanigiri (NLR), *CAB* 6619 (MH); Medarmetta (PKM), *MPG* 158 (DD); Jaggayyapeta (KSN), *CAB* 8213 (MH); Palakonda (VSKP), *KCJ* 17259 (MH); Maripalli (VSKP), *VNS* 4607 (MH); Champabandh Mali (KHD), *HFM* 3186 (RRL-B); Mahendragiri (GJP), *SX* & *MB* 3655, 3736 (RRL-B); Tikarpada, Satkosia Tiger Reserve (AGL), *KCM* 7073 (BSID); Jacum, Karlapat Wildlife Sanctuary (AGL), *AC* 2443, 2452 (BSID); Kuthadya hill (GJM), *VNS* 5906 (MH).

INDIA: Warmer regions of India.

WORLD: Indo-China, China.

Cymbopogon jwarancusa (Jones) Schult., Syst. Veg. 2, Mant. 458. 1824; Bor, Grass. India 128. 1960. *Andropogon jwarancusa* Jones, Asiat. Res. 4: 109. 1875; FBI 7: 203. 1896.

A perennial, tufted grass; culms up to 70 cm high, creeping at base; nodes glabrous. Leaf sheaths glabrous; ligule ciliate membrane; leaf blades 10-25 x 0.1-0.4 cm, linear, flat or convolute, apex acuminate, glabrous. Panicle very variable; branches distant or crowded; racemes 1.5 cm, unequal; joint and pedicels densely villous with long hairs. *Sessile spikelet*: 4 mm, lanceolate, bisexual; lower glume linear-lanceolate, membranous, nerveless, keels scaberulous; upper glume cymbiform, thinly membranous, 3-nerved; lower lemma linear-oblong, obtuse, hyaline, 2-nerved; upper lemma linear, bidentate, hyaline, ciliate; stamens 3; stigmas plumose. *Pedicelled spikelet:* longer than sessile spikelet, lanceolate; lower glume 7-9-nerved, keels scaberulous; upper glume ovate-oblong, hyaline, ciliate; lower lemma broadly oblong, hyaline.

Rare in open forest in Nallamalais and Udayagiri hills. Fl. & Fr.: August–January

Diguvametta (PKM), *MPG* 117 (CAL); Udayagiri (NLR), *BSN* 4733 (BSID).

INDIA: Warmer regions of India.

WORLD: Iran, Iraq, Oman, Arabia, Pakistan, Bhutan, Nepal, China.

Cymbopogon martinii (Roxb.) W.Watson, Gaz. N.W. India 392. 1882; Fischer 3: 1756. 1934; Bor, Grass. India 129. 1960; Matthew, Fl. Tamilnadu Carnatic 3: 1828. 1983. *Andropogon martinii* Roxb., Fl. Ind. 1: 280. 1820. *A. schoenanthus* L. var. *martinii* (Roxb.) Hook. f., Fl.Brit. India 7: 204. 1896.

A perennial, tufted, sweet scented grass; culms up to 2 m high, straw-coloured, glabrous. Leaf sheaths 7-15 cm, glabrous; ligule very long, membranous; leaf blades 10-25 x 1-1.5 cm, flat, linear-lanceolate, base cordate, apex acuminate, midrib prominent. Panicle up to 25 cm long; spathes 3 cm; spatheoles 1.5 cm; lower most pedicel swollen; rachis pilose along the margins; spikelets 2-nate. *Sessile spikelet:* bisexual, oblong, 4 mm; lower glume chartaceous, 2-keeled, slightly bidentate; upper glume boat-shaped, 1-keeled, ciliate; lower lemma narrow-lanceolate, ciliate; upper lemma narrow, deeply bidentate, awned, epaleated; stamens 3, stigmas plumose. *Pedicelled spikelet:* male or barren, oblong; pedicel hairy; lower glume 9-nerved, slightly keeled; upper 2-nerved, hyaline, ciliate; lemmas hyaline, ciliate, epaleate; stamens 3.

Occasional in open forests. Fl. & Fr.: July–December. Tam.: *Moongilpul, Cooraipul.*

Peccheruvu (KNL), *PVP* & *RVR* 2176; Chelama (KNL), *MPG* 105 (CAL); Gundlabrahmeswaram (KNL), *JLE* 16925 (MH); Yerrabalem (KDP), *BDP* 160 (CAL); Sirumalai hills (DGL), *SKS* 409 (SGH); Kolli hills (NMK), *SKS* 581 (SGH); Vandalur (CPT), *sine coll. s.n.* (MH); Moolakadu (SA), *KRM* 77462 (MH); Banigoccha, Satkosia Tiger Reserve (AGL), *KCM* 8356 (BSID); Kollimalai (TP), *S.India Flora* 13011 (MH); Yercaud (SLM), *KCJ* 18068 (MH).

INDIA: Throughout hotter parts of India.

WORLD: Pakistan, Cultivated in the Tropics.

Cymbopogon nardus (L.) Rendle, Cat. Welw. Afr. Pl. 2: 155. 1899.

Key to Varieties

1. Spikelets awned .... var. **confertiflorus**
1. Spikelets unawned .... var. **nardus**

Cymbopogon nardus (L.) Rendle, var. **confertiflorus** (Steud.) Stapf ex Bor, J. Bombay Nat. Hist. Soc. 51: 905. 1853 & Grass India 130. 1960. *Andropogon confertiflorus* Steud., Syn. Pl. Glumac. 1. 385. 1854. *A. nardus* L. subsp. *nilagiricus* Hack. in A. DC. Monog. Phan 6: 604. 1889; FBI 7: 206. 1896. *Cymbopogon confertiflorus* (Steud.) Stapf, Bull. Misc. Inform. Kew 1906: 318. 1906; Fischer 3: 1756. 1934.

An aromatic, perennial; culms up to 80 cm high. Ligule coriaceous; blades 5-10 x 2-4 cm, glaucous below. Panicle 30 cm, erect, dense, interrupted; spathes 1.5 cm, broadly lanceolate, cymbiform, acuminate. *Sessile spikelet:* 5 mm, dorsally flat, glabrous; lower glume slightly depressed; upper lemma bearing a geniculate awn.

Occasional in open forests. Fl. & Fr.: July–December.

Penukonda (ATP), *S.India Flora* 14322 (MH); Kondapalli (KSN), *CAB* 4396 (MH); Rampa hills (EG), *VNS* 462 (MH); Araku valley (VSKP), *GVS* 32884 (MH); Pecheruvu (KNL), *PVS & TRS* 79526 (MH, BSID); Ramanadurg (Ballari), *S.India Flora* 15918 (MH); Sirumalai hills (DGL), *SKS* 318 (SGH); Devarbetta (SLM), *KCJ* 18052 (MH).

INDIA: East to South India.

WORLD: Sri Lanka, Java.

Cymbopogon nardus (L.) Rendle var. **nardus;** Fischer 3: 1756. 1934; Bor, Grass. India 130. 1960. *Andropogon nardus* L., Sp. Pl. 1046. 1753; FBI 7: 205. 1896.

A perennial, tufted grass; culms up to 60 cm high, rootstock stout, leafy. Leaf sheaths *c.* 13 cm, lower ones fibrous at maturity, more or less ribbed; ligule membranous, lacerate with fimbriate apex, brownish yellow; leaf blade linear-lanceolate, 17-46 x 1.4-1.6 cm, base rounded to truncate, apex acuminate. Panicle large, subpyramidal, interrupted; branches loosely or closely packed, drooping; spathes imbricate, lanceolate, 2 cm; spikes with 4-5 pairs of spikelets; joints and pedicels ciliate. *Sessile spikelet:* 5 mm, lanceolate, awnless; lower glume lanceolate, acute, back flat, 2-nerved, 2-keeled, keels narrowly winged, wings with ciliolate or serrulate margins, margins incurved; upper glume oblong-lanceolate, boat-shaped, chartaceous, 3-nerved, 1-keeled, keel narrowly winged from middle above, margins incurved, Florets 2, lower barren, upper bisexual; lower lemma narrowly oblong-lanceolate, hyaline to brownish, margins infolded, ciliate towards apex, lower palea absent; upper lemma narrowly oblong, 2-lobed at apex, awn from sinus, geniculate, upper palea oblong, membranous, hyaline, nerveless, lodicules 2; stamens 3. Caryopsis oblong-lanceolate. *Pedicelled spikelet:* narrowly oblong-elliptic, acute at apex, unawned, lower glume ovate-lanceolate to oblong-lanceolate, brownish, prominently 11-nerved, 2-keeled, keels narrowly wingd, wings serrulate, margins incurved, upper glume oblong-lanceolate, 3-nerved, margins incurved, ciliate towards apex. Florets 2, both epaleate, lower barren, upper bisexual; lower lemma narrowly oblong-lanceolate, nerveless, margins ciliate towards apex; upper lemma narrowly oblong, hyaline, sometimes reduced; stamens 2.

Only known in cultivation.

Jenabil Range, Similipal B.R. (MBJ), *AC* 2301, 2348 (BSID).

INDIA: Throughout hotter parts of India.

WORLD: Sri Lanka, China. Introduced elsewhere as crop plant.

Cymbopogon pendulus (Nees ex Steud.) W. Watson, Gaz. N.W.Ind. 392.1882; Bor, Grass. India 131. 1960; Moulik, Grass. & Bamb. India 1: 217. 1996. *Andropogon pendulus* Nees ex Steud., Syn. Pl Glum. 1: 388. 1854.

Perennials, erect, tufted, to 2 m high. Leaf sheaths glabrous; ligules membranous; leaf blades 50-60 x 0.6 cm, linear-lanceolate, acute, glabrous, aromatic. Inflorescence composed of racemes, subtended by a spatheole; racemes 2, paired; spikelets 2-nate, one sessile and the other pedicelled. *Sessile spikelets*: *c.* 4.3 x 0.9 mm, linear-lanceolate, awn *c.* 4.5 mm; lower glume *c.* 4.2 x 0.8 mm, oblong-lanceolate 2-keeled, keels winged above, nerveless grooved on the dorsal surface in the lower half; upper glume *c.* 3.5 x 0.8 mm, lanceolate, 2-keeled, margins ciliate; lower lemma 3 x 0.5 mm, 2-keeled, keels ciliate at top; upper lemma *c.* 1.6 x 0.4 mm, 2-cleft, ciliate, awned, awn *c.* 4.5 mm long; palea reduced; ovary *c.* 0.5 x 0.2 mm, oblong; styles 0.9 mm lng, stigmas *c.* 1 mm long. *Pedicelled spikelet*: *c.* 3 x 0.7 mm, linear-lanceolate, shorter than sessile spikelets, purplish or reddish coloured; lower glume *c.* 3 x 0.7 mm, lanceolate, 7-nerved, keeled, acute, scabrid; upper glume *c.* 2.6 x 0.6 mm, ovate-lanceolate, 3-nerved, upper half ciliate; lower lemma absent; upper lemma 2.5 x 0.6 mm, 2-keeled, margins ciliate, hyaline; stamens 3, *c.* 2 x 0.4 mm.

Occasional on hill tops in Anantapuram district (B.R.P.Rao *et al.*, 2012). Fl. & Fr.: December.

Gounivariplli-Samrajyamkonda (ATP), *BR* & *KVS* 29719 (SKU).

INDIA: E Himalaya, E. & NE India.

WORLD: Bhutan, Nepal, China.

CYNODON Richard *nom. cons.*

1. Lemmas longer than the upper glume:
   2. Stoloniferous; ligule a ring of hairs; spikes 2-6, up to 6 cm long; hairs on the lemma not clavellate .... **C. dactylon**
   2. Not stoloniferous; ligule membranous; spikes numerous, up to 10 cm long; hairs on the lemma some times clavellate........ **C. radiatus**
1. Lemmas equal or slightly longer than the upper glume .... **C. barberi**

Cynodon barberi Rang. & Tadul., J. Bombay Nat. Hist. Soc. 24: 846. 1916; Fischer 3: 1836. 1934; Bor, Grass. India 469. 1960.

A perennial, tufted grass; culms up to 30 cm high; nodes glabrous. Leaf sheaths 2-4 cm long, glabrous; ligule ciliate; leaf blade short, 2-6 x 0.5-1.5 cm, flat, narrow, base truncate. Spikes 3-5, up to 6 cm long, digitate; spikelets 3 mm long, closely appressed to the rachis; lower glume lanceolate; upper glume longer than the lemmas; lemma narrowly boat-shaped, keels with clavellate hairs; palea linear, as long as the lemmas; stamens 3; stigmas plumose. Caryopsis oblong.

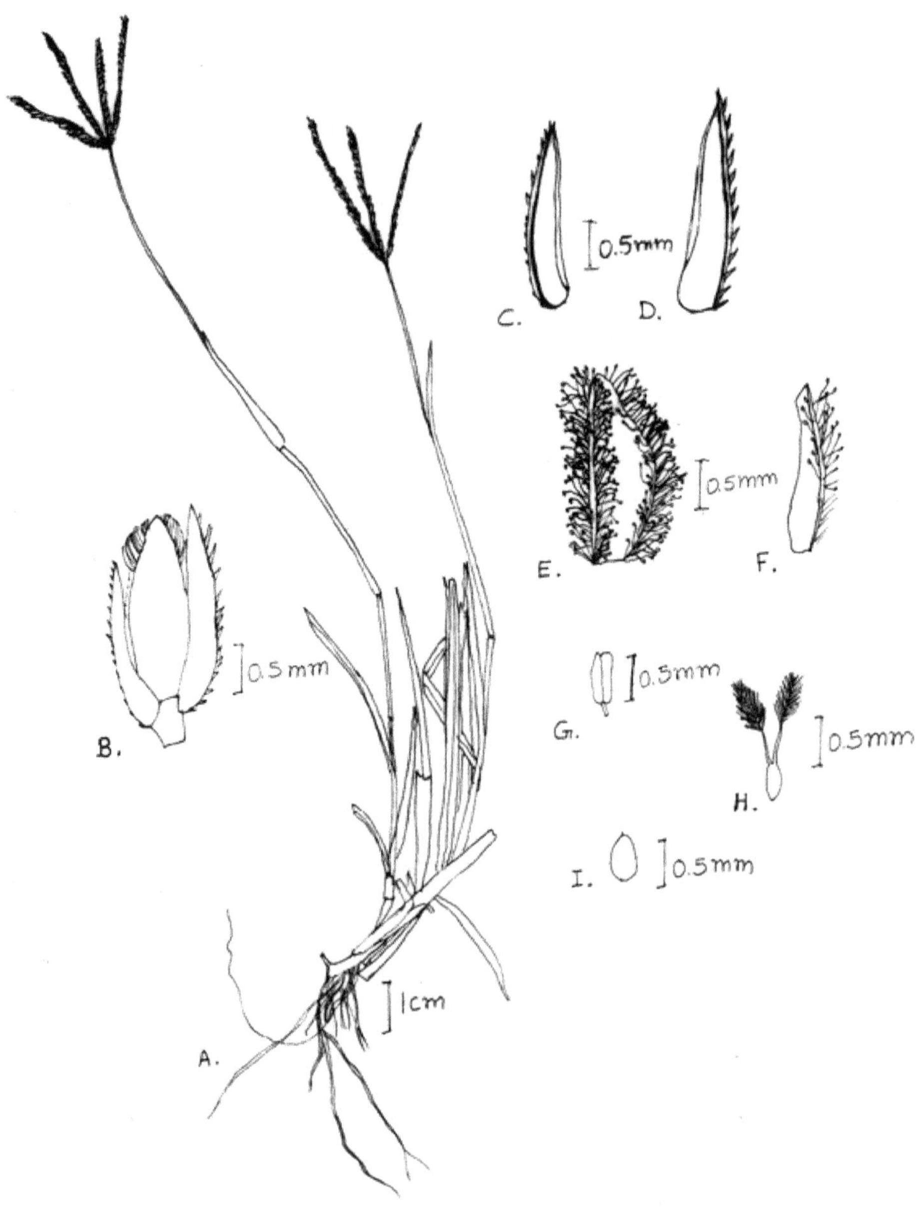

Figure 30. **Cynodon barberi** Rang. & Tadul.

A. Twig, B. Spikelet, C. Lower glume, D. Upper glume, E. Lemma, F. Palea, G. Stamen, H. Pistil, I. Caryopsis.

Occasional in open forests and waste lands. Fl. & Fr.: August–October.

Gani RF (KNL), *RVR* 1528; Gudipala (CTR), *KCJ* 429 (MH); Panapakkam (CTR), *CAB* 15789 (MH); Ethamukkala (PKM), *K.Ramamurthy* & *KR* 88854 (BSID); Maripilli (VSKP), *VNS* 4603 (MH); Red hills (CPT); *S.India Flora* 14188 (MH); Uragadam (CPT), *S.India Flora* 14075 (MH); Krishnagiri (SLM), *S.India Flora* 13870 (MH); Gudianthum (SA) *KCJ* 635 (MH); Krishnagiri (SLM), *CAB* 13870, 14906 (MH); Balipadar (GJM), *AC* 10222 (BSID).

INDIA: Rajasthan, Uttar Pradesh, Bihar, Odisha, West Bengal, Andhra Pradesh, Tamil Nadu, Karnataka, Kerala.

WORLD: Myanmar.

Note: This grass may be confused with *Cynodon dactylon*, but can be easily distinguished by the absence of stoloniferous stems, acute leaves, upper glume always equal to or longer than the lemma, presence of clavellate hairs on the keels and margins of the lemma and on the keels of the palea.

Cynodon dactylon (L.) Pers., Syn. Pl. 1: 85. 1805; FBI 7: 288. 1960; Fischer 3: 1835. 1934; Bor, Grass. India 469. 1960; Matthew, Fl. Tamilnadu Carnatic 3: 1829. 1983. *Panicum dactylon* L., Sp. Pl. 58. 1753.

A perennial, stoloniferous grass; culms creeping by scaly rhizomes or by flat stolons forming matted tufts, decumbent, later erect; nodes glabrous. Leaf sheaths 0.3-3 cm long, glabrous; ligule ciliate; leaf blades 1-10 x 0.2-0.3 cm distichous, linear, base rounded, apex acuminate, softly puberulous. Spikes 4 cm long, 3-4, digitate; spikelets sub sessile, 3 mm long, laterally compressed; glumes sub equal, lanceolate, 1-nerved; lemma oblong, aristate, keels ciliate; palea linear, as long as the lemma; stamens 3, stigmas plumose. Caryopsis linear-oblong.

Very common in plains and lower hills in all hill ranges. Fl. & Fr.: August–October. Tel.: *Garika gaddi*: Tam.: *Arugampul*.

Banaganapalli (KNL), *RVR* 3060; Peapuly (KNL), *PVP* 2658; Thummalapalle (KDP), *KRKS* & *KRS* 33693; Ramgiri (CTR), *BR* & *P.Priyadarsini* 37118; Krishna River bed (GNT), *VRK* 6745; Pathakothacheruvu (ATP), *JSG* 20847 (MH); Horsley hills (CTR), *S.India Flora* 15470 (MH); Vayalpad (NLR), *CAB* 15486 (MH); Diguvametta (PKM); *JLE* 42156 (MH); Polavaram agency (WG), *DCSR* 153 (CAL); Gobanapalem (WG), *KMS* 5156 (CAL); Thummagaddilanka (EG), *CAB* 4348 (MH); Anantagiri (VSKP), *GVS* 19465 (CAL); Gajapathinagaram (VJN), VNS 4592 (MH); on the way to Puliyur from Kuttathur (NA), *KS* 6529 (MH); Inner Javadi RF (NA), *KS* 6103 (MH); Vemmundi, Palakuppam (SA), *VNS* 4147 (MH); Belghar, Kotgarh Wildlife Sanctuary, (KDM), *AC* 2408 (BSID); Badrama Wildlife Sanctuary (SBP), *AC* 10240 (BSID); Charmariya hill base (GJM), *VNS* 4666 (MH); Burmaria hill (GJM), *VNS* 4666 (MH).

INDIA: Throughout India.

WORLD: Tropical and Warm Temperate regions of the World.

Cynodon radiatus Roth ex Roem. & Schult. Syst. Veg. 2: 411. 1817. *C. arcuatus* J.S. Presl in Presl, Reliq. Haenk. 1: 290. 1830; Bor, Grass. India 489. 1960; Saxena & Brahmam, Fl. Orissa 4: 2310. 1996; Moulik, Grass. Bamb. India 2: 565. 1997. *C. dactylon* (L.) Pers. var. *intermedius* (Rang. & Tadul.) Fischer, Fl. Madras 3: 1836. 1934. *C. intermedius* Rang. & Tadul., J. Bombay Nat. Hist. Soc. 26: 309. 1918.

A perennial, stoloniferous grass; culms straggling, up to 30 cm high, roots near the nodes; nodes glabrous. Leaf sheaths 2-5 cm, glabrous; ligule membranous; leaf blade 6-12 x 0.3-0.6 cm, lanceolate, base cordate, margin finely serrulate, apex acute, glabrous. Digitate racemes 4, terminal; racemes 4-10 cm long; spikelets subsessile, 25 mm, elliptic, acuminate; lower glume boat-shaped, narrowly ovate, 1-nerved, 1-keeled, keel scabrous, apex acute; upper glume membranous, boat-shaped, narrowly ovate, 1-nerved, 1-keeled, keels scabrid, apex acute; lemma hyaline, curved, 3-nerved, 1-keeled, keels covered with clavate hairs, apex obtuse; palea involute, 2-nerved, scaberulous back on nerve; lodicules 2; stamens 3, stigmas plumose. Caryopsis elliptic.

Occasional in forests and plains. Fl. & Fr.: August–October.

Rampa (EG), *VNS* 395 (CAL); Godavari, *BO* 3484 (CAL); Chinturu, Maredumilli (EG), *MM* 101399 (MH); Mahendragiri (GJP), *Srivastava & Party* 38323 (LWG); Gunpur (KHD), *HFM* 3541 (DD); Daspalla, Satkosia Tiger Reserve (AGL), *KCM* 8339 (BSID).

INDIA: Uttar Pradesh, Bihar, West Bengal, Assam, Odisha, Madhya Pradesh, Maharashtra, Andhra Pradesh, South India.

WORLD: Nepal, Myanmar, SE Asia, Malaya.

CYRTOCOCCUM Stapf

1. Spikelets shortly pedicelled, rarely as long as the spikelets:
    2. Leaves 2.5-3 cm long, linear-lanceolate; spikelets hispid, few .... **C. trigonum**
    2. Leaves up to 10 cm long, elliptic-oblong; spikelets glabrous, numerous .... **C. oxyphyllum**
1. Spikelets on capillary, longer pedicels:
    3. Spikelets 1.75-2 mm long; glumes not cuspidate .... **C. deccanense**
    3. Spikelets 1.5 mm or less, if longer glumes cuspidate:
        4. Lower lemma obtuse .... **C. patens**
        4. Lower lemma and glume cuspidate .... **C. longipes**

Cyrtococcum deccanense Bor, Kew Bull. 1956: 255. 1956 & Grass. India 291. 1960.

Perennial, creeping or decumbent; culms up to 70 cm, nodes glabrous. Leaf sheath 4-8 x 0.3-0.5 cm; ligule 1 mm long, membranous; leaf blade 5-10 x 0.5-0.8 cm, elliptic or elliptic-lanceolate, apex acuminate, base narrowed. Panicle 10-20 cm long, lax; spikelets with long pedicels, alternate on the rachis, 1.7 -2 x 1 mm, ovate or obovate; lower glume ovate; upper glume boat-shaped; lower floret empty; palea rudimentary; upper floret bisexual; upper lemma deltoid, 1.4-2 x 1-1.5 mm, 5-nerved; palea ovate-oblong, 1-15 x 0.6-0.9 mm, 2-nerved; stamens 3; styles 1.5 mm long, stigmas 1 mm long. Caryopsis ovate.

Rare in moist places of forests in Visakhapatnam district (A.M.Reddy *et al.*, 2004). Fl. & Fr.: July–November.

Chintapalli (VSKP), *BR* & *AMR* 24590 (SKU).

INDIA: South India.

WORLD: Sri Lanka.

Cyrtococcum longipes (Wight & Arn. ex Hook.f.) Camus, Bull. Mus. Hist. Nat. (Paris) 27: 118. 1921; Fischer 3: 1786. 1934; Bor, Grass. India 291. 1960. *Panicum longipes* Wight & Arn. ex Hook. f., Fl. Brit. India 7: 58. 1896.

A perennial, weak grass; culms up to 30 cm high, slender, creeping, decumbent. Leaf sheaths nearly glabrous; ligule rounded; blades 8-20 x 0.5-1.5 cm, narrow, flat, apex acuminate, sparsely hairy at base, puberulous below at apex. Panicle up to 12 cm long, effuse, fastigiately branched; spikelets 2 mm, pale brown; pedicels longer than the spikelet; glumes glabrous, acute or cuspidate, keeled; lower lemma obtuse, apex ciliate; upper lemma semi-ovate, with a linear or oblong, apical pale callus; palea having a small round apical callus; stamens 3, stigmas plumose. Caryopsis small, obovoid, free with the hardened lemma and palea.

Occasional at Bison hill. Fl. & Fr.: April–August

Bhupatipalem (EG), *TP* & *MSG* 12248; Bison hill (EG), *CAB* 5094 (MH); Maredumilli (EG), *MM* 101261 (MH); Purnakote, Satkosia Wildlife Sanctuary (AGL), *AC* 2374 (BSID); Jenabil Range, Similipal B.R. (MBJ), *AC* 2335 (BSID)

INDIA: South India.

WORLD: Indo-China.

Cyrtococcum oxyphyllum (Steud.) Stapf. in Hooker's Icon. Pl. sub. f. 3096. 1922; Fischer 3: 1786. 1934; Bor, Grass. India 291. 1960; Matthew, Fl. Tamilnadu Carnatic 3: 1830. 1983. *Panicum oxyphyllum* Hochst. ex Steud., Syn. Pl. Glumac. 1: 65. 1854. *P. pilipes* Nees & Arn. ex Buese in Miq. Pl. Jungh. 3: 376. 1854; FBI 7: 57. 1896.

A perennial, stoloniferous grass; culms creeping; nodes glabrous. Leaf sheaths 1-2 cm long, glabrous; ligule membranous; blades 3-6 x 0.5-0.7 cm, lanceolate, base rounded, apex acuminate, glabrous. Panicle up to 7 cm long, contracted; pedicels long, 0.3 mm; spikelets 2 mm, globose, glabrous; glumes elliptic-lanceolate, 1-nerved, upper boat-shaped, scaberulous on back side midnerve; lower lemma obovate, ciliate tip; palea broadly ovate, 2-nerved, ciliate at base, neuter, upper lemma boat-shaped; palea boat-shaped, scaberulous near tip. Bisexual floret: stamens 3, linear; stigmas plumose. Caryopsis globose.

Occasional in moist and shady places in forests. Fl. & Fr.: April–August.

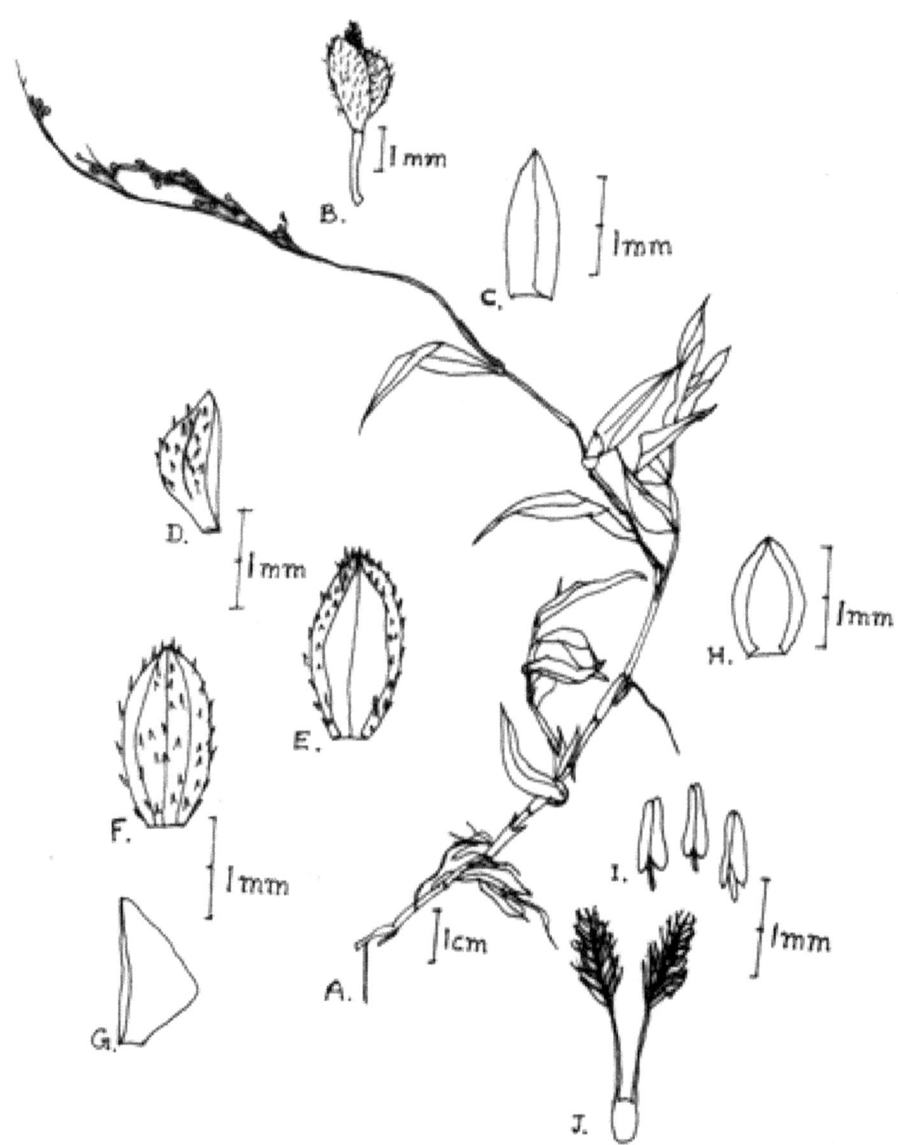

Figure 31. **Cyrtococcum longipes** (Wight & Arn. ex Hook. f.) Camus.

A. Twig, B. Spikelet, C. Lower glume, D. Upper glume, E & F. Lower lemma front and back views, G. Upper lemma, H. Palea, I. Stamens, J. Pistil.

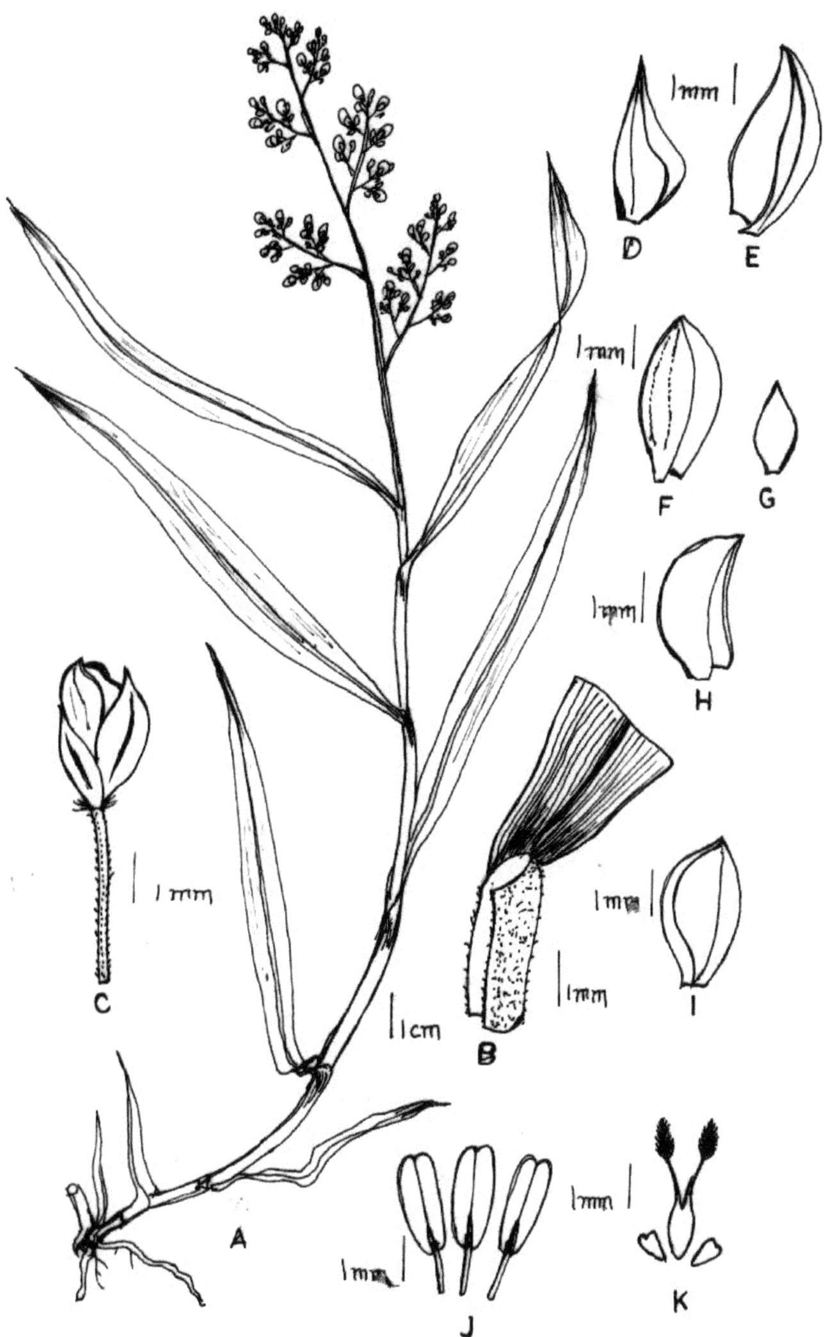

Figure 32. **Cyrtococcum oxyphyllum** (Steud.) Stapf.

A. Plant, B. Collar, C. Spikelet, D. Upper gluma, E. Lower glume, F. Upper lemma, G. Lower lemma, H. Outer palea, I. Inner palea, J. Stamens, K. Ovary and lodicules.

Mogalipenta (KDP), *MVS* & *MB* 38625; Tirumala hills (CTR), *MSG* 12171; Penchalakona (NLR), *MSG* 12195; Maredumilli (EG), *TP* & *MSG* 12263, *GVS* 68508 (MH); Bakirimanchela- way to Ramanapenta (KNL), *JLE* 32646 (MH); Dummakonda (EG), *VNS* 195 (CAL); Bison hill (EG), *CAB* 5055 (MH); Nilavarampadu (VSKP), *VNS* 542 (CAL); Anantagiri (VSKP), *GVS* 2725 (MH); Sirumalai hills (DGL), *SKS* 617 (SGH); Mahendragiri (GJP), *SX* & *MB* 3899; Similipahar (MBJ), *SX* & *MB* 4250, 4 (RRL-B); Chatkei, Purunakote section, Satkosia Tigr Reserve (AGL), *KCM* 5242 (BSID); Barakamara, Similipal Biosphere Reserve (MBJ), *ARKS* & *GPS* 12382 (BSID); Ghat Road, Yercaud (SLM), *DBD* 31437 (MH); Manjakuttai–Yercaud (SLM), *DBD* 31408 (MH).

INDIA: Sikkim, Terai to Nilgiris.

WORLD: Sri Lanka, Myanmar, China and Malesia.

Cyrtococcum patens (L.) A. Camus, Bull. Mus. Natl. Hist. Nat. 27: 118. 1921. var. **latifolium** (Honda) Ohwi, Acta Phytotax. Geobot. 11: 47. 1942 *C. accrescens* Trin. ex Stapf in Hook. Icon. Pl. Sub. Tab. 3096. 1922; Bor, Grass. Ind. 291. 1960; Saxena & Brahmam, Fl. Orissa 4: 2315. 1996.

Perennial weak herbs. Leaves narrow, flat. Inflorescence 10-20 cm long, in lax panicles and longer than spikelets; spikelets obovate, much laterally compressed; glumes thinly membranous; lemmas dissimilar; pedicels long, slightly swollen and cup-shaped at apex. Caryopsis small, obovoid, free within the hardened lemma and palea.

Occurs in moist valleys and also on higher altitudes near stream banks in Northern Eastern Ghats. Fl. & Fr.: September–December.

Similipahar (MBJ), *SX* & *MB* 4629 (RRL-B); Tulka section, Satkosia Tiger Reserve (AGL), *KCM* 5624 (BSID); Mahendragiri hills (GJP), *AC* 4207 (BSID).

INDIA: Throughout India.

WORLD: SE Asia.

Cyrtococcum trigonum (Retz.) A. Camus, Bull. Mus. Hist. Nat. Paris 27: 118. 1921; Fischer 3: 1785. 1934; Bor, Grass. India 292. 1960; Matthew, Fl. Tamilnadu Carnatic 3: 1831. 1983. *Panicum trigonum* Retz., Obs. Bot. 3.9. 1783; FBI 7: 56. 1896.

A perennial, stoloniferous grass; culms creeping, roots at nodes, branches erect, up to 10 cm long; nodes glabrous. Leaf sheaths 0.5-1 cm long, ciliate near the mouth; blades 2.5-4.5 x 0.3-0.5 cm, narrowly lanceolate, base rounded, apex acute, glabrous. Panicle up to 4 cm long; pedicels short or subsessile; spikelets 15 cm long, sub-globose; glumes membranous, glabrous; upper boat-shaped; lemmas obovate-oblong, puberulous; lower floret neuter; upper lemma lanceolate; stamens 3; stigmas plumose. Caryopsis sub-globose.

Common in the dense shade in moist places. Fl. & Fr.: July–March.

Horsley hills (CTR), *PVP* 4664, *MSG* 12215; Mangapuram (CTR), *RVR* & *GO* 2445; Kailasanathakona (CTR), *BR* & *P.Priyadarsini* 37136 (BSID); Bhupatipalem

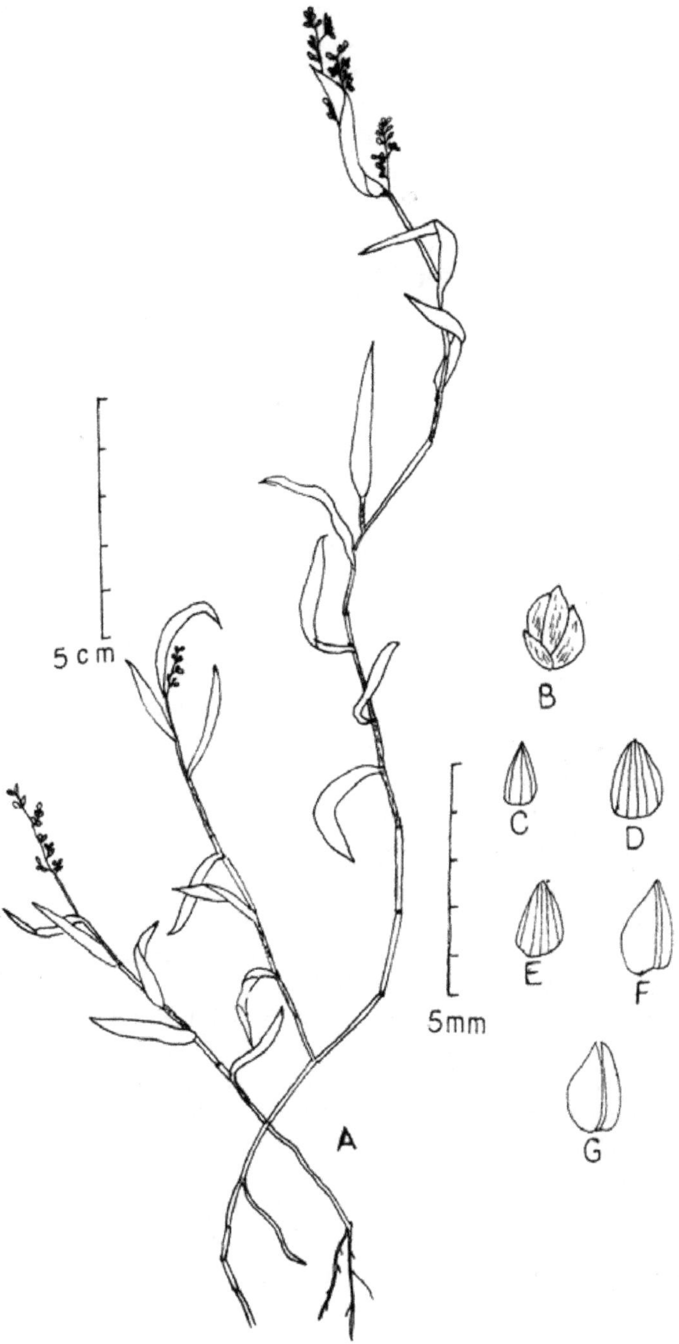

Figure 33. **Cyrtococcum trigonum** (Retz.) Camus.

A. Twig, B. Spikelet, C. Lower glume, D. Upper glume, E. Lower lemma, F. Upper floret, G. Upper lemma.

(EG), *TP & MSG* 12247 & 12242; Simhachalam (VSKP), *TP & EC* 7314; Papanasanam (CTR), *KS* 6953 (MH); Polavaram agency (WG), *DCSR* 202 (CAL); Rayapalli (EG), *VNS* 19 (CAL); Gokavaram (EG), *CAB* 8260 (MH); Santhamarandahalli (SLM), *S.India Flora* 9743 (MH); Sirumalai hills (DGL), *SKS* 412 (SGH); Alagar hills (DGL), *K. Althaf Ahamed Kabeer* 113276 (MH); Jenabil Range, Similipal B.R. (MBJ), *AC* 2321 (BSID).

INDIA: Deccan Peninsula, Odisha, West Bengal.

WORLD: SE Asia.

DACTYLOCTENIUM Willdenow

1. Stoloniferous; spikes 2-6 .... **D. aegyptium**
1. Non-stoloniferous; spikes solitary .... **D. aristatum**

Dactyloctenium aegyptium (L.) P. Beauv., Ess. Agrost. Expl. Pl. 15.1812; Fischer 3:1840.1934; Bor, Grass. India 489. 1960; Matthew, Fl. Tamilnadu Carnatic 3: 1831. 1983; Moulik, Grass. Bamb. India 2:585.1997. *Eleusine aegyptia* (L.) Desf., Fl. Atlant. 1: 85. 1798; FBI 7:295. 1896. *Cynosurus aegyptius* L., Sp. Pl. 72.1753.

Annuals, culms stoloniferous, to 70 cm, erect or ascending, rooting at nodes; nodes glabrous. Leaf sheaths glabrous; ligules membranous; leaf blades linear-lanceolate, acuminate, glabrous, sparsely ciliate on the margins. Spikes digitate; spikelets oblong, 3-5-flowered; lower glume ovate, boat-shaped, 1-nerved, keeled, scabrid; upper glume obovate, 1-nerved, mucronate, scabrid; lemma ovate-lanceolate, mucronate, keeled; palea lanceolate, acuminate, 2-keeled; stamens 3. Caryopsis obovate, transversely rugose.

Common in forest plains, margins and found as a weed in crop fields and waste lands throughout. Fl. & Fr.: June–September.

Kalasamudram (ATP), *BR & AMR* 21898; Kammavaripalli (KNL), *TP & PVP* 2780; Kambagiri konda (KNL), *BR & SS* 20010; Palakonda (KDP), *CS & DAM* 7708; Thummalapalli (KDP), *KRKS & KRS* 36412 (BSID); Krishnapuram (CTR), *AL & BSS* 28567; Leguntapadu (NLR), *BR & PMR* 16432; Macherla (GNT), *VRK* 3242; Lambasangi (VSKP), *AMR* 25630; Tatipudi (VJN), *AMR* 24526; Diguvametta (PKM), *JLE* 42181 (MH); Tirupati hills (CTR), *KS* 6875 (MH); Kondaveedu fort (GNT), *CAB* 4704 (MH); Polavaram agency (WG), *DCSR* 251 (CAL); Rampa (EG), *VNS* 393 (CAL); Nattavaram (VSKP), *CAB* 1894 (MH); Galikonda (VSKP), *GVS* 19587 (MH); Gajapathinagaram (VJN), *VNS* 4591 (MH); Salur (VJN), *NPBK* 957 (CAL); Borapuram (MBNR), *BSS & SKB* 29567, 32291; Mamillagudem (KMM), *RR* 105934 (BSID); Ramanagutta (KMM), *RR* 107993 (BSID); Pennagaram (SLM), *EV* 20775 (MH); Poonamalle (CPT), *S.India Flora* 14847 (MH); Red hills (CPT), *CAB* 5512 (MH); Palur (NA), *VNS* 4105 (MH); Tippukadu RF (NA), *KRM* 17923 (MH); Palakuppam (SA), *VNS* 4040 (MH); Red hills (CPT), *CAB* 5512 (MH).

INDIA: Throughout India.

WORLD: Tropical and Warm Temperate regions of the World.

Dactyloctenium aristatum Link, Hort. Berol. 1: 59. 1827; Bor, Grass. India 489. 1960; Moulik, Grass. Bamb. India 2: 587. 1997. *Eleusine aristata* Hook. f., Fl. Brit. India 7: 296.1896.

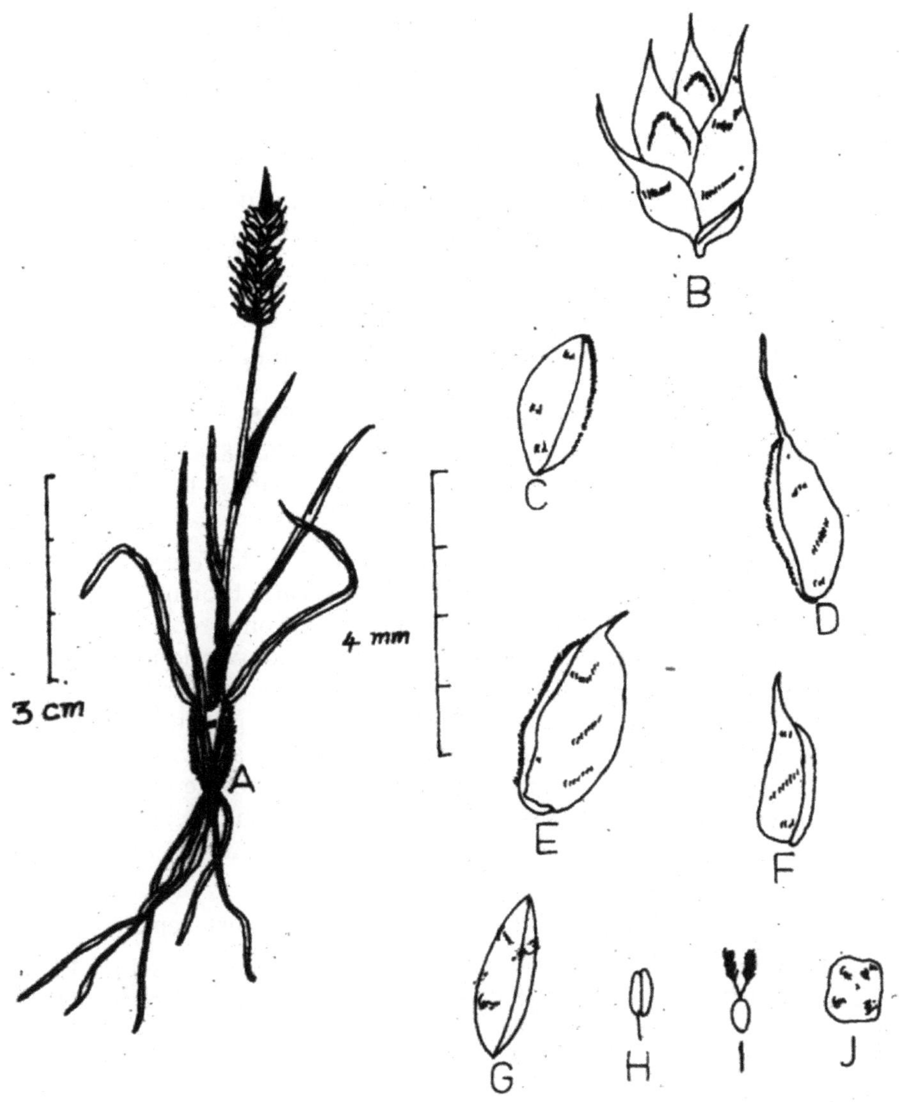

Figure 34. **Dactyloctenium aristatum** Link

A. Twig, B. Spikelet, C. Empty lemma, D. Lower glume, E. Upper glume, F. Fertile lower lemma, G. Palea, H. Stamen, I. Grain.

Annuals, culms erect or geniculately ascending, tufted, to 10 cm, rooting at nodes; nodes glabrous. Leaf sheaths hairy; ligules fimbriate, membranous; leaf blades linear-lanceolate, base cordate or rounded, acuminate, glabrous. Spike solitary; spikelets ovate, 2-3-flowered; lower glume ovate, 1-nerved, keeled, chartaceous; upper glume ovate-elliptic, aristate, scabrid, chartaceous; lower lemma ovate-lanceolate, membranous, keeled, acute; upper lemma akin to lower lemma; palea ovate-lanceolate, chartaceous, keeled; stamens 3. Caryopsis globose, rugose.

Rare in open forests and peripheries in Southern Eastern Ghats. Fl. & Fr.: June–November.

Somulavaripalli (ATP), *KRKS* & *BSS* 40180; Peaupuly (KNL), *PVP* & *RVR* 2675, 3174; Rudrakod (KNL), *SS* & *AMR* 20197; Hanimireddi palli (ATP), *JNP* & *AL* 26892; Nagalooty (KNL), *BR* & *BSS* 30149; Velgode (KNL), *BSS* & *SKB* 30230 (SKU).

INDIA: NW and South India.

WORLD: Kenya north to Sudan and eastwards through Arabia to India.

DENDROCALAMUS Nees

Dendrocalamus strictus (Roxb.) Nees, Linnaea 9: 476. 1834; FBI 7: 404. 1896; Fischer 3: 1858. 1934; Matthew, Fl. Tamilnadu Carnatic 3: 1832. 1983. *Bambusa stricta* Roxb., Pl. Coromandel t. 80. 1798.

A perennial, unarmed, tufted grass; culms up to 15 m high, erect from branched root stock, solid or only with a small cavity; culm sheaths brown, glabrous-hairy, triangular, 5-10 x 6-14 cm; nodes swollen, the lower node often rooting. Leaf sheaths 3-15 cm, striate, glabrous; auricle short, ciliate with few waxy deciduous hairs; ligule narrow, serrate; leaf blades 5-15 x 1.5-3 cm, linear, lanceolate, erect, stiff; shortly petiole. Inflorescence a compound panicle, spikelets in congested globose heads in long spikes, heads 1-5-3 cm wide; spikelets 1 cm long, spiny, 1-3-flowered, all bisexual; glumes ovate, spinescent, many-nerved; lemmas ovate, ending in a sharp spine, surrounded by ciliate tufts of hairs; palea ovate-obovate, emarginate, the lower ones 2-keeled, the upper most keelless; stigmas plumose. Caryopsis ovoid-subglobose.

Very common in the forest up to 800 m. Fl. & Fr.: August–September.

Kalasamudram RF (ATP), *NY* 398 & *PVP* 2601; Peapuly (KNL), *RVR* 2498; Peccheruvu RF (KNL), *PVP* & *RVR* 2177; Tirumala hills (CTR), *RVR* & *GO* 2431, *MSG* 12169; Bhupatipalem (EG), *TP* & *MSG* 12251; Pamuleru (EG), *TP* & *MSG* 12268 A & B; Way to Ahobilam (KNL), *JLE* 32572 (MH); Gazulapalli (KNL), *JLE* 18038 (MH); Kadapa (KDP), *EKK* 78533 (DD); Chintapalli (WG), *DCSR* 68 (CAL); Kuntada RF, Maredumilli (EG), *MM* 100735 (BSID); Karaka (VSKP), *CAB* (MH); On the way to Sapparla peak (VSKP), *GVS* 29698 (CAL); Woddapatti RF (SLM), *EV* 23602 (MH); Kapilas hill (Dhenkanal), *Sine coll.* 11206 (BSID); Jacum, Karlapat Willdlife Sanctuary (KHD), *AC* 2455 (BSID); Magudi (GJM), *CAB* 1431 (MH).

INDIA: In dry hills almost throughout India.

WORLD: Pakistan, Myanmar, Singapore, Indonesia.

DESMOSTACHYA Stapf

Desmostachya bipinnata (L.) Stapf in Dyer. Fl. Cap. 7: 632. 1900; Fischer 3:1819.1934; Bor, Grass. India 491. 1960; Moulik, Grass. Bamb. India 2:588.1997; Matthew, Fl. Tamilnadu Carnatic 3: 1832. 1983. *Briza bipinnata* L., Syst. Nat. ed. 10: 875. 1759. *Eragrostis cynosuroides* (Retz.) P. Beauv., Ess. Agrost. 162. 1812; FBI 7: 324. 1896.

Perennials, stoloniferous, culms erect, to 90 cm, branches from base. Leaf sheaths glabrous; ligule a ring of hairs; leaf blades linear, margin hispid, acuminate. Panicle narrow; spikelets ovate-lanceolate, sessile, secund, 2-seriate, many-flowered; lower glume ovate, small, 1-nerved, membranous, scabrid; upper glume ovate-lanceolate, 1-nerved, membranous; lemmas ovate, acute, glabrous, faintly 3-nerved, chartaceous; palea linear, keeled, shorter than the lemma, scabrid. Caryopsis ovoid.

Common in marshy habitats, in plains and ghats. Fl. & Fr.: Very infrequent.

Kanekal (ATP), *BR* & *KVS* 28052; Hottabetta (ATP), *JNP, AL* & *KVS* 28690; Kurellagudem (WG), *KCJ* 95426 (MH); Nattavaram (VSKP), *CAB* 1866 (MH).

INDIA: Plains of India.

WORLD: Throughout Middele East to Indo-China, Tropical Africa.

DICHANTHIUM Willement

1. Racemes solitary .... **D. foveolatum**
1. Racemes 2-many:
    2. Nodes with short hairs; ligule a short ciliate membrane
        3. Culm to 50 cm high; pedicelled spikelet unawned. .... **D. foulkesii**
        3. Culm above 80 cm high; pedicelled spikelet awned. .... **D. caricosum**
    2. Nodes with long hairs; ligule with long hairs .... **D. annulatum**

Dichanthium annulatum (Forssk.) Stapf in Prain, Fl. Trop. Afr. 9: 178. 1917; Fischer 3: 1741. 1934; Bor, Grass. India 133. 1960; Deshp. in Fasc. Fl. India 15: 5. 1984; Matthew, Fl. Tamilnadu Carnatic 3: 1833. 1983. *Andropogon annulatus* Forssk., Fl. Aegypt.-Arab. 173. 1775; FBI 7: 196. 1896.

A perennial, tufted grass; culms up to 70 cm high, erect or ascending, nodes bearded. Leaf sheaths 5-10 cm long, glabrous, ciliate near the mouth; ligule membranous; blades 6-20 x 0.3-0.5 cm, linear, base rounded, apex acuminate. Racemes 6 cm long, sub digitately fascicled; peduncles glabrous; spikelets 4 mm, 2-nate. *Sessile spikelet:* 4 mm, bisexual, elliptic-oblong, subacute, ciliate; lower lemma as long as the upper glume, nerveless; upper lemma narrow, with a scabrid, slender awn; stamens 3; stigmas plumose. Caryopsis oblong- slightly compressed. *Pedicelled spikelet:*4 mm, male or barren; lower glume elliptic-oblong, 7-11-nerved, keeled; upper glume narrow, 3-nerved; lower lemma ciliate; upper lemma small or obsolete.

Common in plains and hills. Fl. & Fr.: All the year round.

Gani RF (KNL), *RVR* 1539 & 1527; Vaddemanu (KNL), *PVP* 2532; Madhavaram RF (KNL), *TP* & *RVR* 1343; Thummalapalle (KDP), *KRKS* & *KRS* 33683; Horsley hills (CTR), *MSG* & *AA* 12214; Ischonum (WG), *JSG* 15779 (CAL); On the way to Chintapalli & Kannapuram (WG), *DCSR* 92 (CAL); Koruturu (WG), *DCSR* 501 (CAL); Chinnaudekarru (EG), *VNS* 60 (CAL); Valamuru, Maredumilli (EG), *MM* 101353 (BSID); Palakonda (VSKP), *KCJ* 17261 (MH); Krishnapura (VSKP), *CAB* 1990 (MH); Lasari, Kotgarh Wildlife Sanctuary (KDM), *AC* 2437 (BSID); Kuthadya hill (GJM), *VNS* 5926 (MH).

INDIA: Throughout India.

WORLD: Kenya, Tanzania and Senegal, through the Middle East to Indonesia.

Dichanthium caricosum (L.) A. Camus, Bull. Mus. Hist. Nat. Paris 27: 549. 182; Fischer 3: 1741. 1934; Bor, Grass. India 134. 1960; Matthew, Fl. Tamilnadu Carnatic 3: 1834. 1983; Deshp. in Fasc. Fl. India 15: 7. 1986. *Andropogon caricosus* L., Sp. Pl. 2: 1480. 1763; FBI 7: 196. 1896. *Dichanthium aristatum* (Poir.) C.E. Hubb., Kew Bull. 1939. 654. 1939; Bor, Grass. India 134. 1960. *Andropogon aristatus* Poir. in Lam., Encycl. 1: 585. 1810. *Dichanthium pallidum* (Hook.f.) Stapf ex Fischer, Fl. Madras 3: 1741. 1934; Bor, Grass. India 135. 1960; Moulik, Grass. Bamb. India 1: 273. 1997. *Apocopis pallida* Hook.f., Fl. Brit. India 7: 143. 1896.

A perennial, tufted grass; culms up to 1 m high, ascending; nodes bearded or with short hairs. Leaf sheaths 3-10 cm long, glabrous; ligule shortly ciliate membrane; blades 4-20 x 0.3-0.5 cm, linear, base rounded, apex acuminate, glabrous-sparsely pubescent. Racemes 1-6, up to 8 cm long, both terminal and axillary; joints and pedicels of racemes with short hairs; spikelets 4 mm. 2-nate. *Sessile spikelet:* 2 mm, oblong, bisexual; lower glume oblong-obovate, 4-nerved, truncate, pilose; upper glume elliptic, 3-nerved, hairy; lower lemma lanceolate, nerveless, hyaline; upper narrow, awned; stamens 3, stigma plumose. Caryopsis obtuse, slightly compressed. *Pedicelled spikelet:* 2 mm, male or barren; lower glume obovate, many-nerved, pilose, upper glume and lemmas similar to sessile spikelet.

Common in damp places and cultivated fields. Fl. & Fr.: July–December.

Pennahobilam (ATP), *PVP* 2651; Hussenapuram (KNL), *RVR* 3122; Veldurthi RF (KNL), *RVR* 1726 & 1744; Kappaladoruvu (NLR), *MSG* 10897; Rampachodavaram (EG), *TP* & *MSG* 12275; Vayalpadu (CTR), *sine coll.* 56784 (MH); Ramathirtham (NLR), *MPG* 150 (CAL); Kovvur (WG), *CAB* 4344 (MH); Valamuru, Maredumilli (EG), *MM* 101353 (MH); Devarapudi (EG), *CAB* 4303 (MH); Palakonda (VSKP), *KCJ* 17261 (MH); Panjapalli (SLM), *S.India Flora* 9903 (MH); Metigiri (SLM), *KCJ s.n.* (MH); Tulka section, Satkosia tiger Reserve (AGL), *KCM* 6199 (BSID); Chunabatti, Satkosia Willife Sanctuary (AGL), *AC* 2383 (BSID); Satkosia Wildlife Sanctuary (AGL), *AC* 2391 (BSID).

INDIA: Throughout India.

WORLD: Sri Lanka, Myanmar, China, Thailand, Laos, Indonesia, Africa.

Note: Deshpande (1984) reduced *Dichanthiun aristatum* (Poir.) C.E.Hubb. to the synonymy of *D. caricosum* (L.) A.Camus.

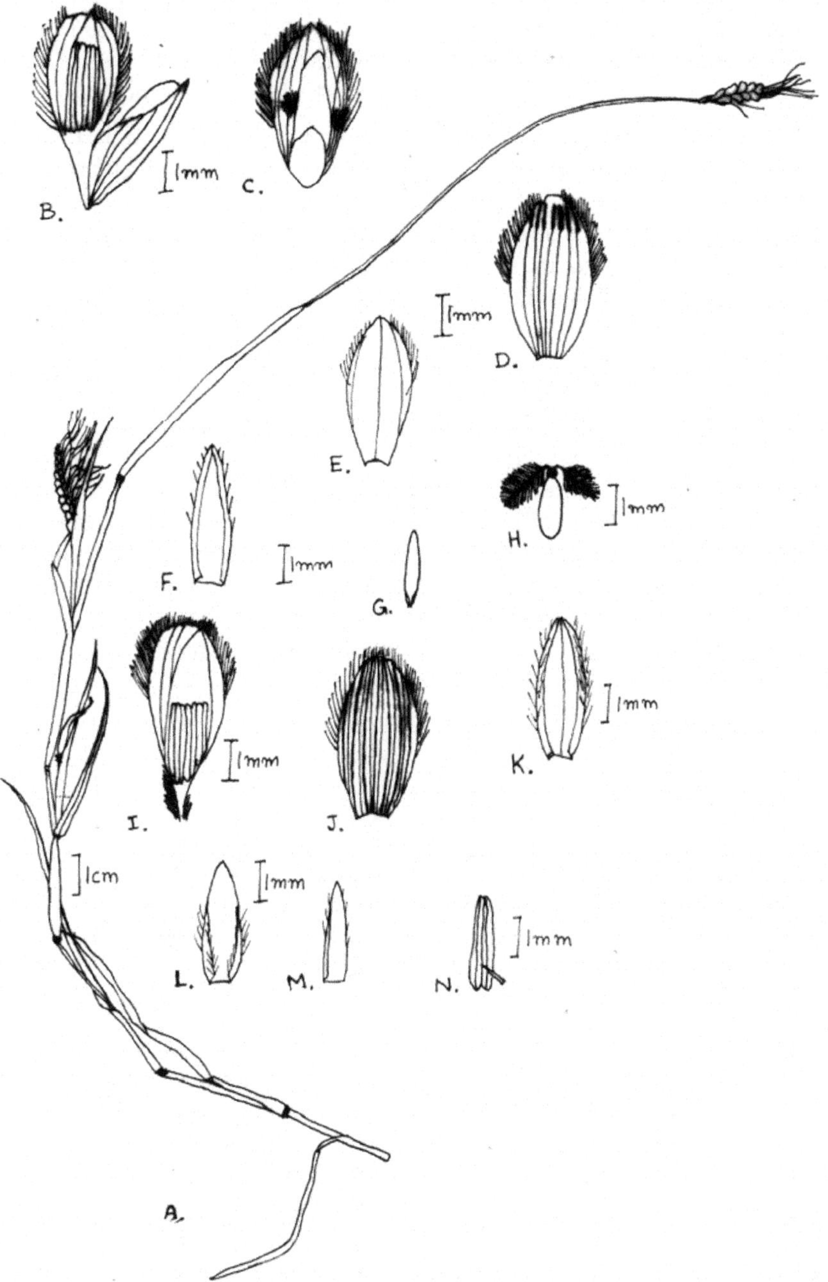

Figure 35. **Dicanthium caricosum** (L.) Camus.
[Syn.: *Dichanthium aristatum* (Poir.) C.E. Hubb]

A. Twig, B. Spikelets, C. Pedicelled spikelet, D. Lower glume, E & F. Upper lemma, G. Palea, H. Ovary, I. Sessile spikelet, J. Upper lemma, K & L. Lower lemma, M. Palea, N. Stamens.

Dicanthium foulkesii (Hook.f.) S.K.Jain & Deshp. in Bull. Bot. Surv. India 20: 134. 1979. *Andropogon foulkesii* Hook.f., Fl.Brit. India 7: 174. 1896. *Amphilopsis foulkesii* (Hook.f.) C.E.C.Fischer, Fl. Madras Pres. 3: 1732. 1934. *Bothriochloa foulkesii* (Hook.f.) Henr., Blumea 3: 457. 1940; Bor, Grass. India 107. 1960.

Tufted perennial grass, culm erect to geniculate, to 50 cm high. Leaves linear, 8-20 x 0.1-0.3 cm, acuminate at apex, often involute, hispid to glabrous, midvein prominent; ligule *c.* 2 mm long, membranous, fimbriate at apex. Racemes 3-8, 10 cm long, subdigitate to distant on a rachis, axis villous; joint pedicel linear, *c.* 2 mm long, flat, margins hairy. *Sessile spikelets*: elliptic to oblong lanceolate, *c.* 5 mm long; callus shortly bearded. Lower glume lanceolate to oblong-lanceolate, dorsally depressed, *c.* 4.5 mm long, acute at apex, chartaceous, faintly 7-nerved. Upper glume oblong-lanceolate, *c.* 4.8 mm long, acute to acuminate at apex, 3-nerved. Florets 2, both epaleate; lower barren, upper bisexual. Lower lemma oblong-laceolate, *c.* 3 mm long, hyaline, faintly 3-5-nerved. Upper lemma stipitate, 2-3 mm long, hyaline base of awn; awn *c.* 2 mm long; lodicules 2, more or less 3-horned at apex; stamens 3; anthers 1-2 mm long; ovary oblong, 0.3-0.5 mm long; stigma 1-2 mm long, plumose. *Pedicelled spikelets*: elliptic-lanceolate, *c.* 5.5. mm long, unawned, slightly longer and narrower than sessile; pedicels *c.* 2.5 mm long, villous; lower glume elliptic-lanceolate, *c.* 5 mm long, chartaceous, 7-9-nerved, margin infolded, ciliate towards apex; upper glume lanceolate, c. 5 mm long, chartaceous, faintly 3-nerved; lower floret similar to sessile spikelet; upper lemma lanceolate, 3-4 mm long, hyaline; stamens 3; anthers 1-2 mm long.

Rare in marshy areas. Fl. & Fr.: September–March.

Valamuru (EG), *MM* 100867A (MH).

INDIA: Karnataka, Kerala, Tamil Nadu. Endemic.

Dicanthium foveolatum (Del.) Roberty, Monogr. Syst. Andropog. Du Globe 163. 1960; Matthew, Fl. Tamilnadu Carnatic 3: 1834. 1983. *Andropogon foveolatus* Del., Fl. d' Egypte. 16. t. 2. 1812; FBI 7: 168. 1896. *Eremopogon foveolatus* (Del.) Stapf in Prain, Fl. Trop. Afr. 9: 183. 1917; Fischer 3: 1751. 1934; Bor, Grass. India 148. 1960.

A perennial, tufted grass; culms up to 80 cm high, shortly rhizomatous below, slender; nodes bearded. Leaf sheaths 4-6 cm, scabrid-glabrous; ligule ciliate, hyaline; blades mostly basal, 5-15 x 0.1-0.3 cm, flat, base rounded, apex acuminate, glabrous. Raceme solitary, up to 5 cm long, subtended by a spatheole; spikelets 6 mm, 2-nate, one sessile, other pedicelled. *Sessile spikelet* 4 mm, bisexual, ovate-oblong; lower glume oblong-ovate, pitted, scabrid on margin; upper glume ovate-lanceolate, hairy near apex; lower lemma linear, scabrid on margin; upper lemma narrow, awned; stamens 3, stigma plumose. Caryopsis obovoid. *Pedicelled spikelet:* 3 mm, oblong-ovate, male or barren; glumes oblong-lanceolate, hairy, lower pitted; lemma oblong, obtuse, scabrid; stamens 3.

Common in open forests, waste lands and cultivated fields. Fl. & Fr.: September-January. Tel.: *Namalachetta, Nopagaddi*

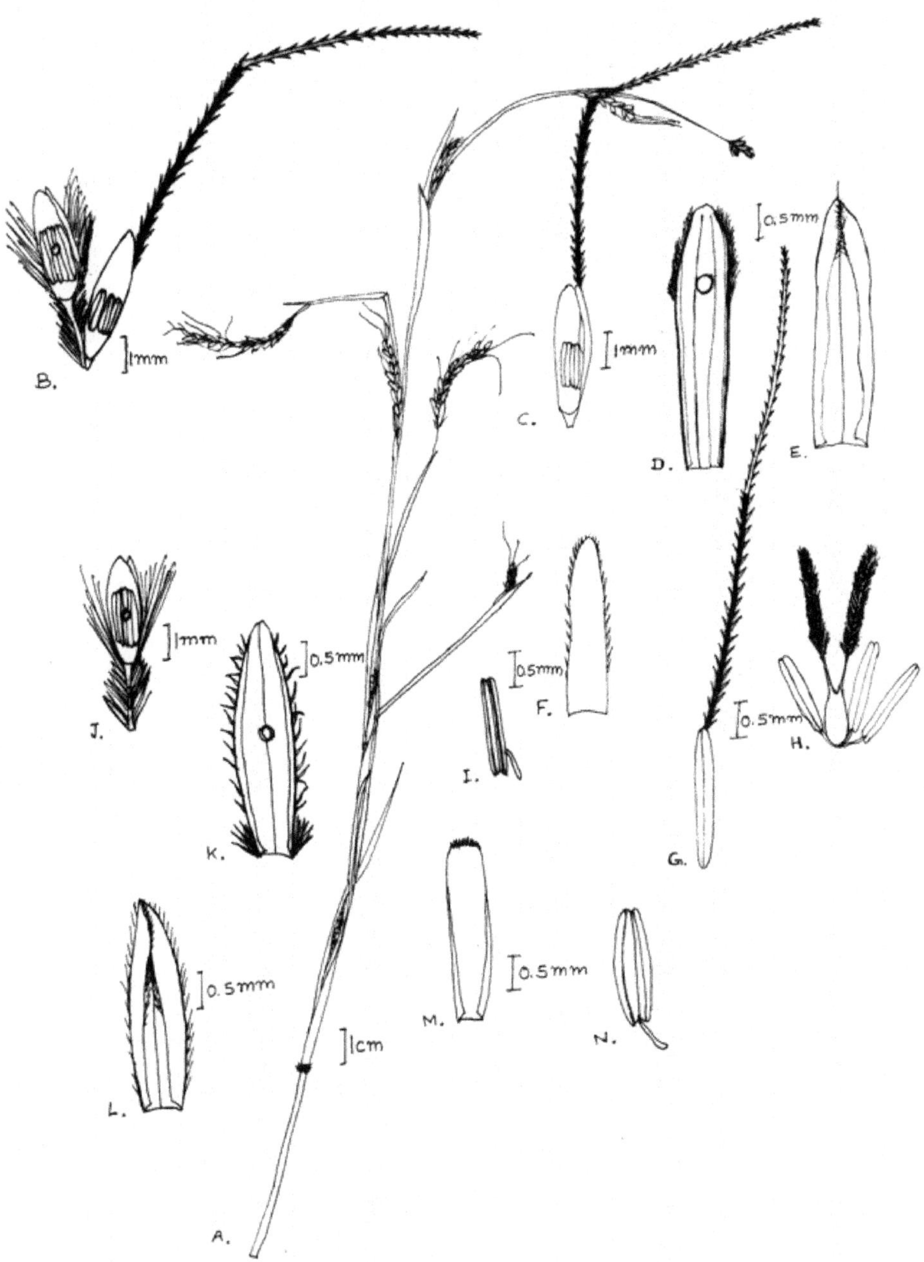

Figure 36. **Dicanthium foveolatum** (Del.) Roberty
[Syn.: *Eremopogon foveolatus* (Del.) Stapf]

A. Twig, B. Spikelet, C. Sessile spikelet, D. Lower glume, E. Upper glume, F. Lower lemma, G. Upper lemma, Pistil with stamens, I. Stamen, J. Pedicelled spikelet, K. Lower glume, L. Upper glume, M. Lemma, N. Stamen.

Muchukota (ATP), *TP* & *NY* 966; Cement nagar valley (KNL), *TP* & *RVR* 1808; Owk RF (KNL), *RVR* 3042; Gundlabrahmeswaram (KNL), *BR* & *BSS* 32971 (BSID); Balapalli RF (KDP), *PVP* 3471; Kalyani dam (CTR), *RVR* & *GO* 2441; Gooty (ATP), *JSG* 20815 (DD); Chelama (KNL), *JLE* 16770 (MH); Badvel (KDP), *JSG* 11207 (MH); Madanapalle (CTR), *JSG* 15051 (CAL); Rollapenta (PKM), *VBH* 83964 (BSID); Kondapalli (KSN), *CAB* 7994 (MH); Kondapudi (EG), *CAB* 5193 (MH); Maripalli hills (VSKP), *VNS* 4608 (MH); Chatrapur (GJM), *CECF* 59 (CAL), *JSG* 2154 (CAL); Mandragiari hill (AGL), *AC* 2370 (BSID).

INDIA: Drier parts of India.

WORLD: Tropical Africa.

DIGITARIA Haller *nom. cons.*

1. Panicle large, effuse; spikelets solitary, scattered; pedicels capillary .... **D. tomentosa**
1. Spike like racemes, digitate, never effuse; spikelets binate or ternate; pedicels stout, small:
    2. Spikelets binate:
        3. Margins of the rachis smooth .... **D. radicosa**
        3. Margins of the rachis serrate:
            4. Spikelets homomorphous:
                5. Spikelets glabrous to sparsely pubescent, lower glume absent, upper glume less than 1/3 as long as the spikelets.......... **D. setigera**
                5. Spikelets various pubescent, lower glume present, upper glume more thin 2/3 or little longer than the spikelet:
                    6. Lower lemma reduced to a triangular scale, nerves smooth, without minute triangular spines on nerves.......... **D. ciliaris**
                    6. Lower lemma well developed, nerved with minute triangular spines on the upper part of the nerves:
                        7. Culm below 30 cm long; upper glume 3-nerved; caryopsis ovoid .... **D. compacta**
                        7. Culm above 30 cm long; upper glume 5-nerved; caryopsis oblong .... **D. sanguinalis**
            4. Spikelets heteromorphous;
                8. Culms bulbously thickened at the base which is covered with densely silky- villous cataphylls.......... **D. nodosa**
                8. Culms not thickened at the base and without villous cataphylls .... **D. bicornis**

2. Spikelets ternate:
   9. Annuals; upper glume distinct:
      10. Hairs on the spikelets verrucose:
         11. Culms creeping and rooting at the nodes; leaves of the innovations rather broad; inflorescence of 2-3 racemes, 1.5-7 cm long; fruit pale to golden yellow:
            12. Rachis flat, broadly winged, wing 0.6-1.2 mm wide .... **D. longiflora**
            12. Rachis triquetrous to capillary, narrowly winged .... **D. griffithii**
         11. Culms erect; leaves all linear; inflorescence of 2-6 racemes, 4-10 cm long; fruit at length purplish .... **D. violascens**
      10. Hairs on the spikelets clavate:
         13. Tips of the pedicels cupuliform with hairs on the rim and below; hairs not exceeding the summit .... **D. ternata**
         13. Tips of pedicels somewhat thickened, without a ring of hairs .... **D. abludens**
   9. Perennials; upper glume represented by a tuft of hairs .... **D. stricta**

Digitaria abludens (Roem. & Schult.) Veldkamp, Blumea 21: 53. 1773. *Panicum abludens* Roem. & Schult., Syst. Veg. 2: 457. 1817. *Paspalum granulare* Trin. ex Spreng., Neue Entd. 2: 47. 1821. *Digitaria granularis* (Spreng.) Henr., Mongr. Digitaria 302. 1950; Bor, Grass. India 301. 1960. *Paspalum pedicellare* Trin. ex Hook. f., Fl. Brit. India 7: 19. 1896. *Digitaria pedicellaris* (Hook. f.) Prain, Bengal Pl. 1182. 1903; Fischer 3: 1765. 1934.

An annual, tufted grass; culms up to 50 cm high, slender, filiform; nodes glabrous. Leaf sheaths 1-3 cm, keeled, glabrous; ligule membranous; blades 3-9 x 0.2 - 0.5 cm, base rounded, apex acuminate, glabrous. Spikes 3-10, 9 cm long, digitate; rachis capillaris, trigonous; pedicels filiform, with clavate hairs; spikelets ternate, 1.5 mm; lower glume absent; upper glume elliptic, 5-nerved, hairy along the margins; lower lemma elliptic-lanceolate, 3-nerved, hairy; upper lemma ovate, subchartaceous, 3-nerved, apiculate; palea similar to upper lemma; stamens 3; stigmas plumose. Caryopsis oblong, tightly enclosed in lemma and palea

Abundant in plains, near water sources. Fl. & Fr.: July–October.

Balugram (KNL), *RVR* 2744; Palakonda hills (KDP), *CS & DAM* 7622; Talakona hills (CTR), *MSG & AA* 12226; B.N.Kandriga (CTR), *BR & P.Priyadarsini* 37340 (SKU & BSID); Isukagundam (PKM), *BR & BSS* 29297 (SKU & BSID); Ramanapenta (KNL), *MH* Acc. No. 54367 (MH); Nagalapuram hills (CTR), *MCB* 45234 (MH); Kothajupudi (GNT), *SCA* 170 (DD); Rampachodavaram (EG), *MM* 102651 (BSID); Palagedda (VSKP), *GVS* 42525 (MH); Forest near Araku (VSKP), *NPBK* 10830 (MH); Aska (GJM), *sine coll. s.n.* (MH); Tulka, Satkosia Tiger Reserve (AGL), *KCM* 8323 (BSID); Jenbil Range, Similipal BR (MBJ), *AC* 2349 (BSID); Pottangi (KPT), *AC* 10234 (BSID).

INDIA: Throughout India.

WORLD: Tropical Asia.

Digitaria bicornis (Lam.) Roem. & Schult., Syst. Veg. 2: 470. 1817; Bor, Grass. India 299. 1960; Matthew, Fl. Tamilnadu Carnatic 3: 1836. 1983. *Paspalum bicorne* Lam., Tabl. Encycl. 1: 176. 1791. *P. heteranthum* Hook. f., Fl. Brit. India 7: 16. 1896. *Digitaria biformis* Willd., Enum. Hort. Berol. 92. 1809; Bor, *loc. cit.*

An annual, tufted grass; culms up to 50 cm high, base geniculately ascending, rooting at the nodes; nodes sparsely bearded, glabrescent. Leaf sheaths 2-8 cm, ligule membranous; blades 2-6 x 0.4-0.6 cm, oblong, base cordate, apex acuminate, glabrous. Racemes 2-9, 10 cm long, digitate; rachis serrulate; spikelets 3 mm, lanceolate, 2-nate, heteromorphous. *Sessile spikelet:* 3 mm, glabrous-pilose; lower glume scaly; upper glume elliptic, 3- nerved, pubescent; lower lemma elliptic, 7-nerved, pubescent at lateral nerves, membranous; upper lemma elliptic-lanceolate, 3- nerved, chartaceous; palea similar to upper lemma; stamens 3; stigmas plumose. Caryopsis ellipsoid, lightly enclosed by hardened lemma and palea.

Very variable grass commonly occurring in plains and foot hills. Fl. & Fr: July-November.

Muchukota RF (ATP), *PVP* 2568; Gani RF (KNL), *RVR* 1562; Lower Ahobilam (KNL), *RVR & GO* 2582; Tirumala hills (CTR), *RVR & GO* 2405; Kailasanathakona (CTR), *BR & P. Priyadarsini* 37137 (SKU & BSID); Beore Chittivelu on way to Rajampeta (NLR), *P. Venu* 110074 (BSID).

INDIA: In dry regions, S & E. India.

WORLD: Africa, Sri Lanka, Myanmar, China, Malaysia, Indonesia, Thailand, Australia.

Digitaria ciliaris (Retz.) Koeler, Descr. Gram 27. 1802; Matthew, Fl. Tamilnadu Carnatic 3: 1837. 1983. *Panicum ciliare* Retz., Obs. Bot. 4: 16. 1786. *Digitaria marginata* Link var. *fimbriata* (Link) Stapf in Prain, Fl. Trop. Afr. 9: 440. 1919; Fischer 3: 1764.1934. *D. adscendens* (Kunth) Henrard., Blumea 1: 92. 1934; Bor, Grass. India 298. 1960. *Panicum adscendens* Kunth, Nov. Gen. Pl. 1:102. 1821. *Digitaria ciliaris* (Retz.) Koeler subsp. *chrysoblephara* (Fig. & de Not.) Henrard, Monogr. Gen. Digitaria 998. 1950; Bor, Grass. India 299. 1960. *D. chrysoblephara* Fig. & de Not., Accad. Rle. Sc. Tor. (ser. 2) 14: 364. t. 27. 1851. *D. marginata* Link, Enum. Hort. Berol. Alt. 1: 102. 1821. *D. sanguinalis* (Link) Scop. var. *extensum* Rang. & Tad., Handb. S. Indian Grass. 56. ff. 74 & 75. 1921.

An annual, tufted grass; culms up to 90 cm high, usually ascending from geniculate or prostrate base; branching from lower nodes; nodes glabrous. Leaf sheaths 4-12 cm long, glabrous; ligule membranous, truncate; blades 4-10 x 0.4-0.6 cm, linear- lanceolate, base cordate, apex acute, glabrous. Racemes 4-9, 10 cm long, digitate; rachis serrate, trigonous; spikelets 3 mm, elliptic-lanceolate, 2-nate, appressed to the rachis, glabrous- sparsely ciliate, homomorphous; lower glume reduced to a triangular scale; upper glume linear-lanceolate, 3-5-nerved, pubescent; lower lemma oblong-lanceolate, 5-nerved, marginal nerves pubescent; upper lemma

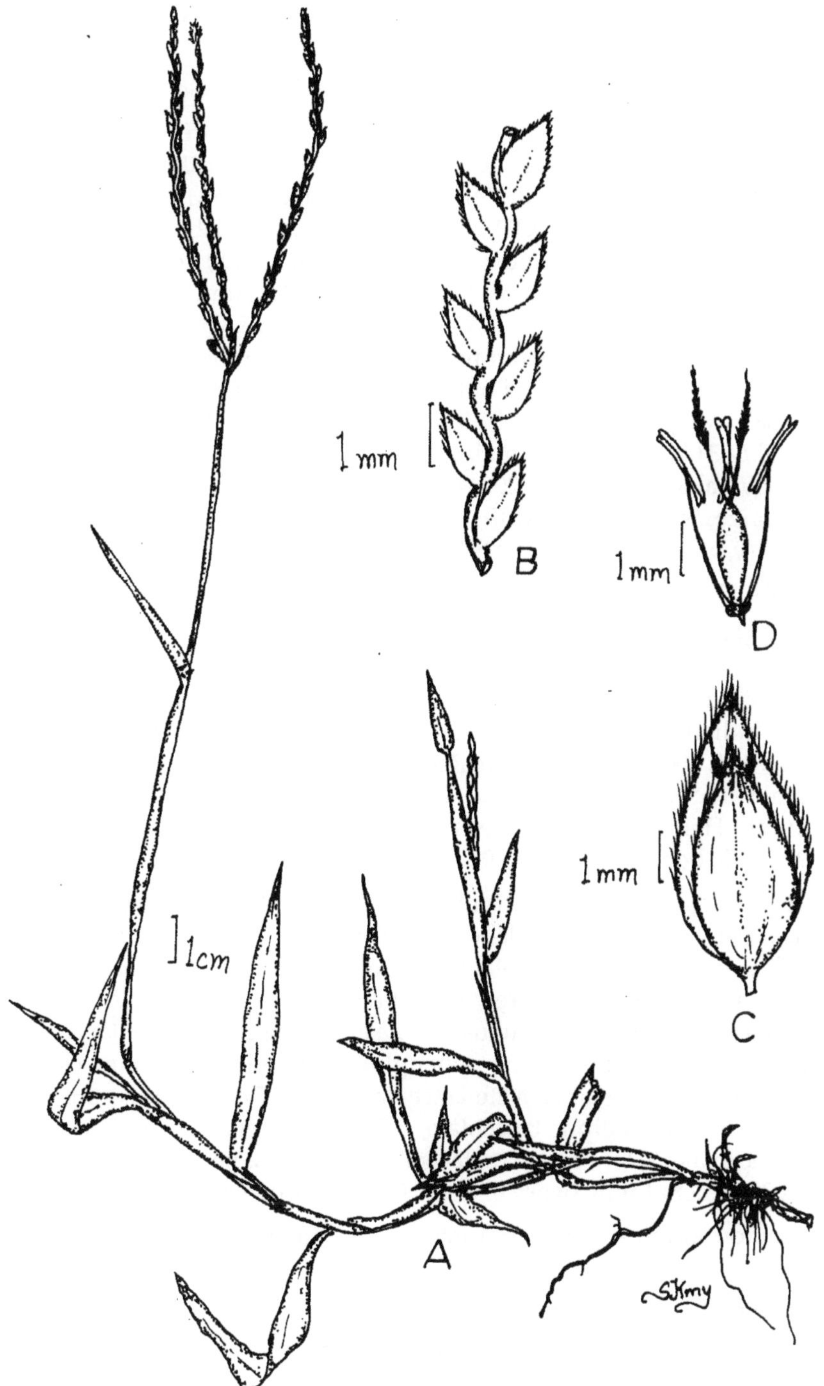

Figure 37. **Digitaria ciliaris** (Retz.) Koel.

A. Habit, B. Spikelet, C. Flower, D. Pistil & stamens.

oblong-lanceolate, obscurely 3-nerved; stamens 3, stigmas plumose. Caryopsis oblong.

Very common in moist plains and hills. Fl. & Fr.:July-October.

Peapuly (KNL), *PVP* 2662; Guvvalacheruvu (KDP), *PVP* 3422; Balarachapalle (KDP), *PVP* 3444; Tirumala hills (CTR), *MSG* 12135; Bhupathaiah kona (CTR), *BR & P.Priydarsini* 37319 (SKU & BSID); Nandipadu (NLR), *PMR* 16469; Macherla RF (GNT), *VRK* 3248; Tadipatri (ATP), *JSG* 21306 (DD); Gundlabrahmeswaram (KNL), *JLE* 16929 (CAL & MH); Mogilipenta (KDP), *JSG* 21324 (DD); Kondapalli (KSN), CAB 7976 (MH); Polavaram agency (WG), *DCSR* 250 (CAL); Araku valley (VSKP), *NPBK* 600 (CAL); Anantagiri (VSKP), *GVS* 19458, 19464 (MH & CAL); Gajapathinagaram (VJN), *VNS* 4573 (MH); Belugam near Parvathipuram (SKLM), *VNS* 4531 (MH); Palur (SA), *CAB* 8350 (MH); Red hills (CPT), *CAB* 5513 (MH); Melkondai RF (SA), *KRM* 60383 (MH); Burgur (SLM), *S.India Flora* 15513 (MH); Ghat road–Yercaud (SLM), *DBD* 31443 (MH); Sirumalai hills (DGL), *SKS* 411 (SGH); Cauvery river bank, Hogainakkal (KGR), *EV* 20644 (MH); Mahendragiri (GJP), *VNS* 5749 (MH); Jenabil Range, Similipal BR (MBJ), *AC* 2341 (BSID); Chahala, Similipal BR (MBJ), *A.R.K.Sastry & G.P.Singh* 12270 (BSID).

INDIA: Throughout India.

WORLD: Throughout much of the Tropics.

Digitaria ciliaris (Retz.) Koeler subsp. **ciliaris** var. **fimbrita** (Link) Jain & Doli Das, Indian Forester 99(9): 572. 1973; A.M.Reddy *et al.*, J.Econ. Taxon. Bot. 33(2): 411. 2009. *Panicum ciliare* Retz. Obs. Bot. 4: 16. 1786. *Paspalum sanguinata* Lam. var. *ciliare* (Retz.) Hook.f., Fl. Brit. India 7: 15. 1896. *Digitaria adscendens* (Kunth) Henr. subsp. *adscendens* var. *criniformis* Henr., Mongr. Gen. Digitaria 999. 1950; Bor, Grass. India 299. 1960.

Annuals, culms erect or decumbent, tufted, to 80 cm, rooting at lower nodes, glabrous. Leaf sheaths 2-10 cm, glabrous, keeled; ligule 2 mm, truncate, membranous; leaf blade 5-10 x 0.2-1 cm, lanceolate or linear-lanceolate, base cordate, acuminate, scabrid or villous. Racemes 2, rarely 3, digitate, rhachis serrate; pedicels triquetrous, scabrid; spikelets 2, homomorphous, sessile and pedicelled, ovate-lanceolate, 3.5 x 2 mm, hairy; lower glume absent; upper glume lanceolate, 3 x 1 mm, chartaceous, 3-nerved, hairy; lower lemma ovate-lanceolate, 3.5 x 1.7 mm, chartaceous, 9-nerved, hairy; upper lemma elliptic-lanceolate, 3.3 x 1 mm, coriaceous, 3-nerved; palea elliptic-lanceolate, 3.2 x 1 mm, crustaceous, 2-keeled, 2-nerved; lodicules 2, ovate; stamens 3, anthers 1 mm; ovary 0.5 mm. Caryopsis ovate-oblong, 1.5 mm.

Common in forest plains and margins in Mahabubnagar (A.M.Reddy *et al.*, 2009).

Fl. & Fr.: August–December.

Appapur (MBNR), *BR & AMR* 25679.

INDIA: Throughout India.

WORLD: Tropical Asia.

Digitaria compacta (Roth ex Roem. & Schult.) Veldkamp, Blumea 21: 71. 1973. *Paspalum compactum* Roth, Nov. Pl. Spec. 36. 1821; FBI 7: 12. 1896; Fischer 3: 1772. 1934; Bor, Grass. India 336. 1960.

An annual grass; culms 8-30 cm, stoloniferous, nodes pilose. Leaf sheath 0.5-2 cm, densely hispid; ligule membranous, ciliate, 0.3 mm; blades ovate-lanceolate, 1.5-4 cm long, 4-8 mm wide, tuberculate hispid along nerves. Racemes 5-15, distant, 0.5-2 cm long, spreading on a common elongate axis; pedicel 0.5-2 mm. Spikelet solitary or paired, ovoid, turgidly plano-covex, 1.5-2 mm, acute; upper glume broadly ovate, to 2 mm, 3-nerved; upper lemma ovate, to 1.7 mm, crustaceous, granular; stamens 3. Caryopsis ovoid.

Occasional on the stream side of southern Eastern Ghats in Tamil Nadu. Fl. & Fr.: September–January. Vern.:Tam.: *Arisipullu.*

On the way to Tanjavanam (VSKP), *GVS* 42626 (MH); Kolli hills (NMK), *SKS* 622 (SGH).

INDIA: Gujarat, Uttar Pradesh, Tamil Nadu.

WORLD: Indonesia.

Note: Addition to Flora of Andhra Pradesh.

Digitaria griffithii (Hook.f.) Henrard, Blumea 1: 100. 1934; Matthew, Fl. Tamilnadu Carnatic 2: 1837. 1983. *Panicum griffithii* Arn. ex Hook.f., Fl Brit. India 7: 15. 1896. *Paspalum sanguinale* var. *griffithii* Hook.f., Fl. Brit. India 7: 15. 1896.

Perennial grass, culms erect or geniculately ascending, 50-100 cm long. Leaf sheaths glabrus or hirsute; ligule a ciliate membrane, 3 mm long; blades 10-20 x 1.0-1.5 cm, flaccid, scabrous, rough on both sides, glabrous or pilose, apex acuminate. Racemes 5-10, digitate, 6-13 cm long; spikelets in pair; fertile spikelets pedicelled; spikelets lanceolate, compressed; upper glume ovate.

Occasional in Southern Eastern Ghats. Fl. & Fr.: July - February

Sirumalai hills (DGL), *SKS* 419 (SGH); Kolli hills (NMK), *SKS* 585 (SGH); Ghat road, Yercaud (SLM), *DBD* 31441 (MH).

INDIA: Southern India.

WORLD: Sri Lanka.

Digitaria longiflora (Retz.) Pers., Syn. Pl. 1: 85. 1805; Fischer 3: 1765. 1934; Bor, Grass. India 302. 1960; Matthew, Fl. Tamilnadu Carnatic 3: 1838. 1983. *Paspalum longiflorum* Retz., Observ. Bot. 4: 5. 1786/87; FBI 7:17. 1896.

An annual, tufted grass; culms up to 60 cm high, ascending, rooting at the nodes; nodes glabrous. Leaf sheaths 2-4 cm long, glabrous; ligule membranous; blades 2-7 x 0.2-0.5 cm, linear-lanceolate, base rounded, apex acuminate, glabrous. Racemes 2-5.7 cm long, digitate; rachis serrate; spikelets 2 mm long, elliptic, homomorphous, sparsely pubescent; lower glume absent; upper glume oblong-ovate, 3-5-nerved, hairy; lower lemma similar to upper glume; upper lemma

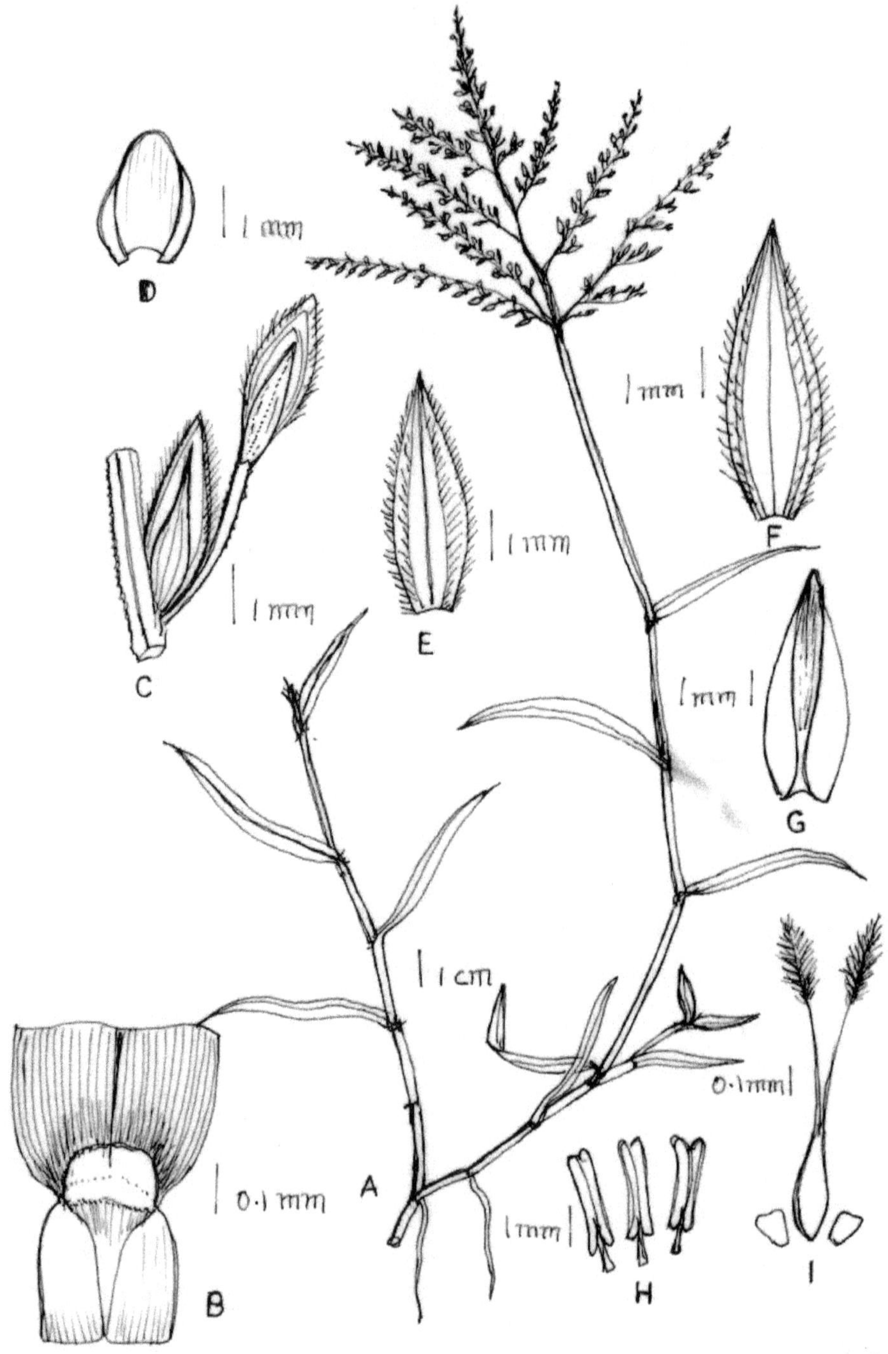

Figure 38. **Digitaria griffithii** (Hook.) Henrard.

A. Plant, B. Collar, C. Spikelets, D. Upper glume, E. Lower glume, F. Lemma, G. Palea, H. Stamens, I. Pistil and lodicules.

ovate-oblong, subchartaceous, nerveless; palea similar to upper lemma; stamens 3, stigmas plumose. Caryopsis ellipsoid.

Occasional in the plains of southern Eastern Ghats. Fl. & Fr.:July-October.

Kalasamudram RF (ATP), *PVP* 2612; Balapalli RF (KDP), *PVP* 3472; Horsley hills (CTR), *PVP* 4669, *MSG* & *AA* 12223; Ramagiri (CTR), *P.Priyadarsini* & *BR* 37115 (SKU & BSID); Chelama (KNL), *MPG* 102 (CAL); Way to Balugram (KNL), *JLE* 25538 (CAL & MH); Gazulapalle (KNL), *KCJ* 83135 (MH); Ambaka (CTR), *MCB* 45193 (MH); Diguvametta (PKM), *JLE* 42156 (MH); Chinturu, Maredumilli (EG), *MM* 101399 (BSID); Red hills (CPT), *CAB* 5515 (MH); Kambakkam hills (CPT), *sine coll.* 8897 (MH); Gudiyattam (NA), *S.India Flora* 15510 (MH); Metigiri (SLM), *S.India Flora* 15440 (MH); Krishnagiri (KGR), *S.India Flora* 14937 (MH); Tikarpada, Satkosia Tiger Reserve (AGL), *KCM* 7620 (BSID); Beghar, Kotgarh Wildlife Sanctuary (KDM), *AC* 2411 (BSID).

INDIA: Throughout India.

WORLD: Tropical and Subtropical regions of the Old World.

Digitaria nodosa Parl., Plant. Nov. 39. 1842; Bor, Grass. India 303. 1960; Moulik, Grass. Bamb. India 1: 94. 1997. *Panicum pabulare* Aitch., J. Linn. Soc. 19: 190. 1882. *Paspalum sanguinale* (L.) Lamk. var. *pabulare* (Aitch.) Hook.f., Fl. Brit. India 7: 15. 1896.

Perennials; culms erect, to 120 cm high, rhizomatous, bulbously thickened at the base with dense silky hairs, nodes glabrous. Leaf sheaths 4-9 cm, glabrous; ligules 1.5 mm, membranous, truncate. Leaf blades 5-20 x 0.2-0.4 cm, linear-lanceolate, acute or acuminate, rounded at base, smooth or glabrous. Racemes digitate, 2-3; pedicels winged, 1 mm wide, truncate; spikelets paired, heteromorphous, sessile one slightly pubescent, pedicelled ones densely pubescent; spikelets elliptic, 3.5x 1.2 mm, acute, lower glume distinct, triangular, 1.7 x 0.7 mm; upper glume well developed, as long as spikelet, lanceolate, 3.2 x 1.3 mm, chartaceous, 5-nerved, silky pubescent; lower lemma oblong-lanceolate, 2.8 x 1.2 mm, chartaceous, 3-nerved, glabrous, upper lemma elliptic-lanceolate, 2.7 x 1 mm, acute, crustaceous; stamens 3, anthers 1.3 mm; ovary ovate, 1.2 mm; styles 2 mm, stigmas 1 mm; lodicues 2, truncate, 0-4 mm.

Rare in forests in sandy soils in Nallmalais. Fl. & Fr.: December–July.

Peccheruvu (KNL), *AMR* & *BR* 24507 (SKU).

INDIA: Arid regions of NW India, Andhra Pradesh.

WORLD: Tropical Africa, West Asia, Arabia.

Digitaria radicosa (J. Presl) Miq., Fl. Ned. Ind. 3: 437 1857; Matthew, Fl. Tamilnadu Carnatic 3: 1838. 1983. *Panicum radicosum* J. Presl, Rel. Haenk. 1: 297. 1830. *Digitaria timorensis* (Kunth) Hal., J. Bot. (Morot) 4: 138. 1890; Bor, Grass. India 306. 1960. *Panicum timorense* Kunth Enum. Pl. 1: 83. 1833. *Paspalum sanguinale* Lam. var. *debile* auct. non Hook. f., Fl. Brit. India 7: 16. 1896.

An annual, tufted grass; culms up to 30 cm high, decumbent or ascending at base; nodes glabrous. Leaf sheaths 2-5 cm long, loose, glabrous; ligule membranous; blades 3-5 x 0.2-0.4 cm, linear-oblong, base narrow, apex acute, sparsely pilose. Racemes 2 or 3, 7 cm long; rachis smooth; spikelets narrowly lanceolate, 3 mm, homomorphous, 2-nate, pubescent; lower glume reduced to a scale; upper glume lanceolate; lower lemma lanceolate-oblong, marginal nerves pubescent; upper lemma elliptic-lanceolate, prominent; palea similar to upper lemma; stamens 3; stigmas plumose. Caryopsis oblong.

Occasional grass in plains and stream side of southern Eastern Ghats. Fl. & Fr.: July-December.

Mangapuram (CTR), *RVR & GO* 2449 (SKU).

INDIA: E and NE India, Andhra Pradesh and Tamil Nadu.

WORLD: Myanmar, Nepal, China, Taiwan, Philippines, Thailand, Indonesia, Australia.

Note: This species in closely related to *Digitaria ciliaris* but it can be distinguished by the slender, weak, trailing culms, few and slender racemes, smooth edged rachis and narrow lanceolate spikelets.

Digitaria sanguinalis (L.) Scop., Fl. Cam. ed. 2.1, 552. 1772; Bor, Grass. India 304 1960; T.A. Cope in Nasir & Ali, Fl. Pak. 143. 231. 1982. *Panicum sanguinalis* L., Sp. Pl. 57. 1753. *Paspalum sanguinale* (L.) Lam., Tabl. Encycl. Meth. 1: 176. 1791; FBI 7: 13. 1896. *Digitaria sanguinalis* (L.) Scop. subsp. *aegyptica* var. *frumentacea* Henr., Monogr. *Digitaria* 985. 1950; Bor, Grass. India 304. 1960. *Digitaria sanguinalis* subsp. *vulgaris* var. *rottleriana* Henr. *loc.cit.* 986; Bor, Grass. India 304. 1960.

An annual, tufted grass; culms up to 40 cm high, erect or ascending from a creeping, branching base. Leaf sheaths glabrous, hairy near the mouth; blades 2.5-18 x 0.4-0.8 cm, linear. Spikes slender, 3-10, 8 cm long, digitate; spikelets 2-nate on abbreviated peduncles, imbricate; lower glume 3-5-nerved, a tuft of hairs as long as the glume on back, usually silky; upper glume 5-nerved, lateral nerves marginal; lower lemma lanceolate or oblong-lanceolate, smooth. Caryopsis oblong, whitish.

A very variable taxon, commonly occurring along the foot hills and plains of northern Eastern Ghats. Fl. & Fr.: August- November.

INDIA: Through out India.

WORLD: Warm temperate and upland subtropical regions throughout the world.

Digitaria setigera Roth ex Roem. & Schult. in L., Syst. Veg. 2: 474. 1817; Bor, Grass. India 305. 1960; Matthew, Fl. Tamilnadu Carnatic 3: 1839. 1983. *Paspalum sanguinale* Lam var. *tensum* Hook. f., Fl. Brit. India 7: 15. 1896; Fischer 3: 1772. 1934.

An annual, stoloniferous grass; culms up to 50 cm, creeping, rooting at the lower nodes; nodes glabrous. Leaf sheaths 2-6 cm long, loose, glabrous; ligule membranous;

leaf blades 3-10 x 0.2-0.6 cm, linear-oblong, base rounded, apex acuminate, glabrous. Racemes 3-6, 10 cm long, digitate; rachis serrate, basally with stiff hairs; hairy near joints and pedicels; lower glume obsolete; upper glume triangular in sessile, linear in pedicelled spikelets, pubescent; lower lemma lanceolate, pubescent on marginal nerves; upper lemma elliptic-lanceolate; stamens 3; stigmas plumose. Caryopsis ellipsoid.

Occasional grass in open mloist places. Fl. & Fr.: July-October.

Choutkoor (CTR), *TP & MSG* 11917; Horsley hills (CTR), *MSG & AA* 12212; Tirumala hills (CTR), *RVR & GO* 2415; Ettukkuru, Maredumilli (EG), *MM* 102613 (MH, BSID); Near Kuttathur (NA), *KS* 6537 (MH); Shevaroys (SLM), *S.India Flora* 14810 (MH); Jenabil Range, Similipal BR (MBJ), *AC* 2318 (BSID).

INDIA: Throughout India.

WORLD: Bangladesh, Myanmar, Bhutan, Nepal, China, Indonesia, Japan, Malaysia, Thailand, Vietnam, E. Africa, Madagascar, Australia.

Digitaria stricta Roth, Syst. Veg. 2: 474. 1817 var. **denudata** (Link) Henr., Monogr. Digitaria 175. 1950; Bor, Grass. India 306. 1960. D. *denudata* Link, Hort. Berol. 1: 222. 1827. *Paspalum royleanum* Nees ex Hook. f., Fl. Brit. India 7: 18. 1896. *Digitaria royleana* (Nees ex Hook. f.) Prain, Bengal Pl. 1182. 1903; Fischer 3: 1765. 1934.

A perennial, tufted grass, culms up to 50 cm high; nodes glabrous. Leaf sheath 2-5 cm, appressed hairy; ligule membranous; blades 6-16 x 0.2-0.4 cm, linear, flat, sparsely hairy along the margins. Racemes 2-10, erect, 10 cm long; rachis filiform, trigonous; spikelets 1.6 mm, broadly ellipsoid, secund, tomentose with minute clavellete hairs. Lower glume absent; upper glume reduced to the tuft of hairs; lower lemma broadly elliptic, thinly membranous, 5-nerved; upper lemma ovate, coriaceous, acute with inflexed margins; stamens 3. Caryopsis oblong.

Rare grass in open grass lands. Fl. & Fr.: August-January.

Bhupathipalem, Rampa hills (EG), *TP & MSG* 12243; Gundlabrahmeswaram RF (KNL), *JLE* 22183 (MH); Diguvametta (PKM), *JLE* 22183 (MH); Near Sileru approach (VSKP), *GVS* 42719 (MH); Gurguria, Similipahar (MBJ), *SX* & *MB* 5183 (RRL-B); Guma (GJM), *SX* & *MB* 3852 (RRL-B); Paniganja (GJM), *SX* & *MB* 1971 (RRL-B); Pottangi (KPT), *AC* 10236 (BSID).

INDIA: Hilly regions of India.

WORLD: Tropical Asia and Africa.

Digitaria ternata (A.Rich.) Stapf ex Dyer, Fl. Cap. 7: 376. 1898; Bor, Grass. India 305. 1960; Matthew, Fl. Tamilnadu Carnatic 3: 1840. 1983; Moulik, Grass. Bamb. India 1: 97. 1997; A.M.Reddy *et al.*, J. Econ. Taxon. Bot. 33: 411. 2009. *Cynodon ternatus* A. Rich., Tent. Fl. Abyss. 2: 405. 1851. *Paspalum ternatum* (A.Rich.) Hook.f., Fl. Brit. India 7: 17. 1896.

Annuals; culms erect or ascending, to 120 cm, nodes glabrous. Leaf sheaths 5-14 cm, glabrous, ciliate at mouth. Ligules hairy, membranous, 0.5 mm; leaf blades 5-15 x 0.5-1 cm, linear-lanceolate, base rounded. Rachis villous; pedicels 2.4 mm, scabrid. Spikelets ovate, 2.5-3 x1.3 mm, clavate, hairy; lower glume absent; upper glume linear-lanceolate, 1.7 x 0.6 mm, hyaline; lower lemma elliptic or oblong-lanceolate, 2.8 x 1.2 mm, chartaceous, scabrid, 7-nerved; upper lemma oblong-lanceolate, 2.5 x 1 mm, crustaceous, keeled; lodicules 2, ovate; stamens 3, anthers 0.6 mm; ovary elliptic, styles 0.5 mm, stigmas 0.8 mm. Caryopsis oblong.

Common in forest plains and margins in Nallmalais (A.M.Reddy *et al.*, 2009). Fl. & Fr.: August–December.

Srisailam (KNL), *AMR* 26802; Appapur (MBNR), *BR* & *AMR* 25677 (SKU).

INDIA: Peninsular and NE India.

WORLD: Tropical Africa, Bhutan, Nepal, China, Malaysia, Indonesia, Thailand, Philippines. Introduced in America and Australia.

Digitaria tomentosa (Willd.) Henr., Blumea 1: 100. 1934; Bor, Grass. India 306. 1960; Matthew, Fl. Tamilnadu Carnatic 3: 1840. 1983. *Milium tomentosum* Koen. ex Willd., Ges. Naturf. Freunde Berlin Neue Schriften 4: 220. 1803. *Panicum subeglume* Trin., Mem. Acad. Imp. Sci. St. Petersb. Ser. 6, Sci. Math. 3: 292. 1834; FBI 7: 51. 1896; Fischer 3: 1783. 1934.

A perennial, tufted grass, culms up to 80 cm high, geniculately ascending at base; nodes sparsely hairy. Leaf sheaths 4-9 cm, loose, sparsely pubescent-pilose; ligule membranous; leaf blades 7-15 x 0.3-0.7 cm, oblong, flat, base rounded, margins ciliate, apex acuminate, sparsely pubescent to tuberculate based hairs. Panicle lax, large, up to 30 cm long, spreading; rachis capillary, pilose at branch axis; spikelets 2 mm, narrowly lanceolate, solitary, pubescent; lower glume obsolete; upper glume elliptic-lanceolate, pubescent; upper lemma elliptic-lanceolate, pilose; upper lemma elliptic-oblong, coriaceous; palea similar to upper lemma; stamens 3, stigmas plumose. Caryopsis oblong.

Common grass in all localities. Fl. & Fr.:July - October

Bugga (ATP), *TP* 453 (SKU & MH); Madhavaram RF (KNL), *RVR* 1667; Palakonda hills (KDP), *CS & DAM* 7720 & 7627; Penchalakona (NLR), *MSG* 12188; Ethipothala (GNT), *VRK* 5791; Kottakota (ATP), *JSG* 20883 (MH & DD); Way to Pathalaganga, Srisailam (KNL), *JLE* 22090 (MH); Modugumamisala (KNL), *JLE* 42200 (MH); Guvvalacheruvu (KDP), *KMS* 6837 (CAL & MH); Tummanabailu (KDP), *MH* Acc. No. 55033 (MH); Kondapalle (KSN), *CAB* 4564 (MH); Kadapa, *EKK* 444; Kondagarulapenta (GNT), *MPG* 118 (CAL); Kondapalli (KSN), *CAB* 4564 (MH); Salur forest (VJN), *NPBK* 1120 (CAL); Kambakkam hills (NA), *sine coll. s.n.* (MH); Vandalur hill (CPT), *sine coll. s.n.* (MH); Sithanur (CPT), *sine coll. s.n.* (MH);

INDIA: Peninsular India. Endemic.

Digitaria violascens Link, Hort. Berol. 1: 229. 1827; Bor, Grass. India 307. 1960; Mishra *et al.*, Indian J. For. 6: 295. 1983. D. *chinensis* sensu Fischer 3: 1765. 1934, non Hornem 1819. *Paspalum longiflorum* sensu Hook.f., Fl. Brit. India 7: 17. 1896. *p.p.* non Retz. 1786.

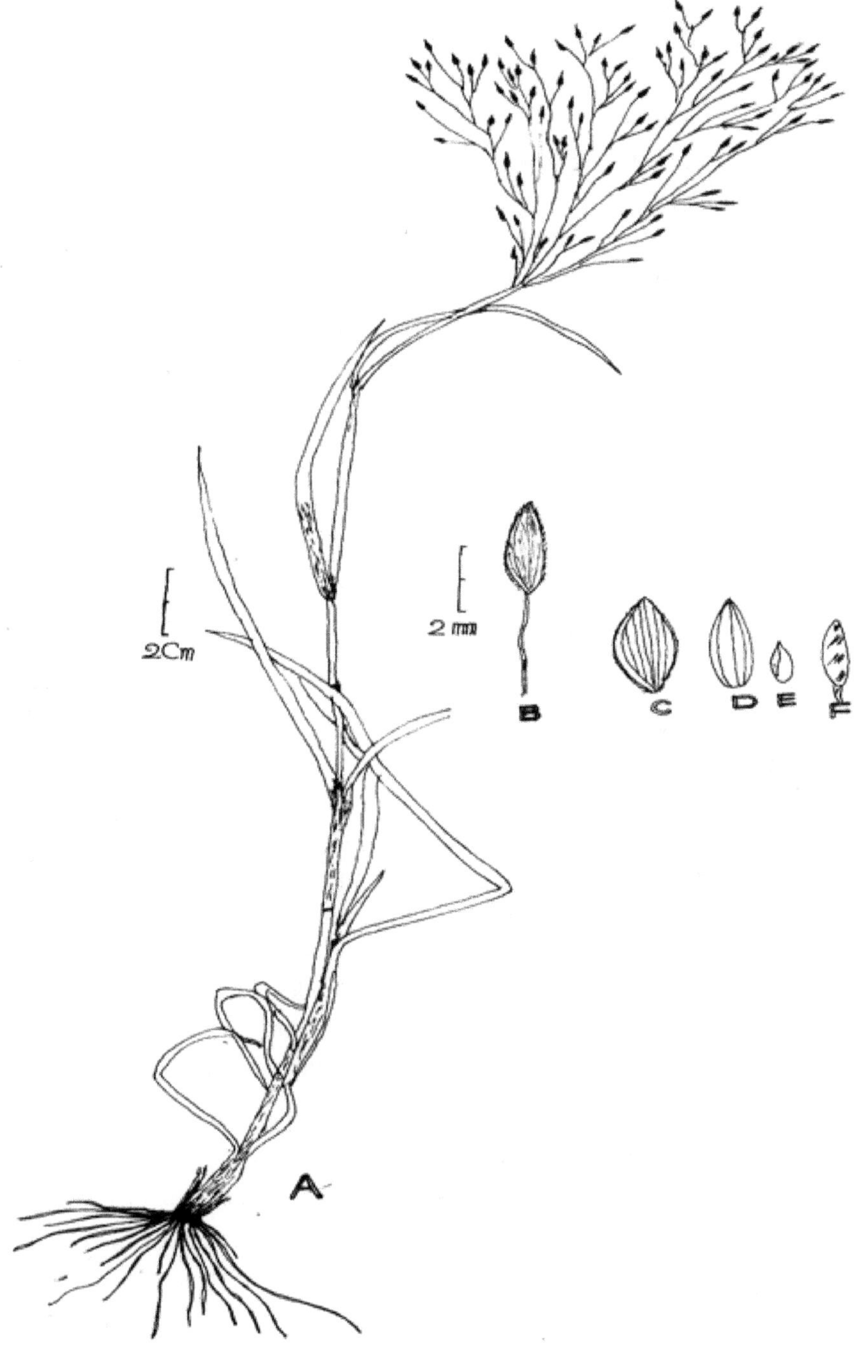

Figure 39. **Digitaria tomentosa** (Willd.) Henr.

A. Twig, B. Spikelet, C. Lower lemma, D. Upper lemma, E. Palea, F. Grain.

Annual, tufted, erect grass, 5-50 cm high, base not creeping. Leaves linear, 2.5-10 cm long. Racemes 2-6, digitate or when more than 3, 2 or 3 from lower level, 4-10 cm long; rachis straight, flat, margins winged; spikelets 1.5-2 mm long, with short verrucose hairs; fruiting lemmas dark brown or black.

Occasional in Northern Eastern Ghats in Odisha.

Mahendragiri (GJP), *VNS* 5746 (MH); Ganjam, *JSG* 21629 (CAL); Deomali hill (KPT), *AC* 10226 (BSID).

INDIA: Assam, Odisha, Tamil Nadu.

WORLD: Tropics and subtropics of Asia, Australia and America.

DIMERIA R.Brown

1. Upper glume awned .... **D. avenacea**
1. Upper glume unawned:
    2. Perennials:
        3. Racemes 2, rarely 3; lower glume keeled at he tip, not winged; wing of the upper glume papery throughout; awn of upper lemma 1.4 cm long .... **D. lehmanii**
        3. Racemes solitary; lower glume keeled all along the back and narrowly winged towards the tip; wing of upper glume papery towards the tip and corky below; awn of upper lemma 1 cm long .... **D. mahendragiriensis**
    2. Annuals:
        4. Upper glume winged:
            5. Upper glume winged all along or at least at the tip:
                6. Upper glume narrowly winged all along the keel; rachis narrow, 0.75 mm wide, racemes rarely divergent .... **D. connivens**
                6. Upper glume broadly winged on the keel; rachis broader, 1-1.25 mm wide, racemes usually divergent:
                    7. Spikelet 7.5 mm long, strongly compressed .... **D. mooneyi**
                    7. Spikelet 3 mm long, less compressed.................... **D. bialata**
            5. Upper glume winged just below the tip only:
                8. Wings of upper glume mucronate .... **D. fuscescens**
                8. Wings of upper glume not mucronate........................ **D. orissae**
        4. Upper glume not winged:
            9. Rachis angular or trigonus........................ **D. ornithopoda**
            9. Rachis flat:

10. Spikelets less than 3.5 mm long; rachis margin narrowly winged.. **D. kanjirapalliana**

10. Spikelets 4-4.5 mm long; rachis margin wingless.. **D. kollimalayana**

Dimeria avenacea (Retz.) C.E.C.Fischer, Kew Bull. 1932: 72. 1932 & Fl. Madras 3: 1713. 1934; Bor, Grass. India 139. 1960; Kapoor, J. Bombay Nat. Hist. Soc. 61: 368. 1964. *Anthoxanthum avenaceum* Retz., Observ. Bot. 3: 8. 1783. *Dimeria pusilla* Thwaites, Enum. Pl. Zeyl. 369. 1864; FBI 7: 103. 1896. *D. acutipes* Bor, Kew Bull. 1952: 560. 1953.

An annual tufted grass; culms up to 40 cm high, filiform. Leaves erect, linear, 2.5-5 cm x 2-2.5 mm, apex acuminate, sparsely ciliate; sheaths keeled, smooth and glabrous, striate; ligule membranous, ciliate. Racemes solitary, 2-4.5 cm long, rachis very narrow, trigonous, scabrid on margins. Spikelets alternate, pedicelled, 5 mm; glumes coriaceous, sparsely ciliate, upper winged, keels ciliate, awned; lemmas hyaline, upper awned; palea small, bisexual. Caryopsis linear, compressed, free within the lemmas.

Rare in Nellore district in Andhra Pradesh (Bourne), Mahendragiri in Ganjam, Odisha (Kapoor, *loc cit.*).

INDIA: Deccan Peninsula.

WORLD: Sri Lanka.

Dimeria bialata C.E.C. Fisch., Bull. Misc. Inform. Kew 1933: 351. 1933; Bor, Grass. India 140. 1960; Kumar *et al.*, Rheedea 26: 162. 2016.

Culms erect, 25–30 cm high; nodes subglabrous. Leaf blades linear-lanceolate, 3–6 x 0.2–0.3 cm, tuberculate-ciliate at margins, pilose with tubercle-based hairs on both sides; ligules *c.* 1 mm long, ciliolate-membrane; sheaths keeled, pilose with tubercle-based hairs. Racemes 2, divergent, straight, unilateral, to 4 cm long; rachis *c.* 1.5 mm wide, flattened, winged, scabrous on margins. Spikelets densely packed, 2-rowed, strongly appressed; fertile spikelets pedicelled, elliptic, 3.5–4 mm long, laterally compressed, falling entire; callus *c.* 0.3 mm long, obtuse at base; pedicels reduced to a stump, flattened; glumes dissimilar, exceeding the floret length; lower glume lanceolate, 3–3.25 mm long, almost to first length of spikelet, coriaceous, much thinner on margins, 1-keeled; keel narrowly winged all along, ciliate at margins, acute at apex, upper glume elliptic, *c.* 3.5 mm to as long as spikelet, acute at apex, 1-keeled; keel broadly winged all along, ciliate at margins, acute at apex; florets 2, lower barren, upper bisexual; basal sterile florets barren, without palea; lemma of lower sterile floret oblong, hyaline; fertile lemma elliptic, *c.* 2.5 mm long, hyaline, without keel, 1-veined; lemma apex dentate, 2-fid, 1-awned; awn from a sinus, geniculate, 9–11 mm long including twisted column; column 2.5–4 mm long, glabrous; palea minute, *c.* 1 mm long; anthers 2, *c.* 1 mm long; stigma plumose. Caryopses cylindrical, *c.* 2 mm long.

Rare in the open habitats of moist deciduous forests in Rampa hills. Fl. & Fr.: September–December.

Sivakasipuram, near Y. Ramavaram (EG), *BR* & *M. Anil Kumar* 49203 (SKU).

INDIA: Andhra Pradesh, Karnataka, Kerala. Endemic.

Dimeria connivens Hack. in DC., Monog. Phan. 6: 689. 1889; FBI 7: 104. 1896; Haines, Bot. Bihar & Orissa 1016. 1924; Mooney, Suppl. Bot. Bihar & Orissa 192. 1950; Bor, Grass. India. 140. 1960.

Annual grass, culms tufted, erect or geniculate, up to 30 cm long, terete, smooth; nodes shortly and very sparsely bearded. Leaves linear or narrowly linear-lanceolate, ascending, 2.5-5.5 cm, mostly basal, apex acuminate, margins with sparse tubercled based hairs, midrib well marked on the lower surface; sheath slightly keeled, broadly hyaline on the margins, leaf sheath longer then the lower internodes, shorter than the upper, upper sheath close to the culm, lower one rather loose and slipping from the culms, smooth and glabrous, striate; ligule ovate, membranous, shortly ciliate, less than 1 mm long, racemes 2, erect, shortly divergent, 4.2 cm long, lower florets empty, upper florets bisexual, rachis flat, tough, 0.75 mm wide, narrowly winged, ciliate along margins; spikelets alternate, oblong or oblanceolate, greenish-yellow, 3.5-4 mm long, callus base bearded, pedicels very short, flat, lower glume 2.5 mm long, excluding the callus, chartaceous, apex acute with a sharp point, keels covered all along the beak with forwardly directed cilia; upper glume lanceolate or elliptic, 3.5 mm long, acute, winged all along the keel, wings ciliate, apical cilia 1 mm long, margin hyaline, lower lemma oblanceolate to oblong, hyaline, 1.5-2 mm long, ciliate towards apex; upper lemma linear-elliptic, acute, 2-2.5 mm long, hyaline, faintly nerved arista or awn 7 mm long; short ciliate, column dark brown, 3 mm long; stamens 2, anthers 1.5-2 mm long, bright yellow; ovary elliptic, *c.* 0.3 mm long; styles 0.5 mm, stigmas plumose, 0.6 mm long. Caryopsis linear, slightly curved.

Rare in grasslands in Northern Eastern Ghats. Fl. & Fr.: September–December.

Way to Y. Ramavaram (EG), *TP* & *MSG* 12286; near Tiger camp, Maredumilli (EG), *MM* 102585 (BSID); Sirkagutta (KJR), *HFM* 2794 (DD); Sonabera (KRT), *HFM* 3658 (DD); Khadgaon (SBP), *HFM* 2922 (DD); Punakote, Satkosia Willife Sanctuary (AGL), *AC* 2387 (BSID); Dimli, Kotgarh Willife Sanctuary, Kandhamal (KDM), *AC* 2423 (BSID).

INDIA: Bihar, Odisha, Andhra Pradesh, Kerala. Endemic.

Key to the varieties of *Dimeria connivens*

1. Rachis *c.* 0.5 mm wide; lower glume narrowly ovate; lower lemma narrowly obovate, 1–1.3 mm long .... **D. connivens** var. **connivens**

1. Rachis *c.* 1 mm wide; lower glume linear-lanceolate; lower lemma linear-lanceolate, 1.8–1.9 mm long .... **D. connivens** var. **roxburghiana**

Dimeria connivens Hack. var. **roxburghiana** K.C. Mohan & Prasanna, Bangladesh J. Plant Taxon. 24: 237. 2017.

Annuals, culms erect, 40–50 cm high; nodes hairy, clothed with leaf sheath. Leaf sheath terete, margins pilose with tubercle based hairs, 2.0–2.5 cm long, loose, upper most spathiform; ligule membranous, ciliate at apex, 0.5–0.6 cm long; leaf blade linear-lanceolate, 7–9 x 0.2–0.3 cm, glabrous on both sides and tubercle based hairy along margins, margins more or less wavy, acuminate at apex. Racemes 2 or 3, erect,eventually divergent, 4.2–7.0 cm long. Rachis narrowly winged, triquetrous, ciliate along margins, 0.9–1.0 mm wide, zig-zag. Spikelets solitary, up to 4.1 x 1.2 mm, falling entire at maturity; pedicels thick, 0.3–0.4 mm long, glabrous; raceme internodes 1 mm long; callus oblong, minute, hairy. Lower glume linear-lanceolate, 3.5–3.8 x 0.2–0.3 mm long, acute to acuminate at apex, sub-coriaceous, 1-keeled, pilose along keel, margins hyaline, ciliate on outer surface; upper glume elliptic-oblong, 4.1–4.3 x 0.9–1 mm, acute at apex, sub coriaceous, 1-keeled, winged all along the keel, pilose, margins hyaline, ciliate on outer surface; florets 2; lower barren, epaleate, upper bisexual; lower lemma lanceolate, 1.8–1.9 mm, hyaline, acute, margins ciliate towards the apex; upper lemma elliptic, 2.8-3.0 x 0.8-1.0 mm, 2-lobed at apex, awned from the sinus, 1-nerved; awn geniculate, 10-11 mm long; stamens 2; anthers 1 mm long; palea very narrow. Caryopsis lanceolate, 2.0–2.4 x 0.2–0.3 mm, brown.

It grows in open rocky slopes in deciduous forests in association of *Dimeria mooneyi* and *Striga angustifolia.*

Fl. & Fr.: September–October.

Odisha, Nayagarh, Satkosia Tiger Reserve (AGL), *Chandramohan* 8354 (*Holotype*: CAL; *Isotype*: BSID).

Dimeria fuscescens Trin., Mem. Acad. Imp. Sci. St.-Petersbourg, Ser. 6, Sci. Math. 2: 335. 1832; FBI 7: 105. 1896; Bor, Grass. India 140. 1960; Matthew, Fl. Tamilnadu Carnatic 3: 1841. 1983.

Perennial tufted grass; culm to 40 cm, with bearded nodes. Leaf sheath 2-4 cm; ligule to 1 mm; blades oblong, 5-10 cm, flat or inrolled. Racemes 2 or 3, digitate, 4.5-8.5 cm, to 3 mm wide; rachis 0.5 mm wide, undulate, glabrous. Spikelets oblong-linear, 5-5.5 mm, reddish-brown, pilose, laterally flattened; glumes winged throughout, reddish, folded; lower glume to 5 x 0.8 mm, ciliate, winged at base, stiff-pilose; upper glume 5.5 mm; wing produced into a short mucro, to 0.2 mm; lower lemma 3 mm, membranous, 1-nerved; upper lemma to 4 mm, hyaline; palea to 0.8 mm; stamens 2. Caryopsis ellipsoid.

Moist places in high altitudes of Shervarayan hills. Fl. & Fr.: August–September.

Shervarayan hills, Kaaka shola (SM), *SKS* 594 (SGH); *KMM & NPV* 25643 (RHT, K).

INDIA: Tamil Nadu.

WORLD: Nepal and Sri Lanka.

Dimeria kanjirapalliana K.C. Jacob, J. Bombay Nat. Hist. Soc. 47: 48. 1947; Manilal & Sivaraj. Fl. Calicut 231. 1982; Sreekumar & V.J. Nair, Fl. Kerala Grasses 91. 1991; Gayathri & Pullaiah, J. Bombay Nat. Hist. Soc. 93: 122.1996.

An annual, tufted grass, culms up to 25 cm high, erect, occasionally branched near the base; nodes sparsely bearded or glabrous. Leaf sheaths 3-4 cm long, glabrous, keeled; ligule membranous, ovate, acute; blades linear, 2-6 x 0.2-0.4 cm, rounded at base, bulbous based hairs along the margins, acuminate at apex. Spikes 2, 3-8 cm long, divaricate; rachis flat, zig-zag, margins sparsely ciliate, narrowly winged; spikelets sessile, 4.5 mm, oblong-lanceolate; callus villous; lower glume linear- lanceolate, 2.5 mm, chartaceous, sparsely hairy at the apex, acuminate; upper glume lanceolate, 3 mm, chartaceous, hairy at the apex, margins hyaline, lower lemma hyaline, 2 mm, oblong, 1-nerved; upper lemma hyaline, 3 mm (excluding the awn), bifid, awned at the sinus, awn 1 cm long; stamens 2, anthers oblong, 0.9 mm, filament 0. 5 mm; pistil 1.75 mm. Caryopsis linear.

Occasional in the humus soil and rocky areas of East Godavari district. Fl. & Fr.: October-January.

INDIA: Kerala, Andhra Pradesh. Endemic.

Dimeria kollimalayana M. Mohanan & S.V.N. Rao, J. Bombay Nat. Hist. Soc. 80: 617. 1984; V.J. Nair in A.N. Henry *et al.*, Fl. Tamil Nadu Anal. 3: 109. 1989; K.A.A. Kabeer & V.J. Nair, Fl. Tamil Nadu Grasses 417. 2009.

Annual, erect grass, culms up to 20 cm high; nodes sparsely bearded. Leaf sheath *c.* 5 mm, keeled; ligule membranous; leaf balde lanceolate, 2-6 x 0.4 cm, rounded at base, villous with tuberculate hairs, midrib prominent; lower ones longer than the upper, sparsely tuberculate hairy in upper half. Racemes 2.2-3 cm long, divergent; rachis *c.* 0.7 mm wide, flat, zigzag, margins scabrid; callus hairs 0.5-1 mm long; spikelets oblong-lanceolate, 4-4.5 mm long, acuminate at apex, villous; lower glume narrowly lanceolate, 3.5 - 4 x 0.5 mm, acuminate at apex, scabrid; upper glume oblong-lanceolate, as long as spikelet, acuminate and winged at apex, densely villous and few hairs at tip, marign hyaline; florets 2; lower barren, upper bisexual; lower lemma oblanceolate, 1.5-2 x 0.4 mm, membranous, hyaline, faintly 1-nerved, margins ciliate in upper half; upper lemma 2.5-3 x 1 mm, 2-lobed at apex, awned from sinus, lobes acuminate, delicate, faintly 1-nerved, awn geniculate, 8-10 mm long, column *c.* 3 mm long, brownish, twisted, bristle barbellate; stamens 2; anthers *c.* 0.7 mm long; filaments short; ovary elliptic, *c.* 0.2 mm long; stigma plumose, *c.* 0.8 mm long. Caryopsis oblong, *c.* 2 mm long.

Along slopes of hills, usually in exposed areas of Kollimalais.

Kolli hills, Seekuparai (NK), *SKS* 1311 (SGH).

INDIA: Tamil Nadu. Endemic.

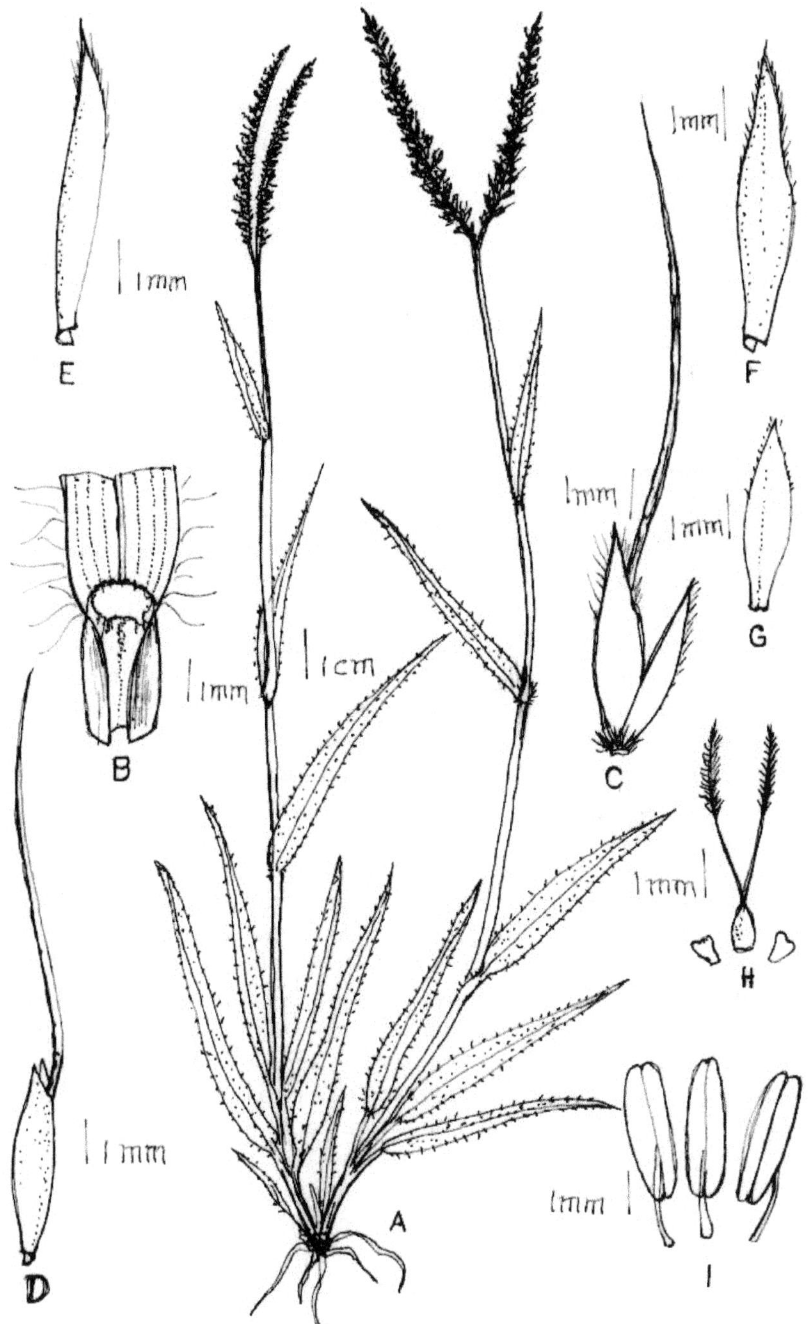

Figure 40. **Dimeria kollimalayana** M. Mohanan & A.V.M. Rao.

A. Plant, B. Collar, C. Spikelet, D. Upper glume, E. Lower glume, F. Lemma, G. Palea, H. Ovary and lodicules, I. Stamens.

Dimeria lehmanii (Nees ex Steud.) Hack. in A. DC. Monog. Phan. 6: 82. 1889; FBI 7: 104. 1896; Fischer 3: 1713. 1934. *Pterygostachyum lehmanii* Nees ex Steud., Syn. Pl. Glum. 1: 413. 1853 *p.p.*

Perennial, erect grass, culms up to 60 cm high, terete, smooth, glabrous. Leaves 4-15 x 0.2-0.4 cm, acuminate, minutely scabrid on both sides. Racemes 2, rarely 3, divergent, 4-8 cm long; spikelets 4-4.5 mm long, firmly compressed; lower glume linear, 3-3.6 mm long, acute; lower lemma 2 mm long; upper lemma elliptic, 3 mm long, acute, hyaline. Caryopsis compressed, ca 3 mm long.

Rare in forests in Northern Eastern Ghats. Fl. & Fr.: September–January.

Rampa hills (EG), *K.N.Reddy* & *CSR* 2255 (FRLHT).

INDIA: Odisha, Andhra Pradesh, Tamil Nadu.

WORLD: Sri Lanka.

Dimeria mahendragiriensis Ravi *et al.*, Rheedea 5: 142. 1995; Saxena & Brahmam, Fl. Orissa 4: 2337. 1996.

Perennial grass; culms slender and branched, lower portion subterranean and rooting at nodes, nodes ciliate. Leaves all along the culm; sheath up to 6 cm long, lower ones longer than the internodes and slipping from the culm, keeled on the back, hyaline on the margins, sparsely ciliate with bulbous-based hairs to glabrous; blade linear-lanceolate to elliptic, acuminate at apex, up to 10 cm x 4 mm, base more or less rounded, flattened on the midrib above and keeled below, ciliate with bulbous based hairs towards the base; prominently so on the margins and towards the margins on the upper surface and scaberulous on the margins towards the tip; ligule membranous, truncate at apex, 1-1.5 mm long. Raceme solitary, rachis triquetrous, flattened on the back, *c.* 7.5 cm wide, prominently winged on the margins, wings ciliate and prominently ridged on the face; pedicels *c.* 0.5 mm long, concave at the apex, thickly ciliate on the outside and sparsely so on the inside; spikelets elliptic, 4-5 mm long, callus *c.* 0.5 mm long, callus hairs up to 0.75 mm long; lower glume oblong-elliptic, acute, aristate at apex, 3-3.5 x 0.5-0.6 mm (when folded), strongly keeled on the back, keel narrowly winged in the upper half, ciliate on the keel except towards the scaberulous apex, and sparsely short-ciliate on the sides; upper glume subcoriaceous, oblong-elliptic, acute-aristate at apex 4-4.5 x 1 mm (when folded), broadly winged on the keel; wing narrowed to the base, papery towards the apex and corky below, densely ciliate on the back except towards the scaberulous apex with a few longer hairs on the sides in the upper half; sides subcoriaceous, sparsely ciliate, margins hyaline, shortly ciliate in the middle; lower lemma oblanceolate, acute at apex, cuneate at base, hyaline, 2-2.5 mm long, sparsely ciliate on the margins in the upper half; upper lemma elliptic, bifid at apex with acute lobes, 2.5-3 x 0.5 mm (when folded), awned from the sinus, awn up to 10 mm long with a 3 mm long column; lodicules 2, cuneate at base; palea linear-elliptic, *c.* 0.75 mm long, papillate in the upper half and long ciliate towards the apex; stamens 2, anthers 1.5 mm long; ovary *c.* 0.5 mm long, styles 2, *c.* 1 mm long, stigmas 2, *c.* 1 mm long. Grains oblong-flattened, acute to subacute at apex, 1.75-2 mm long.

In Mahendragiri hills in Northern Eastern Ghats (Saxena and Brahmam, 1996).

INDIA: Odisha, Endemic.

Dimeria mooneyi Raizada ex Mooney, Suppl. Bot. Bihar & Orissa 263. 1950; Bor, Kew Bull. 1952: 569. 1953; Saxena & Brahmam, Fl. Orissa 4: 2339. 1996.

Annual grass, culms erect, stiff, smooth, glabrous. Leaves linear, 2-6 x 0.2-0.3 mm, acuminate, base rounded, margins scabrous. Racemes 2, pale silky, 5 cm long; spikelets compressed, 7.5 mm long; lower glume narrowly oblong, 5 mm long; upper glume oblong, 6.2 mm long, strongly compressed; lower lemma oblong, hyaline, 3.5 mm long, ciliate on the margins; upper lemma hyaline, 4.3 mm long, apex cleft, 2-lobed, awned, awn 1 cm long, arising from the sinus; palea very narrow, 1.5-2 mm long.

Rare in crevices of massive quarzite outcrops of Northern Eastern Ghats (Subba Rao and Kumari, 2003, 2008). Fl. & Fr.: October–January.

Sonabera (KRT), *HFM* 3652, 3657 (DD); Banogoccha (Nayagarh), *KCM* 8358, 8363 (BSID); Raisili, Laxmipur (KPT), *AC* 4224 (BSID); Badrama Wildlife Sanctuary (SBP), *AC* 10250, 10251 (BSID).

INDIA: Odisha, Andhra Pradesh, Endemic.

Dimeria orissae Bor, Kew Bull. 7: 579. 1952 & Grass. India 142. 1960; Matthew, Fl. Tamilnadu Carnatic 3: 1842. 1983.

Annual, slender grass, up to 22 cm high. Leaf sheaths keeled in the upper half, rounded below, glabrous and smooth, striate; leaf blades linear 3-7 x 0.2-0.3 cm, apex acuminate, erect or spreading, somewhat glaucous on the lower surface, scabrid on both surfaces. Racemes 2, terminating the culms, 3-5.5 cm long, spikelet 4-4.25 mm long, firmly compressed, oblong, awned, not winged, callus shortly bearded; lower glume linear, 3 mm long, upper glume 4 mm long; lower lemma a very small hyaline scale, 2 mm long, oblanceolate; upper lemma elliptic, 2 mm long, awn 1 cm long.

Subgregarious in sal forest under shade, and on the thin soil caps of the hills in Koraput district (Bor, 1960) and Kolli hills (Matthew, 1983). Fl.: October–November.

Mahalkotta forest, Pottangi (KRT), *HFM* 4039 (DD); Devgiri hills, Kuldhia W.L.S. (BLR), *AC* 2361 (BSID); Deomali hill (KPT), *AC* 4229(BSID); Bargakhol, Rajgangpur, Sundargarh, *AC* 10259 (BSID); Badrama Wildlife Sanctuary (SBP), *AC* 10248 (BSID); Dimli, Kotgarh Wildlife Sanctuary (KDM), *AC* 10206 (BSID).

INDIA: Odisha, Tamil Nadu. Endemic.

Dimeria ornithopoda Trin., Fund. Agrost. 167. t. 14. 1820; FBI 7: 104. 1896; Fischer 3: 1713. 1934; Bor, Grass. India 142. 1960; Matthew, Fl. Tamilnadu Carnatic 3: 1842. 1983.var. **ornithopoda**

An annual, tufted grass; culms up to 40 cm long. Leaf sheaths 2.5 cm, keeled, tuberculate pubescent; ligule membranous; blades 4-6 x 0.3-0.8 cm, lanceolate, densely pubescent. Racemes 2 or 3, to 6 cm long, terminal digitate; rachis selnder, flexuous, trigonous; spikelets 2 mm, oblong, alternate, sessile or subsessile; glumes hairy, upper mucronate, short, lower glume linear, a little shorter than the spikelet; lower lemma 0; upper lemma awned, awn 4-5 mm long. Caryopsis ellipsoid.

Occasional in forests. Fl. & Fr.: August-January.

Srisailam (KNL), *BSS, SKB* & *BR* 30691; Irlapenta (MBNR), *BR* & *BSS* 32290 (BSID); Gandhamardan (BL), *SX* & *MB* 6141, 6258 (RRL-B); Mali (KHD), *MB* & *Dhal* 7617 (RRL-B); Garpati forest (SBP), *HFM* 2918 (DD); Tikarpada section, Satkosiaa Tiger Reserve (AGL), *KCM* 6104 (BSID); Punakote, Satkosia Willife Sanctuary (AGL), *AC* 2386 (BSID); Dimili, Kotgarh Wildlife Sanctuary (KDM), *AC* 2422 (BSID); Betkote, Karlapat Wildlife Sanctuary (KHD), *AC* 2468 (BSID).

INDIA: Throughout India.

WORLD: Tropics of Asia extending to Australia.

DINEBRA Jacquin

Dinebra retroflexa (Vahl) Panz., Denkschr., Acad. Wiss. Muenchen. 270. t. 12. 1814; Fischer 3:1841. 1934; Bor, Grass. India 491.1960; Matthew, Fl. Tamilnadu Carnatic 3: 1843. 1983; Moulik, Grass. Bamb. India 2:589.1997. *Cynosurus retroflexus* Vahl, Symb. Bot. 2:20.1791. *Dinebra arabica* Jacq., Fragm. 77. t. 121. f. 1.1809; FBI 7: 297. 1896.

Annuals, culms erect or ascending, tufted, leafy, to 1.2 m; nodes slender, glabrous. Leaf sheaths glabrous; ligules membranous; leaf blades linear-lanceolate, base cordate, apex acuminate, glabrous. Inflorescence elliptic-oblong to pyramidal; spikelets 1-3-flowered; lower glume elliptic-lanceolate, 1-nerved, keeled, coriaceous, scaberulous, awned; upper glume elliptic-lanceolate, 1-nerved, keeled, awned or cuspidate, persistent; lemmas narrow, ovate, hyaline, sub-acute or obtuse, 1-nerved; paleas ovate, hyaline, 2-keeled, sparsely ciliate on the keels; stamens 3. Caryopsis ovoid, trigonous.

Occasional in forests and common in crop fields. Fl. & Fr.: August-September.

Garldinne (ATP), *TP* & *NY* 315; Kalasamudram (ATP), *BR* & *AMR* 21890; Pamidi (ATP), *KVS* 28015; Kammavaripalli (KNL), *TP & PVP* 2786; Gundumallappa kona (KNL), *SS* & *AMR* 20285; Gundlabrahmeswaram (KNL), *BSS* & *SKB* 32972; Kadapa (KDP), *AL* 26579; Jyothi (KDP), *BR* & *BSS* 30859; Jagadevipet (NLR), *BR* & *PMR* 19521; Macherla RF (GNT), *VRK* 5897; Kondapudi (EG), *CAB* 5192 (MH); Achaaryapur–Koilkonda (MBNR), *SRS* 109766 (BSID); Garlapadu RF (MBNR) *SRS* 109085 (BSID); Manbhang, Gandhamardan (BL), *SX* & *MB* 5996 (RRL-B).

INDIA: Rajasthan, Madhya Pradesh, Uttar Pradesh, Maharashtra, Andhra Pradesh, Bihar, Odisha and Tamil Nadu.

WORLD: S. Africa through Tropical Africa to Egypt, Iraq to India.

Note: This grass looks like some species *oi Eragrostis,* but it can be distinguished by the lax, loose, grey panicles, larger spikelets and lobed as well as awned lemmas with villous nerves.

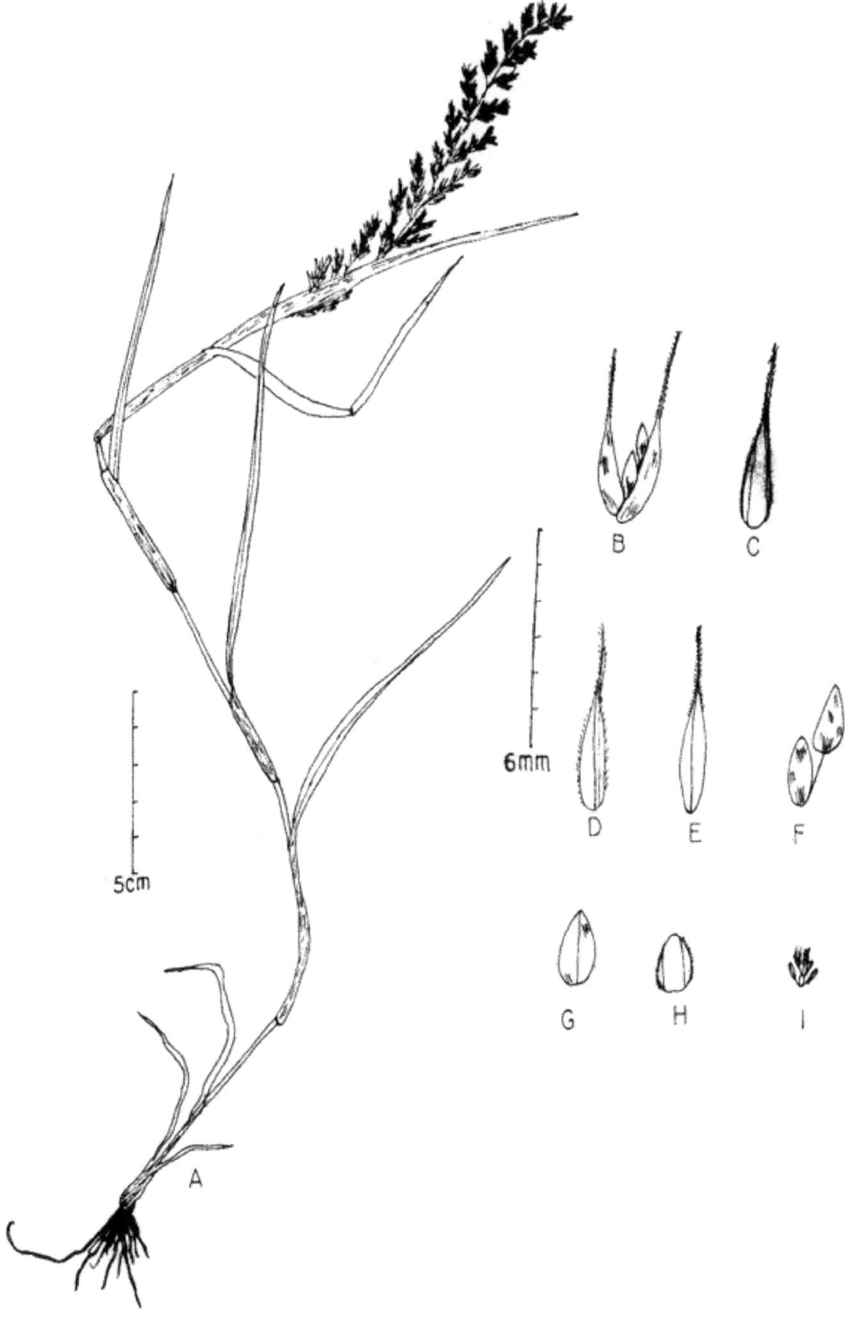

Figure 41. **Dinebra retroflexa** (Vahl) Panz.

A. Twig, B. Spikelet, C. Lower glume (lateral view), D. Back view, E. Upper glume, F. Florets, G. Lemma, H. Lemma, I. Pistil and stamens.

ECHINOCHLOA P. Beauvois *nom. cons.*

1. Ligule present:
   2. Culms very long, up to a meter; lower lemma awned to 4 mm long .... **E. stagnina**
   2. Culms short, to 40 cm long; lower lemma awned to 2 mm long .... **E. picta**
1. Ligule absent:
   3. Spikelets awned .... **E. crusgalli**
   3. Spikelets unawned:
      4. Racemes appressed to axis, 4-rowed .... **E. colona**
      4. Racemes not appressed to axis, many-rowed:
         5. Raceme curved at maturity; spikelet not deciduous.... **E. frumentacea**
         5. Raceme straight at maturity; spikelet deciduous .... **E. glabrescens**

Echinochloa colona (L.) Link, Hort. Berol. 2: 209. 1833; Fischer 3: 1776. 1934; Bor, Grass. India 308. 1960; Matthew, Fl. Tamilnadu Carnatic 3: 1845. 1983. *Panicum colonum* L., Syst. Nat. (ed. 10) 810. 1759; FBI 7: 32. 1896.

An annual, tufted grass; culms up to 70 cm high, branching from lower nodes; nodes glabrous. Leaf sheaths 5-15 cm long, glabrous; ligule obsolete; blade 5-25 x 0.3 - 0.6 cm, linear-lanceolate, base cordate, apex acuminate, glabrous. Spike like panicle, racemes 8-20, appressed to the rachis; rachis angular, scaberulous; spikelets 2.5 cm, broadly ovate, acute or sub- cuspidate, crowded in rows, secund; glumes unequal; lower glume half the length of the upper glume, broadly ovate-orbicular; upper glume broadly ovate, mucronate, puberulous; lower lemma broadly ovate, cuspidate; upper lemma broadly ovate, obtuse; palea oblong; stamens 3; stigmas plumose. Caryopsis broadly ellipsoid.

Very frequent in cultivated fields and near water courses. Fl. & Fr.: Almost round the year. Tam.: *Pullampayaru*.

Dharmapuri RF (ATP), *NY* 1002 (SKU & MH); Pennahobilam (ATP), *PVP* 2652; Gulemallebad RF (KNL), *RVR* 1406; Balarachepalli (KDP), *PVP* 3443, 3477 & 3470; Mabbuchintalapalli (KDP), *TP* & *SSR* 33665 (SKU & BSID); Tirumala hills (CTR), *MSG* 12176; Way to Koppaladoruvu (NLR), *MSG* 10900; Krishna river bed (GNT), *VRK* 6131; Lakkavarapu reservoir (VSKP), *MHR* & *MCK* 14879; Chelama (KNL), *JLE* 16797 (CAL); Kadapa, *EKK* 78530 (DD); way to Mamandur (CTR), *KS* 6918 (MH); Chandragiri (CTR), *S. India Flora* 10111 (MH); Kondapalli (KSN), *sin coll. s.n.* (MH); Kolleru lake (WG), *KS* 5092 (CAL); Zengareddy gudem (WG), *DCSR* 283 (CAL); Gokavaram (EG), *CAB* 8257 (MH); Rampachodvaram (EG), *MM* 102650 (BSID); Araku valley (VSKP), *NPBK* 518 (CAL); Garlapdu RF (MBNR), *SRS* 109090 (BSID); Billyguttalu (KMM), *RR* 10669, 105996 (BSID); Rathamhutta hills (KMM), *RCS* 104233 (MH); Manuguru RF (KMM), *RCS* 104259 (MH); Purunakote, Satkosia Tiger Reserve (AGL), *KCM* 5165, 5263 (BSID); Sirumalai hills (DGL), *SKS* 420 (SGH);

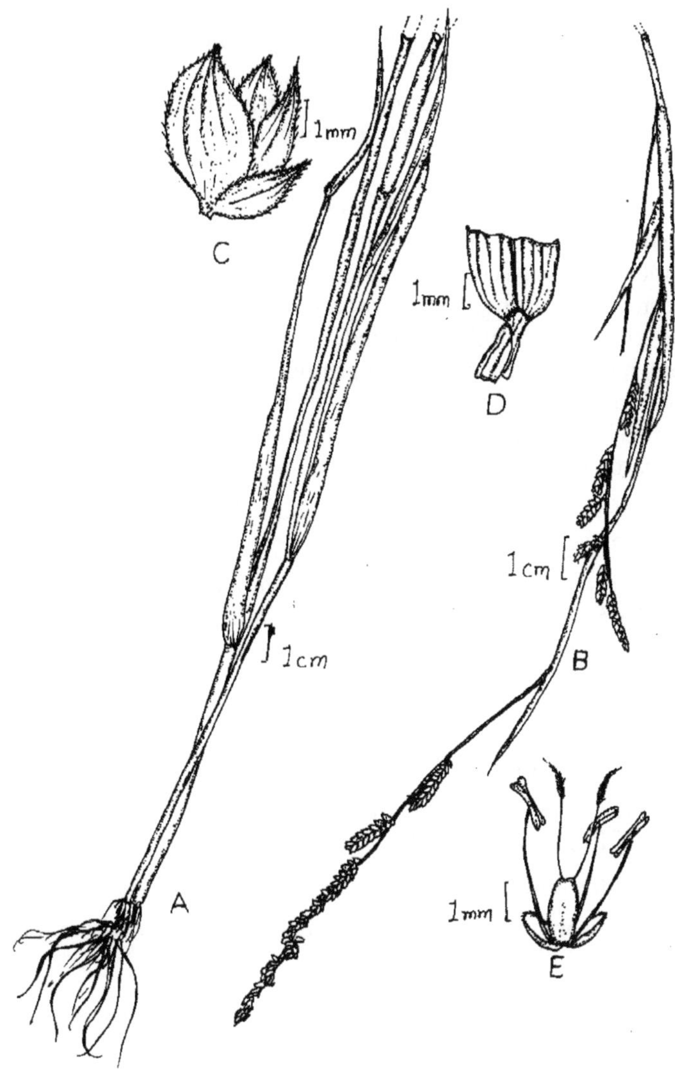

Figure 42. **Echinochloa colona** (L.) Link

A. Habit, B. Inflorescence, C. Flower, D. Ligule, E. Pistil, stamens & lodicules.

Hogainakkal RF (KGR), *EV* 20659 (MH); Gandigam lake, Pennagaram (SLM), *EV* 22450 (MH); Sonanchavadi (SA), *KRM* 64167 (MH); Red hills (CPT), *CAB* 5518 (MH); Kottur RF, Yelagiri hills (NA), *MBV* 806 (MH); Palur (SA), *CAB* 8340 (MH); Tindivanam to Gingee (SA), *KRM* 60297 (MH).

INDIA: Throughout India.

WORLD: Throughout the Tropics and subtropics.

Echinochloa crus-galli (L.) P. Beauv., Ess. Agrost. 53. 161. 1812; Fischer 3: 1777. 1934; Bor, Grass. India 310. 1960. *Panicum crus-galli* L., Sp. Pl. 56. 1753; FBI 7: 30. 1896.

An annual, tufted grass; culms 70 cm high, branching from lower nodes; nodes glabrous. Leaf sheaths 5-25 cm, glabrous; ligule obsolete; leaf blades 5-20 x 0.4-0.8 cm, linear-lanceolate, base cordate, apex acuminate, glabrous. Spikelike panicle 10 cm long, racemes spreading; rachis triquetrous, scabrid; spikelets broadly ovoid, 5 mm, crowded, acute-cuspidate, awned; lower glume broadly ovoid or orbicular, margin ciliate, half as long as the upper glume; upper glume broadly ovate-oblong, concave, cuspidate, hispid on back; lower lemma ovate-oblong, flat or depressed on the back, awned, hispid at apex; palea ovate-oblong; stamens 3; stigmas plumose. Caryopsis broadly ellipsoid.

Common in rice fields and wet places. Fl. & Fr.: Almost through the year especially in rainy season.

Kalasamudram RF (ATP), *NY* 1172; Gulemallebad RF (KNL), *RVR* 1410; Veldurthi (KNL), *RVR* 1742; Maddileti stream (KNL), *GO* 3353; Balapalle (KDP), *PVP* 3451; Balarachapalli (KDP), *PVP* 3441; Gopalpuram (NLR), *MSG* 10889; Vinukonda RF (GNT), *VRK* 6860; Palamaneru (CTR), *KCJ* 468 (MH); (Nagari (CTR), *S.India Flora* 14757 (MH); Nakirekal (GNT), *CAB* 4722 (MH); Gokavaram (EG), *CAB* 8256 (MH); Way to Komattiyur from Kavalur (NA), *KS* 6464 (MH); Priya Ayyampalayam (NA), *EV* 55610 (MH); Bhuvanagiri to Kurinjipadi (SA), *KRM* 60113 (MH); Near Gomukhi dam (SA), *KRM* 52824 (MH); Vannipaka Kannarpalayam (CPT), *DN* 505 (MH); Allikuli (CPT), *DN* 419 (MH).

INDIA: Throughout India.

WORLD: Subtropics of Africa, Afghanistan, Sri Lanka, Myanmar, Malaya, SE Asia.

Echinochloa frumentacea Link, Hort. Berol. 1: 204. 1827; Bor, Grass. India 311.1960; Matthew, Fl. Tamilnadu Carnatic 3: 1845. 1983. *Panicum crusgalli* var. *frumentaceum* Hook. f., Fl. Brit. India 7: 31. 1896. *Echinochloa colona* Link var. *frumentacea* Ridl., Fl. Malay Penin. 5: 223. 1925; Fischer 3: 1777. 1934.

An annual, robust grass; culms up to 1 m. Leaf sheaths 12 cm; ligule absent; leaf blades to 20 x 2 cm. Racemes 18 cm, several seriate; spikelets 4 mm, ovoid-broadly ellipsoid, hispid; glumes subequal, upper with a short mucronate apex; lower lemma mucronate, stiff- hispid outside; upper lemma crustaceous, short mucronate; palea similar to its lemma. Caryopsis broadly ellipsoid.

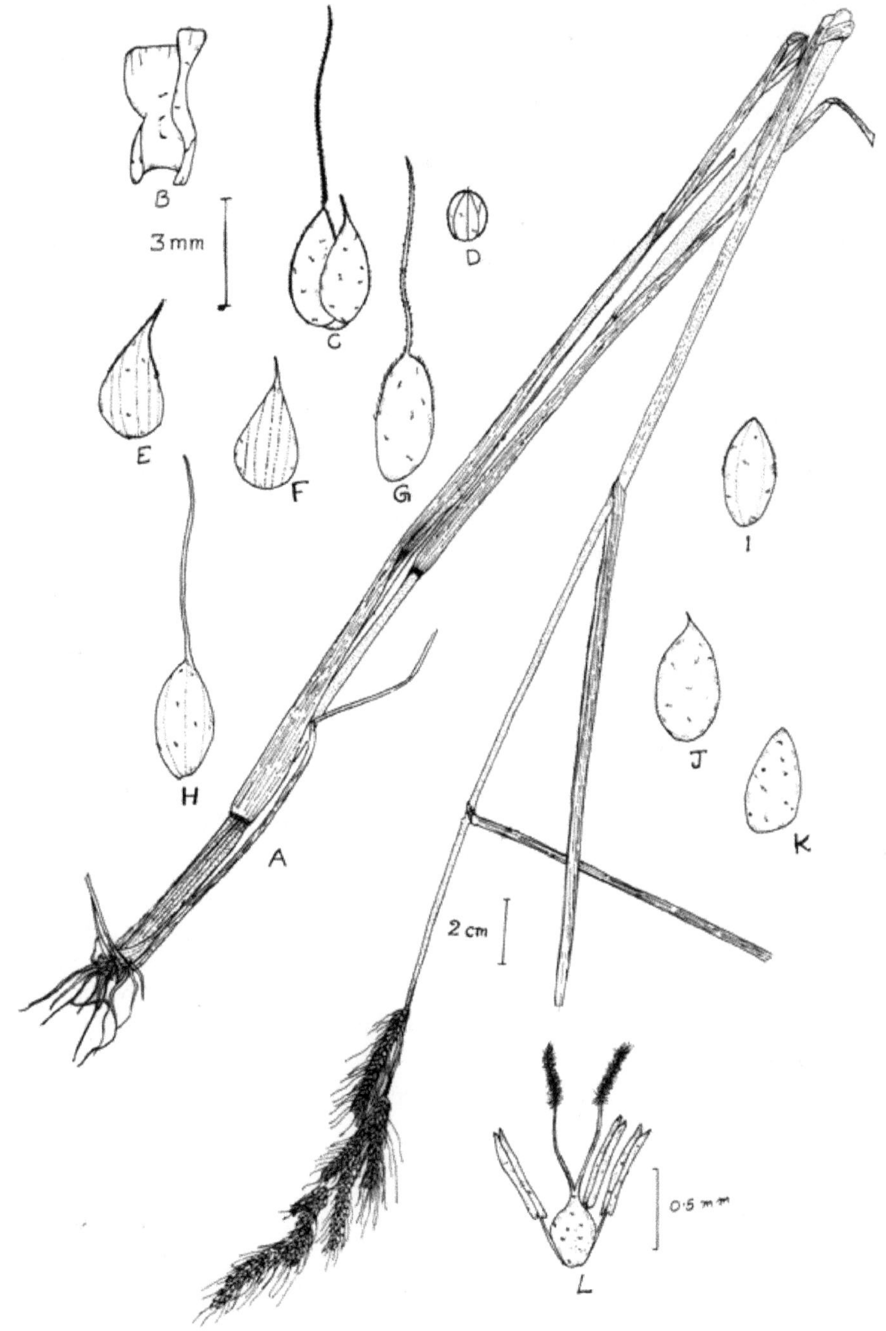

Figure 43. **Echinochloa crusgalli** (L.) P.Beauv.

A. Habit, B. Ligule, C. Spikelet, D. Lower glume, E & F. Upper glume dorsal and ventral view, I. Palea, J. Upper lemma, K. Palea, L. Pistil and stamens.

Cultivated for its grains. Fl. & Fr.: Through the year. Tel.: *Bonta shama, Pala ordalu, Sawa, Hamalu;*Tam.: *Kuthiraivali.*

Chennampalli (ATP), *BR* & *DV* 41149; Kuderu (ATP), *BR* & *DV* 41106; B.N.Kandriga (CTR), *BR* & *P. Priyadarsini* 37338 (SKU & BSID); Darsi (PKM), *MCK* 22828; Chintaladevi (NLR), *VNS* 82179 (MH); Tadepalli (EG), *NRR* & *DN* 84327 (BSID); Salur (VJN), *NPBK s.n.* (MH); Red Hills (CPT), *CAB* 5516 (MH); Hogainakkal (KGR), *EV* 23519 (MH).

INDIA: Cultivated all over India.

WORLD: Hotter parts of Asia and Africa.

Echinochloa glabrescens (Munro) Hook.f., Fl. Brit. India 7: 31. 1896; Bor, Grass. India 311. 1960; Matthew, Fl. Tamilnadu Carnatic 2: 1846. 1983.

Annual erect grass; culm 40-80 cm, nodes glabrous. Leaf sheath to 15 cm, glabrous; ligule absent; blades 15-30 **x** 0.4-0.8 cm. Raceme 5-13 cm, 2 or more rowed, to 3 cm; spikelets ovoid, to 4.5 mm, hispid; lower glume to 2.5 mm, acuminate; upper glume to 4 mm, long-acuminate; lower lemma to 3.8 mm, indurated, acuminate, setose without; upper lemma to 3 mm, aristate; stamens 3. Caryopsis ellipsoid.

Common weed of cultivated fields of Eastern Ghats of Tamil Nadu. Fl. & Fr.: July–October.

Kolli hills (NK), *SKS* 348 (SGH).

INDIA: Tamil Nadu.

WORLD: Pakistan.

Echinochloa picta (J.Koenig) P.W. Michael, Philipp. J. Weed Sci. 5: 18.1978; Matthew, Fl. Tamilnadu Carnatic 2: 1847. 1983. *Panicum pictum* J.Koenig, Naturforscher 23: 204. 1788; FBI 7:32. 1896 (pro. syn.)

Culms 50 cm to 1 m high. Leaf sheaths to 20 cm; ligule present; blades 6-25 x 0.8-1 cm. Inflorescence 7-20 cm; racemes irregularly 2-rowed, 2-5 cm. Spikelets to 4 mm, tuberculate-hispid. Lower glume to 1.5 mm, with a short mucro; upper glume to 4 mm, aristate; lower lemma to 3.5 mm; awn to 2 mm; upper lemma to 4 mm, acuminate.

Marshy or partly submerged in hills in Tamilnadu Carnatic (Matthew, 1983).

INDIA: All the states of India.

WORLD: Asia to New Guinea and Cape York Peninsula in Australia, also in E.Africa.

Echinochloa stagnina (Retz.) P. Beauv., Ess. Agrost. 53. 161 & 171. 1812; Fischer 3: 1777. 1934; Bor, Grass. India 311. 1960. *Panicum stagninum* Retz., Observ. Bot. 5: 17. 1789. *P. crusgalli* sensu Hook. f., FBI 7: 30. 1896 *p.p.* non L. 1753.

An annual stoloniferous grass; culms up to 50 cm high, rooting at the nodes; nodes glabrous. Leaf sheaths 5-10 cm long, loose, glabrous; ligule a fringe of stiff hairs; blades 4-15 x 0.4-0.8 cm, linear-lanceolate, base cordate, apex acuminate,

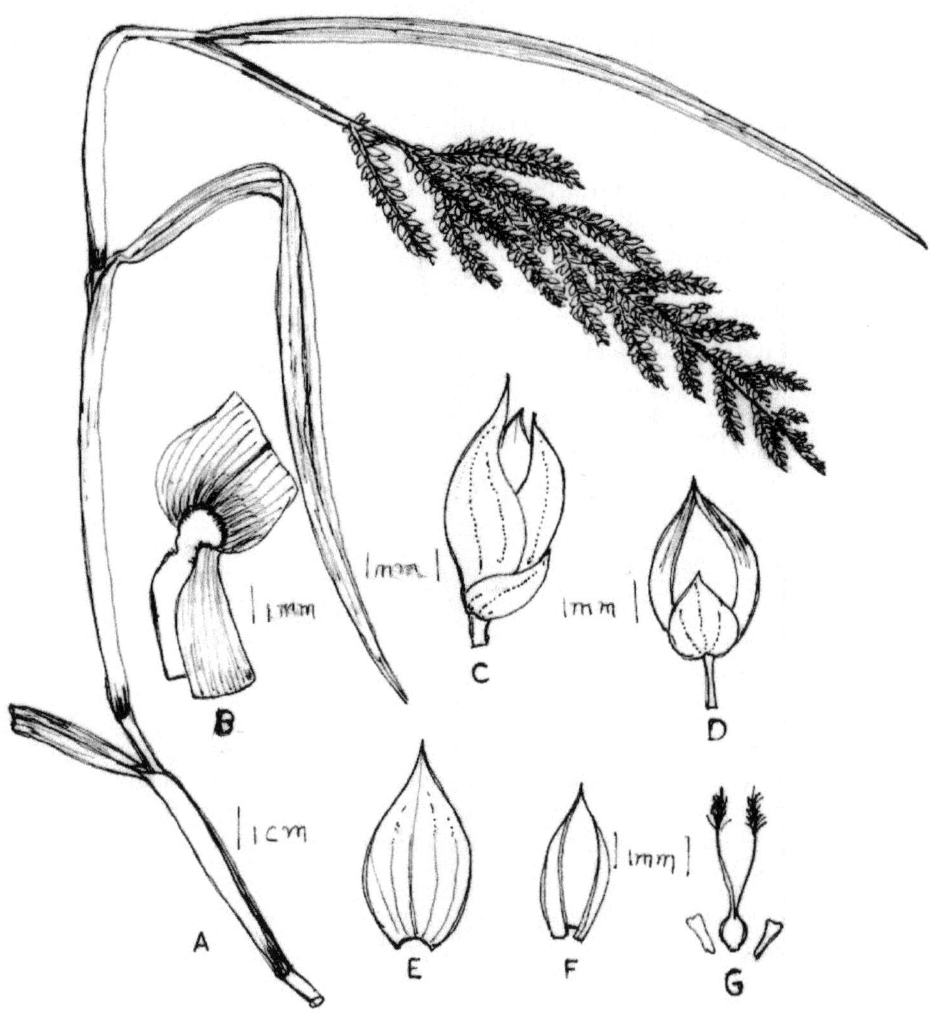

Figure 44. **Echinochloa glabrescens** Munro ex Hook.f.

A. Plant, B. Collar, C & D. Spikelets, E. Upper glume, F. Lemma, G. Ovary and lodicules.

midrib prominent. Spikelike panicle 5-10 cm long, racemes interrupted; rachis angular, scabrid; spikelets ovate, crowded, acutely acuminate, awned, secund, pubescent; lower glume broadly ovate, bidentate, pubescent; upper glume broadly ovate, acuminate, hairy at apex and margins; lower lemma ovate, hairy and scabrid at apex, awned; palea oblong-ovate, scabrid margin, male; upper lemma ovate, scabrid along the margins, acute, chartaceous; palea similar to upper lemma, bisexual; stamens 3, stigmas plumose. Caryopsis broadly ellipsoid.

Occasional in rice fields and marshy areas. Fl. & Fr.: Throughout the year.

Kalasamudram RF (ATP), *NY* 1172 (SKU & MH); Somayajulapalle RF (KNL), *RVR* 1872; Omkaram (KNL), *BSS* & *BR* 29255 (SKU & BSID); Way to Y. Ramavaram (EG), *TP* & *MSG* 12285; Gudur (NLR), *KSJ* 18397 (MH); Kolleru lake (WG), *KS* 5097 (MH & CAL); Palagudem (WG), *VNS* 44940 (MH); Mandapeta (EG), *CAB* 4283 (CAL); Dornacheruvu, Mannanur (MBNR), *SRS* 107458 (BSID); Erloo to Mahabubnagar (MBNR), *SRS* 104509 (BSID); Vandalur (CPT), *S. Indian Flora* 11472 (MH); Inetigiri (SLM), *S.India Flora* 15447 (MH); Bhuvanagiri (SA), *KRM* 53575 (MH); Kala sanapur (GJM), *VNS* 4660 (MH).

INDIA: More or less throughout India.

WORLD: Tropical Asia and Africa.

ELEUSINE Gaertner

1. Spikes slender, narrow, straight; wild species .... **E. indica**
1. Spikes stout, broad, curved at apex; cultivated species .... **E. coracana**

Eleusine coracana (L.) Gaertn., Fruct. Pl. 1: 8. 1789; FBI 7: 294. 1896; Fischer 3:1839. 1934; Bor, Grass. India 492. 1960; Matthew, Fl. Tamilnadu Carnatic 3: 1848. 1983. *Cynosurus coracanus* L., Syst. Nat. (ed. 10) 875. 1759.

Annuals, erect, tufted, to 1 m high; nodes glabrous. Leaf sheaths glabrous; ligule hairy; leaf blades linear, convolute, overlapping the culm, acuminate, base truncate, glabrous. Spikes 4-7; spikelets ovoid, glabrous; glumes oblong, margins scaberulous, acute; lemma broadly ovate; palea ovate; stamens 3. Caryopsis globose, dark red brown, often black.

Cultivated, often run wild. Fl. & Fr.: July–September. Tel.: *Arikelu.* Tam.: *Kelvaragu.*

Dayanidhiguda (KPT), *T.K.Paul* 1402 (BSID); Belghar, Kotgarh Wildlife sanctuary (KDM), *AC* 2412 (BSID).

INDIA: Cultivated all over India.

WORLD: Cultivated in Old World Tropics.

Eleusine indica (L.) Gaertni. Fruct. 1:8.1789; FBI 7: 293.1896; Fischer 3: 1839. 1934; Bor, Grass. India 493. 1960; Matthew, Fl. Tamilnadu Carnatic 3: 1848. 1983; Moulik, Grass. Bamb. India 2: 590.1997. *Cynosurus indicus* L., Sp. Pl. 72.1753.

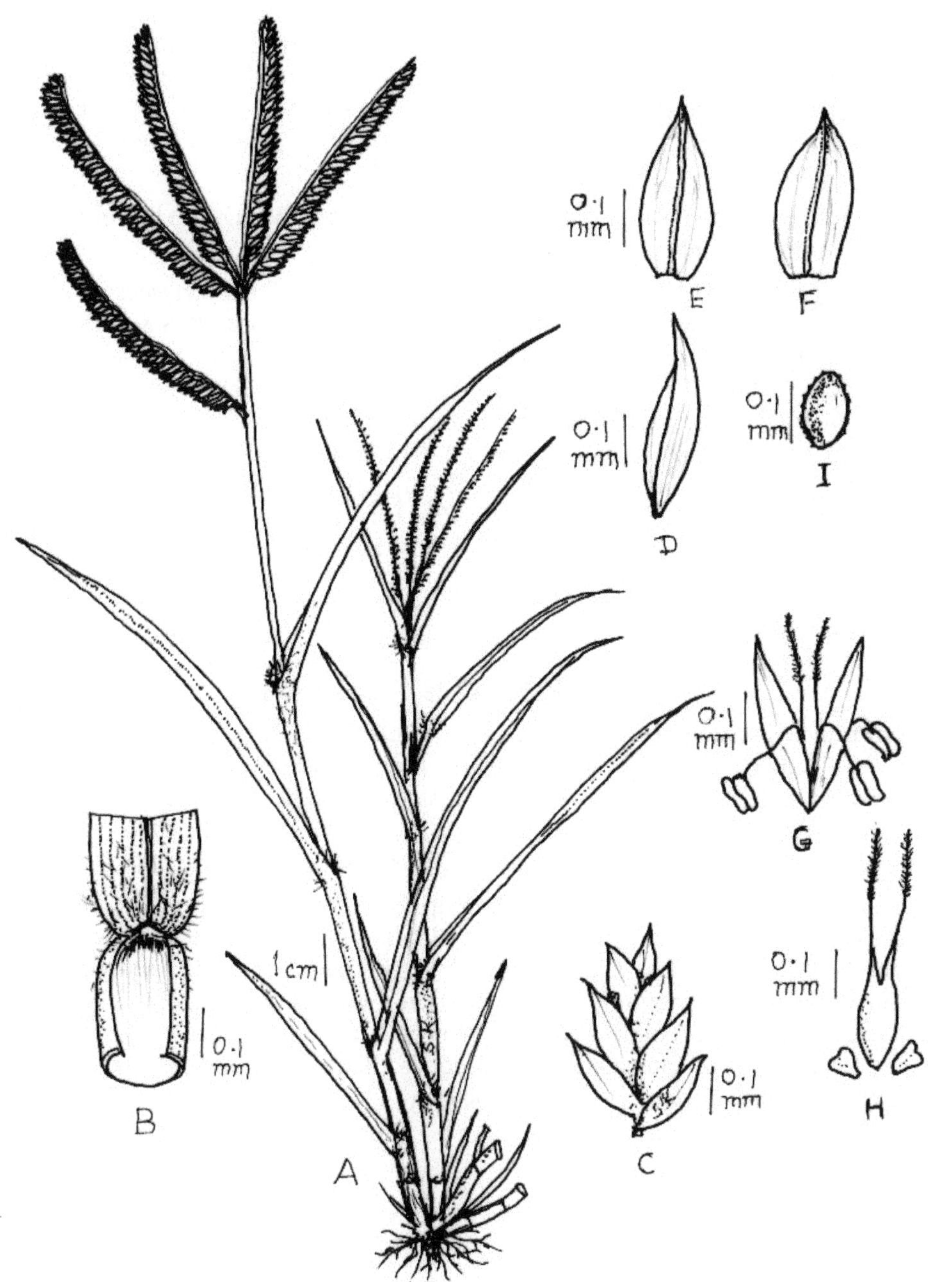

Figure 45. **Eleusine indica** (L.) Gaertn.

A. Habit, B. Ligule, C. Spike, D. Lower glume, E. Upper glume, F. Lemma, G. Spikelet, H. Lodicules and pistil, I. Caryopsis.

Annuals or perennials, culms erect, tufted, to 90 cm high; nodes glabrous. Leaf sheaths glabrous; ligule membranous; leaf blade linear or linear-lanceolate, base cordate, tip blunt. Spikes digitate or sub-digitate; spikelets 2-5-flowered; lower glume oblong-lanceolate, chartaceous, keeled, 1-3-nerved, acute, scabrid on the keels; upper glume ovate-oblong, chartaceous, keeled, 5-7-nerved; lemmas ovate-oblong, chartaceous, 3-nerved; paleas elliptic-oblong, hyaline, 2-keeled, 2-nerved; stamens 3. Caryopsis ovate.

Common in open places usually near water courses throughout Eastern Ghats. Fl. & Fr.: July–December.

Gooty hills (ATP), *MSG* 12238; Penukonda (ATP), *BR* & *P.Priyadarsini* 37169; Kalvabugga (KNL), *RVR* 3104; Gundlabrahmeswaram (KNL), *BSS* & *SKB* 32232; Mydukur (KDP), *AL* 26574; Tirumala hills (CTR), *MSG* 12139 & 12186, *RVR* & *GO* 2418, *BR* & *AMR* 21801; Talakona (CTR), *PVP* 2319; Near Padmapuram gardens (VSKP), *MHR* & *MCK* 14837; Polavarm agency (WG), *DCSR* 223 (CAL); Kottravada, Maredumilli (EG), *MM* 102550; Galikonda (VSKP), *GVS* 19585 (CAL); Anantagiri (VSKP), *GVS* 19499 (MH); Amangal to Shadnagar (MBNR), *SRS* 104574 (BSID); Sirumalai hills (DGL), *SKS* 321 (SGH); Red hills tank (CPT), *CAB* 5511 (MH); Tikarpada, Satkosia Tiger Reserve (AGL), *KCM* & *Swamy* 7637 (BSID).

INDIA: Throughout India in the plains and at lower elevations.

WORLD: Tropical and Subtropical regions throughout the world.

ELYTROPHORUS P. Beauvois

Elytrophorus spicatus (Willd.) A. Camus in Lecomte, Fl. Gen. de 1' Indo-china 7:547. 1923; Fischer 3:1843. 1934; Bor, Grass. India 493. 1960; Matthew, Fl. Tamilnadu Carnatic 3: 1849. 1983. *Dactylis spicata* Willd. in Ges. Naturf. Freunde Berlin Neue Schrift. 3: 416. 1801. *Elytrophorus articulatus* P. Beauv. Essai Agrost. 67. t.14. f.2. 1812; FBI 7: 306. 1896.

An annual, decumbent grass; culms up to 30 cm high; nodes glabrous. Leaf sheaths short, overlapping, glabrous, loose; ligule fringed hairs; leaf blades narrow, 5-15 x 0.2-0.7 cm, glabrous. Catkin like spikes with globose cluster arranged on the rachis of the inflorescence; spikelets short, 5 mm, 4-6-flowered, laterally compressed; glumes subequal, lanceolate, hyaline, 1-nerved, aristate; lemmas ovate-lanceolate, awned; palea broad, truncately 3-lobed, lobes ciliate, keels winged; stamens 1-3; stigmas plumose. Caryopsis narrowly oblong.

Occasional in ditches and wet places. Fl. & Fr: August- January.

Lathipada (GJM), *SX* & *MB* 3272 (RRL-B); Serango (GJM), *MB* 3008 (RRL-B); Indravathi (KHD), *MB* & *Dhal* 7648; Hanumanthpur (KRT), *MB* & *Rout* 7010 (RRL-B).

INDIA: Assam, Nagaland, West Bengal, Deccan Peninsula.

WORLD: Tropical regions of the Old World.

ENNEAPOGON Desvaux ex P. Beauvois

Enneapogon persicus Boiss., Diagn. Pl. Orient. 5: 71. 1844. *E. schimperianus* (Rich.) Renvoize, Kew Bull. 22: 400. 1968; Matthew, Fl. Tamilnadu Carnatic 2: 1849. 1983; Pullaiah, Fl. Andhra Pradesh 5: 2171. 2018. *Pappophorum schimperianum* Hochst. ex Rich., Tent. Fl. Abyss. 2: 403 1854. *Enneapogon elegans* (Steud.) Stapf, Bull. Misc. Inform. Kew 1907: 224. 1907; Fischer 3: 1841. 1934; Bor, Grass. India 610. 1960. *Pappophorum elegans* Nees ex Steud., Syn. Pl. Glum. 1: 199. 1854; FBI 7:301. 1896.

A perennial, tufted grass; culms up to 30 cm high, erect or ascending from the thickened base; nodes glabrous. Leaf sheaths 3-7 cm long, puberulous; ligule a membranous ring; blades 1.5-4.5 x 0.1-0.3 cm, linear, involute, base truncate, apex acute, glabrous. Panicle 9 cm, contracted with very short branches; rachis short; spikelets 5 mm, alternate, softly tomentose, 3-flowered; glumes persistent, keeled, lanceolate, membranous, margins ciliate; lemma broadly concave, 9-nerved, 9-awned, base pilose; palea elliptic, 2-toothed, 2-keeled, lower contains bisexual florets; stamens 3; stigmas 3. Caryopsis ovoid.

Occasional in foot hills and scrub jungles of southern Eastern Ghats. Fl. & Fr: October–December.

Gooty konda (ATP), *MSG* 12238; Kalasamudram RF (ATP), *PVP* 2607, *BR* & *P.Priyadarsini* 37080 (BSID); Gani RF (KNL), *RVR* 1559; Balarachapalli RF (KDP), *PVP* 3445; Panappakkam (CTR), *S.India Flora* 15769 (MH); Kondagorulapenta (GNT), *HDP* 125 (DD); Macherla RF (GNT), *MPG* 126 (DD); Kadapa, *EKK* 78524 (DD): Narasaraopeta (GNT), *CAB* 4547 (MH); Sirumalai hills (DGL), *SKS* 417 (SGH); Kolli hills (NMK), *SKS* 583 (SGH); Santha Marandahalli (SLM), *S.India Flora* 9746 (MH); Burgur (SLM), *S.India Flora* 15502 (MH); Hogainakkal (SLM), *EV* 20716 (MH).

INDIA: S & N.W. India.

WORLD: Egypt and Arabia northwards to Turkmenistan and eastwards to India.

ENTEROPOGON Nees

1. Spike solitary, spikelets 1-4-flowered .... **E. monostachyus**
1. Spike digitate, spikelets 1-2-flowered .... **E. dolichostachys**

Enteropogon dolichostachyus (Lag.) Keng., Clav. Gen. Sp. Gram. Prim. Sin. 197. 1957. *Chloris dolichostachya* Lag., Gen. Sp. Pl. 5. 1816; Fischer 3: 1838. 1934; Bor, Grass. India 466. 1960; Matthew, Fl. Tamilnadu Carnatic 2: 1820. 1983; Saxena & Brahmam, Fl. Orissa 4: 2289. 1996; Pullaiah, Fl. Andhra Pradesh 5: 2123. 2018. *C. incompleta* Roth, Nov. Pl. Sp. 60, 1821; FBI 7: 290. 1896.

A perennial, tufted grass; culms up to 1 m high, procumbent, rooting at base; nodes glabrous. Leaf sheaths 3-7 cm long, sparsely hairy; ligule obscure; blade 0.7-1.5 x 0.4-0.7 cm, base narrow, margin scaberulous, apex acuminate, scabrid, midrib prominent. Spikes arranged digitately, drooping, terminal; spikelets narrowly oblong. 2-flowered; glumes lanceolate, with prominent midrib, aristate; fertile

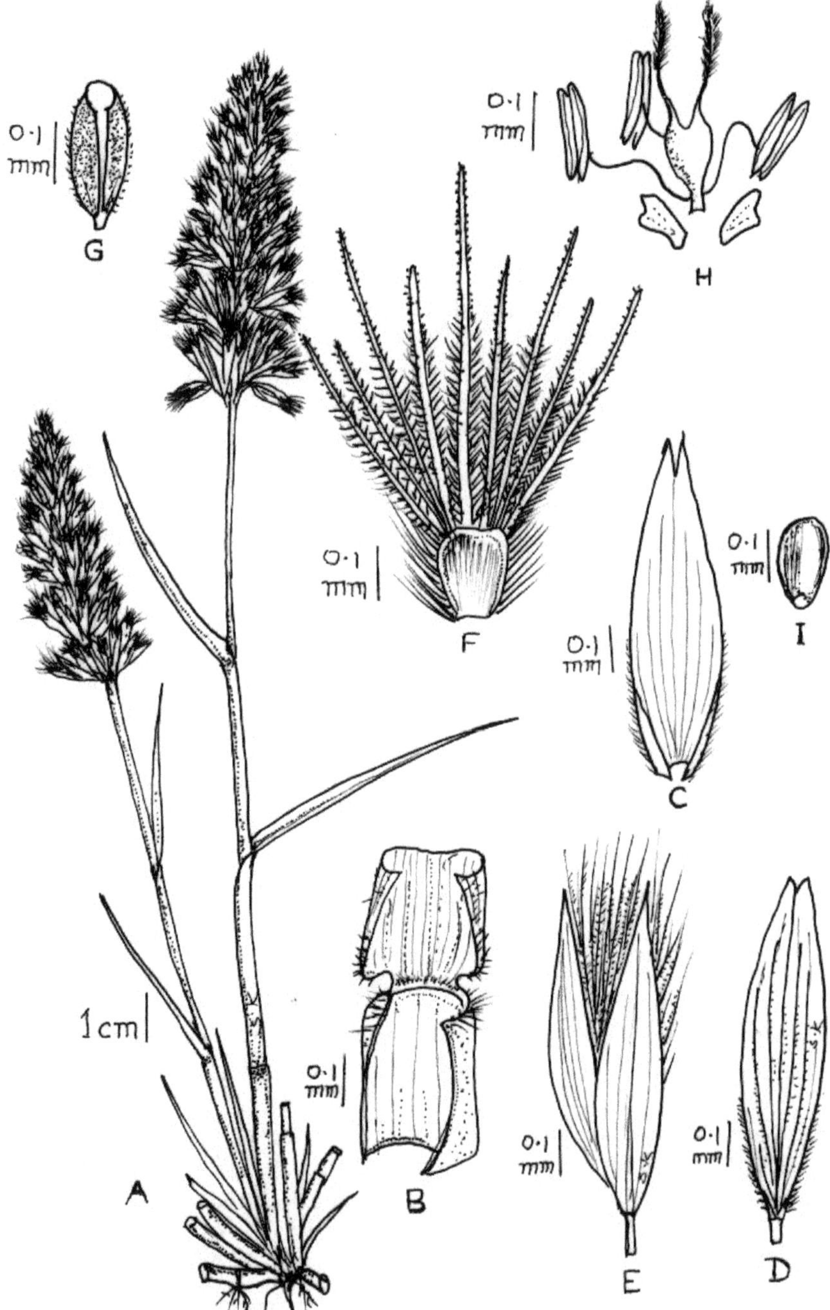

Figure 46. **Enneaopogon persicus** Boiss.

A. Habit, B. Ligule, C. Lower glume, D. Upper glume, E. Spikelet, F. Lemma, G. Palea, H. Lodicules and pistil, I. Caryopsis.

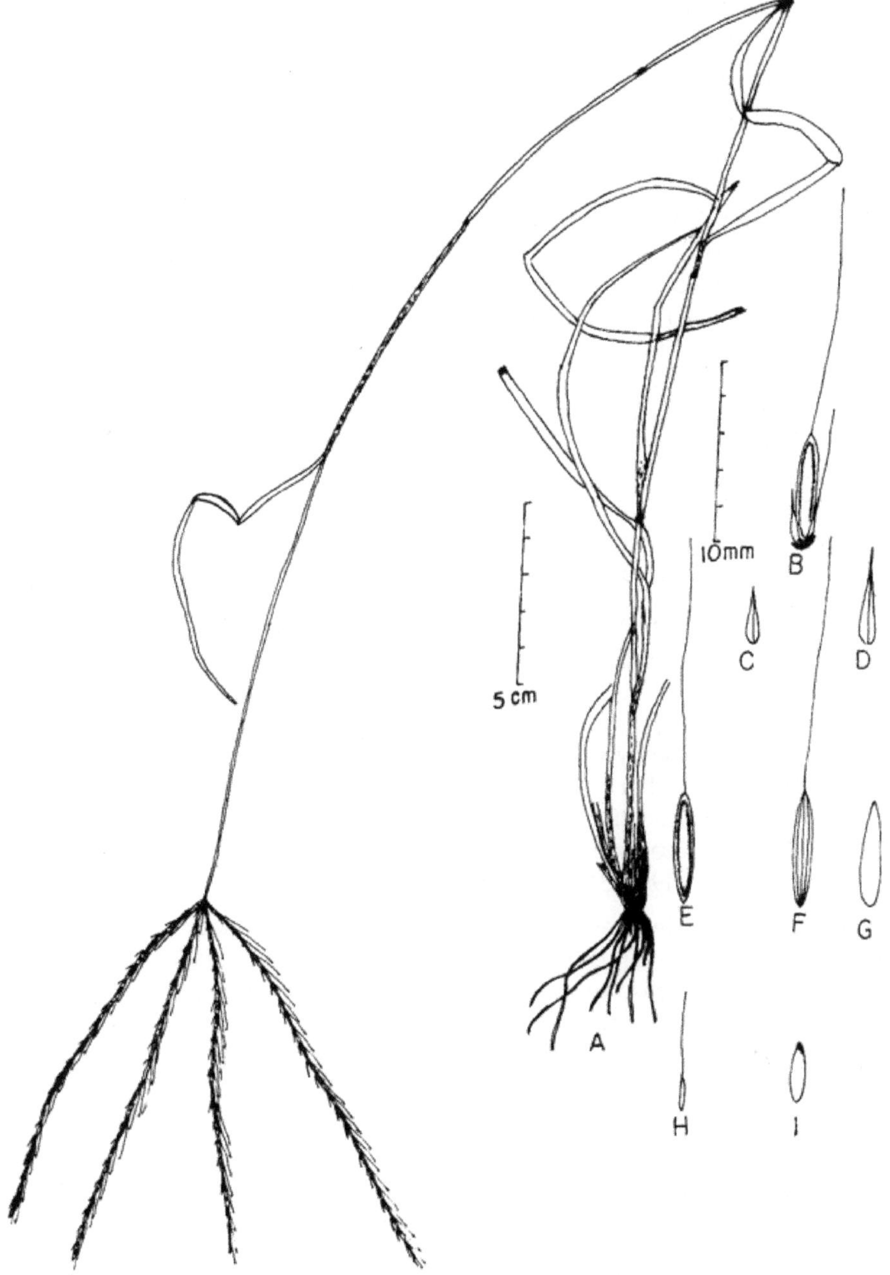

Figure 47. **Enteropogon dolichostachyus** (Lag.) Keng.
(Syn.: *Chloris dolichostachya* Lagasca).

A. Twig, B. Spikelet, C. Lower glume, D. Upper glume, E. Dorsal view of fertile lemma, F. Ventral view, G. Palea, H. Sterile lemma, I. Caryopsis.

lemma oblong-lanceolate, ciliate on the back, bidentate apex, awned, nerved; palea oblong, keels scaberulous, callus sparsely ciliate; sterile lemma reduced to an awn; stamens 3. Caryopsis oblong.

Common grass along the moist hills and occasional in the plains. Fl. & Fr.: November -February

Peddacheruvu (KNL), *PVP* & *RVR* 2166; Lankamalai RF (KDP), *SRSR* 13059; Rachakuntapalli (KDP), *KRKS* 41814; Tirumala hills (CTR), *MSG* 12140: Talakona (CTR), *PVP* 2506; Penchalakona (NLR), *MSG* 12189; Vinukonda RF (GNT), *VRK* 6852; Thimmanabailu (KDP), *MH* Acc. No. 57769 (MH); Chittoor, *MH* Acc. No. 57768 (MH); Penchalakona (NLR), *S.India Flora* 12792 (MH); way to Diguvametta (PKM), JLE 22164 (MH); Kondapalli hill (KSN), *CAB* 7999 (MH); Sirivaka (EG), *BO* 3445 (MH); Bison hill (EG), *CAB* 5095 (MH); Maripilli (VSKP), *VNS* 4599 (MH); Palasa (SKLM), *BDP* 182 (CAL & DD); Nungambakkam (CPT), *sine coll. s.n.* (MH); Vandalur (CPT), *S.India Flora* 11550 (MH); Parigam RF (SA), *S.India Flora* 17883 (MH); Hogainakkal (SLM), *EV* 21980 (MH); Betta Mogilam, Melagiri hills (SLM), *S.India Flora* 9836 (MH); Alagar hills (DGL), *Althaf Kabeer* 113277 (MH); Mahendragiri (GJP), *SX* & *MB* 3593, 3607 (RRL-B); Taptapani (GJM), *SX* 1454 (RRL-B); Karlapat (KHD), *HFM* 1190 (DD); Tulka section, Satkosia Tiger Reserve (AGL), *KCM* 5674 (BSID); Danerguda, Karlapat Wildlife Sanctuary (KHD), *AC* 2456 (BSID).

INDIA: Plains of India.

WORLD: Afghanistan, Pakistan, China to Southeast Asia and northern Australia.

Enteropogon monostachyus (Vahl) K. Schum. ex Engl., Abh. Preuss. Akad. Wiss. 17. 1894; Fischer 3: 1834. 1934; Bor, Grass. India 472. 1960; Matthew, Fl. Tamilnadu Carnatic 3: 1850. 1983. *Cynosurus monostachyus* Vahl, Symb. Hot. 2: 20. 1791. *Enteropogon melicoides* (K.D.Koen. ex Rottb.) Nees in Lindl., Introd. Nat. Syst. ed. 2. 449. 1836; FBI 7: 284. 1896.

A perennial, slender grass; culms up to 1 m high; nodes glabrous. Leaf sheaths 10-25 cm long, glabrous; ligule membranous with hairy apex; blades 10-25 x 0.2-0.4 cm, linear, convolute, base rounded, apex acuminate, glabrous. Spike solitary, 20 cm long; rachis trigonous or flat; spikelets 0.7 mm, narrowly lanceolate, all alike, subsessile, 1-4-flowered; glumes unequal, lanceolate; hyaline; lower lemma oblong, bidentate, awned, scaberulous keeled; palea oblong, 2-keeled, scabrid; upper lemma lanceolate, chartaceous; palea narrowly lanceolate, keeled enclosing a bisexual floret; stamens 3; stigmas plumose. Caryopsis narrowly ellipsoid.

Occasional in southern Eastern Ghats. Fl.: & Fr.: August-December.

Kalasamudram (ATP), *BR* & *KVS* 28994; Alampur (ATP), *BR* & *P.Priyadarsini* 37190; Guvvalacheruvu (KDP), *PVP* 3415; Rachakuntapalli (KDP), *KRKS* 37924; Kadapa, *EKK* 78517 (DD); Arumanipenta (KDP), *MH* Acc. No. 57611 (MH); Sirumalai hills (DGL), *SKS* 418 (SGH).

INDIA: S. India.

WORLD: Tropical Africa, Sri Lanka, Myanmar, Indo-China.

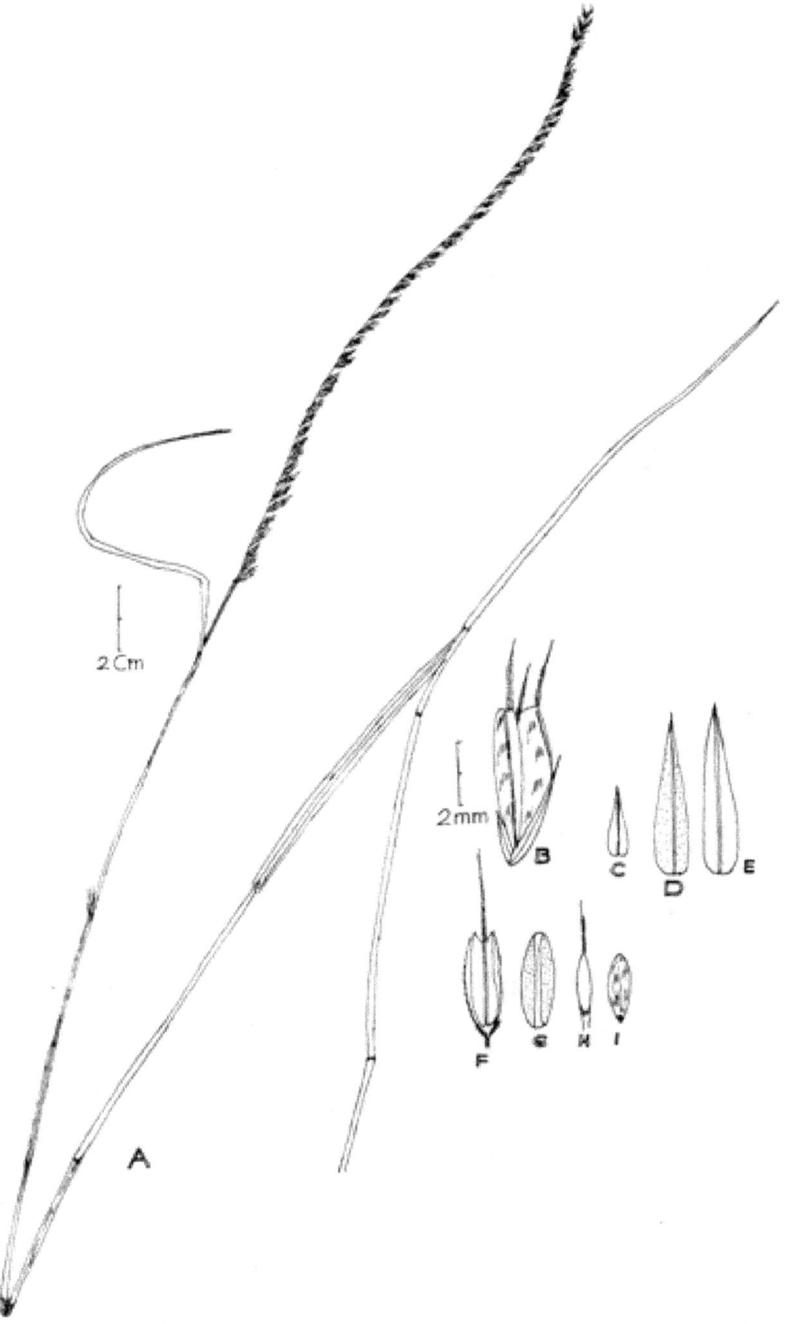

Figiure 48. **Enteropogon monostachyus** (Vahl) K. Schum.

A. Twig, B. Spikelet, C. Lower glume, D. Upper glume front view, E. Back view, F. Lower lemma, G. Palea, H. Upper lemma, I. Grain.

ERAGROSTIELLA Bor

1. Leaves from the basal sheaths reaching to the base of the inflorescence:
   2. Leaves flat; lemmas 2.5 mm .... **E. bifaria** var. **walkeri**
   2. Leaves convolute, filiform, with long white hairs; lemmas below 2 mm .... **E. nardoides**
1. Leaves from the basal sheaths very much shorter, never reaching the base of the inflorescence:
   3. Keels of the palea broadly-winged .... **E. bifaria** var. **bifaria**
   3. Keels of the palea narrowly-winged.. **E. brachyphylla**

Eragrostiella bifaria (Vahl) Bor, Indian Forester 66: 270. 1940 & Grass. India 494. 1960; Matthew, Fl. Tamilnadu Carnatic 3: 1851. 1983; Moulik, Grass. Bamb. India 2: 594. 1997. var. **bifaria.** *Poa bifaria* Vahl, Symb. Bot. 2:19.1791. *Erogrostis bifaria* (Vahl) Wight ex Steud., Nom. Bot. Ed. 2.1:562.1840; FBI 7: 325.1896; Fischer 1828.1934.

Annuals or perennials, culms erect, tufted, to 90 cm; nodes glabrous. Leaf sheaths glabrous; ligules membranous or fimbriate; leaf blades linear or acicular, often rolled, base narrow, acute, glabrous. Inflorescence spikes; spikelets slightly compressed, ovate-lanceolate or oblong-lanceolate, 5-30-flowered; lower glume oblong–lanceolate chartaceous, 1-nerved, keel scabrid; upper glume ovate–lanceolate, chartaceous, 1-nerved; lemmas broadly ovate, chartaceous, 3-nerved; paleas elliptic–ovate, hyaline, 2-keeled, winged, ciliate; stamens 3. Caryopsis globose.

Common in grass lands, plains and open forests. Fl. & Fr.: October–December.

Bukkapatnam (ATP), *NY* 623; Kalasamudram (ATP), *EC* 2895; Ramallakota (KNL), *RVR* 1593; Lower Ahobilam (KNL), *TP* & *RVR* 2730; Horsley hills (CTR), *MSG* & *AA* 12225; Tirumala hills (CTR), *BR* & *AMR* 21862; Penchalakona (NLR), *MSG* 12194; Omkaram (KNL), *MSR* & *BSS* 29349; Sunnipenta (KNL), *JLE* 22141 (CAL); Balapalle (CDP), *CAB* 15116, 15416 (MH); Emmuluru hills (NLR), *PV* 110124 (BSID); Diguvametta (PKM), *PCN* 125 (DD); Kondgorulapenta (GNT), *MPG* 12 (CAL); Nagarjunakonda Sacred grove (GNT), *J.Swamy* & *S.Nagaraju* 14167 (BSID); Kondavedu fort (GNT), *CAB* 17493 (MH); Polavaram agency (WG), *DCSR* 439 (CAL); Rampa hills (EG), *MSR* 1605 (CAL); Rampa hills (EG), *VNS* 202 (MH); Dharakonda (VSKP), *CAB* 17083 (MH); Nattavaram (VSKP), *CAB* 1835 (MH); Salur (VJN), *VNS* 82401 (MH); Itillock (SKLM), *JLE* 32714 (CAL); Koriguttalu (KMM), *RR* 106045 (BSID); Sithanur (CPT), *CAB* 11199 (MH); Thippukadu R.F. (NA), *KRM* 17692 (MH); Dhanei (GJM), *SX* & *MB* 4070 (RRL-B); Sorada (GJM), *SX* 2890 (RRL-B); Sirumalai hills (DGL), *SKS* 326 (SGH); Badrama Wildlife Sanctuary (SBP), *AC* 10242, 10249 (BSID); Jenabil Range, Similipal B.R. (MBJ), *AC* 2342 (BSID).

INDIA: Throughout India.

WORLD: Myanmar, Sri Lanka.

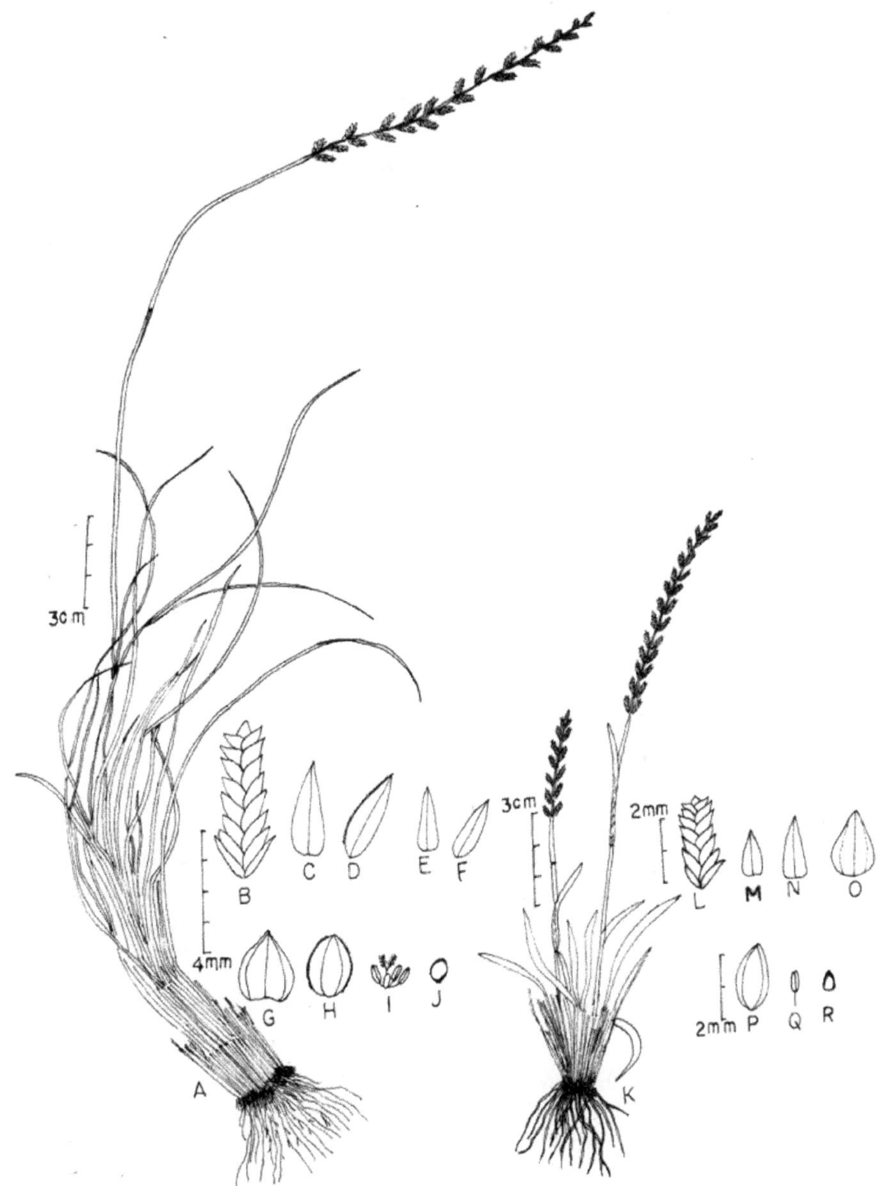

Figure 49. A-J. **Eragrostiella bifaria** (Vahl) Bor

A. Habit, B. Spikelet, C & D. Lower glume front and lateral view, E & F. Upper glume front and lateral view, G. Lemma, H. Palea, I. Pistil and stamens, J. Grain.

K-R. **Eragrostiella brachyphylla** (Stapf) Bor

K. Habit, L. Spikelet, M. Lower glume, N. Upper glume, O. Lemma, P. Palea, O. Stamen, R. Grain.

Eragrostiella bifaria (Vahl) Bor var. **walkeri** (Stapf) Lazarides, Contr. Herb. Austral. 22: 7. 1976. *Eragrostis walkeri* Stapf in Trin., Fl. Ceyl. 5:298. 1900; Fischer 3:1828. 1934. *Eragrostiella walkeri* (Stapf) Bor, Indian Forester 66: 270. 1940; Bor, Grass. India 495. 1960; Moulik, Grass. Bamb. India 2:594. 1997.

Perennials, culms erect, tufted, to 80 cm high, tuft of leaf sheaths at the base; nodes glabrous. Leaf sheaths glabrous; ligules ciliate, membranous; leaf blades filiform, convolute, base rounded, glabrous, margin scabrous, acuminate. Inflorescence terminal, solitary racemes; rhachis flattened; spikelets secund, ovate, subsessile, margin scaberulous, 2-8-flowered; lower glume linear-lanceolate, acute, 1-nerved; upper glume boat-shaped, keeled, acute, 1-nerved; lemmas broadly ovate, margin hyaline, 3-nerved; palea spatheolate, keels ciliate, margin inflexed, hyaline; stamens 3. Caryopsis ellipsoid.

Common in forest plains and on open slopes. Fl. & Fr.: July-December.

Kalasamudram (ATP), *KRKS* & *BSS* 39696; Penukonda (ATP), *JNP* 26751; Vajrakarur (ATP), *AL* 28571; Pamidi (ATP), *BR* & *KVS* 28023; Gundumallappa sacred grove (KNL), *BR* & *SS* 20287; Srisailam (KNL), *AMR* 25688; G. Madugula (VSKP), *BR* & *AMR* 25607; Amrabad (MBNR), *BSS* & *SKB* 32210 (SKU).

INDIA: South India.

Eragrostiella brachyphylla (Stapf) Bor, Indian Forester 66:270. 1940 & Grass. India 494. 1960; Moulik, Grass. Bamb. India 2: 594. 1997; Matthew, Fl. Tamilnadu Carnatic 3: 1852. 1983; *Eragrostis brachyphylla* Stapf in Hook. f., Fl. Brit. India 7:327. 1896; Fischer 3: 1828. 1934.

Perennials, culms erect, tufted, to 50 cm high, slender; nodes glabrous. Leaf sheaths glabrous; ligule a ciliate rim; leaf blades linear, base rounded, apex acute, glabrous. Inflorescence solitary raceme; spikelets ovoid-oblong, secund, 5-18-flowered; lower glume linear-lanceolate, acute, 1-nerved; upper glume lanceolate, acute, 1-nerved; lemma broadly ovate, 2-nerved, keeled, margins hyaline; palea obovate, keels ciliate, margins inflexed, hyaline; stamens 3. Caryopsis ellipsoid.

Common in waste lands, forest plains and on hill slopes. Fl. & Fr.: July-November.

Gundumala RF (ATP), *TP* & *NY* 724; Kalasamudram (ATP), *PVP* 2613; Upper Ahobilam (KNL), *SS* & *AMR* 22692; Rollapenta (PKM), *BR* & *BSS* 29396; Palakonda hills (KDP), *CS* 8282; Balapalli RF (KDP), *PVP* 3460, *JLE* 14255 (MH); Tirumala hills (CTR), *KS* 6880 (MH); Guggamkonda RF (KDP), *KS* 6337 (MH); Chelama (KNL), *JLE* 16737 (CAL); Basavapuram (PKM), *VBH* 83991; Kondaveedu (GNT), *CAB* 4703 (MH); Anantagiri (VSKP), *MRS* 305 (DD); Salur (VJN), *NPBK* 1137 (CAL); Manuguru RF (KMM), *RCS* 104255 (BSID); Bandirevu RF (KMM), *RCS* 104300 (BSID); Tulka, Satkosia Tiger Reserve (AGL), *KCM* 8317 (BSID); Kundali (KPT), *AC* 10231 (BSID).

INDIA: West Bengal, Bihar, Madhya Pradesh, Odisha, South India.

Eragrostiella nardoides (Trin.) Bor, Indian Forester 66: 270. 1940. *Eragrostis nardoides* Trin., Mem. Acad. Imp. Sci. St. Petersbourg 1: 415. 1830; FBI 7: 326. 1896.

Perennial, culms erect, tufted, 30-70 cm high. Leaf blades filiform, 15-23 cm long, convolute, with very delicate, long hairs. Spikes long, slender, 15-35 cm long with smooth, terete rachis and many close-set, erectopatent spikelets; spikelets 7.5-12.5 x 1.5-2 mm; glumes subequal, acutely keeled; lemmas turgid, 1.8-2 mm long; keels of palea narrowly wingled, ciliate. Caryopsis ellipsoid-oblong with oblique base.

Rare in open moist grasslands in Northern Eastern Ghats. Fl. & Fr.: August-December.

Laksmipuram (VSKP), *CAB* 1810 (MH); Adava (GJM), *SX* & *MB* 2026 (RRL-B); Simlipahar, *SX* & *MB* 4920 (RRL-B); Malayagiri (DK), *MB* & *Dhal* 8386 (RRL-B).

INDIA: Himalayas, Bihar, Odisha,

WORLD: Nepal.

ERAGROSTIS Wolf

1. Spikelets breaking up from above downwards; rachis fragile:
   2. Lemma ciliate on the margins:
      3. Panicles clavate, dense, shining; keels of the palea narrowly winged .... **E. deccanensis**
      3. Panicle cylindric, compact, dull; keels of the palea not winged .... **E. ciliata**
   2. Lemma not ciliate on the margins:
      4. Keels of the palea ciliate:
         5. Panicle spiciform or compact and dense:
            6. Lemma ciliate on the keels; annuals .... **E. ciliaris** var. **ciliaris**
            6. Lemma not ciliate on the keels; perennials:
               7. Panicle very dense, cylindric; lemma rounded at apex .... **E. riparia**
               7. Panicle more or less dense, interrupted; lemma acute at apex .... **E. coarctata**
         5. Panicle effuse:
            8. Lemma ciliate on the keels .... **E. ciliaris** var. **clarkei**
            8. Lemma not ciliate on the keels:
               9. Plants viscous; panicles often leaden grey .... **E. viscosa**
               9. Plants not viscous .... **E. amabilis**
      4. Keels of the palea not ciliate:
         10. Panicle lax, open thyrsiform .... **E. aspera**
         10. Panicles not thyrsiform .... **E. japonica**

1. Spikelets breaking up from below upwards; rachis tough:
    11. Annuals:
        12. Lemmas up to 1.5 mm long:
            13. Plants glandular:
                14. Lowest branches of the panicle fascicled or sub whorled:
                    15. Plants up to 61 cm high, ligules fimbriate, membranous; spikelets lanceolate or oblong-lanceolate; florets *c.* 17-flowered........ **E. pilosa**
                    15. Plants up to 25 cm high; ligules villous or long hairs; spikelets linear-oblong or elliptic; florests *c.* 8-flowered........ **E. nairii**
                14. Lowest branches not whorled:
                    16. Spikelets less than 2 mm wide; leaf margins eglandular; caryopsis truncate at both ends........ **E. maderaspatana**
                    16. Spikelets more than 2 mm wide; leaf margins glandular; caryopsis globose .... **E. minor**
            13. Plants glandular:
                17. Lemmas to 1.5 mm long .... **E. gangetica**
                17. Lemmas over 1.5 mm long........ **E. macilenta**
        12. Lemmas over 1.5 mm long:
            18. Spikelets 2 mm or more wide, ovate or ovate-oblong:
                19. Plants with or without glands; upper glume 3-nerved, paleas persistent........ **E. cilianensis**
                19. Plants eglandular; upper glume 1-nerved, paleas deciduous:
                    20. Florets more than 70 in spikelets, rachilla not visible; paleas keels narrowly winged; anthers 2........ **E. unioloides**
                    20. Florets less than 40 in spikelets, rachilla visible; paleas not winged; anthers 3........ **E. collinensis**
            18. Spikelets 0.75-2 mm wide, lanceolate........ **E. tremula**
    11. Perennials:
        21. Lemmas up to 1.8 mm long:
            22. Spikelets 1.7-2.5 mm wide, paleas deciduous........ **E. atrovirens**
            22. Spikelets 1-1.2 mm wide; paleas persistent:
                23. Panicle contracted; spikelets crowded; straight, tip of lemma overlapping base of next lemma........ **E. nutans**

23. Panicles wide spreading; spikelets loose, slightly curved; tip of lemma just touching the base of next lemma.. **E. papposa**

21. Lemmas 2mm or longer:

24. Pedicels of the spikelets with a glandular band .... **E. patula**

24. Pedicels of the spikelets devoid of glandular band .... **E. nigra**

Eragrostis amabilis (L.) Wight & Arn., Cat. Indian Pl. 2: 105. 1834. *Poa amabilis* L., Sp. Pl. 68. 1753. *Eragrostis tenella* (L.) P. Beauv. ex Roem. & Schult., Syst. Veg. 2: 576. 1871; Bor, Grass. India 513.1960; Moulik, Grass. Bamb. India 2: 608.1997; Pullaiah, Fl. Andhra Pradesh 2186. 2018. *Poa tenella* L., Sp. Pl. 69. 1753. *Eragrostis plumosa* (Retz.) Link, Hort. Berol. 1:192.1827; Fischer 3: 1826. 1934. *E. tenella* (L.) P. Beauv. ex Roem. & Schult. var. *plumosa* (Retz.) Stapf in Hook. f., Fl. Brit. India. 7:315.1896.

Annuals, culms erect, tufted, to 80 cm high, slender; nodes glabrous. Leaf sheaths hairs at mouth; ligules membranous; leaf blades linear-acuminate, base rounded, acuminate. Panicles 20 cm long, effuse; spikelets oblong-ovate, 5-9-flowered; lower glume lanceolate, chartaceous, 1-nerved; upper glume ovate-lanceolate, chartaceous, 1-nerved; lemmas oblong-ovate, chartaceous, 3-nerved; paleas lanceolate or elliptic-lanceolate, 2-keeled, 2-nerved, keels ciliate; stamens 3. Caryopsis ovate-elliptic.

Common in hill slopes, waste places and plains. Fl. & Fr.:April–November.

Kalasamudram (ATP), *BR* & *AMR* 20040, 24238; Chendrayuni kona (ATP), *BR* & *JNP* 26736; Sunnipenta (KNL), *BR* & *BSS* 32978; Gajulapalle (KNL), *KCJ* 83130 (MH); Thummalapalle (KDP), *KRKS* & *KRS* 33677; Talakona (CTR), *MSG* 12224; Kailasanathakona (CTR), *BR* & *P.Priyadarsini* 37132 (BSID); Udayagiri (NLR), *BSN* 4223 (BSID); B.K.Palen (GNT), *VRK* 3247; Chinturu (EG), *MM* 101389 (BSID); Karepally (KMM), *RR* 105970 (BSID); Saleswaram (MBNR), *BSS* & *SKB* 34801; Shervarayan hills (SM), *SKS* 584 (SGH); Chinnar (DMP), *P.Venu* 111718 (BSID); Ramjolu river bank (Dhenkanal), *ARKS* 11050 (BSID); Punakote, Satkosia Wildlife Sanctuary (AGL), *AC* 2389 (BSID); Tikarpada, Satkosia Wildlife sanctuary (AGL), *AC* 2400 (BSID).

INDIA: Throughout India.

WORLD: Widely distributed in the Tropics of the Old World, introduced into America.

Eragrostis aspera (Jacq.) Nees, Fl. Afr. Austral. Ill. 408. 1841; FBI 7: 314.1896; Fischer 3: 1825.1934; Bor, Grass. India 501. 1960; Matthew, Fl. Tamilnadu Carnatic 3: 1853. 1983; Moulik, Grass. Bamb. India 2:597. 1997. *Poa aspera* Jacq., Hort. Bot. Vindob 332. t. 56. 1776.

Annuals, culms erect, tufted, to 70 cm. Leaf sheaths glabrous; ligules ring of hairs; leaf blades 7-20 x 0.2-0.8 cm, linear-lanceolate, acuminate. Panicles lax, branches numerous; spikelets linear-oblong, 4-15-flowered; lower glume ovate-lanceolate, chartaceous, acute, 1-nerved; upper glume ovate-lanceolate, acute,

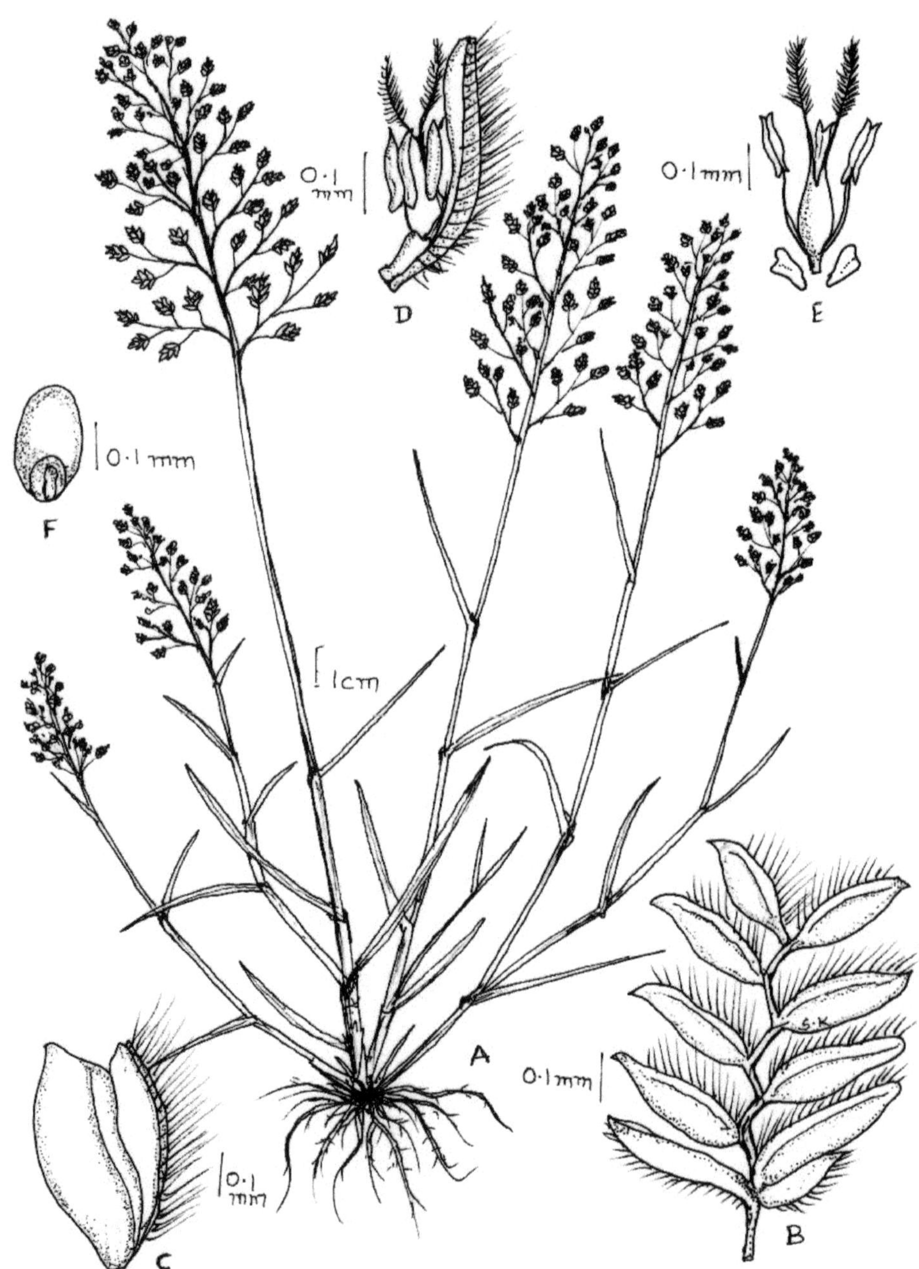

Figure 50. **Eragrostis amabilis** (L.) Wight & Arn.

A. Habit, B. Spikelet, C. Upper glume, D. Glume with pistil and stamens, E. Lodicules, pistil and stamens, F. Caryopsis.

chartaceous, bifid, purple at apex, 1-nerved; lemmas broadly ovate, markedly truncate, not ciliate on margins; paleas obovate, keeled, scabrid, hyaline, 2-nerved. Caryopsis sub-globose.

Occasional in forest plains, waste lands and fields. Fl. & Fr.: July–November.

Amagondapalem RF (ATP), *NY* 1046; Pathakothacheruvu (ATP), *BR* & *JNP* 26631; Gani RF (KNL), *RVR* 1515; Bhajaranga kona (KDP), *KVS* & *AL* 29115; Annavaram (EG), *CAB* 8187 (MH) Mahendragiri (GJP), *SX* & *MB* 2204 (RRL-B); Gangabada (GJM), *SX* & *MB* 3523 (RRL-B); Sirumalai hills (DGL), *SKS* 417 (SGH); Krishnagiri (SLM), *CAB* 13873 (MH); Gomuhi–Vellimala Ghat (SA), *KRM & V. Chandrasekaran* 77344 (MH).

INDIA: Peninsular to North India.

WORLD: Africa and Mascarene Islands.

Eragrostis atrovirens (Desf.) Trin. ex Steud., Nom. Bot. ed. 2.1:562.1840; Bor, Grass. India 503. 1960; Moulik, Grass. Bamb. India 2:597.1997; Matthew, Fl. Tamilnadu Carnatic 3: 1854. 1983. *Poa atrovirens* Desf., Fl. Atlant. 1:73. t.14. 1798.

Perennials, culms erect, tufted, to 1.3 m; nodes glabrous. Leaf sheaths glabrous; ligules membranous; leaf blades 10-30 x 0.3-0.4 cm, linear–lanceolate, base rounded or cordate, apex acuminate. Panicles contracted, often lax; spikelets short-pedicelled, ovate, 3-30-flowered; lower glume lanceolate, chartaceous, 1-nerved; upper glume ovate–lanceolate, chartaceous, 1-nerved; lemmas ovate, chartaceous, acute, 3-nerved; paleas lanceoloate, keeled, hyaline, keels ciliate, 2-nerved; stamens 3. Caryopsis ellipsoid.

Common along stream banks and in marshy areas. Fl. & Fr.:July - February. Tam.: *Injipul.*

Kalasamudram (ATP), *BR* & *AMR* 24227; Batrepalli (ATP), *KVS* & *MSB* 28114; Sunnipenta (KNL), *BR* & *BSS* 29309; Velgode (KNL), *BR* & *BSS* 30231; Palakonda hills (KDP), *CS* & *DAM* 7698; Tirumala hills (CTR), *AMR* 21849, 21863; Talakona (CTR), *MSG* & *AA* 12232; Rollapenta (PKM), *BSS* & *SKB* 29392; Isukagundam (PKM), *BSS* & *SKB* 32983; Macherla (GNT), *VRK* 5883; Appapur (MBNR), *AMR* 25700; Mallelatheertham (MBNR), *SRS* 110705, 111414 (BSID); Araku valley (VSKP), *NPBK* 530 (CAL); Hoggenekal (DMP), *P.Venu* 111616 (BSID); Niyamgiri hill (KHD), *AC* 2471 (BSID); Jenabil Range, Similipal B.R. (MBJ), *AC* 2310 (BSID).

INDIA: Throughout India.

WORLD: NW Africa through tropical Asia to South Africa, Malaysia.

Eragrostis cilianensis (Alt.) Vignolo-Lutati ex Janchen, Mitt. Naturwiss. Vereins Univ. Wien n.s. 5: 110. 1907; Fischer 3:1827.1934; Bor, Grass. India 503.1960; Moulik, Grass. Bamb. India 2:598.1997. *Poa cilianensis* Alt., Fl. Pedem. 2:246. t. 91. f. 2. 1785. *Eragrostis major* (L.) Host., Gram. Austr. 4:14 & 24.1809; FBI 7: 320.1896.

Annuals, culms erect or ascending, tufted, to 70 cm, leafy; nodes glabrous. Leaf sheath hairy at mouth; ligule ciliate; leaf blades 5-12 x 0.4-0.6 cm, linear-lanceolate, base cordate, margins glandular, apex acuminate. Panicles contracted, often lax, green; spikelets oblong, 10-25-flowered; lower glume oblong–lanceolate, chartaceous, 1-nerved; upper glume ovate-lanceolate, acute, chartaceous, 3-nerved; lemmas ovate, chartaceous, 3-nerved, glands on the keels; paleas oblong, 2-keeled, 2-nerved, keels ciliate; stamens 3. Caryopsis sub-globose.

Occasional along the forest margins in shady localities. Fl. & Fr.: December–June.

Kalasamudram (ATP), *BR* & *AMR* 24237; Penukonda (ATP), *JNP* 26763; Guntakal (ATP), *JNP* 26634; Gundumallappa kona (KNL), *SS* & *AMR* 20288; Pathikonda (KNL), *KVS* & *AL* 29141; Ahobilam (KNL), *BR* & *BSS* 29307, 29319; Peddavegi (WG), *VNS* 4516 (MH); Annavaram (EG), *CAB* 8186 (MH); Madapalli (EG), *BO* 3392 (CAL); Devipatnam (EG), *BO* 15777 (CAL); Chadavur (EG), *CAB* 4961 (MH); Kondapudi (EG), *CAB* 5196 (MH); Shammikulam (SA), *CAB* 873 (MH).

INDIA: Throughout India.

WORLD: Widely distributed in the warmer regions.

Eragrostis ciliaris (L.) R. Br. in Tuckey, Narr. Exp. Congo, App. 478.1818; FBI 7: 314.1896; Fischer 3: 1825.1934; Bor, Grass. India 506.1960; Moulik, Grass. Bamb. India 2: 598. 1997; var. **ciliaris.** *Poa ciliaris* L., Syst. Nat. ed. 10. 2:875.1759.

Annuals, culms erect, tufted, to 60 cm, slender; nodes glabrous. Leaf sheaths ciliate at mouth; ligules short silky hairy; leaf blades 2-5 x 0.2-0.4 cm, linear, base rounded, acuminate, glabrous. Panicles spiciform or compact; spikelets ovate; lower glume ovate-acute, chartaceous, 1-nerved; upper glume ovate-lanceolate, chartaceous, scabrid, 1-nerved; lemmas broadly ovate, chartaceous, 3-nerved, ciliate; palea obovate, truncate, soft ciliate, 2-keeled, 2-nerved; stamens 3. Caryopsis ovoid.

Common in plains and forest margins. Fl. & Fr.: August - November.

Chennekothapalli (ATP), *BR* & *JNP* 27339; Guntakal (ATP), *JNP* & *AL* 26605; Hottabetta (ATP), *BR* & *KVS* 28085; Tirumala hills (CTR), *BR* & *AMR* 21808; Sitampeta (SKLM), *AMR* 25653; Sirivaka (WG), *CAB* 5256 (MH); Annavaram (EG), *CAB* 1864 (MH); Jenabil Range, Similipal Biosphere Reserve (MBJ), *AC* 2346 (BSID); Purunkote, Satkosia Wildlife Sanctuary (AGL), *AC* 2373 (BSID).

INDIA: Plains of India.

WORLD: Tropical and South Africa extending through Arabia and the Mascarene Islands to India, Tropical America.

Eragrostis ciliaris (L.) R.Br. var. **clarkei** Stapf ex Hook.f., Fl. Brit. Ind. 7:315 1896.

Annuals, culms erect, tufted, to 30 cm; nodes glabrous. Leaf sheaths 2-4 cm, ciliate at mouth; ligule ciliate; leaf blades lanceolate, 4-10.5 mm, flat, base rounded, glabrous, acuminate. Panicles to 15 cm effuse. Spikelets ovate, 1.5-2 x 1-2 mm, 4-6-flowered; lower glume ovate-acute, chartaceous, 1-nerved, scabrid on upper side; upper glume ovate-lanceolate, chartaceous, 1-nerved, scabrid on upper side;

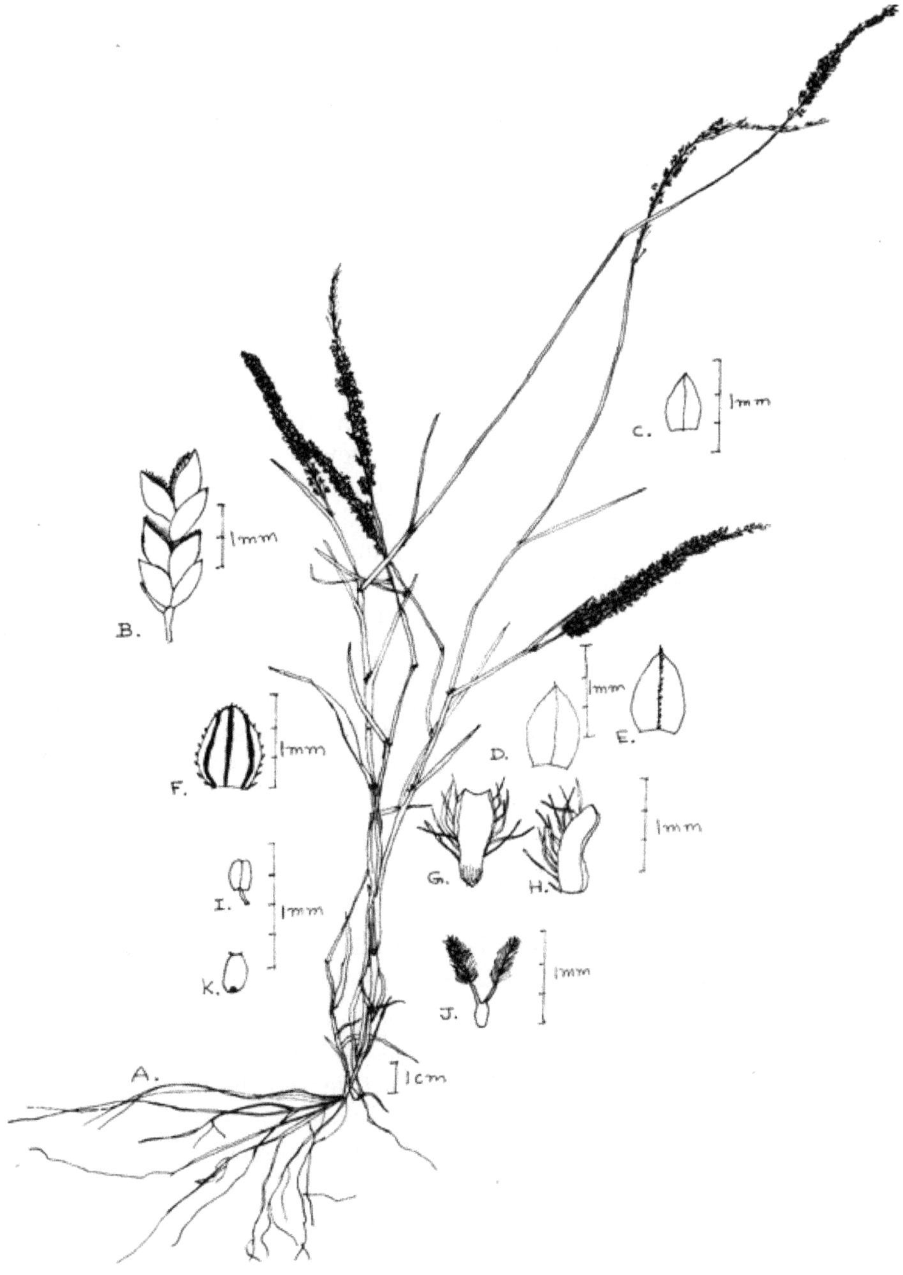

Figure 51. **Eragrostis ciliaris** R.Br. var. **ciliaris**.

A. Twig, B. Spikelet, C. Lower glume, D & E. Upper glume, F. Lemma, G & H. Palea front and side views, I. Palea, J. Pistil, K. Caryopsis.

lemmas broadly ovate, 3-nerved, 2-keeled, ciliate on the keels; paleas oblong, 2-keeled, 2-nerved, long ciliate on the keels. Caryopsis ovoid, 0.3 mm.

Rare in Nallamalais. Fl. & Fr.: August - November.

Gundlabrahmeswaram (KNL), *BR* & *BSS* 32964 (SKU).

INDIA: Delhi, Haryana, Andhra Pradesh. Endemic.

Note: There are three varieties of *Eragrostis ciliaris* in India (Bor, 1960): *ciliaris, brachystachya* and *clarkei,* the latter variety differs from others in having effused panicle. This variety is so far known only from Delhi through collections of C.B.Clarke (Bor, 1960; Sadasivaiah *et al*., 2013).

Eragrostis ciliata (Roxb.) Nees, Agrost. Bras. 512.1829; FBI 7: 373.1896; Fischer 3:1825.1934; Bor, Grass. India 506. 1960. *Poa ciliata* Roxb., Fl. Ind. 2: 256. 1820.

Annuals or perennials, tufted, erect, to 80 cm high, geniculate; nodes glabrous. Leaf sheaths glabrous, bearded at mouth; ligule a few hairs; leaf blades 5-15 x 0.2-0.4 cm, linear-lanceolate, flat or convolute, apex acuminate, glabrous. Panicles compact, cylindrical; spikelets compressed, ovate-lanceolate, 6-12-flowered; lower glume ovate, chartaceous, 1-nerved, acute; upper glume ovate, chartaceous, 1-nerved, acute; lemma obovate, chartaceous, 3-nerved, ciliate; palea ovate-lanceolate, hyaline, 2-keeled, 2-nerved, long ciliate. Caryopsis ovate-elliptic.

Occasional in open places. Fl. & Fr.: May-September.

Gundlabrahmeswaram (KNL), *BR* & *BSS* 32963; Betludatha konda (PKM), *BSS* & *SKB* 32281; Gangavaram (EG), *TP* & *MSG* 12279; Visakhapatnam, *VNS* 663 (CAL); Palasa (SKLM), *MPG* 193 (CAL) & 194 (DD); Punakote, Satkosia Wildlife Sanctuary (AGL), AC 2392 (BSID).

INDIA: Peninsular to E. India.

WORLD: Myanmar to Indo-China.

Eragrostis coarctata Stapf in Hook. f., Fl. Brit. India. 7:313.1896; Fischer 3:1825.1934; Bor, Grass. India 507. 1960; Moulik, Grass. Bamb. India 2: 601.1997.

Perennials, culms erect, tufted, to 1.1 m high, geniculate; nodes glabrous. Leaf sheaths ciliate at mouth; ligules hairy; leaf blades 4-12 x 0.2-0.5 cm, linear–lanceolate, base rounded, apex acuminate. Panicles compact, dense, cylindrical; spikelets short pediceled, oblong, grey colour; lower glume lanceolate, chartaceous, 1-nerved; upper glume ovate-lanceolate, chartaceous, 1-nerved; lemmas broadly ovate, chartaceous, 3-nerved, ciliate on the margins; paleas oblong, 2-keeled, 2-nerved, ciliate on the keels; stamens 3. Caryopsis ovoid.

Occasional in stream beds and marshy areas in forests. Fl. & Fr.: December - July.

Bukkapatnam (ATP), *BR* & *JNP* 27349; Hottabetta (ATP), *BR* & *KVS* 28100; Rangapuram RF (KNL), *RVR* 1498, 1698; Gundlabrahmeswaram (KNL), *PVP* & *RVR* 2497; Kailasanathakona (CTR), *BR* & *P.Priyadarsini* 37130; Kasumuru (NLR), *BR* & *PMR* 22334; Isukagundam (PKM), *BR* & *BSS* 29291; Perayya kunta (PKM), *BSS* & *SKB* 32953; Vinukonda RF (GNT), *VRK* 6862; Punyagiri (VJN), *BR* & *AMR* 24538;

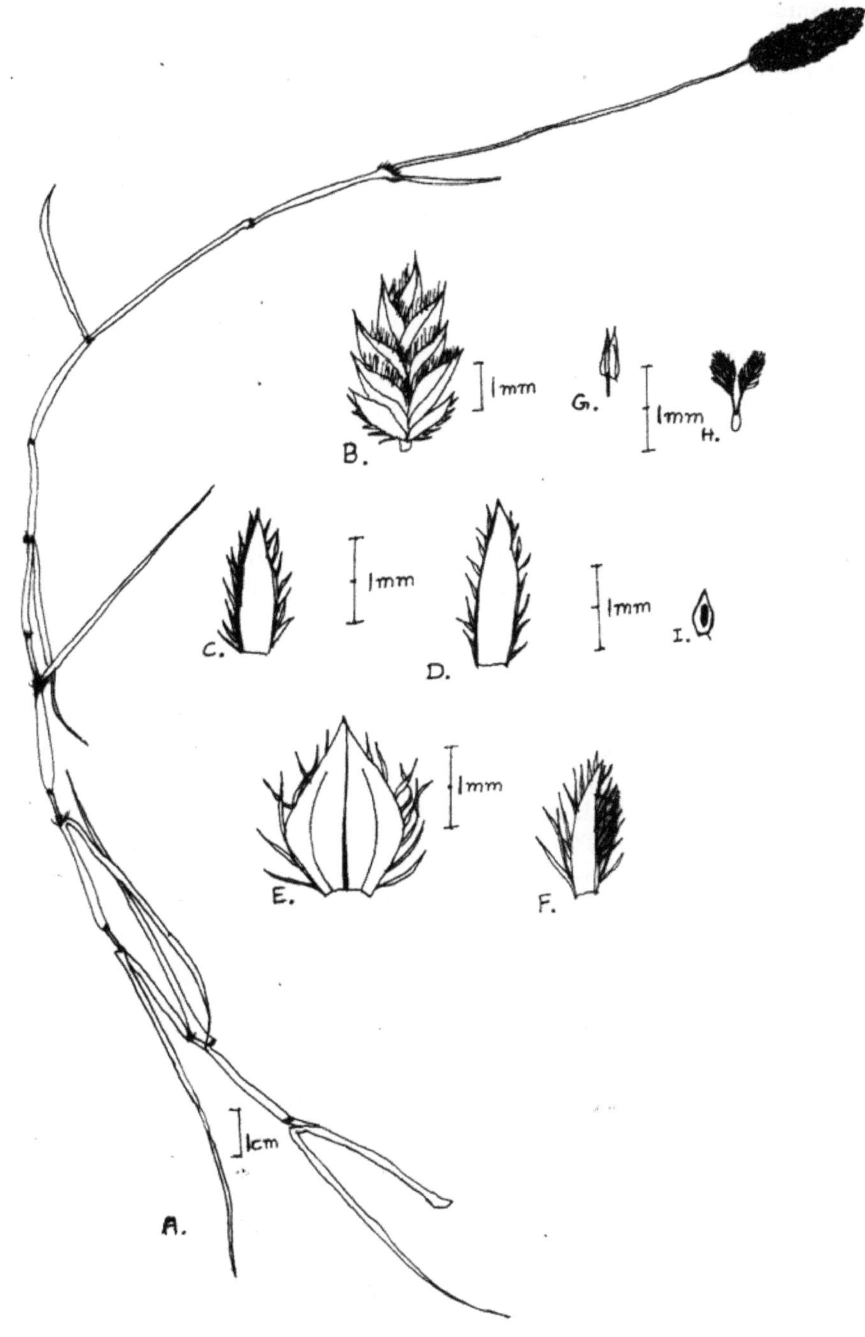

Figure 52. **Eragrostis ciliata** (Roxb.) Nees.

A. Twig, B. Spikelet, C. Lower glume, D. Upper glume, E. Lemma, F. Palea, G. Stamen, H. Pistil, I. Caryopsis.

Kailasanathakona (CTR), *BR* & *P. Priyadarsini* 37130 (BSID); Macherla (GNT), *BDP* 129 (CAL); Ankannapuram (WG), *DCSR* 233 (CAL); Sirivaka (WG), *BO* 3404 (CAL); Araku valley (VSKP), *NPBK* 587 (CAL); Palasa (SKLM), *BDP* 186 (CAL) & *MPG* 174 (CAL); Farahabad (MBNR), *MVS* & *MB* 38026; Lakshmipuram (KMM), *RCS* 98910 (BSID); Rathamhutta hills (KMM) *RCS* 99059 (BSID); Tikarpada, Satkosia Tiger Reserve, *KCM* & *J. Swamy* 7632 (BSID); Deaogaon, ca 15 miles on way to Dhenkenal (Dhenkenal), *sine coll.* 11280 (BSID); Ramyalu river bank, N.W. of Kamakhyanagar (Dhenkenal), *ARKS* 11032 (BSID); Betkote, Karlapat Wildlife Sanctuary (KHD), *AC* 2463 (BSID).

INDIA: Throughout India.

WORLD: Bangladesh, Myanmar.

Eragrostis collinensis Vivek, G.V.S. Murthy & V.J. Nair, Ind. J. Forestry 36(3): 401. 2013.

Culms tufted, erect, 10- 38 cm high, nodes brownish; internodes 2-3.5 cm long. Leaves linear to lanceolate, apex acute to acuminate, base subcordate, margins eglandular, upper surface scabrid to sparsely ciliate towards base; leaf sheaths c.4 cm long, mouth bearded, margins glabrous or finely ciliate; ligule *c.* 0.2 mm long, membranous with fine ciliate rim. Panicle ovate, open, 5-9 x 1.5-4 cm, branches alternate, angled, scabrous along angles, eglandular, base and axis glabrous; peduncle *c.* 5 cm long, eglandular. Spikelets ovate to oblong or elliptic; serrate in appearance, 4-10 x 3-4 mm, purplish to dark pink or chocolate brown in colour; lateral pedicels up to 3 mm long, eglandular, angled, minutely scabrous along angles. Glumes deciduous, unequal, lanceolate, apex acute, margins entire, 1-nerved, keeled, minutely scabrous along keel above middle; lower glume 1.3-1.7 x 0.3 mm; upper glume 1.9-2.1 x 0.4 mm. Florets 3-39, breaking from below upwards, reducing in size towards apex, purplish or dark pink; rachilla zigzag, internode up to 1 mm long. Lemma deciduous, lanceolate to elliptic, 1.9-2 x 0.4-0.8 mm, chartaceous, apex acute, margins entire, 1-keeled, minutely scabrid along keel towards apex, 3-nerved, lateral nerves straight. Paleas deciduous, elliptic to lanceolate, 1.5-1.9 x 0.4-0.8 mm, apex acute to obtuse, margins entire, 2-keeled, scaberulous along keels, 2-nerved; lodicules 2, *c.* 0.2 mm long, apex truncate, white; stamens 3; filaments c.0.5 mm long, white; anthers 0.5-0.8 mm long, deep purple; ovary ovoid, *c.* 0.2 mm long, pale green; style *c.* 0.6 mm long, white; stigma *c.* 0.5 mm long, white or cream coloured. Caryopsis obovoid to narrowly ellipsoid, up to 0.7 x 0.4 mm, deep brown.

Hills above 1000 m in Shevaroy hills. Fl. & Fr.: Throughout the year.

Balmadies, Yercaud (SLM), *A.V.N. Rao* 26970 (MH); Kaceri peak, Yercaud (SLM), *DBD* 31301 (CAL & MH).

INDIA: Tamil Nadu and Kerala. Endemic.

Note: This taxon is closely allied to *Eragrostis uniloides* but differs mainly by its deciduous florets disarticulating from below upwards; scabrous pale keels and obovoid caryopsis. It is distributed mainly above 1000 m in Anamalais, Nilgiris and Shevaroy hills (Vivek *et al.*, 2013).

Eragrostis deccanensis Bor, Grass. India 507.1960; Moulik, Grass, Bamb. India 2: 602. 1997. *E. spicata* Jedwaln., Bot. Arch. 5:185.1924, non Varsey 1891; Fischer 3:1825.1934. *E. phleoides* Stapf in Hook. f., Fl. Brit. India 7: 313.1896.

Annuals, culms erect or geniculate, tufted, to 75 cm high. Leaf sheaths glabrous, ciliate at mouth; ligule ciliate; leaf blades 2-10 x 0.2-0.6 cm, linear-lanceolate, flat, base rounded, apex acuminate, glabrous. Panicle clavellate, dense, interrupted at base, shining; branches very short, villous at base; spikelets compressed; glumes sub-equal, obtuse, chartaceous, 1-nerved; lemma ovate, membranous, minutely scaberulous, 3-nerved, nerves prominent; palea keels narrowly winged, long ciliate, hyaline, 2-nerved; stamens 3. Caryopsis ellipsoid-globose.

Occasional in forest plains and margins. Fl. & Fr.: March-June.

Tirumala hills (CTR), *KCJ* 471 (MH).

INDIA: South India. Endemic.

Eragrostis gangetica (Roxb.) Steud., Syn. Pl. Glum. 1: 266. 1854; Fischer 3: 1826.1934; Bor, Grass. India 508. 1960; Matthew, Fl. Tamilnadu Carnatic 3: 1854. 1983; Moulik, Grass. Bamb. India 2:602.1997. *Poa gangetica* Roxb., Fl. Ind. 1:341.1820. *Eragrostis stenophylla* Hochst. ex Miq., Ver. Konink. Nederl. Inst. 3, 4, 39. 1851; FBI 7: 318. 1896.

Annuals, culms erect, tufted, to 80 cm high, slender; nodes glabrous. Leaf sheaths hairy at mouth; ligules membranous; leaf blades 5-15 x 0.2-0.4 cm, linear-lanceolate or linear, base cordate, apex acute or acuminate. Panicles lax, branches slender; spikelets ovate-oblong, 6-15-flowered, grey; lower glume ovate, chartaceous, 1-nerved; upper glume oblong-lanceolate, chartaceous, 1-nerved; lemmas ovate-lanceolate, chartaceous, 3-nerved; paleas elliptic-lanceolate, 2-keeled, 2-nerved, keels ciliate; stamens 3. Caryopsis orbicular.

Common in forest plains. Fl. & Fr.: July–January.

Hanimireddipalli (ATP), *JNP* & *KVS* 27368; Gundlabrahmeswaram (KNL), *SS* & *AMR* 22436, *JLE* 23838 (MH); Gundumallappa kona (KNL), *BR* & *SS* 20093; Bhajaranga kona (KDP), *KVS* & *BSS* 29119; Gangutla RF (NLR), *BR* & *BSS* 29304; Perayya kunta (PKM), *BSS* & *SKB* 32952; Sunnipenta (KNL), *JLE* 22130 (MH); Balapalle (KDP), *JLE* 14721 (MH); Potangi (CDP), *CAB* 15373 (MH); Seethalam river bed (CTR), *M. Chandrabose* 45101 (MH); Horesleykonda (CTR), *South India Flora* 12643 (MH); Diguvametta (PKM), *JLE* 42180 (MH); Chinturu, Maredumilli (EG), *MM* 102503; Forest near Anigevi (EG), *GVS* 68552 (MH); Anantagiri (VSKP), *GVS* 21759 (CAL); Chatigona water falls (VSKP), *KCJ* 453 (MH); Saleswaram (MBNR), *BSS* & *SKB* 30584; Appapur (MBNR), *BSS* & *SKB* 32297; Sirumalai hills (DGL), *SKS* 325 (SGH); Hogainakkal (SLM), *EV* 20658 (MH); Hogainakkal (SLM), *CAB* 11619, 17994 (MH); Yercaud (SLM), *DBD* 31415 (MH); Purunakote section, Satkosia Tiger Reserve (AGL), *KCM* 5199 (BSID); Tikarpada, Satkostia Tiger Reserve (AGL), *KCM* & *Swamy* 7658 (BSID); Belghar, Kotgarh Willdlife Sanctuary (KDM), *AC* 2415 (BSID); Mahendragiri (GLM), *VSN* 5743 (MH).

INDIA: Throughout India.

WORLD: Tropical Asia and Africa.

Eragrostis japonica (Thunb.) Trin., Mem. Acad. Imp. St. Petersbourg., Ser. 6. Sci. Math. 1:405.1830; Fischer 3: 1826.1934; Bor, Grass. India 509.1960; Matthew, Fl. Tamilnadu Carnatic 3: 1855. 1983; Moulik, Grass. Bamb. India 2: 603. 1997. *Poa japonica* Thunb., Fl. Jap. 51. 1784. *Eragrostis interrupta* sensu Stapf in Hook. f., Fl. Brit. India. 7: 316. 1896. incl. vars., non (R.Br.) P. Beauv. 1812. *E. diarrhena* (Schult.) Steud., Syn. Pl. Glumac. 1: 266. 1854; Fischer 3: 1826. 1934; Bor, *loc. cit. Poa diarrhena* Schult., Mant. 2: 616. 1824. *Eragrostis diplochoides* Steud., Syn. Pl. Glumac. 1: 268. 1854; Bor, *loc. cit.* 508.

Perennials, culms erect, tufted, to 1.3 m high, ascending; nodes glabrous. Leaf sheaths glabrous; ligules membranous; leaf blades 10-20 x 0.2-0.4 cm, linear-lanceolate, involute, base rounded, apex acuminate, glaucous. Panicles contracted, dense, spiciform; spikelets oblong-lanceolate, 5-15-flowered; lower glume ovate-lanceolate, chartaceous, 1-nerved; upper glume obovate, chartaceous, 1-nerved; lemmas obovate, chartaceous, 3-nerved; paleas oblong-lanceolate, 2-keeled, 2-nerved, keels ciliate; stamens 3. Caryopsis ellipsoid.

Occasional in stream beds and moist places of forests. Fl. & Fr.: October-June.

Garugudu kona (ATP), *TP* 957; Kammavaripalli (KNL), *TP* & *PVP* 2794; Nandalur (KDP), *AL* & *BSS* 26568; Chaparlapalli (KDP), *BSS* & *SKB* 29306; Udayagiri fort (NLR), *BR* & *BSS* 29301; Rollapenta (PKM), *BR* & *BSS* 29386; Gundlamotu Reservoir (PKM), *BSS* & *SKB* 32933; Nagarjunakonda (GNT), *VRK* 6713; East of Nerabylu (CTR), *GVS* 46974 (MH); Gobanapalem (WG), *KS* 5140 (CAL); Polavarm agency (WG), *DCSR* 142 (CAL); Talapudi (EG), *BO* 3296 (CAL); Uttanoor (MBNR), *MVS* & *DV* 40722; Tekulagudem, Opp. to Tupakulagudem, Godavari bank (KMM), *NRR* & *TRS* 86105 (BSID); Lanjigarh (KHD), *MB* & *Dhal* 7398, 7404 (RRL-B); Tikarpada, Satkosia Tiger Reserve (AGL), *KCM* & *Swamy* 7617 (BSID).

INDIA: Throughout India.

WORLD: South and SE Asia to Tropical and South Africa, Australia.

Eragrostis macilenta (A.Rich.) Steud., Syn. Pl. Glum. 1: 268. 1854; Bor, Grass. India 509. 1960; B.R.P.Rao *et al.*, Grass. Arid Zone Peninsular India 116. 2017. *Poa macilenta* A.Rich., Tent. Fl. Abyss. 2: 428. 1851.

Annuals; culms slender, tufted, up to 70 cm high. Spikelets oblong,10-flowered. Lower glume 1-nerved, oblong-ovate, apex lobed; upper glume ovate, obscurely 1-nerved or nerve absent, keeled. Lemmas ovate-lanceolate, keeled, acute, paleas oblong-ovate, scabrid.

Occasional in drier parts on hill tops (B.R.P.Rao *et al.*, 2017). Fl. & Fr.: October–February.

Samrajyam konda, Gorantla (ATP), *KVS* & *ANS* 29710 (SKU).

INDIA: Tamil Nadu, Andhra Pradesh.

WORLD: East Africa.

Eragrostis maderaspatana Bor, Grass. India 509. 1960. *E. willdenowiana* Nees ex Stapf in Hook.f., Fl. Brit. India 7: 322. 1896, non Nees ex Hook. & Arn. 1838; Fischer 3:1827. 1934.

Annuals, culms erect or geniculately ascending, tufted, to 40 cm high; nodes glabrous. Leaf sheaths glabrous; ligule a ring of hairs; leaf blades 1-1.5 x 0.1-0.3 cm, linear, margins eglandular, acuminate, glabrous. Panicle oblong, lax, open; spikelets ovate-laneolate, 4-18-flowered; lower glume oblong-lanceolate, 1-nerved, chartaceous; upper glume ovate-lanceolate, 1-nerved, chartaceous; lemmas broadly elliptic, 3-nerved; paleas lanceolate, 2-keeled, keels scabrid, persistent. Caryopsis oblong.

Rare in open areas and near streams. Fl. & Fr.: March-September.

Gundlabrahmeswaram (KNL), *BR* & *BSS* 32966; Narasaraopet (GNT), *CAB* 4549 (MH).

INDIA: South India.

WORLD: Sri Lanka.

Eragrostis minor Host., Icon. Descr. Austr. 4. 15. 1809; FBI 7: 321. 1896. *E. poaeoides* P. Beauv., Ess. Agrost. 162. 1812; Fischer. 3:1827. 1934; Bor, Grass. India 512. 1960; Matthew, Fl. Tamilnadu Carnatic 3: 1855. 1983.

Annuals, culms erect, tufted, to 45 cm high, leafy; nodes glabrous. Leaf sheaths ciliate at mouth, glabrous; ligule a pubescent ridge; leaf blades 7-15 x 0.1-0.2 cm, linear, base narrow, usually glandular along the margin, apex acuminate, glabrous. Panicle lax, open, branches filiform, false whorled at base; spikelets ovate-oblong, 12-flowered; glumes subequal, ovate, acute, 1-nerved; lemma broadly ovate, acute, 3-nerved; palea obovate-oblong, keels scabrid or slightly ciliate; stamens 3. Caryopsis broadly oblong, dorsally rounded or slightly flattened.

Rare in crop fields and waste lands in northern Eastern Ghats. Fl. & Fr.: July–December.

Mandasa mendu (EG), *VNS* 4624 (MH); Sirumalai hills (DGL), *SKS* 419 (SGH).

INDIA: Throughout India.

WORLD: Pakistan, Warm temperate and subtropical regions of the Old World.

Eragrstis nairii Kalidass, J.Econ. Taxon. Bot. 39: 126. 2015.

Annuals; culms densely tufted, branched at the base, up to 25 cm high, 1-1.5 mm wide, nodes glabrous. Leaf sheats 1-4.5 cm, ligulate, ligule a membranous rim, villous; leaf blades linear, 3-8 x 0.1-0.15 cm, apex filiform, base truncate. Panicless lanceolate or elliptic, 10-15 cm long, purple to violet; pedicel 0.4-1.2 cm; spikelets lanceolate or elliptic, 2-3.5 x 0.1-0.8 mm, pink to violet, breaking up from the base, *c.* 8-florets, rachilla branched, persistent, opposite florests slightly overlapping; lower glume ovate-lanceolate, to 0.3 mm, acute at apex, cuneate at base, glabrous, chartaceous, minutely ciliated, 1-nerved; upper glume ovate-lanceolate, up to 1.0 mm, tips acuminate, obtuse at base, chartaceous, 1-nerved; lemma broadly ovate, to 1.1 mm, prominent, membranus, tip acuminate, 3-nerved; palea obovate, to 0.7 mm long, 2-keeled, minutely ciliate along the keels; lodicules 2, oblong, *c.* 0.05 mm; stamens 2, anthers elliptic, purple, *c.* 0.1 mm long, basifixed, glabrous, filaments *c.* 0.03 mm long; ovary ovate to obovate, *c.* 0.04 mm, style *c.* 0.04 mm long, divided into two filament with ovary, stigmas plumose, *c.* 0.2 mm long. Caryopsis ovate or globose, *c.* 0.1-0.3 x 0.05 mm.

Found in the riparian banks of semi-evergreen forest of Similipal Biosphere reserve in Odisha. Fl. & Fr.: July–December.

Similipal Biosphere reserve (MBJ), *Kalidass* 9402 (Holotype MH; Isotype CAL).

INDIA: Odisha. Endemic.

Eragrostis nigra Nees ex Steud., Syn. Pl. Glum. 1: 267. 1854; FBI 7: 324.1896; Fischer 3: 1827. 1934; Bor, Grass. 511. 1960; Matthew, Fl. Tamilnadu Carnatic 3: 1856. 1983; Moulik, Grass. Bamb. India 2: 605.1997.

Perennials, culm erect, tufted, to 75 cm high; nodes glabrous. Leaf sheaths ciliate at mouth; ligule ciliate; leaf blades 8-37 x 0.2-0.5 cm, linear, convolute, sparsely pilose above. Panicle effuse; spikelets ovate-lanceolate, 8-flowered; pedicels glandular; glumes ovate-lanceolate, subequal; lemmas broadly ovate, eglandular, straight on the back; palea obtuse, denticulate, persistent; stamens 3. Caryopsis smaller, subquadrate, dorsally slightly grooved.

Common in forest plains and peripheries. Fl. & Fr.: July–December.

Tirumala hills (CTR), *MSG* 12136 & 12179; Anantagiri (VSKP), *GVS* 19470 (CAL).

INDIA: Throughout India.

WORLD: Bhutan, Nepal, China, Sri Lanka, Myanmar, SE Asia.

Eragrostis nutans (Retz.) Nees ex Steud., Nom. Bot. ed. 2. 563. 1840; Fischer 3: 1826. 1934; Bor, Grass. India 511.1960; Matthew, Fl. Tamilnadu Carnatic 3: 1856. 1983; Moulik, Grass. Bamb. India 2: 605.1997. *Poa nutans* Retz., Observ. Bot. 4:19.1786. *Eragrostis elegantula* Nees ex Steud., Syn. Pl. Glum. 1: 266.1854; FBI 7: 318. 1896.

Perennials, culms erect, tufted, to 90 cm high, slender; nodes glabrous. Leaf sheaths glabrous; ligules hairy; leaf blades 6-15 x 0.3-0.5 cm, linear-lanceolate, base rounded, apex acuminate. Panicles contracted, black or grey colour; spikelets oblong-lanceolate, 6-15-flowered; lower glume ovate, chartaceous, 1-nerved; upper glume oblong, chartaceous, 1-nerved; lemmas ovate, chartaceous, 3-nerved; palea, obovate, 2-keeled, 2-nerved, ciliate on the keels; stamens 3. Caryopsis ellipsoid.

Common in moist habitats throughout. Fl. & Fr.: June–April.

Kalasamudram RF (ATP), *NY* 1184; Penukonda (ATP), *BR* & *P.Priyadarsini* 37174 (BSID); Gundumallappa kona (KNL), *BR* & *SS* 20100; Palakondalu (KDP), *PVP* & *GO* 3435; Tirumala hills (CTR), *MSG* 12143, *BR* & *AMR* 21847, 21871, 21860, 21866, 21878; Perayya kunta (PKM), *BSS* 32951; Garnipenta (PKM), *BSS* & *SKB* 32959; Rampa hills (EG), *TP* & *MSG* 12284; G. Madugula (VSKP), *AMR* 24554; Mahendragiri

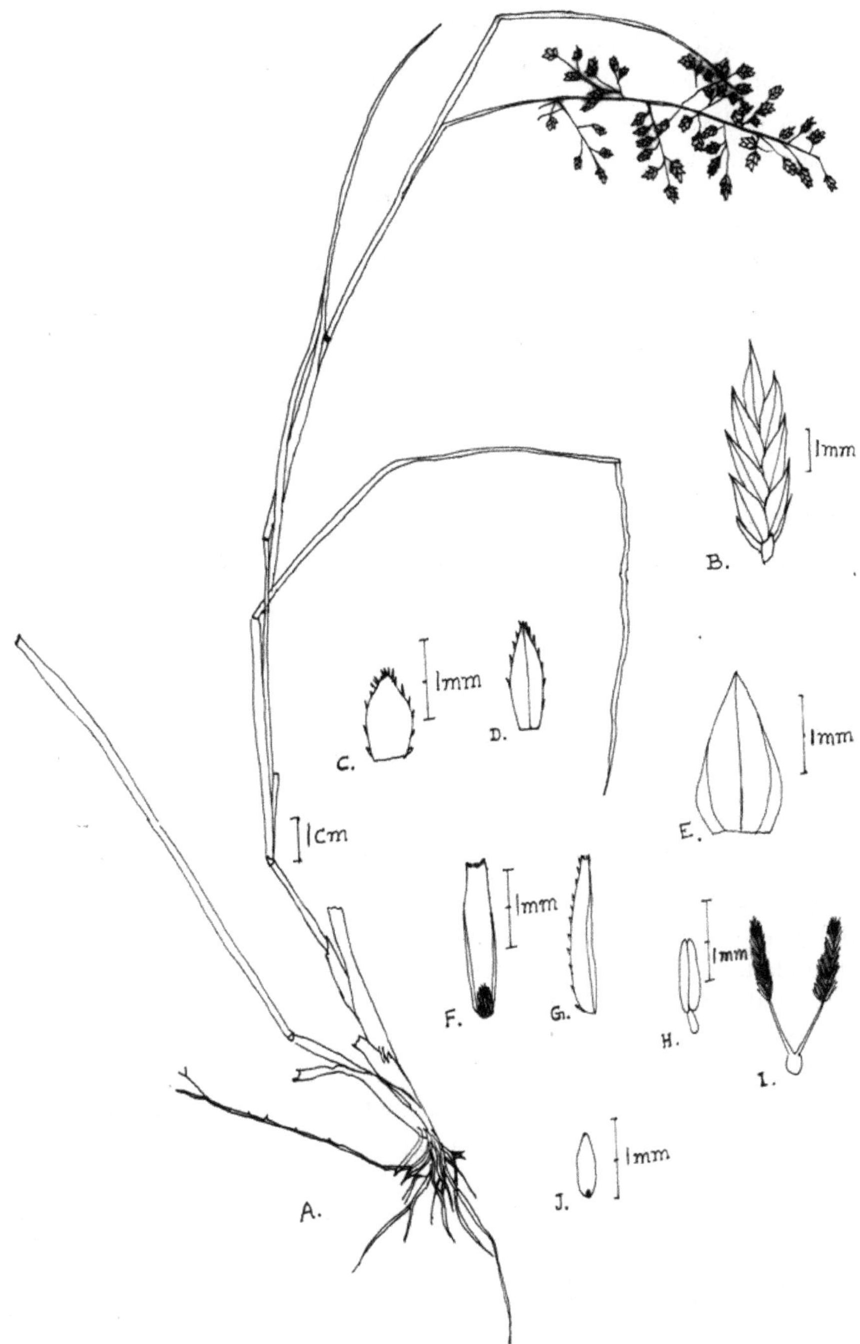

Figure 53. **Eragrostis nigra** Steud.

A. Twig, B. Spikelet, C. Lower glume, D. Upper glume, E. Lemma, F & G. Palea front and side views, H. Stamen, I. Pistil, J. Caryopsis.

(SKLM), *AMR* 25661; Palasa (SKLM), *MPG* 195 (DD & CAL); Tikarpada, Satkosia Tiger Reserve (AGL), *KCM* & *Swamy* 7633 (BSID); Noana vicinity, Similipal Biosphere Reserve (MBJ), *ARKS* & *GPS* 12295 (BSID); Balipadar (GJM), *AC* 10216 (BSID).

INDIA: S. & E. India.

WORLD: S. China, Malesia.

Eragrostis papposa (Roem. & Schult.) Steud., Syn. Pl. Glum. 1: 263.1854; Bor, Grass. India 511.1960. *Megastachya papposa* Roem. & Schult., Syst. Veg. 2: 585.1817.

Perennials, culms erect, tufted, to 50 cm high; nodes glabrous. Leaf sheaths glabrous, ciliate at mouth; ligule ciliate; leaf blades short, stiff, linear, convolute. Panicle lax, open, filiform; spikelets linear, 7-23-flowered; glumes subequal, hyaline, lower nerveless, upper 1-nerved; lemma broadly ovate, hyaline; palea persistent, obtuse, keels scabrid; stamens 3. Caryopsis obovoid, dorsally grooved.

Occasional in streams and forest shade in Rampa hills. Fl. & Fr.: November-May.

Peddabeerampalli, Rampa hills (EG), *TP* & *MSG* 12278 (SKU).

INDIA: N.W.India, Andhra Pradesh.

WORLD: Spain and North Africa through the Middle East to India and Sudan to Arabia.

Eragrostis patula (Kunth) Steud., Nomencl. Bot. Ed. 2, 1: 564. 1840. *Poa patula* Kunth, Nov. Gen. Sp. 1: 158. 1816. *Eragrostis tenuifolia* (A.Rich.) Hochst. ex Steud., Syn. Pl. Glum. 1: 268. 1854; FBI 7: 322. 1896; Fischer 3:1827. 1934; Bor, Grass. India. 514.1960; Moulik, Grass. Bamb. India 2:609.1997; Matthew, Fl. Tamilnadu Carnatic 2: 1858. 1983; Pullaiah, Fl. Andhra Pradesh 5: 2187. 2018. *Poa tenuifolia* A. Rich., Tent. Fl. Abyss. 2: 425. 1850.

Perennials, culms erect, slender, tufted, to 60 cm high; nodes glabrous. Leaf sheaths glandular, sparsely hairy; ligules ciliate; leaf blades 4-7 x 0.1-0.3 cm, linear, base rounded, apex acuminate, glabrous. Panicles lax; spikelets linear-oblong, pedicels glandular; lower glume minute, ovate, linear, 1-nerved, lobed; upper glume ovate-oblong, 1-nerved, 2-3-lobed; lemmas ovate-elliptic, 3-nerved, mucronate; palea oblanceolate, 2-nerved, 2-keeled, keels scabrid, hyaline, persistent. Caryopsis oblong.

Rare in open dry places and along road sides. Fl. & Fr.: July–October.

Muniswaruni kunta (ATP), *PVP* 2617; Bugga (ATP), *BR* & *JNP* 27389; Kalasamudram (ATP), *BR* & *P.Priyadarsini* 37094 (BSID); Gani (KNL), *TP* & *AA* 1928; B.K.Palem (GNT), *VRK* 3247; Tirumala hills (CTR), *KCJ* 475 (MH); Araku valley (VSKP), *DDSR* 21375 (MH); Anantagiri (VSKP), *GVS* 19470 (MH); Melagiris (SLM), *CAB* 15436 (MH); Pulluhallli (SLM), *CAB* 9861 (MH); Shevaroy hill SLM), *AVNR* 26771 (MH); Yercaud (SLM), *S. Karthikeyan* 28295 (MH).

INDIA: Peninsular India.

WORLD: Tropical Asia, Tropical Africa, Tropical America.

Eragrostis pilosa (L.) P. Beauv., Ess. Agrost. 71: 162.175.1812; FBI 7:323.1896; Fischer 3:1827.1934; Bor, Grass. India 512.1960; Matthew, Fl. Tamilnadu Carnatic 3: 1856. 1983; Moulik, Grass. Bamb. India 2:606. 1997. *Poa pilosa* L., Sp. Pl. 68.1753.

Annuals, culms erect, tufted, to 1.2 m high, slender; nodes glabrous. Leaf sheaths glabrous; ligules fimbriate, membranous; leaf blades linear-lanceolate, base rounded, acuminate. Panicles lax, filiform; spikelets ovate-lanceolate or linear-oblong, 4-10-flowered; lower glume ovate-lanceolate, chartaceous, 1-nerved; upper glume lanceolate, chartaceous, 1-nerved; lemmas oblong-lanceolate, chartaceous, 3-nerved; paleas ovate, keeled, 2-nerved, keels scabrid; stamens 3. Caryopsis ovate.

Occasional in forest margins and plains. Fl. & Fr.:August–December.

Kalasamudram (ATP), *EC* 2903, *BR* & *AMR* 24219; Hanimireddi palli (ATP), *JNP* & *AL* 27379, 27801; Kambagiri konda (KNL), *BR* & *SS* 20039; Peapuly (KNL), *PVP* 2667, 3164; Balapalii RF (KDP), *PVP* 3461; Guvvalacheruvu (KDP), *AMR* 25606; Gundlamotu Reservoir (PKM), *BSS* & *SKB* 32934; Palamber R.F. (CTR), *KCJ* 465 (MH); Chandargiri (CTR), *CAB* 9997 (MH); Talapudi (WG), *BO* 3302 (CAL); Appapur (MBNR), *AMR* 25699; Sirumalai hills (DGL), *SKS* 346 (SGH); Pelakuppam (SA), *VNS* 4047 (MH); Gudiyatham (SA), *KCJ* 6353 (MH); Polur (NA), *KCJ* 6371 (MH); Kollimalai (NMK), *CAB* 347 (MH); Peddamugilalam, Melagiri (DMP), *CAB* 9830 (MH); Shevaroys (SLM), *sn. coll.* 95543 (MH); Punakote, Satkosia Wildlife Sanctuary (AGL), *AC* 2390 (BSID); Barga Khol, Rajgangapur (Sundargarh), *AC* 10256 (BSID); Jacum, Karlapat Wildlife Sanctuary (KHD), *AC* 2449 (BSID).

INDIA: Throughout India.

WORLD: Tropical and Warm Temperate regions of the Old World.

Eragrostis riparia (Willd.) Nees, Agrost. Bras. 512.1829; Fischer 3: 1825.1934; Bor, Grass. India 513.1960; Matthew, Fl. Tamilnadu Carnatic 3: 1857. 1983; Moulik, Grass. Bamb. India 2: 606.1997. *Poa riparia* Willd. in Ges., Natur. Freunde Berlin, Neue, Schrift. 4: 185.1803. *Eragrostis tenella* P.Beauv. var. *riparia* Stapf in Hook. f., Fl. Brit. India 7: 315.1896.

Perennials, culms erect, tufted, to 90 cm high, ascending; nodes glabrous. Leaf sheaths glabrous, hairy at mouth; ligules a row of hairs; leaf blades 2-4 x 0.2-0.4 cm, linear-lanceolate, base rounded, apex acute, hairy. Panicles spiciform, dense; spikelets ovate, oblong, 5-10-flowered; lower glume ovate-linear, chartaceous, 1-nerved; upper glume oblong, chartaceous, 1-nerved; lemmas oblong, chartaceous, 3-nerved; paleas oblong, 2-keeled, 2-nerved, keels ciliate; stamens 3. Caryopsis elliptic.

Occasional in marshy areas especially near sandy moist soils. Fl. & Fr.: May-November.

Kalasamudram (ATP), *NY* 1166, *BR* & *AMR* 24256; Ahobilum-Mattamcheruvu (KNL), *MSR* & *BSS* 29315; Tirumala hills (CTR), *MSG* 1241, *RVR & GO* 2413, *AMR* 21842; Rollapenta (PKM), *BR* & *BSS* 29384; Marripadu (NLR), *PMR* 16513; Gundlamotu reservoir (PKM), *BSS* & *SKB* 32935; Bhupathipalem, Rampa hills

(EG), *TP* & *MSG* 12244; Ramannapenta (KNL), *JSG* 12683 (MH); Mamandur valley (CTR), *KS* 6905 (MH); Tirumala hills (CTR), *KCJ* 472 (MH); Srivaka (EG), *CAB* 5256 (MH); Galikonda (VSKP), *GVS* 19584 (CAL); Nupuraganga (MDU), *KS* 3355 (MH); Hogainakkal (DMP), *EV* 20543 (MH); Polur (SA), *KCJ* 6370 (MH).

INDIA: Bengal southwards.

WORLD: Sri Lanka.

Eragrostis tremula Hochst. ex Steud., Syn. Pl. Glum. 1: 269. 1854; FBI 7: 320.1896; Fischer 3: 1827.1934; Bor, Grass. India 514.1960; Moulik, Grass. Bamb. India 2: 609. 1997.

Annuals or perennials, culms erect, tufted, to 90 cm high, ascending; nodes glabrous. Leaf sheaths bearded at mouth; ligules fimbriate; leaf blades 5-15 x 0.2-0.5 cm, linear, flat, base cordate, apex acuminate, glbrous. Panicles up to 20 cm long, effuse, filiform; spikelets oblong, 15-50-flowered; lower glume oblong, chartaceous, 1-nerved; upper glume oblong, chartaceous 1-nerved; lemmas ovate-oblong, chartaceous, 3-nerved; paleas elliptic-oblong, 2-keeled, 2-nerved, keels scabrid; stamens 3. Caryopsis globose.

Occasional in forest plains throughout. Fl. & Fr.: June - November.

Kalasamudram RF (ATP), *NY* 393, *BR* & *AMR* 24239; Baireni (KNL), *BR* & *BSS* 32246; Chelama (KNL), *BSS* & *SKB* 32948; North Dhone RF (KNL), *RVR* 1631; Bhogeswaram (KNL), *BR* & *SS* 22679; Tirumala hills (CTR), *BR* & *AMR* 21863; Chintapalli (VSKP), *AMR* 24577; Gazulapalle (KNL), *KCJ* 83171 (MH); Ankannapalem (WG), *DCSR* 246 (CAL); Kondapudi (WG), *CAB* 5189 (MH); Erapuntapalli (EG), *JSG* 18559 (DD); Jummilam (EG), *CAB* 5281 (MH); Guddaeguda-Koilkonda (MBNR), *SRS* 110797 (BSID); Tikarpada, Satkosia Tiger Reserve (AGL), *KCM* & *Swamy* 7636 (BSID).

INDIA: Throughout India.

WORLD: Myanmar to Tropical Africa.

Eragrostis unioloides (Retz.) Nees ex Steud., Syn. Pl. Glum. 1:264. 1854; Fischer 3: 1826.1934; Bor, Grass. India 515.1960; Matthew, Fl. Tamilnadu Carnatic 3: 1858. 1983; Moulik, Grass. Bamb. India 2:609.1997. *Poa unioloides* Retz., Observ. Bot. 5:19.1789. *Eragrostis amabilis* auct. non Wight & Arn. ex Nees 1838: FBI 7: 317.1896.

Annuals or perennials, culms erect or ascending, tufted, to 90 cm high; nodes glabrous. Leaf sheaths glabrous; ligules membranous, leaf blades 5-10 x 0.2-0.3 cm, linear-oblong, base cordate, acuminate. Panicles lax, purplish; spikelets ovate, 10-30-flowered, purplish; lower glume ovate-oblong, chartaceous, 1-nerved; upper glume ovate-oblong, chartaceous, 1-nerved; lemmas broadly ovate, chartaceous, 3-nerved; paleas oblong-elliptic, 2-keeled, 2-nerved, keels scabrid; stamens 3. Caryopsis obovoid.

Common in wet lands, streams and in moist habitats throughout Eastern Ghats. Fl. & Fr.: July–April.

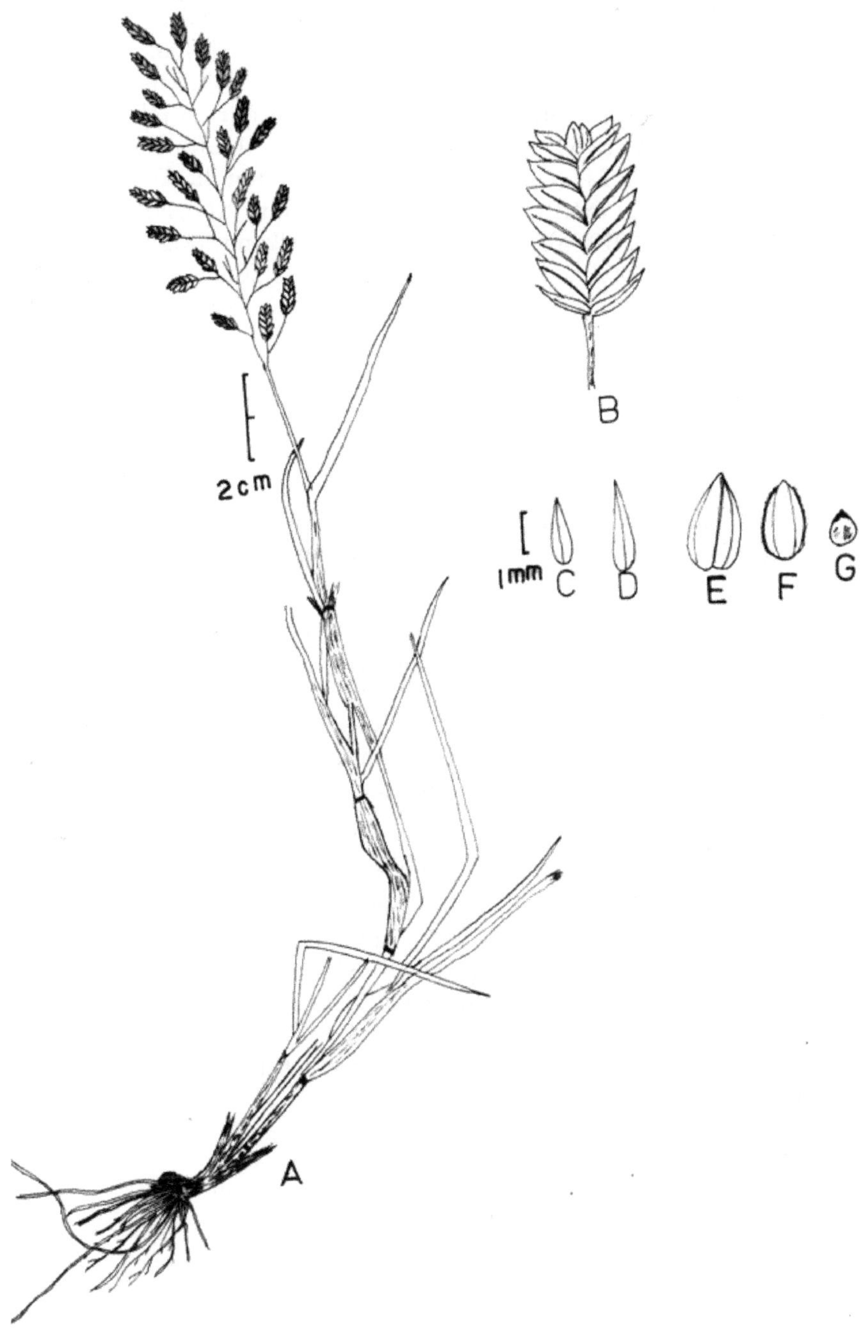

Figure 54. **Eragrostis unioloides** (Retz.) Nees ex Steud.

A. Twig, B. Spikelet, C. Lower glume, D. Upper glume, E. Lemma, F. Palea, G. Grain.

Garugudukona (ATP), *TP* 951; Kadiri (ATP), *JNP* & *AL* 27342; Peapuly (KNL), *PVP* & *RVR* 3191; Kambagirikonda (KNL), *BR* & *SS* 20042; Namalagundu (KDP), *KRKS* & *BSS* 40175; Bhajaranga kona (KDP), *KVS* 29108; Tirumala hills (CTR), *RVR* & *GO* 2409; Kambakkam (CTR), *AMR* 24283; Chintapalli (VSKP), *AMR* 24579; Lambasangi (VSKP), *AMR* 25632; Punyagiri (VJN), *BR* & *AMR* 24537; Modavagulvanka (KNL), *JLE* 18014 (MH); Kota (NLR), *VNS* 519 (CAL); Bhupatipalem (EG), *VNS* 247 (CAL); Kottravada, Maredumilli (EG), *MM* 102551; Way to Cherukoda peak (VSKP), *GVS* 42728 (MH); Salur (VJN), *NPBK* 951 (CAL); Bandirevu (KMM), *RCS* 99100 (BSID); Guddaeguda–Koilkonda (MBNR), *SRS* 116774 (BSID); Sirumalai hills (DGL), *CAB* 9083 (MH), *SKS* 420 (SGH); Swamymalai (SA), *CAB* 17920 (MH); Jenabil range, Similipal B.R. (MBJ), *AC* 2309 (BSID).

INDIA: Throughout India.

WORLD: SE Asia, Africa.

Eragrostis viscosa (Retz.) Trin., Mem. Acad. Sci. Petersb. Ser. 6.1: 397. 1830; Fischer 3: 1826.1934; Bor, Grass. India 515. 1960; Matthew, Fl. Tamilnadu Carnatic 3: 1859. 1983; Moulik, Grass. Bamb. India 2: 611.1997. *Poa viscosa* Retz., Observ. Bot. 4:20.1786. *Eragrostis tenella* var. *viscosa* Stapf in Hook.f., Fl. Brit. India 7:315.1896.

Annuals, culms erect or ascending, tufted viscous below, to 70 cm high; nodes glabrous. Leaf sheaths glabrous; ligules membranous; leaf blades 3-10 x 0.2-0.4 cm, linear, convolute, base rounded, apex acuminate. Panicles up to 20 cm, narrowly cylindric, glandular; spikelets oblong-lanceolate, 4-20-flowered; lower glume linear, chartaceous, 1-nerved; upper glume oblong, chartaceous, 1-nerved; lemmas ovate, chartaceous, 3-nerved; paleas linear, ciliate on the keels, 2-keeled, 2-nerved; stamens 3. Caryopsis ovoid.

Common in wastelands and forest plains throughout. Fl. & Fr.: April–November.

Gani RF (KNL), *RVR* 1515; Kottala cheruvu (KNL), *BR* & *SS* 21283; Papavinasanam (CTR), *BSS* & *SKB* 34822; Ramagiri (CTR), *BR* & *P.Priyadarsini* 37112 (BSID); Leguntapadu (NLR), *BR* & *PMR* 16437; Variguntapadu (NLR), *PMR* 16475; Macherla (GNT), *VRK* 5883; Guntur (GNT), *SCA* 587 (DD); Varadayapalem, Kambakkam (CTR), *MCB* 45157 (MH); Ramavaram (KMM), *RR* 112584 (BSID); Uragadam (CPT), *S.India Flora* 14080 (MH); Chettuppattu to Punnathur (SA), *KRM* 60318 (MH).

INDIA: Throughout India.

WORLD: Tropical and South Africa, Thailand, Philippines.

EREMOCHLOA Buese

Eremochloa muricata (Retz.) Hack. in A.DC. & C.DC., Monogr. Phan. 6: 262. 1889; FBI 7: 142. 1896; Fischer 3: 1726. 1934; Bor, Grass. India 146. t.5(1). 1960. *Aegilops muricata* Retz., Observ. Bot. 2: 27. 1781.

Perennial grass with woody stonoliferous base. Culms geniculate, slightly ribbed, 21-45 cm high; nodes brownish. Leaves linear-lanceolate, 4-12 x 0.5-0.8 cm, acute to obstuse at apex, infold towards base, margins tuberculate, hairy below, serrulate above; ligule 1-1.2 mm, truncate at apex, membranous, hyaline to brownish. Racemes long, spikate, solitary, exerted; peduncle 10-22 cm long, more or less ribbed; pedicel short, 0.2-0.4 mm, represents as pedicelled spike. Sessile spikelet broadly elliptic, *c.* 6.3 x 2.8 mm, dorsally compressed, acute at apex, greenish yellow to brown, comb like with lateral spines on both sides; callus glabrous. Lower glume broadly ovate to elliptic-lanceolate, *c.* 6.3 x 2.6 mm, acuminate at apex, wings broad, *c.* 1.5 mm across, fimbriate above, chartaceous or subcoriaceous, greenish yellow, 7-9-nerved. Upper glume ovate to elliptic-laceolate, *c.* 5.3 x 1.8 mm, acuminate at apex, chartaceous to subcoriaceous, greenish yellow, prominently 3-nerved. Florets 2; lower male, upper bisexual; lower lemma oblong to elliptic, *c.* 4.7 x 1.2 mm, acute to obtuse at apex, ciliolate on upper margins; lower palea oblanceolate to elliptic, *c.* 4.7 x 1.2 mm, acute at apex, membranous to chartaceous, hyaline, nerveless, margins infold towards apex; upper palea ovate-laceolate, *c.* 4.2 x 0.8 mm, acute at apex, otherwise similar to lemma; stamens 3; anthers 2.8-3 mm, brown or yellow; filaments *c.* 1.3 mm long, hyaline; ovary oblong, *c.* 1.2 x 0.4 mm, acute; style *c.* 0.7 mm long; stigma plumose, purplish brown.

Rare in wet places. Fl. & Fr.: August–March.

Nungambakkam (CPT), *sine coll. s.n.* (MH); Avadi (CPT), *S. India Flora* 14879 (MH); Velacheri (CPT), *sine coll. s.n.* (MH); Ponneri (CPT), *S. India Flora* 14037 (MH); Palakuppam (SA), *VNS* 4148 (MH); Perungudinedu (CPT), *sine coll s.n.* (MH).

INIDA: Guajarat, Maharashtra, Tamil Nadu.

WORLD: Sri Lanka, Myanmar, Thailand, Australia.

ERIOCHLOA Kunth

Eriochloa procera (Retz.) C.E. Hubb., Kew Bull. 1930: 256. 1930; Fischer 3: 1767. 1934; Bor, Grass. India 312. 1960; Matthew, Fl. Tamilnadu Carnatic 3: 1859. 1983. *Agrostis procera* Retz., Observ. Bot. 4: 19. 1786. *Eriochloa polystachya* Hook. f., Fl. Brit. India 7: 20. 1896.

A perennial, tufted grass; culms up to 50 cm high, creeping at base; nodes glabrous. Leaf sheaths 4-15 cm long, loose, glabrous; ligule fimbriate; blades 5-20 x 0.4-0.7 cm, linear, flat, base rounded, apex acuminate, glabrous. Panicle up to 10 cm long, lax; spikelets 3 mm, ellipsoid; lower glume obscure; upper glume elliptic, equal to spikelet, pilose; lower lemma lanceolate- elliptic, equal to spikelet; pilose, male or empty; upper lemma crustaceous, ovate, apical barbellate mucro at apex; palea crustaceous, ovate; stamens 3, stigmas plumose. Caryopsis oblong- ellipsoid, dorsally compressed.

Common grass near stream and marshy places. Fl. & Fr.:August- October. Tam.: *Karunkanipul*.

Pennahobilam (ATP), *PVP* 2648; Kalasamudram RF (ATP), *KRKS* 36470 (SKU & BSID); Malyala (KNL), *PVP* 2553; Balapalli RF (KDP), *PVP* 3468; Kalyanidam (CTR), *RVR & GO* 2439; Moodavagulavanka, Chelama (KNL), *JLE* 18013 (MH); Krishnanandi (KNL), *JLE* 42320 (MH); Guvvalacheruvu (KDP), *KS* 6430 & 7794 (MH); Palakonda RF (KDP), *KS* 6430 (MH); Satyavedu to Madras (CTR), *MCB*

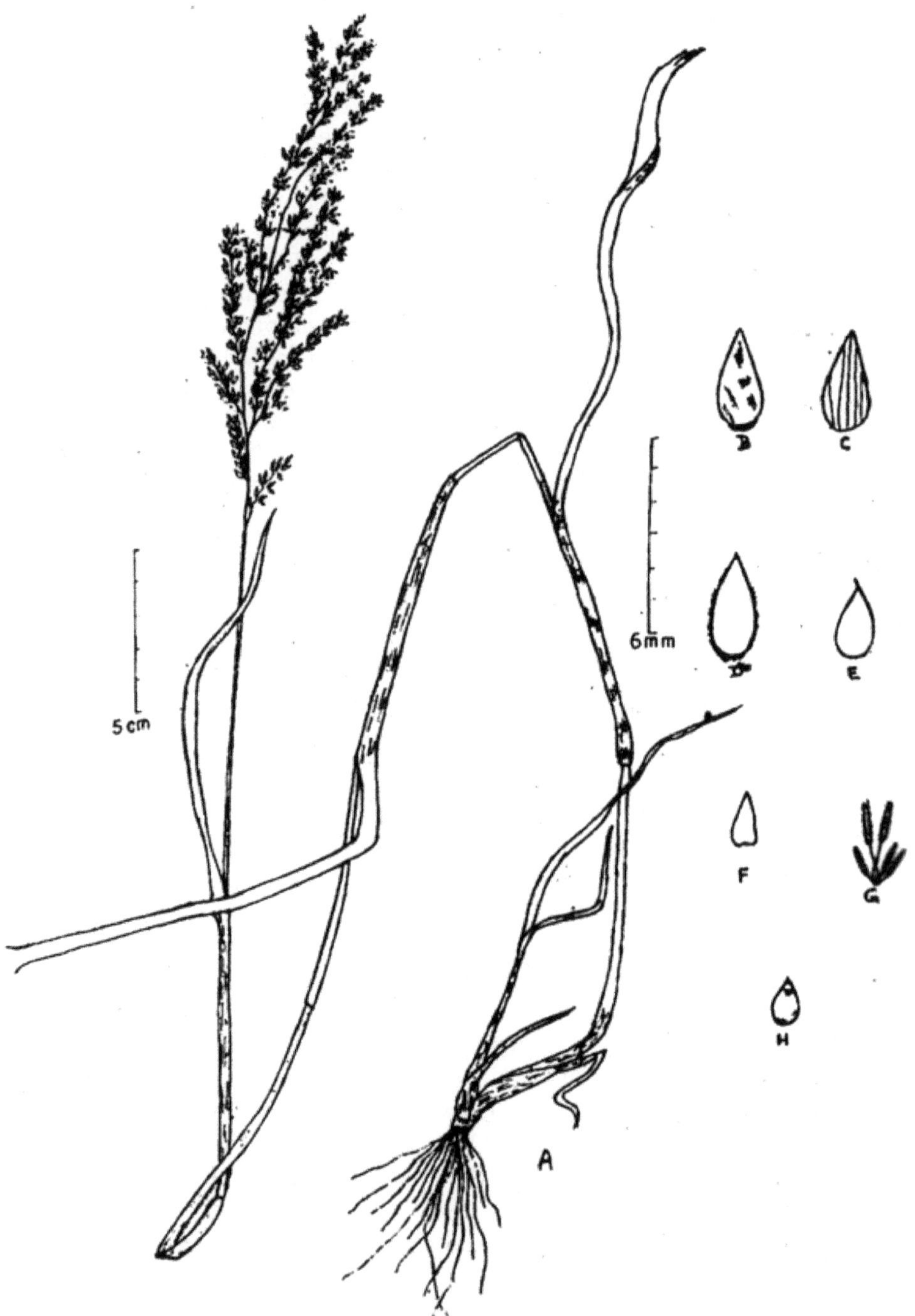

Figure 55. **Eriochloa procera** (Retz.) C.E. Hubb.

A. Twig, B. Spikelet, C. Upper glume, D. Lower lemma, E. Upper lemma, F. Palea, G. Pistil and stamens, H. Grain.

43251 (MH); Macherla RF (GNT), *SCA* 138 (DD); Kondaveedu fort (GNT), *CAB* 4708 (MH); Kondapalli (KSN), *CAB* 7974 (MH); Amangal RF (MBNR), *SRS* 108695 (BSID); Vandalur (CPT), *S.India Flora* 11552 (MH); Gudiyattam (NA), *S.India Flora* 15522 (MH); Tippukadu RF (NA), *KRM* 17697 (MH).

INDIA: Plains of India.

WORLD: SE Asia, Tropical Africa; introduced in Tropical America.

EULALIA Kunth

1. Basal sheaths covered with brownish or white wool:
    2. Basal sheaths covered with brownish wool .... **E. phaeothrix**
    2. Basal sheaths covered with white wool:
        3. Spikelet sessile; upper lemma with 5 mm longer awned .... **E. leschenaultinana**
        3. Spikelet pedicelled; upper lemma with 2 mm long awned .... **E. quadrinervis**
1. Basal sheaths glabrous:
    4. Spikelets paired pedicelled; upper glume awned .... **E. contorta**
    4. Spikelets not paired, sessile; upper glime unawned:
        5. Sessile spikelets 4-7 mm long; lower glume dorsally concave, 4-nerved .... **E. villosa**
        5. Sessile spikelets 2-3.5 mm long; lower glume narrowly elliptic, 2-nerved .... **E. trispicata**

Eulalia contorta (Brongn.) Kuntze, Revis, Gen. Pl. 2: 775. 1891. *Pogonatherum contortum* Brongn. in Duperr., Voy Coq. Bot. 90 & 17. 1831. *Pseudopogonatherum contortum* (Brongn.) A. Camus in Lecomte, Gen. de Indochine 7: 255. 1922; Fischer 3: 1716. 1934; Bor, Grass. India 204. 1960. *Pollinia articulata* Trin., Bull. Sci. Acad. Petersb. 1: 71. 1836; FBI 7: 109. 1896. *P. articulata* var. *pedicellata* Hack. ex Hook. f., Fl. Brit. India 7: 100. 1896.

A perennial, tufted grass; culms up to 25 cm high; nodes glabrous. Leaf blades 15-40 x 0.2-0.3 cm, linear-lanceolate, broad, apex acuminate, hairy below, glabrous above; ligule a tuft of hairs. Spiciform racemes 6 cm long, clusters, supported by spathe; spikelet 2 mm, paired, 2-flowered; lower glume narrowly truncate, keels villous below with white hairs; upper glume glabrous; lower lemma ovate, obtuse; upper narrowly oblong, bi-dentate, awned, awn very long ciliate; palea short, broad; stamens 3. Caryopsis free.

Rare in norther Eastern Ghats. Fl. & Fr.: November-January.

Kodagundi RF (VSKP), *KCJ* 17282 (MH).

INDIA: Assam, East Himalaya, West Himalaya.

WORLD: Nepal, China, Myanmar, Australia, Philippines, Thailand.

Eulalia leschenaultiana (Decne) Ohwi, Bull. Tokyo Sci. Mus. 18: 2. 1947; Bor, Grass. India 156. 1960; Saxena & Brahmam, Fl. Orissa 4: 2374. 1996. *Andropogon leschenaultianus* Decne, Herb. Timor. Descr. 29. 1835. *Pollinia cumingii* Nees in Hook., Kew J. Bot. 2: 98. 1850; FBI 7: 114. 1896.

Perennial stoloniferous grass. Leaf sheaths usually covered with white wool or with stiff white villae; blades narrow, flat, 7.5-10 cm x 2.5-4.3 mm, glabrous, ligule short, truncate, glbrous. Racemes usually 2, rarely 3-4, brown or yellow-brown, slender, 2.5-8.7 cm long, rachis and pedicels covered with brown hairs on the edges. Sessile spikelets 2.5-3 mm long, awned, callus short, blunt, hairy; lower glume involute, oblong, truncate, brown hairy; upper glume slightly longer, broadly truncate, less hairy; lower lemma 0; upper lemma reduced to a long awn 2-3 times the length of the spikelet or lemma developed, 2-lobed.

Occasional in Eastern Ghats of Odisha. Fl. & Fr.: October–November.

Parhaipad (KJR), *HFM* 2831 (DD); Chahal, Similipahar (MBJ), *SX* & *MB* 5303 (RRL-B); Jenabil range, Similipal B.R. (GJM), *AC* 2305 (BSID).

INDIA: Plains of N., E. & S. India

WORLD: E. Asia, Malaya, Australia.

Eulalia phaeothrix (Hack.) Kuntze, Revis. Gen. Pl. 2; 775. 1891; Fischer 3: 1715. 1934; Bor, Grass. India 156. 1960. *Pollinia phaeothrix* Hack, in DC. Mongr. Phan. 6: 168. 1889; FBI 7: 112. 1896.

A perennial, tufted grass; culms up to 2 m high, slender; nodes sparsely bearded. Leaf sheaths 5-10 cm, lower sheaths rusty, reddish brown tomentose; ligule bearded; blades 10-20 x 0.3-0.6 cm, linear. Racemes digitate, 2-10, densely silky under the panicle; joints and pedicels long villous; spikelets 4-7 mm long, linear-oblong, 2-nate, similar, dorsally compressed; pedicel long pilose. *Sessile spikelets:* 4 mm, lanceolate-oblong; lower glume oblong-lanceolate, pilose; upper glume lanceolate; hairy and scabrid outside; lower lemma oblong, acuminate, hairy near apex; upper lemma linear, bidentate, awned; stamens 3, stigma plumose. Caryopsis oblong. *Pedicelled spikelets:* Similar to sessile spikelet.

Common in Nallamalais, Tirumala hills, Nallakonda, Palakonda and Sirumalai hills. Fl. & Fr.: September-December.

Gundlabrahmeswaram (KNL), *PVP & RVR* 2400; Peddacheruvu RF (KNL), *PVP & RVR* 2174; Ramanapenta, Nallamalais (KNL), *JLE* 32627 (MH); Tirumala hills (CTR), *MSG* 12150; Isukagundam (PKM), *BR* & *BSS* 29296 (SKU & BSID); Dummakonda, Rampa (EG), *VNS* 643 (MH); Chintapalli (VSKP), *JLE* 37102 (MH); Sirumalai hills (DGL), *SKS* 349 (SGH).

INDIA: South India.

WORLD: Sri Lanka, China, Thailand, Vietnam.

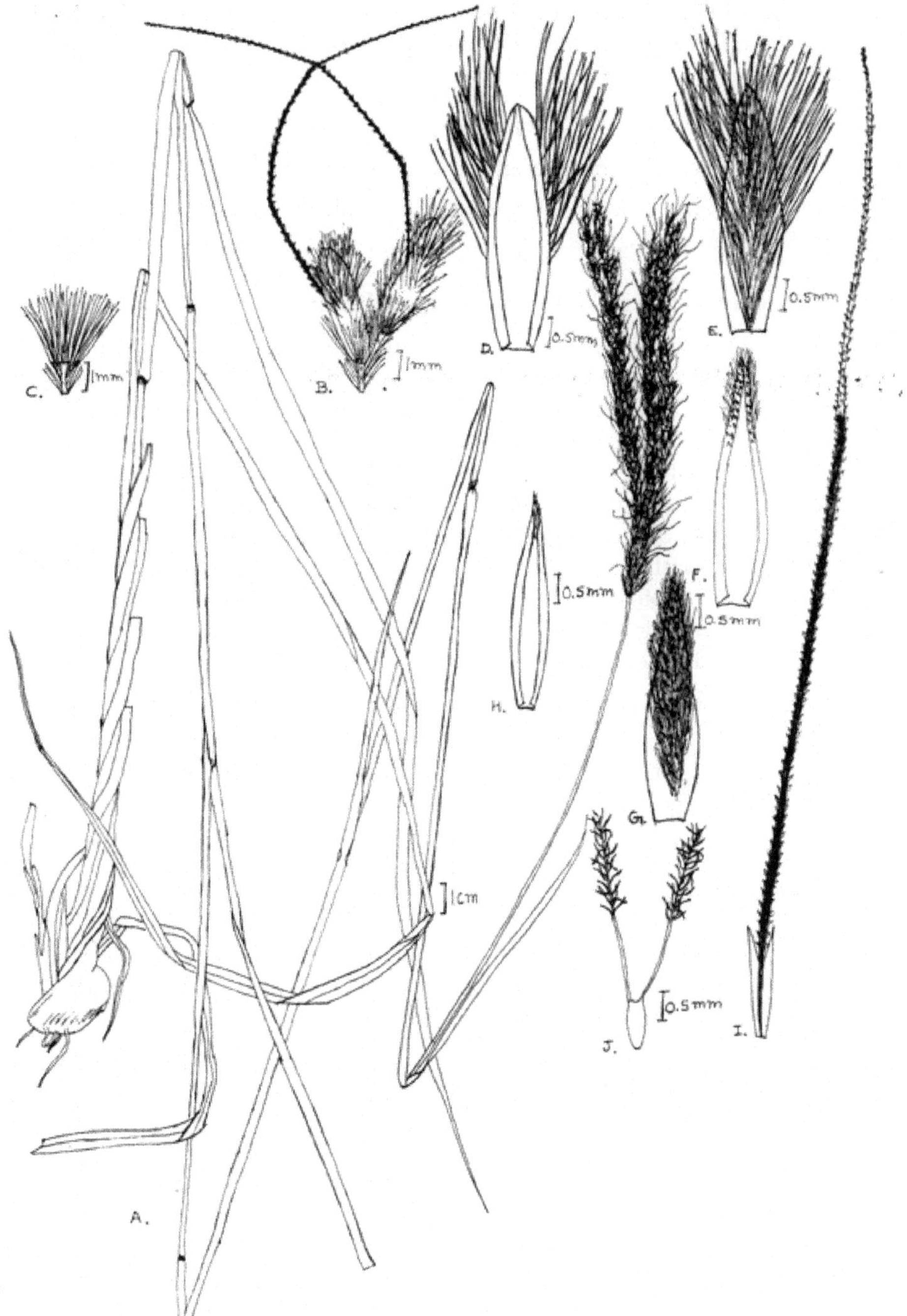

Figure 56. **Eulalia phaeothrix** (Hack.) Kuntze.

A. Twig, B. Spikelets, C. Pedicel, D & E. Lower glume front and back views, H. Lower lemma, I. Upper lemma, J. Pistil.

Eulalia quadrinervis (Hack.) Kuntze, Revis, Gen. Pl. 2: 775. 1891; Bor, Grass. India 156. 1960. *Pollinia quadrinervis* Hack., Monogr. Phan. 6: 158. 1889; FBI 7: 110. 1896.

Key to the Varieties

1. Lower leaves crowded; lower glumes of sessile spikelet larger...... var. **wightii**
1. Lower leaves not crowded; lower glumes of sessile spikelet smaller .... var. **quadrinervis**

Eulalia quadrinervis (Hack.) Kuntze var. **quadrinervis**

Perennial tufted grass, culms 0.6-1 m high. Leaves subrigid, spreading, linear, 10-20 cm x 3-9 mm, apex acuminate, hairy; sheaths hairy; ligule hairy. Racemes 3-6, 7-15 cm long, rachis and pedicels with golden brown hairs; sessile spikelets 4-6.5 mm long, oblong-lanceolate, subacute, awned, pale brown, callus hairs long, white; lower glume oblong or oblong-lanceolate, 5-6 mm long, flat, coriaceous, hyaline at apex, obscurely truncate, back depressed, scabrous, margins with narrow incurved flaps, keels hispid above, base with long, soft hairs, 4-nerved; upper glume lanceolate, acute,without an arista, keels glabrous, obscurely 3-nerved; lower lemma hyaline, oblong-lanceolate, acute, 2-nerved, ciliate; upper lemma broadly cuneiform, oblong, hyaline, 2.5-3 mm long, ciliate towards the top, deeply bifid, awn 1.7-2 cm; palea oblong, short. Pedicelled spikelets with 1.4 cm long awn.

Frequent on open grassy hill slopes in Northern Eastern Ghats in Odisha. Fl. & Fr.: Oct.- Nov.

Mahendragiri hills (GJP), *SX* & *MB* 3651, 3943 (RRL-B), *AC* 4231 (BSID); Deomali (KPT), *SX* & *MB* 6620 (RRL-B).

INDIA: Subtropical Himalaya, Odisha.

WORLD: Myanmar, China, Laos, Thailand.

Eulalia quadrinervis (Hack.) Kuntze var. **wightii** (Hook.f.) Fischer, Fl. Madras 3: 1715. 1934. *Pollinia quadrinervis* Hack. var. *wightii* Hook.f., Fl. Brit. India 7: 110. 1896. *Eulalia wightii* (Hook.f.) Bor, Grass. India 158. 1960.

Differs from the type variety in the very crowded lower leaves and lower glume of sessile spikelets being larger and narrowly elliptic-acute with broad flaps.

Common in Northern Eastern Ghats in Odisha. Fl.: August.

Kuthadya hill (GJM), *VNS* 5910 (MH); Pottangi (KPT), *HFM* 4066 (DD).

INDIA: Tamil Nadu, Odisha

Eulalia trispicata (Schult.) Henr., Blumea 3: 453. 1940; Bor, Grass. India 157. 1960; Matthew, Fl. Tamilnadu Carnatic 3: 1861. 1983. *Andropogon trispicatus* Schult. Syst. Veg. 2, Mant. 452. 1824. *A. tristachyus* Roxb., Fl. Ind, 1: 261. 1820 non Kunth 1816. *Eulalia tristachya* (Roxb.) Kuntze, Revis. Gen. Pl. 2: 775. 1891; Fischer 3: 1715. 1934. *Pollinia argentea* Trin., Bull. Sci. Acad. Imp. Sci. Petersbourg. 1: 71. 1836; FBI 7: 111. 1896.

A perennial, tufted grass; culms up to 1.5 m high; nodes glabrous. Leaf sheaths 15 cm long, glabrous; ligule membranous, hairy; blades 10-25 x 0.3-0.5 cm, linear-lanceolate, glabrous-sparsely hairy. Racemes 2-12, 15 cm long, sub-digitate, capillary; rachis and pedicels densely villous with whitish silvery hairs; spikelets 4 mm, 2-nate, oblong, compressed. *Sessile spikelets:* Glumes subequal, subchartaceous; lower glume narrowly elliptic, 2-nerved, 2-keeled, hairy below the middle; upper glume boat-shaped, 1-nerved, 1-keeled, ciliate along the margin; lemmas hyaline, unequal; lower oblong epaleated, male; upper stipiform, bi-dentate, awned, bisexual; stamens 3; stigmas plumose. Caryopsis oblong. *Pedicelled spikelet:* 4 mm, similar to sessile spikelet.

Occasional in dry open forests. Fl. & Fr.: September–December.

Chelama (KNL), *JLE* 16857 (CAL); Satyavedu (CTR), *MVS* & *BR* 32161 (SKU & BSD); Kambakkam hills (CTR), *S.India Flora* 10274 (MH); Similipahar (MBJ), *V. Naik* Acc. No. 3891, 3901 (RRL-B); Niyamgiri hill (KHD), *AC* 2475 (BSID).

INDIA: Throughout India.

WORLD: Pakistan, Sri Lanka, Nepal, Myanmar, China, Malesia, Australia.

Eulalia villosa (Spreng.) Nees, Fl. Afr. Austral. Ill. 91. 1841. *Pollinia villosa* Spreng., Syst. Veg. 1: 288. 1824. *Eulalia wighii* (Hook, f.) Bor, Grass. India 158, 1960. *Pollinia quadrinervis* Hack. var. *wightii* Hook. f., Fl. Brit. India 7: 110. 1896. *P. quadrinervis* Kuntze. var. *wightii* Hook. f., Fl. Brit. India 7: 110. 1896.

A perennial, tufted grass; culms up to 50 cm high, leafy; nodes glabrous. Leaf sheaths hirsute, apex acuminate; ligule membranous, oblong; blades linear, slightly hairy on both surfaces. Spikes 3-5, 15 cm long, erect; spikelets 6 mm, oblong-lanceolate, subacute; lower glume dorsally concave, 4-nerved, keels hispid, truncate, margins incurved; upper glume acute, minutely apiculate, 3-nerved, margins ciliate; lower lemma linear-oblong, long ciliate above; upper lemma reduced to an awn, bi-dentate; palea lanceolate ciliate.

Rampa hills at 1400 m (Narayanaswamy).

INDIA: South India, Endemic.

EULALIOPSIS Honda

Eulaliopsis binata (Retz.) C. E. Hubb. in Hook. Ic. Pl. sub. t. 3262. 1935; Bor, Grass. India 158. 1960. *Andropogon binatus* Retz., Obs. Bot. 5: 21. 1789. *Pollinidium binatum* (Retz.) C.E. Hubb., Kew Bull. 1932. 72. 1932; Fischer 3: 1717. 1934. *Ischaemum angustifolium* (Trin.) Hack. in DC., Monogr. Phan. 6: 241. 1889; FBI 7: 129. 1896.

A perennial, tufted grass; culms up to 90 cm high, slender, woolly at base. Leaf sheaths glabrous; ligule a ring of short hairs; blades 20-50 x 3-5 cm, linear, base broad, margins scaberulous, apex acuminate, glabrous. Racemes 2-4, 5 cm long, distant or compact, axillary and terminal, golden or rusty villous; joints and pedicels ciliate. *Sessile spikelet:* 4 mm, lanceolate; callus hairy; lower glume elliptic-oblong, chartaceous, 2-3-dentate, densely hairy; upper glume ovate-lanceolate, membranous, bi-dentate; lower lemma oblong, obtuse, bidentate, hyaline; palea linear, bi-dentate; upper lemma linear, hyaline, bi-dentate or acute, awned; palea small, *Pedicelled spikelet:* Similar to sessile spikelet.

Rare in moist waste places in hilly regions of northern Eastern Ghats. Fl.: February-June, Fr.: May-July.

Itennavaram (WG), MH *s.n.* (MH); Annavaram (EG), *S.India Flora* 12579 (MH); Rajavommangi (EG), *S.India Flora* 12665 (MH); Barnakonda RF (VSKP), *KCJ* 17119 (MH); Lakshmipuram (VSKP), *KCJ* 1768 (MH); Karakakonda (VSKP), *CAB* 1715 (MH); Barmakonda (VSKP), *KCJ* 17119 (MH); Tulka section, Satkosia Tiger Reserve (AGL), *KCM* 6197 (BSID).

INDIA: Throughout India.

WORLD: China, Thailand, Philippines.

GARNOTIA Brongniart

1. Annual grass; nodes bearded; blades hirsute; spikelets up to 3 mm long .... **G. tenella**
1. Perennial grass; nodes glabrous; blades glabrous; spikelets up to 6 mm long .... **G. elata**

Garnotia elata (Arn. ex Miq.) Janowski, Feddes Repert. 17: 86. 1921; Bor, Grass. India 567. 1960. *Berghausia elata* Arn. ex Miq., Verh. Konink. Nederl. Inst. 3, 4: 32. 1851. *Garnotia scoparia* Stapf ex Hook. f., Fl. Brit. India 7: 242. 1896 non Thw. 1864; Fischer 3: 1812. 1934. *G. tenuiglumis* Stapf ex Hook. f. *loc. cit.* 242.

A stout, perennial, tufted grass; culms up to 1 m high; nodes glabrous. Leaf sheaths 4-12 cm long, glabrous; ligule ciliate membrane; blades 10-30 x 0.2-0.4 cm, linear, flat, base rounded, scaberulous margin, apex acuminate, glabrous. Panicle 40-70 cm long, contracted; branches fastigiate, filiform; spikelets 5 mm long, awned; glumes equal, linear, margin scaberulous, 3-nerved; lemma lanceolate, 3-nerved, apex bidentate, awned; palea hyaline, 2-nerved; stamens 3; stigmas plumose. Caryopsis linear.

Rare grass in hills of Chittoor district. FI. & Fr.: July-December.

Samrajyam konda, Gorantla (ATP), *KVS & MVS* 29190; Tirumala hills (CTR), *MSG* 12180; Tirumala–Talakona (CTR), *BR, BSS & P. Priyadarsini* 30534 (SKU & BSID); Horsley hills (CTR), *MSG & AA* 12210; *DAM & PVP* 3498 & *JSG* 21308 (MH); Devaribetta (SLM), *KCJ* 18054 (MH).

INDIA: South India. Endemic to Peninsular India.

Garnotia tenella (Arn. ex Miq.) Janowski, Feddes Repert. 17: 86. 1921; Matthew, Fl. Tamilnadu Carnatic 3: 1861. 1983. *Berghausia tenella* Arn. ex Miq., Verh. Konink. Nederl. Inst. 3, 4: 34. 1851. *Garnotia stricta* sensu Hook. f., FBI 7: 243. 1896; Fischer 3: 1812. 1934; Bor, Grass. India 569. 1960.

An annual, tufted grass; culms up to 60 cm high, geniculately ascending and branched at base; nodes bearded. Leaf-sheaths 4-9 cm long, glabrous; ligule a narrow

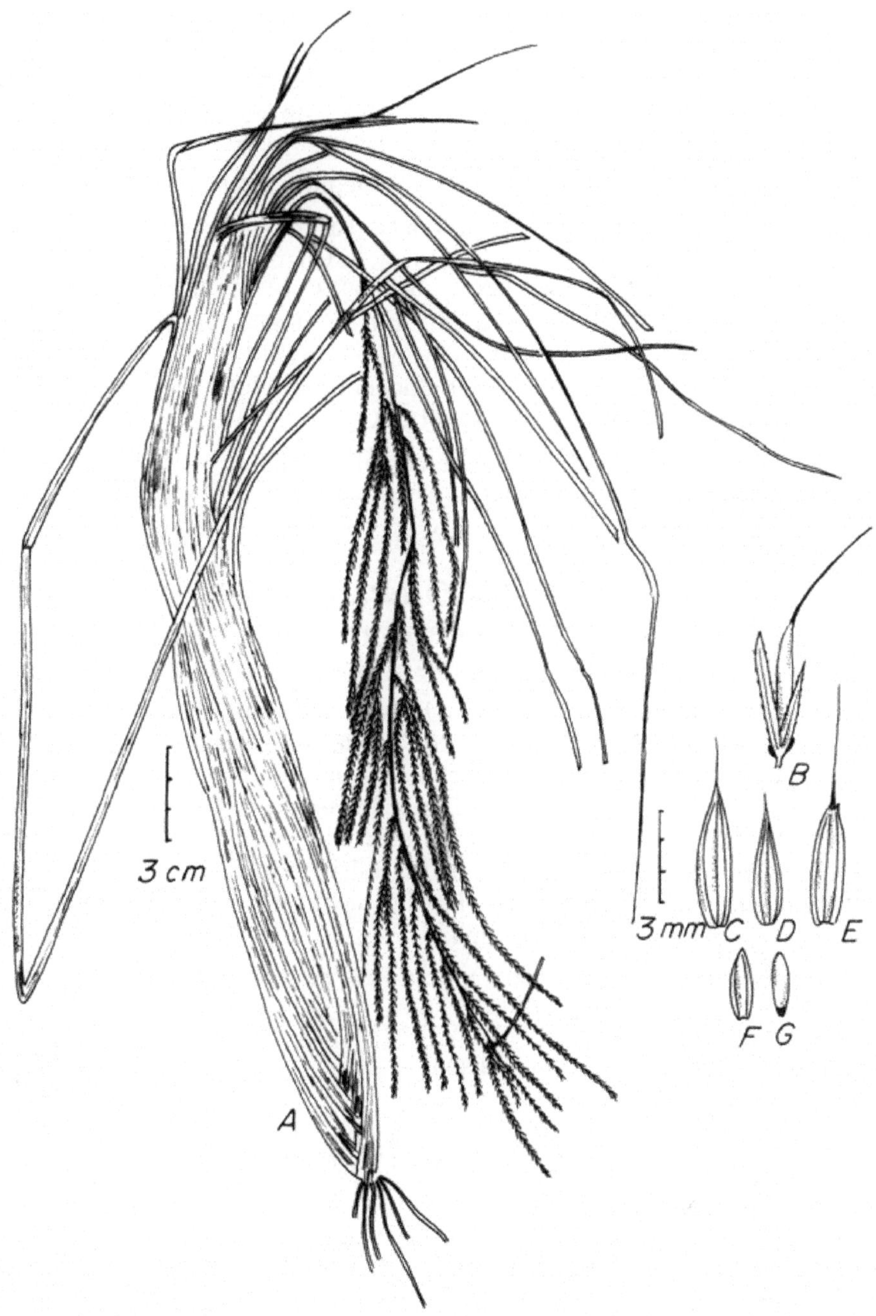

Figure 57. **Garnotia elata** (Arn. ex Miq.) Jan.

A. Twig, B. Spikelet, C. Lower glume, D. Upper glume, E. Lemma, F. Palea, G. Caryopsis.

ciliate membrane, blades 4-15 x 0.3-0.5, linear-lanceolate, flat, base rounded, apex acuminate, hirsute above. Panicle up to 15 cm long; branches distant at base, close at top, in dense clusters, appressed to the rachis; spikelets 4 mm, oblong-linear, paired, sparsely hairy at the base; glumes ovate-lanceolate, 3-nerved, apex acuminate, hairy; lemma lanceolate, 3-nerved, apex bidentate, awned; palea oblong; stamens 3, stigmas plumose. Caryopsis linear, dorsally compressed.

Occassional grass in the hills. Fl.& Fr.: September-January.

Rollapenta (KNL), *PVP & RVR* 2342; Way to Tirumala hills (CTR), *KS* 7849 (MH); Sunkarimetta (VSKP), *NPBK* 117 (CAL); way to Cherukonda peak (VSKP), *GVS* 42724 (MH); Umamaheswaram (MBNR), *SRS* 109137, 111376 (BSID); Mahendragiri (GJP), *SX* & *MB* 3641 (RRL-B); Chingudia fall, Similipahar (MBJ), *SX* & *MB* 4460 (RRL-B); Joranda, Similipahar (MBJ), *SX* & *MB* 5236 (RRL-B); Tikarpada, Satkosia Tiger Reserve (AGL), *KCM* 7071 (BSID); Mahendragiri hills (GJP), *AC* 4206, 4212 (BSID); Jenabil Range, Similipal B.R. (MBJ), *AC* 2336, 2339 (BSID).

INDIA: Subtropical and Temperate Himalayas, Bihar, West Bengal, Odisha, Western Ghats, Gujarat, South India.

WORLD: Widely distributed in SE Asia.

GYMNOPOGON P. Beauvois

Gymnopogon delicatulus (C.B.Clarke) Bor, Grass. Burma, Ceylon, India Pakistan 472. 1960. *Chloris delicatula* C.B.Clarke ex Hook.f., Fl. Brit. India 7: 290. 1896.

Slender, erect grass, 10-30 cm high. Leaves spreading, concave or convolute, 0.75-2.5 cm long. Racemes 5-10, at the top of the culm, slender, erect, 3.7-6.2 cm long, with closely crowded spikelets, rachis capillary, scabrid; spikelets 2.5 mm long including the subaristate tip of upper glume; lower glume 2 mm long, acuminate, not awned; lemma 1.5 mm log, delicately hairy on the back, tip 2-lobed, awn very fine, 7.5 mm long; palea narrow, glabrous; rhachilla produced beyond base of lemma to half its length or rather more, and bearing at its tip 2, rarely 3, very fine awns nearly as long as the awn of the lemma.

Occasional in Eastern Ghats of Odisha. Fl. & Fr.: October–November.

Badrama forest, Bamra (SBP), *HFM* 3722 (DD).

INDIA: Central India, Bihar, Odisha.

WORLD: Myanmar.

HACKELOCHLOA Kuntze

Hackelochloa granularis (L.) Kuntze, Revis. Gen. Pl. 2. 776. 1891; Fischer 3: 1758.1934; Bor, Grass. India 159. 1960; Matthew, Fl. Tamilnadu Carnatic 3: 1862. 1983. *Cenchrus granularis* L., Mant. 2, App. 575. 1771. *Mnesithea granularis* (L.) Koning & Sosef, Blumea 31: 295. 1986. *Manisuris granularis* (L.) L.f., Nov. Gram. Gen. 40. 1779; FBI 7: 159. 1896.

An annual, tufted grass; culms up to 40 cm high, tuberculate pilose; nodes bearded. Leaf sheaths 2-6 cm long; ligule membranous; blades 4-10 x 0.4-1 cm,

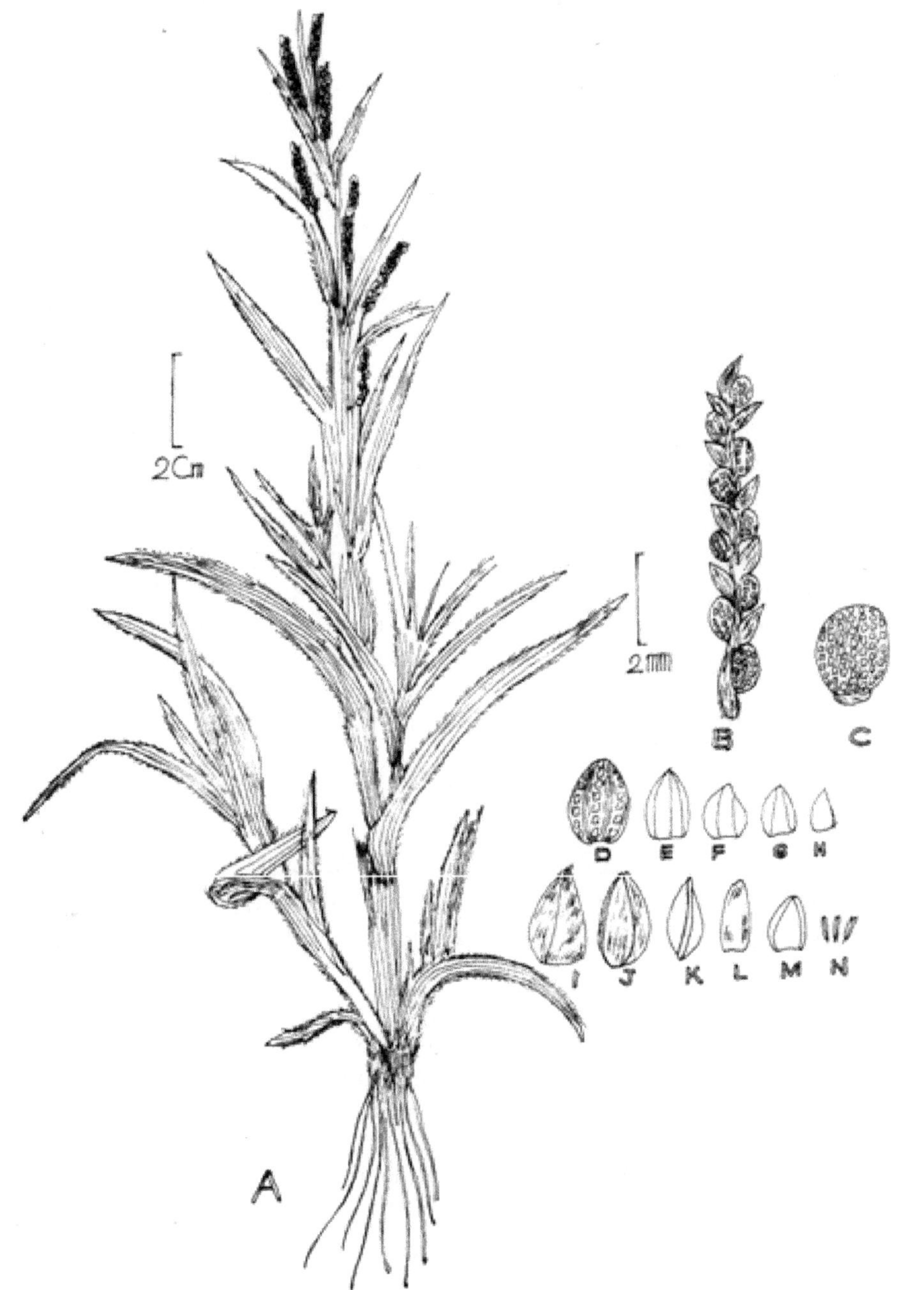

Figure 58. **Hackelochloa granularis** (L.) Kuntze

A. Twig, B. Raceme, C. Sessile spikelet, D. Lower glume, E. Upper glume, F. Lower lemma, G. Upper lemma, H. Palea, I. Pedicelled spikelet, J. Lower glume, K. Upper glume, L. Upper lemma, M. Palea, N. Stamens.

lanceolate, flat, base rounded, apex acute. Racemes up to 4 cm long, spiciform, both terminal and axillary; rachis fragile and hollow; spikelets 3 mm, globose, 2-nate. *Sessile spikelet:* globose, 2 mm; lower glume globose, crustaceous, pitted and tuberculate outside; upper glume broadly ovate, chartaceous, concave; lower lemma ovate-obtuse; upper lemma broadly ovate; palea ovate; stamens 3, stigmas plumose. Caryopsis ovoid. *Pedicelled spikelet:* 3 mm, narrowly globose; lower glume obovate, incurved, winged; upper glume navicular, keeled on back; lemma often wanting, when present upper lemma either male or empty; palea ovate; stamens 3.

Common in the hills and weed in cultivated fields. Fl. & Fr.: July-September.

Amagondapalem RF (ATP), *NY* 1043; Ahobilam RF (KNL), *TP* 2719, *TP & GNVF* 7060; Ramallakota (KNL), *RVR* 1847; Guvvalacheruvu (KDP), *PVP* 3421 & *RVR* 8013; Horsley hills (CTR), *MSG & AA* 12220; Talakona RF (CTR), *MSG & AA* 12227; Bhupathipalem, Rampa hills (EG), *TP & MSG* 12246; Balugram (KNL), *JLE* 32598 (CLA); Chelama (KNL), *JLE* 22005 (CAL & MH); Nerabylu (CTR), *GVS* 46039 (MH); Kadapa, *EKK* 78514 (DD); Mettakota (WG), *DCSR* 386 (CAL); way to Tadepalli (EG), *NRR* & *DN* 84302 (BSID); Adativalsa (VSKP), *GVS* 21588 (CAL); Salur (VJN), *NPBK* 953 (CAL); Sirumalai hills (DGL), *SKS* 517 (SGH); Vandalur (CPT), *S.India Flora* 11117 (MH); Tippukadu RF (NA), *KRM* 17931 (MH); Chatkei beat, Purunakotee section, Satkosia Tiger Reserve, *KCM* 5253 (BSID); Chahala, Similipal Biosphere Reserve (MBJ), *ARKS* & *GPS* 12263 (BSID); Belghar, Kotgir Wildlife Sanctuary (KDM), *AC* 2407 (BSID); Badrama Wildlife Sanctuary (SBP), *AC* 10245 (BSID); Jenabil range, Similipal Biosphere Reserve (MBJ), *AC* 2341 (BSID); Charunaroya Hill slopes (GJM), *VNS* 4668 (MH).

INDIA: Throughout India.

WORLD: Pantropic.

HEMARTHRIA R. Brown

Hemarthria compressa (L.f.) R. Br., Prodr. 207. 1810; Fischer 3: 1762. 1934; Bor, Grass. India 161. 1960. *Rottboellia compressa* L. f., Suppl. Pl. 114. 1781; FBI 7: 153. 1896.

A perennial, tufted grass; culms up to 1.5 m high, decumbent from rooting base, leafy. Leaf sheaths below the internodes compressed, loose, glabrous; ligule membranous, ciliate; blades 3-15 x 3-5 cm, linear, flat, base rounded, margin scaberulous, apex acuminate, glabrous. Racemes few-many, 15 cm long. *Sessile spikelets:* 5 mm, lower glume coriaceous, 7-9-nerved; upper glume membranous, linear-oblong, 3-nerved; lemmas subequal, lanceolate, nerveless, hyaline; palea linear. *Pedicelled spikelet;* 6 mm, similar to sessile spikelet.

Occasional in wet locations and near water courses. Fl. & Fr.: July-November. Vern.: Tel.: *Shervu panuku.*

Tummalapenta (NLR), *JSG* 12788 (DD); Kankipadu (KSN), *CAB* 8228 (MH); Jenabil Range, Similipal Biosphere Reserve (MBJ), *AC* 2316 (BSID).

INDIA: Throughout the hotter parts of India.

WORLD: Sri Lanka, Myanmar, Malaysia.

HEMISORGHUM Clayton

Hemisorghum venustum (Thw.) Clayton, Kew Bull. 27: 448 1972. *Andropogon venustus* Thw., Enum. Pl. Zeyl. 367 1864. *Bothriochloa venusta* (Thw.) A. Camus, Ann. Soc. Linn. Lyon n.s., 76: 165 1930 (1931). *Capillipedium venustum* (Thw.) Bor, Grass. India 113. 1960. *Sorghum venustum* (Thw.) Kuntze, Revis. Gen. Pl. 2: 792 1891. *Vetiveria venusta* (Thw.) Willis, Ann. Roy. Bot. Gard. (Peradeniya) 5: 88 1911.

Perennial grass, caespitose, clothed with basally aggregated pubescent sheaths; culms up to 85 cm high; internodes solid, terete. Leaf-blade linear-lanceolate, up to 65 x 2 cm, upper truncate at base, lower narrowed, convolute, pale green, scaberulous on ventral surface, dorsal surface dark green, glabrous with prominent *c.* 2 mm wide midrib, acuminate at apex; ligule a fringe of hairs, up to 6 mm long, white. Panicle elliptic-lanceolate to ovate, up to 13 x 5 cm, open, dull purplish green in colour; primary panicle branches whorled at most nodes, up to 6 cm long, hairy at base, hairs white, up to 2 mm long. Racemes up to 2 cm long; joints of rachis linear, sub-angular, *c.* 3.5 mm long, ciliate below middle on both sides, discoid at apex. Spikelets two, one sessile, hermaphrodite, the other pediceled, staminate. *Sessile spikelets*: linear-lanceolate, *c.* 5 mm long, awned, yellowish green to dull purplish in colour, lower floret empty, upper floret bisexual. Lower glume linear-lanceolate to lanceolate, *c.* 5 x 1 mm, chartaceous to subcoriaceous, 2-keeled, glabrous, slightly convex, nerveless, keels scabrid above the middle, with two short cusps at hyaline apex, margins incurved below middle. Upper glume elliptic, boat-shaped in profile, *c.* 5 x 1 mm, cuspidate at apex, chartaceous, 1-nerved, 1- keeled, keel scabrid, margins incurved, ciliate. Lower floret epaleate, barren; lower lemma lanceolate, *c.* 3 x 1 mm, subhyaline, 2-keeled, 2- nerved, margin infolded along the nerves, obtuse, lacerate at apex. Upper floret epaleate, bisexual; upper lemma oblong, *c.* 3 x 0.5 mm, subhyaline, lobed at apex, 1 to 3-nerved, lobes acute at apex, awned in sinus; awn *c.* 11 mm long, twisted, geniculate, scabrid, with *c.* 4 mm long column, golden brown in colour. Lodicules 2, denticulate at apex, *c.* 0.5 x 0.5 mm, fleshy, glabrous; stamens 3; anthers *c.* 2 mm long, yellow; ovary elliptic, *c.* 1 mm long; style *c.* 0.5 mm long; stigma plumose, *c.* 1.5 mm long. *Pedicelled spikelets*: elliptic, unawned, longer than the spssile spikelets; pedicel *c.* 3 mm long, linear, subangular, discoid at apex, hairy below the middle on both sides, hairs *c.* 1.5 mm long, white; lower glume elliptic to elliptic-lanceolate, *c.* 6 x 1.5 mm, hairy at base, chartaceous, 7- nerved, 3-towards the margin on each side, 2- keeled, scabrid along keels, acuminate at apex. Upper glume elliptic-lanceolate, *c.* 5 x 1.5 mm, chartaceous, 5-nerved, 2-keeled, margins hyaline, ciliate above the middle, acuminate-aristate at apex. Lower floret epaleate, barren; lower lemma oblong-lanceolate, *c.* 4 x 1 mm, hyaline, delicate, margins winged, wing infolded, ciliate. Upper floret epaleate, staminate: Upper lemma elliptic-lanceolate, *c.* 5 x 1.5 mm, chartaceous, 3-nerved, 2-keeled, winged on margins above middle, acuminate at apex; wings hyaline, delicate, ciliate, acuminate. Stamens 3; anthers *c.* 2 mm long, yellow.

Rare in Odisha. Fl. & Fr.: January–October.

Chahala–Daldali, Similipal Biosphere Reserve (MBJ), *ARKS* & *GPS* 12215, 12237 (BSID).

INDIA: Karnataka, Maharashtra, Odisha.

WORLD: Sri Lanka, Myanmar.

Note: Clayton (1972) while placing the species under *Hemisorghum* observed that in *Sorghum* ciliate nature is seen in pedicels, internodes and lodicules whereas these characters are absent in the former genus. The authentic specimen of this species for India collected from Kodachardi hills of south Canara of Karnataka (Bhat and Nagendran, 2001). Later it was recorded from Maharastra of northern Western Ghats by Gad and Janarthanam (2007). Maharashtra specimens observed with the pedicels and intemodes ciliate in the upper half and lodicules are glabrous, thus, indicating that these characters are variable and further studies are required to ascertain its taxonomic position (Gad and Janarthanam, 2007).

HETEROPOGON Persoon

1. Annual grass.. **H. melanocarpus**
1. Perennial grass:
    2. Spikelets pubescent.. **H. contortus**
    2. Spikelets glabrous .... **H. fischerianus**

Heteropogon contortus (L.) P. Beauv. ex Roem. & Schult., Syst. Veg. 2: 836. 1817; Fischer 3: 1743. 1934; Bor, Grass. India 163. 1960; Matthew, Fl. Tamilnadu Carnatic 3: 1862. 1983. *Andropogon contortus* L., Sp. Pl. 1045. 1753; FBI 7: 199. 1896.

Key to Varieties

1. Culms up to 60 cm high.. var. **contortus**
1. Culms 10-15 cm high.. var. **polystachyos**

Heteropogon contortus (L.) Beauv. ex Roem. & Schult. var. **contortus**.

An perennial, tufted grass; culms up to 60 cm high, erect or decumbent below; nodes glabrous. Leaf sheaths 5-10 cm, glabrous; ligule membranous, fimbriate; blades 4-10 x 0.2-0.4 cm, lanceolate, flat, base rounded, apex acute, scaberulous. Racemes solitary, 6 cm, awns forming a twisted spire; lower spikelets homogamous, unawned, either neuter or male; upper spikelets heterogamous, awned, 2-nate, sessile bisexual, pedicelled male or barren; pedicel glabrous; callus pungent, bearded with brown hairs. *Sessile spikelets:* 7 mm, linear, hispidulous, female; callus bearded with brown hairs; lower glume linear-oblong, hispidulous; upper glume linear, obtuse, margin hyaline, dark brown, hispidulous; lower lemma oblong, truncate, epaleate; upper lemma linear, stipitiform, awned, epaleate; stamens 3; stigma plumose. Caryopsis linear. *Pedicelled spikelet:* oblong, hispidulous; lower glume lanceolate, dorsally hispidulous with long bulbous based hairs, winged, serrulate; upper glume oblong-lanceolate, margin hyaline; lemmas hyaline, nerveless, hairy.

Very common in plains and waste lands throughout Eastern Ghats. Fl. & Fr.:July - November. Tel.: *Kaserigaddi, Eddigaddi.* Tam.: *Oosipul.*

Tadipatri-Bugga (ATP), *TP* 436; Dhone RF (KNL), *RVR* 1626; Gundlabrahmeswaram (KNL), *PVP & RVR* 2374 & 2370; Palakonda hills (KDP), *CS* 8229; Guvvalacheruvu (KDP), *PVP* 3414, *KS* 6838 (MH); Tirumala hills (CTR), *MSG* 12138 & 12145; Nagarjuna konda (GNT), *VRK* 6709, *J.Swamy* & *S.Nagaraju* 7237 (BSID); Chelama (KNL), *JLE* 22036 (MH); Balapalle (KDP), *JSG* 11167 (MH); Kondagorulapenta (GNT), *MPG* 124 (DD) & *PCN* 121 (CAL); Kondapalli (KSN), sine coll. s.n. (MH); North Bison hills, Sirivaka (EG), *BO* 3394 (CAL); Balabuthrapuram, Godavari delta (EG), *BO* 3227 (CAL); Adativalasa (VSKP), *GVS* 21605 (CAL); Kothuru (VSKP), *NPBK* 867 (CAL); Salur (VJN), *NPBK* 1138 (CAL); Umamaheswaram top (MBNR), *SRS* 111327 (BSID); Sirumalai hills (DGL), *SKS* 421 (SGH); Guthirayan RF (SLM), *CAB* 6037 (MH); Ghat Road, Pennagaram RF (SLM), *EV* 22413 (MH); Kollimalai (Trichy), *S.India Flora* 11371 (MH); Tirumani (CPT), *S. India Flora* 14053 (MH); Gingee to Chittupattu (SA), *KRM* 60304 (MH); Tippukadu RF (NA), *KRM* 17698 (MH); Kambukudi R.H. (NA), *KS* 6506 (MH); Tulka section, Satkosia Tiger Reserve (AGL), *KCM* 5697 (BSID); Raigada, Satkosia Tiger Reserve (AGL), *KC* 6536 (BSID); Jenabil Rabge, Similipal B.R. (MBJ), *AC* 2345 (BSID); Mandargiri hill (AGL), *AC* 2366 (BSID); Chahahla–Daldali, Similipal, B.R. (MBJ), *ARKS* & *GPS* 12240 (BSID); Badrama Wildlife Sanctuary (SBP), *AC* 10253 (BSID); Tickapalli to Linepada (GJM), *CAB* 1213 (MH).

INDIA: Throughout India.

WORLD: Tropical and warm temperate regions.

Heteropogon contortus (L.) P. Beauv. ex Roem. & Schult. var. **polystachyos** (Roxb.) Deshpande, Bull. Bot. Surv. India 30: 121. 1990. *Andropogon polystachyos* Roxb., Fl. Ind. ed. Carey 1: 261. 1832.

Perennial, branched grass, culms slender, 10-15 cm high. Leaves 2.5-7 cm long, flat, glabrous or puberulous, ciliate at base. Racemes 1.2-6 cm long; lower spikelets sessile; fertile spikelets rufous hairy on one side; upper glumes not channeled; lemmas hyaline. Caryopsis sublinear, terete.

Very rare, in hills of dry deciduous forests. Fl. & Fr.: August–November.

It is known from a single collection only from Mogilikuppa in Kadapa district (C.S.Reddy *et al.*, 2001).

INDIA: Central India.

WORLD: Tropical Asia.

Heteropon fischerianus Bor, Kew Bull. 1951: 170. 1951 & in Grass. India 165. 1960; Moulik, Grass. Bamb. India 1: 227. 1997. *H. contortus* (L.) P. Beauv. ex Roem. & Schult. var. *distichus* Fischer,Fl. Madras 3: 1743. 1934.

Perennials; culms erect, tufted, to 80 cm high; nodes glabrous. Leaf sheaths 3-4.5 x 0.2 cm, glabrous; ligules ciliate rim, 1 mm; leaf blades 8-13 x 0.2 cm, slightly hairy at the base, base rounded, margin terete, distichous. Racemes solitary, 2-5 cm, glabrous; spikelets paired, sessile and pedicelled, lower pair homogamous, upper heterogamous; lower sessile spikelet 5-5.6 x 1.5 mm, male; lower glume 5.4 x 1.5 mm, crustaceous, glabrous, winged, 9-nerved; upper glume 5-5.6 x 1.5 mm, glabrous, membranous, 3-nerved, winged; lower lemmas 5 x 1 mm, hairy; upper lemma 4.5 x 1mm, hairy; stamens 3, anthers 1.5 mm; lodicules 2, 0.5 mm, truncate; pedicelled spikelet 7 x 1.5 mm, pedicel 1.5 mm, male; lower glume 5.2-5.5 x 1.5 mm, crustaceous, 9-nerved, winged one side; upper glume 5.6-5.8 x 1.8 mm, membranous, 3-nerved, winged upper side; lower lemma 3.8 x 1 mm, hyaline; stamens 3; anthers 1 mm; lodicules 2, 0.3 mm, truncate; upper sessile spikelet, 8-8.5 x 1 mm, bisexual, awned; lower glume 6.2-6.5 x 1 mm, brown, hispid, oblong, crustaceous; upper glume 6-6.2 x 1 mm, centre coriaceous; lemma reduced to a hyaline base of the awn, awn 5.7 cm; lodicules 2, truncate.

Occasional on slopes and among rocks in southern Eastern Ghats, (A.M.Reddy *et al.*, 2009). Fl. & Fr.: July–December.

Batrepalli (ATP), *KRKS* 40333; Bhumaiahgaripalli (KDP), *BSS & KRKS* 37610; Pecheruvu (KNL), *BR* & *AMR* 24505; Samrajyamkonda (ATP), *KVS* & *MSB* 29196, *BR* & *P.Priyadarsini* 37354 (SKU).

INDIA: Andhra Pradesh and Tamil Nadu, Endemic.

Note: The taxon differs from its allied taxon *H. contortus* (L.) P. Beauv. ex Roem. & Schult. in having glabrous spkelets.

Heteropogon melanocarpus (Elliott) Benth., J. Linn. Soc. Bot. 19: 71. 1881; Mooney, Suppl. Bot. Bihar & Orissa 191. 1950; Bor, Grass. India 165. 1960. *Andropogon melanocarpus* Elliott, Sketch. Bot. S. Carolna 1: 146. 1816.

Tall annual grass, up to 2.1 m high, culms 6.2 mm thick below; fastigiately branched above. Leaves linear or linear-lanceolate, 30-60 x 0.6-1.2 cm, setaceously acuminate, scabrid above, margins glabrous, scaberulous or with long hairs at the base; sheaths compressed, glabrous; ligule rounded, glabrous. Racemes 1.2-2.5 cm long (excluding awns); spathe 7.5-12.5 cm long; lower 1-3 pairs of spikelets male or neuter with glabrous joints, upper with bearded fragile joints. Female or bisexual spikelets sessile, 7.5-15 mm long, callus bearded; lower glume coriaceous, linear-oblong, 3-nerved, keel hispid; lemma shorter, oblong, nerveless, awns stout, column hirsute. Pedicelled spikelets 7.5-12.5 mm long, obliquely lanceolate, acuminate; pedicel very short; lower glume many-nerved, glabrous, keels ciliate, serrulate, with a row of glands down the centre; upper glume lanceolate, acute; lemma much shorter, nerveless, ciliate; palea minute, linear.

Taprang, Kalahandi, not common in grassy glades in sal forest (Mooney *loc.cit.*).

INDIA: Central India and Upper Gangetic Plain.

WORLD: Probably native to the Eastern Hemisphere, but is now found in tropical regions throughout the world.

HYGRORYZA Nees

Hygroryza aristata (Retz.) Nees ex Wight & Arn., Edinburgh New Philos. J. 15: 380. 1833; FBI 7: 95. 1896; Fischer 3: 1846. 1934; Bor, Grass. India 597. 1960. *Pharus aristatus* Retz., Obs. Bot. 5: 23. 1789.

Glabrous, floating, aquatic grass, culms 30 cm or more long, spongy below, emitting capillary roots at the nodes. Leaves ovate, ovate-lanceolate or oblong, 1.8-7.5 x 0.5-1.75 cm, subcoriaceous, base rounded or subcordate, apex obtuse, rounded or rarely acute, usually shortly petioled; sheaths much inflated, sometimes pubescent, mouth auricled; ligule 0. Panicle 2.5-7.5 cm long, with few spreading branches; spikelets few, 7.5 mm long (without awn); lemma narrowly lanceolate, 5.75-7.5 mm long, strongly 5-nerved, awn 6.2-13.7 cm long, scabrid; palea 6.2-7.5 mm long, subaristate-acuminate, with white margins.

Common in tanks, lakes and slow moving streams. Fl.: October.

INDIA: Upper and Lower Gangetic plains Assam, Bengal, Bihar, Odisha, Deccan.

WORLD: SE Asia.

HYMENACHNE P. Beauvois

Hymenache amplexicaulis (Rudge) Nees, Fl. Bras. Enum. Pl. 2: 276. 1829. *Panicum amplexicaule* Rudge, Pl. Guian 21. 1805. *Hymenachne acutigluma* (Steud.) Gilliland, Gard. Bull. Singapore 20: 314. 1963; Saxena & Brahmam, Fl. Orissa 4: 2390. 1996. *Panicum acutiglumum* Steud., Syn. Pl. Gram. 66. 1854. *Hymenachne pseudointerrupta* C. Muell., Bot. Zietung (Berlin) 19: 333. 1861; Fischer 3: 1784. 1934; Bor, Grass. India 313. 1960; Matthew, Fl. Tamilnadu Carnatic 3: 1863. 1983. *Panicum myuros* sensu Hook. f., FBI 7: 39. 1896, non Lam. 1791.

An annual, tufted grass; culms up to 1 m high, root stock spongy, prostrate, leafy. Leaf sheaths 15 cm long, glabrous, ligule membranous; blades 16-25 x 1-2 cm, linear, flat, base cordate, margins serrulate, apex acuminate. Panicle 35 cm long, dense, contracted; branches appressed to the rachis, winged, scabrid; spikelets 6 mm, oblong, secund, grouped; glumes aristate; lower glume broadly ovate, keels ciliate; upper glume narrowly lanceolate, aristate, ciliate; lower lemma longer than upper glume, lanceolate, awned, ciliate on nerves; upper lemma small, linear; palea having bisexual floret, 2-keeled; stamens 3; styles distinct. Caryopsis ellipsoid-obovoid.

On sands and near water courses of Northern Eastern Ghats. Fl. & Fr.: September-January

D. Mallapuram, Rampa hills (EG), *TP & MSG* 12281 (SKU).

INDIA: Assam, West Bengal, Uttar Pradesh, Odisha, Andhra Pradesh.

WORLD: Nepal, Bangladesh, Myanmar, Thailand, Indo-China, China, Malesia, Polynesia.

Note: It looks like *Sacciolepis interruptea* in the field, because of its interrupted, spiciform panicles and creeping or trailing, spongy culms. But it can be easily distinguished from *Sacciolepis* by the rigid culms, lanceolate, acuminate spikelets, glumes and lemmas acuminate or aristate.

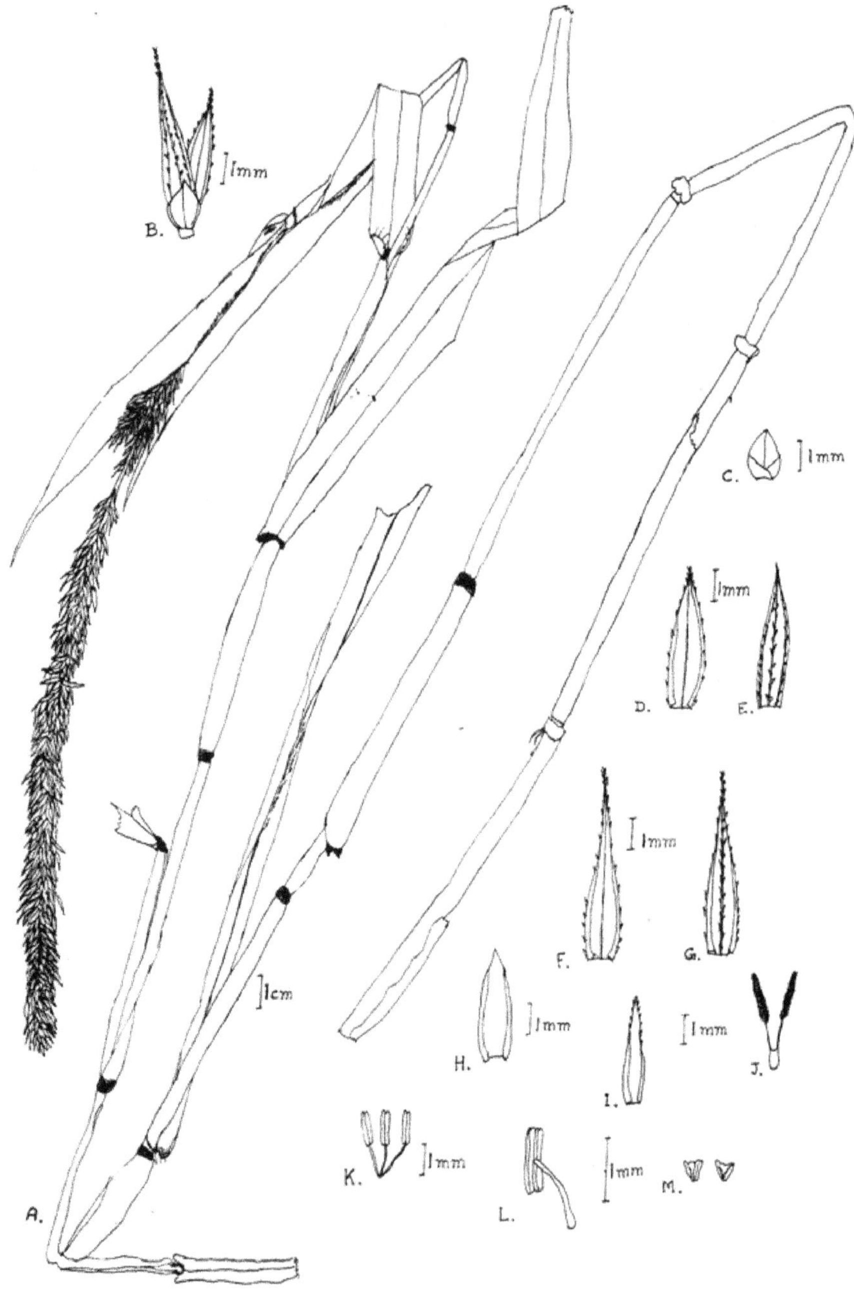

Figure 59. **Hymenachne amplexicaulis** (Rudge) Nees.
(Syn.: *Hymenachne pseudointerrupta* P.Beauv.).

A. Twig, B. Spikelet, C. Lower glume, D & E. Upper glume, F & G. Lower lemma, H. Upper lemma, I. Palea, J. Pistil, K & L. Stamens front and back views, M. Lodicules.

ICHNANTHUS P. Beauvois

Ichnanthus pallens (Sw.) Munro ex Benth., Fl. Hongk. 414. 1861. var. **majus** (Nees) Stieber, Syst. Bot. 12: 207. 1847. *Panicum pallens* Sw., Prodr. Veg. Ind. Occ. 23. 1788. var. *majus* Nees, Fl. Bras. Enum. Pl. 2: 138. 1829. *Ichananthus vicinus* (Bailey) Merr., Philipp. J. Sci. 20: 367. 1922; Bor, Grass. India 314. 1960. *Panicum vicinum* Bailey, Syn. Queensl. Fl. Suppl. 3: 82. 1890. *Ichananthus pallens* sensu Hook. f., FBI 7: 60. 1896; non Munro 1861.

A perennial, tufted grass; culms up to 40 cm high, branched and creeping in the basal parts, often with long and very slender aerial roots, terete, in groove above beds with row of hairs, otherwise glabrous. Leaf sheaths softly hairy along outer margin; ligule truncate, apex acute, glabrous or softly hairy below. Panicle 2.5-15 cm long, terminal, usually solitary or binate in 1-3 upper leaf axils, rather long-peduncled, erect or nodding; primary branches of obliquely erect and often secund, long hairy at base; spikelets 4-5 mm; pedicels 1-3 mm long; glumes ovate-oblong, acute or subulate-acuminate, thinly membranous, margins long hairy or glabrous; lower 3-nerved; upper as long as the spikelets, straight-backed upwards; lower lemma boat-shaped, 7- nerved; palea shorter. Caryopsis oblong.

On the hills and shady humid places of Northern Eastern Ghats. Fl.& Fr.: October–January.

Way to Chintapalli (VSKP), *JLE* 37135 (MH); Way to Sileru, Chintapalli (VSKP), *JLE* 37167 (MH); Bhalupodar vlley, Pottangi (KRT), *HFM* 3871 (DD).

INDIA: Peninsular India, Plains and hills of N.E.India, West Bengal, Odisha.

WORLD: Polynesia to the Philippines and Queensland in Australia.

IMPERATA Cyrillo

Imperata cylindrica (L.) Raeusch, Nom. Bot. ed. 3: 10. 1797; var. **cylindrica.** *Lagurus cylindricus* L.,Syst. Nat. ed. 10. 878. 1759. *Imperata arundinacea* Cyr., Pl. Rar. Neep. 2: 26, t. 11. 1792; FBI 7: 160. 1896. *I. cylindrica* var. *major* (Nees) C.E. Hubb. ex C. E. Hubb. & Vaughan Grass Maur. 96. 1940; Matthew, Fl. Tamilnadu Carnatic 3: 1864. 1983. *I. cylindrica* var. *koenigii* (Retz.) T. A. Durands Sching, Consp. Fl. Afric. 5: 694; Fischer 3: 1707. 1934.

A perennial, tufted grass; culms up to 70 cm high; nodes bearded. Leaf sheaths overlapping, glabrous, ciliate near mouth; ligule membranous; blades mostly basal, 6-20 x 0.2-0.5 cm, lanceolate, base rounded, apex acuminate, glabrous. Spike like panicle, up to 15 cm long, very dense; callus bearded; spikelets 3 mm, 2-nate, all alike, unawned, covered with long silky hairs; lower glume lanceolate, hyaline, long pilose outside, apex bidentate; upper glume lanceolate, hyaline, apex fringed; lower lemma oblong, margin scaberulous, hyaline; upper lemma lanceolate, margin scaberulous, hyaline; palea truncate, inrolled; stamens 2-3, stigmas plumose. Caryopsis oblong.

Common in hedges of field bunds and waste lands. Fl. & Fr.: July-November. Vem.: *Dharbagaddi.* Tam.: *Tharbaipul.*

Gani RF (KNL), *RVR* 1535; Orampadu (KDP), *PVP* 3484; Tirumala hills (CTR), *MSG* 12178; B.N.Kandriga (CTR), *BR* & *P.Priyadarsini* 37332 (SKU &

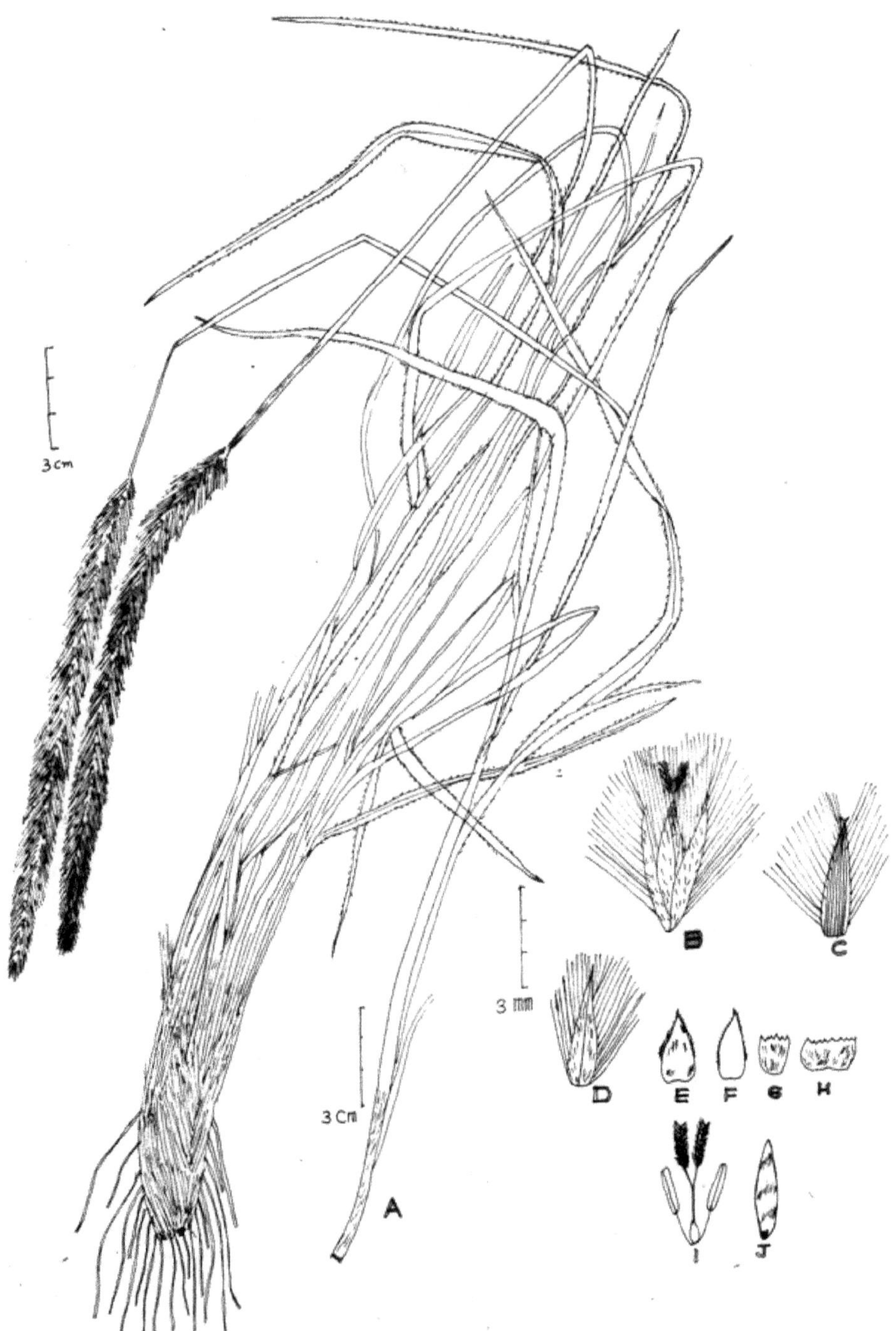

Figure 60. **Imperata cylindrica** (L.) Raeusch.

A. Twig, B. Spikelet, C. Lower glume, D. Upper glume, E. Lower lemma, F. Upper lemma, G. Palea external view, H. Palea spread open, I. Pistil and stamens, J. Caryopsis.

BSID); Madhuravada tank (VSKP), *MHR* & *MCK* 14958; Krishnanandi (KNL), *VBH* 86693 (BSID); Ramanapenta, Nallamalais (KNL), *S.India Flora* 14982 (MH); Gundlabrahmeswaram (KNL), *JLE* 16916, 16922 (MH); Balapalle (KDP), *JLE* 14909 (MH); Ramathirtham (NLR), *SCA* 151 (CAL); Kutravada (EG), *GVS* 27284 (CAL, BSID); Rajavommangi (EG), *S.India Flora* 12659 (MH); Anantagiri (VSKP), *GVS* 19488 (MH); Sirumalai hills (DGL), *SKS* 422 (SGH).

INDIA: Throughout India.

WORLD: Mediterranean region, extending to Iraq, Iran, Afghanistan and also in Chile.

ISACHNE R. Brown

1. Spiklets not more than 1.5 mm long .... **I. pulchella**
1. Spikelets over 1.5 mm long:
   2. Florets equal in size or nearly so and of same texture .... **I. albens**
   2. Florets unequal in size and of different texture, the lower more delicate:
      3. Culm nodes glabrous; ligule a fimbriate rim:
         4. Lower glume 7-nerved .... **I. miliacea**
         4. Lower glume 11-nerved .... **I. dispar**
      3. Culm nodes sparsely villous; ligule a tuft of long white hairs:
         5. Glumes 7-nerved; caryopsis ellipsoid .... **I. globosa**
         5. Glumes 7-9-nerved; caryopsis ovoid .... **I. kunthiana**

Isachne albens Trin., Sp. Gram. 1: t 85. 1828; FBI 7: 22. 1897; Bor, Grass. India 579. 1960; Moulik, Grass. Bamb. India 542. 1997 var. **albens**

Perennials, culms erect or ascending, to 130 cm, sometimes rooting at nodes, nodes glabrous. Leaf sheath 3-8 cm, smooth, loose, striate; ligule a ridge of stiff hairs, 1.5 mm; leaf blades 4-15 x 0.5-1.2 cm, linear-lanceolate, stiff, margins cartilaginous, acute-acuminate, rough on both sides. Panicles decompound, 7-30 x 10-15 cm, racemes rounded with many spikelets, rhachis trigonous; pedicels eglandular; spikelets solitary, globose, 1.4-1.5 x 1.4 mm, florets 2, equal in size and same in texture; lower glume orbicular, glabrous, obovate, 1.5 x 1 mm, 7-nerved, chartaceous; upper glume ovate-obovate, chartaceous, 1.3 x 0.9 mm, 7-nerved; lemma orbicular, obtuse, pubescent on the dorsal surface, margins infolded, cartilaginous, 1 x 0.7 mm; palea elliptic-oblong, 1 x 0.6 mm, 2-keeled, 2-nerved, membranous; stamens 3, anthers 0.8 mm; ovary oblong, 0.7 mm; styles 0.5 mm, stigmas 0.7 mm; lodicules truncate, 0.3 mm. Caryopsis obovate.

Occasionally found in marshy areas in forests in Northern Eastern Ghats. Fl. & Fr.: August–December.

Galikonda (VSKP), *BR* & *AMR* 25645 (SKU).

INDIA: Temperate and Subtropical Himalaya, N.E.India.

WORLD: Widely distributed in SE Asia.

Isachne dispar Trin., Sp. Gram. Stip. 1: t.86. 1828; FBI 7: 26. 1896; Fischer 3: 1797. 1934; Bor, Grass. India 580. 1960; Matthew, Fl. Tamilnadu Carnatic 3: 1865. 1983.

Annual grass; culms to 30 cm, decumbent, glabrous at nodes. Leaf sheath 1.5-3.5 cm; ligule to 3 mm; blades lanceolate, 2.6 x 0.5-0.8 cm, flat, tuberculate-pilose along nerves. Panicles to 4 cm; pedicel 1-5 mm. Spikelets ovoid-globose, to 2 mm; florets unequal; glumes boat-shaped, scaberulous; lower glume to 2.5 mm, 11-nerved, acute; upper glume to 2 mm, 7-9-nerved; lower lemma to 2 mm, 5-nerved, glabrous, paleate; upper lemma to 1.8 mm, subcrustaceous, granular, ciliate; stamens 3. Caryopsis elliptic.

Occasional in marshy fields of southern Eastern Ghats. Fl. & Fr.: August–October.

Shervarayan hills (SM), *SKS* 587 (SGH); Graigmore estate–Yercaud (SLM), *S.Karthikayan* 26991 (MH).

INDIA: Throughout hotter parts.

WORLD: Myanmar, South-East Asia extends up to Australia.

Isachne globosa (Thunb.) Kuntze, Revis. Gen. Pl. 2: 478. 1891; Bor, Grass. India 580. 1960; Matthew, Fl. Tamilnadu Carnatic 3: 1866. 1983. *Milium globosum* Thunb., Fl. Jap. 49. 1784. *Isachne australis* R. Br., Prodr. 196. 1810; FBI 7: 24. 1896. *I. dispar auct. non* Trin. 1828; Fischer 3: 1797. 1934; Bor, Grass. India 580. *p.p.* 1960; FBI 7: 26. 1896. *p.p.*

An annual, tufted, marshy grass; culms up to 25 cm high, slender, often ascending from a root stock, leafy, branches below; nodes sparsely villous. Leaf sheaths 2-5 cm, margins densely ciliate; ligule a tuft of long white hairs, blades 2-5 x 0.2-0.3 cm, lanceolate, base rounded, scabrid on both surfaces. Panicle 8 cm long; rachis angular, smooth; spikelets 2 mm, globose, sessile or pedicelled; glumes similar, ovate, boat-shaped, membranous, scabrid, 7-nerved; lemmas unequal, lower ovate, obscurely nerved, glabrous; stamens 3; upper crustaceous, 7-nerved, ciliate; palea glabrous. Caryopsis ellipsoid.

Common in marshy areas, rice fields etc. Fl. & Fr.: May-December.

Maredumilli (EG), *GVS* 27404 (CAL & MH), *MM* 101322 (MH); Tadepalli (EG), *DN* 85547 (BSID); Near Anantagiri (VSKP), *NPBK* 11007 (CAL & MH); Khotnigedda area (VSKP), *GVS* 19520 (CAL); Barheipani, Similipahar (MBJ), *SX* & *MB* 5196 (RRL-B); Ektali (DK), *MB* & *Dhal* 8281 (RRL-B); Manaskonda (KRT), *SX* & *MB* 6805, 6814, 6771 (RRL-B); Lasari, Kotgarh Wild Life Sanctuary (KDM), *AC* 2430 (BSID); Jenabil Reange, Similipal BR (MBJ), *AC* 482 (BSID).

INDIA: Hotter parts of India.

WORLD: Sri Lanka, Myanmar, China, Japan, Malesia, Australia, New Zealand.

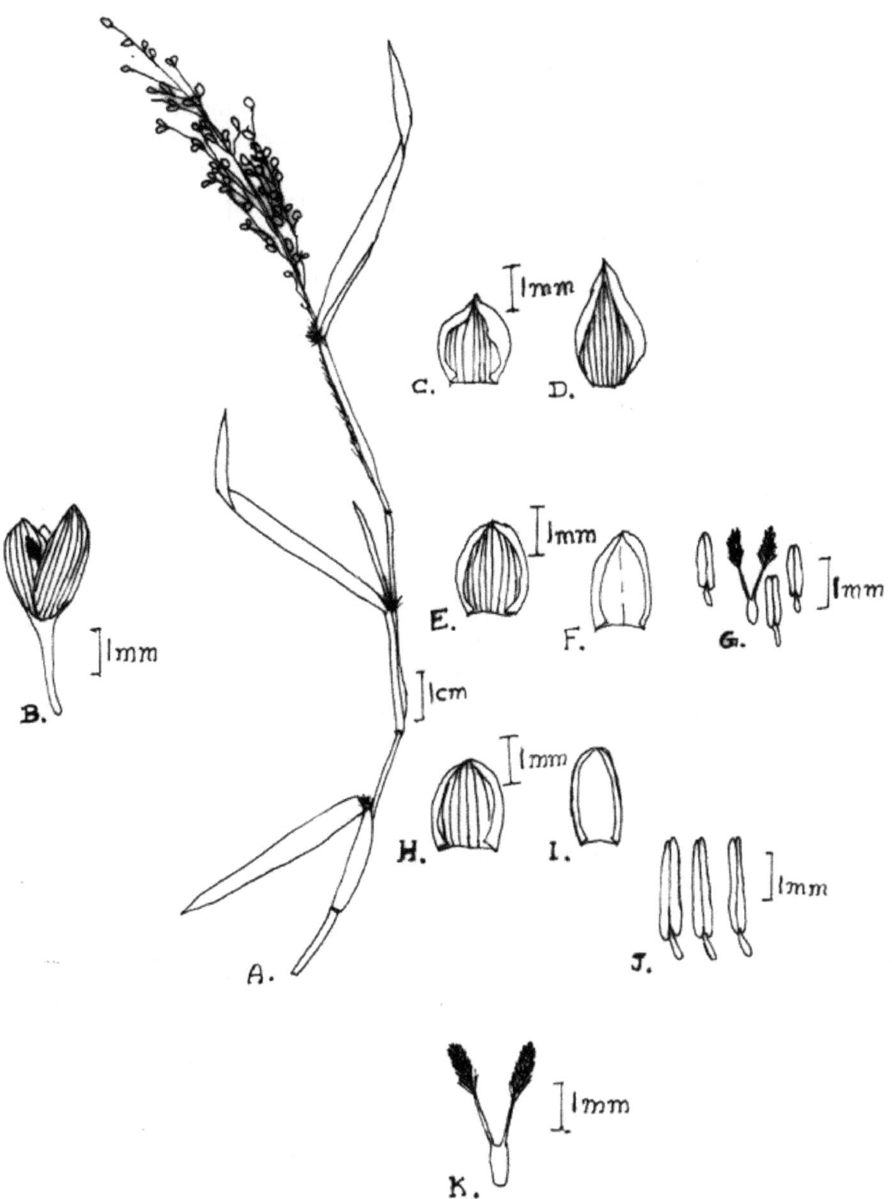

Figure 61. **Isachne globosa** (Thunb.) Kuntze

A. Twig, B. Spikelet, C. Lower glume, D. Upper glume, E. Lower lemma, F. Palea, G. Pistil with stamens, H. Upper lemma, I. Palea, J. Stamens, K. Pistil.

Isachne kunthiana (Steud.) Miq., Fl. Ned. Ind. 3: 460. 1857; FBI 7: 21. 1896; Fischer 3: 1796. 1934; Bor, Grass. India 581. 1960; Matthew, Fl. Tamilnadu Carnatic 3: 1866. 1983. *Panicum kunthianum* Wight & Arn. ex Steud., Syn. Pl. Glum. 1: 96. 1854.

Annual grass; culms to 50 cm, nodes bearded. Leaf sheath 1.5-4.5 cm, tuberculate-pilose; ligule to 1 mm; blades lanceolate, 1.5-5.5 x 0.8-1.5 cm, flat. Spikelets ovoid, to 3 mm; florets nearly equal, of the sme texture; glumes exceeding florets, setose in the upper half, lanceolate, cuspidate; lower glumes to 3 mm, 7-9-nerved; upper glume similar to lower; lower lemma to 2 mm, ciliate; upper lemma somewhat puberulous; stamens 3. Caryopsis ovoid.

Slopes of Shervarayan hills and Sirumalai hills at marshy places. Fl. & Fr.: Septermber–December.

Sirumalai hills (DGL), *SKS* 423 (SGH).

INDIA: Peninsula.

WORLD: Southeast Asia and Sri Lanka.

Isachne miliacea Roth, Nov.Pl. Sp. 58. 1821; FBI 7: 25. 1896; Fischer 3: 1797. 1934; Bor, Grass. India 582. 1960; Matthew, Fl. Tamilnadu Carnatic 3: 1867. 1983.

An annual tufted grass; culms up to 25 cm high, slender, prostrate, branching below; nodes glabrous. Leaf sheaths 1-4 cm; ligule a fimbriate rim; blades 2-5 x 0.2-0.3 cm, lanceolate, flat, base rounded, apex acute, sparsely scabrid. Panicle 5 cm long, small, many-flowered; spikelets 2 mm, globose, few, distant; glumes orbicular-oblong, membranous, 7-nerved, scaberulous; lower lemma ovate-oblong, membranous, obscurely 5-nerved, male; stamens 3; upper lemma small, ovate, 3-nerved, ciliate, coriaceous; palea female glabrous. Caryopsis ellipsoid.

Rare in swampy places. Fl. & Fr.: September–October.

INDIA: Widely distributed in India.

WORLD: SE Asia to China, Malesia.

Isachne pulchella Roth ex Roem. & Schult., Syst. Veg. 2: 476. 1817; Bor, Grass. India 582. 1960; Matthew, Fl. Tamilnadu Carnatic 3: 1867. 1983; Chorghe *et al.*, J. Econ Taxon. Bot. 39: 81. 2015.

An annual, tufted, marshy grass; culms up to 40 cm high; nodes pilose. Leaf sheaths 1-2 cm, tuberculale-pilose; ligule fimbriate membrane; blades 1.5-4 x 0.5-0.7 cm, ovate-lanceolate, flat. Panicle 6 cm long, contracted; branches spreading; pedicels glandular; spikelets 2 mm, obovoid, 2-flowered, unequal; glumes nearly equal to the florets, membranous, glabrous; lower lemma ovate-lanceolate, male; upper lemma similar but smaller than lower lemma, female; stamens 3. Caryopsis ellipsoid.

On the hills and marshy areas. Fl. & Fr.: August-December.

Similipal Biosphere Reserve, *AC* 2329 (BSID); Shervaroy hills (SLM), *KMM* 1795, 10189 (RHT, K).

INDIA: Tamil Nadu, Andhra Pradesh, Odisha, Maharashtra, Madhya Pradesh,West Bengal, Assam.

WORLD: Bangladessh, Nepal, China, Taiwan, Malaysia, Thailand, Vietnam.

ISCHAEMUM Linnaeus

1. Margins of the lower glume of the sessile spikelet expanded below the middle:
   2. Upper lemma of pedicelled spikelet awned.......................... **I. polystachyum**
   2. Upper lemma of pedicelled spikelet awnless...................................... **I. ciliare**
1. Margins of the lower glume of the sessile spikelet narrowly inflexed from base to apex:
   3. Leaf blade oblong, narrow into a petiole at base; lower glume of pedicelled spikelets very unequally winged .... **I. commutatum**
   3. Leaf blades linear, rounded, or sub cordate at base;
      4. Base of lower blades sub cordate or rounded, sessile;
         5. Perennial, rhizomatous grass; lower glume of the spikelets not rugose with transverse ridges; dorsal surface densely pilose or ciliate.. **I. afrum**
         5. Annual tufted grass; lower glume of the spikelets rugose with transverse ridges; dorsal surface sparsely ciliate.. **I. rugosum**
      4. Base of the lower blades sagittate or deeply cordate, with distant petiole:
         6. Lower glume of pedicelled spikelet very broadly winged on one margin... **I. fischeri**
         6. Lower glume of pedicelled spikelet without a very broad wing .... **I. semisagittatum**

Ischaemum afrum (J.F.Gmel.) Dandy, Fl. Pl. Sudan 3: 476. 1956. *Andropogon afer* J.F.Gmel., Syst. Nat. 166. 1791. *Ischaemum pilosum* (Klein ex Willd.) Wight, Madras Lit. Sci. J. 3: 138. 1835; FBI 7: 130 1896; Fischer 3: 1723. 1934; Bor, Grass. India 183. 1960; Pullaiah, Fl. Andhra Pradesh, 5: 2208. 2018. *Andropogon pilosus* Klein ex Willd. Sp. Pl. 4: 920. 1806.

A perennial, stoloniferous grass; culms up to 90 cm high, erect; nodes glabrous. Leaf sheaths 5-20 cm long, glabrous; ligule membranous; blades 7-15 x 0.4-0.7 cm, linear, base rounded, apex acuminate, veins prominent, glabrous. Racemes 6, up to 15 cm long, pilose; rachis hairy; joints and pedicels slender, compressed, sparsely ciliate; callus bearded with long hairs; spikelet 9 mm, 2-nate. *Sessile spikelet:* 6 mm long, narrowly lanceolate; glumes lanceolate, chartaceous, laterally compressed, dorsally hairy in lower; lower lemma linear-oblong, ciliate above; palea nerveless, male; upper lemma cleft to the middle into acute, ciliate lobes; palea lanceolate,

Figure 62. **Ischaemum afrum** (J.F.Gmel.) Dandy.
[Syn.: *Ischaemum pilosum* (Klein ex Willd.) Wight].

A. Twig, B. Sessile spikelet, C. Lower glume, D. Upper glume, E. Lemma, F. Palea, G. Anther, H. Pistil, I. Pedicelled spikelet, J. Joint.

nerveless; stamens 3, stigma plumose. Caryopsis lanceolate, *Pedicelled spikelet:* Similar to sessile spikelet; upper lemma contains short awn.

Common weed in black cotton soils Eastern Ghats. Fl. & Fr.: October–March. Tel.: *Kundaranattu.*

Tadipatri (ATP), *TP & NY* 1122; Perusomula (KNL), *TP & PVP* 3365; Ramallakota RF (KNL), *RVR* 1838; Chinnakudala (KDP), *KRKS* 40464; Kadapa, *EKK* 79525 (DD); Narasaraopet (GNT), *CAB* 4717 (MH); Mylavaram (KSN), *CAB* 8143 (MH); Toranagallu (Ballari), *sine coll. s.n.* (MH).

INDIA: Peninsular and Central India. Endemic.

Ischaemum ciliare Retz., Obs. Bot. 6. 36. 1791; FBI 7: 133. 1896; V.S.Raju & Kumar, J. Threat. Taxa 10(7): 11969. 2018. *Polytrias indica* auct non (Houtt.) Veldkamp 1991. *Ischaemum indicum* auct (Houtt.) Merr. 1938; Bor, Grass. India 180. 1960; Pullaiah, Fl. Andhra Pradessh 2206. 2018. *I. aristatum* auct. non. L., Fischer 3: 1721. 1934.

A perennial, tufted grass; culms up to 60 cm high, decumbent at base, branched above; nodes glabrous. Leaf sheaths 3-9 cm long, loose, glabrous auricled; ligule membranous; blades 3-12 x 0.3-0.5 cm, linear-lanceolate, base rounded, hairy, apex acuminate, midrib prominent. Racemes 2, 5 cm, stout or slender; spikelets 6 mm, 2-nate. *Sessile spikelet:* 4 mm, oblong; lower glume broadly oblong, 2-winged at apex, cuspidate, scabrid; upper glume lanceolate, hyaline, bidentate, slightly winged; lower lemma ovate-lanceolate, hyaline, margin scaberulous; palea lanceolate, involute, hairy at apex, upper lemma hyaline, bidentate, awned; stamens 3, stigmas plumose. Caryopsis oblong. *Pedicelled spikelet:* Similar to sessile spikelet; upper lemma awnless or with short mucro.

Common near the water sources. Fl.& Fr.: August-January.

Ramanapenta RF (KNL), *PVP & RVR* 2377; Sunnipenta (KNL), *BSS & SKB* 32982 (BSID): Tirumala hills (CTR), *MSG* 12147 & *PVP* 3495; Pamuleru, Rampa hills (EG), *TP & MSG* 12272; Pinnintiraminaidu valasa (EG), *PCN* 198 (CAL); Anantagiri (VSKP), *NPBK* 286 (CAL); Arakuvalley (VSKP), *NPBK* 6191 (CAL); Adativalasa (VSKP), *GVS* 21609 (CAL); Balugam near Parvathipuram (SKLM), *VNS* 4551 (MH); Yercaud, near Shervarayan Temple (SLM), *P.Venu s.n* (BSID); Balmadies to Kakasholai, Yercaud (SLM), *KS* 7564 (MH); Tulka section, Satkosia Tiger Reserve (AGL), *KCM* 6193 (BSID); Purnakote, Satkosia Wildlife Sanctuary (AGL), *AC* 2376 (BSID); Tulka section, Satkosia Tiger Reserve (AGL), *KCM* 5951 (BSID); Dimili, Kotgarh Wildlife Sanctuary (KDM), *AC* 2420 (BSID); Niyamgiri hill (KHD), *AC* 2474 (BSID).

INDIA: Throughout India.

WORLD: SE Asia.

Note: V.S.Raju and Kumar (2018) discussed in detail the nomenclatural problems of this taxon and its distribution. The following account is from their paper. *Phleum indicum* was described by Houttuyn (Nat. Hist. 2(13): 198, t.90. f.2) in 1782 of a collection from Java. Houttuyn's exotic herbaria were from the

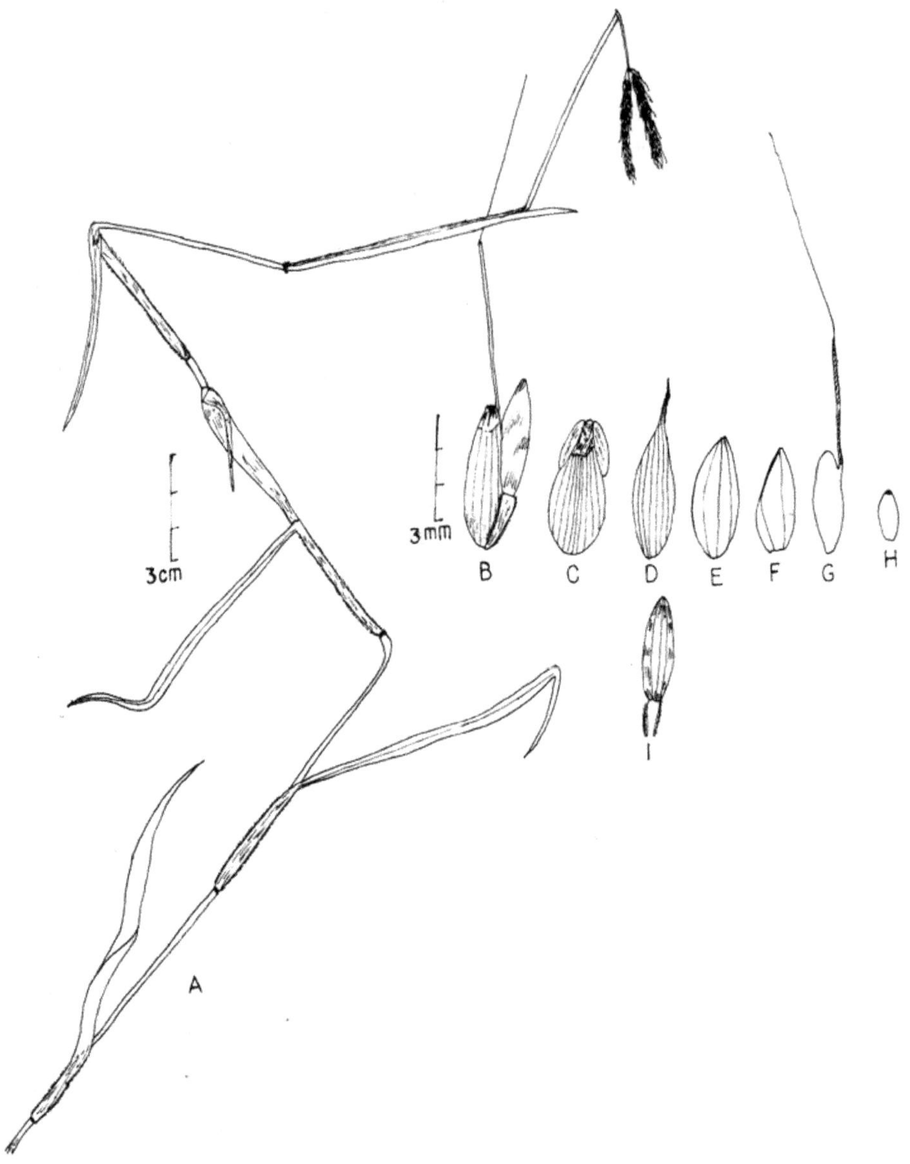

Figure 63. **Ischamum ciliare** Retz., (*Ischaemum indicum auct.* (Houtt.) Merr.)

A. Twig, B. Sessile spikelet, C. Lower glume, D. Upper glume, E. Lower lemma, F. Palea, G. Upper lemma, H. Grain, I. Pedicelled spikelet.

Cape of Good Hope (South Africa), Ceylon (Sri Lanka), Java (Indonesia), and Japan. Houttuyn did not go abroad to collect these plants (Wijnands *et al.*, 2017). The type of *Phleum indicum* was deposited in Herb. Houttuyn present in Geneva Herbarium (G), under Pre-Linnaean collections. Later, this name became the basionym for two different entities–*Ischaemum indicum* (Houtt.) Merr. (J. Arnold Arbor. 19: 320. 1938) and *Polytrias indica* (Houtt.) Veldkamp (Blumea 36: 180. 1991). Since Merrill (1938) conceived *Phleum indicum* Houtt. to be an *Ischaemum* and made the combination under the genus, it suppressed the name *I. ciliare* Retz. (1791). Hackel (1889) considered *I. ciliare* Retz. to be polymorphic and recognized in it three varieties and four subvarieties. The name "*Ischaemum indicum* sensu Merr." (Bor 1960), however, continues to be used in all the Indian floras and the recent generic accounts of *Ischaemum* (Sur 2001: 419; Srivastava and Nair 2010:75) in India, despite the fact that Veldkamp (1991) clearly established that the holotype (specimen) of *Phleum indicum* is not an *Ischaemum* but a *Polytrias* by virtue of its solitary racemes and ternate flowers bearing one pedicellateand two sessile spikelets. Therefore, the name *I.ciliare* Retz. has to be used in lieu of the replacement name "*Ishaemum indicum* (Houtt.) Merr.". One can also know this fact from the Merrill's (1938: 320) own making. Realizing Houttuyn's name, Merrill made the observation that Houttuyn's material was from Java and his species was clearly indicated as new (he used '*mihi*' for the binomials coined by himself - cf. Wijnands *et al.*, 2017: 158), following the short Latin diagnosis. To Merrill, Houttuyn's illustration is a reasonably good one for the grass currently known as *Ischaemum ciliare* Retz., which is common in Java, and more definitely Houttuyn's species seems to be the same as *Ischaemum ciliare* Retz. var. *genuinum* Hack. In fact, Merrill mentioned *I. ciliare* Retz. as a synonym under the combination. Above all, Veldkamp (1991: 180) actuality pleaded for the reinstatement *I. ciliare* Retz. for the *Ischaemum indicum* auct. non (Houtt.) Merr. though it was not taken into the cognizance by the subsequent workers on Indian grasses. When the 'www.theplantlist.org' mentions the name *Ischaemum indicum* (Houtt.) Merr. as a synonym of *Polytrias indica* (Houtt.) Veldkamp, many grass workers perhaps simply adopted that name. Efloraofindia (Google sites) is no exception in this regard. *Polytrias indica* (Houtt.) Veldkamp is the name of Java Lawn Grass misapplied to another grass from Java found very common in India or elsewhere which is to be rightly called *Ischaemum ciliare* Retz. Although Veldkamp (1991) clearly brought out the distinction of *Polytrias indica* of its ternate flowers, the confusion continues in Indian grass literature. It pertains to the subtribe Sacchrinae rather than Ischaeminae of Andropogoneae. The 2013 IUCN Red List assessment of *Ischaemum indicum* sensu Merr. (i.e., *I. ciliare* Retz.) was attributed to *Polytrias indica* as a result of misapplication of the name. Although *Polytrias indica* is stated to be invasive outside its native range, it is yet to be confirmed as such in India. Thus far, it is known from Meghalaya and West Bengal of India.

Ischaemum commutatum Hack. in DC., Monogr. Phan. 6: 209. 1889; FBI 7: 131. 1896; Fischer 3: 1722. 1934; Bor, Grass. India 178. 1960; Matthew, Fl. Tamilnadu Carnatic 3: 1868. 1983; Moulik, Grass. Bamb. India 1: 291. 1997; Reddy *et al.*, J. Econ. Taxon. Bot. 38: 22. 2014

Perennials, culms tufted, to 90 cm high, nodes villous. Leaf sheaths 9 cm, sparsely tuberculate hairy, keeled towards apex. Ligule 3 mm, membranous. Leaf blades narrowly oblong to elliptic-lanceolate, 5-15 x 0.3-0.8 cm, margins cartilaginous, scabrid, base cordate, ciliate on both surfaces. Racemes 2, paired, 3-7 cm, rachis cupiliform. Spikelets 2-nate, sessile and pedicelled, florets 2, lower male, upper hermaphrodite; sessile spikelet 7 x 1.5 mm, oblong, acute, crustaceous, callus bearded; lower glume 7 x 1.5 mm, oblong, acute, 11-13-nerved, 2-keeled, keels narrowly winged, wings finely serrulate; upper glume 7 x 1.5 mm, oblong to ovate-lanceolate, acuminate with an arista, subcoriaceous, faintly 3-5-nerved 2-keeled towards apex, ciliate on keels, margins slightly infolded, lower lemma 5.2 x 1.2 mm, ovate-lanceolate, acute, chartaceous, 3-nerved, margin infolded, ciliate, lower palea 5.2 x 1.2 mm, elliptic-lanceolate, acute, chartaceous, 2-nerved, 2-keeled, margin infolded; upper lemma 4.8 x 0.8 mm, ovate to elliptic-lanceolate, acute, deeply 2-lobed, awned at sinus, membranous, hyaline, nerveless; upper pales 3.8 x 0.6 mm, oblong, acute-obtuse, membranous, hyaline, nerveless, margin infolded, stamens 3, anthers 1 x 0.3 mm, filaments 2 mm, ovary 1 x 0.3 mm, styles 1.5 mm, stigmas 1.2 mm, plumose. Caryopsis 2.2 x 0.4 mm, oblong or oblanceolate. Pedicelled spikelet oblong-elliptic, 4.8 x 1.8 mm, acute, chartaceous; lower glume 4.6 x 1.8 mm, oblong-elliptic, acute, chartaceous, 9-11-nerved, 2-keeled, keels broadly winged on one side throughout and narrowly winged below middle on other side, wings serrulate towards apex, margins infolded, upper glume 4.5 x 1.2 mm, elliptic-lanceolate, acute, chartaceous,3-5-nerved, margin infolded, lower lemma 3.6 x 1.0 mm lanceolate, acute-obtuse, lower palea 3.2 x 0.8 mm lanceolate, acute-obtuse, membranous, hyaline, nerveless, upper lemma 4.0 x 0.8 mm, ovate-lanceolate, acute, membranous, hyaline, nerveless, upper palea 3.5 x 0.8 mm, elliptic, acute-obtuse, membranous, hyaline, nerveless, margin infolded. Caryopsis 2 x 0.4 mm, elliptic, acute.

In moist shady places like fringes of forest and banks of rivers, usually in hilly areas. Fl. & Fr.: June–April

Galikonda (VSKP), *BR* & *AMR* 25640; Shervarayan hills (SM), *SKS* 548 (SGH); Manjakuttai (SLM), *DBD* 31406 (MH).

INDIA: Maharashtra, Andhra Pradesh, Kerala and Tamil Nadu.

WORLD: Tropical Asia and Sri Lanka

Ischaemum fischeri Ravi & Kiran Raj, Bot. Bull. Acad. Sin. 42: 227. 2001. *I. rangacharianum* C.E.C. Fischer, Kew Bull. 1933, 1933: *p.p.* 352 & 353. *p.p. quoad specim.* (*Paratypus*) V. Narayanaswamy 1743 (CAL) & Fl. Madras 3:1820. 1934; Bor, Grass. India 516. 1960.

Perennial; culms creeping, rooting at nodes and mat-forming with the flowering culms up to 75 cm high; nodes sparsely ciliate, glabrous at length. Leaves all along the culm; sheath up to 8 cm long, keeled on the back, striate and very sparsely bulbous-pilose closer to the blade, glabrous otherwise; ligule subcoriaceous, triangular-ovate, 2.5-5 mm long, 2-keeled on the back and subacute to obtuse-truncate at the apex; blades elliptic to lanceolate, up to 15 x 1 cm, lower ones acuminate to tapering with the lowest ones tapering into a short or long 'pseudopetiole' at the base and acute to acuminate at the apex, upper ones cuneately rounded at the base towards the culm apex with the margins shortly undulate, prominently so on one side and scaberulous towards the apex and sparsely bulbous-pilose, glabrous at length and with a small tuft of hairs at the base behind the ligule. Inflorescence mostly solitary, sometimes 2, terminal on lateral, erect shoots; racemes 2, up to 10 cm long; joints of rhachis oblong-clavate, 3-5 mm long, triquetrous, tridentate at the apex and ciliate on the outer angle. *Sessile spikelets*: oblong to oblong-elliptic, 5-7 mm long including the 0.75-1 mm long, ciliate callus; lower glume oblong-elliptic, 4-6 x *c.* 1.25 mm, thickly coriaceous to subcrustaceous below and coriaceous above, laterally keeled and incurved on the margins with the keels thickened and 2 or 3 (or 4) nodulose in the lower 1/3-1/2 and laterally flattened and winged unequally above the middle with scrabrid margins, acute to subacute, sometimes apiculate and hyaline-tipped, often ultimately unevenly notched at the apex, 11-13-nerved, and minutely and sparsely scaberulous outside; upper glume coriaceous, lanceolate in profile, boat-shaped, 4.5-6.5 x *c.* 1 mm (when folded), keeled on the back and the keel very shortly and narrowly winged or not towards the acuminate to aristate apex, incurved on margins, 5-7-nerved and scaberulous outside in the upper half and also longitudinally along the middle on the sides; lower lemma hyaline, elliptic to elliptic-lanceolate, 3.5-5 mm long, acute at the apex, 3-nerved, incurved on the sides and sparsely short-pilose outside closer to the margins, glabrous otherwise; lower palea hyaline, elliptic to elliptic-lanceolate, 3.5-5 mm long, subacute to shortly obtuse at the apex, 2-nerved and incurved along the nerves with the nerves keeled and very narrowly winged; upper lemma hyaline, oblong to oblong-elliptic in profile, 3-4.5 mm long, deeply cleft to below the middle with acute to acuminate lobes and awned from the sinus; awn geniculate, up to 15 mm long with a column up to 6.5 mm long; upper palea hyaline, oblong-elliptic, 3.5-5 mm long, 2-nerved, incurved on the sides and subacute at the apex. *Pedicelled spikelets*: Pedicel 1-2 mm long, triquetrous and ciliate on the outer angle; lower glume obliquely oblong-elliptic, 4-6.5 x 1.5-2 mm, thickly coriaceous below and coriaceous above, laterally keeled and the keels broadly winged all along on one side with a few hairs on the margin at the extreme base and narrowly winged towards the apex and slightly thickened and somewhat rounded below and 1-3 nodulose or not towards the base on the otherside with the wings scabrid on the margins, subacute, sometimes apiculate and hyaline-tipped, often ultimately unequally notched at the apex, 11-13-nerved; upper glume, lower lemma, lower palea and upper palea similar to those of the sessile spikelet, sometimes slightly smaller; upper lemma hyaline, oblong to oblong-elliptic, 3-4 mm long, sometimes very reduced, mostly entire and acute to mucronate at the apex, sometimes shortly bifid with subacute to obtuse lobes and aristate or imperfectly awned or deeply cleft and geniculately awned; when awned, awn usually shorter than that of the sessile spikelet; lodicules 2, 0.5-0.75 mm long, cuneate at the base and truncately lobed at the apex; stamens 3; anthers 2.5-3 mm long, yellow turning to brown; ovary *c.* 0.5 mm long; styles *c.* 2 mm long; stigmas *c.* 2 mm long, glassy-white, turning to brown. Caryopsis ellipsoid to oblong-ellipsoid, 2-2.5 x 0.75-0.9 mm, pale white.

Rare in Rampa hills. Fl. & Fr.: October -April.

Godavari District, Rampa Agency, *VNS* 1743(CAL).

INDIA: Eastern Ghats, Endemic.

Note: Ravi *et al.* (2001) examined 2 of the specimens of *I. rangacharianum* quoted by Fischer in his original description and deposited at CAL. One of these carried the label "Shornur 23, Nov. 1900, Collector, K. Ranga Acharyar" and the other "Rampa country Godavari District., Sept. 1920, Collector, *V. Narayanaswamy* 1743". Both these specimens were determined as *I. rangacharianum* C.E.C Fischer by Fischer himself. As indicated in the protologue the first sheet is the type of *I. rangacharianum* and the other, one of its paratypes. On critical examination of the above 2 specimens, Ravi *et al.* (2001) became thoroughly convinced that they represented 2 distinct taxa, Ranga Acharyar's specimen *I. rangacharianum*, and *Narayanaswamy* 1743 a hitherto undescribed species, which they described as *Ischaemum fischeri* Ravi & Kiran Raj.

The differentiating characters of the two are tabulated below:

Ischaemum rangacharianum C.E.C.Fischer	*Ischaemum fischeri* Ravi & Kiran Raj
Annual	Perennial
Culms diffuse and more or less erect	Culms creeping, rooting at nodes and mat-forming
Lower leaves rounded to cordate at base with distinct pseudopetioles	Lower leaves acuminate at base, sometimes lowermost ones tapering into pseudopetioles
Peduncles 1-9, often fasciculate	Peduncles solitary or rarely 2
Lower glume of sessile spikelets 4-6-nodulose in the lower half with acutely ridged nodules separated by deeper furrows	Lower glume of sessile spikelets 2 or 3 (or 4)-nodulose with obtusely ridged nodules separated by shallower furrows.
Lower glume of sessile spikelets sparsely long-pilose outside in the upper half.	Lower glume of sessile spikelets never pilose, at the most sparsely scaberulous outside

Ischaemum polystachyum J.Presl., Reliq. Haenk 1: 328. 1830. *I. nilagiricum* Hack, Oesterr. Bot. Z. 51: 150. 1901; Fischer 3: 1722. 1934; Bor, Grass. India 183. 1960. *I. hirtum* sensu Hook. f., Fl. Brit. India 7: 135. 1896, non Hack. 1889; Saxena & Brahmam, Fl. Orissa 4: 2399. 1996.

A perennial, tufted, marshy grass; culms up to 60 cm high, robust; nodes sparsely bearded. Leaf sheaths sparsely hairy; ligule trigonous, ciliate; blades 10-20 x 0.6-2.5 cm, base rounded, apex acuminate, glabrous. Spikes 2-20, 10 cm long, racemes corymbose, silvery hirsute. *Sessile spikelets;* 4 mm, 2-nate; callus long-bearded; lower glume membranous, 6-9-nerved, margins incurved below, bidentate; upper glume boat-shaped, bidentate, 3-5-nerved; lower lemma oblong-lanceolate, 3-nerved, ciliate; palea oblong, wings ciliate, male; upper lemma hyaline, bidentate, ciliate, awned; palea bisexual. *Pedicelled spikelet:* lower glume awned.

Occasional in Eastern Ghats. Fl. & Fr.: August-January.

Hoginakkal (SLM), *S.India Flora* 11617 (MH); Morappur (SLM), *S.India Flora* 11659 (MH); Mahendragiri (GJP), *Fischer* Acc. No. 574580 (CAL); Joranda, Similipahar (MBJ), *SX* & *MB* 5229 (RRL-B).

INDIA: Endemic to Peninsular India (Srivastava and Nair, 2010).

Ischaemum rugosum Salisb., Icon. Stirp. Rar. l, t. 1. 1791; FBI 7:127 1896; Fischer 3: 1722. 1934; Bor, Grass. India 184. 1960; Matthew, Fl. Tamilnadu Carnatic 3: 1869. 1983.

An annual, tufted grass; culms up to 60 cm high, decumbent at base; nodes bearded. Leaf sheaths 4-6 cm long, compressed, loose, glabrous, auricled, membranous; ligule membranous; blades 6-12 x 0.3-0.5 cm, linear-lanceolate, base rounded, apex acuminate. Racemes usually 2.5 cm, digitate, spiciform; rachis trigonous; spikelets 2-nate. *Sessile spikelet:* 5 mm, ovate-oblong; callus villous; lower glume oblong-elliptic, cattilaginous from the base, pale yellow, concave, crossed by 3-6 deep smooth; ridges, the upper portion flat, membranous, incurved, winged: upper glume oblong-lanceolate, keeled, keel ciliate; lower lemma ovate-lanceolate, palea containing male floret; upper lemma deeply cleft into 2 acute, lanceolate lobes, cuneate; palea lanceolate; stamens 3, stigmas plumose. Caryopsis oblong. *Pedicelled spikelet:* Similar to sessile spikelet.

Common weed in rice fields and in wet marshy grounds. Fl. & Fr.: August-December.

Pennahobilam (ATP), *TP* 1273; Kammavaripalli (KNL), *TP & PVP* 2782; Bhajaranga kona (KDP), *KVS* & *AL* 29118; Buranpur (NLR), *PMR* 16495; Vinukonda RF (GNT), *VRK* 6861; way to Diguvametta (PKM), JLE 22160 (MH); Palakonda RF (VSKP), *KCJ* 17260 (MH); Viriyambakkam (CPT), *sine coll. s.n.* (MH); Similipahar (MBJ), *V. Naik* Acc. No. 3858 (RRL.-B); Chitrakonda (KRT), *SX* & *MB* 6826 (RRL-B); Tulka RF, Satkosia Tiger Reerve (AGL), *KCM* 5970 (BSID); Lasari, Kotgarh Wildlife Sanctuary (KDM), *AC* 2436 (BSID).

INDIA: Throughout India.

WORLD: Tropical Asia, Tropical Africa.

Ischaemum semisagittatum Roxb., Fl. Ind. 1: 322. 1820; FBI 7: 130. 1896; Bor, Grass. India 185. 1960. *I. conjugatum* Roxb. Fl. Ind. 1; 323. 1820; FBI 7: 131. 1896;

An annual, tufted grass; culms up to 60 cm high, decumbent at base, slender; nodes glabrous. Leaf sheaths loose, glabrous, upper most spathiform, higher pilose; blades 5-8 x 1-2 cm, oblong- lanceolate, auricled, base cordate-sagittate, margin scabrid, apex acuminate, lower with a filiform petioles, upper sessile, glabrous-sparsely hairy. Racemes 2, 7 cm long, villous with white or grey hairs, fragile. *Sessile spikelet:* 6 mm; callus bearded. Lower glume ovate, convex, the lower half coriaceous, the upper half chartaceous, ciliate margins; upper glume ovate- oblong, coriaceous; lower lemma ovate; palea linear-oblong, hyaline, male; upper lemma linear, 2-lobed, awned; palea linear-oblong, female. *Pedicelled spikelet:* short and narrower than the

Figure 64. **Ischaemum rugosum** Salisb.

A. Twig, B. Part of the raceme, C. Sessile and pedicelled spikelets, D. Lower glume dorsal view, E. Ventral view, F. Upper glume dorsal view, G. Upper glume ventral view, H. Lower lemma, I. Palea, J. Upper lemma, K. Palea, L. Grain, M. Pedicelled spikelet.

sessile, closely appressed to the sessile spikelet; pedicel broad, lower lemma ovate; upper lemma narrowly oblong, 2- lobed, awned, minute.

On crest of ghats usually in shades of trees in West Godavari district. Fl.& Fr: August-December.

Cultivated fields (WG), *KS* 5110 (CAL).

INDIA: Peninsular India.

WORLD: Sri Lanka.

ISEILEMA Andersson

1. Pedicelled spikelets shorter than and hidden by the involucral spikelets .... **I. venkateswarlui**
1. Pedicelled spikelets projecting beyond the tips of the involucral spikelets:
   2. Glandular tubercles present on the keels of the spatheoles and often on the nerves of the lower glume of the involucral spikelets .... **I. prostratum**
   2. Glandular tubercles not present on the keel of the spatheoles and on the nerves of the lower glume of the involucral spikelets:
      3. Pedicels of the involucral spikelets slender, longer than broad .... **I. laxum**
      3. Pedicels of the involucral spikelets almost as broad as long .... **I. anthephoroides**

Iseilema anthephoroides Hack, in DC., Monogr. Phan. 6: 683. 1889; FBI 7: 219, 1896; Fischer 3: 1748. 1934; Bor, Grass. India 187. 1960; Matthew, Fl. Tamilnadu Carnatic 3: 1870. 1983.

An annual, tufted grass; culms up to 50 cm high, decumbent at base; nodes glabrous. Leaf sheaths 1-4 cm long, glabrous; ligule membranous with ciliate apex; blades 3-10 x 0.1-0.3 cm, linear-lanceolate, flat, base rounded, apex acute, glabrous. Panicle up to 8 cm long; spatheoles 8 mm long, cymbiform, glabrous. *Involucral spikelet:* 4 mm, oblong, enclosing both sessile and pedicelled spikelets, male; glumes lanceolate-oblong, 3-nerved, upper scabrid and pilose on outside; lemma obovate-oblong, nerveless; stamens 3. *Sessile spikelet:* 3 mm, linear, bisexual; lower glume ovate, acuminate, scabrid outside; upper glume oblong, scabrid outside; lower lemma lanceolate; upper lemma linear, awned; stamens 3; stigma plumose. Caryopsis ellipsoid. *Pedicelled spikelet:* Similar to involucral spikelet.

Common along the hedges of the fields and abandoned plains and in lower hills. Fl. & Fr.: June-October.

Tadipatri (ATP), *TP* 412; Batrepalli (ATP), *BR* & *P.Priyadarsini* 37066; Bhumayyagaripalli (KDP), *KRKS* & *KRS* 37609; Way to Kappaladoruvu (NLR), *MSG* 10899; Giddalur (PKM), *BR* & *BSS* 30575 (BSID); Yerrabalem (KDP), *MPG* 163 (CAL); Kadapa, *EKK* 78578 (DD); Kothajupudi (GNT), *SCA* 172 (CAL), *PCN* 173

Figure 65. **Iseilema anthephoroides** Hack.

A. Twig, B. Spatheolate spikelet, C. Pedicelled spikelet, D. Lower glume, E & F. Upper glume front and back views, N. Lemma, O. Palea, P. Pistil, Q. Spatheole.

(CAL & DD); Bhupathaiah palem, way to Maredumilli (EG), *MM* 102683 (BSID); Gajapathinagaram (VJN), *VNS* 4576 (MH); Forest near Srungavarapukota (VJN), *NPBK* 110031 (CAL); Koriguttalu (KMM), *RR* 106034 (BSID); Purunakote section, Satkosia Tiger Reserve (AGL), *KCM* 5183 (BSID); Kuanaria, Satkosia Tiger Reserve (AGL), *KCM* 8370 (BSID); Tikarpada, Satkosia Willife Sanctuary (AGL), *AC* 2398 (BSID); Mandragiri hill (AGL), *AC* 2372 (BSID).

INDIA: Peninsular India. Endemic.

Iseilema laxum Hack., in DC., Monogr. Phan. 6: 682. 1889; FBI 7: 219. 1896; Fischer 3: 1748. 1934; Bor, Grass. India 188. 1960; Matthew, Fl. Tamilnadu Carnatic 3: 1871. 1983.

A perennial, tufted grass; culms up to 30 cm high, decumbent or ascending at base; nodes glabrous. Leaf sheaths 2-4 cm, loose, glabrous; ligule a fimbriate rim; blades cauline, 2-8 x 0.2-0.4 cm, lanceolate, flat, base rounded, margin scaberulous, apex acuminate. Panicle up to 15 cm long, flexuous, interrupted; spatheoles 18 mm plicate, glabrous. *Involucral spikelet:* 3 mm, linear-oblong, whorled, joints pilose; lower glume chartaceous, linear-oblong, pilose near apex at outside; upper glume oblong, acute; lemma linear, hyaline; stamens 3. *Sessile spikelet:* 3 mm, elliptic-lanceolate, bisexual; lower glume chartaceous, lanceolate, bidentate, beak elongate, tip scabrid, sparsely hairy; upper glume lanceolate, acuminate, scabrid on outside; lower lemma lanceolate; upper lemma bidendate, awned; stamens 3, stigmas plumose. Caryopsis ellipsoid. *Pedicelled spikelet:* Similar to involucral spikelet.

Common in waste places and in forests. Fl. & Fr.: August-March.

Gundumala RF (ATP), *TP* & *NY* 718; Kalasamudram (ATP), *NY* 662, *KRKS* & *BSS* 40457; Bugganipalli (KNL), *TP* & *RVR* 1814; Malyala (KNL), *PVP* 2552; Orampadu (KDP), *PVP* 3485; Yerrabalem (KDP), *MPG* 163 (DD); Kondapalli (KSN), *CAB* 4597 (MH); Karkalpadu beat, Balugam near Parvathipuram (SKLM), *VNS* 4541 (MH); Amangal RF (MBNR), *SRS* 10549 (BSID); Chinnar (SLM), *CAB* 6042 (MH); Vandalur (CPT), *S.India Flora* 11549 (MH); Kabrapalli (SBP), *AC* 10238 (BSID); Purnakote, Satkosia Wildlife Sanctuary (AGL), *AC* 2377 (BSID).

INDIA: Peninsular India, Maharashtra.

WORLD: Sri Lanka.

Iseilema prostratum (L.) Andersson, Nov. Act. Soc. Sci. Upsal. 3, 2; 251. 1856; Fischer 3: 1747. 1934; Bor, Grass. India 188. 1960. *Andropogon prostratum* L., Mant. Alt. 304. 1771. *Iseilema wightii* Andersson *loc.cit.* 1856; FBI 7: 218. 1896.

A perennial, tufted grass; culms up to 70 cm high, prostrate and ascending at base; nodes more or less bearded. Leaf-sheaths 3-5 cm, loose, glabrous; ligule membranous; blades 4-12 x 0.2-0.4 cm, linear, flat, base rounded, apex acute, glabrous. Panicle up to 30 cm, lax, spreading; rachis bearded at nodes; spathes linear-lanceolate; margin scarious. *Involucral spikelet:* 4 mm, oblong, male, whorled; lower glume oblong, 5-nerved, sulcate on the back, pilose; upper glume lanceolate, 1-nerved, glabrous; lemma oblong, glabrous; stamens 3. *Sessile spikelet:* 5 mm, lanceolate, bisexual; lower glume ovate, acuminate, scabrid on outside near the tip,

4-nerved; upper glume elliptic, acute, 3- nerved; lower lemma lanceolate; upper lemma, bidentate, awned; stamens 3; stigmas plumose. Caryopsis oblong. *Pedicelled spikelet:* Similar to involucral spikelet.

Common in plains and occasional in forests. Fl. & Fr.: November-June.

Deyyalakuntapalli (ATP), *PVP* 2645; Somulavaripalli (ATP), *KRKS* & *KRS* 37835; Gutturu (ATP), *BR* & *P.Priyadarsini* 37158; Kammavaripalli (KNL), *TP* & *PVP* 2785; Cement Nagar valley (KNL), *TP* & *RVR* 1811; Sangam (NLR), *AMR* 23047; Balapalli (KDP), *JLE* 15025 (MH); Nerabylu (CTR), *GVS* 46024 (MH); Palamaner RF (CTR), *KCJ* 458 (MH); Salur (VJN), *NPBK* 1153 (CAL); Karkalpadu, Amangal RF (MBNR), *SRS* 104549 (BSID); Ponneri (CPT), *S.India Flora* 14183 (MH); Eana, Jenabil, Similipal RF (MBJ), *ARKS* & *GPS* 12314 (BSID).

INDIA: Throughout India.

WORLD: Myanmar.

Iseilema venkateswarlui Satyavathi, J. Bombay Nat. Hist. Soc. 65: 667. 1969.

An annual, prostrate grass; culms 30-40 cm high; rooting at nodes, round and purple or white in colour. *Involucral spikelets:* 6.5-7.5 mm, oblong-acute, markedly umbonate at base; pedicels 1.25 mm, densely ciliate on the outer margins; stamens 3. *Sessile spikelet:* Central 5.5 mm of which 3.5 mm are elliptic and the reminder beak; in some racemes beak longer than lower elliptic portion; glumes ovate-elliptic; lower floret empty; upper floret female; lemma lanceolate; upper lemma bidentate, awned. *Pedicelled spikelet:* present; glumes 5.5 mm long; stamens 3, pedicels 4-4.5 mm long.

Rare in Visakhapatnam district. Fl.& Fr.:June-February.

A.U. Botanical garden from Lam Farm (VSKP), *SVT* 4 (K).

INDIA: Andhra Pradessh, Endemic.

JANSENELLA Bor

Jansenella griffithiana (C.Muell.) Bor, Kew Bull. 1955: 98. 1955 & Grass. India 516. 1960. *Danthonia griffithiana* C.Muell., Bot. Zeit. 14: 347. 1856. *Arundinella avenacea* Munro ex Thw., Enum. Pl. Zeyl. 362. 1864; FBI 7:69. 1896; Fischer 3: 1801. 1934.

An annual, slender grass, up to 25 cm high, often rooting at nodes. Leaf sheath 1.2-2 cm long, margins ciliate; blade 3-5 cm long, ovate-lanceolate, base amplexicaul, apex acute; ligule 0.3-0.5 mm, a membranaceous rim, acute at apex. Spikelets 4-5 mm long, subequal, lower 3-nerved, upper 5-nerved; lower lemma shorter than upper, upper with tufts of white hairs above the middle; awn 8 mm long; stamens 3; ovary elliptic, bead-like, stipitate, greenish. Caryopsis obovate, acute to obtuse at apex, greenish brown, hilum basal.

Rare, restricted to exposed grassy hilly areas over 700 m, growing in rocky crevices. Fl. & Fr.: October–February.

Mahendragiri (GJP), *SX* & *MB* 3676, 3883 (RRL-B), *AC* 4213 (BSID); Pottangi (KRT), *HFM* 4082 (DD); Kolli hills (NMK), *SKS* 472 (SGH).

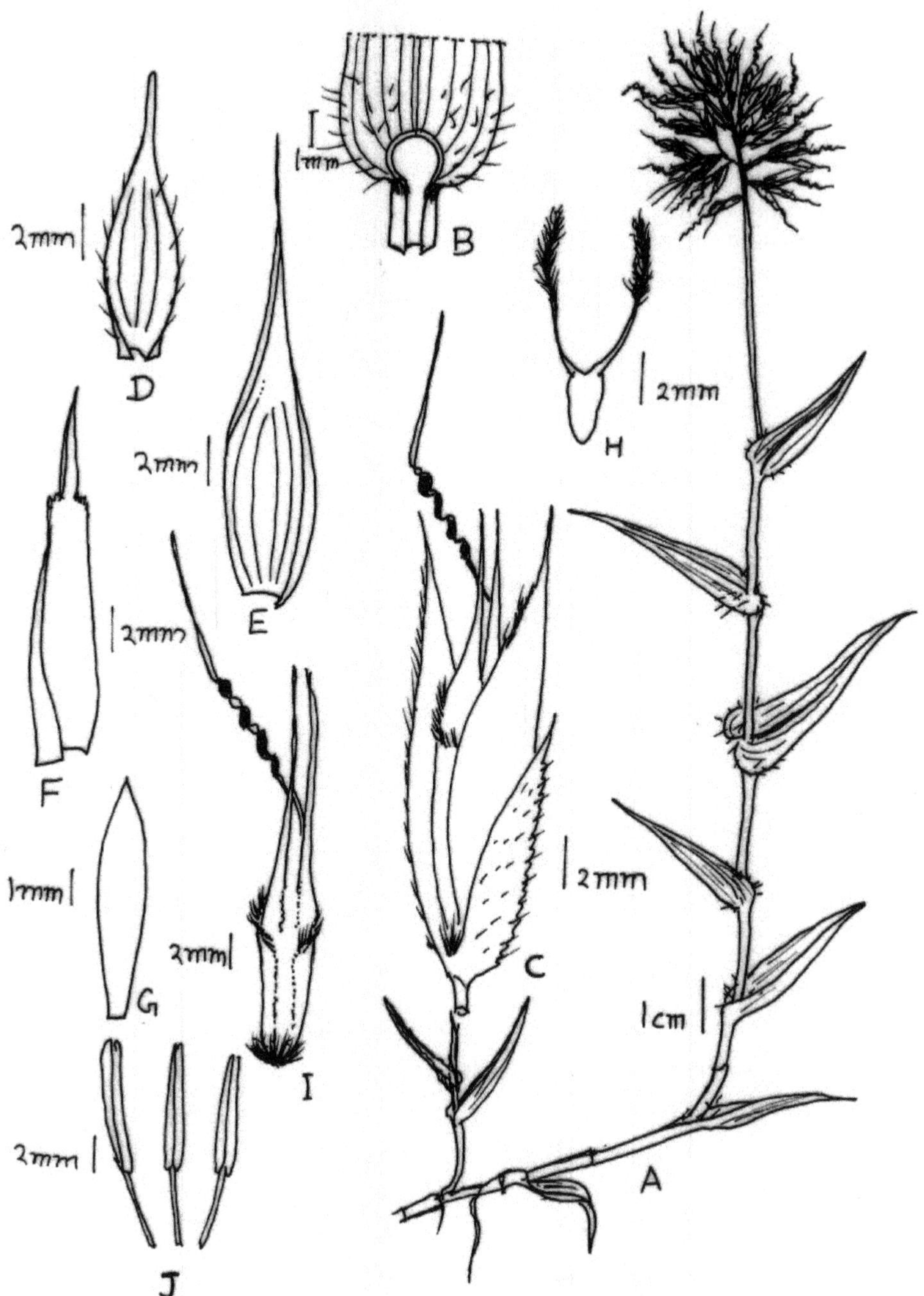

Figure 66. **Janseniella griffithiana** (Muell-Stuttg.) Bor.

A. Habit, B. Ligule, C. Spikelet, D. Upper glume, E. Lover glume, F. Upper lemma, G. lower lemma, H. Pistil, I. Palea, J. Stamens.

INDIA: Peninsular and N.E.India.

WORLD: Sri Lanka, Myanamar.

LEERSIA Solander ex Swartz. *nom. cons*

Leersia hexandra Sw., Prodr. 21. 1788; FBI 7: 94. 1896; Fischer 3: 1845. 1934; Bor, Grass. India 599. 1960; Matthew, Fl. Tamilnadu Carnatic 3: 1871. 1983.

An annual, aquatic, floating grass; culms up to 40 cm high; nodes bearded. Leaf sheaths 4-10 cm long, glabrous; ligule membranous; leaf blades 5-15 x 0.3-0.7 cm, lanceolate, base narrowed, apex acute, scabrid on both surfaces. Panicle up to 7 cm long; pedicel apically cupular, 2-lipped; spikelets 4 mm, solitary, oblong, laterally compressed; glumes obscure; lemma boat-shaped, coriaceous, 3-nerved, ciliate, 1-keeled; palea linear, midrib scabrid, 3-keeled; stamens 6; stigma plumose. Caryopsis oblong.

Common in ponds, tanks, lakes and marshy areas. Fl. & Fr.: September-February.

Ratnachalam Anjaneyaswamy temple (KDP), *KRKS* & *BSS* 40385; Tirumala hills (CTR), *MSG* 12156; Way to Kappaladoruvu (NLR), *MSG* 10898; Darsi (PKM), *MCK* 22826; Nallapadu (GNT), *VRK* 6771; Kambakkam (CTR), *sine coll. s.n.* (MH); Araku valley (VSKP), *DDSR* 21368 (MH); Sirumalai (DGL), *S.India Flora* 9084 (MH); Orchidarium, Shevaroys (SLM), *AVNR* 23081 (MH); Red hills (CPT), *DN* 942 (MH); Tulka ection, Satkosia Tiger Reserve (AGL), *KCM* 5978 (BSID); Jenabil Range, Similipal B.R. (MBJ), *AC* 2328 (BSID).

INDIA: Throughout India.

WORLD: Tropics of Old and New Worlds.

LEPTOCHLOA P. Beauvois

1. Lemmas 3-lobed; racemes loosely speculate:
    2. Spikelets 2-7-flowered .... **L. fusca**
    2. Spikelets 1-flowered:
        3. Leaves membranous, broadly oblong to elliptic, flat........ **L. uniflora**
        3. Leaves not membranous, linear, flat to infolded........ **L. neesii**
1. Lemmas not lobed; racemes densely speculate:
    4. Lemmas acuminate; anthers 0.2-0.5 mm........ **L. panicea**
    4. Lemmas acute-obtuse, anthers 0.5-0.7 mm:
        5. Racemes many; peduncle not exerted, enclosed by upper leaf sheath; spikelets 1-1.2 mm broad........ **L. chinensis**
        5. Racemes few; peduncle well exerted from leaf sheath; spikelets 1.8–2 mm broad........ **L. obtusiflora**

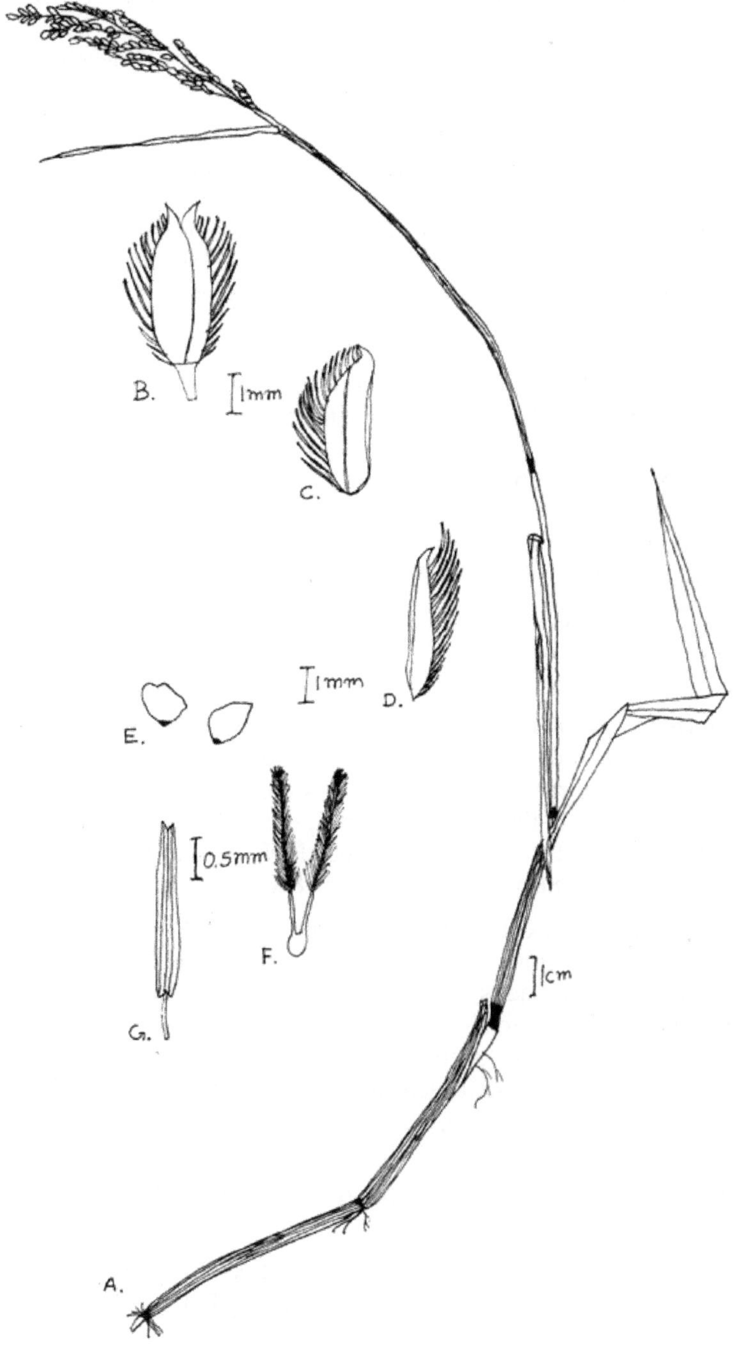

Figure 67. **Leersia hexandra** Sw.

A. Twig, B. Spikelet, C. Lemma, D. Palea, E. Lodicules, F. Pistil, G. Stamen.

Leptochloa chinensis (L.) Nees, Syll. Ratisb. 1: 4. 1824; FBI 7: 299. 1896; Fischer 3: 1820. 1934; Bor, Grass. India 516. 1960; Matthew, Fl. Tamilnadu Carnatic 3: 1872. 1983; Moulik, Grass. Bamb. India 2:613.1997. *Poa chinensis* L., Sp. Pl. 69. 1753.

Annuals, culms erect or ascending, tufted, to 90 cm; nodes glabrous. Leaf sheaths glabrous; ligules membranous; leaf blades 10-20 x 0.3-0.6 cm, linear or linear-lanceolate, base rounded, acuminate, glabrous. Panicles open; spikelets ovate-oblong, 3-7-flowered; lower glume lanceolate, chartaceous, 1-nerved; upper glume ovate-lanceolate, chartaceous, 1-nerved; lemmas ovate, chartaceous, 3-nerved; paleas ovate or oblong, 2-keeled, 2-nerved, keels ciliate; stamens 3. Caryopsis oblong.

Common near water sources throughout. Fl. & Fr.: July-December.

Kalasamudram RF (ATP), *PVP* 2616; Gundlabrahmeswaram (KNL), *BR* & *BSS* 32967; Rajampet (KDP), *AL* & *BSS* 26570; Musalreddypalli (KDP), *BSS* & *SKB*, 30823; Ramagiri (CTR), *BR* & *P.Priydarsini* 37107; Gopalapuram (NLR), *MSG* 10887; Leguntapadu (NLR), *PMR* 16430; Andhra Reservoir (VJN), *AMR* 24544; Kothajupudi (GNT), *SCA* 168 (DD); Palagudem (WG), *VNS* 4475 (MH); Medapenta (EG), *CAB* 4326 (MH); Red hills (CPT), *CAB* 5519 (MH); Sonachavadi (SA), *KRM* 64165 (MH); Tikarpada, Satkosia Tiger Reserve (AGL), *KCM* & *Swamy* 7618 (BSID).

INDIA: Throughout India.

WORLD: SE. Asia.

Leptochloa fusca (L.) Kunth, Revis. Gramin. 1: 91. 1829. *Diplachne fusca* (L.) P.Beauv. ex Roem. & Schult. Syst. Veg. 2. 615. 1817; FBI 7: 329. 1896; Fischer 3:1829.1934; Bor, Grass. India 492. 1960; Matthew, Fl. Tamilnadu Carnatic 2: 1844. 1983; Saxena & Brahmam, Fl. Orissa 4: 2344. 1996; Gayathri & Pullaiah in Pullaiah, Fl. Andhra Pradesh 3: 1183. 1997. *Festuca fusca* L., Sp. Pl. (ed.2) 109.1762. *Poa procera* Roxb., Fl. Ind. 1.344.1820.

A perennial, tufted, coarse grass; culms up to 70 cm high, decument or geniculately ascending below; nodes glabrous. Leaf sheaths 5-10 cm long, glabrous; ligule membranous; leaf blades 5-15 x 0.2-0.4 cm, linear, convolute, base rounded, apex acuminate. Panicle up to 15 cm, erect; rachis scabrid; spikelets 7 mm, oblong, 4-10-flowered; pedicel short, scabrid; glumes oblong, 1-nerved, membranous, scabrid on the keels, mucronate; lemma oblong, 1-nerved, 4-toothed, aristate, scabrid, pilose below; palea oblong, truncate, pilose on the keels; stamens 3; stigmas plumose. Caryopsis oblong.

Common in moist and shady localities of the forests, and in streams and ponds. Fl. & Fr.: October - April.

N.S.P.Kottala (ATP), *KRKS* & *BSS* 40482; Kottala cheruvu (KNL), *SS* & *AMR* 22637; Nagaluty (KNL), *BSS* & *SKB* 30150; Pamidi (ATP), *JSG* 20829 (MH & DD); Atapaka (KSN), *PV* 5349 (MH).

INDIA: Upper Gangetic Plains, West Bengal, Odisha, Andhra Pradesh

WORLD: Tropical and subtropical regions of Old world and Australia.

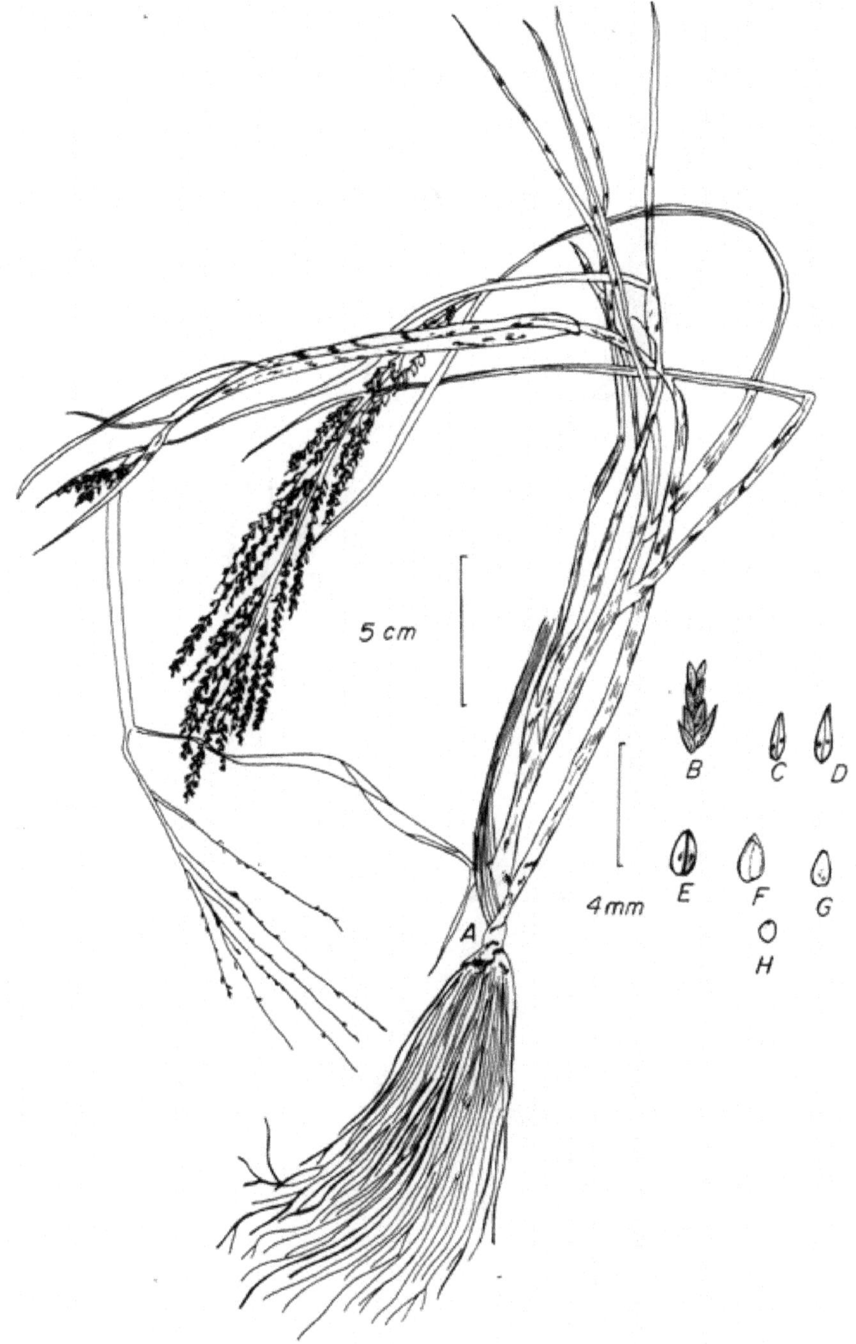

Figure 68. **Leptochloa chinensis** (L.) Nees.

A. Twig, B. Spikelet, C. Lower glume, D. Upper glume, E. Floret, F. Lemma, G. Palea, H. Caryopsis.

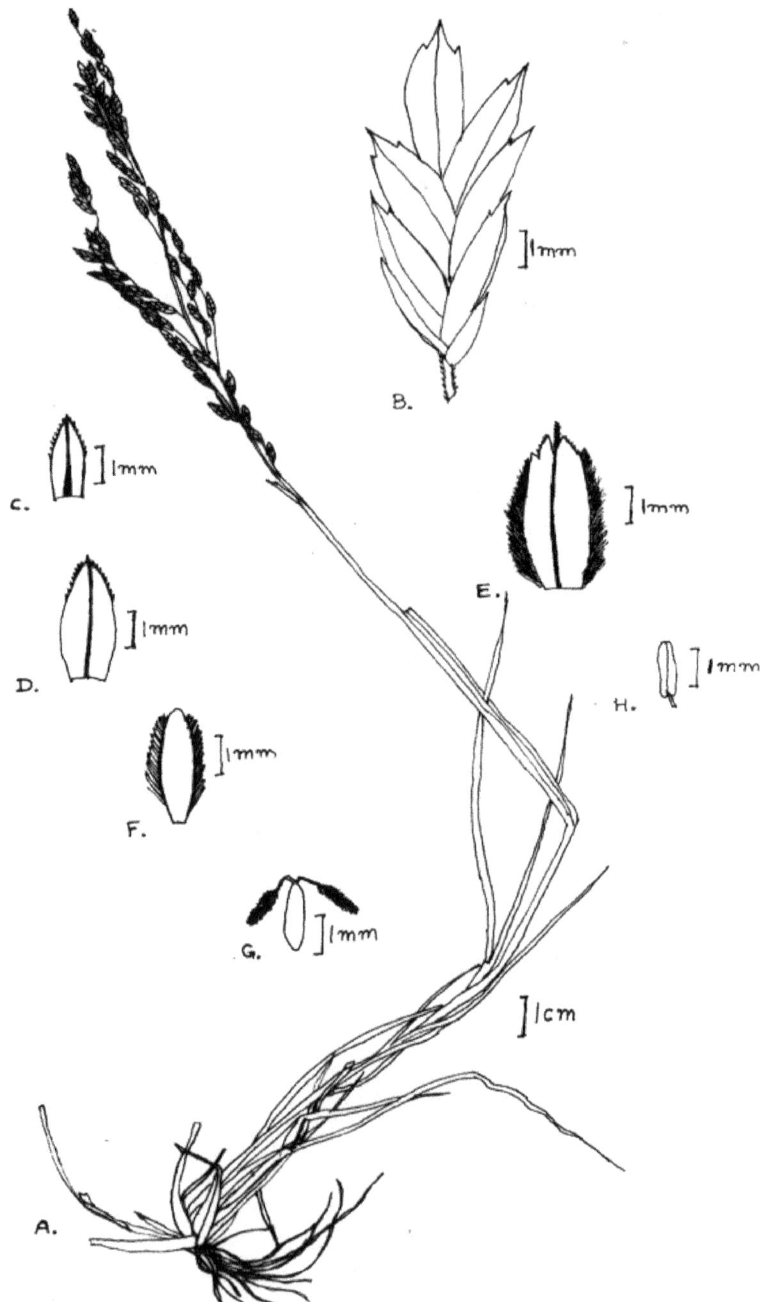

Figure 69. **Leptochloa fusca** (L.) Kunth.
[Syn.: *Diplachne fusca* (L.) P.Beauv.].

A. Twig, B. Spikelet, C. Lower glume, D. Upper glume, E. Lower lemma, F. Upper lemma, G. Pistil, H. Stamen.

Leptochloa neesii (Thw.) Benth., J. Linn. Soc. (Bot.) 19:108.1881; Bor, Grass. India 517.1960; Matthew, Fl. Tamilnadu Carnatic 3: 1873. 1983; Moulik, Grass. Bamb. India. 2:613. 1997. *Cynodon neesii* Thw., Enum. Ceyl. Pl. 3:71.1864. *Leptochloa polystachya* Benth., Fl. Austral 7: 617.1878; FBI 7: 298.1896; Fischer 3: 1820. 1934.

Annuals, culms erect or geniculately ascending, tufted, to 1.1 m high; nodes glabrous. Leaf sheaths glabrous; ligules membranous; leaf blades linear or linear-lanceolate, base cordate, apex acuminate. Spikelets ovate-oblong, 1-flowered; lower glume ovate-lanceolate, chartaceous, 1-nerved; upper glume ovate-lanceolate, chartaceous, 1-nerved; lemmas lanceolate, membranous, keeled, 3-nerved; paleas ovate-oblong, hyaline, keeled, 2-nerved; stamens 3. Caryopsis ovoid.

Occasional in marshy areas. Fl. & Fr.: July–December.

Baireni (KNL), *BSS* & *SKB* 32239; Kambakkam (CTR), *AMR* 24286; Kankipadu (KSN), *CAB* 8227 (MH); Kothajupudi (GNT), *SCA* 169 (CAL); Mandapenta (EG), *CAB* 4709 (MH); Salur (SKLM), *NPBK* 1134 (CAL); Gingee RF (SA), *KRM* 13057 (MH); Vandalur (CPT), *S.India Flora* 11489 (MH).

INDIA: South India.

WORLD: Sri Lanka, Australia, Malesia.

Leptochloa obtusiflora Hochst. in Flora 38: 203. 1855; FBI 7: 299. 1896; Fischer 3: 1820. 1934; Bor, Grass. India 517. 1960.

Annual tufted grass; culms erect to geniculate, rooting at lower nodes, 90 cm high; nodes glabrous, less ribbed. Leaf sheath more or less ribbed, 9-12 cm; leaf blade linear-lanceolate, 14-24 x 0.5-0.6 cm, apex acuminate to attenuate, base truncate to subcordate, margin serrulate, midnerve prominent; ligule acute at apex, 1.6 mm, membranaceous with long fimbriate hairs at apex. Spikelets alternate, subsessile, oblong, acute at apex, greenish to yellow brown; glumes persistent; lower glume ovate-lanceolate, boat-shaped, 1.3-1.5 x 0.8-1.2 mm, acute to subacute at apex, chartaceous, 1-nerved; upper glume broadly ovate-lanceolate, 1.8 x 0.8 mm, subacute to acute and minutely mucronate, 1-nerved; lemma broadly ovate to elliptic-lanceolae, prominently 3-nerved; palea obong-elliptic, 1.7 x 0.8 mm, retuse at apex, 2-nerved; stamens 3. Caryopsis oblong to elliptic, turgid, flat and shallowly concave dorsally.

Occasional in southern Eastern Ghats of Tamil Nadu. Fl. & Fr.: May–December.

Sirumalai hills (DGL), *SKS* 425 (SGH); Kolli hills (NMK), *SKS* 583 (SGH).

INDIA: Peninsula.

WORLD: Arabia, E. Africa.

Leptochloa panicea (Retz.) Ohwi, Bot. Mag. Tokyo 55:311. 1941; Bor, Grass. India 517. 1960; Matthew, Fl. Tamilnadu Carnatic 3: 1873. 1983. *Poa panicea* Retz., Obs. Bot. 3:11. 1783. *Leptochloa filformis* sensu Hook. f., FBI 7: 298.1896 non Beauv. 1812; Fischer. 3: 1820. 1934.

Annuals, erect, tufted grass, to 25 cm high; nodes glabrous. Leaf sheaths tuberculate-hispid; ligule membranous; leaf blades linear-lanceolate, glabrous.

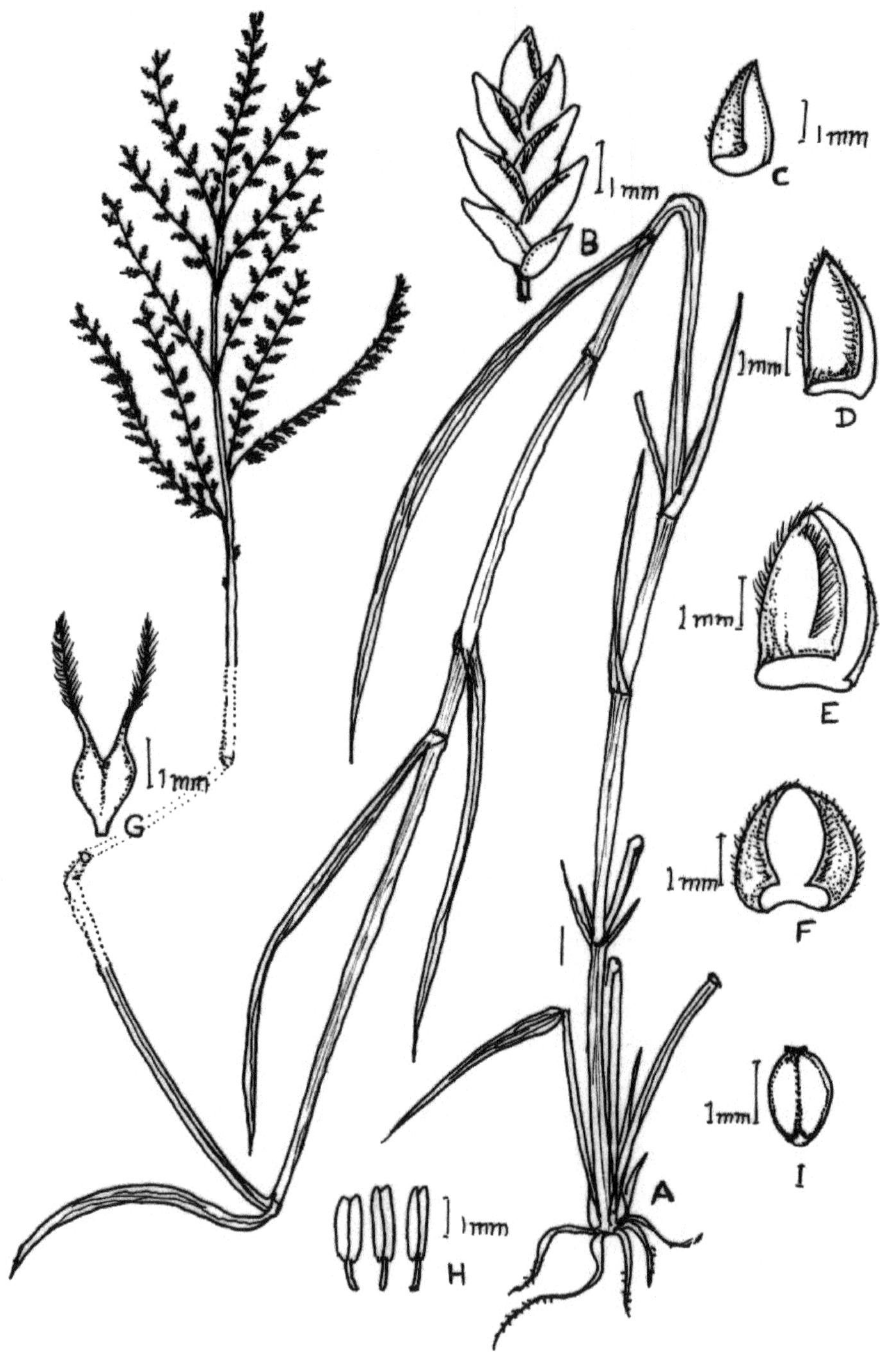

Figure 70. **Leptochloa obtusiflora** Hochst.

A. Habit, B. Spikelet, C. Lower glume, D. Upper glume, E. Lemma, F. Palea, G. Pistil, H. Stamens, I. Caryopsis.

Panicle interrupted at base; spikelets linear, 3-flowered; glumes subequal, persistent, membranous; lemmas membranous, keeled. Caryopsis broadly ellipsoid.

Rare in wet lands in northern Eastern Ghats. Fl. & Fr.: August - November.

Samalkota (EG), *S.India Flora* 12486 (MH); Anakapalle (VSKP), *sine coll. s.n.* (MH); Gajapathinagaram (VJN), *VNS* 8241 (MH).

INDIA: Throughout India.

WORLD: Tropical Asia and Africa.

Leptochloa uniflora Hochst. ex A. Rich., Tent. Fl. Abyss. 2: 409. 1850: FBI 7: 297. 1896; Fischer, 3: 1820. 1934; Bor, Grass India 517. 1960; Althaf & V.J. Nair, Fl. Tamil Nadu, Grasses 162. 2009.

Annual tufted grass; culm geniculate, 60 cm high. Leaves elliptic-lanceolate, 6.3-11.5 x 1.3-1.6 cm, acuminate or acute at ends, margins cartilaginous, scabrid, midrib white; ligule ovate, *c.* 2 mm, membranous, lacerate, hyaline. Panicle 18-28 cm long, lax on maturity; central axis more or less ribbed, scabrid on ribs; peduncle *c.*11 cm long. Racemes many, alternate or subwhorled, 4.5-8 cm long. Spikelets subsessile, secund, narrowly lanceolate, 2-2.4 x 0.5 mm, acuminate at apex, acute at base; pedicels *c.* 0.4 mm, flat, scabrid on margins; glumes longer than floret; lower glume oblong-lanceolate, 2-2.3 x 0.4-0.5 mm, acuminate, chartaceous, yellow, prominently 1-nerved, keeled, scabrid, margins finely serrulate towards atpex; upper glume 2-2.3 x 0.4 mm, similar to lower glume, divergent at maturity. Floret single, bisexual. Lemma ovate to oblong, *c.*1.5 x 0.8 mm, acute, membranous to chartaceous, greenish, faintly 3-nerved, minutely hairy along nerves, margins sparsely ciliate towards apex. Palea elliptic, *c.* 1.3 x 0.6 mm, acute at apex, narrowly truncate at base, membranous to chartaceous, greenish, 2-nerved; lodicules 2, 0.2-0.4 mm, truncate with raised hook like projections on one side, hyaline; stamens 3; anthers 0.6-1 mm, yellowish; filaments 0.3-0.5 mm long, hyaline; ovary obovate to elliptic, *c.* 0.3 mm long, yellowish green; style *c.* 03 mm long; stigma plumose, c. 0.6 mm long, yellowish green. Caryopsis oblong-lanceolate, acute, brownish, smooth, deeply grooved on one side.

Rare in exposed lands. Fl. & Fr.: October–February.

Pennagaram RF (SLM), *EV* 21992 (MH).

INDIA: Kerala, Tamil Nadu.

WORLD: Sri Lanka, Tropical Africa.

LEPTURUS R.Brown

Lepturus radicans (Steud.) A. Camus, Ann. Soc. Linn. Lyon, n.s. 69: 87. 1923; Bor, Grass. 585. 1960; Kumar & Rao, Indian J. Forestry 41: 307. 2018.

Annual or short lived perennial. Culms stoloniferous, up to 75 cm long, at times mat forming. Leaf sheaths glabrous; ligules distinct, ciliate membranes; leaf blades lanceolate, acute to acuminate at apex. Inflorescence a solitary, spike like raceme. Spikelets sunken in rachis, 1-flowered, with or without rachis extension; lower glume absent; upper glume well developed, oblong-lanceolate, 3-3.4 mm long, chartaceous, as long as or longer than the length of floret, exceeding the cavity of rachis, 5-7-nerved, apex acuminate; nerves prominent; lemmas lanceolate-elliptic, 2.8-3 mm long, membranous-hyaline, slightly curved, acute-acuminate at apex, 3-nerved, lateral nerves close to the margins; pales oblong, hyaline, as long as or slightly longer than lemma, 2-nerved; keels wingless or narrowly winged; wings ciliolate, 2.9-3.2 mm long; stamens 3; anthers 0.6 mm long, stigmas plumose, 2.

Rare, usually under moist shades, in Coffee plantation near Kothaveedhi colony. Fl & Fr.: July–December.

Coffee plantation, Kothaveedhi colony (VSKP), *BR* & *M.Anil Kumar* 55527 (SKU).

INDIA: Andhra Pradesh, Karnataka, Kerala and Maharashtra.

WORLD: Africa, India, Malesia, Sri Lanka.

LOPHOPOGON Hackel

1. Annual grass, nodes silky; ligule a rim of white hairs .... **L. kingii**
1. Perennial grass, nodes glabrous; ligule prominently ciliate .... **L. tridentatus**

Lophopogon kingii Hook.f., Fl. Brit. Ind. 7: 149. 1896; Haines, Bot. Bihar & Orissa: 1025. 1924; Bor, Grass. India 190. 1960; Chandramohan *et al.*, Int. J. Adv. Res. Sci. Tech. 5: 604. 2016.

Annuals; culms tufted, up to 35 cm high, silky tomentose at base. Leaf sheath 2.5 - 4 cm long; ligule a rim of white hairs; leaf blades narrowly linear, 4-15 x 0.06-0.1 cm, acuminate at apex, incurved when mature, gradually smaller upwards and sheathing the peduncle; nodes glabrous; base of the leaves and mouth of the sheaths softly villous. Inflorescence spatheolate, spatheous sheath narrowly lanceolate, 2.5-3 cm long, caudate to acuminate. Racemes 2, closely appressed together, 1.5-2 cm long, yellowish brown, exerted from spatheole, up to 6.5 cm long. Spikelets densely imbricate, fragile, lower glumes of the lowest spikelets closely nerved, with tufts of hair on the margins; lower sessile and upper 2 pedicelled spikelets, dissimilar, bisexual; sessile and pedicelled spikelets obovate - oblong, base of the joints and callus bearded with 1.5-2 mm long brown hairs. Sessile spikelets 4.2-4.5 mm long, lower glume oblong, 3.5-3.8 x 0.8-0.9 mm, truncate with two lateral blunt teeth and 2-3 shorter ones between them, narrowed at base, yellowish brown, prominently 7-9 nerved, tufts of hairs between nerves in the middle forms continuous transverse band, margins incurved, glabrous. Upper glume oblong - elliptic, awned, *c.* 5.5 mm long, brown to yellow, 3- nerved, glabrous, margins broadly hyaline at the base, ciliate with long brown hairs on upper lateral sides, glabrous inside, awn as long as glume, barbellate. Florets 2, lower male or barren; upper bisexual. Lower lemma oblong - lanceolate, 4.1 x 0.9 mm, slightly two-lobed, acute and hairy at apex, membranous, hyaline, nerveless. Upper lemma oblong - lanceolate *c.* 4.4 mm long, faintly 1- nerved, glabrous below the middle, ciliate or pubescent above the middle, acute, awned between two acuminate scabrid lobes, awn geniculate, column *c.* 4 mm long, twisted, margins sparsely barbellate; bristle *c.* 9 mm long, barbellate; palea rectangular, 2.2 x 1 mm, membranous, hyaline, nerveless, undulate at apex, ciliolate, margins infolded. Pedicelled spikelets 5-6 mm long, pedicels densely brown hairy at base; similar to sessile spikelets but upper glume 5-7-nerved with tufts of hairs, forming a continuous transverse band in the middle, awned, awn *c.*10 mm long, upper lemma awn longer than the awn of upper lemma of sessile spikelet; sometimes awn reduced as short arista between two acuminate scabrid lobes; stamens 2; filaments *c.* 1.6 mm long; anthers *c.* 2.1 mm long, light yellow; style *c.* 1.5 mm long; stigma plumose, up to 3 mm long, purple, exerted. Caryopsis narrow, *c.* 1.8 mm long, oblique at base, tapering below, bulged above the middle.

Grows usually in the rocky crevices on hills in dry deciduous forests. Fl. & Fr.: September-October.

Satkosia Wildlife Sanctuary (AGL), *KCM* 7061 (BSID).

INDIA: Bihar, Odisha.

Lophopogon tridentatus (Roxb.) Hack. in Engl. & Prantl, Pflanzen. 2 (2): 26. 1887; FBI 7: 149 1896; Fischer 3: 1724. 1934; Bor, Grass. India 190. 1960; Matthew, Fl. Tamilnadu Carnatic 3: 1874. 1983. *Andropogon tridentatus* Roxb., Fl. India 1: 261. 1820.

A perennial, tufted grass; culms up to 30 cm high, decumbent; nodes glabrous. Leaf sheaths 1-4 mm, glabrous; ligule hyaline, ciliate; leaf blades 2-5 x 0.1-0.3 cm, linear, flat or convolute, base narrowed, apex acuminate, glabrous. Binate spiciform racemes up to 3 cm, solitary, closely appressed, enclosed within ventricose sheaths or exserted; spikelets 4 mm, 2-nate. *Sessile spikelet:* oblong, 4 mm, dorsally compressed; lower glume oblong, 5-nerved, brownish tuft hairs above the middle, margin ciliate, apex truncate with lateral teeth; upper glume oblong, tufted hairs on outer side, apex bidentate, narrowed into a fine arista; lower lemma linear, 1-nerved, keeled palea linear, male or empty; upper lemma lanceolate, apex bidentate, hyaline, awned; palea bisexual; stamens 3; stigmas plumose. *Pedicelled spikelets:* Similar to sessile spikelet.

Occurs on hill slopes and abandoned plains of southern Eastern Ghats. Fl. & Fr.:July–October.

Gutturkona (ATP), *NY* 990, *BR* & *P.Priyadarsini* 37142; Gooty fort (ATP), *MSG* 12240 & *PVP* 3672; Rangapuram RF (KNL), *RVR* 1704; Krishna River (GNT), *VRK* 6753; Panapakkam (CTR), *S.India Flora* 15780 (MH); Madinikonda RF, near Madanapalle (CTR), *KCJ* 469 (MH); Chintaladevi (NLR), *MH* Acc. No. 56344 (MH); Way to Mysamma temple (MBNR), *SRS* 108628 (BSID); Umamaheswaram (MBNR), *SRS* 108812 (BSID).

INDIA: Maharashtra, Madhya Pradesh, Andhra Pradesh, Tamil Nadu. Endemic.

MANISURIS Linnaeus

Manisuris myurus L., Mant. 2: 300. 1771; Fischer 3: 1760. 1934; Bor, Grass. India 192.1960. *Rottboellia myurus* (L.) Benth., J. Linn. Soc. (Bot.) 19: 68. 1881; FBI 7: 154. 1896.

A perennial, tufted grass; culms up to 50 cm high, decumbent or ascending at base; nodes sparsely bearded. Leaf sheaths 2-10 cm long, pubescent basal sheaths, upper glabrous; ligule a fringe of hairs; leaf blades 5-10 x 0.2-0.4 cm, linear, flat or

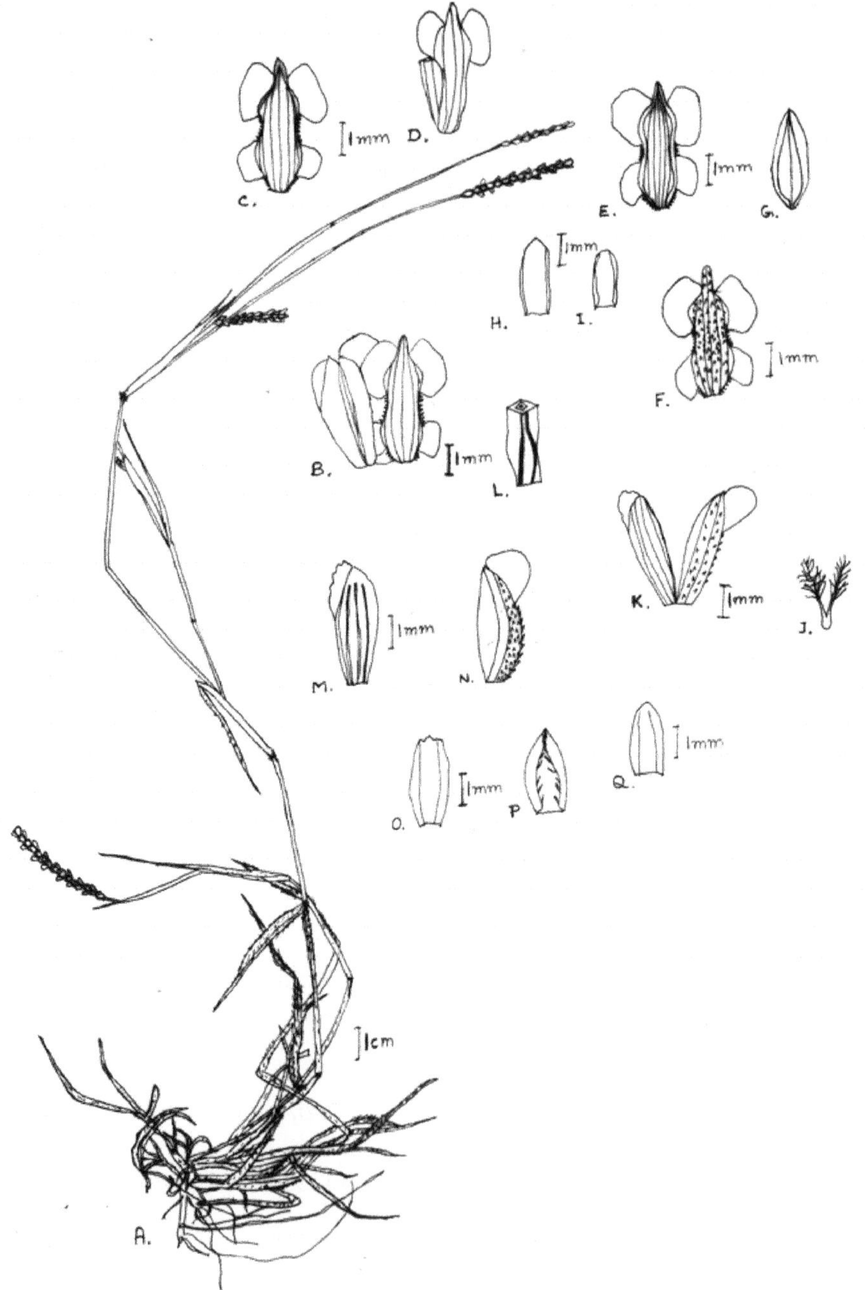

Figure 71. **Manisuris myurus** L.

A. Twig, B. Spikelets, C & D. Sessile spikelet front and side views, E & G. Lower glume front and back views, G. Upper glume, H. Lower lemma, I. Upper lemma, J. Pistil, K. & L. Pedicelled spikelet and its pedicel, M. Lower glume, N. Upper glume, O. Lower lemma, P. Palea, Q. Upper lemma.

convolute, base rounded, apex acuminate, pubescent. Racemes 7 cm long, solitary, both terminal and axillary, compressed; spikelet 4 mm, 2-nate; pedicel truncate, hollowed on the innerside to receive the sessile spikelet. *Sessile spikelet:* 4 mm, lower glume ovate-oblong, margins winged, back of the glume transversely grooved, scabrid; upper glume oblong, not winged, curved, hairy near the apex; lower lemma oblong, nerveless; upper lemma lanceolate, paleated; stamens 3, stigma plumose. Caryopsis oblong. *Pedicelled spikelet:* 4 mm; lower glume oblong, winged on one side, 3 strong-nerved; upper glume boat-shaped, winged at apex, scabrid on back; lemmas oblong, hyaline, obtuse, 2-nerved; palea oblong, acute, sparsely hairy; stamens 3.

Occasional on stream banks and wet places in southern Eastern Ghats. Fl. & Fr.: August-February. Tel.: *Nallapanuku.*

Peccheruvu–Rollapenta (KNL), *BSS* & *SKB* 30165; Bhupathayyakona (CTR), *BR* & *P.Priydarssini* 37320; Kambakkam hills (CTR), *MCB* 45082 (MH); Kothamarikuppam (CTR), *MCB* 45238 (MH); Nellore, *JSG* 12785 (MH); Ethamukkala (PKM), *PVS* & *KR* 88866 (BSID).

INDIA: South India, Endemic.

MELANOCENCHRIS Nees

1. A tiny, annual grass; rachilla glabrous; lemmas bi-dentate .... **M. jacquemontii**
1. A perennial grass; rachilla puberulous; lemmas acute, apiculate or aristate .... **M. monoica**

Melanocenchris jacquemontii Jaub. & Spach, Illus. Pl. Orient. 4: 36. t. 325. 1851; Bor, Grass. India 473. 1960. *M. royleana* Nees ex Steud., Syn. Pl. Glum. 1: 218 1854; Fischer 3: 1831. 1934. *Gracilea royleana* Hook. f., Fl. Brit. India 7: 284. 1896.

A tiny annual, tufted grass; culms up to 20 cm high, decumbent; nodes glabrous. Leaf sheaths 1-2 cm long, sparsely hairy; ligule hairy; blades 2-4 x 0.1-0.5 cm, linear, convolute, base rounded, apex acuminate, glabrous. Spikes 6 cm long, filiform, recurved, flexuous; spikelets 8 mm, clustered on a simple axis, 2-flowered; lower glume linear, coriaceous, densely ciliate on margin, awned; upper glume ovate, densely ciliate on margin, awned; lower lemma and palea ovate, membranous, bidentate, sparsely ciliate, awned; stamens 3; stigmas plumose. Caryopsis elliptic, obtuse. Upper lemma ovate-oblong, bidentate, sparsely ciliate, awned; upper palea broadly ovate, bidentate, scabrid, male; stamens 3.

Common in dry sandy soils and waste lands. Fl. & Fr.: July–September.

Gundumala RF (ATP), *TP & NY* 725; Kalasamudram RF (ATP), *TP* & *EC* 2811; Peapuly (KNL), *PVP* 2671; Madhavaram RF (KNL), *RVR* 1661; Gani RF (KNL), *RVR* 1779; Rachakuntapalli (KDP), *KRKS* & *KRS* 37644; Agastheswarakona, Lingapuram (KDP), *J.Swamy* & *S.Nagaraju* 7996 (BSID); Varikuntapadu (NLR), *PMR* 16474; Ethipothala (GNT), *VRK* 5190; Guvvalacheruvu (KDP), *KS* 6839 (CAL & MH); Chelama (KNL), *JLE* 22061 (CAL & MH); Rollamadugu, Sechachalam hills (CDP), *CAB* 15372 (MH); Kondaveedu fort (GNT), *CAB* 4707 (MH); Kondapalle fort (KSN), *P.Venkanna* 5632 (MH); Guttakindipalle (MBNR), *SRS* 107482 (BSID); Mudwin,

Amamgal RF (MBNR), *SRS* 108735 (BSID); Perungudimedu (CPT), *JSG* 88717 (MH); Bhaludungri (SPR), *HFM* 2403 (DD).

INDIA: Throughout India.

WORLD: Pakistan.

Melanocenchris monoica (Rottl.) C.E.C. Fischer, Fl. Madras 1831. 1934; Bor, Grass. India 473. 1960; Matthew, Fl. Tamilnadu Carnatic 3: 1874. 1983. *Pommereulla monoica* Rottl., Ges. Natur. Frunde Berlin, Neue Schr. 4: 218. 1803. *Gracilea nutans* Koen. ex Rottl. *loc.cit.;* FBI 7: 283. 1896.

A perennial, tufted grass; culms up to 60 cm high; nodes glabrous. Leaf sheaths 2-5 cm long, sparsely tubercled based hairs; ligule fimbriate membrane; blades 2-10 x 0.2-0.5 cm, linear-lanceolate, flat or convolute, base cordate, apex acuminate, sparsely tubercled based hairs. Spike up to 15 cm long; rachilla puberulous. Spikelets 6 mm, clustered, subsessile, secund, 2-nate, lower floret bisexual, upper male or neuter; lower glume linear, coriaceous, pilose, awned; upper glume oblong, pilose, margin hyaline; lower lemma broadly ovate, membranous, sparsely ciliate, aristate; palea oblong-ovate, obtuse, sparsely ciliate; stamens 3; stigmas plumose. Caryopsis oblong. Upper lemma oblong, sparsely aristate, ciliate; palea oblong-ovate, scabrid margins.

Rare grass in southern Eastern Ghats. Fl. & Fr.:August-December.

Penukonda (ATP), *BR & P.Priyadarsini* 37175; Batrepalli (ATP), *BR & P.Priyadarsini* 37067; Mellavagu (GNT), *VRK* 5713; Diguvametta (PKM), *JLE* 32467 (CAL & MH); Kamalapuram (KDP), *JSG* 11208 (MH); Bairanidokku (KNL), *JLE* 42227(MH); Nallamalai (KNL), *JLE* 32487 (MH); Kambakkam hills (CTR), *S.India Flora* 10266 (MH); Keradi RF (NLR), *PV* 112116 (BSID); Pollur (NLR), *CAB* 12728 (MH); Phirangipuram (GNT), *CAB* 4610 (MH); Narasarapet (GNT), *JSG* 57578 (MH);

Palakuppam (SA), *VNS* 4078 (MH); Arkonam panchayat forest (NA), *KCJ* 85428 (MH); Sidevi RF (SA), *KRM* 51164 (MH); Acharapakam (CPT), *S.India Flora* 11234 (MH); Vandalur RF (CPT), *ANH* 45553 (MH).

INDIA: South India.

WORLD: Sri Lanka.

MELINIS P. Beauvois

Melinis repens (Willd.) Zizka, Biblioth. Bot. 38: 55. 1988. *Saccharum repens* Willd., Sp. Pl. 1; 372 1798. *Rhynchelytrum repens* (Willd.) C.E. Hubb., Kew Bull. 1934. 110: 1934; Bor, Grass. India 355. 1960; Matthew, Fl. Tamilnadu Carnatic 2: 1893. 1983; Pullaiah, Fl. Andhra Pradesh 5: 2261. 2018. *Tricholaena wightii* Nees & Arn. ex Steud., Syn. Pl. Glum. 1: 98. 1854; FBI 7: 65. 1896. *Rhynchelytrum villosum* (Parl.) Chiov., Annuario Reale Ist. Bot. Roma 8: 310. 1908; Fischer 3: 1791. 1934; Bor, *loc. cit. Monachyron villosum* Parl. in Hook., Niger Fl.191. 1849.

An annual, tufted grass; culms up to 60 cm high; nodes pubescent. Leaf sheaths 2-4 cm long, tuberculately pilose; ligule ciliate; blades 3-15 x 0.2-0.5 cm, lanceolate, base truncate, apex acuminate, pubescent - glabrous on both surfaces. Panicle up to 10 cm long, contracted; branches filiform; pedicels capillary, pilose; spikelets

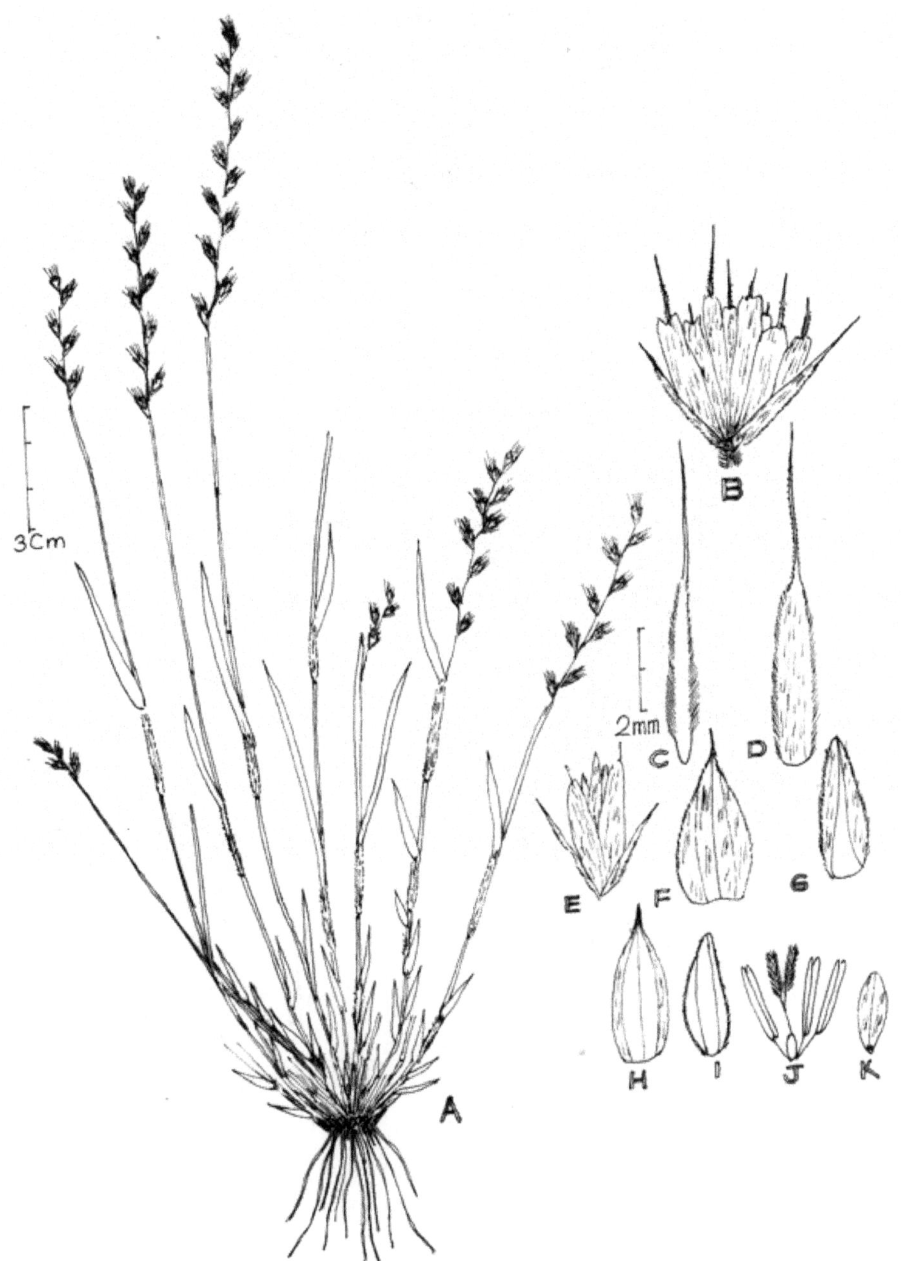

Figure 72. **Melanocenchris monoica** (Rottl.) Fischer

A. Twig, B. Cluster of spikelet, C. Lower glume, D. Upper glume, E. Florets, F. Lower lemma, G. Lower palea, H. Upper lemma, I. Palea, J. Pistil and anthers, K. Grain.

Figure 73. **Melinis repens** (Willd.) Zizka.
[Syn.: *Rhynchelytrum repens* (Willd.) C.E. Hubb.]

A. Twig, B. Spikelet, C. Lower glume, D. Upper glume, E. Lower lemma, F. Palea, G. Anther of lower floret, H. Pistil, I. Anther of upper floret.

5 mm, ovate-obovate, laterally compressed, 2-flowered, villous; callus bearded; lower glume linear, pilose; upper glume ovate, acuminate, 5-nerved, densely pilose, awned; lower lemma oblong, densely pilose, aristate; palea linear, male; upper lemma retuse, arista absent; palea hyaline, bisexual; stamens 3; stigmas plumose. Caryopsis obovoid.

Occasional in hills and open forests of southern Eastern Ghats. Fl. & Fr.:August-December.

Kekati RF (ATP), *TP & NY* 783; Maddileti stream (KNL), *GO & PVR* 3352; Dudemmakonda (KNL), *BR* & *BSS* 29251 (SKU & BSID); Namalagundu (KDP), *KRKS* & *BSS* 39661; Bhupathaiah kona (CTR), *PP* & *BR* 37321 (SKU & BSID); Hampi (Ballari), *S. India Flora* 16002 (MH); Kolli hills, on the road sides (NMK), *SKS* 472 (SGH); Shevaroy Bauxite hills–Yercaud (SLM), *AVNR* 26770 (MH); Ghat road, Yercaud (SLM), *DBD* 31433 (MH).

INDIA: Introduced. Drier parts of India.

WORLD: Native to Africa. Pantropical weed.

MICROCHLOA R. Brown

1. Annual, up to 20 cm high .... **M. indica**
1. Perennial, over 25 cm high .... **M. kunthii**

Microchloa indica (L.f.) P. Beauv., Ess. Agrost. 20. 1812; Fischer 3: 1830. 1934; Bor, Grass. India 473. 1960; Matthew, Fl. Tamilnadu Carnatic 3: 1875. 1983. *Nardus indica* L.f., Suppl. Pl. 105. 1781. *Microchloa setacea* R. Br., Prodr. 208. 1810; FBI 7: 283. 1896.

An annual, tufted grass; culms up to 20 cm high, slender. Leaf sheaths sparsely hairy; ligule a minutely ciliate rim; leaf blade 1.5-4 cm long, linear, apex acute, glabrous. Spikes solitary, 8 cm long, curved; rachilla minutely ciliate; spikelets 2 mm, 1-flowered, dorsally compressed; glumes subequal, lanceolate; lemmas lanceolate-oblong, ciliate along the nerves; stamens 3, stigmas plumose. Caryopsis oblong, subfusiform.

Occasional grass Eastern Ghats. Fl. & Fr.: August-November.

Gazulapalli (KNL), *BSS* & *SKB* 32254; Sithanur (CPT), *S.India Flora* 11210 (MH); Palasa (SKLM), *PCN* 195 (CAL); Motijharan (SBP), *HFM* 3575, 3587 (DD); Angul (AGL), *HFM* 3574 (DD); Depisahi, Satkosia Tiger Reserve (AGL), *KCM* 9406 (BSID), Talkudar, Rajgangapur, Sundargarh, *AC* 10263 (BSID); Raisili, Laxmipur (KPT), *AC* 4222 (BSID); Kundali (KPT), *AC* 10229 (BSID).

INDIA: Tamil Nadu, Andhra Pradesh, Odisha, Central India and N.E.India.

WORLD: Malaya, East and West Africa, S. America.

Microchloa kunthii Desv., Opusc. Sci. Phys. Nat. 75. 1831; Bor, Grass. India 474. 1960; Saxena & Brahmam, Fl.Orissa 4: 2413. 1996.

A perennial, tufted grass; up to 80 cm high; nodes glabrous. Leaf sheaths 3-5 cm long, glabrous; ligule a narrow membranous ring; blades 5-6 x 0.1-0.2 cm, narrowly linear, convolute, base rounded, margin cartilaginous, apex acute. Spike solitary, up to 8 cm long, curved; spikelets 3 mm, lanceolate, acute, uniseriate.

Rare in northern Eastern Ghats. Fl. & Fr.:August-November.

Sonabera (KRT) *HFM* 3659 (DD).

INDIA: South and North-East India.

WORLD: Myanmar, China, Malaya, Tropics of Africa and S. America.

MICROSTEGIUM Nees

1. Upper lemma well developed, split to the middle into 2, long, glabrous, triangualr lobes .... **M. petiolare**
1. Upper lemma small, very shortly 2-lobed or entire, 1-2 mm long .... **M. ciliatum**

Microstegium ciliatum (Trin.) A. Camus, Ann. Soc. Linn. Lyon 68: 201. 1921; Fischer 3: 1716. 1934; Bor, Grass. India 193. 1960. *Pollinia ciliata* Trin., Mem. Acad. Imp. Sci. St. Petersbourg. Ser. 6, Sci. Math. 2: 306. 1832; FBI 7: 116. 1896.

A perennial, tufted grass; culms up to 50 cm high, straggling. Leaf blades 2.5-12 x 0.2-0.5 cm, lanceolate-elliptic, base narrowed, apex acuminate, some times petioled. Racemes 2-10, 8 cm long, fascicled; joints and pedicels densely ciliate; lower glume lanceolate, minutely scaberulous; upper glume boat-shaped, keeled; lower lemma absent; upper lemma lanceolate, bidentate, awned; palea minute.

Rare in northern Eastern Ghats.

South east of Gudem Camp (VSKP), *VNS* 592 (CAL); Kapilas (DK), *SX* 2604 (RRL-B); Meghasani, Similipahar (MBJ), *SX* & *MB* 4174 (RRL-B); Harishankar (BG), *SX* & *MB* 5854 (RRL-B); Betkote, Karlapat Wildlife Sanctuary (KHD), *AC* 2465 (BSID).

INDIA: Throughout India except N.W.Himalaya.

WORLD: Sri Lanka, Bhutan, Nepal, China, Myanmar, Malaysia, Thailand.

Microstegium petiolare (Trin.) Bor, Indian Forest Rec., Bot. n.s., 1: 87 1938 & Grass. India 194. 1960; Saxena & Brahmam, Fl. Orissa 4: 2415. 1996. *Spodiopogon petiolare* Trin., Mem. Acad. Imp. Sci. St.-Peterbourg, ser. 6. Sci. Math. 2: 301. 1832. *Ischameum petiolare* (Trin.) Hack. in A. & C.DC., Mong. Phan. 6: 238. 1889.

Culms 60-120 cm high, stout; nodes pubescent. Leaf sheaths terete, margins ciliate; ligule linear-oblong, 4-9 mm long, truncate, fulvous in the middle, back often pubescent, decurrent on the margins of sheaths; leaf blades 20-30 cm x 2-2.5 mm, rigid, spreading, setaceous-acuminate, both surfaces with appressed tubercle-based hairs, margins scabrid. Inflorescence of 8-20 racemes, forming an oblong panicle. 6-10 cm long, simple or divided; axis glabrous, axils bearded, joints and pedicels subtrigonous, about as long as the spikelets, shortly ciliate. Sessile spikelets 4.5-5 mm long, lanceolate-oblong, callus minutely or sparsely bearded; lower glume dorsally deeply channeled, chartaceous, 2-toothed; margins broadly inflexed from base to tip, 6-8-nerved; keels scabridly ciliate; upper glume chartaceous, keel scabrid,

3-nerved. Lower floret male. Lemma 4.5-6 mm long, oblong, membranaceous-hyaline, 3-nerved, the hyaline margins inflexed above and ciliate; palea equally long, hyaline, acute; stamens 3. Upper floret bisexual; lemma 3-4.5 mm long, ovate-oblong, membranaceous, hyaline, divided into 2 lobes to below the middle, lobes triangular, acute, glabrous, 3-nerved below, awned from the sinus; awn 4-8 mm long; palea as long as the lemma, linear-lanceolate, acute, hyaline, nerveless, glabrous. Pedicelled spikelets rather shorter than the sessile, for the rest similar except that the lower glume is not so deeply channed (description from Bor *loc.cit.*).

Patna, Bolangir (Panig. *et al.*, Bull. Bot. Surv. India 6: 263. 1964; Jain *et al.*, J.Bombay Nat. Hist. Soc. 72: 762. 1975). Fl.: November.

INDIA: Subtropical Himalayas, Khasia hills, Odisha.

WORLD: Nepal, Myanmar.

MNESITHEA Kunth

Mnesithea laevis (Retz.) Kunth, Revis. Gram. 1: 154. 1829; Fischer 3: 1761. 1934; Bor, Grass. India 197. 1960; Matthew, Fl. Tamilnadu Carnatic 3: 1876. 1983. *Rottboellia laevis* Retz., Obs. Bot. 3: 11. 1783. *R. perforata* Roxb., Pl. Coromandel 2: 43. t. 182. 1798; FBI 7: 158. 1896.

An annual, tufted grass. Culms up to 1 m high; nodes glabrous. Leaf sheaths 10 cm long, glabrous; ligule membranous; blades 50 x5 cm, linear-lanceolate, flat. Raceme solitary, 10 cm long, terminal and axillary; rachis fragile, with 2 sessile spikelets at each node, sunk into internode, separated by a narrow pedicel. *Sessile spikelet:* 2-flowered, dorsally compressed, falling entire. Lower glume coriaceous, 3-5-nerved, wingless; upper glume rigidly membranous; lower lemma hyaline, 3-nerved, 2-keeled, epaleated; upper lemma coriaceous, paleated; palea similar to its lemma; stamens 3. *Pedicelled spikelet:* reduced to a small scale or 0.

Rare in open grasslands and wet places. Fl. & Fr.: September-November. Tel.: *Penuku.*

Balapalle (KDP), *JLE* 14916 (CAL & MH); Palamaneru (CTR), *KCJ* 443 (MH); Bhimavaram (WG), *CAB* 4369 (CAL); Peddapuram (EG), *CAB* (MH); Kothuru (VSKP), *NPBK* 866 (CAL); Salur (VJN), *NPBK* 1166 (CAL); Baluga near Parvathipuram (SKLM), *VNS* 4554 (MH); Nungambakkam (CPT), *sine coll. s.n.* (MH); Nawna, Similipahar (MBJ), *SX* 381 (RRL-B); Sorada (GJM), *SX* 2892; Dhanei (GJM), *SX* & *MB* 4071 (RRL-B); Purunakote section, Satkosia Tiger Reserve (AGL), *KCM* 5182 (BSID); Budhakol (GJM), *AC* 10213 (BSID).

INDIA: Throughout India.

WORLD: Afghanistan, Pakistan, Sri Lanka, Myanmar, Malay Islands, Indo-China, Thailand, Indonesia.

OPHIUROS Gaertner

Ophiuros exaltatus (L.) Kuntze, Revis Gen. Pl. 2: 780. 1891; Fischer 3: 1758. 1934; Bor, Grass. India 199. 1960; Matthew, Fl. Tamilnadu Carnatic 3: 1876. 1983. *Aegilops exaltata* L., Mant. 2: 575. 1771. *Rottboellia corymbosa* L.f., Suppl. Pl. 114. 1781. *Ophiuros corymbosus* Gaertn. f., Fruct. 3. 4. t. 181. 1805; FBI 7: 160. 1896.

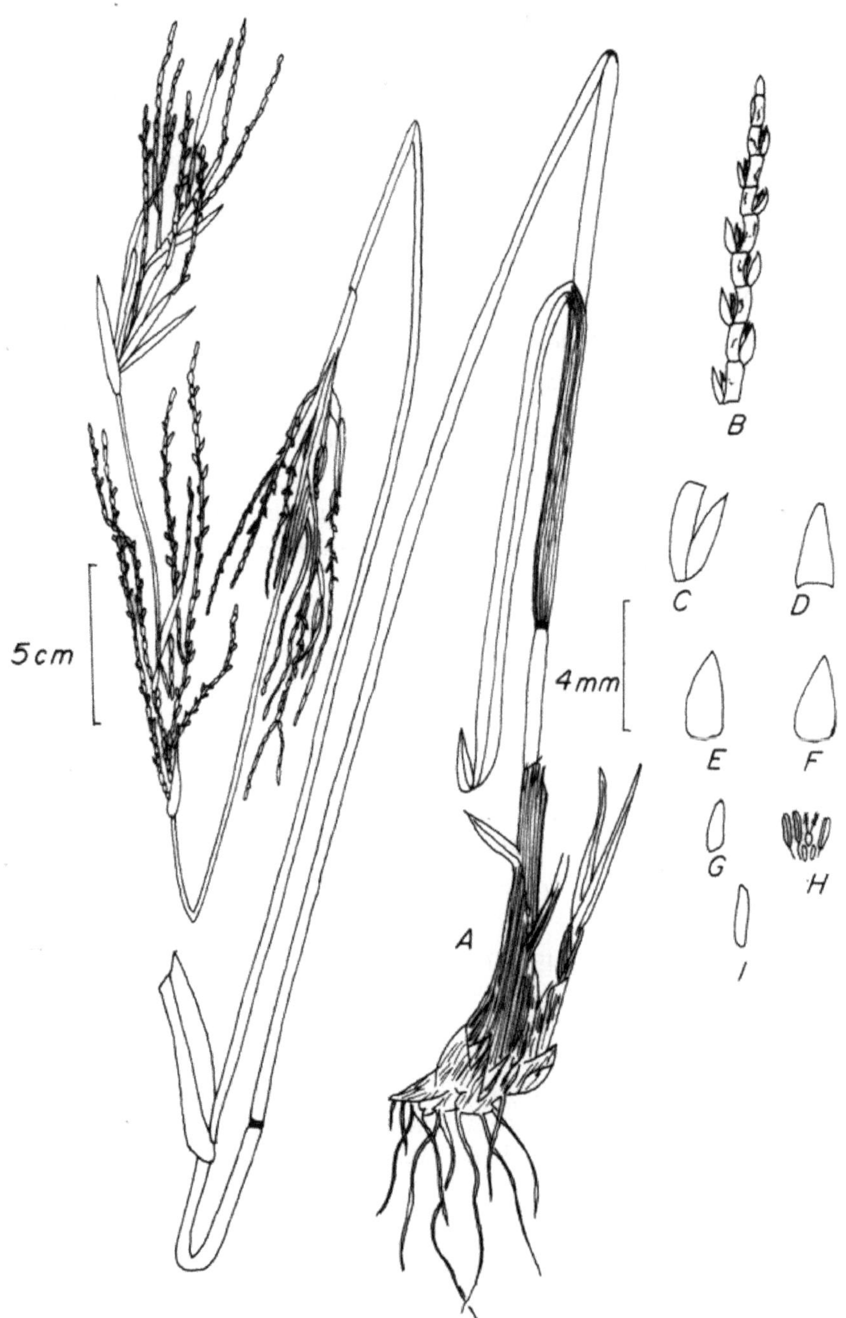

Figure 74. **Ophiuros exaltatus** (L.) Kuntze.

A. Twig, B. Part of the raceme, C. Spikelet, D. Lower glume, E. Upper glume, F. Lower lemma, G. Palea of the lower lemma, H. Lodicules, pistil and stamens.

A perennial, stoloniferous grass; culms up to 70 cm high, erect; nodes glabrous. Leaf sheaths 7-15 cm long, glabrous; ligule membranous; blades 10-20 x 0.5-1 cm, lanceolate, base cordate, margins minutely tubercled at base, apex acute, glabrous, midrib prominent. Spikes corymbose, up to 15 cm long; rachis fragile; joints completely fused with pedicels, hollow; spikelets 3 mm, awnless; lower glume oblong, coriaceous; upper glume oblong, boat-shaped, membranous; lower lemma elliptic, hyaline; palea lanceolate, hyaline; upper lemma lanceolate; palea linear; stamens 3, stigmas plumose. Caryopsis oblong slightly compressed.

Occasional grass particularly in black cotton soils. Fl. & Fr.: July-November. Vern.: Tel.: *Peddapanuku.*

Kammavaripalle (KNL), *TP & PVP* 3363; Vaddemanu (KNL), *PVP & RVR* 2539 & 2538; Pacherla (KNL), *BR* & *BSS* 30563; Sangam (NLR), *PMR* & *AMR* 23048; Kothajupudi (GNT), *SCA* 178 (DD); Kondapudi (EG), *CAB* 5195 (MH).

INDIA: Drier parts of Peninsular & N.E.India.

WORLD: Sri Lanka, China, Laos, Malaysia, Thailand, Vietnam, Australia.

OPLISMENUS P. Beauvois *nom. cons*

1. Perennial, diffuse grass; leaf blades elliptic–lanceolate; awns reddish, stout, smooth .... **O. compositus**
1. Annual, slender branching grass; leaf blades linear-lanceolate; awns pale, capillary, minutely scaberulous .... **O. burmanii**

Oplismenus burmannii (Retz.) P. Beauv. Ess. Agrost. 54, 168, 169. 1812; FBI 7: 68. 1896; Fischer 3: 1778. 1934; Bor, Grass. India 317. 1960. *Panicum burmannii* Retz., Observ. Bot. 3: 10. 1783.

An annual, slender grass; culms up to 50 cm high, creeping, diffusely branched at base; nodes glabrous-sparsely hairy. Leaf sheaths 15 cm long, below the nodes, margins densely ciliate; ligule a ridge of hairs; blades 2-5 x 0.5-1 cm, ovate-lanceolate, base amplexicaul, apex acuminate, sparsely pilose on both sides. Spiciform secund racemes up to 6 cm long, terminal; rachis slender, trigonous; branches distant; spikelets elliptic-ovate, 3 mm, solitary, 2-nate, awned; glumes ovate, hairy, awned, lower 3-nerved, upper 5-nerved; lower lemma ovate, margins curved, hairy on margins, aristate; upper lemma ovate, margins curved; palea linear; stamens 3; stigmas plumose. Caryopsis linear.

Common grass in cultivated fields and shades of high trees in the forests. Fl. & Fr.: August-December.

Alampur (ATP), *BR* & *P.Priyadarsini* 37176 (SKU & BSID); Upper Ahobilam (KNL), *TP* & *RVR* 2600, *TP & GNVF* 7064; Jayanthipuram (GNT), *VRK* 6952; Panapakkam (CTR), *S. India Flora* 15772 (MH); Chelama (KNL), *JLE* 22025 (CAL & MH); Diguvametta (PKM), *JLE* 32414 (MH); Bison hills (EG), *CAB* 5099 (MH); Towards Chinturu from Maredumilli (EG), *MM* 102502 (BSID); Simhachalam hills (VSKP), *GVS* 21498 (MH); Palasa (SKLM), *SCA* 179 (CAL & DD); Amrabad (MBNR), *BR* & *BSS* 32205 (SKU & BSID); Mukundapur (GJM), *VNS* 4682 (MH); Ramanadrug (Ballari), *S. India Flora* 15915 (MH).

INDIA: Plains of India.

WORLD: Widespread in the tropics of both the spheres.

Oplismenus compositus (L.) P. Beauv., Ess. Agrost. 54, 168, 169. 1812; FBI 7: 66 1896; Fischer 3: 1778. 1934; Bor, Grass. India 317. 1960; Matthew, Fl. Tamilnadu Carnatic 3: 1877. 1983. *Panicum compositum* L., Sp. Pl. 57. 1753.

A perennial, stoloniferous grass; culms up to 60 cm high, creeping, rooting at nodes; nodes glabrous. Leaf sheaths 2-9 cm long, margin densely ciliate; ligule fimbriate membrane; blades 3-10 x 1-2 cm, elliptic-lanceolate, base amplexicaul, apex acuminate, sparsely pubescent. Spiciform racemes up to 20 cm long, terminal; branches distant, long; spikelets 4 mm, elliptic, 2-flowered, awned; lower glume ovate, membranous, scaberulous above, awned; upper glume oblong, membranous, aristate; lower lemma ovate-oblong, membranous, ciliate on margin, shortly aristate; palea oblong; upper lemma ovate; palea ovate; stamens 3, stigmas plumose. Caryopsis linear.

Common in damp, shady places in the forests. Fl. & Fr.: November-January. Tel.: *Kodijuttugaddi.* Tam.: *Kozhi arugampul.*

Batrepalli (ATP), *KRKS* & *BSS* 40350; Upper Ahobilam RF (KNL), *TP & RVR* 2769, *TP & GNVF* 7065; Rollapenta (KNL), *PVP & RVR* 2349; Guvvalacheruvu (KDP), *R.V. Reddy* 8053; Talakona RF (CTR), *PVP* 2317; Paladhara Panchadhara (KNL), *JLE* 22150 (CAL & MH); Balapalle (KDP), *S. India Flora* 15418 (MH); Thummalabailu (KDP), *MH,* Acc. No. 55310 (MH); Panapakkam (CTR), *MH* Acc. No. 55304 (MH); Bison hills (EG), *CAB* 5036, 5096 (MH); Jaddangi (EG), *S. India Flora* 12638 (MH); Araku valley (VSKP), *NPBK* 571 (CAL); Umamaheswaram (MBNR), *SRS* 111365 (BSID); Mallelatheertham (MBNR), *SRS* 108853, 111401 (BSID); Tulka section, Satkosia Tiger Reserve, *KCM* 5625 (BSID); Jenabil Range, Sililipal B.R. (MBJ), *AC* 2333 (BSID); Sirumalai hills (DGL), *SKS* 426 (SGH); Kolli hills (NMK), *SKS* 362 (SGH); Vandalur (CPT), *S. India Flora* 11486 (MH); Kalpadai (SA), *S. India Flora* 17852 (MH); Pattaraikadu RF, Javadi hills (NA), *MBV* 1346 (MH); Alagar hills (DGL), *K. Althaf Ahmed Kabeer* 113275 (MH); Orchidarium, Shevaroys (SLM), *AVNR* 18286 (MH).

INDIA: Throughout India.

WORLD: Tropical regions of the Old and New Worlds.

OROPETIUM Trinius

1. Spikelets 1-flowered:
    2. Lemmas glabrous .... **O. thomaeum**
    2. Lemmas villous .... **O. villosulum**
1. Spikelets 1-2-flowered .... **O. roxburghianum**

Oropetium roxburghianum (Steud.) S.M. Phillips, Kew Bull. 36: 469. 1975. *Lepturus roxburghianus* Steud., Syn. Pl. Glum. 1:357. 1854; FBI 7: 365. 1896. *Tripogon roxburghianus* (Steud.) Bhide, J. Proc. Asiat. Soc. Beng. 1911 (n.s.) 7: 515. 1912; Fischer. 3: 1834. 1934; Bor, Grass. India 524. 1960.

Figure 75. **Oplismenus compositus** (L.) P. Beauv.

A. Twig, B. Spikelet, C. Lower glume, D. Upper glume, E. Lower lemma, F. Upper lemma, G. Palea, H. Grain.

Annual, culms erect, tufted, slender, to 15 cm high; nodes glabrous. Leaf sheaths glabrous; ligules membranous; leaf blades 5-8 cm long, filiform, margin ciliate. Spike solitary; spikelets 1-2-flowered; lower glume oblique, hyaline, 2-lobed, 1-nerved; upper glume coriaceous, 3-nerved; lower lemma membranous, bidentate, awned in the cleft, ciliate in lower part, callus bearded; upper lemma bisexual or imperfect or 0; palea 2-keeled, scabrid; stamens 3. Caryopsis terete, oblong-ellipsoid, pointed at one side.

Rare in moist rock crevices in southern Eastern Ghats. Fl. & Fr.: July–December.

Gooty fort (ATP), *BR* & *JTC* 46657; Talakona (CTR), *BR* & *BSS* 32929 (SKU & BSID); Puttalur (NLR), *CECF* 4136 (CAL); Karempudi (GNT), *South India Flora* 17451 (MH).

INDIA: Peninsular India, Endemic.

Oropetium thomaeum (L.f.) Trin., Fund. Agrost. 98. t.3. 1820; FBI 7:336.1896; Fischer 3:1830.1934; Bor, Grass. India 474. 1960; Matthew, Fl. Tamilnadu Carnatic 3: 1878. 1983; Moulik, Grass. Bamb. India 2:617.1997. *Nardus thomaea* L.f., Suppl. 105.1781. *Rottboellia thomaea* (L.f.) Koenig, Naturforscher 23: 210. 1788.

Annuals, culms erect, tufted, to 6 cm high; nodes glabrous. Leaf sheaths small, compressed; ligules membranous; leaf blades 1.5-2.5 cm long, linear, filiform, base rounded, acute, ciliate on margins. Spikes solitary, terminal, slender; spikelets sessile, lanceolate, solitary, sunk in the alternatingly distichous hollows in the rhachis, 1-4-flowered; lower glume minute; upper glume linear-lanceolate; lemmas ovate-oblong, hyaline, 1-3-nerved; paleas lanceolate, 2-keeled; stamens 3. Caryopsis oblong.

Common in forest plains in dry places, throughout. Fl. & Fr.:July-November.

Batrepalli (ATP), *KRKS* & *BSS* 39610; Chendrayunikona (ATP), *BR* & *P.Priyadarsini* 37061; Devaragattu (KNL), *SS* & *AMR* 22671; Rapur (NLR), *AMR* 24563; Thummalabailu (PKM), *BSS* & *SKB* 29383; Pachipenta (VJN), *AMR* 24544; Gooty (ATP), *JSG* 20839 (CAL & MH); Guvvlacheruvu (KDP), *KS* 6410 (MH); Agasthyeswarakona, Lingapuram (KDP), *J.Swamy* & *S.Nagaraju* 14707 (BSID); Thippukadu (NA), *KRM* 17650 (MH); Uragadam (CTP), *South India Flora* 14073 (MH); Sithanur (CTP), *South India Flora* 11216 (MH); Hogainakkal (SLM), *EV* 21985 (MH); Talkudar, Rajgangpur (Sundargarh), *AC* 10264 (BSID).

INDIA: Throughput India.

WORLD: Tropical Africa, Arabia, Indo-China.

Oropetium villosulum Stapf ex Bor, Kew Bull. 1949: 571. 1950 & Grass. India 474. 1960; Mooney, Suppl. Bot. Bihar & Orissa 261. 1950; Saxena & Brahmam, Fl. Orissa 2421. 1996.

Perennial, tufted grass, 1.2-3.7 cm high, culms scabrous, rather dichotomously branched, leafy in upper part. Leaves rigid, linear, up to 2.5 cm long, acute; sheaths striate, glabrous, with hyaline margins, ligule very short. Spikes simple, terminal,

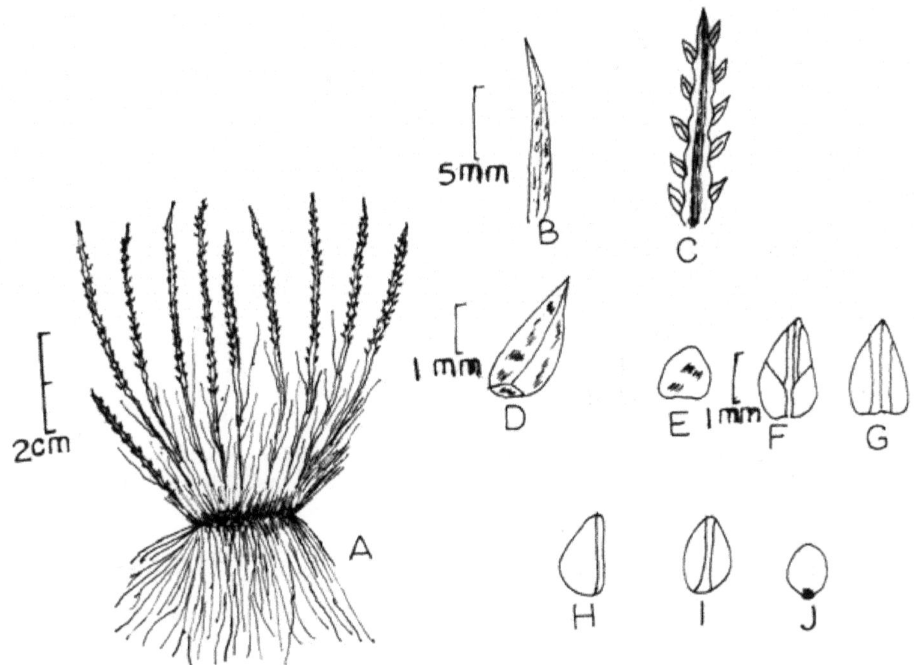

Figure 76. **Oropetium thomaeum** (L.f.) Trin.

A. Twig, B. Leaf blade, C. Part of the raceme, D. Spikelet, E. Lower glume, F & G. Upper glume front and back view, H. Lemma, I. Palea, J. Grain.

slender, to 1.2 cm long; spikelets solitary, sessile, 2-2.5 mm long, 1-flowered, embedded in the depressions in the rachis; lower glume minute, under 1 mm or 0, hyaline, hemispherical, nerveless; upper glume rigid, dorsal aspect oblong-obtuse, 2 mm long, coriaceous, 3-nerved; lemmas elliptic, 1.5 mm long, as broad, truncate, 3-nerved, villous on its dorsal surface and margins; palea as long as the lemma, 2-keeled 2-nerved. Terminal spikelet 3.7 mm long; glumes equal, long, coriaceous, lanceolate, acuminate with recurved apices. Caryopsis 1.5 mm long.

Rare in rock crevices in drier habitats. Fl.: July - September.

Pathikonda (KNL), *BR* & *KVS* 29126; Sonabera (KRT), *HFM* 3660 (DD).

INDIA: Andhra Pradesh, Bihar, Madhya Pradesh, Odisha, Maharashtra, Karnataka. Endemic.

ORYZA Linnaeus

1. Ligule of the lower leaf blades less than 0.6 cm long:
   2. Fertile lemma awned:
      3. Erect annual grass; leaf sheath purple line inside .... **O. jeyporensis**
      3. Decumbent perennial grass; leaf sheath green inside................ **O. minuta**
   2. Fertile lemma not awned:
      4. Panicle spiciform, branches 3-10 cm long, surface of the lemma and palea granulate.. **O. meyeriana**
      4. Panicle lax, branches spreading, 10-30 cm long, surface of the lemma and palea smooth, not granulate........ **O. officinalis**
1. Ligule of the lower leaf blades 1-5 cm long:
   5. Spikelets persistent, unawned, cultivated .... **O. sativa** var. **sativa**
   5. Spikelets caducous, awned, wild .... **O. rufipogon**

Oryza jeyporensis Govindasw. & K.H. Krishnam., Sci. & Cult. 24: 236. 1958; Bor, Grass. India 606. 1960; Saxena & Brahmam, Fl. Orissa 4: 2423. 1996.

Erect annual grass, up to 1 m high, internodes 20-25 cm long, greenish yellow with purple lines, septum purple. Leaves broadly lanceolate, 30-35.5 x 1.5-1.8 cm, deep green, hairy, margin entire, apex acute; leaf sheaths green, purple lined inside with deep purple axils; ligule white, triangular, 3-5 mm long. Panicles erect and lax, 22-26 cm long; spikelets broadly oblong, 7-8.8 mm long, 2-2.5 mm broad, with long awns that are pigmented purple, awns 3-6.8 cm long; sterile glumes green, 0.9-1 cm; sterile lemma acute, considerably longer than the fertile lemma, 5-nerved, hairy and deep purple apiculas; palea 3-nerved, hairy. Mature spikelets hard, dull brown and non-shattering. Caryopsis non-glutionous and dirty whte.

Rare in Jeypore in Eastern Ghats of Odisha.

INDIA: Odisha, Endemic.

Oryza meyeriana (Zollin. & Mor.) Baill., Hist. Pl 12: 166. 1894. *O. myeriana* subsp. *granulata* Tateoka, Hot. Mag. Tokyo 75: 460. 1962; Fischer 3: 1845. 1934. *O. granulata* Nees & Arn. ex Watt., Dict. Econ. Prod. Ind. 5: 500. 1891; FBI 7: 96. 1896; Bor, Grass. India 604. 1960.

A perennial, tufted grass; culms up to 90 cm high. Leaf sheaths 5- 7 cm long, glabrous; ligule short, membranous; blades 7-15 x 1- 1.5 cm, lanceolate, glabrous. Racemes 8 cm long, very slender; spikelets 6 m long, obtuse, unawned; lower glume absent; upper glume linear; lemma densely irregularly granulate, glabrous; palea elliptic-oblong, granulate, glabrous; stamens 6. Caryopsis ellipsoid or narrowly oblong.

Occasional grass in streams in northern and central Eastern Ghats. Fl. & Fr.: July-December.

Talakona RF (CTR), *MSG &. AA* 12231; Gundlabrahmeswaram RF (KNL), *JLE* 16953 (MH); Chintapalli (VSKP), *GVS* 30087 (CAL); Rampa hill (EG), *MSR* 1524 (CAL); Maredumilli - Kakur (EG), *GVS* 67590 (MH); Bison hill (EG), *CAB* 5058 (MH); Bodalanka (EG), *DN* 85559 (BSID); Chintapalli towards Narsipatnam (VSKP), *GVS* 30087 (MH); on the way to Cherukonda (VSKP), *GVS* 42750 (MH); Madgole to Paderu (VSKP), *GVS* 28107 (MH); Tulka section, Satkosia Tiger Reserve (AGL), *KCM* 5692 (BSID); Betkote, Karlapat Wildlife Sanctuary (KHD), *AC* 2464 (BSID).

INDIA: Eastern India, West Bengal, Bihar, Sikkim, Uttar Pradesh, Odisha, South India.

WORLD: Sri Lanka, Myanmar, Thailand, Laos, Cambodia, China, Malesia.

Oryza minuta J. Presl., Reliq. Haenk. 1: 208. 1830; Bor, Grass India 605. 1960; Saxena & Brahmam, Fl. Orissa 4: 2424. 1996. *O. latifolia* auct. non Desv., FBI 7: 92. 1896.

A perennial, caespitose grass; culms geniculately ascending or decumbent, 50–100 cm long. Leaf blades 14–28 x 0.6–1.3 cm wide, scaberulous, rough on both sides, margins scabrous. Inflorescence a panicle, comprising 6–20 fertile spikelets, ovate, 8–17 cm long; primary panicle branches appressed or ascending or spreading, 1–2–nate, whorled at lower nodes, simple, 6–10 cm long, panicle branches scaberulous, spikelets solitary. *Fertile spikelets* pedicelled; pedicels linear, angular, 0.5 long, comprising 2 basal sterile florets and 1 fertile floret; florets without rhachilla; spikelets elliptic or ovate, laterally compressed, 3.8–4 mm long, 1.75 mm wide; glumes both absent or obscure; basal sterile florets similar, barren, without palea; lemma of lower sterile floret lanceolate, 1.5 mm long, 0.3–0.4 length of spikelet, 1–veined, without lateral veins, acuminate; lemma of upper sterile floret lanceolate, 1.5 mm long, 1 length of lower sterile floret; fertile lemma oblong or orbicular, laterally compressed, 3.8–4 mm long, coriaceous, dark brown or black, keeled, 5–veined; lemma midvein spinulose, surface scabrous and reticulate, margins interlocking with palea margins; lemma apex awned; 1–awned; principal lemma awn 3–12 mm long overall; palea elliptic, coriaceous, 3–veined, 1-keeled; palea keels

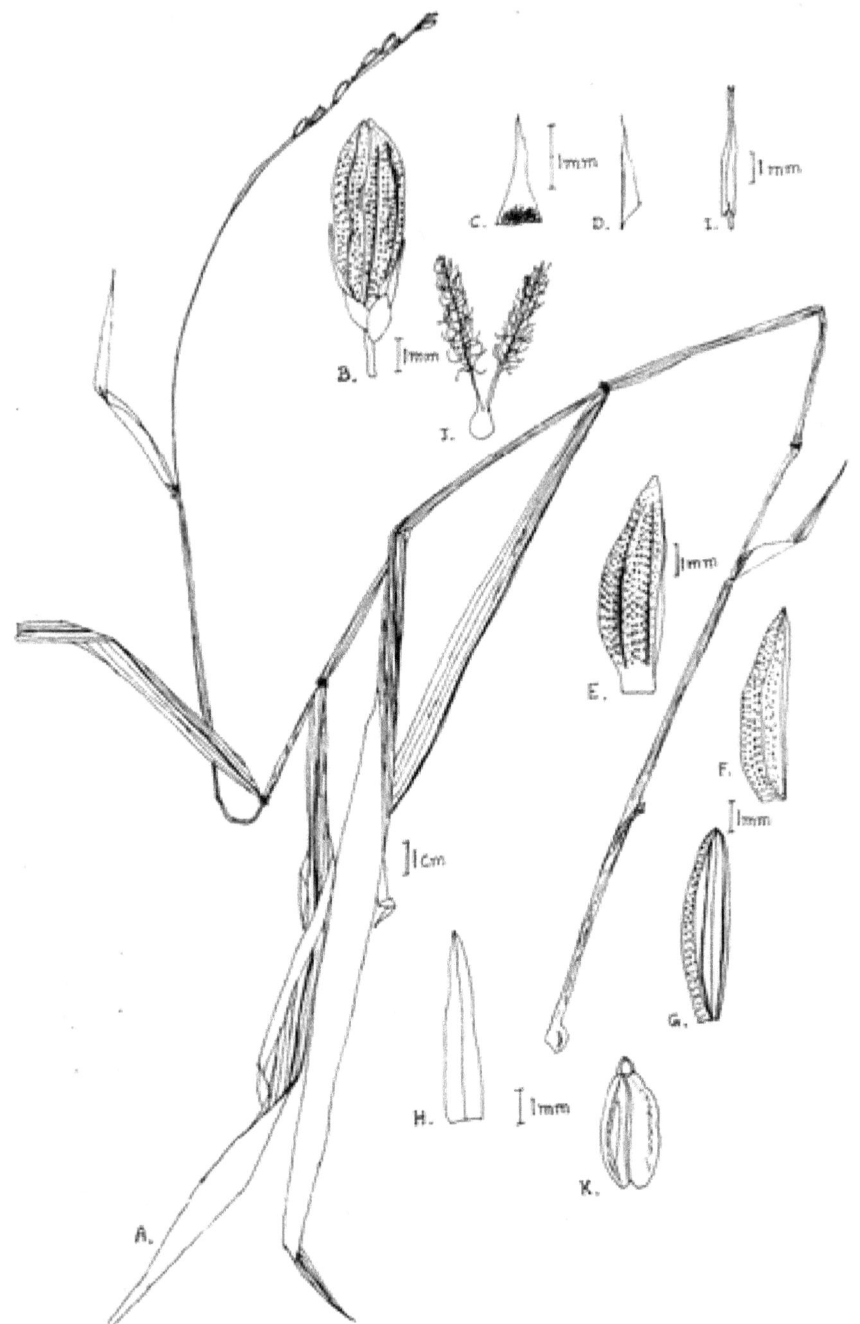

Figure 77. **Oryza meyeriana** (Zollin. & Mor.) Baill. subsp. **granulata** Tateoka.

A. Twig, B. Spikelet, C. Lower glume, D. Upper glume, E. Lemma, F & G. Palea side and front views, H. Sterile lemma, I. Stamen, J. Pistil, K. Caryopsis.

spinulose, palea surface scaberulous, apex acute, awned, awns 1 mm long; lodicules 2, membranous; anthers 6, 2 mm long; stigmas 2. Caryopsis with adherent pericarp.

Khejari hills, Similipahar, in thick Sal forest on stream banks in marshy land; very rare and sporadic (Malik and Dikshit, *loc cit.*). Fl.: January.

Jenabil Range, Similipal B.R. (MBJ), *AC* 2327 (BSID).

INDIA: Andhra Pradesh, Odisha, Maharashtra, Madhya Pradesh, Assam, Sikkim.

WORLD: Asia-tropical: India, Indo-China, Malesia, and Papuasia.

Oryza officinalis Wall. ex Watt, Dict. Econ. Prod. India 5: 501. 1891; Malik & Dikshit, J. Econ. Taxon. Bot. 14: 605. 1990. *O. officinalis* Wall. ex Watt, subsp. *malampuzhaensis* (Krish. & Chandr.) Tateoka, Bot. Mag. Tokyo 75: 422. 1962. *O. malampuzhaensis* Krish. & Chandr., Madras Agric. J. 45: 471. 1958; Bor, Grass. India 606. 1960.

A perennial, stoloniferous grass; culms up to 50 cm high, trailing, branching at base; nodes glabrous. Leaf sheaths overlapping, keeled, glabrous; ligule membranous with a fringe of hairs at apex; leaf blades 20-50 x 1-2 cm, linear-lanceolate, flat, base rounded, apex acuminate, glabrous, midrib prominent. Panicle lax, up to 10 cm long; rachis and branches scaberulous; spikelet 5 mm, oblong, deciduous, awned; glumes triangular, 1-nerved, scaberulous, deciduous; lemma boat-shaped, apex acute, sparsely hairy over the outer surface; palea narrowly lanceolate, sparsely hairy over the outer surface; stamens 6, style plumose. Caryopsis oblong, brownish white.

Rare in Nallamalais in Andhra Pradesh and Khejari hills in Similipal forest in Odisha along the margins of the river. Fl. & Fr.: September-January.

Gundlabrahmeswaram (KNL), *PVP* & *RVR* 2391, *BR* & *BSS* 32274, 322276, *JLE* 22158, 32665 (MH).

INDIA: Assam, Sikkim, Odisha, Andhra Pradesh, Madhya Pradesh and Maharashtra,

WORLD: S. and SE. Asia.

Oryza rufipogon Griff., Notul. 3: 5. 1851; Bor, Grass. India 605. 1960; Moulik, Grass. Bamb. India 1: 4. 1997.

A perennial, stoloniferous, aquatic grass; culms up to 1 m high, rooting at the nodes, spongy below; nodes glabrous. Leaf sheaths overlapping, glabrous; ligule membranous; leaf blades 20-30 x 0.2-0.6 cm, linear, flat, base rounded, apex acuminate, scaberulous. Panicle up to 15 cm long, drooping; spikelets linear-oblong, 7 mm, awned; glumes triangular, acuminate, 1-nerved, scaberulous; lemma boat-shaped, 3-nerved, scabrid on the outer surface, awned, coriaceous; palea linear, scabrid on the outer surface, coriaceous; stamens 6; stigmas plumose. Caryopsis narrowly oblong, free with in the lemma and palea.

Occasional aquatic grass found near the pools and ditches. Fl. & Fr.:August-December.

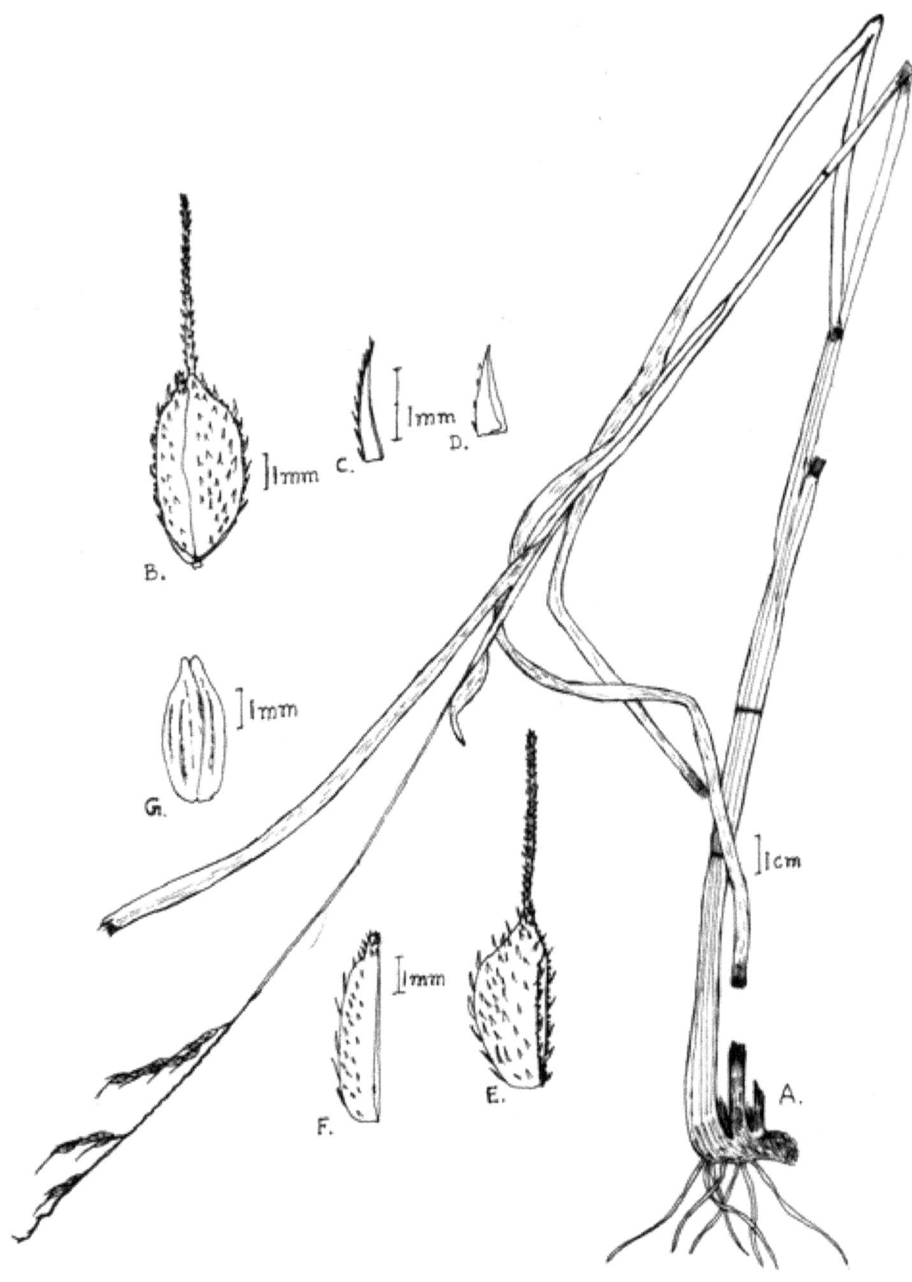

Figure 78. **Oryza officinalis** Wall. ex Watt,

A. Twig, B. Spikelet, C. Lower glume, D. Upper glume, E. Lemma, F. Palea, G. Caryopsis.

Neelithotti Kandriga, Satyavedu (CTR), *MCB* 45242 (MH); Parlakhemudi (GJM), *SX* & *MB* 3862 (RRL-B); Malkngiri - Chitrakonda (KPT), *SX* & *MB* 6824, 6825 (RRL-B); Thakurmunda (MBJ), *SX* & *MB* 5373 (RRL-B); Purunakote section, Satkosia Tiger Reserve (AGL), *KCM* 5251 (BSID).

INDIA: Peninsular, E. and NE. India.

WORLD: Sri Lanka, Bangladesh, Myanmar, China, Cambodia, Indonesia, Malaysia, Philippines, Thailand, Vietnam, Australia.

Oryza sativa L., Sp. Pl. 333. 1753; FBI 7: 92. 1896; Fischer 3: 1844. 1934; Bor, Grass. India 605. 1960; Matthew, Fl. Tamilnadu Carnatic 3: 1878. 1983. var. **sativa**

An annual, cultivated grass; culms up to 1 m high, creeping or floating at base; nodes glabrous. Leaf sheaths 15-25 cm long, glabrous; ligule membranous, 2-partite; blades 20-50 x 4-7 cm, linear, flat, base rounded, margin scabrid, apex acuminate, glabrous, midrib prominent. Panicle up to 30 cm long, lax, compound; spikelets 5 mm, linear-oblong, numerous, unawned; glumes oblong, scaly, 1-nerved; lemma boat-shaped, coriaceous, punctate, dorsally ciliate, 3-nerved; palea linear, scabrid on the outer surface; stamens 6; stigma plumose. Caryopsis narrowly oblong, compressed.

Widely cultivated for its grains, often found as an escape. Fl. & Fr.: August-February. Tel.: *Vadlu.*

Chinnakunta (ATP), *PVP* 2637; Namalagundu (KDP), *KRKS* 40166; Krishnnandi (KNL), *VBH* 86694 (BSID); Kolleru lake (WG), *KS* 5082 (MH); Leelanagar (KSN), *PV* 5524 (MH); Srungavarapu Kota (VJN), *GVS* 21816 (MH).

INDIA: Cultivated throughout.

WORLD: Native of SE Asia, now cultivated all over the World in Tropical to Mediterranean region.

OTTOCHLOA Dandy

Otochloa nodosa (Kunth) Dandy, J. Bot. 69: 55 1931; Fischer 3: 1778.1934; Bor, Grass. India 318. 1960; Kabeer & Nair, Fl. Tamil Nadu Grass. 265. 2009; Kumar *et al.*, J. Threatened Taxa 9: 10974. 2017.

Annual or perennial mat forming grass; culms up to 60 cm long, decumbent, rooting from lower nodes. Leaf sheaths glabrous to sparsely pubescent on surface, densely ciliate on one margin, 2.5–4.5 x 0.2 cm; ligules eciliate, membranes; leaf blades elliptic - ovate or lanceolate, base unequal or shallowly cordate, apex acuminate, surface glabrous, margins scabrous, 5–10 x 0.8–1.5 cm. Inflorescence of racemes, branches appressed to raceme axis or effuse; spikelets single or 2–3 in a cluster on secondary branches; spikelets 2, lower barren, upper fertile; glumes 2, lower ovate, membranous, margins ciliate,1.6-1.8 mm long, prominently 5 nerved; upper ovate - lanceolate, membranous, margins ciliate, *c.* 2 mm long, 7 - nerved. Lower floret barren, without significant palea; lemma of lower sterile floret akin to upper glume, first length of fertile lemma, membranous, 2.9–3.1 mm long, 7–nerved; upper fertile lemma coriaceous, margins thin, involute, 2.8–3 mm long, 5 - nerved;

palea coriaceous, glabrous, involute, sub equal to its lemma, 2.5–2.8 mm long, 2–nerved; stamen 3, anthers 1.2 mm long; stigma plumose.

Occasional; along perennial water bodies in the interior of the forest. Fl. & Fr.: July–December.

Maredumilli forest (EG), *BR* & *M. Anil Kumar* 49680 (SKU).

INDIA: Andhra Pradesh, Assam, Karnataka, Kerala, Manipur, Meghalaya, Nagaland, Sikkim, Tamil Nadu.

WORLD: West Africa.

Note: *Ottochola* is closely allied to the genus *Panicum*, but is distinguished from the later by its shorter upper glumes. The leaf sheaths of *Ottochloa nodosa* are conspicuously ciliate on one margin and are a key vegetative character to identify the species.

PANICUM Linnaeus

1. Leaf blades linear-lanceolate:
    2. Upper lemma transversely rugose:
        3. Panicle to 45 cm long, effuse; nodes glabrous .... **P. auritum**
        3. Panicle to 25 cm long, lax; nodes bearded .... **P. maximum**
    2. Upper lemma smooth:
        4. Lower glume more than half as long as the spikelet:
            5. An annual grass; leaf blades base rounded or truncate; caryopsis ovoid:
                6. Cultivated; nodes bearded; spikelet not gaping at maturity:
                    7. Lower glume cuspidate or cuspidate-acuminate; spikelets 1.5-1.7 mm long, slightly compressed, gibbous .... **P. sparsicomum**
                    7. Lower glume and spikelets not as above .... **P. miliaceum**
                6. Wild; nodes glabrous; spikelets gaping at maturity .... **P. curviflorum**
            5. A perennial grass; leaf blades base amplexicaul; caryopsis ellipsoid .... **P. notatum**
        4. Lower glume less than half as long as the spikelet:
            8. Culms spongy, lower glume nerveless .... **P. paludosum**
            8. Culms not spongy, lower glume prominently nerved:
                9. Leaf-sheaths margins densely ciliate; lower glume suborbicular, apex truncate:
                    10. Stoloniferous grass; leaf seath margin densely ciliate; spikelet 3 mm long .... **P. repens**

10. Erect grass; leaf sheath mrging not ciliate; spikelet 1.5 mm long .... **P. humidorum**

9. Leaf-sheath margins glabrous; lower glume ovate, apex acute or acuminate:

11. Panicle effuse; caryopsis ellipsoid, wild:

12. An annual; leaf blades linear-lanceolate, upper glume 3 or 10-nerved:

13. Lower glume 5-nerved; upper glume 10-nerved .... **P. fischeri**

13. Lower glume 3-nerved; upper glume 5-nerved .... **P. humile**

12. A perennial, tall grass; leaf blades elliptic-lanceolate; upper glume 9-nerved.......... **P. sumatrense**

11. Panicle contracted, caryopsis ovoid, cultivated...... **P. antidotale**

1. Leaf-blades ovate or ovate-lanceolate.......... **P. brevifolium**

Panicum antidotale Retz., Observ. Bot. 4: 17: 1786. *P. miliare* Lam.,Tabl. Encycl. 1: 173. 1791; FBI 7: 46. 1896; Fischer 3: 1782. 1934; Bor, Grass. India 329. 1960.

An annual, tufted grass; culms up to 1 m high; nodes glabrous. Leaf sheaths 4-10 cm, glabrous; ligule membranous; leaf blades 7-15 x 0.5-1 cm, flat, base rounded. Panicle 25 cm long, thyrsiform or contracted; lower branches alternate; spikelets 5 mm, ellipsoid, glabrous; lower glume broadly ovate, 3-5-nerved, obtuse; upper glume ovate-lanceolate, membranous, 9-13 nerved, obtuse; lower lemma 9-nerved; palea equal to the length of its lemma, neuter; upper lemma elliptic-oblong, 3-5-nerved, glabrous; palea similar to its lemma; stamens 3. Caryopsis ovoid.

Occasionally cultivated grass. Fl. & Fr.: October-December. Tel.: *Gangasamulu, Konda samalu;* Tam.: *Saamai.*

Pinjari konda (EG), *TP & MSG* 12288; Kuppam (CTR), *sine coll. s.n.* (MH); Tadepalli (EG), *NRR & DN* 84326 (BSID); Mandasa mendu (VSKP), *VNS* 4626 (MH); Tadikonnda (VJN), *NRR* 83633 (BSID); Kolli hills (NMK), *SKS* 602 (SGH); Pudur Nadu, Javadi hills (NA), *MBV* 1186 (MH); Kuthadya hill (GJM), *VNS* 5895 (MH).

INDIA: N.W. India, Upper Gangetic plains, Maharashtra, Andhra Pradesh, Odisha, Tamil Nadu.

WORLD: Afghanistan, Iran.

Panicum auritum Presl ex Nees, Agrost. Bras. 176. 1829; FBI 7: 40. 1896; Haines, Bot. Bihar & Orissa 3: 996. 1924; Bor, Grass. India 324. 1960.

Perennial, erect grass, nodes glabrous. Leaves broadly linear, 20-35 x 1.8-3 cm, glabrous or sparsely hairy beneath, base broadly cordate; sheaths glabrous or sparsely hairy with villous mouth; ligule very short, membranous. Panicle green, compound, 20-45 cm long, contracted or more or less effuse, much fastigiately branched; spikelets numerous, lanceolate, 1.8-2.25 mm long, sessile or shortly pedicelled, strongly nerved, acute or subacute; lower glume ovate, one third the length of the spikelet, acute or obtuse, 3-5-nerved; upper glume acute or acuminate, 5-nerved; lower lemma neuter with small palea; upper lemma thinly coriaceous, white, as long as the lower lemma.

Occasional in Eastern Ghats of Odisha.

INDIA: NE & E. India.

WORLD: Tropics of SE Asia.

Panicum brevifolium L., Sp. Pl. 59. 1753; Fischer 3: 1783. 1934; Bor, Grass. India 324. 1960. *P. ovalifolium* Poir. in Lam., Encycl. Suppl. 4: 279. 1816; FBI 7: 44. 1896.

A perennial, tufted grass; culms up to 40 cm high, weak, decumbent and rooting at base. Leaf sheaths slightly glabrous; ligule very short, ciliate; leaf blades 2.5-8 x 0.5-1.5 cm, ovate-lanceolate, base amplexicaul, apex acute-acuminate. Panicle effuse, ovoid; spikelets 3 mm, solitary, sparsely hairy; glumes unequal, puberulous; upper glume gibbose; lower lemma equal to the length of upper glume, puberulous; palea broad; upper lemma oblong, shorter than the lower lemma, margins incurved.

Common near water courses in the plains and along the hill slopes. Fl. & Fr.: November-February.

Tirumala hill (CTR), *MSG* 12172; Kambakkam hills (CTR), *sine coll. s.n.* (MH); Bhupatipalem, Rampa hills (EG), *TP & MSG* 12250; Near about Sesharayi (EG), *VNS* 183 (CAL); Bison hills (EG), *CAB* 5093 (MH); Maredumilli (EG), *MM* 101305 (MH & BSID); Koinpur (GJM), *SX* 1224 (RRL-B); Parlakhemudi (GJM), *SX* & *MB* 3846; Harishankar (BL), *SX* & *MB* 5789; Meghasani, Similipahar (MBJ), *SX* & *MB* 4256 (RRL-B); Raigoda, Satkosia Tiger Reserve (AGL), *KCM* 7747 (BSID); Jenabil Range, Similipal BR (MBJ), *AC* 456 (BSID).

INDIA: Throughout Tropical and subtropical forests of India.

WORLD: Tropical Africa, Tropical Asia.

Panicum curviflorum Hornem., Hort. Bot. Hafn. Suppl. 116. 1819. *P. trypheron* Schult., Syst. Veg. 2. Mant. 244. 1824; FBI 7: 47. 1896; Fischer 3: 1783. 1934; Bor, Grass. India 331. 1960; Matthew, Fl. Tamilnadu Carnatic 2: 1883. 1983; Saxena & Brahmam, Fl. Orissa 4: 2437. 1996.

An annual, tufted grass; culms up to 60 cm high; nodes glabrous. Leaf sheaths 2-8 cm long, sparsely ciliate-glabrous; ligule membranous, with long pilose tufts; blades 2-12 x 0.2-0.6 cm, linear-narrowly lanceolate, base truncate, apex acuminate, pilose. Panicle 15 cm long, open; branches alternate; spikelets 3 mm, ellipsoid, gaping at maturity; glumes ovate, acuminate-cuspidate; lower glume 3-5-nerved, scabrid; upper glume 7-9- nerved; lower lemma ovate, acute; palea male or empty; upper lemma pallid, smooth; palea similar to its lemma; stamens 3; stigmas plumose. Caryopsis ovoid.

Common grass in moist plains and hill slopes. Fl. & Fr.: October–March.

Figure 79. **Panicum brevifolium** L.

A. Twig, B. Spikelet, C & D. Lower glume front and back views, E. Upper glume, F. Lower lemma, G. Palea, H. Upper lemma, I. Palea, J. Stamens, K. Pistil, L. Caryopsis.

Kailasanathakona (CTR), *BR* & *P.Priyadarsini* 37139 (SKU & BSID); Udayagiri (NLR), *BR* & *BSS* 29303 (SKU & BSID); Gajulapalle (KNL), *KCJ* 83136 (MH); Diguvametta (KNL), *CAB* 8114 (MH); Macherla (GNT), *SCA* 131 (DD); Kadapa, *EKK* 78521 (DD) & *KS* 6315 (CAL); Medraitta (KDP), *MPG* 159 (CAL); Dongalacheruvu (KDP), *KS* 6315 (MH & CAL); Balapalle (KDP) *MH Acc.* No. 95119 (MH); Rajugunta (CTR), *MCB* 45202 (MH); Kavali range (NLR), *CAB* 6611 (MH); Phirangipuram (GNT), *CAB* 4606 (MH); Aditavalasa (VSKP), *GVS* 21597 (CAL); Parvathipuram (VSKP), *VNS* 4571 (MH); Jacum, Karlapat Wild Life Sanctuary (KHD), *AC* 2439 (BSID); Inchigiri (SLM), *sine coll. s.n.* (MH); Gudiyattam (NA), *S.India Flora* 15514 (MH); Gingee (SA), *S.India Flora* 17961 (MH); Sithanur (CPT), *S.India Flora* 11206 (MH).

INDIA: Gangetic plains to East India and Southwards.

WORLD: SE. Asia and China.

Panicum fischeri Bor, Kew Bull. 1956; 257. 1956 & Grass. India 324. 1960; Pullaiah, Fl. Andhra Pradesh 5: 2234. 2018.

An annual, tufted grass; culms up to 1 m high; nodes glabrous. Leaf sheaths 10 cm long, glabrous; ligule a densely ciliate rim; blades 15-45 x 0.4-0.6 cm, linear-lanceolate, base-narrowed. Panicle 30 cm long; branches long, filiform; spikelets 5 mm, pedicelled; lower glume broadly ovate, 5-nerved; upper glume lanceolate, equal to the length of the spikelet, 10-nerved; lower lemma similar to upper glume, epaleated; upper lemma crustaceous, smooth; palea similar to its lemma; stamens 3, stigmas thickly plumose.

Rare grass in southern Eastern Ghats. Fl. & Fr.:August- October.

Batrepalli (ATP), *KRKS* & *BSS* 40341; Bhajarangakona (KDP), *BR* & *P.Priyadarsini* 37359; Near Papavinasanam falls (CTR), *KS* 6962 (MH); Kambakam hills (CTR), *MCB* 45084 (MH).

INDIA: South India, Endemic.

Panicum humidorum Buch.-Ham. ex Hook.f., Fl. Brit. India 7: 53. 1897; Bor, Grass. India 326. 1960; Mondal & Mukh., J. Bombay Nat. Hist. Soc. 88: 307. 1991.

Perennial slender grass. Leaves linear, 8-10 mm wide, base shortly cordate. Panicle large, pedicels capillary, smooth; spikelets ellipsoid, 0.8-1.5 mm long, obtuse, glabrous; lower glume obtuse, 3-nerved; upper glume 5-nerved; lemma obscurely 5-nerved; lower sterile, epaleate; upper lemma bisexual, ovate, 1-1.2 mm long, chartaceous, smooth; palea as long as the lemma.

Rare in Esterm Ghats of Odisha. Fl. & Fr.: October–December.

INDIA: NE India, Odisha.

WORLD: Bangladesh, Malaya.

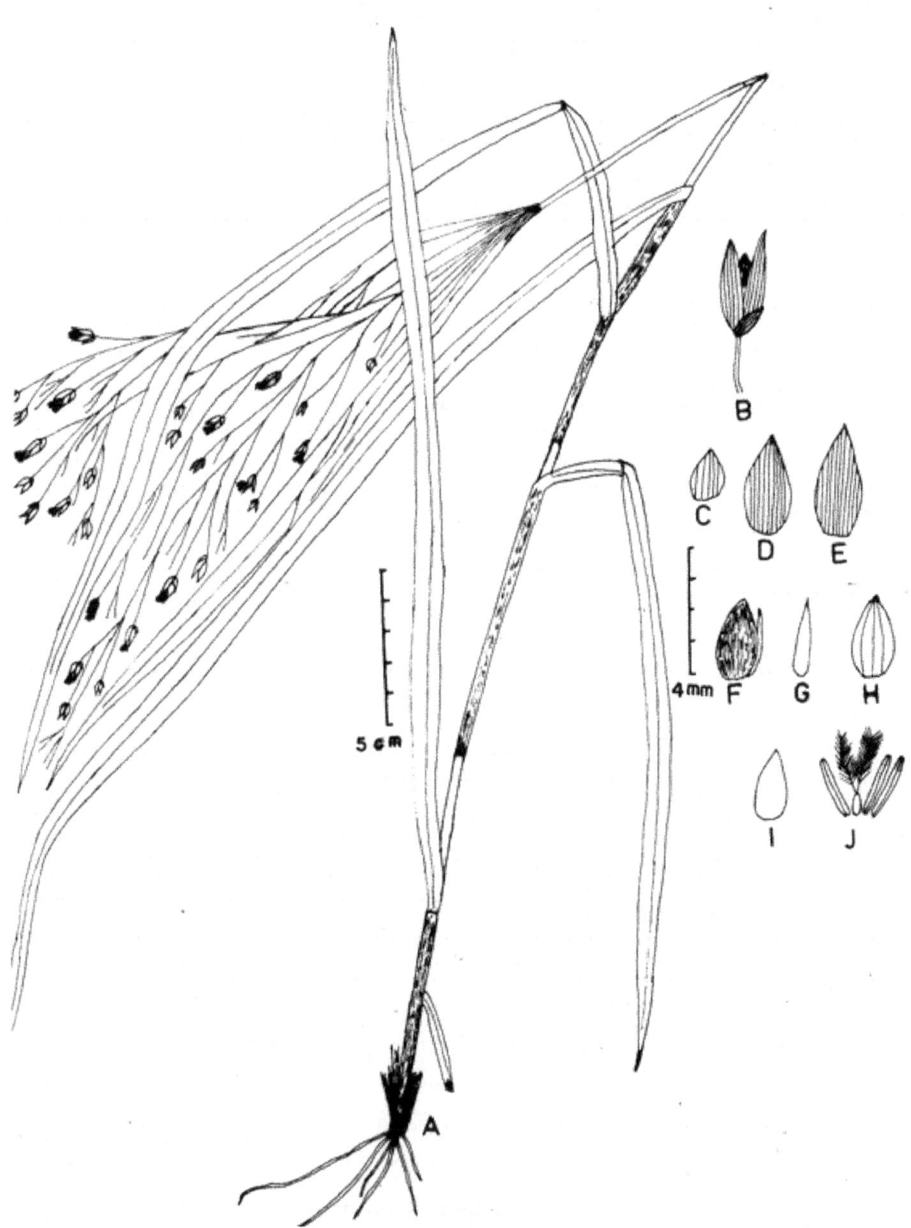

Figure 80. **Panicum fischeri** Bor.

A. Twig, B. Spikelet, C. Lower glume, D. Upper glume, E. Lower lemma, F. Upper floret, G. Membranous scale, H. Upper lemma, I. Palea, J. Pistil and stamens.

Panicum humile Steud. in Syn. Pl. Glumac. 1: 84 1854. *P. walense* Mez, Bot. Jahrb. Syst. 34: 146 1904; Pullaiah, Fl. Andhra Pradesh 2240. 2018.

Annual grass; culms erect, branched, 20–70 cm tall. Leaves mostly cauline; leaf sheaths glabrous, occasionally ciliate on the margins; leaf blades linear, flat, 7–20 x 0.2–0.6 cm, glabrous, base tapering, apex acute or acuminate; ligule *c.* 1 mm, a ciliolate membrane. Panicles axillary or terminal, ovate or oblong in outline, usually scarcely exserted from the uppermost sheath, 5–15 cm, moderately branched. Spikelets ovate, 1.5–2 mm, glabrous, upper glume and lower lemma divergent at maturity; lower glume ovate, 0.5-0.8 mm long, 3-veined, acuminate or cuspidate; upper glume ovate, as long as spikelet, 3–5-veined acuminate; lower floret barren, lemma similar to upper glume, palea present; upper floret pale yellow, smooth, shiny.

Rare in moist areas and hill slopes. Fl. & Fr.: August–December.

Kondagarulapenta (GNT), *MPG* 123 (DD & CAL); Parvatipuram (VSKP), *VNS* 4570 (MH); Palasamendu (SKLM), *VNS* 46290 (MH); Harishankar (BL), *SX* & *MB* 62546 (RRL-B); Pathar, Gandhamardan (BL), *SX* & *MB* 6254 (RRL-B); Kundali (KPT), *AC* 10232 (BSID).

INDIA: Odisha, Madhya Pradesh, Kerala, Andhra Pradesh, West Bengal.

WORLD: Malaysia, Philippines, Sri Lanka, China, Thailand, Vietnam, Tropical Africa.

Panicum maximum Jacq., Coll. Bot. 1: 76. 1786; FBI 7: 49 1896; Fischer 3: 1783. 1934; Bor, Grass. India 327. 1960; Matthew, Fl. Tamilnadu Carnatic 3: 1879. 1983.

A perennial, tufted grass; culms up to 1 m high, stout, branching from the base; nodes bearded. Leaf sheaths 10-25 cm long, ciliate near the mouth; ligule membranous, usually dense hairs from behind it; leaf blades 10-60 x 4-10 cm, linear, flat, base cordate, margin scaberulous, apex acuminate, glabrous, midrib prominent. Panicle up to 25 cm, lax; spikelets 4 mm, linear-oblong, dense; lower glume orbicular, half the length of spikelet, 1-nerved; upper glume broadly oblong, margins incurved, 5-nerved; lower lemma oblong, 5-nerved; palea oblong, 2-nerved, male; upper lemma ovate-oblong, 3-nerved, transversely rugose, pallid; palea similar to upper lemma, nerveless, bisexual; stamens 3; stigma plumose. Caryopsis narrowly ellipsoid.

Rarely cultivated and as escape. Fl. & Fr.: July-October.

Ettukkuru, Maredumilli (EG), *MM* 102626 (BSID); Anantagiri (VSKP), *GVS* 19503, 21715 (MH); Sirumalai hills (DGL), *SKS* 427 (SGH); Kancheepuram (CPT), DN 92 (MH); Alagarkoil RF (MDU), *EV* 25274 (MH).

INDIA: Cultivated and Naturalized.

WORLD: A tropical African grass cultivated for its valuable fodder and occasionally found as an escape.

Panicum miliaceum L., Sp. Pl. 58. 1753; FBI 7: 45. 1896; Fischer 3: 1782. 1934; Bor, Grass. India 327. 1960; Matthew, Fl. Tamilnadu Carnatic 3: 1880. 1983.

An annual, tufted grass; culms up to 1 m high, sparingly branched; nodes pilose. Leaf sheaths 7 cm long, loose, hirsute with tuberculate based hairs; ligule a ciliate rim; leaf blades 5-20 x 0.4-1 cm, lanceolate, base rounded, margins involute, apex acuminate, sparsely hairy on both the surfaces. Panicle 20 cm long, narrowly oblong; branches alternate, filiform, scabrid; spikelets 5 mm, ovate-oblong, apiculate; glumes persistent, unequal; lower broadly ovate, 5-nerved; upper ovate-oblong, 11-nerved; lower lemma oblong; palea ovate-oblong; upper lemma elliptic-oblong, crustaceous; palea similar to its lemma, bisexual; stamens 3; stigmas plumose. Caryopsis ovoid.

Occasionally cultivated for its grains and often as an escape in black cotton soils. Fl. & Fr.:July-November. Tel.: *Warigalu, Barigalu.* Tam.: *Mani varagu samai.*

Kammavaripalli (KNL), *TP & PVP* 3371; Guntur (GNT), *JSG*18610 (MH); Kollapur (MBNR), *BSS & SKB* 32915 (BSID); Belghar, Kotgarh Wild Life Sanctuary, (KDM), *AC* 2413 (BSID); Purunakote RF, Satkosia Tiger Reserve (AGL), *KCM* 5177 (BSID); Sirumalai hills (DGL), *SKS* 428 (SGH).

INDIA: Throughout the hotter parts of India, cultivated and naturalized.

WORLD: Pakistan. Sri Lanka. Introduced to parts of Africa, Europe, N. & S. America, Australia, Central and E. Asia.

Panicum notatum Retz., Observ. Bot. 4: 18. 1786; Bor, Grass. India 701. 1960; Matthew, Fl. Tamilnadu Carnatic 3: 1881. 1983. *P. montanum* Roxb. Fl. Ind. 1:315. 1820; FBI 7: 53. 1896; Fischer 3: 1783. 1934; Bor, Grass. India 329. 1960.

A perennial, slender grass; culms up to 1 m high, erect from woody stock, sparsely branched: nodes glabrous-sparsely pilose. Leaf sheaths 3-5 cm, ciliate on margin; ligule of soft long hairs; blade 5-12 x 1-2 cm, linear-oblong, base amplexicaul, margin with tubercled based hairs, apex acuminate, sparsely pubescent on both surfaces. Panicle up to 25 cm long, lax, open; branches spreading, filiform; spikelets 2.5 mm, oblong, few, not gaping at maturity; pedicel very long, 2 cm, filiform; lower glume ovate, 3-5-nerved, pilose 3/4 length of the spikelets; upper glume broadly ovate, 5-nerved; lower lemma ovate, 5-nerved, sparsely hairy on outside, epaleated; upper lemma ovate, crustaceous, indurated; palea coriaceous with involute margins; stamens 3, stigmas plumose. Caryopsis ellipsoid, dorsally compressed.

Occasional grass in the hill slopes and under the shades of big trees. Fl. & Fr.: September–February.

Rollapenta (KNL), *PVP & RVR* 2336; Tirumala hills (CTR), *MSG* 12153, 12177; Horsley hills (CTR), *MSG & AA* 12177; Papavinasanam (CTR), *JHFB* 117488 (BSID); Bhupatipalem, Rampa hills (EG), *TP & MSG* 12252; Araku valley (VSKP), *NPBK* 591 (CAL); Donubai (VSKP), *KCJ* 17218 (MH); Galikonda (VSKP), *GVS* 19589 (MH); Srungavarapukota (VJN), *GVS* 21817 (CAL); Chahala, Similipal BR (MBJ), *A.R.K.Sastry & G.P.Singh* 12236 (BSID); Kapila Hill (Dhenkenal), *sine coll.* 11221 (BSID); Jenabil Range, Similipal B.R. (MBJ), *AC* 2304 (BSID); Sirumalai hills (DGL), *SKS* 449 (SGH); Kolli hills (NMK), *SKS* 617 (SGH); Kongampallm–Yercaud (SLM),

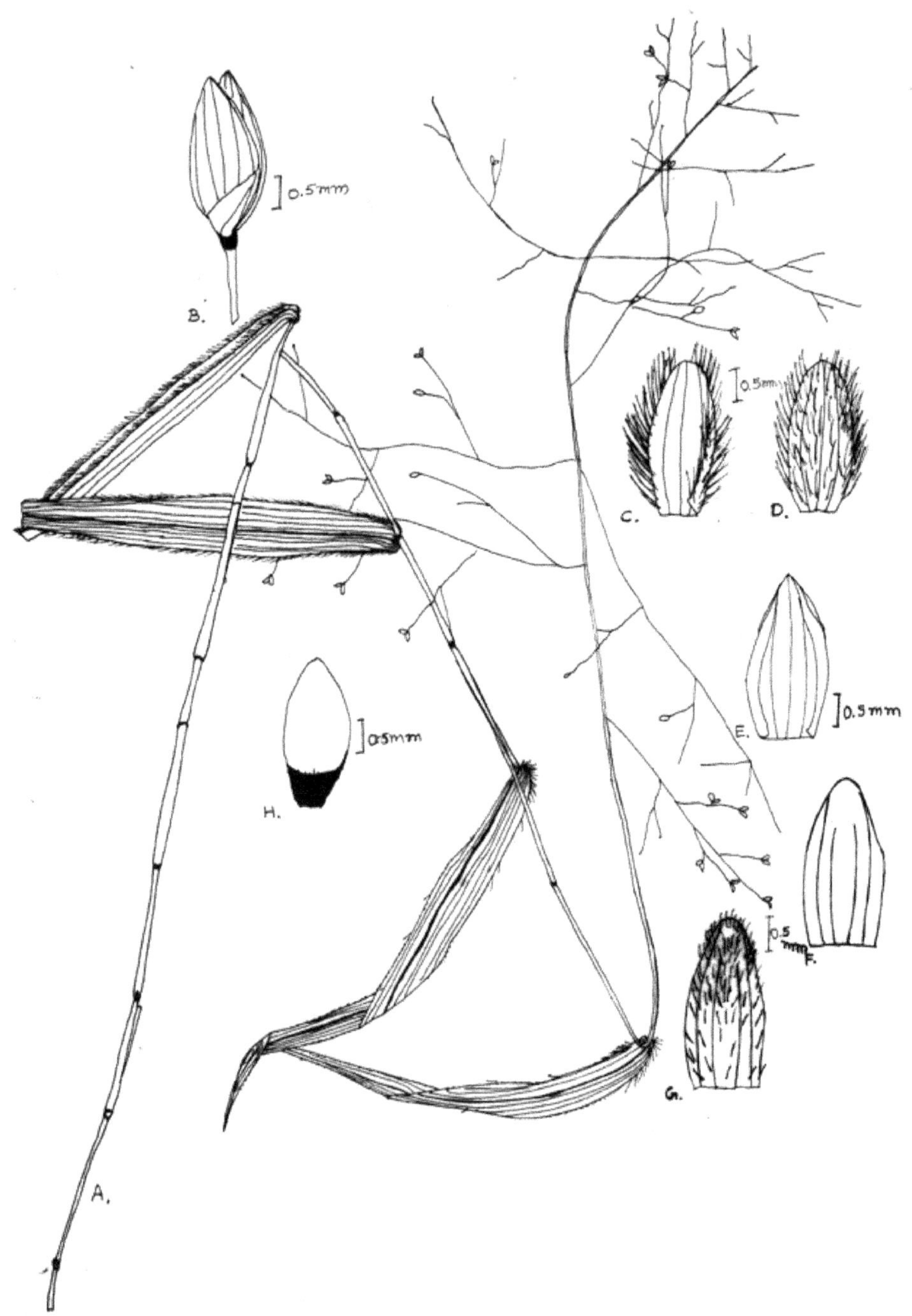

Figure 81. **Panicum notatum** Retz.

A. Twig, B. Spikelet, C & D. Lower glume front and back views, E. Upper glume, F & G. Lower lemma front and back views. H. Caryopsis.

DBD 31422 (MH); Orchidarium–Shevaroys (SLM), *AVNR* 18285 (MH); way to Kambukkudi (NA), *KS* 6499 (MH).

INDIA: Throughout India, especially in the hills.

WORLD: Tropics of SE. Asia.

Panicum paludosum Roxb., Fl. Ind. 1:310. 1820; Fischer 3: 1783. 1934; Bor, Grass. India 329. 1960; Matthew, Fl. Tamilnadu Carnatic 3: 1881. 1983. *P. proliferum* sensu Hook. f., FBI 7: 50 1896. non Lam. 1797.

A perennial, tufted, aquatic grass; culms up to 60 cm high, branched, spongy; nodes glabrous. Leaf sheaths 4-12 cm long; ligule a fimbriate membrane; blades 10-20 x 0.3-0.6 cm, oblong-lanceolate, scabrid. Panicle 20 cm long, ovate-oblong, spreading; branches alternate; spikelets 4 mm, narrowly ellipsoid-lanceolate, shortly acute-acuminate; lower glume orbicular, less than half the length of the spikelet; upper glume lanceolate, 7-9-nerved; lower lemma 9-11-nerved, acuminate; palea male or empty; upper lemma pallid, smooth, 3-nerved; palea similar to its lemma; stamens 3, stigmas plumose. Caryopsis ellipsoid.

Occasional near marshy areas. Fl. & Fr.: September-March

Pennahobilam (ATP), *TP* & *NY* 347; Alampur (ATP), *BR* & *P.Priyadarsini* 37195 (SKU & BSID); Idupulapaya (KDP), *KRKS* & *BSS* 40304; Veyinuthulakona (KDP), *MHR* 13312; Tirumala hills (CTR), *MSG* 12162; Kadapa, *JSG* 21217 (MH & DD); Araku (VSKP), *GVS* 19732 (MH).

INDIA: Throughout India, in the plains and at lower elevations.

WORLD: Tropics of Asia,Tropical Australia.

Note: It is often confused with *Panicum repens,* but can be distinguished by the spongy culms, broad, flat leaf blades, lax panicle and narrowly lanceolate spikelets which are appressed to the branches. *P. paludosum* does not possess rhizomes unlike the other ones.

Panicum repens L., Sp. Pl. ed. 2. 87. 1762; FBI 7: 49. 1896; Fischer 3: 1783. 1934; Bor, Grass. India 330. 1960; Matthew, Fl. Tamilnadu Carnatic 3: 1882. 1983.

A perennial, stoloniferous grass; culms up to 45 cm long, creeping at base, rooting from the nodes; nodes glabrous. Leaf sheaths 2-6 cm, densely ciliate on margin; ligule membranous; leaf blades 5-15 x 0.4-0.8 cm, linear-lanceolate, flat, base rounded, apex acute, sparsely pubescent. Panicle up to 10 cm long; spikelets 3 mm, ovate-lanceolate; lower glume suborbicular, apex truncate, 3-nerved; upper glume lanceolate, 9-nerved; lower lemma lanceolate, 10-nerved, paleated, male; upper lemma ovate, pallid, indurated; palea similar to upper lemma; stamens 3; stigma plumose. Caryopsis ellipsoid.

Common in fields and waste lands. Fl. & Fr.: March–December.

Garugudu kona (ATP), *TP* 946; Yellutla (ATP), *PVP* 2644; Gulemallebad (KNL), *RVR* 1411; Veldurthi (KNL), *RVR* 1743; Gadimalla (KNL), *AA* 2035; Balarachapalli (KDP), *PVP* 3442; Aadavulapalle (KDP), *JSG* 21250 (MH); Chandragiri (CTR), *S.India*

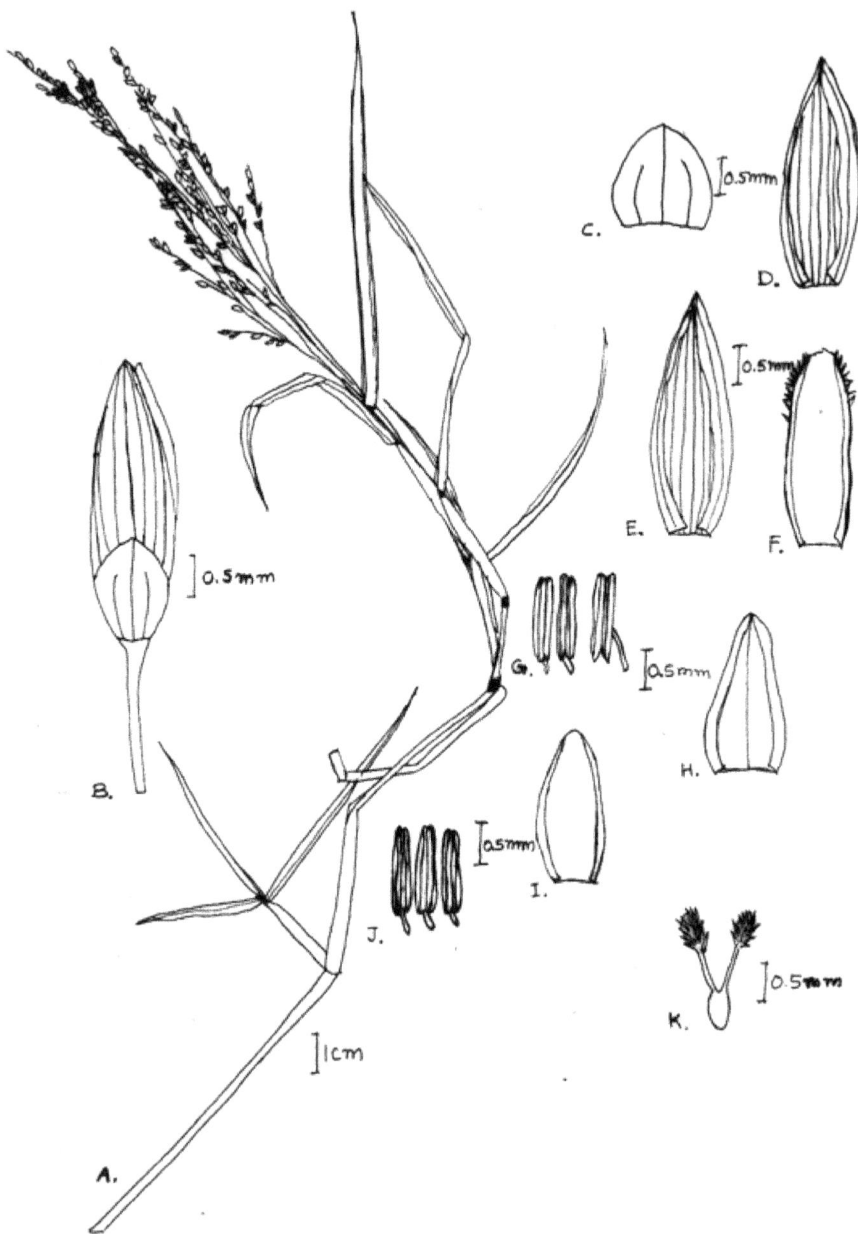

Figure 82. **Panicum paludosum** Roxb.

A. Twig, B. Spikelet, C. Lower glume, D. Upper glume, E. Lower lemma, F. Palea, G. Stamens, H. Upper lemma back view, I. Upper lemma front view; J. Stamens, K. Pistil.

Flora 10110 (MH); Ramatheertham (NLR), *MPG* 156 (CAL); Palasa (VSKP), *VNS* 4638 (MH); Tulka section, Satkosia Tiger Reserve (AGL), *KCM* 5688 (BSID); Palakuppam (SA), *VNS* 4037 (MH); Red hills (CPT), *CAB* 5517 (MH).

INDIA: Throughout the hotter parts of India.

WORLD: Tropical and subtropical regions of both the hemispheres.

Panicum sparsicomum Nees ex Steud., Syn. Pl. Glum. 1: 83. 1854; FBI 7: 58. 1896; Bor, Grass. India 330. 1960; Moulik, Grass. Bamb. India 1: 123. 1997. *Cyrtococcum sparsicomum* (Nees ex Steud.) A. Camus, Bull. Mus. Hist. Nat. Paris. 27: 118. 1921; Fischer 3: 1786. 1934.

Annuals, culms erect or ascending, to 30 cm, nodes glabrous. Leaf sheaths 2-3 x 0.3 cm, glabrous; ligules 0.3 mm, fimbriate; leaf blades linear-lanceolate, 3-10 x 0.4 cm, acute, base rounded, glabrous or sparsely villous above. Panicles open, 5-10 x 2-3 cm, branches whorled, rachis glabrous, pedicels 3-6 mm; spikelets solitary, 1.5-1.7 x 0.7 mm, ovate-lanceolate or elliptic-oblong, acute, glabrous, purplish, deciduous, slightly compressed; lower glume cuspidate-acuminate, 1.5 x 0.1 mm, chartaceous, 3-nerved, glabrous; upper glume ovate-lanceolate, acute, 1.5 x 0.7 mm, chartaceous, 5-nerved; lower lemma ovate, acute, 1.4 x 0.6 mm, chartaceous, 5-nerved; lower palea oblong-lanceolate, 1 x 0.5 mm, hyaline, keeled; upper lemma ovate-oblong, 1 x 0.5 mm, coriaceous, 3-nerved, upper palea ovate-lanceolate, 0.1 x 0.5 mm, obtuse, subcoriaceous; stamens 3, anthers 1 mm; ovary ovate, 0.5 mm, styles 0.3 mm, stigmas 0.5 mm; lodicules 2, obvate, 0.3 mm, truncate.

Rare along the dry slopes and open forest in Srikakulam district (B.R.P. Rao and A.M. Reddy, 2010) and Alagar hills in Tamil Nadu. Fl. & Fr.: May–December.

Alagar hills (DGL), *K.Althaf Ahamed Kabeer* 113276 (MH).

INDIA: South India.

WORLD: Sri Lanka.

Panicum sumatrense Roth ex Roem. & Schult.,Syst . Veg. 2;434. 1817; Bor, Grass. India 701. 1960. *P. psilopodium* Trin., Gram. Panic. 217. 1826; FBI 7: 46. 1896; Fischer 3: 1782. 1934; Bor, Grass. India 329. 1960; Matthew, Fl. Tamilnadu Carnatic 2: 1882. 1983; Saxena & Brahmam, Fl. Orissa 4: 2434. 1996.

A perennial tufted grass; culms up to 50 cm high; nodes glabrous. Leaf sheaths 4-8 cm long, glabrous; ligule a fimbriate membrane; blades 4-10 x 0.3-0.9 cm, elliptic-lanceolate, base cordate, apex acuminate, glabrous. Panicle up to 12 cm long, ovate-oblong, effuse; branches numerous, alternate; spikelets 3 mm, elliptic-oblong, deciduous; lower glume less than half the length of the spikelet, ovate, obscurely 3-nerved; upper glume broadly ovate, acute, 9-nerved, margins incurved; lower lemma broadly ovate, 10-nerved; palea oblong, obtuse, scabrid on margin, male; upper lemma oblong, acute; palea oblong, obtuse, bisexual; stamens 3; stigmas plumose. Caryopsis ellipsoid.

Common in shady and wet or marshy areas. Fl. & Fr.:July-September.

Kalasamudram RF (ATP), *NY* 1171, *PVP* 2611; Ramgiri (CTR), *BR* & *P.Priyadarsini* 37114 (SKU & BSID); Bukkapatnam RF (ATP), *JSG* 20947 (DD); Mogilipenta (KDP), *JSG* 2136 (DD); Balapalle (KDP), *JLE* 14939 (MH); Horsley konda (CTR), *JSG* 20990 (DD); Varadayapalem to Kambakkam (CTR), MCB 45155 (MH); Palagudem (WG), *VNS* 4483 (MH); Tadepalli (EG) *NRR* & *DN* 84326 (BSID); Araku valley (VSKP), *NPBK* 10778 (MH); Gajapathinagaram (VJN), *VNS* 4581 (MH); Way to Vadabai, Tadikonda (VJN), *NRR* 83633 (BSID); Ramanagutta (KMM), *RR* 108009 (BSID); Sirumalai hills (DGL), *SKS* 513 (SGH); Hogainakkal RF (KGR), *EV* 20569 (MH); Gudiyattam (NA), *sine coll. s.n.* (MH); Mahendragiri (GJM), *VNS* 5744 (MH).

INDIA: Throughout hotter parts of India.

WORLD: Pakistan, Sri Lanka, Myanmar, Malesia, SE. Asia.

PARAHYPARRHENIA A. Camus

Parahyparrhenia bellariensis (Hack.) W.D. Clayton, Kew Bull. 27: 448. 1972. *Andropogon bellariensis* Hack., Flora 68; 124. 1885; FBI 7: 201. 1896. *Heteropogon bellariensis* (Hack.) C.E.C. Fischer, Fl. Pres. Madras 3: 1744. 1934; Bor, Grass. India 163. 1960.

An annual, tufted grass; culms up to 60 cm high, slender, terete, glabrous. Leaf sheaths terete, glabrous; blades 7.5-20 x 0.3-0.4 cm, involute, filiform, glabrous. Racemes solitary, 5-8 cm long, slender, partially sheathed; joints ciliate. *Sessile spikelets:* 4-7 cm, glabrous; callus bearded; lower glume linear-oblong, bidentate, 2-keeled, at the apex with a deep median groove on the dorsal surface; upper glume narrowly oblong- lanceolate; lower lemma ovate-oblong; upper lemma reduced to a hyaline base of an awn, 2-lobed. Caryopsis ovoid. *Pedicelled spikelet:* oblong, glabrous; callus short; lower glume linear- lanceolate; upper glume linear.

Extremely rare on Gooty fort of Anantapuram district; Known only from type collection.

INDIA: Andhra Pradesh. Endemic, probably extinct.

Note: Hackel (1885) description of *Andropogon bellariensis* was based on Wight 2321 collected from Gooty fort hill, Anantapuram (Andhra Pradesh). No specimen at CAL, MH, DD, Isotype is at Kew. Fischer mentions of another collection by Campbell as well from the same locality. Efforts made by us for its collection now proved futile.

PASPALIDIUM Stapf

1. Annual; ligule absent; spikelets gibbose .... **P. flavidum**
1. Perennial; ligule present; spikelets not gibbose:
    2. Upper glume broadly oblong; upper lemma rugulose............. **P. punctatum**
    2. Upper glume broadly ovate; upper lemma granulose .... **P. geminatum**

Paspalidium flavidum (Retz.) A. Camus in Lecomte., Fl. Gen. de 1′ Indo-china 7:419. 1922; Fischer 3: 1744. 1934; Bor, Grass. India 333. 1960; Matthew, Fl. Tamilnadu Carnatic 3: 1884. 1983. *Panicum flavidum* Retz., Observ. Bot. 4. 15. 1786; FBI 7: 28. 1896.

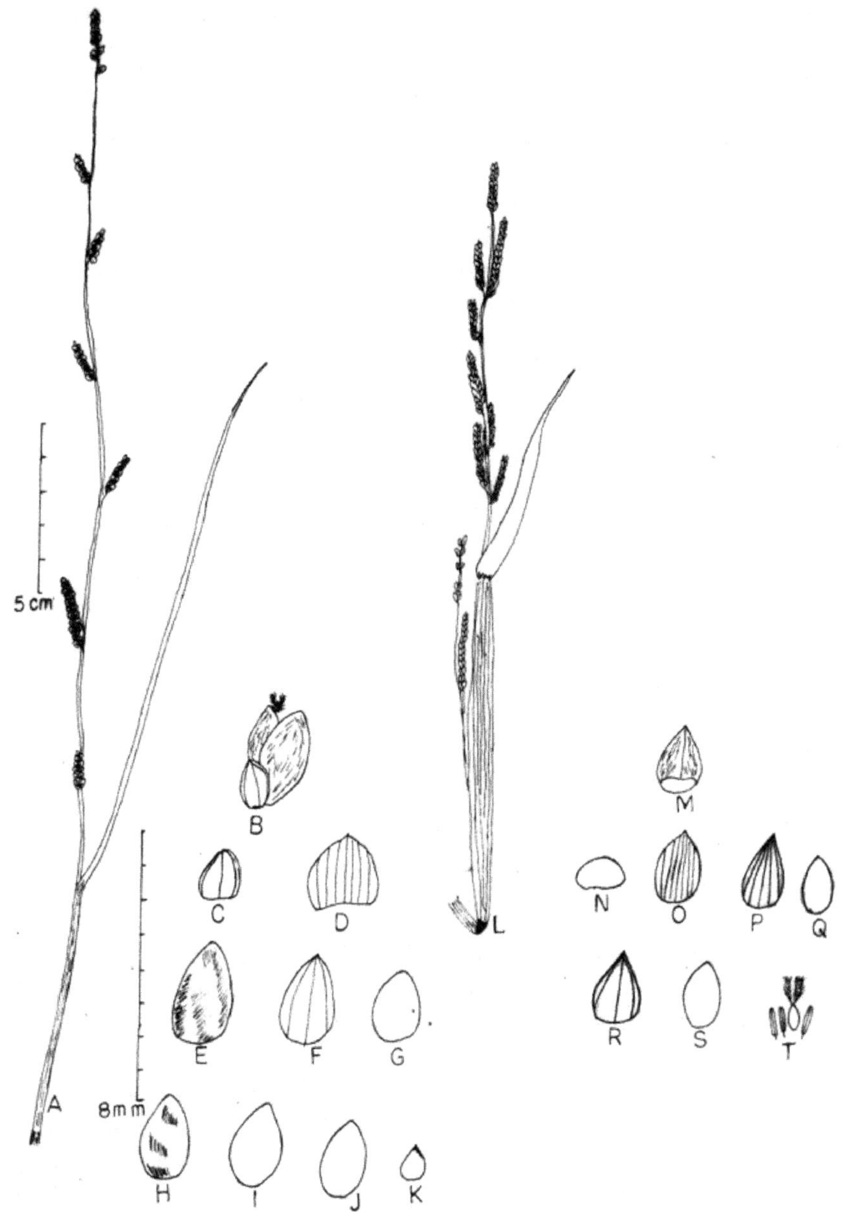

Figure 83. A-K. **Paspalidium flavidum** (Retz.) A. Camus

A. Twig, B. Spikelet, C. Lower glume, D. Upper glume, E. Lower floret, F. Lower lemma, G. Palea, H. Upper floret, I. Upper lemma, J. Palea, K. Grain.

L-T. **Paspalidium geminatum** (Forssk.) Stapf.

L. Twig, M. Spikelet, N. Lower glume, O. Upper glume, P. Lower lemma, Q. Palea, R. Upper lemma, S. Palea, T. Anthers and ovary.

An annual, tufted grass; culms up to 70 cm high, geniculately ascending; nodes glabrous. Leaf sheaths 6-15 cm, glabrous; ligule fringed, membranous; blades 5-20 x 0.5-0.8 cm, linear-lanceolate, base rounded, margins scaberulous, apex acuminate, slightly pubescent. Paniculate racemes up to 20 cm long; branches distant, secund; spikelets 3 mm, broadly ovate, secund in two rows, 2 florets, lower usually neutral or sometime male, upper bisexual; lower glume orbicular, 3-nerved, keeled; upper glume broadly ovate, 7-nerved, keeled; lower lemma broadly ovate, 5-nerved, paleated, empty; upper lemma broadly ovate or orbicular, convex, coriaceous, mucronate; palea similar to its lemma; stamens 3; stigmas plumose. Caryopsis ovate-oblong, tightly enclosed in the hardened lemma and palea.

Common in cultivated fields and waste lands. Fl. & Fr.:July-October. Tel.: *Ubagaddi.*

Kalasamudram (ATP), *KRKS* 40127; Gounivaripalli (ATP), *BR* & *P.Priyadarsini* 37310; Rollapenta (KNL), *PVP & RVR* 2343; Balapalli (KDP), *PVP* 3458, *JLE* 15017 (MH); Palakonda hills (KDP), *CS & DAM* 7702; Rachakuntapalli (KDP), *KRKS* & *KRS* 37910; Horsley hills (CTR), *MSG &. AA* 12211; Bhupathayyakona (CTR), *BR* & *P.Priyadarsini* 37324; Eeduru (NLR), *PMR* 22325; Chelama (KNL), *MPG* 104 (DD), *JLE* 16754, 22063 (MH); Mogilipenta (KDP), *JSG* 21323(DD); Yerrabalem (KDP), *MPG* 161 (DD); Mamandur valley (CTR), KS 6896 (MH); Maredumilli (EG), *GVS* 68543 (MH); Dharakonda (VSKP), *sine coll., s.n.* (MH); Anantagiri (VSKP), *GVS* 19486 (MH); Koriguttalu (KMM), *RR* 106036 (BSID); Tikarpada, Satkosia Tiger Reserve (AGL), *KCM* 7084 (BSID); Pampasar RF, Satkosia Tiger Reserve, *KCM* 5103 (BSID); Inchigiri (SLM), *S.Indian Flora* 15444 (MH); Gudiyattam (NA), *S.Indian Flora* 15515 (MH); Kuthadya hills (GJM), *VNS* 58910 (MH); Melpat (SA), *CAB* 1072 (MH); Urigam forest (DMP), *EV* 57951 (MH).

INDIA: Throughout India.

WORLD: Tropical Asia.

Paspalidium geminatum (Forssk.) Stapf in Prain, Fl. Trop. Afr. 9: 583 1920; Fischer 3: 1774. 1934; Bor, Grass. India 333. 1960; Matthew, Fl. Tamilnadu Carnatic 3: 1884. 1983. *Panicum geminatum* Forssk., Fl. Aegypt.-Arab. 18. 1715. *P. paspaloides* Pers., Syn.1, 81. 1805; FBI 7: 30. 1896.

A perennial, marshy grass; culms up to 50 cm high, spongy, creeping below; branches from lower nodes; nodes glabrous. Leaf sheaths 4-10 cm, loose, glabrous; ligule a ciliate rim; leaf blades 5-15 x 0.3-0.6 cm, linear, base rounded, apex acuminate, glabrous. Spike like panicle up to 15 cm long, racemes 2 cm, 10- 20, alternate on the rachis, compressed; spikelets 2 mm, ovoid, 2 florets, lower male or neutral, upper bisexual; lower glume orbicular, nerveless; upper glume ovate, 5-7-nerved, concave; lower lemma ovate-oblong, membranous, 5-7-nerved, paleated, male; upper lemma coriaceous, broadly ovate, mucronate; palea similar to its lemma, margins inflexed; stamens 3, stigma plumose. Caryopsis ellipsoid.

Occasional grass near water courses, usually near marshy areas. Fl. & Fr.: December-July.

Tirumala hills (CTR), *MSG* 12157; Kalyani dam (CTR), *RVR & GO* 2435; Simhachalam (VSKP), *MHR* & *MCK* 14953; Duggeru (VJN), *MCK* 25245; Kinthali (SKLM), *MCK* 18872; Way to Guvvalacheruvu (KDP), *KS* 7785 (CAL); Nagari (NA) *sin. coll. s.n.* (MH); Red hills (CPT), *DN* 939 (MH); Komattiyur (NA), *KS* 6471 (MH).

INDIA: Throughout plains of India.

WORLD: Tropical Africa and Asia.

Paspalidium punctatum (Burm. f.) A. Camus in Lecomte, Fl. Gen. de 1' Indo-chine 7: 41. 1922; Fischer 3: 1774.1934; Bor, Grass. India 333. 1960; Matthew, Fl. Tamilnadu Carnatic 3: 1885. 1983. *Panicum punctatum* Burm. f., Fl. Ind. 26. 1768; FBI 7: 29. 1896.

A perennial, tufted grass; culms up to 1 m high, prostrate, rooting at lower nodes, spongy. Leaf sheaths 10 cm long; ligule a ridge of hairs; blades 10-15 x 0.5-1 cm, linear, flat or convolute, apex acuminate, glabrous. Spike like panicle 30 cm long; racemes 20 cm long, distant, sessile, compressed, appressed to the rachis; spikelets 3 mm long, 18-40 per raceme, ovoid; lower glume truncate; upper glume broadly oblong; lower lemma acuminate; palea empty; upper lemma punctate, apiculate; palea coriaceous, margins incurved; stamens 3; stigmas plumose. Caryopsis ovoid.

Common in wet places, ponds and along margins or streams. Fl. & Fr.:December-July.

Dorigallu (ATP), *KRKS* 40473; Mallemkonda temple (KDP), *BSS* & *SKB* 30524; Kolleru lake (WG), *KS* 5083 (MH); Mandapeta (EG), *CAB* 4296 (MH); Annamalainagar (SA), *KMS* 5222 (MH).

INDIA: Throughout India.

WORLD: N. Africa, Tropical Asia.

PASPALUM Linnaeus

1. Stoloniferous grass; upper glume pubescent
   2. Spikelets fringed with fine white hairs from the margins of the upper glume .... **P. conjugatum**
   2. Spikelets glabrous or pubescent, not ciliate on the margins .... **P. distichum**
1. Tufted grass, non-stoloniferous; upper glume glabrous:
   3. Annual; cultivated; erect; leaf sheaths glabrous, blades linear-lanceolate, glabrous .... **P. scrobiculatum**
   3. Perennial; wild; decumbent at base; leaf sheaths densely hairy, blades linear or elliptic-lanceolate, sparsely hairy on both surfaces:
      4. Panicle with 6-13 racemes on axis; racemes 0.7-2 cm long.......... **P. canarae**
      4. Panicle with 2-6 racemes on axis; racemes more than 2 cm long.......... **P. dilatatum**

Paspalum canarae (Steud.) Veldk., Blumea 21: 72. 1973. *Panicum canarae* Steud., Syn. Pl. Glumac. 1: 58. 1853. *Paspalum compactum* auct. non Roth 1821; FBI 7: 12 1896; Fischer 3: 1772. 1934; Bor, Grass. India 336. 1960.

A perennial, tufted grass; culms up to 30 cm high, decumbent and branched below, leafy, hairy. Leaf sheaths densely hairy with fine tuberculate-based hairs; ligule very short; leaf blades 3-8 x 0.8-1.5 cm, elliptic-lanceolate, apex acute, sparsely hairy on both sides. Spikes numerous, pyramidal; racemes 10 cm long; rachis triquetrous; spikelets 2 mm, orbicular; lower glume absent; upper glume broadly elliptic-orbicular; lower lemma narrower; upper lemma subcoriaceous, concave, orbicular; palea obscurely auricled; stamens 3, stigmas laterally exserted near the tip of the floret.

Occasional in marshy areas of Northern Eastern Ghats. Fl. & Fr.:August-November.

On the way to Tajavi (VSKP), *GVS* 42626 (MH); Mahendragiri (GJP), *SX* & *MB* 3680 (RRL-B); Karlapat (KHD), *MB* & *Rout* 7645 (RRL-B); Deomali hill (KPT), *AC* 4230 (BSID).

INDIA: Goa, Karnataka, Maharashtra, Andhra Pradesh, Odisha, Tamil Nadu. Endemic.

Paspalum conjugatum P.J. Bergius, Act. Helv. Phys.-Math. 7: 129. t. 8. 1772; Bor, Grass. India 336. 1960.

Perennial, stoloniferous grass, rooting at nodes; culm to 1 m, surface smooth; nodes glabrous. Leaf sheaths 6-20 x 0.6-1 cm, surface glabrous, ciliate along one margin, compressed, keeled; ligule rounded, to 1 mm, membranous; leaf blade 17-25 x 0.7-1 cm, linear-lanceolate, surface glabrous, apex acuminate, base narrow, margin ciliate. Racemes 2, conjugate, 6-18 cm, rachis triquetrous, glabrous; spikelets subsessile, 2-flowered, solitary, alternate on the rachis, 1.8-2 x 1-1.5 mm, ovate or broadly elliptic, or orbicular, acute, hairy, greenish-yellow; lower glume absent; upper glume ovate or orbicular, 1.5-1.9 x 1-1.3 mm, membranous, 2-nerved, long ciliate along the margins; lower lemma barren, ovate or orbicular, acute, 1.6-1.8 x 1mm, membranous, 2-nerved; upper lemma ovate or orbicular, 1.5-1.7 x 1 mm, crustaceous, 2-keeled; palea ovate, 1.5 x 1 mm, crustaceous, 2-keeled; stamens 3, anther 0.5 mm, deep yellow; pistil 1 mm, stigmas white; lodicules 0.3 mm. Caryopsis elliptic-ovate.

Occasional in streams and moist localities of Chintapalli forest in Visakhapatnm district (A.M. Reddy *et al.*, 2005). Fl. & Fr.: July–November.

Chintapalli (VSKP), *BR* & *AMR* 24576 (SKU).

INDIA: Peninsular & E. India.

WORLD: Tropics and subtropics through out the World.

Paspalum dilatatum Poir. in Lam. & Poir. Encycl. 5: 35. 1804; Fischer 3: 1772. 1934; Bor, Grss. India 338. 1960.

Tufted perennial grass; culm erect, geniculate, 60-150 cm high, nodes slightly swollen. Leaves linear-lanceolate, 11-27 x 0.6-1.3 cm, acuminate, sparsely tubeculate hairy near base, midrib prominent, margins deflexed towards apex; ligule ovate, acute, 3-6 mm, membranous, hyaline to brownish, lacerate. Panicle with 4-6 alternate racemes, 8-21 cm long, ciliate, greenish, nodes long villous haity; peduncle 18-65 cm long. Racemes 3-12 x 0.6 cm, with 4 rows of spikelets. Spikelets secund, paired with short and long pedicels, similar, ovate-laceolate, 2.8-4 x 2-2.3 mm, acute, sparsely hairy with ciliate margins; pedicels 0.4 -1.7 mm, greenish to purplish, serrulate throughout. Lower glume absent. Upper glume ovate-lanceolate, 3.2-4 x 1.8-2 mm, acute, chartaceous, hairy. Florets 2; lower barren and epaleate, upper bisexual; lower lemma ovate-lanceolate, 2.8-3.5 x 2 mm, acute, chartaceous, 5-nerved, sparsely ciliate towards apex. Upper lemma ovate to elliptic, 2-2.2 x 1.8 mm, obtuse at apex, narrowly truncate at base, subcoriaceous, smooth, faintly 2-nerved; lodicules 2, 0.4-0.4 mm long, truncate at apex, membranous, hyaline; stamens 3; anthers 1.2-1.3 mm long, purplish; filaments 0.8-1.2 mm long, hyaline; ovary elliptic, 0.6-0.3 mm; style 0.3-1.2 mm long, hyaline; stigma plumose, *c.* 1.3 mm long, purplish.

A weed in shaded places in hilly ares. Fl. & Fr.: May–March.

Melagiri (SLM), *S.India Flora* 15445 (MH).

INDIA: Assam, Himachal Pradesh, Kerala, Maharashtra, Sikkim, Uttar Pradesh, West Bengal, Tamil Nadu.

WORLD: Native to S. America, now established almost all over tropical countries.

Paspalum distichum L., Syst. Nat. ed. 10. 2: 855. 1759; FBI 7: 12. 1896; Bor, Grass. India 338. 1960. *Paspalum paspalodes* (Michx.) Scribner, Mem. Torr. Bot. Cl. 5: 29. 1894. (*'paspaloides'*); Bor, Grass. India App. 11. 1973. *Digitaria paspalodes* Michx., Fl. Bor. Amer. 1: 46. 1803.

A perennial, stoloniferous grass; culms up to 50 cm, ascending from creeping base, rooting at nodes, branches from lower nodes; nodes glabrous. Leaf sheaths 4-10 cm, glabrous; ligule membranous, divaricate; blades 5-14 x 0.4-0.8 cm, linear-lanceolate, base rounded, acute, glabrous. Spikes up to 9 cm long, axis flattened; spikelets 3 mm, elliptic-oblong, subsessile, dorsally compressed, 2-flowered; lower glume suppressed; upper glume elliptic-oblong, membranous, pubescent; lower lemma similar to upper glume; upper lemma elliptic-concave, apex acuminate, glabrous; palea elliptic; stamens 3; stigmas plumose. Caryopsis ovoid-ellipsoid.

Occasional grass in marshy areas. Fl. & Fr.:August-November.

Pennahobilam (ATP), *TP* & *NY* 359; Madhavaram RF (KNL), *RVR* 1645; Vaddemanu (KNL), *PVP & RVR* 2529 & 2527; Kodavalur (NLR), *PMR* & *AMR* 23012; Red hills (CPT), *DN* & *PVS* 943 (MH); Sonanchavadi (SA), *KRM* 64170 (MH).

INDIA: NW to E India and southwards.

WORLD: Tropics and subtropics of the world.

Figure 84. **Paspalum distichum** L. [Syn.: *Paspalum paspalodes* (Michx.) Scribner].

A. Twig, B. Spikelet, C. Lower glume, D. Upper glume, E. Lemma, F. Palea, G. Stamens and Pistil, M. Grain.

Paspalum scrobiculatum L., Mant. Pl. 29. 1767; FBI 7: 10. 1890; Fischer 3: 1772. 1934; Matthew, Fl. Tamilnadu Carnatic 3: 1887. 1983. *P commersonii* Lam.,Tabl. Encycl. 1: 175. t. 43. f. 1. 1791; Bor, Grass. India 335. 1960. *P. orbiculare* Forst. f., Insul. Austr. Prodr. 7. 1786; Fischer 3: 1772. 1934.

An annual, cultivated, sometimes escaping, tufted grass; culms up to 70 cm high, leafy; nodes glabrous. Leaf sheaths 5-15 cm long, glabrous; ligule a ciliate rim; blades 10-40 x 0.3-0.7 cm, linear, base rounded, apex acute, midrib prominent, glabrous. Racemes 2-5, up to 10 cm long, axis flattened; spikelets 2 mm, globose, in two rows; lower glume obscure; upper glume ovate; lower lemma ovate, membranous; upper lemma ovate, crustaceous, 3-nerved; palea orbicular; stamens 3; stigmas plumose. Caryopsis plano-convex, free within the lemma and palea.

Occasionally cultivated in black cotton soils and rarely as an escape. Fl. & Fr.:Throughout the year. Tel.: *Arikelu.*

Thalaricheruvu (ATP), *TP & NY* 1127; Kalasamudram (ATP), *KRKS* 40112; Cement nagar valley (KNL), *TP & RVR* 1817; Palakonda hills (KDP), *CS* 8285, 7718; Tirumala hills (CTR), *MSG* 12170; Papavinasanam (CTR), *PVP* 3680; Galikonda (VSKP), *DAM* 6511 (MH); Chelama (KNL), *JLE* 17969 (MH); Gobanapalem (WG), *KS* 5153 (MH); Near Cheedipalem (EG), *GVS* 68620 (MH); Anantagiri (VSKP), *GVS* 19551 (MH); Balugam, Parvathipuram (SKLM), *VNS* 4539 (MH); Mallelatheertham (MBNR), *BSS* & *BKS* 32211 (BSID); Red hills (CPT), *CAB* 5514 (MH); Kurumapuram RF (SA), *KRM* 60272 (MH); (MH); Ammapetai (SA), *VNS* 4173 (MH); Sirumalai hills (DGL), *SKS* 430 (SGH); Selur, Kollimalai (NMK), *S.India Flora* 12981 (MH); Kaveri peak–Yercaud (SLM), *DBD* 31299 (MH); Mahendragiri (GJP), *VNS* 5745 (MH); Bhimanthu, Similipal BR (MBJ), *Bahali* & *Agrawala* 414 (BSID); Buguda, Satkosia Tiger Reserve (AGL), *KCM* 8330 (BSID); Tikarpada RF, Satkosia Tiger Reserve (AGL), *KCM* & *J.Swamy* 7621 (BSID); Jenabil Range, Sililipal B.R. (MBJ), *AC* 2311 (BSID).

INDIA: Throughout the hotter parts of India, wild and cultivated.

WORLD: Tropics of the Old World.

PENNISETUM Richard ex Persoon

1. Anther cells not bearded at the apex:
   2. A low, copiously branched perennial, forming mats with numerous stolons with short internodes, culms very short .... **P. clandestinum**
   2. Annual or perennial, erect, culms very tall:
      3. Involucral bristles scabrid; lemmas homomorphous, not 3-lobed:
         4. An annual; upper glume broadly ovate, upper lemma transversely rugose; leaf blades flat........ **P. glaucum**
         4. A perennial; upper glume elliptic-oblong, upper lemma not transversely rugose; leaf blades convolute .... **P. hohenackeri**

3. Involucral bristles densely villous or ciliale below the middle; lemmas heteromorphous, 3-lobed:
   5. Spikelets solitary, pedicellcd or in groups of 2-5, with one sessile and other pedicelled, lower glume at least half the length of the lower lemma.................. **P. pedicellatum**
   5. Spikelets solitary, sessile, lower glume small rarely if so half as long as the lower lemma:
      6. Involucral bristles densely plumose; annual grass..**P. polystachyon**
      6. Involucral bristles outer not ciliate, the inner long with long silky hairs below the middle; perennial grass ....**P. setosum**

1. Anther cells bearded at the apex:
   7. Wild, perennial; spikelets subsessile, not gaping..................... **P. purpureum**
   7. Cultivated, annual; spikelets pedicelled, gaping.......................... **P. glaucum**

Pennisetum clandestinum Hochst. ex Chiov., Ann. Reale 1st. Bet. Roma 8: 41. t. 5 f. 2. 1903; Fischer 3: 1793. 1934; Bor, Grass. India 344. 1960.

A perennial, stoloniferous grass; culms very short, up to 10 cm high, spread on the floor, like mats; nodes glabrous. Leaf sheaths overlapping, terete, glabrous; ligule a fringe of hairs; leaf blades 5-15 x 0.2-0.5 cm, linear, convolute, base rounded, apex obtuse.

Excellent tropical African grass, cultivated to improve the grazing.

Palasa (SKLM), *BDM* 183 (CAL).

INDIA: Introduced.

WORLD: Native of E. Africa, introduced into many countries and has become invasive weed.

Pennisetum glaucum (L.) R.Br., Prodr. Fl. Nov. Holl. 195. 1810. *Panicum glaucum* L., Sp. Pl. 56. 1753. *Pennisetum americanum* (L.) Leeke, Z. Naturwiss. 79: 52. 1907; Bor in Rech. f., Fl. Iran. 70: 503. 1970. *Panicum americanum* L.,Sp. Pl. 56. 1753. *Alopecurus typhoides* Burm. f., Fl. Ind. 27. 1768. *Pennisetum typhoideum* Rich., in Pers., Syn. Pl. 1: 72. 1805; FBI 7: 82. 1896. *P. typhoides* (Burm. f.) Stapf & C. E. Hubb., Kew Bull. 1933: 271. 1933; Fischer 3: 1792. 1934; Bor, Grass. India 350. 1960. *Setaria glauca* (L.) P. Beauv. Ess. Agrost. 51. 109. 178. 1812; FBI 7: 78. 1896; Bor, Grass. India 360. 1960.

An annual, cultivated, tufted grass; culms up to 2 m high; nodes bearded. Leaf sheaths overlapping, sparsely tubercled based hairs; ligule a ciliate rim; leaf blades 30-90 x 0.5-5 cm, linear-lanceolate, base rounded, apex acuminate, pilose below, midrib prominent. Panicle up to 25 cm long, spiciform, cylindric, dense; rachis stout, villous below the inflorescence; involucral bristles 4 mm, often purplish-coloured; spikelet 4 mm, oblong; lower glume absent; upper glume minute, truncate, 3-nerved;

lower lemma oblong, obtuse, 5-nerved, epaleated; upper lemma ovate-oblong, membranous, 2-nerved; palea very broad, truncate, ciliate at the tip; stamens 3, apex penicillate, stigmas plumose. Caryopsis oblong-obovoid or fusiform; glabrous, top exposed.

Widely cultivated for its grains. Fl. & Fr.: August-December. Vern.: Tel.: *Sajjalu, Pitta gantee;* Ori.: *Gantiya, Bajri.*

Alurukona (ATP), *TP* & *NY* 1128; Kailasanathakona (CTR), *BR* & *P.Priyadarsini* 37131 (SKU & BSID); Puttagandhilanka, Tadepalli (EG), *NRR* & *DN* 84349 (BSID); Salur (VJN), *NPBK* 1046 (CAL); Visakhapatnam (VSKP), *NPBK* 585 (CAL); Salur (VJN), *NPBK* 958, 1046 (CAL).

INDIA: Cultivated and escape.

WORLD: Tropical Africa.

Pennisetum hohenackeri Hochst. ex Steud., Syn. Pl. Glumac. 1: 103. 1854; Fischer 3: 1792. 1934; Bor, Grass. India 344. 1960; Matthew, Fl. Tamilnadu Carnatic 3: 1888. 1983. *P. alopecuros* Steud., Syn. Pl. Glumac. 1: 102. 1854; FBI 7: 84. 1896.

A perennial, densely tufted grass; culms up to 70 cm high; nodes glabrous. Leaf sheaths overlapping, glabrous; ligule a small hairy ring; blades 15-40 x 0.2-0.5 cm, linear, coriaceous, convolute, base truncate, margins hairy near base, apex acuminate, glabrous. Spikelike raceme up to 12 cm long, solitary; rachis below the inflorescence pilose; involucral bristels 1.5 cm; spikelet 8 mm, elliptic-lanceolate; lower glume sub-orbicular, hyaline, 3-nerved; upper glume elliptic-oblong, acuminate, 5-nerved; lower lemma elliptic-oblong, margins incurved, 9-nerved; upper lemma lanceolate, acuminate, 5-nerved; stamens 3; stigma plumose. Caryopsis narrowly oblong.

Occasional in sandy soils near streams and bunds of rice fields. Fl. & Fr.: August- October.

Amagondapalem RF (ATP), *NY* 1021; Nallamada (ATP), *NY* 1209; Talakona Road (CTR), *PVP* 2507; Kuppam (CTR), *S. India Flora* 14271 (MH); Araku (VSKP), *GVS* 21570 (CAL); Gudem (VSKP), *VNS* 632 (CAL); Araku valley (VSKP), *NPBK* 588 (CAL); on the way to Tanjavanam (VSKP), *GVS* 42630 (MH); Mannanur (MBNR), *MRS* 54 (DD); Melpat (SA), *CAB* 1062 (MH); Komatiyur (NA), *KS* 6468 (MH); Chinnamettappanur, Mangalam RF, Yelagiri hills (NA), *MBV* 656 (MH); Kaveri peak, Yercaud (SLM), *DBD* 31246 (MH); Pottangi (KPT), *AC* 10233 (BSID).

INDIA: Peninsular and N.W.India.

WORLD: Tropical Africa, Madagascar, Pakistan.

Pennisetum hordeoides (Lam.) Steud., Syn. Pl. Glum. 1: 103. 1854; Bor, Grass. India 345. 1960; Saxena & Brahmam, Fl. Orissa 4: 2451. 1996.

Erect and rather slender grass; culms and rachis glabrous. Leaves suberect, 10-25 cm x 4.2-7.5 mm, base and mouth of sheath ciliate. Spike narrow, erect, tapering, purple, 7.5-12.5 cm long, to 7.5 cm diam. Spikelets sessile, solitary, 3-3.5 mm long, close but not clustered on the undulate-toothed angles of the rachis; bristles all

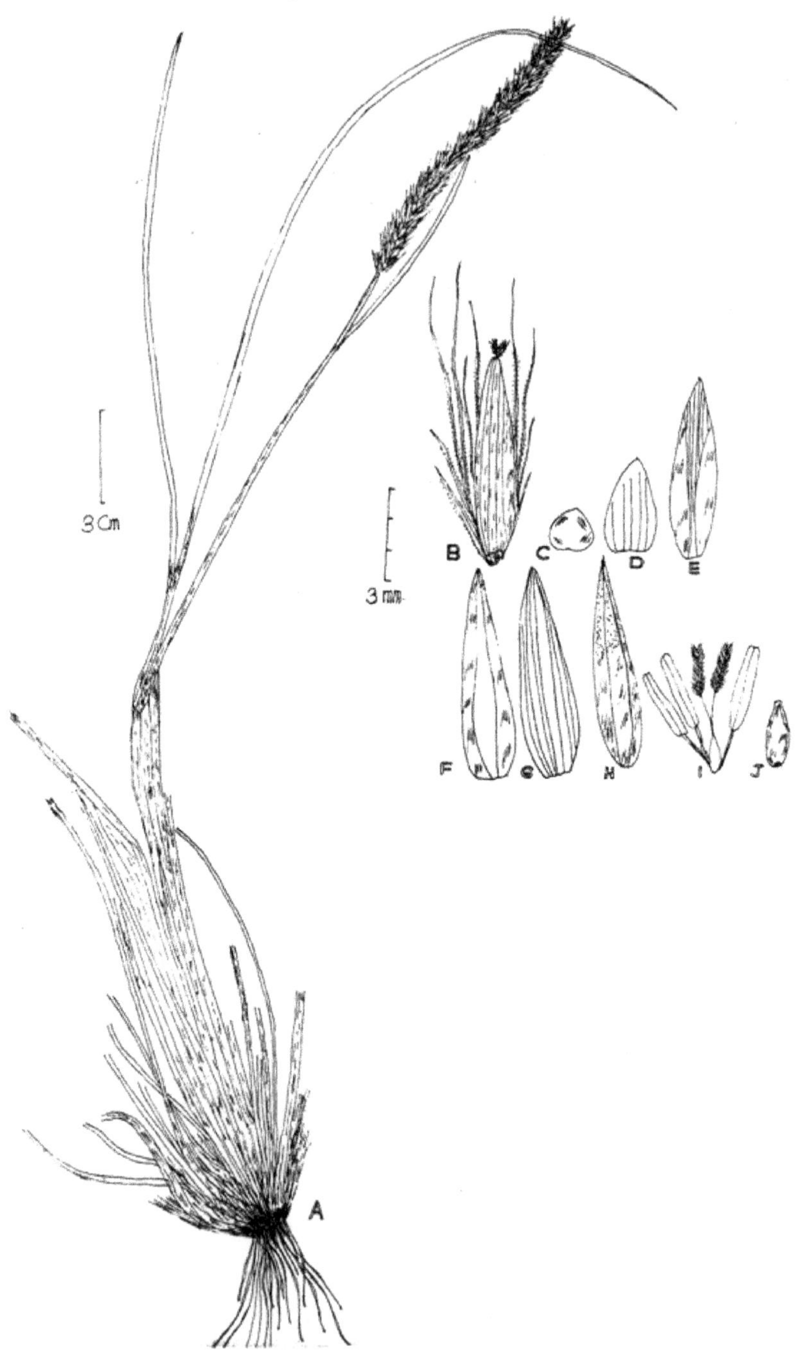

Figure 85. **Pennisetum hohenackeri** Hochst. ex Steud.

A. Twig, B. Spikelet, C. Lower glume, D. Upper glume, E. Lower lemma, F & G. Upper lemma front and back view, H. Upper palea, I. Pistil and stamens, J. Grain.

free to the base, unbranched, glabrous, scaberulous, without any cilia, some up to 1.5 cm long; lower glume minute, suborbicular, upper glume equal to the spikelet, 5-nerved, truncate, cuspidate; lower lemma 5-nerved, truncate, toothed, paleatee, male. Upper floret readily disarticulating, upper lemma obtuse, nerveless, shining.

Malda, Sambalpur (Panigr. *et al.*, Bull. Bot. Surv. India 6: 263. 1964). Fl.: November.

INDIA: Rajasthan, Bihar, Odisha.

WORLD: W. Africa.

Pennisetum pedicellatum Trin., Mem. Acad. Imp. Sci. St. Petersbourg Ser. 6. Sci. Math., Seconde Pt. Sci. Nat. 3: 184. 1934; FBI 7: 86. 1896; Fischer 3: 1792. 1934; Bor, Grass. India 346. 1960.

An annual, tufted grass; culms up to 90 cm high, branches from the base, leafy. Leaf sheaths glabrous; ligule a shortly ciliate membrane; leaf blades 10-20 x 3-9 cm, linear, flat, glabrous-sparsely hairy. Racemes 13 cm long, cylindric, densely-flowered; involucral bristles outer few, slender, short, 3 mm, inner long, 1 cm, numerous, densely villous below the middle, unequal, free from the base; spikelets 4 mm, solitary in the involucral, sub sessile; lower glume very small, woolly; upper glume oblong- lanceolate, hyaline, apiculate; lower lemma oblong, truncate, 3-lobed, hyaline; upper lemma ovate-oblong, obtuse, apex fimbriate ciliate, coriaceous, shining; palea lanceolate, toothed.

Common in plains and hill slopes. Fl. & Fr.: October-November.

Kalvabugga (KNL), *RVR* 1884; Rachakuntapalle (KDP), *KRKS* 39042; Tirumala (CTR), *BR* & *BSS* 30490 (SKU & BSID); Mahanandi (KNL), *VBH* 86673 (BSID); Chiduru (EG), *CAB* 4976 (MH); Jolaput (VSKP), *GVS* 32899 (MH); Parvathamalai RF (NA), *MBV* 1333 (MH); Jacum, Karlapat Wildlife Sanctuary (KHD), *AC* 2442 (BSID).

INDIA: Rajasthan, Uttar Pradesh, Madhya Pradesh, West Bengal, Odisha, Andhra Pradesh, Maharashtra, Tamil Nadu.

WORLD: West Africa.

Note: It can be distinguished from all the other species of this genus by its fluppy - white or purplish panicles, densely woolly, plumose involucral bristles and pedicelled spikelets. This is similar to *P. polystachyon*, but differs by the same character mentioned above.

Pennisetum polystachion (L.) Schult., Mant. 2: 146. 1824; Fischer 3: 1792. 1934; Bor, Grass. India 346. 1960. *Panicum polystachion* L., Syst. Nat. ed. 10. 870. 1759. *Pennisetum setosum* (Sw.) Rich. in Pers., Syn. Pl. 1: 72. 1805; FBI 7: 87. 1896; Bor, Grass. India 340. 1960. *Cenchrus setosus* Sw., Prodr. Veg. Ind. Occ. 26. 1788.

A perennial, tufted grass; culms up to 80 cm high, often fastigiately branched at the nodes. Leaf sheaths glabrous; ligule with fringed long, soft hairs; leaf blades 10-20 x 5-10 cm, linear, apex acuminate, slightly hairy. Racemes 10 cm long, purplish brown; involucral bristles unequal, the outer not ciliate, short, 3 mm, the inner longer, long silky hairy below the middle, 1.3 cm; spikelet 4 mm, solitary within the

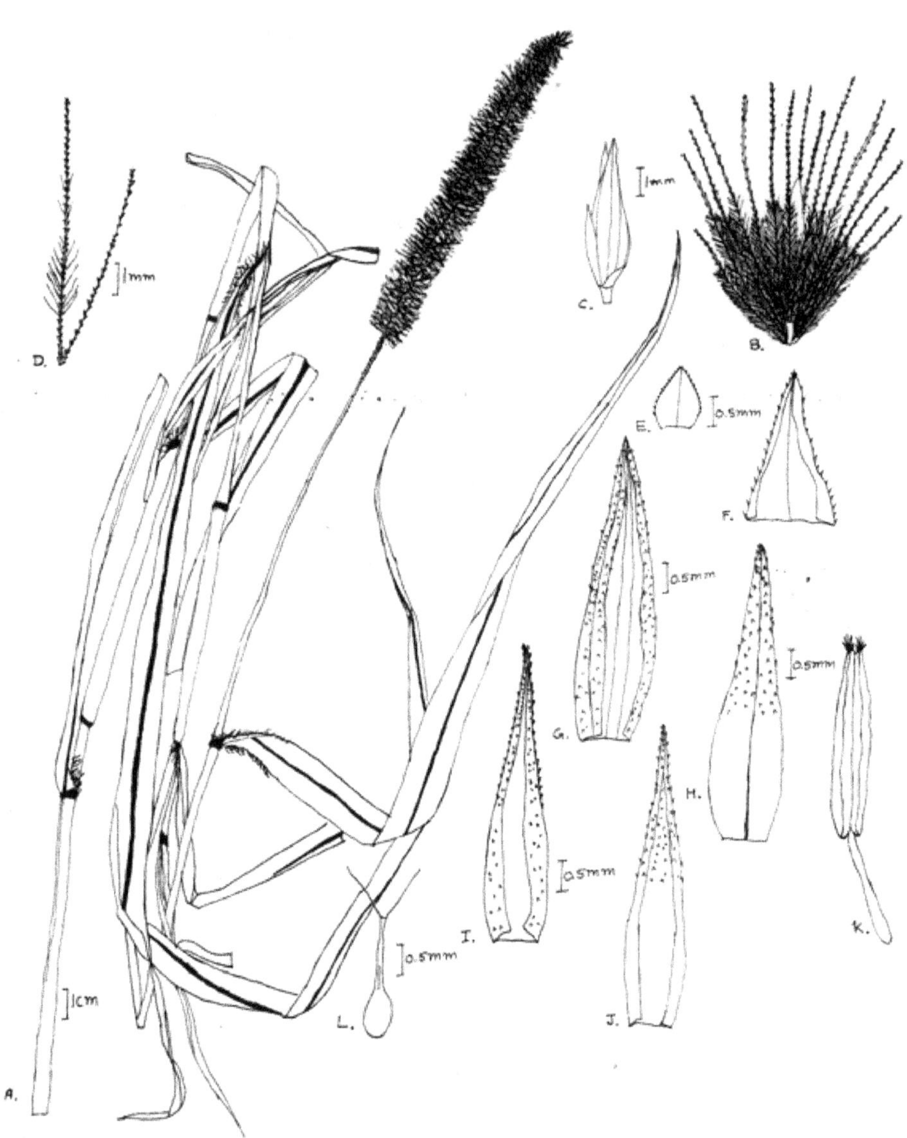

Figure 86. **Pennisetum pedicellatum** Trin.

A. Twig, B. Cluster of spikelets, C. Lower glume, D. Upper glume, E. Lower lemma, F. Upper lemma, G. Lower palea, H. Stamens, I. Upper palea, J. Pistil, K. Involucral bristles.

involucral; lower glume minute or completely absent; upper glume ovate-oblong, cuspidate, hyaline; lower lemma oblong, obtuse, bidentate, palea narrowly oblong, hyaline, male; upper lemma ovate-oblong, truncate, coriaceous, shining; palea oblong, truncate, toothed or ciliate at the apex; stamens 3.

Common in the plains and hills. Fl. & Fr.: July-January.

Talakona top (KDP), *BR* & *BSS* 32926 (SKU & BSID); Kanjarampugha (KNL), *KMS* 22834 (CAL); Papavinasanam (CTR), *JHFB* 117059 (MH, BSID); Sirivaka (WG), *CAB* 5272 (MH); (CAL); Araku valley (VSKP), *GVS* 32885 (CAL); Majhipada section, Satkosia Tiger Reserve (AGL), *KCM* 6133 (BSID).

INDIA: Peninsular, C. & E. India.

WORLD: SE USA, Mexico, South America, Tropical Africa, Arabia, China, Indo-China, Malesia, Australia.

Pennisetum purpureum Schumach., Beskr. Guin. Pl. 44. 1827; Bor, Grass. India 348. 1960; Matthew, Fl. Tamilnadu Carnatic 3: 1889. 1983.

A perennial, stoloniferous grass; culms up to 2 m high, erect, branching from lower nodes; nodes bearded. Leaf sheaths 10-15 cm long, glabrous; ligule a dense fringe of hairs; blades 30-60 x 0.7-1.5 cm, linear - lanceolate, base rounded, margins scaberulous, apex acuminate, scabrid, midrib prominent. Spikelike raceme up to 18 cm long, solitary, cylindric, yellowish or purplish; involucral bristles numerous of unequal length, one usually very much longer, 8 mm, scabrid, ciliate, inner most scabrid, 6 mm, sparingly plumose towards the base; spikelet solitary, sessile, if in fascicles 2-4, lateral pedicelled, all lanceolate, 5 mm; lower lemma absent or if present ovate- lanceolate, truncate, scabrid; upper glume triangular, margin incurved, 3-nerved, scabrid; lower lemma oblong - lanceolate, scabrid on outside surface, 5-nerved, male or barren; upper lemma lanceolate, rolled, scabrid at apex, 2-nerved; stamens 3, peniciliate at apex; style united throughout; stigmas plumose. Caryopsis ovoid.

Occasional in hill slopes and forest areas. Fl. & Fr.: July–January.

Tirumala hills (CTR), *MSG* 12173; Diguvametta (PKM), *BR* & *BSS* 30347 (SKU & BSID); Vinukonda RF (GNT), *VRK* 6854; Anantagiri (VSKP), *GVS* 21558 (MH); Chahala, Similipal RF (MBJ), *A.R.K.Sastry* & *G.P.Singh* 12272 (BSID).

INDIA: Introduced.

WORLD: Native of Tropicl Africa, introduced in many other tropical countries.

Note: The 'Napeur grass' is a native of Tropical Africa. But now introduced into many tropical countries including India as a valuble fodder grass. Like *Themeda cymbaria,* it is also known as 'elephant grass.

PEROTIS Alton

Perotis indica (L.) Kuntze, Revis. Gen. Pl. 2:787. 1891; Fischer 3: 1814. 1934; Bor, Grass. India 611. 1960; Matthew, Fl. Tamilnadu Carnatic 3: 1889. 1983. *Anthoxanthum indicum* L., Sp. Pl. 28. 1753. *Perotis latifolia* Ait., Hort. Kew Bull. 1: 96. 1789; FBI 7: 98. 1896.

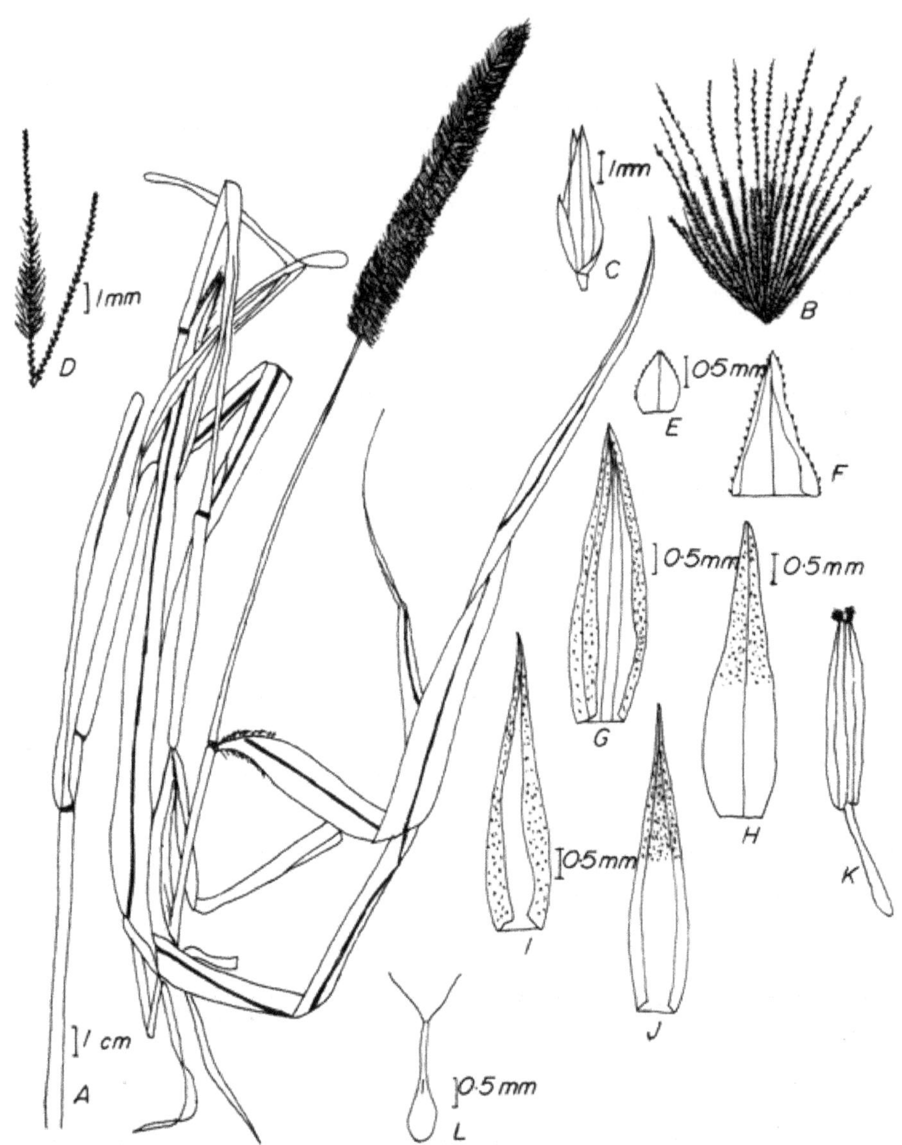

Figure 87. **Pennisetum purpureum** Schumch.

A. Twig, B. Cluster of spikelets, C. Spikelet, D. Involucral bristles, E. Lower glume, F. Upper glume, G & H. Lower lemma front and back views, I & J. Upper lemma front and back views, K. Stamens, L. Pistil.

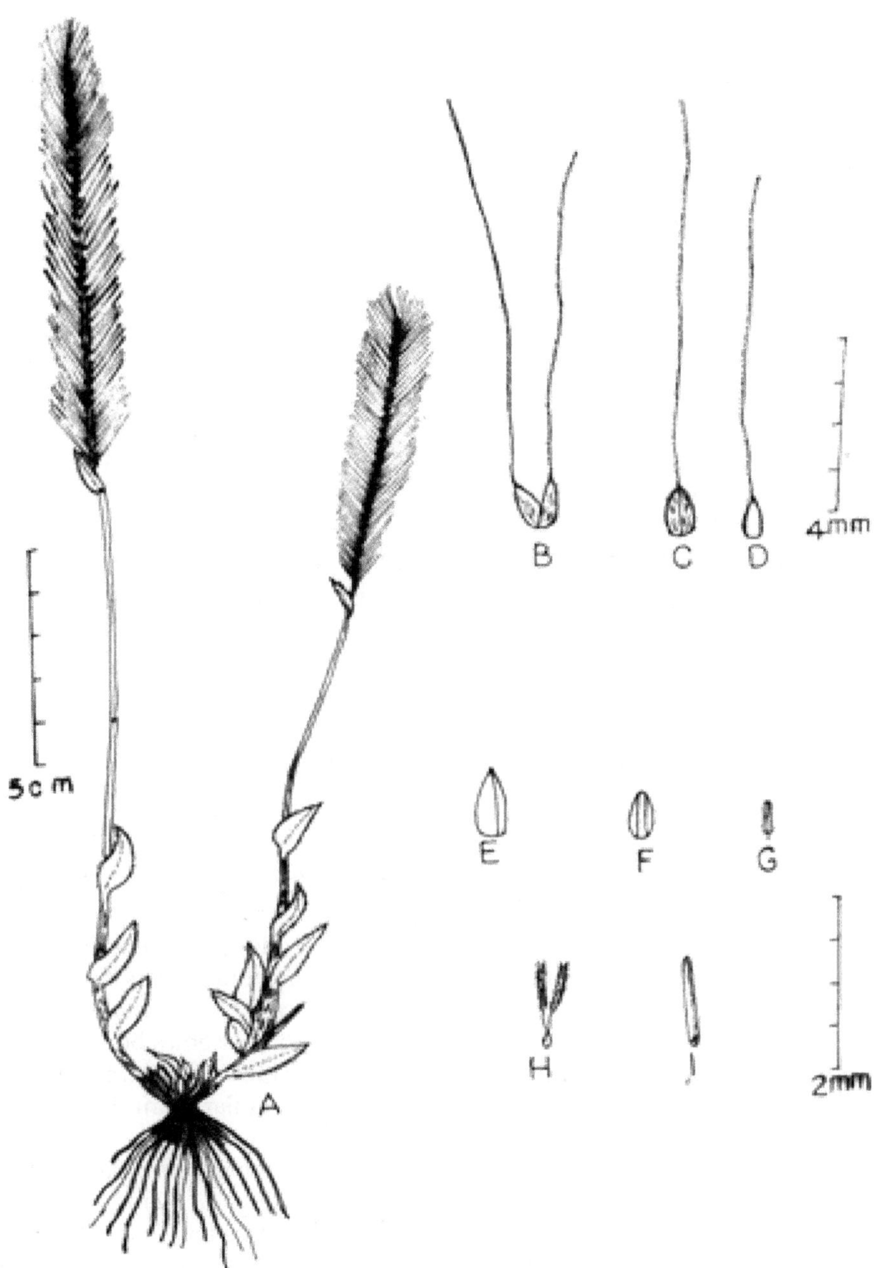

Figure 88. **Perotis indica** (L.) Kuntze.

A. Twig, B. Spikelet, C. Lower glume, D. Upper glume, E. Lemma, F. Palea, G. Anther, H. Pistil, I. Caryopsis.

A perennial, tufted grass; culms up to 50 cm high, geniculately ascending from base; nodes glabrous. Leaf sheaths 1-7 cm long, glabrous; ligule absent; blades 1-4 x 0.3-0.6 cm, ovate- lanceolate, base amplexicaul, margin cartilaginous, spinulose, apex acute, glabrous. Spikelike raceme up to 15 cm long, terminal; spikelets 2 mm, linear, subsessile; glumes subequal, linear, 1-nerved, margins scabrid in lower, awned; lemma linear, hyaline, 1-nerved; palea narrow, 2-nerved; stamens 3; stigmas plumose. Caryopsis linear.

Very common in sandy tracts and plains. Fl. & Fr.: July-January.

Thalaricheruvu (ATP), *TP* 918; Kalasamudram (ATP), *KRKS* & *KRS* 36464 (SKU & BSID); North Dhone RF (KNL), *PVP* 2694; Palakonda hills (KDP), *CS* 8275: Ramagiri (CTR), *BR* & *P.Priyadarsini* 37117 (SKU & BSID); Penchalakona (NLR), *MSG* 12192; Narapareddikunta (KNL), *VBH* 86649 (BSID); Balapalle (KDP), *JLE* 14962; (CAL & MH); Seshachalam hills (KDP), *MH* Acc. No. 5844 (MH); Nagrjunakonda Sacred Grove (GNT), *J.Swamy* & *S.Nagaraju* 7251 (BSID); Kondapalli (KSN), *CAB* 4600 (MH); Polavaram agency (WG), *DCSR* 208 (CAL); Nellore, *JSG* 12816 (CAL); Dolphin's nose hill (VSKP), *ARKS* 7137 (CAL); Palasamendu (VSKP), *VNS* 4630 (MH); Salur (VJN), *NPBK* 1003 (CAL); Sirumalai hills (DGL), *SKS* 329 (SGH); Illalur RF, Tirupporur (CPT), *ANH* 45584 (MH); Tippukadu RF (NA), *KRM* 17696 (MH); Pennagaram RF (SLM), *EV* 20747, 22405 (MH); Purunakote, Satkosia Tiger Reserve (AGL), *KCM* 5207 (BSID); Deomali hill(KPT), *AC* 10228 (BSID).

INDIA: Plains of India.

WORLD: Nepal, Sri Lanka, Myanmar, Malaya, SE. Asia.

PHACELURUS Grisebach

Phacelurus zea (C.B.Clarke) Clayton, Kew Bull. 33: 177. 1978. *Rottboellia zea* C.B.Clarke, J. Linn. Soc. Bot. 25: 86. 1889; Hook.f., Fl. Brit. India 7: 152. 1896. *Thyrsa zea* (C.B.Clarke) Stapf, Hook.f. Icon. 31: p. 3078. 1922; Bor, Grass. India 255. 960.

Perennial grass, culms 150-200 cm high, culm nodes pubescent. Leaf sheath keeled, glabrous, 20-40 cm long; ligule a ciliolate membrane; leaf blades flat or involute, 50-100 cm x 1-2 mm, midrib widened, surface scabrous, rough adaxially, margins scabrous. Inflorescence composed of numerous racemes borne along the central axis, 5-15 cm long, central axis 25-40 cm long; rachis semiterete, glabrous. Spikelets in pairs, one sessile and other pedicelled, similar, joint 3.4 x 1.7 mm, pedicels 2 x 1.2 mm, club-shaped; spikelets ovate, dorsally compressed, 4 mm long, deciduous, callus base truncate, attached transversely; glumes dissimilar, exceeding the florets, firmer than fertile lemma; lower glumes oblong, 3.8 x 2 mm, coriaceous, 2-keeled, winged narrowly all along, 1 to 3-nerved, surface flat, scabrid, obtuse at apex; upper glumes ovate, chartaceous, 1-keeled, 3-nerved, 3.9 x 1.2 mm, acute at apex. Florets 2, basal florets barren; lower lemma lanceolate, 3.4 x 1 mm, hyaline, 2-nerved; upper lemma lanceolate, 3 x 1 mm,hyaline, 2-veined; palea hyaline, 2-nerved, 2-3 x 0.9 mm; anthers, 2 mm long

Occsional near marshes in Similipal Biosphere reserve. Fl. & Fr.: September–November.

Similipal Biosphere reserve (MBJ), *AC* 2454 (BSID); Hatigarh, Similipal B.R. (MBJ), *AC* 2354 (BSID);

INDIA: Assam, West Bengal and Odisha.

WORLD: Bhutan, Myanmar, Nepal, Thailand and Vietnam.

PHALARIS Linnaeus

Phalaris minor Retz., Obs. Bot. 3: 8. 1783; FBI 7: 221. 1896; Bor, Grass. India 616. 1960.

An annual, tufted grass; culms up to 60 cm high; nodes glabrous. Leaf sheaths 3-9 cm; glabrous, loose; ligule membranous; blades 5- 15 x 0.3-0.5 cm, linear, flat, base sub-cordate, apex acuminate, glabrous. Spikelike panicle up to 5 cm long, dense, ovate-oblong; spikelets all alike, 5 mm, solitary, sterile florets present; pedicels scabrid; glumes equal, 5 mm, membranous, boat-shaped, broadly winged, the wing margin denticulate, prominently 3-nerved, enclosing the floret; sterile lemma 2 mm, glabrous; fertile lemma 3 mm, pubescent, broadly ovate, 1-nerved; palea narrowly linear, bisexual; stamens 3; stigmas plumose. Caryopsis ovoid, pubescent.

Occasional on waste lands in Tirumala hills. Fl. & Fr.: October-March

Tirumala hills (CTR), *MSG* 12160.

INDIA: Introduced with Mexican wheat, naturalized in NW. to C. Himalayas.

WORLD: Indigenous to Mediterranean region, introduced into Africa, Australia, Pakistan, India, Latin America and many other countries.

PHRAGMITES Adanson

Phragmites karka (Retz.) Trin. ex Steud. Nom. Bot. ed. 2. 2: 324. 1841; FBI 7: 304. 1896; Fischer 3: 1807. 1934; Bor, Grass. India 416. 1960; Matthew, Fl. Tamilnadu Carnatic 3: 1888. 1983; var. **karka.** *Arundo karka* Retz., Obs. Bot. 4: 21. 1786.

A perennial shrub; culms up to 5 m high, stout woody below, stoloniferous, creeping; nodes glabrous. Leaf sheaths 1.5-20 cm long, glabrous; ligule a ciliate rim; blades 15-60 x 0.7-4 cm, linear-lanceolate, stiff, base rounded, apex acuminate, glabrous. Panicle up to 60 cm long, lax, open, decompound; branches filiform, scabrid; spikelets 1.3 cm long, lanceolate, 4-flowered, laterally compressed, pedicelled; callus pilose, it covers the spikelet; lower glume oblong, 3-nerved; upper glume linear-lanceolate, apex acuminate; lemma linear, 1-3-nerved, apex acuminate, lowest sterile, upper fertile; palea oblong, obtuse, 2-nerved; stamens 3; stigmas plumose. Caryopsis oblong, terete.

Occasional robust grass near the water courses. Fl. & Fr.: November-March. Tel.: *Kapiri gaddi.* Tam.: *Tharbaipul.*

Gutturu (ATP), *BR & P. Priyadarsini* 37153 (BSID); Palakonda hills (KDP), *CS* 8295; Idupulapaya (KDP), *KRKS & BSS* 41988; Loyapalle (GNT), *VRK* 3882; Diguvametta (PKM), *JLE* 32468 (CAL & MH); Kadapa, *JSG* 1134 (CAL); Kambakkam hills (CTR), *S. India Flora* 16440 (MH); Isukagundlu, GB WLS (PKM), *J. Swamy & L. Rasingam* 7590 (BSID); Polavaram agency (WG), *DCSR* 135 (CAL); Maredumilli (EG), *MM* 100828 (BSID); Donkarayi near Mangapadu (VSKP), *GVS* 29675 (CAL);

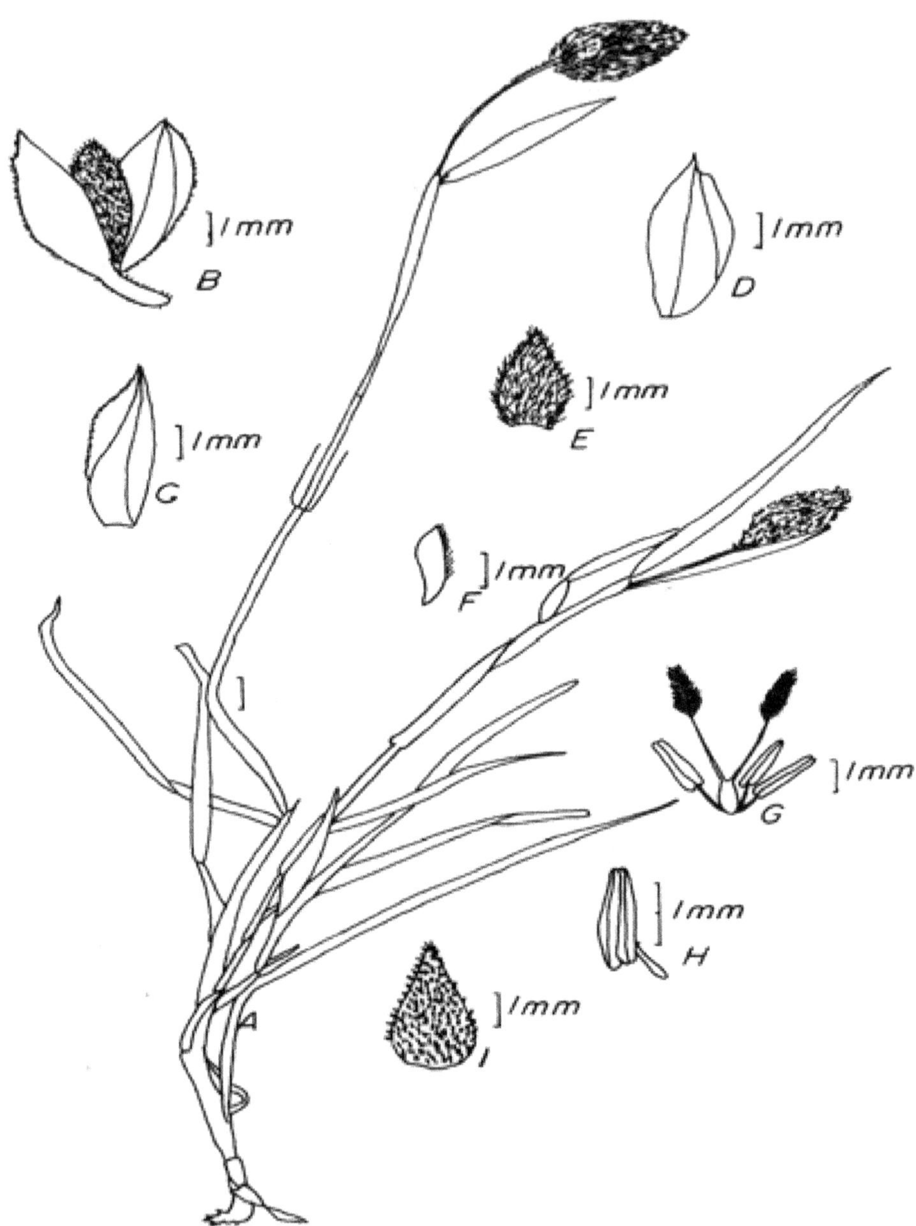

Figure 89. **Phalaris minor** Retz.

A. Twig, B. Spikelet, C. Lower glume, D. Upper glume, E. Lemma, F. Palea, G. Pistil with stamens, H. Stamen, I. Caryopsis.

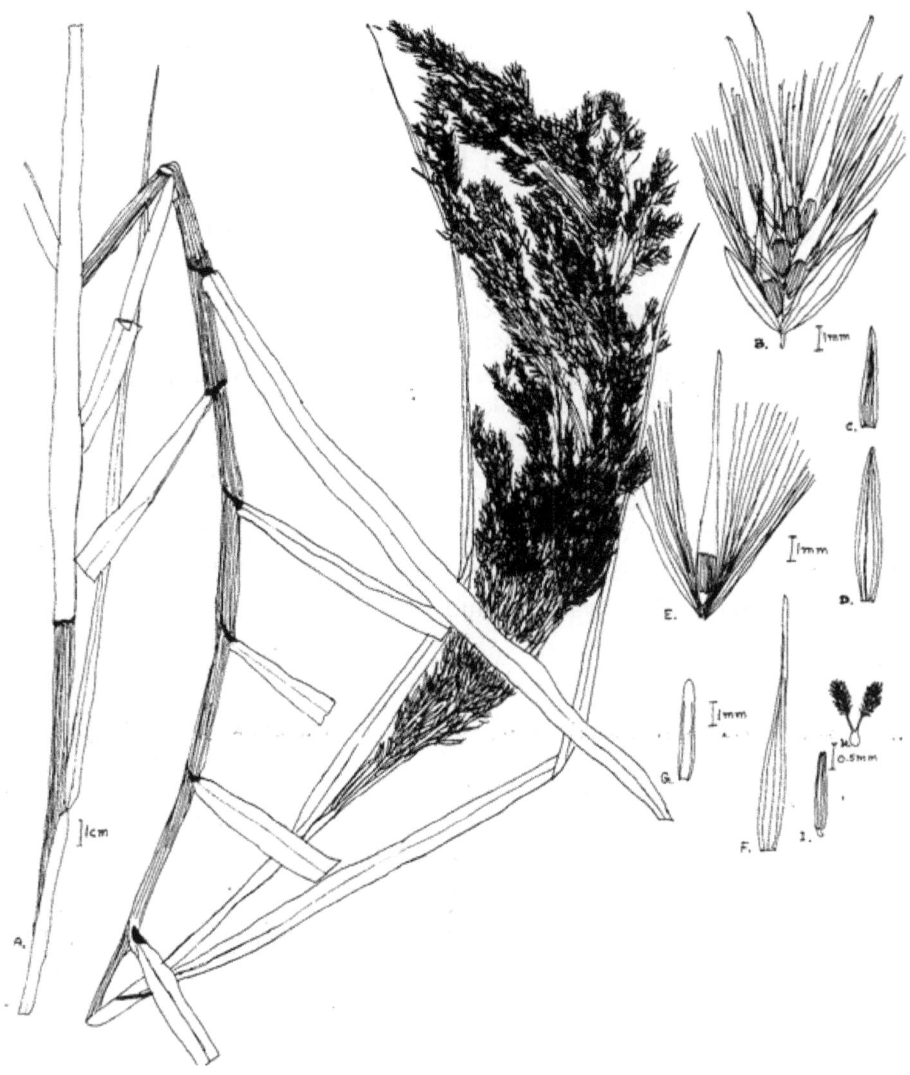

Figure 90. **Phragmites karka** Steud.

A. Twig, B. Spikelet, C. Lower glume, D. Upper glume, E. Floret, F. Lemma, G. Palea, H. Pistil, I. Stamen.

Hogainakkal (SLM), *KCJ* 17999 (MH), *EV* 23575 (MH); Similipahar (MBJ), *V. Naik* Acc. No. 3817 (RRL-B); Tulka section, Satkosia Tiger Reserve (AGL), *KCM* 6194 (BSID); Jenbil range, Similipal B.R. (MBJ), *AC* 2325 (BSID).

INDIA: Throughout India.

WORLD: Tropical Africa, Tropical Asia, Polynesia and North Australia.

Note: It is very similar to *Arundo donax,* but can be distinguished by the narrower leaves, glabrous lemma and densely villous spikelets. This grass is said to be poisonous to cattle.

POA Linnaeus

Poa annua L., Sp. Pl. 68. 1753; FBI 7: 345. 1896; Fischer 3: 1850. 1934; Bor, Grass. India 555. 1960; Matthew, Fl. Tamilnadu Carnatic 2: 1891. 1983; B.R.P.Rao *et al.*, Indian J. Forestry 41: 113. 2018.

Annual tufted grass; culms to 10 cm, decumbent, rooting at lower nodes, nodes pilose. Leaf sheath 1.5-2 cm, glabrous; ligule membranous, 2 mm; blades lanceolate, 1.5-4 x 0.1-0.2 cm, flat or folded, scaberulous. Panicles 3-8 cm, lax or contracted; branches paired. Spikelet ovoid or oblong, 2-4 mm, laterally compressed, 3-flowered; rachilla jointed at base and between lemmas. Glumes unequal, thinly membranous; lower glume ovate-lanceolate, to 1.5 mm, 1-nerved; upper glume 1.5-2 mm, 3-nerved, glabrous. Florets disarticulate in between rachilla internodes, nodes of rachilla 0.8 mm, lanceolate; lemma oblong, to 2 mm, 3-5-nerved, ciliate on keel and nerves; stamens 3. Caryopsis terete, to 1 mm, grooved, surface reticulate.

Occasional in Shervarayan (Yercaud) hills and Nallamalais. Fl. & Fr.: August–December.

Shervarayan hills, near Yercaud lake area (SM), *SKS* 609 (SGH); Upper Ahobilam (KNL), *BR* & *BSS* 29307 (SKU).

INDIA: Throughout.

WORLD: Cosmopolitan.

POGONATHERUM P. Beauvois

1. Hairs of the callus longer than the spikelets; glumes nerveless .... **P. crinitum**
1. Hairs of the callus shorter than the spikelet; glumes 1-4- nerved .... **P. paniceum**

Pogonatherum crinitum (Thunb.) Kunth, Enum. Pl. 1: 478. 1833; FBI 7: 141. 1896; Bor, Grass. India 200. 1960; Matthew, Fl. Tamilnadu Carnatic 3: 1891. 1983. *Andropogon crinitus* Thunb., Fl. Jap. 40. t. 7. 1784. *Pogonatherum paniceum* sensu Fischer, Fl. Madras 3: 1714. 1934; non (Lam.) Hack. 1706.

A perennial, densely tufted grass; culms up to 30 cm high; nodes sparsely bearded. Leaf sheaths 1.5 cm long, lax, scaberulous; ligule of long hairs: blades 1.5-5 x 0.2-0.5 cm, lanceolate, flat, minutely scaberulous. Racemes 2.5 cm long, slender, spiciform, spathiolate; spikelets 2.5 mm, 2-nate. *Sessile spikelet;* callus bearded; glumes nerveless, hyaline; lower oblong, truncate, incurved margins, setose;

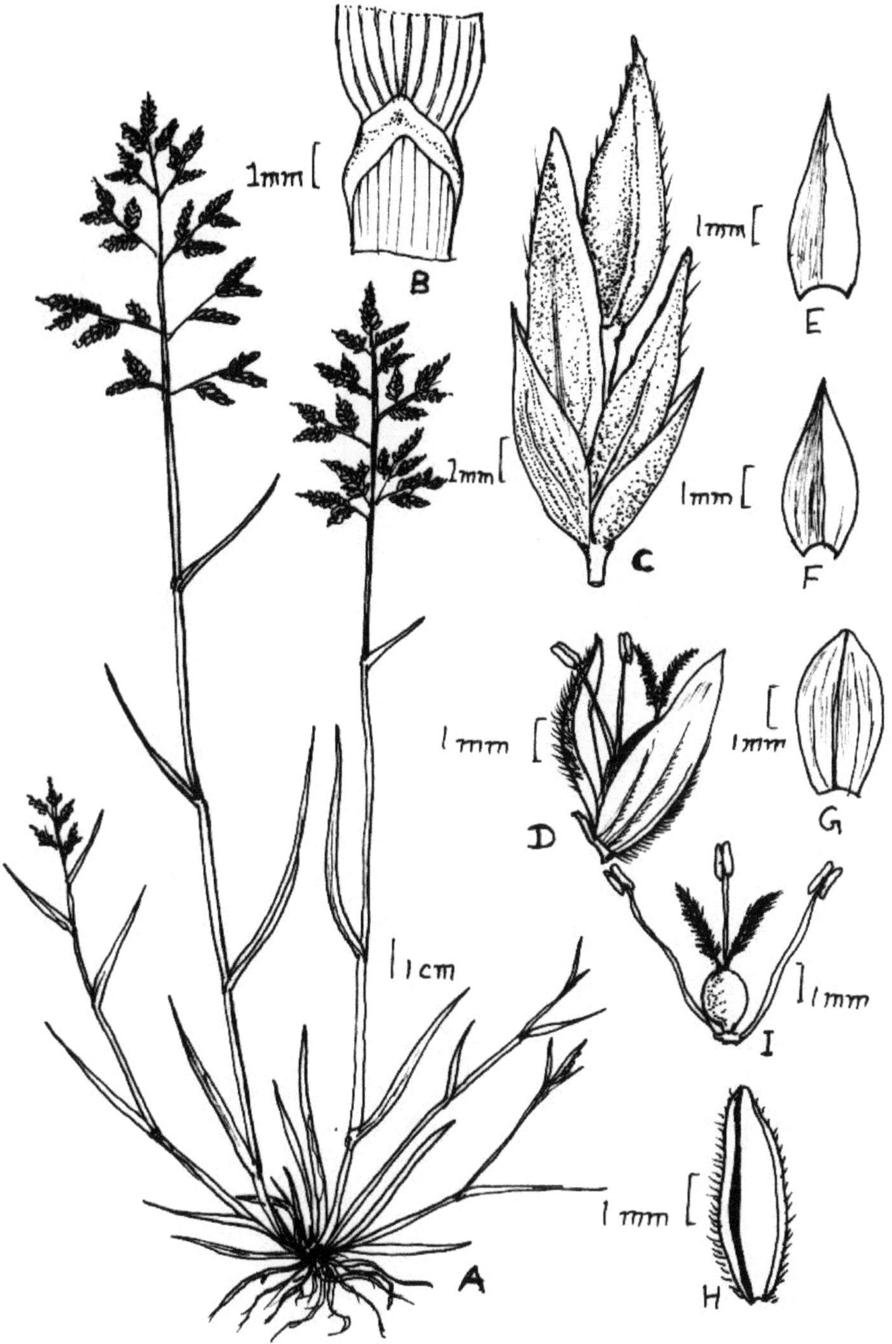

Figure 91. **Poa annua** L.

A. Habit, B. Ligule, C. Spike, D. Spikelet, E. Upper glume, F. Lower glume, F. Lemma, H. Palea, I. Stamens with pistil.

Figure 92. **Pogonatherum crinitum** (Thunb.) Kunth

A. Twig, B. Spikelets, C. Pedicelled spikelet, D. Lower glume, E. Upper glume, F. Lower lemma, G. Upper lemma, H. Stamen, I. Pistil.

upper glume ovate, keeled, bidentate, lobes ciliate, awned; lower lemma linear or some times absent; upper lemma ovate, deeply bidentate, awned; stamen(s) 1 or 2. *Pedicelled spikelet:* similar to sessile spikelet.

Common near water courses. Fl. & Fr.:August - January.

Pachagundala (KNL), *BR & SKB* 29277; Talakona water falls (CTR), *MSG & AA* 12234; Papavinasanam, Tirumala (CTR), *MHR* 13341; Near water falls, Penchalakona (NLR), *MSG* 12191; Near Padmapuraam gardens (VSKP), *MHR & MCK* 14846; Gundlakamma (KNL), *JLE* 23832 (MH & CAL); Kailasakona (CTR), *GVS* 32030 (CAL & MH); Rajavommangi (EG), *S.India Flora* 12645A (MH); Valamuru, Maredumilli (EG), *MM* 101354 (MH); Araku valley (VSKP), *DDSR* 95873 (MH); Anantagiri (VSKP), *GVS* 19532 (MH); Marapalam–Nagalur Road, Yercaud (SLM), *AVNR* 27463 (MH); Kanibori (KHD), *MB & Dhal* 7447 (RRL-B); Raigada, Satkosia Tiger Reserve (AGL), *KCM & Swamy* 7714 (BSID); Betkote, Karlapat Wildlife Sanctuary (KHD), *AC* 2466 (BSID).

INDIA: Throughout India.

WORLD: Afghanistan, Pakistan, China, Malaya, Japan, Solomon Islands.

Note: This species is closely related to and often confused with *P. paniceum,* but differs by the smaller spikelets, empty first floret, solitary stamens of the upper floret and longer callus hairs and awns. It is very difficult to distinguish these two species in the field. Several specimens in MH were wrongly identified and kept under *P. paniceum,* a rare plant and probably introduced. *P. crinitum* is rather common and less robust.

Pogonatherum paniceum (Lam.) Hack., Allg. Bot. 2.12: 178; Fischer 3: 1714. 1934; Bor, Grass. India 202. I960. *Saccharum paniceum* Lam., Encycl. 1: 595. 1785. *Pogonatherum saccharoideum* P. Beauv., Ess. Agrost. 176. 177. 1812; FBI 7: 141. 1896.

A perennial, densely tufted grass; culms up to 60 cm high, leafy, woody; nodes on the culms sparsely hairy. Leaf shealhs bearded; leaf blades 2.5-7 x 0.25-1 cm, linear, broad, apex acuminate. Spikes 5 cm long, spathiolate, terminal; rachis compressed, bearded; spikelets 2 mm long, 2-nate. *Sessile spikelet:* lower glume narrow-oblong, apex ciliate; upper glume conduplicate, apex densely ciliate, awned; lower lemma sometimes absent; palea of the upper lemma broadly ovate-oblong; ovary minute. *Pedicelled spikelets:* similar to the sessile spikelet.

Fairly common on moist hill sides among the baks of streams etc. Fl. & Fr.: September-April.

Gundlakamma, GBM, Nallamalais (KNL), *JLE* 23832 (MH); Puthur, Kailasakona (CTR), *GVS* 32030 (MH); Rajavommangi (EG), *S.India Flora* 12645 (MH); Araku valley (VSKP), *NPBK* 592 (CAL); Mallelatheertham (MBNR), *SRS* 110703 (BSID).

INDIA: Throughout the hilly parts of India.

WORLD: Sri Lanka, Myanmar, Malaya, China.

Note: This grass is extensively cultivated as an ornamental in gardens due to its peculiar tillering and foliage.

POLYPOGON Desfontaines

Polypogon monspeliensis (L.) Desf., Fl. Atlant. 1: 66. 1798; FBI 7: 245. 1896; Haines, Bot. Bihar & Orissa 3: 976.1924; Fischer 3: 1807. 1934; Bor, Grass. India 403. 1960.

Annual, tufted grass, 15-40 cm high. Leaves linear, 7.5-15 x 0.3-0.8 cm, gradually tapering to a point, glabrous, scabrid on the close nerves; sheath striate, scaberulous; ligule membranous, lacerate, ciliolate. Panicle cylindrical or oblong, 1.2-7.5 cm long, 0.75–2.3 cm diam.; spikelets 1.5-2.5 mm long, minutely pubescent; glumes subequal, narrowly oblong, scaberulous, obtuse or notched at the tip, with awn from the sinus, 5-9.75 mm long, 2-3 times as long as the glumes; lemmas 1-1.2 mm long, lower oblong, bifid, easily detachable with its palea and grain, with minute, deciduous awn; palea 2-toothed, about as long as the lemma. Caryopsis subterete, grooved, about as long as the lemma.

Rare in Eastern Ghats of Odisha. Fl.: March–April.

Rajpal, Similipahar (MBJ), *MB* 5499 (RRL-B).

INDIA: Cooler parts of NW Himalayas, Odisha.

WORLD: Sri Lanka, Pakistan, Europe.

POMMEREULLA Linnaeus f

Pommereulla cornucopiae L.f., Diss. Nov. Gram. 31. 1779; FBI 7: 300. 1896; Bor, Grass. India 620. 1960.

A perennial, stoloniferous grass; culms up to 30 cm high. Leaf sheaths broad, flattened; ligule a ciliate ridge; blades 2.5-8 x 0.2-0.3 cm, linear, margin sparsely ciliate, apex obtuse. Racemes 25 cm long, partially enclosed in the upper most sheaths; spikelet 7 mm long, subsessile; glumes much shorter than the spikelet, keeled, persistent, lower 1-nerved, upper 3-nerved; lemmas 5-6; 1, 2 lemmas narrowed into tomentose stipes; 3,4 lemmas 4-lobed, short awned, disk coriaceous, margins and lobes hyaline; 5, 6 lemmas cuneately obovate, 3-lobed.

Rare. Fl. & Fr.:October-January.

Vijayapuri Road (KSN), *CAB* 8245 (MH); Gajapatinagaram (VSKP). *VNS* 4578 (MH).

INDIA: South India.

WORLD: Sri Lanka.

PSEUDANTHISTIRIA (Hackel) Hooker f.

Pseudanthistiria umbellata (Hack.) Hook. f., Fl. Brit. India 7: 220. 1896; Fischer 3:1749. 1934; Bor, Grass. India 204. 1960. *Andropogon umbellatus* Hack. in A. DC., Monog. Phan. 6: 401. 1889.

An annual, tufted grass; culms up to 60 cm high, slender, prostrate, rooting at the nodes. Leaf sheaths tuberculate base hairy; ligule emerging 2-short, herbaceous, fimbriate auricle; blades 2-5 cm long, linear-oblong, sessile or shortly petioled, base rounded, apex acuminate, sparsely hairy on both surfaces. Panicle 20 cm long, leafy, lax, interrupted; racemes 1.2 cm long, enveloped by a spathe, the raceme sub-umbellate, the umbel subtended by a common spathe; spikes 3-6 in a fascicle, proper spathe 1.2 cm, glabrous, tuberculate base hairs on the margin. *Sessile spikelet:* 3 mm, linear, scabrid; lower glume dorsally concave, chartaceous, truncate, margin incurved; upper glume linear-lanceolate; lower lemma small, quadrate; upper lemma represented by a hyaline base, awn geniculate; stamens 3. Caryopsis small. *Pedicelled spikelet:* linear-lanceolate, awned.

Rare in Bison hills. Fl. & Fr.: November-December.

Bison hills (EG), *CAB* 5097 (MH).

INDIA: Deccan, W. Peninsula, Andhra Pradesh.

WORLD: Sri Lanka.

PSEUDECHINOLAENA Stapf

Pseudechinolaena polystachya (Kunth) Stapf in Oliver, Fl. Trop. Africa 9: 495-1919; Fischer 3: 1766. 1934; Bor, Grass. India 352. 1960; Matthew, Fl. Tamilnadu Carnatic 2: 1892. 1983. *Echinolaena polystachya* Kunth, Nov. Gen. Pl. 1: 119. 1816.

Slender annuals; branches to 20 cm long, nodes appressed pubescent. Leaves ovate-lanceolate, 1.5-4.5 x 0.5-1 cm, base oblique, apex acute, pilose. Inflorescences of distant racemes. Spikelet ovoid, solitary, dorsally compressed; lower glume broadly ovate, 3-nerved, aristate; upper glume boat-shaped, 7-nerved, with rows of hooked bristles in between; lower lemma ovate, 3-nerved; palea convolute. Upper lemma crustaceous, 5-nerved, paleate; stamens 3. Caryopsis oblong, gibbous on the back.

Hills above 1200 m, sparse on shola floor in moist places, in Tamilnadu Carnatic (Matthew, 1983).

Kaduguthadi, Thandikudi hills (DGL), *K.Ravikimar* 92312 (BSID).

INDIA: Assam, Western Ghats, north-eastern states.

WORLD: Tropical regions of Old and New Worlds.

PSEUDORAPHIS Griffith

1. Spikelets 4-6 mm long .... **P. spinescens**
1. Spikelets 8-10 mm long .... **P. brunoniana**

Pseudoraphis brunoniana (Griff.) Pilg., Notizbl. Bot. Gart. Berlin-Dahlem 10: 209, 210. 1928. *Panicum brunonianum* Griff., J. Asiat. Soc. Bengal., 5: 547. 1836.

Aquatic grass, often floating, culms up to 90 cm long, very branchy, striate. Leaves narrowly linear-lanceolate, 6.2-18.7 cm long; sheaths lax, pubescent at the throat, margins eciliate, ligule hairy. Panicles pyramidal, 5-7.5 cm long, axis and branches angular, furrowed, scabrid-hispid, flexuous; lower branches 2.5 cm long, terminated by bristle, 1.5-1.8 cm long; pedicels short, thick, obconic. Spikelets usually 1, rarely 2 on the branches of the panicle, lanceolate, 8-10 mm long; lower glume very small, whitish, orbicular or truncate, upper glume oblong, 7.5-10 mm long,

long-acuminate or awned, scabrid on the convex back and especially towards the apex, smooth on the sides. Lower floret male or neuter; lemma 5-7.5 mm long, finely acuminate, many-nerved, margins thin, clasping the palea; palea half as long as the lemma, hyaline, margins infolded. Upper floret bisexual; lemma papyraceous, linear-oblong, 2.5 mm long, acute, embracing the palea; palea slightly shorter, oblong, thin. Caryopsis ovate-oblong, to 1.2 mm long.

Frequent in tanks, margins of lakes, ditches and marshes in Eastern Ghats of Odisha. Fl. & Fr.: August–November.

Burkhat pass (GJM), *SX* & *MB* 3863 (RRL-B).

INDIA: North-eastern to eastern states.

WORLD: Sri Lanka, China, Malaysia, Indonesia to Australia.

Pseudoraphis spinescens (R.Br.) Vickery, Proc. Roy. Soc. Queens. 62, n. 7. 69. 1952; Bor, Grass. India 353. 1960. *Panicum spinescens* R. Br., Prodr. 193. 1810. *Chamaeraphis spinescens* (R.Br.) Poir. in Lam., Encycl. 2: 189. 1811; FBI 7: 62. 1896. *Pseudoraphis aspera* (Koen.) Pilger, Notiabl. Bot. Gart. Berlin 10: 210. 1928; Fischer 3: 1790. 1934.

A perennial, marshy, floating grass; culms up to 60 cm high, decumbent, much branched; nodes woolly. Leaf sheaths 4-8 cm long, loose, glabrous; ligule membranous; blades 5-10 x 0.2-0.5 cm, linear, flat or convolute, base truncate, margins spinulose, apex acuminate, tuberculate pubescent above, glabrous below. Panicle up to 14 cm long, effuse or contracted; branches filiform, scabrid; spikelets narrowly linear, 6 mm, 2-flowered; lower glume ovate, truncate, membranous, nerveless; upper glume ovate, acuminate, aristate, 11-nerved, scabrid along the margins; lower lemma ovate-oblong, 7-nerved; palea hyaline, male; upper lemma oblong, 2-nerved; palea linear, hyaline, bisexual; stamens 3, stigma plumose. Caryopsis linear-oblong.

Occasional grass in ponds and marshy areas. Fl. & Fr.: October-March.

Gogarbam dam (CTR), *MSG* 12161; Horsley konda (CTR), *JSG* 21172 (MH, DD); Ramgiri (CTR), *BR* & *P. Priyadarsini* 37108 (BSID); Kottapalli (KSN), *CAB* 8270 (MH); Purunakote section, Satkosia Tiger Reserve (AGL), *KCM* 5179 (BSID); Dunaei dam (GJM), *AC* 10209 (BSID); Puliyur tank (NA), *KS* 6520 (MH); Chambarampakam (CPT), *S.India Flora* 14840 (MH); Perumbedu (CPT), *DN* 514 (MH).

INDIA: Throughout India.

WORLD: Sri Lanka, Myanmar, China, Taiwan, Japan, Indo-China, Malesia, northern Australia.

PSEUDOXYTENANTHERA Soderstr. & R.P. Ellis

1. Culms up to 4 m high; internodes 30-45 cm long; leaf blade 20-30 cm long .... **P. monadelpha**
1. Culms up to 9 m high; internodes 15-30 cm long; leaf blade 10-20 cm long .... **P. stocksii**

Pseudoxytenanthera monadelpha (Thw.) Soderstr. & R.P. Ellis, Smith. Contr. Bot. 72: 52. 1988. *Dendrocalamus monadelphus* Thw., Enum. Pl. Zeyl. 376. 1864. *Oxytenanthera monadelpha* (Thw.) Alston in Trimen., Hanb. Fl. Ceylon 6 (Suppl.) 342. 1931; Fischer 3: 1861. 1934. *Pseudotenanthera monadelpha* (Thw.) R. Majumdar in Karthik. *et al.*, Fl. Ind. Enum. Monocot. 280. 1989; Gayathri & Pullaiah in Pullaiah, Fl. Andhra Pradesh 3: 1245. 1997. *Oxytenanthera thwaitesii* Munro, Trans. Linn. Soc. 26: 126. 1868; FBI 7: 402. 1896.

Unarmed trees, culms up to 4 m, gregarious, straggling, hollow; branchlets whorled and sheathed, climbing; nodes prominent; internodes 30-45 cm long, rough; culms sheaths 15 x 7.5-10 cm, glabrous when young, hairy at maturity. Leaf sheaths keeled, coriaceous, hirsute, bristly near mouth; ligule long; blades 20-30 x 2-4 cm, sparsely hairy beneath, pellucid glands on the midrib. Panicle of spicate heads of few-many spikelets; spikelets 8-11 cm, dense heads, 1-flowered; bracts ovate, acute, pale, 1-fid, rachis smooth; glumes 2-3, ovate, mucronate, many-nerved, margins ciliate; lemmas ovate, cuspidate, acute, pubescent above; palea convolute; stamens 6, anthers tips hairy; styles pubescent, stigmas 3. Caryopsis elongate, grooved.

Rare, gregarious, scandent bamboo in Kurnool forests.

Kurnool forest (KNL), *CBDM s.n.* (MH).

INDIA: Andhra Pradesh, Karnataka, Kerala, Tamil Nadu.

WORLD: Sri Lanka.

Pseudoxytenanthera stocksii (Munro) T.Q.Nguyen, Bot. Zhurn. (Moscow & Leningr.) 76: 993. 1991. *Oxytenanthera stocksii* Munro, Trans. Linn. Soc. Lond. 26: 130. *Pseudoxytenanthera stocksii* (Munro) Naithani, J. Bombay Nat. Hist. Soc. 87: 440. 1990. *Pseudotenanthera stocksii* (Munro) R. Majumdar in Kartik. *et al.*, Fl. Ind. Enum. Monocot. 280. 1989.

A medium-sized bamboo; culms up to 9 m high, glabrous, when young covered with dense white or grey, deciduous tomentum; internodes 15-30 x 2.5-3.7 cm; branches few at the node; culm sheaths 15-22 x 7-17 cm, striate, silvery shining within, covered outside with matted, deciduous, reddish hairs mixed with small, white hairs with bulbous black bases, tapering gradually upwards and somewhat concavely truncate at apex; imperfect blade subulate-acuminate, rounded at the base and again expanded into a rounded, waved, long fringed auricle on the top of the sheath; ligule 7 mm, fimbriate. Leaves 10-20 x 1-2 cm, linear-lanceolate, base rounded or attenuate into a 2 mm long petiole, apex ending in a setaceous point, glabrous above, except near the margins where the veins are scabrous, margins scabrous; main vein narrow, pale shining on under surface, secondary veins 5-6, intermediate 6-7; sheath striate, glabrous or at first pubescent; ligule long, dentate. Inflorescence large panicle of spicate heads with many closely-packed spinous spikelets; heads about 2.5 cm in diameter supported by round, chaffy bracts, rhachis striate, smooth; spikelets 1-1.2 cm long, narrow, glabrous, mucronate, fertile floret mixed up with a few sterile; glumes ovate, mucronate, 5-7-nerved; two bisexual florets; lemmas ovate, sub-acute, mucronate at the the back; paleas of the lower floret as long as the lemmas, 2-keeled, ciliate on the keels, 5-nerved between, blunt, that of upper floret concave, convolute, blunt; stamens long exserted; anthers short, acute, ovary ovoid, hairy, style long, stigma one, plumose.

Included here on the basis of a collection made by Madhava Raja, from Araku valley, Visakhapatnam district deposited in the herbarium of Forest Research Insitute, Dehra Dun (DD) (Naithani, 1993).

INDIA: Maharashtra, Goa, Karnataka, Kerala, Andhra Pradesh. Endemic.

PSEUDOSORGHUM A. Camus

Pseudosorghum fasciculare (Roxb.) A. Camus, Bull. Mus. Hist. Nat. (Paris) 26: 662. 1920; Fischer 3: 1734. 1934; Bor, Grass. India 205. 1960. *Andropogon fascicularis* Roxb., Fl. Ind.1: 269. 1820; FBI 7: 177. 1896.

An annual, tufted grass; culms up to 75 cm high, slender, decumbent below; nodes villous. Leaf blades 3-20 x 0.2-0.8 cm, linear, sparsely tuberculate-based hairs. Panicle 25 cm long, dense, compound; branches erect, verticillate; rachis fragile; joint and pedicels hairy; spikelets 2-nate. *Sessile spikelet:* 6 cm long, bisexual; callus bearded; glumes subequal, chartaceous; lower 2-keeled; upper boat-shaped; lemmas hyaline, lower similar to upper, narrow, short, bi-dentate, awned, paleated; stamens 3. *Pedicelled spikelet:* narrower than sessile spikelet, male, similar to sessile spikelet.

Rare in hills. Fl. & Fr.: October-January.

Rollapenta (KNL), *PVP* & *RVR* 2338; Bison hills (EG), *CAB* 5179 (MH). Tikarpada section, Satkosia Tiger Reserve, *KCM* 5998 (BSID); Purunakote, Satkosia Wildlife Sanctuary (AGL), *AC* 2375 (BSID); Jacum, Karlaapat Wildlife Sanctuary (KHD), *AC* 2454 (BSID).

INDIA: Throughout India.

WORLD: China, Myanmar, Philippines, Thailand.

PSEUDOSTACHYUM Munro

Pseudostachyum polymorphum Munro, Trans. Linn. Soc. London 26: 142. 1868. *Schizostachyum polymorphum* (Munro) R.B.Majumdar in Karthikeyan *et al.*, Fl. Ind. Enum. Monocot 282. 1989; Naithani *et al.*, Indian Forester 138: 96. 2012.

A large shrubby or semi-arborescenct bamboo with single culm from a long creeping, jointed rhizome; culms 7 m high, often supported by adjoining trees, and so appears scandent; wall 5 mm thick; nodes not swollen; internodes to 22 x 3.5 cm, glaucous at first, afterwards dark green, whitish below the nodes, somewhat scabrous; culm sheaths shorter than internodes, triangular–truncate in out line, very shortly auricled with a tuft of short bristles, covered out side with appressed dark brown hairs; imperfect blade caducous, long–acuminate in young culms, shorter and triangular on older ones, striate and furnished with purplish transverse nerves; ligule short, slightly dentate. Leaves oblong-lanceolate, 10-35 x 2.5-5 cm, unequally rounded at the base into a short petiole, attenuate at the top into a long twisted point, surface smooth, margins scabrid; main vein pale, conspicuous, secondary veins 7-11, intermediate 5-6, transverse, veinlets many, oblique; sheaths faintly white-pubescent; glabrescent; ligule short. Inflorescence very large, leafy panicle, composed of branches fascicled at the nodes, pendulous or curved spikelets; 5 mm long with one fertile floret, rhachilla produced bearing glumes or incomplete floret; glume one, broad, mucronate, 7-nerved; lemmas similar to the glumes, finely ciliate above; paleas thin, much convolute, 2-keeled, ciliate on the keels; lodicules 3-5, usually 4, large, acute, rounded or truncate at apex, ciliate, persistent; stamens 6, free, filaments short, anthers apiculate, ovary narrow, linear-oblong, ending in a rigid style; stigmas 2, hairy. Caryopsis globose-depressed, surmounted by the base of the style; pericarp crustaceous.

Occurs in damp and moist places in plains and lower hills near Valamuru, Maredumilli in East Godavari district. Tel.: *Bongu veduru, Puttu veduru.*

Valamuru, Maredumilli (EG), *H.B.Naithani* & *P.V.Chalapathi Rao* 5112 (DD).

INDIA: E. Himalaya, NE. India, Andhra Pradesh

WORLD: Myanmar, Indo-China, Thailand, Vietnam.

ROTTBOELLIA Linnaeus f. *nom. cons.*

Rottboellia cochinchinensis (Lour.) Clayton, Kew Bull. 35; 817.1981. *Stegosia cochinchinensis* Lour., Fl. Cochin. 1: 51. 1790. *Rottboellia exaltata* L.f., Suppl. Pl. 114. 1781. non (L.) L. f. 1779; FBI 7: 156. 1896; Fischer 3: 1759. 1934; Bor, Grass. India 206. 1966; Matthew, Fl. Tamilnadu Carnatic 3: 1894. 1983.

A perennial, coarse grass; culms up to 3 m high, supported by stilt roots, branching from the base; nodes glabrous. Leaf sheaths up to 12 cm long, loose, glabrous; ligule a short ciliate rim; leaf blades 10-35 x 0.2-0.7 cm, linear, flat; base rounded, margins scabrid, apex acuminate, glabrous, midrib prominent. Spiciform racemes up to 12 cm long, solitary, cylindric; rachis fragile; spathe clasping the racemes at base; spikelets 4 mm, 2-nate, one sessile and other pedicelled. *Sessile spikelet*: 2 mm, ovoid; glumes subequal, crustaceous, keeled; lower ovate, close to the cavity; upper boat-shaped; lemmas hyaline, oblong-ovate, paleated; lower male, upper bisexual; paleas hyaline; stamens 3; stigmas plumose. Caryopsis ellipsoid. *Pedicelled spikelet:* 4 mm, ovate; glume ovate, crustaceous; lemmas hyaline, ovate-oblong, upper palea red; palea oblong, empty or male.

Occasional grass in plains and hills. Fl. & Fr.:August-December. Tel.: *Pendinekku gaddi.*

Batrepalli (ATP), *KRKS* & *BSS* 37987; Kammavariplli (KNL), *TP & PVP* 3382; Veldurthi RF (KNL), *RVR* 1725; Tirumala hills (CTR), *MSG* 12149; Peddamallapuram, Rampa hills (EG), *TP & MSG* 12283; Ramanapenta - Nallamalais (KNL), *JLE* 2621 (CAL & MH); Dugarajpatnam (NLR), *JSG* 12785 (DD); Kondapalli (KSN), *CAB* 4566 (MH); Near Tiger camp, Maredumilli (EG), *MM* 102582 (BSID); Dummakonda (EG), GVS 68594 (MH); Forest near Anantagiri (VSKP), *NPBK* 11004 (CAL); Gajapathingaram (VSKP), *VNS* 4593 (MH); Salur (VJN), *NPBK* 961 (CAL); Lanjigargh (KHD), *MB & Dhal* 7419; Similpahar (MBJ), *V.Naik* Acc. No. 3823 (RRL-B); Harishankar (BL), *SX* & *MB* 6246 (RRL-B); Kwadoli, Raigoda, Satkosia Tiger Reserve (AGL), *KCM* 6718 (BSID); Gurandia, Similipal B.R., *AC* 2330 (BSID); Dimli, Kotgarh Wildlife Sanctuary (KDM), *AC* 2417 (BSID).

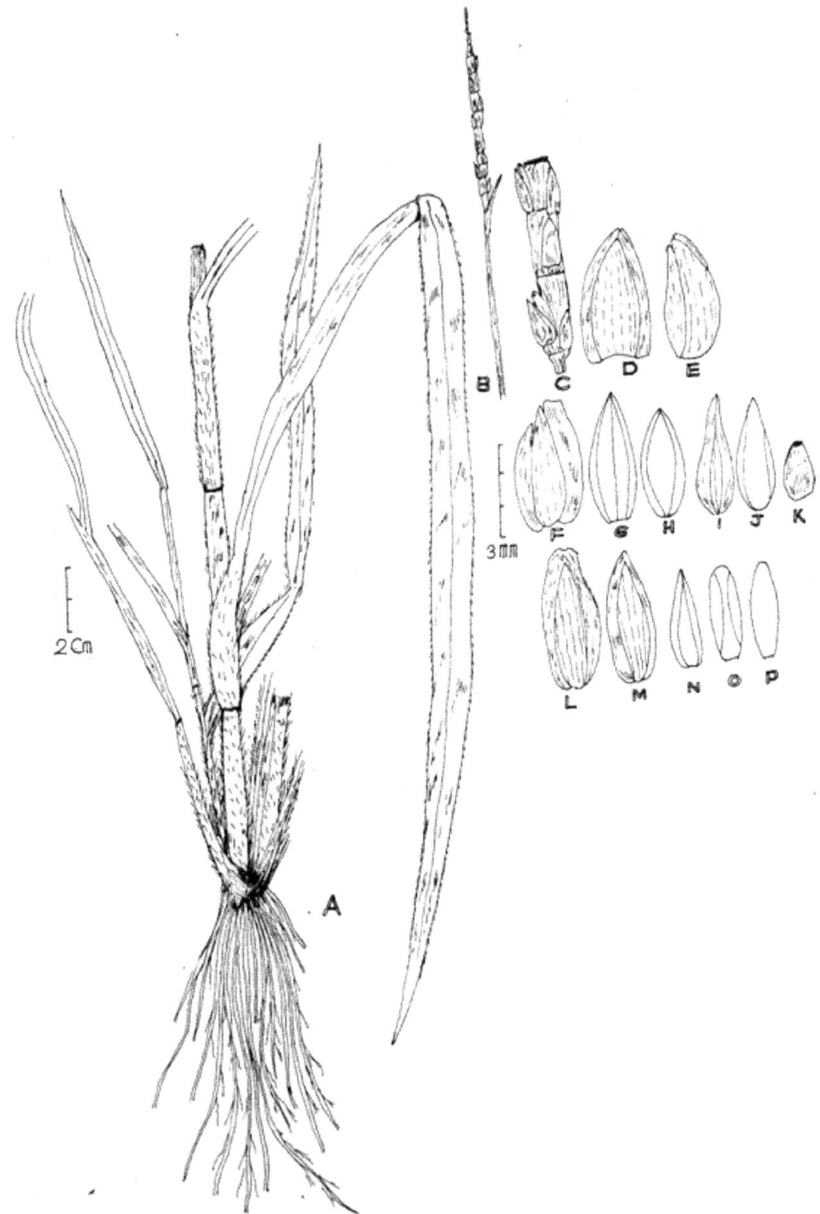

Figure 93. **Rottboellia cochinchinensis** (Lour.) W.D.Clayton

A. Twig, B. Raceme, C. Enlarged part of the raceme, D-K. ***Sessile spikelet***: D. Lower glume, E. Upper glume side view, F. Front view, G. Lower lemma, H. Lower palea, I. Upper lemma, J. Upper palea, K. Grain. L-P. Pedicelled spikelet: L. Lower glume, M. Upper glume, N. Lower lemma, O. Upper lemma, P. Palea.

INDIA: Throughout hotter parts of India.

WORLD: Throughout the Old World Tropics.

SACCHARUM Linnaeus

1. Callus hairs half to as long as spikelet .... **S. narenga**
1. Callus hairs much longer than spikelet:
   2. Lower glume villous; peduncle glabrous .... **S. arundinaceum**
   2. Lower glume glabrous; peducle hairy towards apex:
      3. Cultivated; peduncle glabrous below the panicle; glumes glabrous .... **S. officinarum**
      3. Wild; peduncle hairy below the panicle; glumes ciliate:
         4. Culm stout, nodes waxy; leaf sheaths up to 20 cm long .... **S. spontaneum**
         4. Culm not stout, nodes rusty tomentose; leaf sheaths up to 10 cm long:
            5. Leaves glabrous or hirsute; ligule membranous; panicle to 60 cm long, golden brown .... **S. fallax**
            5. Leaves glaucous; ligule coriaceous, densely villous; panicle to 90 cm long, slivery grey .... **S. bengalense**

Saccharum arundinaceum Retz., Obs. Bot. 4: 14. 1786; FBI 7: 119. 1896, *p.p.*; Fischer 3: 1709.1934; Bor, Grass. India 211. 1960; Saxena & Brahmam, Fl. Orissa 4: 2470.1996.

Very large, tufted grass, up to 4 m high, leafy, glabrous, smooth; internodes long. Leaves linear, up to 1.8 m x 2.5 cm, tapering to a fine point, green, villous at base only, midrib stout, margins very scabrid, upper leaves much reduced, folded and filiform; sheaths glabrous, striate; ligule truncate, membranous, with a ring or tuft of silky hairs 6-25 mm away from its base. Panicle 60-120 cm long, silvery or grey silky, not very loose, axis smooth, glabrous, main branches tufted on the axis, tufts alternate or subverticillate; peduncle not hairy, joints of rachis 3.5-5 mm long, white-villous. Spikelets unawned, 2.5-4 mm long, pedicels 2-2.5 mm long, white villous; callus hairs not dense, as long as the spikelet. Sessile spikelet: Glumes subequal, 3 mm long, acute; lower glume with long hairs; upper glume glabrous; lemma hyaline, lower lemma slightly shorter than the lower glume; upper lemma together with its short cusp equal to the lower lemma, sometimes ciliolate. Pedicelled spikelet: Glumes dorsally villous.

Occasional along borders of fields in Eastern Ghats of Odisha. Fl. & Fr.: November–January.

Parlakhemudi (GJP), *MB* 2946 (RRL-B).

INDIA: Through out the plains and low hills of India.

WORLD: Sri Lanka, Myanmar.

Saccharum bengalense Retz., Obs. Bot. 5: 16. 1789; Bor, Grass. India 211. 1960; Saxena & Brahmam, Fl. Orissa 4: 2471. 1996. *S. muja* Roxb., Fl. India1: 250. 1820. *S. ciliare* Anderss. in Oefvers. Vet. Akad. Forhand. Stockh.155. 1855; Fischer 3: 1709. 1934. *S. arundinaceum* Hook.f., Fl. Brit. India 7: 119. 1896, *p.p.*, non Retz.

Tufted grass, up to 5.4 m high. Leaves 1.5-2.1 m x 1.8-2.5 cm, glaucous, upper cauline leaves tapering to a setiform tip, densely white villous near the ligule. Panicle rather contracted, thyrsoid, more or less lanceolate, 30-90 cm long, yellowish villous, silvery or grey; rachis glabrous, rather angular or sulcate; branches verticillate, compound; rachis of racemes and pedicels with long dense silvery hairs; joints of rachis and pedicels 2.5-3 mm long; peduncle not hairy. Spikelets 4-6 mm long, mostly longer than the joints of the rachis; pedicelled spikelets subequal; pedicel shorter than spikelet; callus hairs much shorter than the spikelet. Sessile spikelet: Lower glume lanceolate, cuspidately acuminate, 2-keeled with faint midnerve, keels scabrid above, dorsally villous in the lower half to three-fourths, upper glume more cymbiform, glabrous or obscurely hispid on keel. Lower lemma lanceolate, 3.2-3.5 mm long, chartaceous, acuminate, sometimes minutely aristulate, obscurely ciliate, upper lemma broader, ciliolate, usually aristulate; palea short, broad, wraping around the linear-oblong, 1.2-1.8 mm long grain. Pedicelled spikelet: Upper glume dorsally villous.

Occasional in open places in Ganjam. Fl. & Fr.: October–January.

Paniganda (GJP), *SX* & *MB* 1972 (RRL-B).

INDIA: N and NW India, Odisha.

WORLD: Pakistan.

Saccharum fallax Balansa, J. Bot. (Morot) 4: 80. 1890; Saxena & Brahmam, Fl. Orissa 4: 2472. 1996. *Narenga fallax* (Balansa) Bor, Kew Bull. 1948. 162. 1948 & Grass. India 198. 1960. *Erianthus chrysothrix* Hack., Ost.Bot. Z. 41: 6. 1891; FBI 7: 125. 1896.

Culms 2.4-3 m and 1.5-1.8 cm diam., rusty-tomentose below the large panicle. Leaves 60-90 x 1-1.5 cm, glabrous or hirsute, margin scabrid, sheaths with tubercle-based hairs; ligule short, membranous. Panicle 30-60 cm long, oblong, erect, contracted, golden or golden-brown, nodes of the rachis golden-brown-bearded and ciliate; branches subverticillate, shortly branched at the base; racemes suberect, slender, usually 8-10 cm long, dense-flowered; pedicels and internodes almost equal to the spikelets or shorter, subclavate at the tips, ferruginous, hairs not exceeding the spikelets. Sessile spikelets 4-5 mm long, linear-oblong, brownish, ferrugineous pubescent. Lower glume chartaceous, narrowly truncate, 2-nerved, dorsally flat and villous, keels ciliate, upper glume truncate, keeled, 3-nerved. Lower lemma shorter, oblong-lanceolate, ciliate; upper lemma half as long as the lower lemma, ovate or linear-oblong, acutely bifid, ciliate, awn 3-5 mm or 0; palea minute, broad, truncate, ciliate with long hairs.

Occasional in Eastern Ghats of Odisha. Fl. & Fr.: August–November.

Dudurchampa- Nawna, Similipahar (MBJ), *SX* & *MB* 4928 (RRL-B); Gurandia, Similipal B.R., *AC* 2331 (BSID).

Saccharum narenga (Nees ex Steud.) Hack., Hack. in A. & C.DC. Monog. Phan. 6: 119. 1889; FBI 7: 120. 1896; Haines, Bot. Bihar & Orissa 3: 1013. 1924; Saxena & Brahmam, Fl. Orissa 4: 2472. 1996. *Eriochrysis narenga* Nees ex Steud., Syn. Pl. Glum. 1: 411. 1855. *Erianthus porphyricoma* Hance ex Trin., J. Bot. 14: 294. 1876. *Narenga porphyrocoma* (Hance ex Trin.) Bor, Indian For. 66: 267. 1940; Mooney, Suppl. Bot. Bihar & Orissa 183. 1950; Bor, Grass. India 198. 1960.

Very large, stout grass, 2.7-3.6 m high, with stout horizontal rootstock; culms pubescent below the panicle, otherwise smooth and glabrous; nodes densely bearded; whole plant often reddish. Leaves flat or folded, lower up to 1.8 m long, 2.5 cm wide, dark green with white midrib, apex long acuminate, margins scabrid, base narrowed, upper surface more or less pubescent; sheaths hirsute or hairy with bulbous-based hairs; ligule very coriaceous, 2-fid, with very long hairs from the base; leaves below the panicle often reduced to a brown hirsute sheath with villous margins and setaceous blade. Panicle purple or greyish purple, 30-60 cm long, dense, lanceolate or oblong, erect, stiff; branches numerous, close, short, erecto-patent or spreading, 1-4-nate, 1.8-7.5 cm long with bearded nodes, lower often compound; rachis with red or purple hairs. Spikelets 2-3 mm long, both similar, usually longer than or nearly equal to the joints of the rachis or sometimes only two thirds as long; callus hairs red or purple, shorter or as long as the spikelet; lower glume dorsally sparsely villous or not, villi not overtopping the glume, brown or purplish, shining coriaceous, 2-keeled, margins ciliolate above, tip narrowly truncate; upper glume similar but 1-keeled, equal, obtuse or minutely apiculate, margins ciliolate above; lower lemma hyaline, very ciliate; upper lemma rather smaller, 1-nerved above, fimbriate or ciliate.

Usually along the water course of hill tracts. Fl. & Fr.: August–December.

Upper Barakampada, Similpahar (MBJ), *SX* & *MB* 4185 (RRL-B); Sijimali hill (KHD), *HFM* 1221 (DD).

INDIA: Tropical Central Himalaya, Punjab, Madhya Pradesh, Odisha, Manipur.

WORLD: Widely distributed in the tropical parts of SE Asia.

Saccharum officinarum L., Sp. Pl. 54. 1753; FBI 7: 118. 1896; Fischer 3: 1709. 1934; Bor, Grass. India 212. 1960; Matthew, Fl. Tamilnadu Carnatic 3: 1894. 1983.

Cultivated, annual, tufted grass; culms up to 7 m high, solid; nodes many, glabrous below the panicle, coated with white wax below the nodes. Leaf sheaths up to 25 cm, terete, glabrous; ligule membranous, ciliate; blades 100-150 x 4-7 cm, linear-lanceolate, midrib stout, margins scabrid. Panicle up to 1 m long, densely branched, silvery; primary branches fragile, glabrous; racemes 12 cm; joints and pedicels filiform, ciliate. *Sessile spikelet*: 5 mm, lanceolate; callus with a tuft of long silky hairs; glumes subequal, lanceolate, chartaceous below and hyaline upwards; lower 4-nerved; upper 1-3-nerved; lower lemma oblong, hyaline, empty; upper lemma ciliate; stamens 3. Caryopsis oblong. *Pedicelled spikelet:* similar to sessile

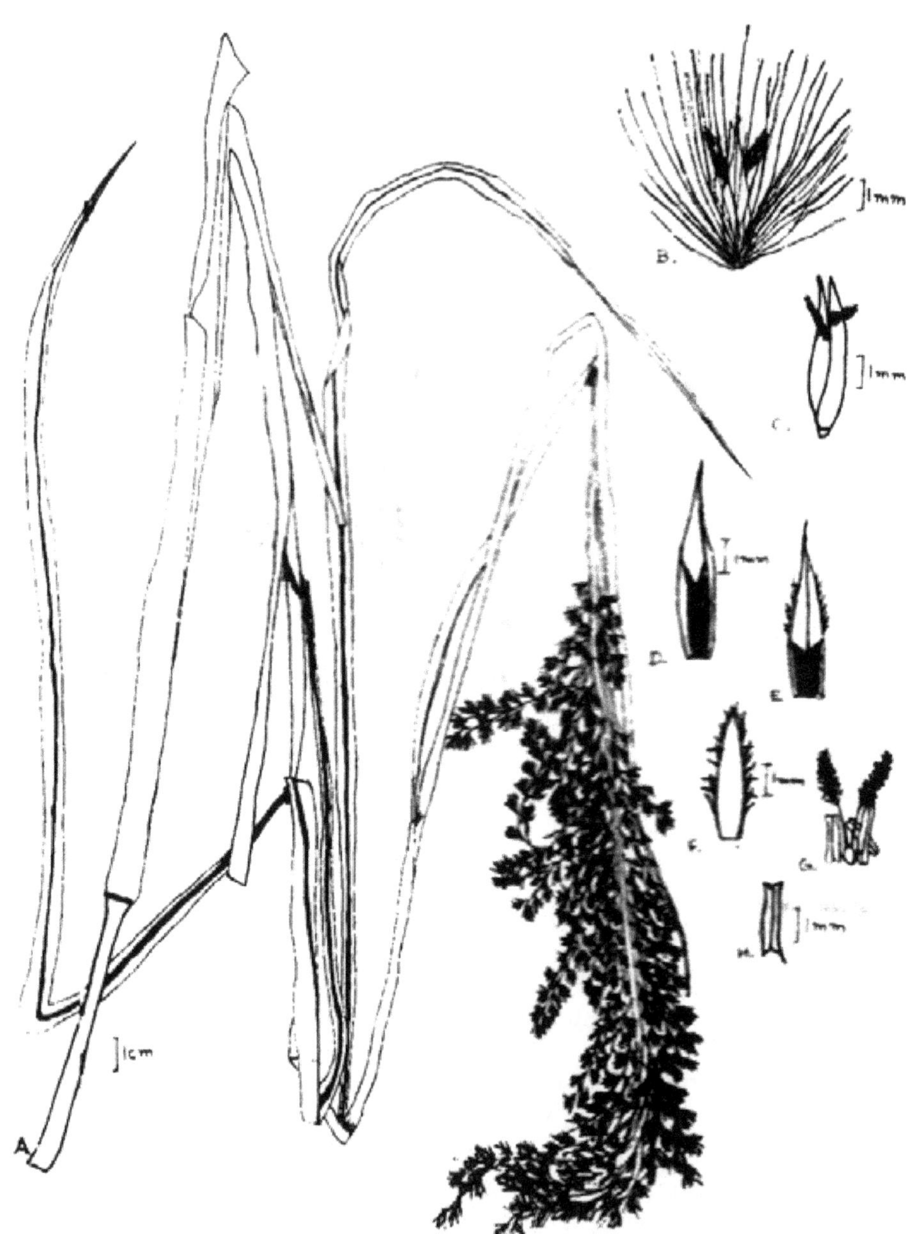

Figure 94. **Saccharum spontaneum** L.

A. Twig, B. Spikelet with hairs, C. Sessile spikelet, D. Lower glume, E. Upper glume, F. Lemma, G. Pistil with stamens, H. Stamen.

spikelet.

Widely cultivated in plains. Fl. & Fr.: October-March. Tel.: *Cheruku.*

INDIA: Cultivated throughout the hotter parts of India.

WORLD: Cultivated in many parts of the tropics.

Saccharum spontaneum L., Mant. Alt. 2: 183. 1771; FBI 7: 118. 1896; Fischer 3: 1709. 1934; Bor, Grass. India 214. 1960; Matthew, Fl. Tamilnadu Carnatic 3: 1895. 1983.

A perennial, tufted grass; culms up to 6 m high, stout, solid, terete; nodes waxy. Leaf sheaths 5-20 cm long, glabrous; ligule membranous with fringe of hairs; blades 15-90 x 0.3-0.8 cm, linear, flat or convolute; base narrowed, margins scaberulous, apex acuminate, midrib prominent, coriaceous, glabrous. Panicle up to 60 cm long, contracted; rachis below the inflorescence villous; spikelets linear, 5 mm, 2-nate; callus with silky hairs. *Sessile spikelets:* 4 mm, linear-lanceolate; glumes equal, lanceolate, chartaceous below, membranous above, acuminate, hairy in the middle, 1-nerved; lower lemma linear, hyaline, ciliate, epaleate; upper lemma linear, ciliate; palea ovate; stamens 3, plumose. Caryopsis oblong. *Pedicelled spikelet:* 4 mm, linear-lanceolate, falling at maturity, similar to sessile spikelet.

Very common along streams and banks of rivers. Fl. & Fr.:October-February. Tel.: *Rellugaddi.*

Amagondapalem RF (ATP), *NY* 1028; Kalvabugga (KNL), *RVR* 1880; Peccheruvu (KNL), *BSS* & *SKB* 30376; Palakonda hills (KDP), *CS* 8246; Kailasakona (CTR), *PVP* 2136; Penchalakona (NLR), *MSG* 12190; Ethipothala falls (GNT), *VRK* 3573; Balapalle (KDP), *JLE* 14904 (MH); Diguvametta, Nallamalais (PKM), *JLE 22182*, 32465 (CAL); Chilakalagadda (VSKP), *GVS* 32791 (CAL); Anantagiri (VSKP); *GVS* 21562 (CAL); S.Kota (VJN), *GVS* 32791 (MH); Salur (VJN), *NPBK* 997 (CAL); Sirumalai hills (DGL), *SKS* 431 (SGH); Kottaiedu to Gavarappattu (SA), *KRM* 60216 (MH); Periya Ramakuppam (SA), *VNS* 5076 (MH); Chinnar river bak - Hogainakkal RF (SLM), *EV* 23574 (MH); Jenabil Range, Similipal B.R., *AC* 2320 (BSID); Raigoda, Satkosia Tiger Reserve (AGL), *KCM* 6737 (BSID); Bhimkote, Karlapat Wildlife Sanctuary (KHD), *AC* 2457 (BSID); Balipadar (GJM), *AC* 10218 (BSID).

INDIA: Throughout the warmer parts of India.

WORLD: Widely distributed in the warmer regions of the Old World.

SACCIOLEPIS Nash

1. Inflorescence with spreading branches; lower glume lanceolate .... **S. curvata**
1. Inflorescence not spreading, with appressed branches; lower glume ovate to ovate-laceolate:
    2. Spikelet ovate-oblong, 1-2 mm long, subglobose, obtuse at apex .... **S. myosuroides**
    2. Spikelet ovate-lanceolate, 2.5-4 mm long, obliquely bulged at base, not subglobose, acute to acuminate at apex:
        3. Culms creeping at base; spikelets in interrupted

crowded panicle, glabrous; lower glume less than half the length of the spikelets.. **S. interrupta**

3. Culms slender; spikelets in continuous crowded panicle, tuberculate-pilose; lower glume half as long as the spikelets .... **S. indica**

Sacciolepis curvata (L.) Chase, Proc. Biol. Soc. Wash. 21: 8. 1908; Fischer 3: 1787. 1934; Bor, Grass. India 357. 1960; Matthew, Fl. Tamilnadu Carnatic 3: 1896. 1983. *Panicum curvatum* L., Syst. Nat. Ex. 12. 2: 732. 1767; FBI 7: 42. 1896.

Annual tufted grass; clum decumbent, 40-48 cm high, finely purple dotted; lower nodes rooting. Leaves linear-lanceolate, 4.3-8.2 x 0.5-0.7 cm, acuminate at apex, rounded to subcordate or truncate at base, midnerve prominent, margins serrulate, sparsely hairy within; ligule 0.2-0.3 mm long, membranous, hyaline to brown, finely fimbriate at apex. Panicle 3-8.5 x 2.5-3 cm, effuse; peduncle 6.5-10 long, more or less ribbed, margins ciliate on one side. Racemes alternate, 2.4-3.6 cm long; spikelets paired, short and long pedicelled, ovate-laceolate, 2.1-2.4 x 1 mm, acute to acuminate at apex, truncate to oblique at base; pedicels 0.3-2.2 mm long, smooth, ends with cup like callus. Glumes unequal. Lower glume lanceolate, 0.4-0.5 x 0.4 mm, acute to subacute at apex, chartaceous, faintly 1-3-nerved. Upper glume ovate-lanceolate, boat-shaped, 2.1-2.4 x 0.8-1 mm, as long as spikelet, acute at apex, promenently 11-13-nerved. Florets 2; lower male or barren; upper bisexual. Lower lemma oblong-lanceolate, 1.8-2.2 x 0.8 mm, acute at apex, prominently 5-7-nerved. Upper palea oblong to narrowly elliptic, 1-1.2 x 0.5 mm, obtuse to subacute at apex, chartaceous to subcoriaceous, 2-nerved; lodicules obovate, 0.2-0.3 mm long, truncate at apex, narrowly acute at base, membranous, hyaline; stamens 3; filaments 1.2 mm long, hyaline; ovary narrowly oblong, 0.4-0.5 x 0.2 mm, acute at apex; style *c.* 0.6 mm long; stigma plumose, to 1 mm long, purple brown. Caryopsis oblong, *c.* 0.5 mm across, shallowly concave at middle, brownish.

In borders of forests and grassy wastelands. Fl. & Fr.: September–February.

Vandalur (CPT), *S.India Flora* 11491 (MH).

INDIA: Andhra Pradesh, Kerala, Tamil Nadu, West Bengal.

WORLD: South Africa East to Sri Lanka.

Sacciolepis indica (L.) Chase, Proc. Biol. Soc. Wash. 21: 8. 1908; Fischer 3: 1787. 1934; Bor, Grass. India 357. 1960. *Aira indica* L., Sp. Pl. 1753 in "Errata" *Panicum indicum* L., Mant. 2: 184. 1771 non Mill. 1768; FBI 7: 41. 1896.

An annual, marshy, tufted grass; culms up to 70 cm high, slender; nodes glabrous. Leaf sheaths 2-5 cm long, glabrous; ligule absent; leaf blades 5-15 x 0.1-0.4 cm, linear, filiform, base truncate, apex acute, glabrous. Spikelike panicle spiciform, up to 7 cm long, contracted, cylindric, dense; spikelets linear, 3.5 mm; lower glume elliptic, 5-nerved; upper glume obovate, keeled, 9-nerved, tuberculate based hairs at apex of outer surface; lower lemma oblong, 7-nerved, keeled, hairs at apex of outer surface; upper lemma ovate, hyaline; palea linear; stamens 3; stigma plumose.

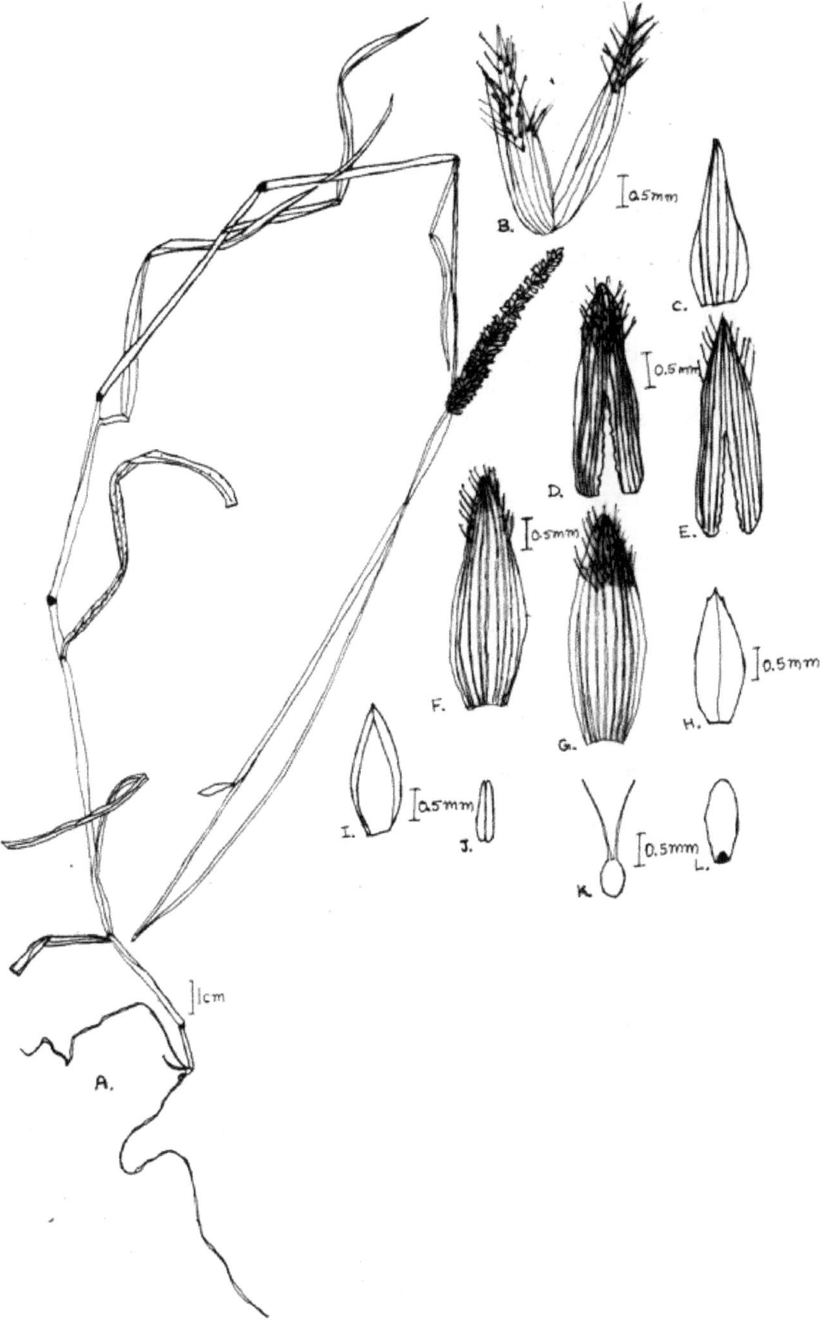

Figure 95. **Sacciolepis indica** (L.) Chase.

A. Twig, B. Spikelet, C. Lower glume, D & E. Upper glume front and back views, F & G. Lower lemma front and back views, H. Upper lemma, I. Palea, J. Stamen, K. Pistil, L. Caryopsis.

Caryopsis linear.

Occasional marshy grass. Fl. & Fr.:August-October.

Talupula (ATP), *KRKS* & *BSS* 41913; Gundlabrahmeswaram (KNL), *BR* & *BSS* 32275; Lankamala (KDP), *BR, SKB* & *BSS* 30508; Tirumala hills (CTR), *MSG* 12164; Talakona RF (CTR), *MSG* & *AA* 12233; East of Nerabylu (CTR), *GVS* 46976 (CAL & MH); Near Tiger camp, Maredumilli (EG), *MM* 102583 (BSID); Vatangi (EG), *S.India Flora* 12675 (MH); Araku (VSKP), *GVS* 21564 (CAL & MH); Yenkur (KMM), *RR* 108570 (BSID); Similipahar (MBJ), *SX* & *MB* 4944, 5156, 5385 (RRL-B); Tulka section, Satkosia Tiger Reserve (AGL), *KCM* 5969 (BSID); Jenabil Range, Similipal B.R. (MBJ), *AC* 2323 (BSID); Baniabasa, Similipal B.R. (MBJ), *Bahali & Agrawala* 652 (BSID); Badrama Wildlife Sanctuary (SBP), *AC* 10241 (BSID); Vandalur (CPT), *S.India Flora* 11490 (MH); Kambakkam hills (CPT), *sine coll. s.n.* (MH); Madancheri (NA), *MBV* 774 (MH); Kongampallam (SLM), *DBD* 31425 (MH).

INDIA: Throughout the Tropical and subtropical India.

WORLD: Tropical Asia, Polynesia, Australia; introduced in Africa and America.

Sacciolepis interrupta (Willd.) Stapf in Prain, Fl. Trop. Afr. 9: 757. 1920; Fischer 3: 1787. 1934; Bor, Grass. India 358. 1960; Matthew, Fl. Tamilnadu Carnatic 3: 1897. 1983. *Panicum interruptum* Willd., Sp. Pl. 1: 341. 1798; FBI 7: 40. 1896.

A perennial, aquatic, tufted grass; culms up to 1 m high, spongy, creeping at base; nodes glabrous, with thick coloured patch. Leaf sheaths 5-15 cm long, loose, glabrous; ligule membranous; blades 10-30 x 0.5-1.5 cm, linear-lanceolate, flat, base truncate, margins scaberulous, apex acuminate, midrib prominent, sparsely hairy. Spikelike panicle up to 20 cm long, cylindric, contracted, interrupted; spikelet up to 6 mm, elliptic-lanceolate; lower glume broadly ovate, 5-9-nerved; upper glume elliptic-lanceolate, 9-11-nerved, scabrid; lower lemma 7-9-nerved, scabrid on outer surface; upper lemma indurated, 5-nerved; stamens 3, stigmas plumose. Caryopsis ovoid.

Occasional grass in marshy and swamp areas. Fl. & Fr.: June-December.

Kalasamudram (ATP), *KVS* & *BSS* 28991; Ahobilam (KNL), *BSS* & *SKB* 32262; Neelithottikandriga, Satyavedu (CTR), *MCB* 45250 (MH); Kilagada (VSKP), *GVS* 21531 (MH); Kalasanapur (GJM), *VNS* 4661 (MH); Graigmore estate–Yercud (SLM), *S. Karthikeyan* 26992 (MH); Manjakuttai lake–Yercaud (SLM), *DBD* 31398 (MH).

INDIA: In Swamps throughout India.

WORLD: Tropics of SE Asia.

Sacciolepis myosuroides (R.Br.) A. Camus in Lecomte, Fl. Indo-Chine 7: 460. 1922; Fischer 3: 1787, 1934; Bor, Grass. India 358. 1960; Matthew, Fl. Tamilnadu Carnatic 3: 1897. 1983. *Panicum myosuroides* R.Br., Prodr. 189. 1810; FBI 7: 42. 1896.

Erect annual grass; culms to 70 cm, nodes pilose. Leaf sheath 12-15 cm; blades oblong-lanceolate, 10-20 x 0.3-0.7 cm, flat, scabrid. Spikelets 2-5 in a cluster, ovoid, 1.5-2 mm, purplish; lower glume ovate, to 1 mm, 3-nerved, acute, scaberulous

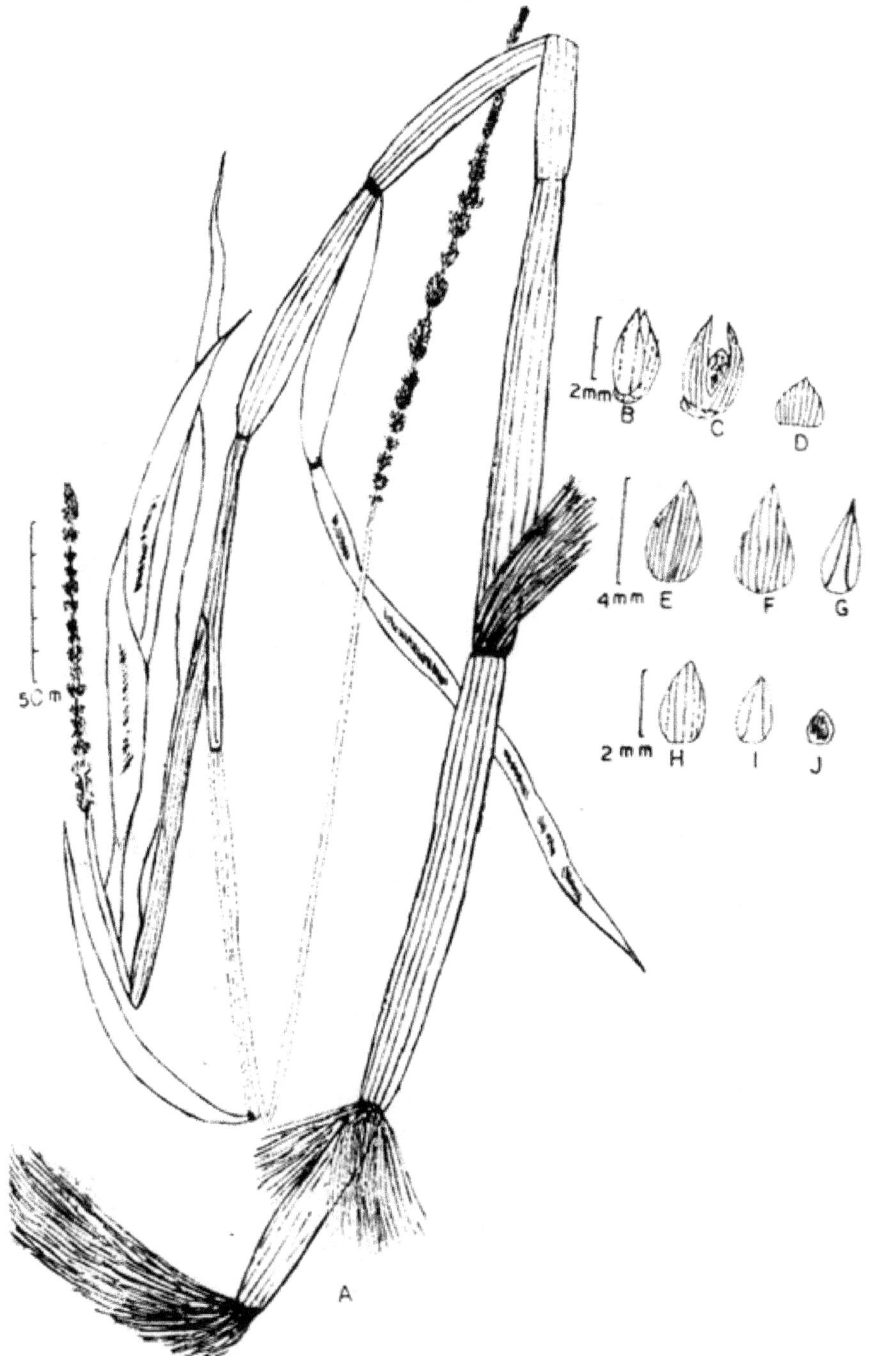

Figure 96. **Sacciolepis interrupta** (Willd.) Stapf

A. Twig, B. Closed spikelet, C. Opened spikelet, D. Lower glume, E. Upper glume, F. Lower lemma, G. Palea, H. Upper lemma, I. Palea, J. Caryopsis.

without; upper glume to 1.5 mm, 9-11-nerved; lower lemma 7-nerved, glabrous; palea a hyaline scale; upper lemma to 1 mm, crustaceous, glossy; palea equal; stamens 3, 1 mm. Caryopsis ellipsoid, turgid.

Occasional in Northern Eastern Ghats and Chitteri hills of Tamil Nadu. Fl. & Fr.: August to December.

Lasari, Kotagarh Wildlife Sanctuary (KDM), *AC* 2435 (BSID); Barga Khol, Rajganjapur, Sundargarh, *AC* 10257 (BSID).

INDIA: Tamil Nadu, Assam, Bihar, Kerala, Madhya Pradesh, Odisha, Sikkim, West Bengal.

WORLD: Tropics of Asia and Australia.

SCHIZACHYRIUM Nees

1. Leaf blades linear, convolute, apex obtuse; racemes 1.5 cm long; very slender plant .... **S. brevifolium**
1. Leaf blades linear, filiform, apex acuminate; racemes 3-6 cm long; robust plant .... **S. exile**

Schizachyrium brevifolium (Sw.) Nees ex Buse in Miq., Pl. Jungh. 359. 1854; Fischer 3: 1752. 1934; Bor, Grass. India 215. 1960; Matthew, Fl. Tamilnadu Carnatic 3: 1898. 1983. *Andropogon brevifolius* Sw., Prodr. 26. 1788; FBI 7: 165. 1896.

An annual, tufted, slender grass; culms up to 60 cm long, weak, ascending from decumbent base, many-noded; nodes glabrous. Leaf sheaths 1.5 cm, glabrous; ligule membranous, fimbriate; blades 1.5-3.5 x 0.3-0.5 cm, linear, convolute, apex obtuse. Racemes spathiolate. 1.5 cm, slender; spikelets 2-nate, both falling together. *Sessile spikelet:* linear-lanceolate, 3 mm, dorsally compressed; lower glume oblong-lanceolate, chartaceous, bidentate, 1- nerved, scaberulous along the keels; upper linear, bidentate, 1-nerved, scaberulous along the keels; lower lemma hyaline, elliptic, nerveless; upper lemma deeply 2-lobed, awned; palea small, hyaline; stamens 3. Caryopsis linear. *Pedicelled spikelet:* reduced to a minute glume.

Rare in grasslands and hill slopes. Fl. & Fr.: September-December.

Gajulapalli (KNL), *BR & BSS* 32253; Rachakuntapalle (KDP), *KRKS* 41811; Bitragunta (NLR), *PMR* 23024; Maredumilli (EG), *TP & MSG* 12265; Bison hills (EG), *CAB 5021* (MH); Near Vedurupalli (EG), *GVS* 42763 (MH, BSID); Majhipada, Satkosia Tiger Reserve (AGL), *KCM* 7055 (BSID); Chunnabatti, Satkosia Wildlife Sanctuary (AGL), *AC* 2382 (BSID); Jacum, Karlapat Wildlife Sanctuary (KHD), *AC* 2450 (BSID).

INDIA: Throughout Tropical India.

WORLD: Tropics of the Old World.

Schizachyrium exile (Hochst.) Pilger in Engl., Bot. Jahrb. 54: 284. 1917; Fischer 3: 1752. 1934; Bor, Grass. India 216. 1960. *Andropogon exilis* Hochst., Flora 27: 241. 1844; FBI 7: 166. 1896.

An annual, tufted, robust grass; culms up to 15 cm high, decumbent; nodes glabrous. Leaf sheaths 1-3 cm long, glabrous; ligule membranous; blades 2- 8 x 0.05-0.1 cm, linear, filiform, base truncate, apex acuminate. Racemes spathiolate, up to 6 cm long, slender, spathe 2 cm long. Spikelets 4 mm, lanceolate, 2- nate. *Sessile spikelet:* 4 mm, dorsally compressed; callus hairy; lower glume lanceolate, keeled, hairy; upper glume linear; lemmas hyaline, upper bidentate, awned; palea hyaline, narrow; stamens 3, stigmas plumose. *Pedicelled spikelet*: 3 mm long, empty.

Rare grass in open dry places. Fl. & Fr.: August-December.

Thalaricheruvu (ATP), *TP* 905; Penukonda (ATP), *S.India Flora* 14325 (MH); Gajulapalli (KNL), *BR* & *BSS* 32253 (BSID); Kambakkam hills (CTR), *S.India Flora* 16442 (MH); Purunakote, Satkosia Wildlife Sanctuary (AGL), *AC* 2385 (BSID); Lasari, Kotgarh Wildlife Sanctuary (KDM), *AC* 2434 (BSID); Talkudar, Rajgangapur, Sundargarh, *AC* 10266 (BSID); Betkote, Karlapat Wildlife Sanctuary (KHD), *AC* 2460 (BSID).

INDIA: Throughout Tropical India.

WORLD: Tropics of the Old World.

SCHOENEFELDIA Kunth

Schoenefeldia gracilis Kunth, Revis, Gramin. 1: 283. Pl. 53. 1829; Bor, Grass. India 474. 1960; Chorghe *et al.*, J. Econ. Taxon. Bot. 39: 84. 2015. *S. pallida* Edgew., J Asiat. Soc. Beng. 21: 183. 1852. *Chloris pallida* (Edgew.) Hook.f., Fl. Brit. India 7: 289. 1896.

Perennials, culms decumbent, 20-60 cm long, nodes glabrous. Leaf blades linear, 5-15 cm x 0.8-3 mm, apex fimbriate, upper surface with tuberculate based long hairs, finely scabrellate along margins; ligules membranous, fimbriate at apex, 0.5 mm long, leaf sheaths glabrous, 5-10 cm long. Inflorescence comprising of 1-2 racemes, 5-10 cm long; peduncle 10-15 cm long; spikelets sessile, compactly arranged on raceme alternately in 2 rows, rachis flat, *c.* 0.7 mm wide, margins hairy; spikelets linear-lanceolate, 3-4 x 0.8-1 mm, laterally compressed; lower glumes linear–lanceolate, 2.7-3 x 0.8 mm, aristate, arista 0.3 mm long, 1-keeled, keels scabrid; upper glumes linear-lanceolate, 3.2–3.5 x 1-1.2 mm, aristate, arista 0.5-1 mm long, 1-keeled, keels scabrid; floret 1, fertile; lemmas linear-lanceolate, 2-2.5 x 0.8-1 mm, hairy, 1-keeled, 1-awned, awns 2-3 cm long; paleas linear-lanceolate, 1.5-2 x 0.4- 0.5 mm, hyaline, 1-keeled; stamens 3; styles 2, stigma feathery; ovary oblong. Caryopsis oblong, 1.2-1.5 x 0.3-0.4 mm.

Rare in northern Eastern Ghats of Odisha.

Fl. & Fr.: Septembder–December.

Talkudar forest (Sundergarh Distr.), *AC* 10267 (BSID).

INDIA: Uttar Pradesh, Madhya Pradesh, Bihar, Odisha, Maharashtra,

WORLD: Pakistan, Tropical Africa, Arabia.

SEHIMA Forsskål

1. Ciliate on one margin only of pedicels; lower glume of sessile spikelets rounded on the back with a deep longitudinal slit-like groove .... **S. sulcatum**

1. Ciliate on both edges of pedicels; lower glume of sessile spikelets flat on the back, without a slit, but grooved deeply or shallowly .... **S. nervosum**

Sehima nervosum (Rottl.) Stapf in Prain, Fl. Trop. Afr. 9: 36. 1917; Fischer 3: 1724. 1934; Bor, Grass. India 218. 1960; Matthew, Fl. Tamilnadu Carnatic 3: 1898. 1983. *Andropogon nervosus* Rottl. apud Willd., Verh. Ges. Naturf. Freund. Berlin. Neue Schr. 4: 218. 1806. *Ischaemum laxum* R. Br., Prodr. Fl. Nov. Holl. 205. 1810; FBI 7: 136. 1896.

A perennial, tufted grass; culms up to 1 m tall; nodes glabrous. Leaf sheaths 5-9 cm long, glabrous; ligule membranous; leaf blades 5-30 x 0.3-0.5 cm, linear, flat, coriaceous, base rounded, apex acuminate, glabrous. Racemes up to 10 cm long, solitary, densely ciliate along the angles; rachis jointed, fragile; spikelets 12 mm, laterally compressed, 2-nate. *Sessile spikelet*: 3 mm, linear; lower glume chartaceous, linear, keeled, apex truncate, winged on the keels with a transluscent longitudinal depression in the centre; upper glume boat-shaped, apex scabrid, awned; lower lemma hyaline, lanceolate; palea lanceolate, margin ciliate; upper lemma hyaline, bidentate, awned; palea linear, apex scabrid; stamens 3, stigmas plumose. Caryopsis oblong. *Pedicelled spikelet:* 6 mm, lanceolate; lower glume lanceolate, chartaceous, winged on the keels, ciliate; upper glume hyaline, lanceolate; lower lemma linear, margins ciliate; palea linear, male; upper lemma oblong, obtuse; palea oblong, margins ciliate; stamens 3.

Common in open forests and occasional in plains. Fl. & Fr.: August-December.

Muchukota RF (ATP), *TP & NY* 965, *PVP* 2564; Palakonda hills (KDP), *CS & DAM* 7637 & *PVP* 3433; Chelama (KNL), *JLE* 22055 (MH); Talakona RF (CTR), *PVP* 2327, *JHFB* 112978 (BSID); Panapakkam (CTR), *MH* Acc. No. 57553 (MH); Chittoor RF (CTR), *KCJ* 83943 (MH); Balapalle (KDP), *JLE* 15712; Seshachalam hills (CTR), *MH*, Acc. No. 56226 (MH); Kurnool hills, *CBDM* s.n. (MH); Gundlabrahmeswaram (KNL), *JLE* 32674 (MH); Mandalakatta (KNL), *JSG* 18669 (DD); Malkondapenta (PKM), *VBH* 83979 (BSID); Nagarjunakonda Sacred Grove (GNT), *J.Swamy* & *S.Nagaraju* 7238 (BSID); Kondaveedu fort (GNT), *CAB* 4702 (MH); Kavali Range (PKM), *CAB* 6610 (MH); Ongole (PKM), *CAB* 7951 (MH); Kondapalli (KSN), *Rangachari s.n.* (MH); Araku (VSKP), *GVS* 21560 (MH); Avadi (CPT), *S.India Flora* 15460a (MH); Usur RF, Vellore (NA), *KCJ* 6358 (MH); Tippukadu RF (NA), *KRM* 17929 (MH); Chettupattu to Pennathur (SA), *KRM* 60317 (MH); Ramanadurg (Ballari), *S.India Flora* 15917 (MH); Majhipada section, Satkosia Tiger Reserve (AGL), *KCM* 6137 (BSID); Purnakote, Satkosia Wildlife Sanctuary (AGL), *AC* 2380 (BSID).

INDIA: Rajasthan, Madhya Pradesh, Bihar, Odisha and South India.

WORLD: E. Africa, SW Asia, Pakistan, Sri Lanka, China, Myanmar, Laos, Indonesia, Thailand, Vietnam, Australia.

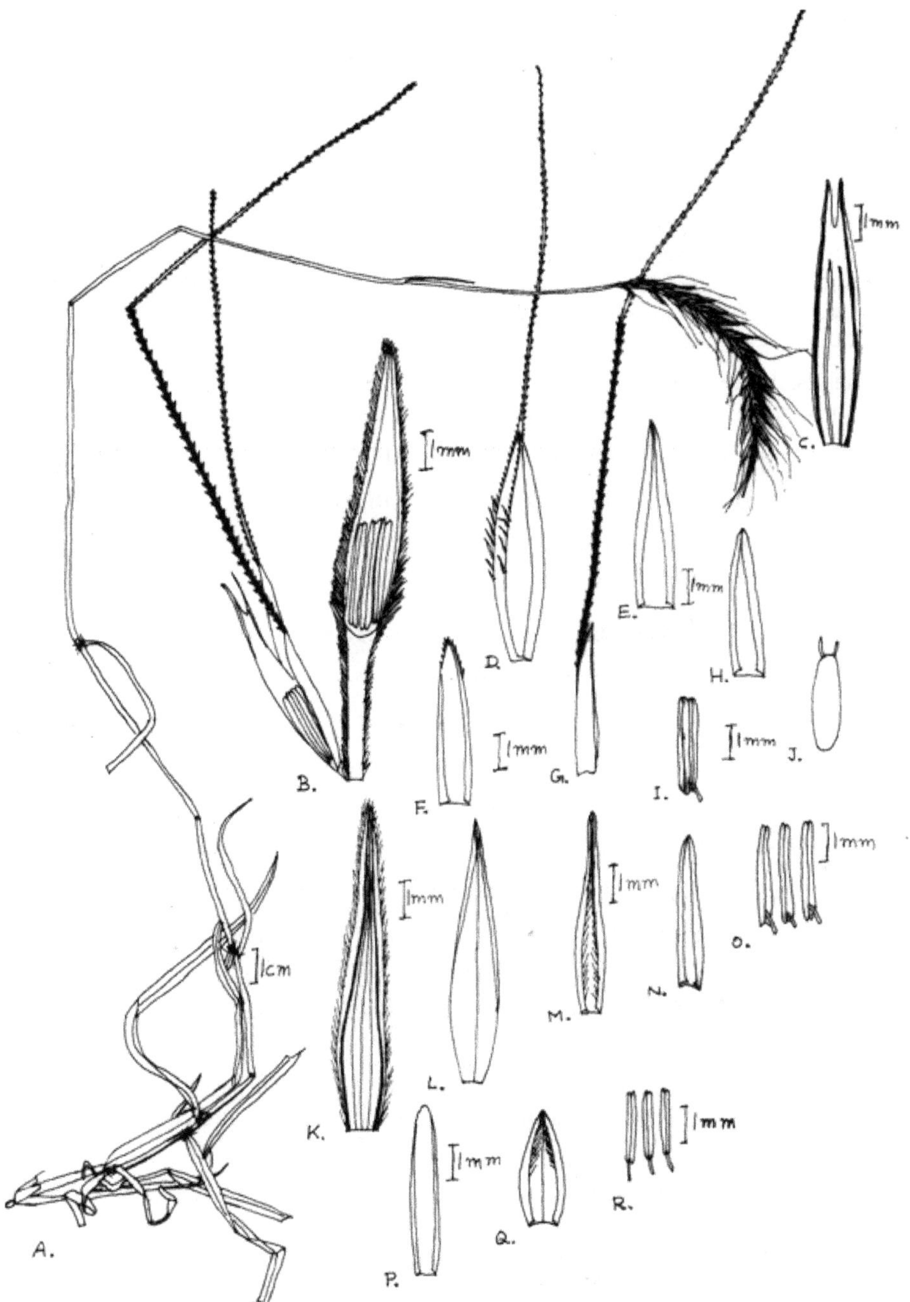

Figure 97. **Sehima nervosum** Stapf

A. Twig, B. Spikelets, C. Sessile spikelet lower glume, D. Upper glume, E. Lower lemma, F. Palea, G. Upper lemma, H. Palea, I. Stamen, J. Pistil, K. Pedicelled spikelet lower glume, L. Upper glume, M. Lower lemma, N. Palea, O. Stamens, P. Upper lemma, Q. Palea, R. Stamens.

Sehima sulcatum (Hack.) A. Camus, Bull. Mus. Hist. Nat. Paris 27: 373. 1921; Fischer 3: 1724. 1934; Bor, Grass. India 219. 1960. *Ischaemum sulcatum* Hack., in DC., Monogr. Phan. 6: 248. 1889; FBI 7: 137. 1896.

A perennial, tufted grass; culms up to 50 cm high, geniculately ascending, much branched below; nodes glabrous. Leaf sheaths lax, glabrous, constricted at the mouth; ligule ciliate membrane; leaf blade 10-20 x 0.3-0.4 cm, linear-lanceolate, apex acuminate, glabrous. Racemes solitary, 8 cm long. *Sessile spikelet:* 1 cm long; callus shortly hairy at the base; lower glume lanceolate, subchartaceous, bidentate, margins incurved; upper glume shorter than the lower, chartaceous, shortly bidentate, apex ciliate, awned; lower lemma linear, hyaline, margins incurved, palea oblong; upper lemma bidentate, awned; palea scaberulous towards apex, *Pedicelled spikelet:* smaller than the sessile, awnless; lower glume lanceolate, bidentate, margin incurved; upper glume unawned; lemmas similar to sessile but unawned.

In Nellore and Krishna districts (Fischer). Fl. & Fr.:August- December

INDIA: Madhya Pradesh, Maharashtra, Andhra Pradesh, Tamil Nadu. Endemic.

Note: No specimens in MH & CAL. In our present study we were unable to find this species in the type locality or elsewhere.

SETARIA P. Beauvois *nom. cons.*

1. Involucral bristels retrorsely barbellate..**S. verticillata**
1. Involucral bristles antrorsely barbellate:
    2. Leaf blades more or less plicate; culms robust:
        3. Racemes up to 2 cm long; annuals.. **S. barbata**
        3. Racemes more than 2 cm long; perennials ....**S. palmifolia**
    2. Leaf blades flat, not plicate; culms slender:
        4. Cultivated; upper lemma faintly transversely rugose, upper glume equal to the length of upper lemma ....**S. italica**
        4. Wild; upper lemma distinctly transversely rugose, upper glume smaller than the upper lemma:
            5. Panicle not continuous, interrupted; involucral bristles 3-4; spikelets up to 2 mm long..............................**S. intermedia**
            5. Panicle continuous, dense; involucral britles 6-11; spikelets more than 2 mm long .... **S. parviflora**

Setaria barbata (Lam.) Kunth, Revis, Gramin. 1: 47. 1829; Bor, Grass. India 360. 1960. *Panicum barbatum* Lam., Tabl. Encycl. 1: 171 1791. *P. costatum* Roxb., *Fl. Ind. 1: 314 1832. P. flavescens* sensu Hook.f., FBI 7: 56. 1986.

Annual grass; culms 20-75 cm long, decumbent, rooting at the lower nodes; nodes sparsely bearded. Leaves 4-15 x 0.5-2 cm, elliptic-lanceolate or lanceolate, base narrow or rounded, apex acuminate, plicate towards base, softly hispid; sheaths to 9 cm long, keeled, villous towards mouth; ligule a row of hairs. Panicles 4-12 cm long, lax; bristles 3-8 mm long, scabrid; spikelets 2-3 mm long, elliptic; lower glume *c.* 1 x 1 mm, ovate; upper glume 1-2 x 1 mm, ovate-oblong, obtuse; lower floret male. Upper floret bisexual, first lemma 2-3 x 1-1.5 mm, elliptic; palea *c.* 2 x 1 mm, elliptic, hyaline, 2-keeled; stamens 3; second lemma 1.5-2.5 x 1-1.5 mm, elliptic or elliptic-lanceolate, acute, crustaceous, rugose; palea 1-2 x 1-1.5 mm, elliptic, crustaceous, rugose, 2-keeled. stamens 3; ovary oblong; stigmas purple.

In moist places like gardens, stream banks, canals and shady places. Fl. & Fr.: December–April.

Valamuru, Maredumilli ((EG), *MM* 105056 (MH & BSID).

INDIA: Kerala, Tamil Nadu, Andhra Pradesh, Madhya Pradesh, Maharashtra, Assam, West Bengal.

WORLD: Tropical Asia and Africa.

Setaria intermedia Roem. & Schult., Syst. Veg. 2. 489. 1817; FBI 7: 79. 1896; Fischer 3: 1789. 1934. *Panicum tomentosum* Roxb., Fl. Ind. 1: 303. 1820. *Setaria tomentosa* (Roxb.) Kunth, Revis. Gramin. 1: 47. 1829; Bor, Grass. India 365. 1960; Matthew, Fl. Tamilnadu Carnatic 2: 1901. 1983.

An annual, tufted grass; culms up to 80 cm high, decumbent; nodes glabrous. Leaf sheaths 4-10 cm long, loose, ciliate at mouth; ligule ciliate; leaf blades 7-15 x 0.4-1 cm, linear-lanceolate, flat, flaccid, base rounded, apex acuminate. Panicle up to 10 cm long, lax, interrupted; involucral bristles antrorsely barbed; glume orbicular; upper glume broadly ovate, hyaline; lower lemma lanceolate-ovate, thinly membranous, paleated, empty; upper lemma broadly ovate, crustaceous, transversely rugose; palea elliptic; crustaceous; stamens 3, stigma plumose. Caryopsis ellipsoid.

Common grass in moist localities of all districts. Fl. & Fr.: July-November.

Yellutla (ATP), *PVP* 2642; Alampur (ATP), *BR* & *P. Priyadarsini* (SKU & BSID); Ramallakota (KNL), *RVR* 1847; Vaddemanu (KNL), *PVP* 2530; Palakonda hills (KDP), *CS & DAM* 7707; Mabbuchintalapalli (KDP), *TP* & *SSR* 33667 (BSID); Tirumala hills (CTR), *MSG* 12159; Chittoor (CTR), *MH* Acc. No. 55650 (MH); Ramanapenta (KNL), *MH* Acc. No. 55654 (MH); Nandyal farm (KNL), *KCJ* 83139 (MH); Mudurlanka, Maredumilli (EG), *MM* 102571 (BSID); way to Vantali (VSKP), *GVS* 42652 (MH); Anantagiri (VSKP), *GVS* 19472 (CAL); Salur (VJN), *NPBK* 1014 (CAL); Mysamma temple (MBNR), *SRS* 108633 (BSID); Sirumalai hills (DGL), *SKS* 432 (SGH); Cherapadi (SA), *KRM* 77447 (MH); way to Kuttathur (NA), *KS* 7447 (MH); Kottur RF, Yelagiri hills (NA), *MBV* 831 (MH); Manchavadi ghat (SLM), *KS* 7581 (MH); Metigiri (SLM), *S.India Flora* 15438 (MH).

INDIA: Throughout India.

WORLD: Pakistan, Sri Lanka, Myanmar extending to Malaysia and Polynesia.

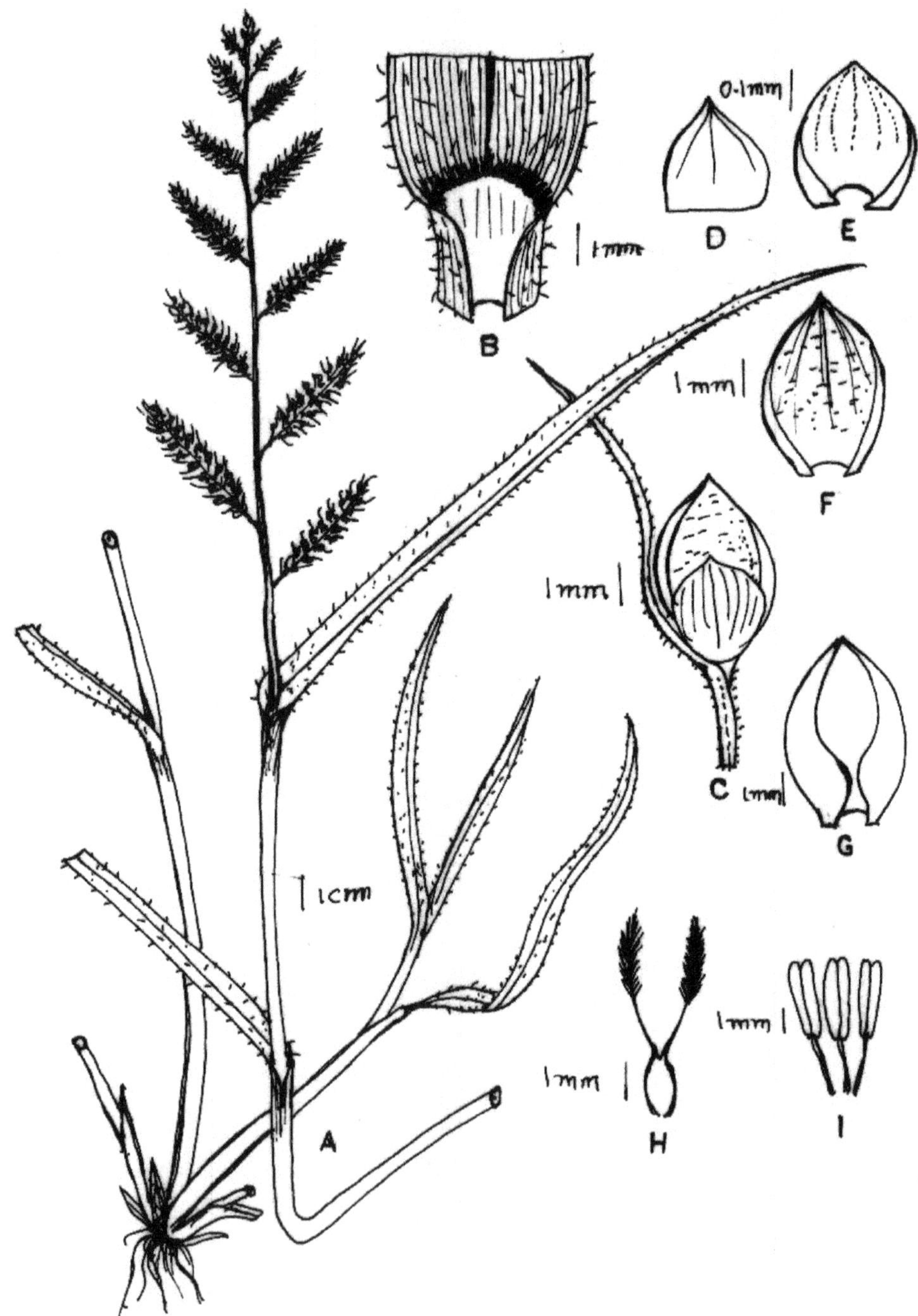

Figure 98. **Setaria intermedia** Roem. & Schult.

A. Plant, B. Collar, C. Spikelet, D. Upper glume, E. Lower glume, F. Lemma, G. Palea, H. Ovary, I. Stamens.

Setaria italica (L.) P. Beauv., Ess. Agrost. 51. 170. 178. 1812; FBI 7: 78. 1896; Fischer 3: 1789. 1934; Bor, Grass. India 362. 1960; Matthew, Fl. Tamilnadu Carnatic 3: 1900. 1983. *Panicum italicum* L., Sp. Pl. 56. 1753.

An annual, tufted grass; culms up to 1 m high; nodes glabrous. Leaf sheaths 3-10 cm long, ciliate on the margins; ligule membranous; blades 10-35 x 0.5-2 cm, linear, flat, base cordate, apex acuminate, minutely scaberulous. Panicle up to 10 cm long, contracted; involucral bristles antrorsely barbed; spikelets 2 mm, ovoid, compressed; lower glume ovate, hyaline; upper glume broadly ovate, hyaline; lower lemma ovate, hyaline and its palea scaly, empty; upper lemma ovate, coriaceous, transversely rugose; palea elliptic, coriaceous; stamens 3, stigmas plumose. Caryopsis ellipsoid.

Cultivated for its grains. Fl. & Fr.: September-November. Tel.: *Korralu;* Tam.: *Thinai.*

Peapuly (KNL), *PVP* 2690; N.S. Canal Site (GNT), *VRK* 3531; Tadepalli (EG), *NRR* & *DN* 84328 (BSID); Bonni (SKLM), *PVS* & *NRR* 76879 (MH); Sirumalai hills (DGL), *SKS* 430 (SGH); Andiappanur (NA), *MBV* 1144 (MH); Belghar, Kotgarh Wildlife Sanctuary (KDM), *AC* 2414 (BSID); Boragharo (GJM), *VNS* 5962 (MH).

INDIA: Cultivated throughout India.

WORLD: Most warm temperate and tropical countries.

Setaria palmifolia (Koen.) Stapf, J.Linn. Soc. 42: 186. 1914; Fischer 3: 1789. 1934; Bor, Grass. India 363. 1960; Matthew, Fl. Tamilnadu Carnatic 3: 1900. 1983. *Panicum palmifolium* J. König, Naturforsch. 23: 208. 1788. *P. plicatum* Willd., Enum. Hort. Berol. 1033. 1809. non Lam 1791; FBI 7: 55. 1896. *Setaria paniculifera* (Steud.) Fourn. ex Hemsl., Biol. Centr. Amer. Hot. 3: 505. 1885; Bor, Grass. India 363. 1960. *Panicum paniculiferum* Steud., Syn. Fl. Glum. 1: 54. 1854.

A perennial, tufted grass; culms up to 2 m high, ascending at base; nodes glabrous. Leaf sheaths up to 12 cm long; ligule hairy; blades 10-30 x 2-4 cm, base cordate, margins scaberulous, apex acuminate, glabrous-sparsely pubescent. Panicle up to 15 cm long; involucral bristles antrorsely barbed; spikelet 3 mm, ellipsoid; lower glume ovate, obtuse; upper glume ovate, membranous; lower lemma ovate, shortly apiculate; palea ovate, hyaline, empty; upper lemma ovate, crustaceous, transversely rugose; palea similar to its lemma; stamens 3; stigmas plumose. Caryopsis ellipsoid.

Occasional on hills. Fl. & Fr.: August-November.

Horsley hills (CTR), *MSG &. AA* 12216; Devarapalli, Maredumilli (EG), *MM* 100799 (BSID); Maredumilli, Cocoa plantation (EG), *GVS* 68527 (MH); Cherukonda (VSKP), *GVS* 28150 (CAL & MH); Sunkarimetta (VSKP), *NPBK* 10911 (CAL & MH); Anantagiri (VSKP), *NPBK* 10999 (MH); Kotnigedda (VSKP), *GVS* 44346 (MH); Jenabil Range, Similipal B.R. (MBJ), *AC* 2340 (BSID); Sirumalai hills (DGL), *SKS* 433 (SGH); Kuthadya hill (GJM), *VNS* 5901 (MH).

INDIA: Peninsular to E. India.

WORLD: Tropics of the Old World, introduced into the New World.

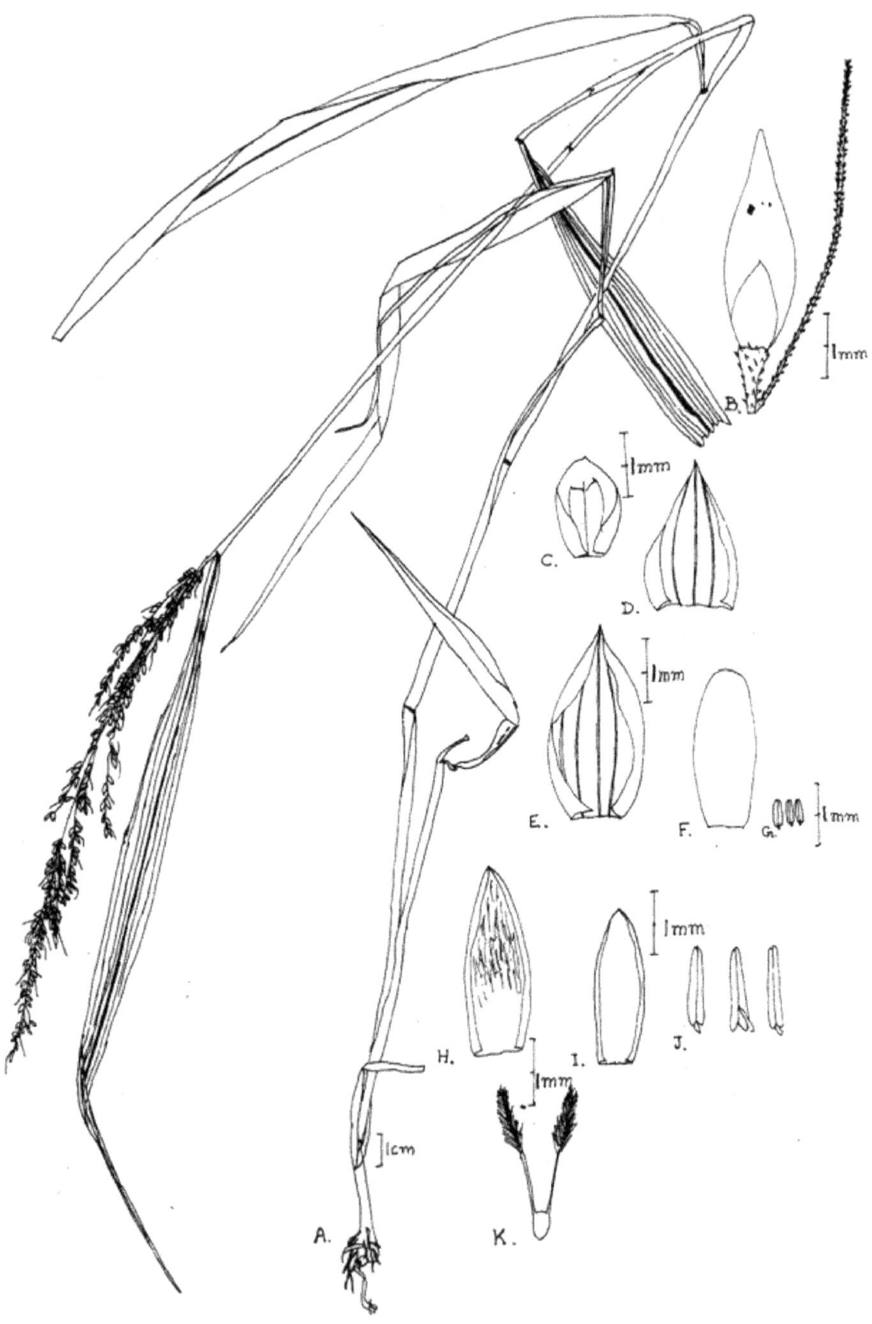

Figure 99. **Setaria palmifolia** Stapf

A. Twig, B. Spikelet, C. Lower glume, D. Upper glume, E. Lower lemma, F. Palea, G. Stamens, H. Upper lemma, I. Palea, J. Stamens, K. Pistil.

Setaria parviflora (Poir.) Kerguelen, Lejeunia 120: 161. 1987. *Cenchrus parviflorus* Poir. in Lam., Encycl. Meth.6: 52. 1804. *Setaria glauca* non (L.) P.Beauv. 1812; FBI 7: 78. 1896 p.p.; Bor, Grasses India 360. 1960. *S. pumila* auct non (Poir.) Roem. & Schult., 1817; Matthew, Fl. Tamilnadu Carnatic 3: 1901. 1983. *Panicum pallidefuscum* Schumach., Beskr. Guin. Pl. 58. 1827. *Setaria pallidefusca* (Schumach.) Stapf & C. E. Hubb., Kew Bull. 1930: 25. 1930; Fischer 3: 1789. 1934; Bor, Grass. India 363. 1960.

An annual, tufted grass; culms up to 60 cm high, ascending; nodes glabrous. Leaf sheaths 2-5 cm long, glabrous; ligule a ridge of hairs; blades 5-15 x 0.2-0.6 cm, linear-lanceolate, flat, base cordate, apex acuminate, flaccid. Panicle spiciform, up to 8 cm long, cylindric, densely-flowered; involucral bristles antrorsely barbed; spikelets 2 mm, ovoid; lower glume ovate; upper glume ovate, hyaline; lower lemma ovate, hyaline, paleated and its palea hyaline, empty; upper lemma ovate, crustaceous, transversely rugose; palea similar to its lemma; stamens 3, stigmas plumose. Caryopsis ovoid-ellipsoid.

Common in plains and hill slopes. Fl. & Fr.: July-November. Tel.: *Nakkathoka gaddi.*

Eswaramala RF (ATP), *TP & NY* 836; Thalaricheruvu (ATP), *TP* 901 & 907; Gani RF (KNL), *TP & RVR* 1785; Bhumayyagaripalli (KDP), *KRKS* & *KRS* 37633; Horsley hills (CTR), *MSG & AA* 12208; Tirumala hills (CTR), *RVR & GO* 2422; Talakona (CTR), *PVP* 2509, 3416; Jayanthipuram (GNT), *VRK* 6965; Papaiah palem (GNT), *VRK* 5844; Kadapa, *EKK* 78544 (DD); Mamandur valley (CTR), *KS* 6911 (CAL); Polavaram agency (WG), *DCSR* 426 (CAL); Chodavaram (EG), *MSR* 1514 (CAL); Rampachodvaram (EG), *MM* 102653 (BSID); Anantagiri (VSKP), *GVS* 19457 (CAL); Salur (VJN), *NPBK* 958 (CAL); Koriguttalu (KMM), *RR* 106047 (BSID); Mysamma temple (MBNR), *SRS* 108630 (BSID); Ramanadurg (Ballari), *S. India Flora* 15927 (MH); Avadi (CPT), *S. India Flora* 14883 (MH); Melpat (SA), *CAB* 1071 (MH); Yelagiri hills (NA), *MBV* 804 (MH); Sirumalai hills (DGL), *SKS* 441 (SGH); Vannipakka Kammarpalayam (CPT), *DN* 141 (MH); Betta Mogilam, Melagiri hills (SLM), *S.India Flora* 9795 (MH); Kelamangalam (SLM), *KCJ* 20108 (MH); Purunakote section, Satkosia Tiger Reserve (AGL), *KCM* 5609 (BSID); Deomali hill (KPT), *AC* 10227 (BSID); Jenabil Range, Similipal B.R. (MBJ), *AC* 2302, 2315, 2317 (MBJ); Belghar, Kotgarh Wildlife Sanctuary (KDM), *AC* 2409 (BSID); Mahendragiri (GJM), *VNS* 5750 (MH); Kuthdya hill (GJM), *VNS* 5907 (MH).

INDIA: Throughout India.

WORLD: Tropical and warm temperate regions of the Old World.

Note: It is a polymorphic species, where the length of leaves and panicles show a gradation in variation. This highly variable grass occurs in all sorts of localities in the plains and hills of the State.

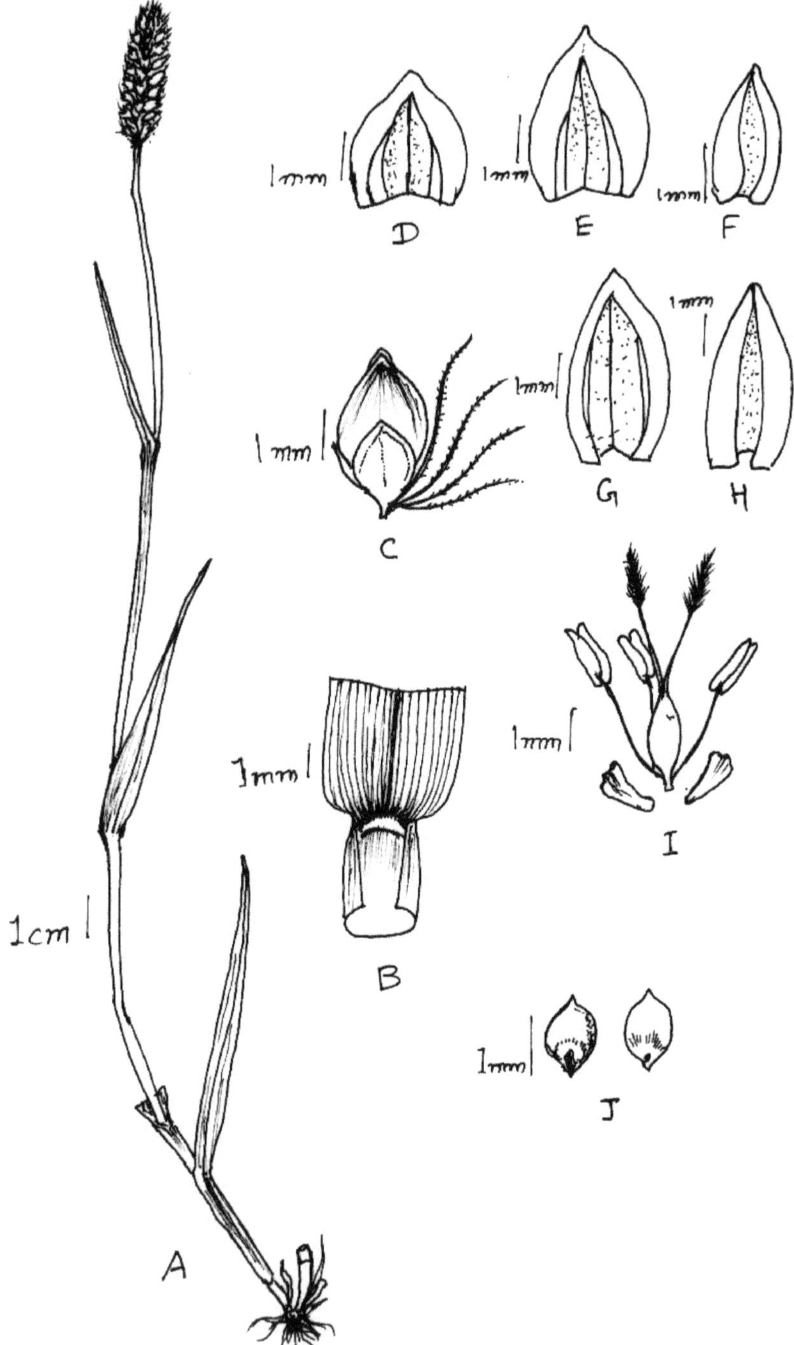

Figure 100. **Setaria parviflora** (Poir.) Kerguelen, (*S. pumila* auct.).

A. Habit, B. Ligule, C. Spikelet, D. Upper glume, E. Lower glume, F. Palea, G. Upper lemma, G. Lower lemma, I Lodicules, stamens and pistil, J. Caryopsis.

Setaria verticillata (L.) P. Beauv., Ess. Agrost. 51, 178. 1812; FBI 7: 80. 1896; Fischer 3; 1789. 1934; Bor, Grass. India 365. 1960; Matthew, Fl. Tamilnadu Carnatic 3: 1902. 1983. *Panicum verticillatum* L., Sp. Pl. ed. 2. 82. 1762.

An annual, tufted grass; culms up to 60 cm high; nodes glabrous. Leaf sheaths 3-7 cm long, glabrous; ligule a fringe of hairs; blades lanceolate, 3-10 x 0.3-1 cm, flat, base rounded, sparsely pubescent; apex acuminate. Panicle up to 10 cm long, spiciform; involucral bristles retrorsely barbed; spikelets oblong-ovate, 2 mm. Lower glume orbicular; upper glume ovate; lower lemma oblong- ovate, shortly apiculate, margins incurved; palea linear; upper lemma oblong, crustaceous, transversely rugose; palea similar to its lemma; stamens 3, stigmas plumose. Caryopsis ellipsoid.

Occasional in moist and shady places. Fl. & Fr.: July-January.

Veldurthi RF (KNL), *RVR* 1719; Tirumala hills (CTR), *RVR & GO* 2404, *MSG* 12184; Nandipadu (NLR), *PMR* 16471; Macherla RF (GNT), *VRK* 3255; Upper Ahobilam (KNL), *JLE* 25585 (CAL & MH); Modugamamisala (KNL), *JLE* 42201 (MH); Guvvalacheruvu (KDP), *KS* 6936 (CAL & MH); Chandragiri (CTR), *S.India Flora* 10114 (MH): Kondapalli (KSN), *CAB* 4393, 7985 (MH); Valamuru, Maredumilli (EG), *MM* 100864 (MH); Appanapally (MBNR), *SRS* 110735 (BSID); Sirumalai hills (DGL), *SKS* 434 (SGH); Hogainakkal RF (SLM), *EV* 21937 (MH); Anchetti (SLM), *S. India Flora* 18046 (MH).

INDIA: Throughout India.

WORLD: Widely distributed in the tropics and temperate regions of the Old World, introduced into America.

SORGHUM Moench *nom. cons.*

1. Panicle branches simple; awn long, well developed; nodes bearded .... **S. nitidum**
1. Panicle branches divided; awn short, poorly developed; nodes glabrous or shortly pilose:
    2. Annual; cultivated; sessile spikelet broadly obovate, without rhizomatous culms .... **S. bicolor**
    2. Perennial; wild; sessile spikelet linear-lanceolate; with well developed creeping rhizomes .... **S. halepense**

Sorghum bicolor (L.) Moench, Meth. Pl. Hort. Hot. Merburg. 207. 1794. var. **bicolor**; Bor, Grass. India 228. 1960. *Holcus sorghum* L., Sp. Pl. 1047. 1753. *Sorghum subglabrescens* (Steud.) Schweinf. & Asch. in Schweinf., Beitr. Fl. Aethio. 302; 306. 1867; var. *rubidium* (Burkill ex Benson) Snowden, Kew Bull. 1935: 253. 1935; Bor, Grass. India 242. 1960. *Andropogon sorghum* var. *mediocris* subvar. *ruber* Burkill ex Benson & Subba Rao, Madras Dept. Agr. Bull. No. 55, 68, 69. 1906. *A. subglabrescens* Steud., Syn. Pl. Glum. 1: 393. 1854.

An annual, robust, tufted grass; culms up to 2 m high, glabrous, polished. Leafblades 10-30 x 2.5-5 cm, broadly lanceolate, flat, apex acuminate, glabrous. Panicle 25 cm long, densely thyrsiform, decompound; branches verticillate; spikelets

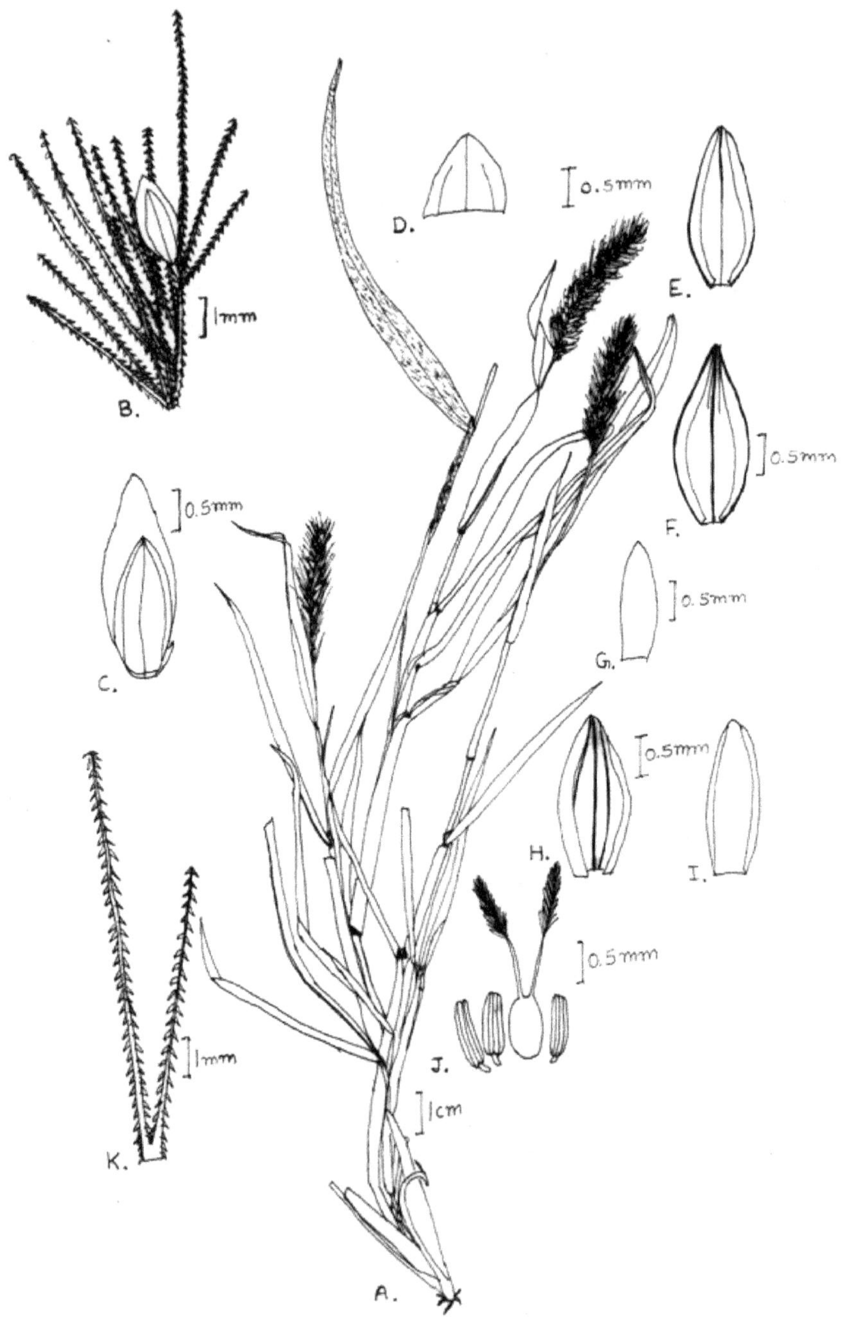

Figure 101. **Setaria verticillata** (L.) P.Beauv.

A. Twig, B. Spikelets with involucral bristles, C. Spikelet, D. Lower glume, E. Upper glume, F. Lower lemma, G. Palea, H. Upper lemma, I. Palea, J. Pistil with stamens, K. Involucral bristles.

5 mm, in groups at the ends of the branches. *Sessile spikelets:* broadly ovate-lanceolate, hairy; glumes coriaceous, nerves obscure; upper lemma awned. Caryopsis broadly obovate.

Cultivated for its grains at some places. Fl. & Fr.: October-March. Tel.: *Jonnalu.*

Anakapalle (VSKP), *S.India Flora* 15455 (MH).

INDIA: Cultivated in S. & N. India.

WORLD: Native to Africa, cultivated in the Tropics.

Sorghum halepense (L.) Pers., Syn. 1: 101. 1805; Fischer 3: 1735. 1934; Bor, Grass. India 222. 1960; Matthew, Fl. Tamilnadu Carnatic 3: 1902. 1983. *Holcus halepensis* L., Sp. Pl. 1047. 1753. *Andropogon halepensis* (L.) Brot., Fl. Lusit, 1: 89. 1804; FBI 7: 182. 1896. *Sorghum miliaceum* (Roxb.) Sw., J. Linn. Soc. Bot. 55: 207. 955 incl. var. *parvispiculum* Sw. *loc.cit.* 209; Bor, Grass. India 223. 1960.

A perennial, tufted grass; culms up to 2 m high, stout, sparsely branched; nodes minutely silky. Leaf sheaths 15-20 cm long, glabrous; ligule short, membranous, ciliate; blades 15-60 x 1-4 cm, linear-lanceolate, base cordate, margins scaberulous, apex acuminate, midrib prominent, glabrous. Panicle effuse, up to 20 cm long, decompound, branches scaberulous; spikelets 2-nate. *Sessile spikelet:* 6 mm, linear-lanceolate, hairy; callus short, bearded; glumes subequal, coriaceous; lower ovate, silky hairs outside; upper lanceolate, acuminate; lower lemma triangular-lanceolate, hyaline, upper hyaline, margins ciliate, 2-lobed, awned; stamens 3, stigmas plumose. Caryopsis ovoid. *Pedicelled spikelet:* similar to sessile spikelet, but narrowly lanceolate, awnless, male.

Occasional near water courses. Fl. & Fr.:Throughout the year. Tel.: *Gaddi jonna;* Tam.: *Karuppanpul.*

Kodavaluru (NLR), *PMR* 23013; Yellavaram, Rampa hills (EG), *TP & MSG* 12280; Diguvametta, Nallamalais (PKM), *JLE* 32479 (CAL & MH); Way to Kolleru (WG), *KS* 5107 (CL); On the way to Kontrukota, (WG), *DCSR* 138 (CAL); Devipatnam (WG), *JSG*15780 (CAL); Beyond Ettakonda (EG), *VNS* 542 (CAL); Mandapeta (EG), *CAB* 4269 (MH); Peddakonda (EG), *VNS* 211 (MH); Adativalasa (VSKP), *GVS* 21613 (MH); Araku valley (VSKP), *NPBK* 623 (CAL); Kodagandi RF (VSKP), *KCJ* 17273 (MH); Srungavarapukota-Chilakalagedda (VJN), *GVS* 32793 (CAL); Kolhusa (KMM), *PVS* 84062 (BSID); Sirumalai hills (DGL), *SKS* 435 (SGH); Ghat road, Salem to Yercaud (SLM), *KS* 7586 (MH); Pudurnadu, Javadi hills (NA), *MBV* 1188 (MH); Belghar, Kotgarh Wildlife Sanctuary (KDM), *AC* 2406 (BSID).

INDIA: Throughout India.

WORLD: Most warm countries, widely cultivated

Note: The introduced 'Johnson grass' is now running wild and rapidly spreading all over India. Snowden (J. Linn. Soc. Bot. 55: 207, 1856) believed that *S.halepense* is not indigenous to India and treated the Indian materials under *S.miliaceum.* But, later on Cope (1982) has merged the later with *S. halepense.*

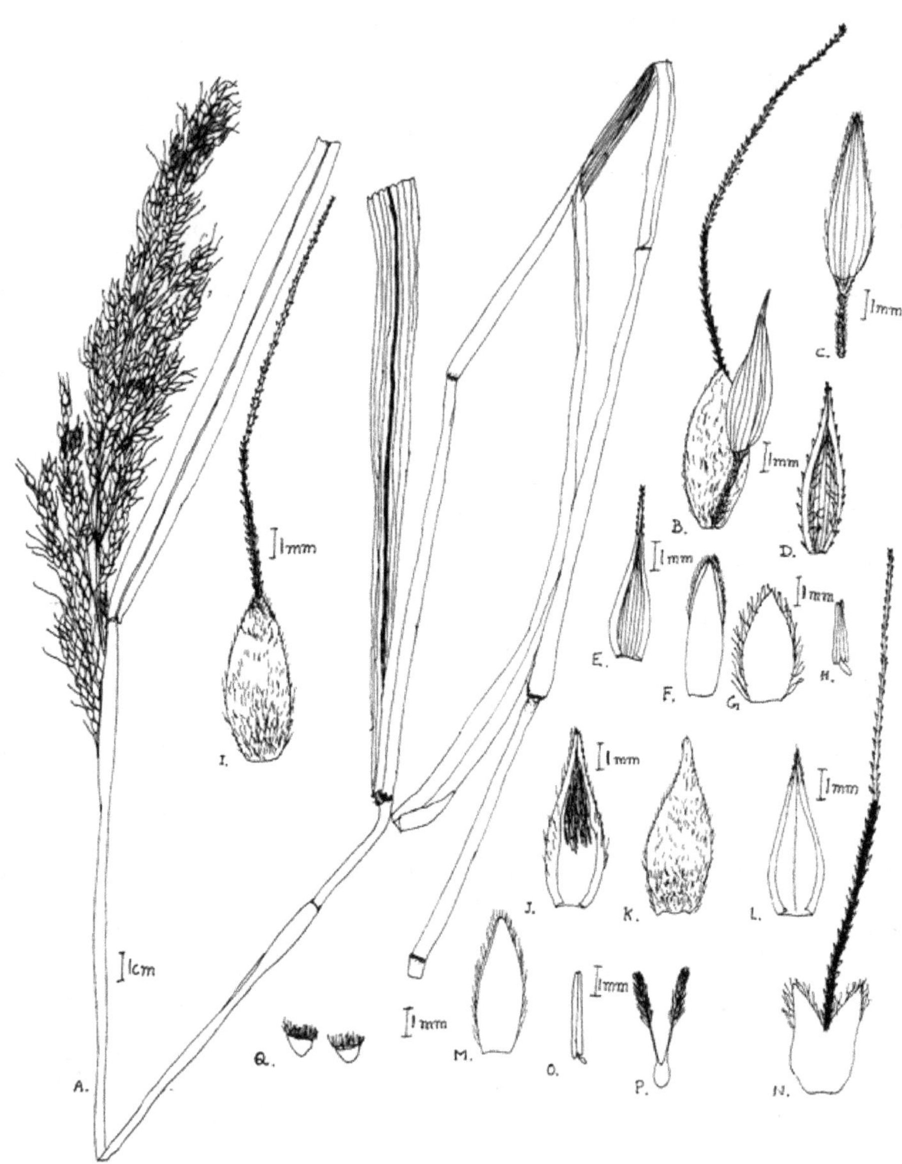

Figure 102. **Sorghum halepense** Pers.

A. Twig, B. Spikelets, C. Pedicelled spikelets, D. Lower glume, E. Upper glume, F. Lower lemma, H. Stamen, I. Sessile spikelet, J & K. Lower glume front and back views, L. Upper glume, M. Lower lemma, N. Upper lemma, O. Stamen, P. Pistil, Q. Lodicules.

Sorghum nitidum (Vahl) Pers., Syn. 1: 101. 1805; Fischer 3: 1735. 1934; Bor, Grass. India 245. 1960. *Holcus nitidum* Vahl, Symb. Bot. 2: 102. 1791. *Andropogon serratus* Thunb., Fl. Jap. 41. 1784; FBI 7: 185. 1896. *Sorghum serratum* (Thunb.) Kuntze, Revis. Gen. Pl. 2: 792. 1891, non Roem. & Schult. 1817.

A perennial, tufted grass; culms up to 4 m high; nodes densely villous. Leaf sheaths terete below, keeled, silky-villous near the mouth; ligule truncate; blades 10-50 x 0.8-2 cm, base rounded, apex acuminate, midrib prominent. Panicle 30 cm long, lax; branches capillary; spikelet margins shortly villous. *Sessile spikelet:* broadly ellipsoid; lower glume broadly oblong-elliptic, coriaceous, keels hispid; upper glume broadly cymbiform, ciliate upwards; lower lemma hyaline, margins incurved, 2-keeled, ciliate; upper lemma linear-oblong, bidentate, awned or unawned. *Pedicelled spikelet:* linear-oblong; lower glume oblong, dorsally depressed, keeled; upper glume narrower, margins incurved, keeled; lower lemma hyaline, linear.

Occasional in hills. Fl. & Fr.:October- March.

Araku (VSKP), *NPBK* 10755 (MH & CAL); Borapuram (MBNR), *BSS* & *SKB* 32288 (BSID); Sirumalai hills (DGL), *SKS* 442 (SGH); Chotkei RF, Purunakote section, Satkosia Tiger Reserve (AGL), *KCM* 5265 (BSID).

INDIA: Throughout India.

WORLD: Sri Lanka, Myanmar, extending through Thailand and Indo-China to Australia.

SPODIOPOGON Trinius

Spodiopogon rhizophorus (Steud.) Pilger in Engl. & Prantl, Pflanzenf. Aufl. 2, 14: 119. 1940; Bor, Grass India 246. 1960; B.R.P.Rao *et al.*, Grasses Arid Zone Peninsular India 180. 2017. *Andropogon rhizophorus* Steud., Syn. Pl. Glumac. 1: 381. 1854.

Perennials; culms tufted, much branched, up to 1.25 m high, rooting at lower nodes. Spikelets lanceolate, densely covered with white hairs. Inflorescence a panicle, at first enclosed in a spathaceous sheath. Spikelets usually one sessile and 1-2 pedicelled, 1-2-flowered, puberulous to villous, usually with whit hairs. Glumes 2, subequal, membranous, many-nerved, densely ciliate, lower glume ovate-lanceolate, upper lanceolat, aristate. Lemmas 2, lower lemma lanceolate, acuminate, keeled, upper lemma bifid, with a long awn from sinus; palea 4-5-nerved, hyaline; upper lemma hyaline, deeply divided at apex inot 2 very acute loes; palea broad, fimbriate.

Common in the shade of trees in forests of Sandur hills in Bellary district of Karnataka (B.R.P.Rao *et al.*, 2017). Fl. & Fr.: October–December.

Swamimalai forest, Sandur (Bellary district), *RS* & *RS* 10469 (JCB).

INDIA: South India, Guajarat, Madhya Pradesh, Maharashtra, Goa, Rajasthan. Endemic.

SPOROBOLUS R.Brown

1. Lower glume shorter than the spikelet:
    2. Upper glume conspicuously shorter than the lemma:
        3. Panicle effuse:
            4. Lemmas more than 1mm long:
                5. Stamens 2; spikelets crowded, slatey-grey in colour, lower glume truncate .... **S. diandrus**
                5. Stamens 3; spikelets scattered, lower glume very short, obtuse.. **S. wallichii**
            4. Lemmas not more than 1 mm long................................. **S. tenuissimus**
        3. Panicle otherwise:
            6. Plants grow near shady or saline soils; panicle spicate .... **S. spicatus**
            6. Plants otherwise; panicle subspiciform or interrupted........... **S. fertilis**
    2. Upper glume as long as or longer than the lemma:
        7. Panicle contracted; branches not whorled:
            8. Stamens 3; perennials.. **S. maderaspatanus**
            8. Stamens 2; annuals .... **S. piliferus**
        7. Panicle effuse; branches truly whorled:
            9. Lower glume half as long as lemma; grain ellipsoid or discoid .... **S. capillaris**
            9. Lower glume minute, less than half the length of the lemma; grain obovate:
                10. Annuals.. **S. coromandelianus**
                10. Perennials .... **S. ioclados**
1. Lower glume as long as the spikelet.. **S. virginicus**

Sprobolus capillaris Miq., Verh. Konink. Nederl. Inst. 3(4): 37. 1851; Bor, Grass India 627. 1960; B.R.P.Rao *et al.*, Grasses Arid Zone Peninsular India 181. 2017. *S. scabrifolius* Bhide, J. Proc. Asiat. Soc. Bengal 8: 312. 1912; Fischer, 3: 1818.1934.

Annuals or perennials, culms erect or ascending up to 65 cm high. Leaf sheath smooth; ligule a fimbriatee rim; leaf blades ovate to lanceolate, 3-10 x 0.5-1.2 cm, acute at apex, rounded to semiamplexicaulis at base, midrib prominent, sparsely tubeculate or ciliate on both sides; ligule 0.3 mm long, membranus, 0.8 cm long fimbriate hairy at apex. Inflorescence panicle, 10-14.5 x 2-6 cm, effuse, pyramid like at maturity; racemes whorled, compact and enclosed by basal sheat, 2.3-4 cm, base pulvinus like; pedicles glandular, greenish to purplish. Spikelets ovate-lanceolate, solitary, paired or in threes, 1.5-1.7 x 0.4-0.7 mm, purple when mature; lower glume ovate-lanceolate, 1 x 0.4 mm, acute at apex, rounded or truncate at base, greenish to yellow, 1-nerved. Upper glume ovate-lanceolate, 1.5-1.7 x 0.8 mm, as long as spikelet, acuminate, chartaceous, purple when mature, 1-nerved. Floret 1, bisexual; lemma ovate-laceolate, boat-shaped, 1.5 x 0.5 mm, acuminate, 3-nerved; palea ovate to elliptic, 1.2 x 0.4 mm, acute, chartaceous, hyaline, 2-nerved; stamens 3; anthers 0.4-0.5 mm, greenish yellow; filaments short; ovary narrowly elliptic, 0.4 x 0.2 mm, acute, greenish. Caryopsis elliptic, obovate or oblong, truncate at apex, dorsally flat or convex, brownish.

Found along roads in moist and shady places. Fl. & Fr.: October–December.

B.T.Dam cross road (ATP), *KVS* & *BR* 29164; Jenabil Range, Similipal B.R. (MBJ), *AC* 2343 (BSID); Kallakurichi (SA), *KCJ* 6373 (MH).

INDIA: South India, Madhya Pradesh, Maharashtra.

WORLD: Myanmar.

Sporobolus coromandelianus (Retz.) Kunth, Revis. Gram. 1: 68.1829; FBI 7: 252. 1896; Fischer 3: 1817. 1934; Bor, Grass. India 627. 1960; Matthew, Fl. Tamilnadu Carnatic 3: 1904. 1983; Moulik, Grass. Bamb. India 2: 477.1997. *Agrostis coromandeliana* Retz., Observ. Bot. 4:19.1786. *Sporobolus commutatus* (Trin.) Kunth, Enum. Pl. 1: 214.1833. *Vilfa commutata* Trin., Diss. Bot. 156.1824.

Perennials, culms erect or decumbent, tufted, to 20 cm high; nodes glabrous. Leaf sheaths glabrous. Ligules membranous. Leaf blades linear-lanceolate with sparsely tubercle based hairy, base rounded or cordate, margins spinulosely toothed, acute. Panicles up to 5 cm, pyramidal, decompound, whorled; spikelets ovate; lower glume oblong, small; upper glume ovate, chartaceous, 1-nerved; lemma ovate, chartaceous, 1-nerved; palea elliptic, 2-nerved; stamens 3. Caryopsis ellipsoid.

Occasional in forest plains and in dry habitats. Fl. & Fr.:August-December.

Thimmammamarrimanu (ATP), *KRKS* 41937; Konapuram (ATP), *JNP* & *AL* 26662; Mantralayam (KNL), *MSG* 12206; Peapuly (KNL), *PVP* 2660; Devaragattu (KNL), *BR* & *SS* 22678; Way to Gollalaguduru (KDP), *KRKS* 41965; Chennur (KDP), *AL* 26577; Gooty (ATP), *JSG* 20840 (MH); Palasa (SKLM), *SCA* 185 (DD); Kinnerasani (KMM), *BR& AMR* 26816; Avadi (CPT); *S.India Flora* 14869(MH).

INDIA: Plains of India.

WORLD: S. Africa, Afghanistan, Pakistan, Sri Lanka, Myanmar.

Sporobolus diandrus (Retz.) P. Beauv., Ess. Agrost. 26.147.178. 1812; FBI 7: 247.1896; Fischer 3:1817.1934; Bor, Grass. India 629.1960; Matthew, Fl. Tamilnadu Carnatic 3: 1905. 1983 *'diander'*; Moulik, Grass. Bamb. India 2:478.1997. *Agrostis diandra* Retz., Obs. Bot. 5:19.1789. *Sporobolus indicus* (L.) R.Br., Prodr. Fl. Nov. Holl. 170. 1810. var. *diander* (Retz.) Jovet & Guedes, Taxon 22: 163. 1973; Gayathri & Pullaiah in Pullaiah, Fl. Andhra Pradesh 3: 1259. 1997.

Annuals, culms erect, tufted, to 90 cm high; nodes glabrous. Leaf sheaths glabrous; ligules membranous; leaf blades 5-12 x 0.1-0.2 cm, linear-lanceolate, base rounded, acuminate, glabrous. Panicles up to 25 cm long, effuse, filiform; spikelets obovoid, lanceolate; lower glume lanceolate, truncate, chartaceous; upper glume ovate, chartaceous, 1-nerved; lemma lanceolate, chartaceous, 1-nerved; palea oblong, elliptic, 2-nerved; stamens 2. Caryopsis obovoid.

Occasional in moist and shady localities of the forests throughout Eastern Ghats. Fl. & Fr.:July–December.

Gundumala RF (ATP), *TP* & *NY* 734; Kalasamudram (ATP), *BR* & *AMR* 21889; Perusomula (KNL), *TP* & *PVP* 3366; Rudrakod (KNL), *BR* & *SS* 20168; Maddiletiswamy temple (KNL), *BR* & *SS* 20266; Balapalli (KDP), *PVP* 3469; Horsley hills (CTR), *PVP* 4661; Tirumala hills (CTR), *RVR* & *GO* 2408; Rollapenta (PKM), *BSS* & *SKB* 29385; Rampa hills (EG), *TP* & *MSG* 12276; Gudem (VSKP), *BR* & *AMR* 24595; Palakonda (SKLM), *AMR* 24250; Forest ear Anigeri village, Addathigala (EG), *GVS* 68550 (MH); Sirumalai hills (DGL), *SKS* 436 (SGH); Tippukadu RF (NA), *KRM* 17927 (MH); Purunakote, Satkosia Tiger Reserve (AGL), *KCM* 5212 (BSID); Dunaei dam (GJM), *AC* 10207 (BSID); Balipadar (GJM), *AC* 10214 (BSID); Jenabil Rang, Similipal B.R. (MBJ), *AC* 2303 (BSID); Badrama Wildlife Sanctuary (SBP), *AC* 10239 (BSID); Belghar, Kotgarh Wildlife sanctuary (KDM), *AC* 2410 (BSID).

INDIA: Throughout India.

WORLD: Pakistan, Sri Lanka, Myanmar extending to Australia.

Note: Jovet and Guedes (1973) treated this taxon as a variety under *S. indicus*. We followed the concept of Clayton and Renovoize (1986).

Sporobolus fertilis (Steud.) Clayton, Kew Bull. 19:291. 1965; Moulik, Grass. Bamb. India 2: 478.1997. *Agrostis fertilis* Steud. Syn. Pl. Glumac. 1: 170. 1854. *Sporobolus indicus* auct non (L.) R.Br.; sensu Hook. f., Fl. Brit. India 7: 247. 1896; Fischer 3:1817. 1934; Bor, Grass. India 630. 1960. *S. indicus* var. *fertilis* (Stud.) Jovet & Guedes, Bull. Centre et Rech. Sc. Biarritz 7: 50. 1968; Saxena & Brahmam, Fl. Odisha 4: 2497. 1996; Gayathri & Pullaiah in Pullaiah, Fl. Andhra Pradesh 3: 1259. 1997.

Perennials, culms erect, tufted, to 1.1 m high; nodes glabrous. Leaf sheaths glabrous; ligules hairy or membranous; leaf blades 15-30 x 0.2-0.5 cm, linear, base rounded or cordate, apex acuminate, glabrous. Panicles to 45 cm long, contracted; spikelets obovoid; lower glume ovate, oblong, chartaceous, obtuse; upper glume ovate, chartaceous, 1-nerved; lemma oblong-ovate, chartaceous, 1-nerved; palea oblong-elliptic, hyaline, 2-nerved; stamens 3. Caryopsis ovate-oblong.

Occasional along road sides and forests plains. Fl. & Fr.: September–December.

Penukonda (ATP), *JNP* 27227; Hottebetta (ATP), *KVS* & *AL* 28082, 28097; Gangutla RF (NLR), *BR* & *BSS* 29305; Althurupadu (NLR), *BSS* & *SKB* 32916; Chintapalli (VSKP), *AMR* 24580; Araku valley (VSKP), *NPBK* 639 (CAL).

INDIA: Throughout India.

WORLD: Sri Lanka, Myanmar.

Note: Jovet and Guedes (1973) treated this taxon as a variety under *S. indicus*. We followed the concept of Clayton and Renovoize (1986).

Sporobolus iocladOS (Nees ex Trin.) Nees, Fl. Afr. Austral. Ill. 161. 1841; Kumar *et al.*, Indian J. Forestry 40: 79. 2017. *Vilfa ioclados* Nees ex Trin., Mem. Acad. Sci. Imp. Sci. St. Petersbourg, Ser. 6. Sci. Maths, 5(2): 65. 1840. *Sporobolus marginatus* Hochst. ex A. Rich., Tent. Fl. Abyss. 2: 397. 1851; Bor, Grass. India 632. 1960; Moulik, Grass. Bamb. India 2: 480. 1997.

Annual or stoloniferous perennials; culms 15-60 cm high, tufted. Leaf sheath glabrous, ligule a hairy rim with fimbriate apex. Leaf blade linear lanceolate, 3-5 x 0.4-0.6 cm, flat, apex acuminate, margins cartilaginous and usually bearing stiff hairs. Panicles 3-20 cm long, pyramidal, usually spreading; peduncle terete; racemes usually spreading, 1.4-4 cm, whorled below, alternate or subwhorled at apex, spikelets pedicelled, oblong-lanceolate, 1.5-2.2 mm long, grey-green to pallid; pedicel 0.5-1 mm long, glabrous; glumes unequal, lower glume oblong, 0.2-0.8 mm long, nerveless, scale like; upper glume oblong-elliptic, chartaceous, 1-nerved. Floret one, bisexual; lemmas oblong-lanceolate, 1.6-2 mm long; palea oblong, nearly as long as lemma, 2-nerved; stamens 3; ovary oblong-elliptic. Caryopsis 0.6-1 cm long, ellipsoid.

Rare in open forests. Fl. & Fr.: September–November.

Ramallakota RF (KNL), *RVR* 1596; Muddanaur (KDP), *BR* & *M. Anil Kumar* 49204 (SKU).

INDIA: NW. India, Upper Gangetic plain, Odisha, Andhra Pradesh.

WORLD: Arabia, Iran, Pakistan.

Sporobolous maderaspatanus Bor, Kew Bull. 12: 234. 1957 & Grass. India 632. 1960; Moulik, Grass. Bamb. India 2: 480. 1997; Matthew, Fl. Tamilnadu Carnatic 3: 1905. 1983. *S. orientalis* Kunth, *nom. illegit*; FBI 7:251.1896; Fischer 3: 1817. 1934.

Perennials, culms erect, to 50 cm high, stoloniferous; nodes glabrous. Leaf sheaths glabrous; ligules membranous; leaf blades 5-15 x 0.2-0.4 cm, linear, base rounded, acuminate, convolute. Panicles 9 cm long, pyramidal; spikelets oblong; lower glume ovate, hyaline; upper glume ovate-lanceolate, acute; lemma ovate, lanceolate, chartaceous, acute; palea ovate-lanceolate, hyaline. Caryopsis ovoid.

Rare in forest shade and plains. Fl. & Fr.: July-November.

Joharapuram (KNL), *AA* 2002; Gooty (ATP), *JSG* 20845 (MH); Mallapukota (EG), *CAB* 8294 (MH); Kollapur (MBNR), *BR* & *BSS* 32917 (SKU).

INDIA: Peninsular India.

WORLD: Sri Lanka.

Sporobolus piliferus (Trin.) Kunth, Enum. Pl. 1:211.1833; FBI 7:251.1896; Fischer 3: 1817. 1934; Bor, Grass. India 632. 1960; Matthew, Fl. Tamilnadu Carnatic 3: 1906. 1983; Moulik, Grass. Bamb. India 2: 480.1997. *Vilfa pilifera* Trin., Dess. Bot. 157. 1824.

Annuals, culms erect, tufted, to 30 cm high; nodes glabrous. Leaf sheaths ciliate at mouth; ligules ring of hairs; leaf blades 2.5-10 x 0.1-0.3 cm, radical, linear, base narrow, margins scabrid, acuminate. Panicles 8 cm long, linear, densely flowered; lower glume ovate-lanceolate, hyaline; upper glume oblong-lanceolate; lemma oblong; palea oblong; lodicules 2, minute; stamens 3; ovary elliptic, styles 2, free, short, stigmas plumose. Caryopsis ellipsoid, reddish-brown.

Occasional on hill slopes and moist habitats. Fl. & Fr.: August–December.

Sunkarimetta, Galikonda area (VSKP), *GVS* 21631 (CAL & MH); Top of Cherukonda peak (VSKP), *GVS* 42747 (MH); Raisili, Laxmipur (KPT), *AC* 4223 (BSID); Deomali hill (KPT), *AC* 10225 (BSID).

INDIA: Temperate places and hills of India.

WORLD: E. Africa, Sri Lanka.

Sporobolus spicatus (Vahl) Kunth, Revis. Gram. 1:67.1829; FBI 7:250.1896; Fischer 3: 1817.1934; Bor, Grass. India 632.1960; Matthew, Fl. Tamilnadu Carnatic 3: 1906. 1983; Moulik, Grass. Bamb. India 2: 481. 1997. *Agrostis spicata* Vahl, Symb. Bot. 1:9.1970.

Perennials, stoloniferous, aquatic, culms erect, tufted, to 30 cm high; nodes glabrous. Leaf sheaths glabrous; ligules membranous; leaf blades 2-6 x 0.05-0.1 cm, oblong, base cordate or truncate, apex acute, glabrous. Panicles spiciform, up to 9 cm long, solitary; spikelets elliptic; lower glume ovate-oblong, truncate, hyaline, 1-nerved; upper glume ovate-oblong, chartaceous, 1-nerved; lemma linear, lanceolate, chartaceous, hyaline; palea oblong, truncate, 2-nerved; stamens 3. Caryopsis obovoid.

Occasional along stream banks and moist places. Fl. & Fr.: April–December.

Peravali (ATP), *TP* & *NY* 505; Pamidi (ATP), *JNP* & *AL* 27365; B.T.dam (ATP), *BR* & *KVS* 29181; Ahobilam (KNL), *MSR* & *BSS* 32264; Peapuly (KNL), *PVP* 2682; Balaraju palli (KDP), *AMR* 24573; Gooty (ATP), *JSG* 20845 (MH).

INDIA: Peninsular India.

WORLD: Tropical Africa, Arabia.

Sporobolus tenuissimus (Schrank) Kuntze, Revis. Gen. Pl. 3: 369. 1893; Bor, Grass. India 633.1960; Moulik, Grass. Bamb. India 2: 481.1997. *Panicum tenuissimum* Schrank in Denkschr., Bot. Ges. Regensb. 2: 26. 1822. *Sporobolus minutiflorus* (Trin.) Link., Hort. Berol. 1: 88. 1827; FBI 7: 248.1896; Fischer 3: 1816. 1934.

Annuals, culms erect, tufted, to 1.2 m high; nodes glabrous. Leaf sheaths glabrous; ligules hairy, membranous; leaf blades 3-6 x 0.05-0.15 cm, linear-lanceolate, base rounded, acuminate. Panicle up to 25 cm long effuse; spikelets ovate; lower glume oblong, hyaline; upper glume ovate, hyaline, 1-nerved; lemma ovate, chartaceous, 1-nerved; palea oblong, hyaline, 2-nerved; stamens 2. Caryopsis obovoid.

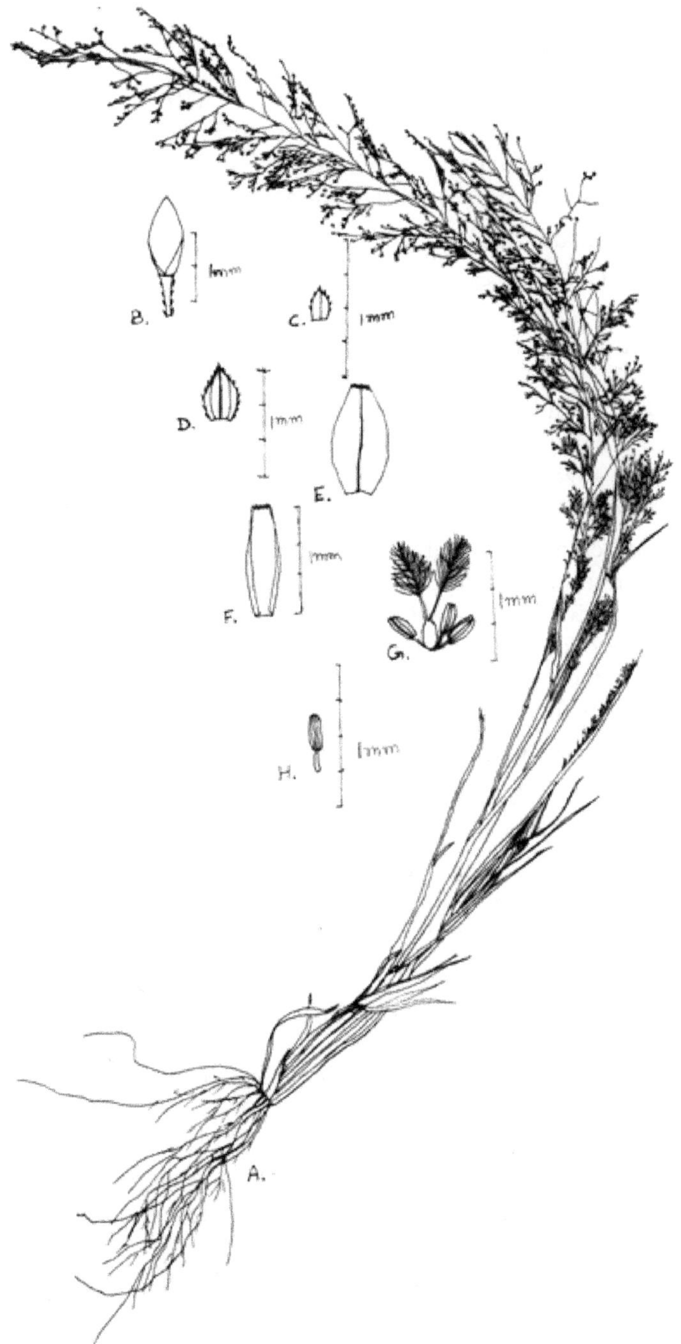

Figure 103. **Sporobolus tenuissimus** Kuntze

A. Twig, B. Spikelet, C. Lower glume, D. Upper glume, E. Lemma, F. Palea, G. Pistil with stamens, H. Stamens.

Occasional near water courses in southern Eastern Ghats. Fl. & Fr.:August–December.

Muchukota (ATP), *JNP* 27397; Upper Ahobilam (KNL), *SS* 22698; Balapalli RF (KDP), *PVP* 3459; Ramathirtham, Tirumala hills (CTR), *MSG* 12175; Sri Kalahasthi (CTR), *AL* 28700; Gangutla RF (NLR), *BR* & *BSS* 29305; Gundlabrahmeswaram (KNL), *JLE* 16919 (MH); Arumanipenta (KDP), *S.India Flora* 15403 (MH); Valamuru, Maredumilli (EG), *MM* 105054 (BSID); Kinnerasani (KMM), *BR* & *AMR* 26820.

INDIA: Peninsular India.

WORLD: Mexico, Mesoamerica, Caribbean, Brazil, Malesia, Arabia.

Sporobolus virginicus (L.) Kunth, Revis. Gram. 1:67. 1829; FBI 7:249. 1896; Fischer 3:1818. 1934; Bor, Grass. India 634. 1960; Matthew, Fl. Tamilnadu Carnatic 3: 1907. 1983; Moulik, Grass. Bamb. India 2:482. 1997. *Agrostis virginica* L., Sp. Pl. ed. 1:63. 1753. *Sporobolus tremulus* (Willd.) Kunth, Revis. Gramin. 1: 67. 1829; FBI 7: 250. 1896; Fischer 3: 1817. 1934; Bor, Grass. India 633. 1960; Moulik, Grass. Bamb. India 2:481. 1997. *Agrostis tremula* Willd., Sp. Pl. 1: 372. 1798.

Perennials, culms erect, tufted, to 10 cm high, stoloniferous, nodes glabrous. Leaf sheaths terete; ligules row of hairs, leaf blades 2.5-7.5 cm long, linear-lanceolate, base rounded, convolute, apex acute. Panicle 10 cm long, sub-spiciform; spikelets lanceolate, crowded; lower glume lanceolate, chartaceous; upper glume ovate-lanceolate, chartaceous; lemma ovate-lanceolate, chartaceous, 1-nerved; palea oblong, elliptic, hyaline; stamens 3. Caryopsis obovoid.

Occasional in river banks, hill slopes and near streams. Fl. & Fr.: July - December.

Pamidi (ATP), *BR* & *JNP* 26617; Penukonda (ATP), *BR* & *JNP* 26749; Velgode (KNL), *BSS* & *SKB* 30233; Devaragattu (KNL), *SS* & *AMR* 22675; Gundlabrahmeswaram (KNL), *BR* & *BSS* 32965; Horsley hills (CTR), *MSG* & *AA* 12209; Tatipudi (VJN), *AMR* 24529; Macherla RF (GNT), *SCA* 139 (DD); Avadi (CPT), *S.India Flora* 14855 (MH); Palakuppam (SA), *VNS* 4032 (MH).

INDIA: Peninsular India.

WORLD: Tropics and Subtropics through out the World.

Sporobolus wallichii Munro ex Trin., J. Bot. 27: 171. 1889; FBI 7: 248. 1896; Fischer 3: 1817. 1934; Bor, Grass. India 634. 1960; Matthew, Fl. Tamilnadu Carnatic 3: 1907. 1983; Moulik, Grass. Bamb. India 2: 482. 1997.

Perennials, culms erect, tufted, to 1.2 m high, slender; nodes glabrous. Leaf sheaths glabrous; ligules membranous; leaf blades 10-25 x 0.2-0.4 cm, linear, convolute, base cordate or truncate, apex acuminate, glabrous. Panicles up to 10 cm long, effuse, flexuous; spikelets oblong-lanceolate; lower glume ovate, chartaceous, truncate; upper glume ovate-lanceolate, acute, 1-nerved; lemma ovate-elliptic, chartaceous, 3-nerved; palea oblong, chartaceous; stamens 3. Caryopsis oblong.

Common in moist localities of the forests in southern Eastern Ghats. Fl. & Fr.: July-December.

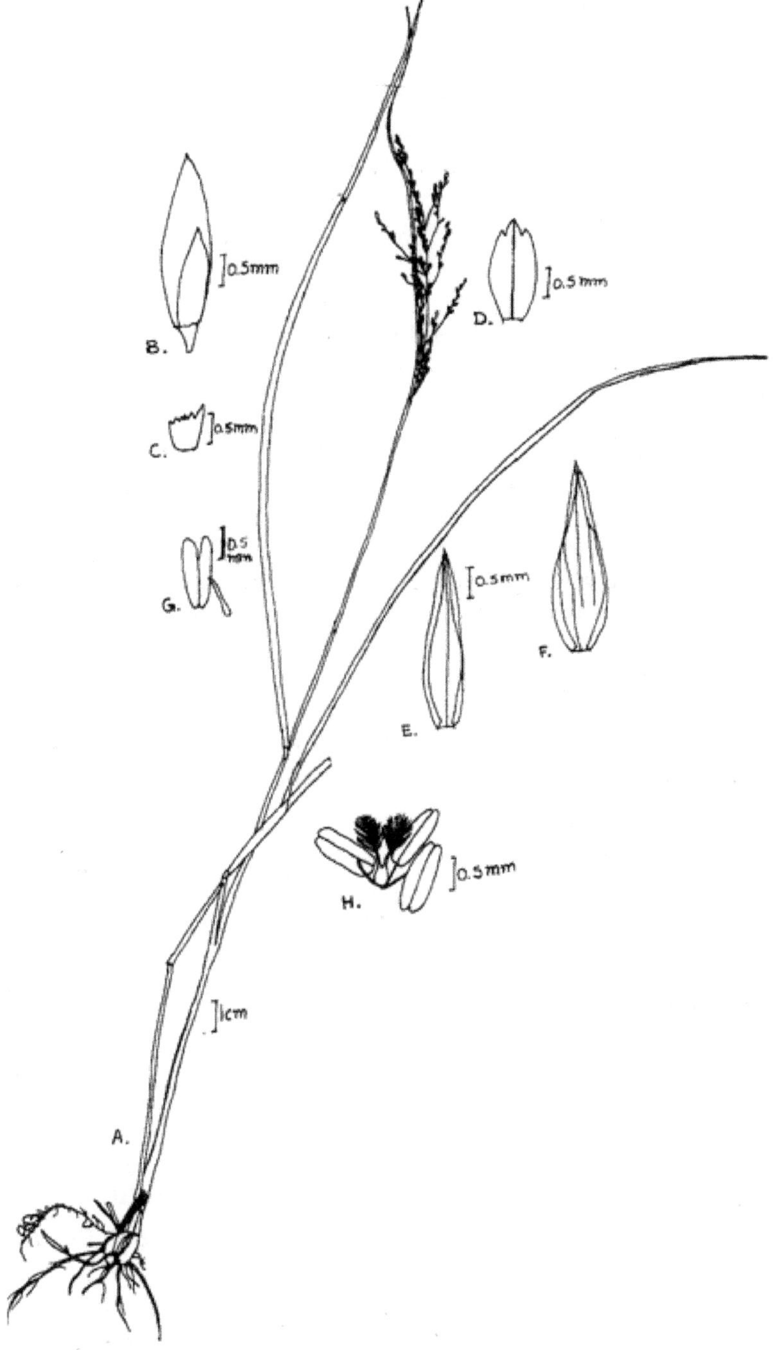

Figure 104. **Sporobolus wallichii** Munro ex Trin.

A. Twig, B. Spikelet, C. Lower glume, D. Upper glume, E. Lemma, E. Palea, G. Stamen, H. Pistil with stamens.

Kalasamudram (ATP), *EC* 2892, *BR* & *AMR* 24230; Nallamada (ATP), *BR* & *JNP* 27337; Kottala cheruvu (KNL), *SS* 22464; Velgode (KNL), *BSS* & *SKB* 30232, 30600; Balapalli RF (KDP), *PVP* 3459; Guvvalacheruvu (KDP), *BR* & *AMR* 26074; Tirumala hills (CTR), *MSG* 12174, *BR* & *AMR* 21841, 21845, 21848; Rollapenta (PKM), *PVP* 2341; way to Diguvametta (PKM), *JLE* 22169 (MH); Mannanur (MBNR), *BSS* & *SKB* 30600, 30601 (SKU, BSID); Sirumalai hills (DGL), *SKS* 437 (SGH); Kuthadya hill (GJM), *VNS* 59240 (MH).

INDIA: Bihar, Odisha, Andhra Pradesh, Tamil Nadu

WORLD: Sri Lanka, Myanmar.

TETRAPOGON Desfontaines

Tetrapogon tenellus (Koen. ex Roxb.) Chiov., Ann. Ist. Bot. Roma 8: 352. 1908; Bor, Grass. India 475. 1960. *Chloris tenella* Koen. ex Roxb., Fl. Ind. 1: 330. 1820; FBI 7: 290. 1896; Fischer 3: 1838. 1934.

An annual, tufted grass; culms up to 60 cm high; nodes glabrous. Leaf sheaths 3-7 cm long, glabrous; ligule membranous; blades 2-8 x 0.5-0.15 cm, linear-lanceolate, base rounded, apex acuminate, glabrous. Racemes up to 7 cm long, solitary, terminal; spikelets 5 mm, cuneate, bifarious, 3-4-awned; perfect florets 4; glumes linear, persistent, strongly 1-nerved; upper broader than lower, mid nerve produced a small arista; all perfect lemmas broadly ovate, strongly 3-nerved, truncate; callus densely hairy, awned (shape is same, but size is different in all 4 perfect lemmas); paleas elliptic, 2-nerved, margins ciliated; stamens 3, stigmas plumose. Caryopsis ellipsoid, compressed. Empty lemma broadly ovate, 3-nerved, awnless.

Occasional grass of sandy tracts and waste lands. Fl. & Fr.: August- September. Tel.: *Eruvangaddi.*

Gooty fort (ATP), *MSG* 12237, *PVP* 3671; Madhavaram RF (KNL), *RVR* 1653; North Dhone RF (KNL), *RVR* 1614; Jyothi chemical area (KDP), *PVP & GO* 3432; Mabbuchintalapalli (KDP), *KRKS* 40065; Pellakuru (NLR), *PMR* 22318; Mellavagu (GNT), *VRK* 5824; VP South (GNT), *VRK* 5778; Yerrabalem (KDP), *SCA* 168 (CAL); Kadapa, *EKK* 78532 (DD); Macherla RF (GNT), *PCN* 137 (DD); Nandyal farm (KNL), *KCJ* 83141 (MH).

INDIA: Peninsular and NW. India.

WORLD: Tropical Africa Eastwards to India and southwards to Rhodesia and Angola.

THELEPOGON Roth ex Roemer & Schultes

Thelepogon elegans Roth ex Roem. & Schult., Syst. Veg. 2: 788. 1817; FBI 7: 148. 1896; Fischer 3: 1892. 1934; Bor, Grass. India 247. 1960.

An annual, tufted grass; culms up to 60 cm high, geniculately ascending, rooting from base; nodes glabrous. Leaf sheaths 3-7 cm, loose, tubercle-based hairs; margins densely ciliate; ligule ciliate; blades 5-15 x 0.7-2 cm, linear-lanceolate, base amplexicaul, margins spinulose, apex acute, tubercle-base pubescent, nerves prominent. Racemes 4-12, up to 15 cm long, digitate, corymbosely fasicled; rachis

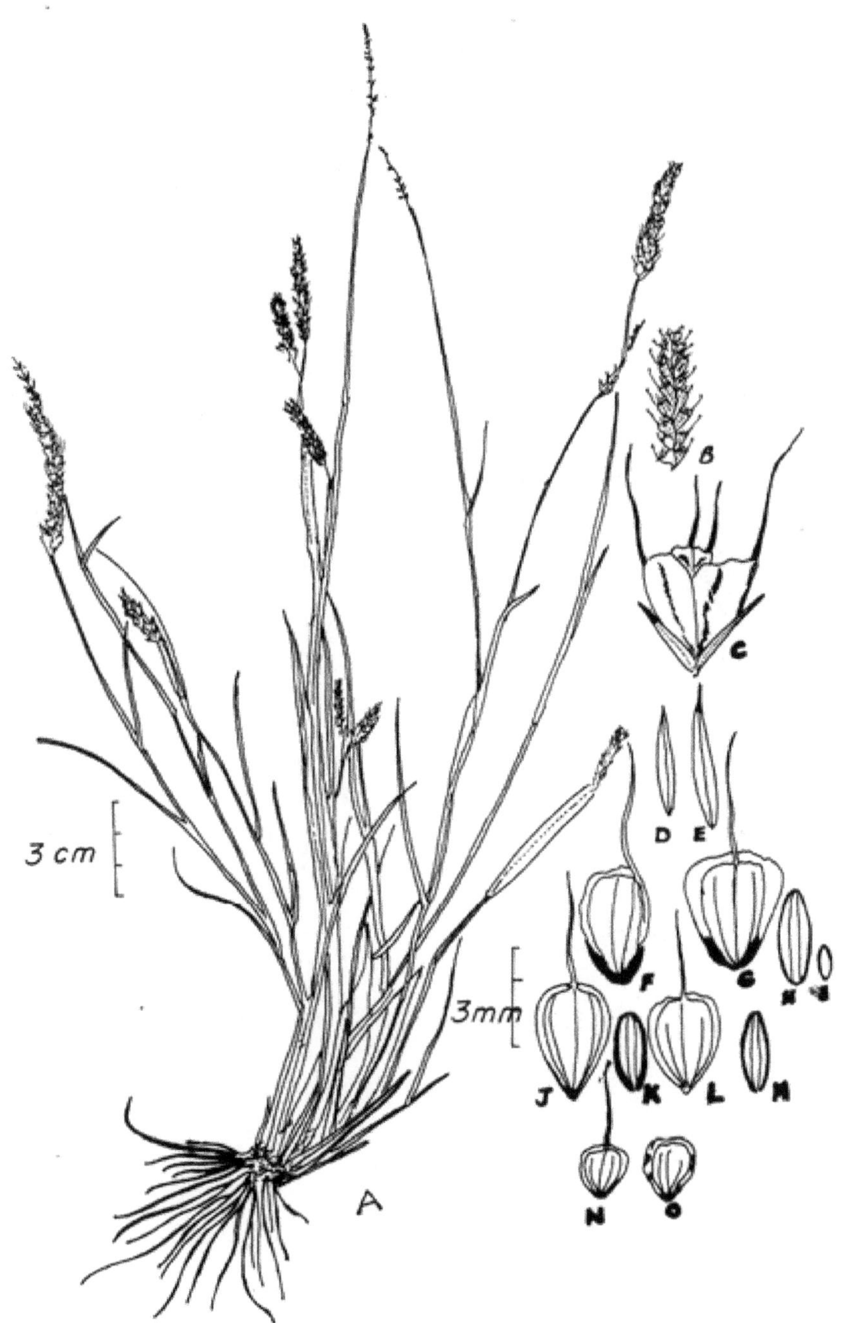

Figure 105. **Tetrapogon tenellus** (Koen. ex Roxb.) Chiov.

A. Twig, B. Raceme, C. Spikelet, D. Lower glume, E. Upper glume, F. Lateral view of the I floret lemma, G. Ventral view, H. Palea, I. Grain, J. II lemma, K. Palea, L. III Lemma, M. Palea, N. IV Lemma, O. Empty lemma.

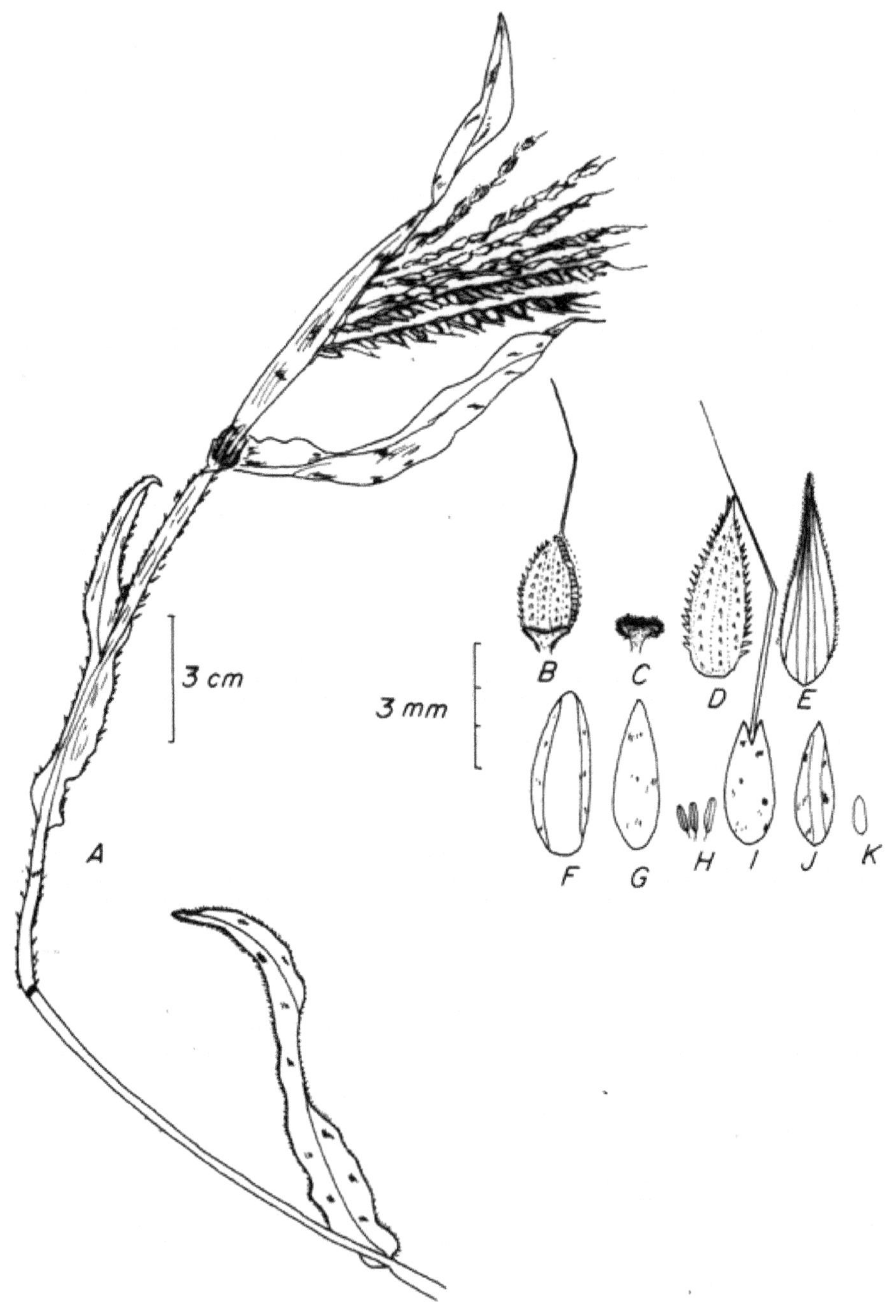

Figure 106. **Thelepogon elegans** Roth ex Roem. & Schult.

A. Twig, B. Spikelet, C. Callus, D. Lower glume, E. Upper glume, F. Lower lemma, G. Palea, H. Stamens, I. Upper lemma, J. Palea, K. Grain.

fragile, compressed; spikelets ovate oblong, 5 mm, 2-nate, pedicelled spikelet reduced. *Sessile spikelet;* ovate-oblong, 5 mm; callus bearded; lower glume ovate, crustaceous, apex 2-lobed, transversely rugose interrupted, strongly ridged, prominent at apex; upper glume ovate, acuminate, coriaceous, keeled, scabrid; lower lemma oblong, hyaline; palea oblong, nerveless, male; upper lemma obovate, 2-lobed, awned; palea narrower, bisexual; stamens 3; stigmas plumose. Caryopsis oblong.

Occasional gregarious grass in Kurnool district. Fl. & Fr.: August-October.

Malyala fields (KNL), *PVP & RVR* 2545.

INDIA: Peninsular to Central India.

WORLD: Tropical Africa, Indo-China, Malesia.

THEMEDA Forsskal

1. Involucral spikelets all on the same level, only one bisexual spikelet in the raceme:
    2. Branching effuse; clusters few, in few small superposed fascicles on filiform branches .... **T. laxa**
    2. Branching short; clusters of inflorescence numerous, in numerous fascicles:
        3. Perennial; culms robust; clusters globose or fan-shaped:
            4. Moderate to robust grass, 30-200 cm high; racemes usually arising from the upper leaf sheaths; fertile spikelets hispidulous or hispid at tip only .... **T. quadrivalvis**
            4. Short grass, 10-30 cm high; racemes usually arising from the basal leaf sheaths and hence culms and inflorescence appear equal in height; fertile spikelets hispidulous or hispid through out the dorsal surface .... **T. helferi**
        3. Annual; culms geniculately ascending; clusters usually narrow, erect .... **T. triandra**
1. Involucral spikelets inserted at different levels, often 2 bisexual spikelets in the raceme:
    5. Involucral spikelets entirely glabrous or puberulous but without tubercle-based hairs:
        6. Involucral spikelets entirely glabrous .... **T. saxicola**
        6. Involucral spikelets puberulous with very short hairs on the lower glumes:
            7. Sessile spikelet awned:
                8. Racemes solitary; spikelet solitary; upper glume 5-nerved, glabrous .... **T. cymbaria**

8. Racemes 2-3 together; central racemes with 2-3 sessile spikelets; terminal raceme with 2-pedicelled spikelets; upper glume 3-neved, scaberulous .... **T. caudata**

7. Sessile spikelet awnless .... **T. villosa**

5. Involucral spikelets covered on lower glume with numerous or sparse tubercle-based hairs:

9. Tall, robust grass with noddling panicle and tubercle based golden hairs on the lower glume of the involucral spikelet; awn stout .... **T. arundinacea**

9. Weak but not wiry, bushy grass or if robust the hairs on the lower glume of the involucral spikelet always grey; awn slender:

10. Hairs on sessile spiklet covered with sparse brownish white hairs .... **T. tremula**

10. Hairs on the lower glumes of invlucral spikelets always grey:

11. Peduncle less than 6 cm long; sessile male spikelets 2 .... **T. mooneyi**

11. Peduncle more than 8 cm long; sessile male spikelets 3 .... **T. odishae**

Themeda arundinacea (Roxb.) Ridl., Trans. Linn. Soc. 2, 3: 401. 1893; Bor, Grass. India 250. 1960. *Anthistiria arundinacea* Roxb., Fl. Ind. 1: 256. 1820. *A. gigantea* Cav. subsp. *arundinacea* (Roxb.) Hack. in DC. Monogr. Phan. 6. 674. 1889; FBI 7: 217. 1896.

A perennial, tufted grass; culms up to 7 m high. Leaves linear, gradually attenuate towards the base, tapering to a subulate tip, rough on both surfaces, very scabrid on the margins, up to 100 x 1-1.5 cm. Panicle compound, dense or lax, terminating the culms, upper branches in short solitary, racemes subtended by spathes. Involucral spikelets: Pairs inserted at different levels, 1.2-1.8 cm long, dorsally hirsute with golden or rusty hairs; lower glumes 2-keeled, closely many-nerved; upper glume a little shorter, 3-nerved, chartaceous, acuminate; lemmas 2, male or suppressed. Sessile spikelets: 1 or 2, 8-11 mm long, linear-lanceolate; lower glume oblong, acute, 2-toothed, thinly hairy below, above hispid or scabrid; upper glume as long as the lower and of the same texture; lower floret empty, lemma hyaline; upper floret bisexual, awned. Caryopsis linear-obovate. Pedicelled spikelets 1-1.5 cm long, similar to the involucral spikelet but much more slender and delicate, covered on the back with numerous long, tubercle-based yellow hairs.

Rare along the hill slopes of northern Eastern Ghats. Fl. & Fr.: August-December.

INDIA: Himalaya and plains of India from Uttar Pradesh eastwards to Khasi and Naga hills.

WORLD: Myanmar, SE. Asia.

Themeda caudata (Nees) A. Camus in Lecomte, Fl. Gen. Indo-Chine 7: 364. 1922; Haines, Bot. Bihar & Orissa 3: 1052. 1924; Mooney, Suppl. Bot. Bihar & Orissa 191.1950; Bor, Grass. India 250.1960; Saxena & Brahmam, Fl. Orissa 2504. 1996. *Anthistiria caudata* Nees in Hook. & Arn., Bot. Beechay Voy. 245. 1838. *Themeda gigantea* (Cav.) Hack. subsp. *caudata* (NNees) Hack. in A. & C.DC., Monog. Phan. 6: 676. 1889, *p.p.*; FBI 7: 217. 1896.

Very large caespitose grass from a stout root stock; culms flattened, 1.2-6 m, greenish yellow or pinkish. Leaves markedly equitant, linear, 0.9-1.5 m x 7.5-18 mm, tapering to a fine point, glabrous or somewhat rough, glaucous beneath, margins scabrid; sheaths very smooth, striate, glabrous; ligule a very narrow ciliate ring, convex upwards. Spathes and their spatheoles forming a very large, leafy and pendulous compound panicle; spatheoles 3.7-5 cm long, finely acuminate, many-nerved, glabrous, on slender peduncles enclosed with others and a branch in a spathe, these with other peduncles and branches in a spathe of a lower order and so on; racemes 2-3 together; peduncle of raceme about half as long as the spatheole, long-pilose upward; central raceme with 2-3 sessile spikelets, the terminal with 2 pedicelled spikelets. Involucral spikelets inserted at different levels, 1-1.5 cm long in somewhat unequal pairs, one pair with longer calli than the others, one of each pair apparently often empty, the other spikelet staminate. Lower glume long, subulate, subaristate, one margin much inflexed and its keel scantily ciliate and scabrid, back flat, puberulous with very short hairs (or scabrellous), 9-nerved; upper glume *c.* three-fourths the lower, 3-nerved and an additional nerve in margins, scaberulous. Lower and upper lemmas sub-similar, ciliate, hyaline; anthers nearly as long as the glume (7.5 mm). Sessile spikelets 1 cm long, densely hirsute with appressed dark brown hairs; awn 4.3 cm long, dark brown, scaberulous but often 0; with oblique, densely brown bearded, ultimately pungent callus, falling with its joint (3 mm long) and pedicelled spikelet; lower glume very coriaceous nearly covering the upper glume, 11-nerved, upper glume with a hirsute keel and involute coriaceous margins; lemmas three-fourths as long, thinly brown hairy; lodicules very large, oblong-lanceolate, 1.5 mm, cuspidate. Caryopsis free, oblong, 3.5 mm long, brown.

Occasional in valleys, forests and plateus in Eastern Ghats of Odisha. Fl. & Fr.: September–December.

Barakamada, Similipahar (MBJ), *SX* & *MB* 4106 (RRL-B); Jenabil Range, Similipal B.R. (MBJ), *AC* 2324 (BSID).

INDIA: Uttar Pradesh, Bihar, West Bengal, Assam, Meghalaya, Odisha.

WORLD: Myanmar.

Themeda cymbaria (Roxb.) Hack. in DC. Monogr. Phan. 6: 69. 1889; Fischer 3: 1746. 1934; Matthew, Fl. Tamilnadu Carnatic 3: 1908. 1983. *Anthistiria cymbaria* Roxb., Fl. Ind. 1: 255. 1820.

Perennial grass, culms up to 3 m high. Leaves 80-90 x 0.7-1 cm, linear-lanceolate, glaucous beneath, margin scabrid. Panicles 30-50 cm long; involucral spikelets 4-6 mm long, in superposed pairs; bisexual spikelets solitary; callus short, bearded;

glumes unequal, glabrous, coriaceous; lemmas dissimilar; lower empty; upper bisexual, stipitiform, with 1.8-2.5 cm long awn. Caryopsis linear, grooved.

Common in hilly areas. Fl. & Fr.: August–January. Tel.: *Konda gaddi.*

Mallelamma temple (KDP), *BR* & *BSS* 30527; Thotamamidi near Devarapalli (EG), *NRR* 76819 (MH, BSID); Sirumalai hills (DGL), *SKS* 343 (SGH).

INDIA: Peninsular India. Endemic.

Themeda helferi Hack. in A.D.C. & C.D.C., Monogr. Phan. 6: 665. 1889; Kumar *et al.*, Rheedea 28: 57. 2018. *Anthistiria ciliata* var. *helferi* (Hack.) Hook.f., Fl. Brit. India 7 (21): 214. 1897. *Themeda ciliata* subsp. *helferi* (Hack.) A. Camus, Bull. Mus. Natl. Hist. Nat. 26: 424. 1920. *Themeda quadrivalvis* var. *helferi* (Hack.) Bor, Grass. India 252. 1960; Karthik. *et al.*, Fl. Ind. Enum. Monocot. 268-269. 1989; Moulik, Grass. Bamb. India. 1: 233. 1997. *Themeda acualis* B.S. Sun & S. Wang, J. Yunnan Univ. (Nat. Sci. Ed.) 21: 96. 1999.

Annuals; culms solitary or caespitose, erect or geniculate-ascending, rooting from lowermost nodes, (15–)25–30 cm high. Leaf sheaths glabrous, laterally compressed, half to longer than their internodes; ligule membranous, truncate, ciliolate on upper margins, (1–)2–3 mm long; blades glabrous, 0.3–0.6 x 15–20 cm, acute to acuminate at apex, a few tubercle-based hair on lower margins. Inflorescence paniculate, composed of fascicled racemes, each raceme compact, subtended by a spathe. Principal spathe lanceolate, caudate acuminate at apex, margins membranous-hyaline, with few tuberculate hairs on margins in the middle, 2–3 x 0.2–0.3 cm long; secondary spatheoles similar to principal, but smaller in size, tubercles absent or very less on margins, 1.5–1.8 x 0.15–0.18 cm long. Spikelets seven in total; of which basal involucral spikelets four, sessile, sterile or male; companion sterile spikelets two, pedicelled, sterile; fertile spikelet one, sessile, bisexual. Involucral spikelets: Glumes 2, sub equal to equal in length, lower glume 4–4.5 mm long, glabrous, with few tuberculate ciliate hairs in upper half, upper glume glabrous. Lemmas hyaline. Paleas minute or absent; stamens 3, anthers 1.7–2 mm long. Companion pedicelled sterile spikelets: reduced to two empty sub equal glumes, 4–4.5(–6) mm long. *Fertile spikelet:* Lower glumes oblong, coriaceous, terete or rounded on dorsal side, 4–5 mm long, densely hispidulous to hispid on the dorsal midline; upper glumes as long as lower glumes, membranous to thinly coriaceous, flat. Lower lemma hyaline; upper lemma reduced to base of awn; principal lemma awn apical, geniculate, column twisted, 3–4 mm long, densely ciliate with stiff hairs; palea minute or absent; stamens 3; stigma plumose. Caryopsis oblong, *c.* 3 mm long.

Found along way sides in the moist deciduous forests and forest margins in Araku valley. Fl. & Fr.: October–March.

Way to Janamguda, near Araku (VSKP), *BR, M. Anil Kumar* & *P. Anjaneyulu* 52576 (SKU); near P. K. Gudem(VSKP), *BR & M. Anil Kumar* 52997 (SKU); Nallamalias, near Thummalabayalu base camp (PKM), *BR, M. Anil Kumar A. Sreenath & P. Anjaneyulu* 53069 (SKU).

INDIA: Andhra Pradesh.

WORLD: China, Myanmar, India.

Themeda laxa (Anderss.) A. Camus, Bull. Mus. Hist. Nat. Paris 26: 423. 1920; Fischer 3: 1746. 1934; Bor, Grass. India 251. 1960. *Anthistiria laxa* Anderss., Nov. Act. Sci. Upsal. 3, 2: 243. 1856; FBI 7: 213. 1896.

A perennial, tufted grass; culms up to 40 cm high. Leaf blades on the main culms up to 30 x 0.7 cm, apex acuminate, upper blades 7.5-12 cm long, filiform. Panicle up to 8 cm long, leafy, few racemes, shortly peduncled outer spathes 2.5 cm; proper spathes glabrous, longer than the spikes. *Involucral spikelet:* 4, all on the same level, 4 mm, male; lower glume glabrous except few bristles near the apex. *Pedicelled spikelets:* similar to involucral spikelet. *Sessile spikelet:* solitary, 4 mm, bisexual, awned; lower glume glabrous, shining below, hispidulous at apex.

Along the hill slopes. Fl. & Fr.: August-December

Upper Ahobilam (KNL), *TP & GNVF* 7062; Jacum, Karlapat Wildlife Sanctuary (KHD), *AC* 2445 (BSID); Hatigarh, Similipal Biosphere Reserve (MBJ), *AC* 2352 (BSID).

INDIA: E. Himalaya, Madhya Pradesh, Odisha, Andhra Pradesh.

WORLD: Bhutan, Nepal.

Themeda mooneyi Bor, Kew Bull. 1951: 451. 1952 & in Grass. India 252. 1960; Moulik, Grass. Bamb. India 1: 231. 1997.

Perennials, culms erect, bushy, to 90 cm, glabrous, branched from all nodes, leafy. Leaf sheaths 4-6 cm, flattened, keeled, glabrous or minutely scabrid, lacerate; leaf blades 20-30 x 0.4 cm, attenuated to long setaceous apex, slightly narrowed to apex. Racemes solitary, peduncles 6 cm, filiform, 2 pairs of involucral spikelets inserted at different levels, lanceolate, 2-5 x 0.5 cm, lower glume elliptic, acute, 10-13 x 1.5 mm, covered with sparsely tubercle based bristles, 2-keeled, keels narrowly winged, margins incurved, many-nerved; upper glume oblong-elliptic, acute, 7-8 x 1 mm, 3-nerved, margins ciliate, hyaline; lower lemma elliptic, oblong, 6-7 x 0.9 mm, acute, hyaline, margins sparsely ciliate; upper lemma obglong-obtuse, 4 x 0.3 mm, hyaline, margins ciliate; lodicules 2, oblong, truncate, 1mm; stamens 3, anthers 2.4 mm. *Sessile spikelets:* oblong, 6-7 x 2 mm, slightly compressed, dark brown; callus obtuse, 1.5 mm, densely pilose; lower glume oblong-truncate, 6 x 1.7 mm, coriaceous, granulate, scabrid towards apex; slightly compressed on dorsal side; upper glume oblong, 5.5 x 1.5 mm; upper lemma hyaline, lanceolate, awn 3-4 cm long; column of awn 1-2 cm, brown, twisted; lodicules 2, oblong, truncate, 1 mm; stamens 3, anthers 2.5 mm; ovary oblong, 1 mm; styles 1 mm, stigmas 1 mm. *Pedicelled spikelet:* lanceolate, acuminate, 15 x 1.5 mm, glabrous, pedicel 2 mm, lower side pilose; lower glume lanceolate-acuminate, 13 x 2 mm, many-nerved, rounded but keeled above; upper glume lanceolate-acute, 7-9 x 1.5 mm, 3-nerved, glabrous; lower lemma oblong, 7 x 1 mm, hyaline; upper lemma narrow, 5 mm; stamens 3, anthers 3 mm; lodicules 2, large, truncate.

Rare in hilly grounds, forest margins and on hill slopes in Visakhapatnam district. Fl. & Fr.: October–December.

Galikonda (VSKP), *AMR* & *BR* 25650.

INDIA: Odisha, Andhra Pradesh, Endemic.

Themeda odishae Chorghe *et al.*, Phytotaxa 245: 183. 2016.

Tufted perennials; culms decumbent, up to 35 cm long; nodes glabrous. Leaf blades linear, attenuated at apex, 5–12 x 0.2–0.25 cm, scabrid on both sides with tubercle based hairs, margins scabrid; ligules membranous, ciliate at apex, 1 mm long; leaf sheath 2–5 cm long, keeled, margins winged, wings hyaline with tubercle based hairs. Inflorescence not branched, subtended by leaf sheath, consisting 3 racemes each one subtended by spatheole; spatheole linear-lanceolate 3–7 cm long, peduncles up to 8 cm long. Racemes 1.5–2 cm long; with 4 involucral male spikelets arranged at different levels; 2 sessile fertile spikelets and 3 pedicelled male spikelets. Involucral spikelets lanceolate, 7–9 x 1.2–1.5 mm, sessile, consist of male floret; lower glumes linear-lanceolate, 7–9 x 1.2–1.5 mm, 7–9-nerved, sparsely tubercule based, white hairy; acute at apex; 2-keeled; keels winged; wings up to 0.3 mm long, scabrid. Upper glumes linear-lanceolate, 7–8.5 x 1–1.2 mm; 3-nerved, chartaceous; acute at apex; florets 2, epaleate; lower lemma linear-lanceolate, 7–7.5 x 1–1.2 mm; hyaline; 3-nerved, 2-keeled; margins infolded, ciliate; upper lemma linear, 6.5–7 x *c.* 0.5 mm, hyaline; lodicules 2, linear, truncate at apex; stamens 3, anthers 3.8–4 mm long. Sessile spikelets linear, 5–5.5 x 1–1.2 mm; fertile, awned; callus *c.* 2 mm long, pungent, bearded with whitish brown hairs; lower glumes linear-lanceolate, 5–5.5 x 1–1.2 mm, coriaceous, whitish-brown, scabrid, glabrous; obtuse at apex; margins scabrid at apex; upper glumes linear-lanceolate, 5–5.5 x 0.8–1 mm, chartaceous; obtuse at apex, 2-keeled; margins infolded, ciliate; florets 2, epaleate; lower lemma linear, 4–4.5 x 0.7–0.8 mm, hyaline; upper lemma linear, 2.5–3 mm, awned; awn geniculate, pubescent 3–3.5 cm long. Fruits not seen. Pedicelled spikelet linear-lanceolate, 8–9 x 1.2–1.5 mm; containing a male floret; pedicel 2 mm long, pubescent; lower glumes linear-lanceolate, 8–9 x 1.2–1.5 mm; coriaceous, multi-nerved, tubercle based white hairs in upper half; apex acute; upper glume linear-lanceolate, 7.5–8.5 x 1–1.2 mm, 3-nerved, chartaceous; florets 2, epaleate; lower lemma linear, 6.5–7 x 0.8–1 mm, hyaline, 3-nerved, acute at apex. Upper lemma linear, 5.4–6 x 0.5–0.7 mm, hyaline, nerveless.

Themeda odishae is currently known from the Mahendragiri hills in the Gajapati district of Odisha state. It grows on rocks at an elevation between 1300 to 1400 m a. s. l., in association with *Arundinella setosa*, *Garnotia tenella* and *Oropetium roxburghianum*. Fl.: September.

Mahendragiri Hills (GJP), *AC & K. Prasad*, 4209 (holotype BSID, isotypes CAL).

Themeda quadrivalvis (L.) Kuntze, Revis. Gen. Pl. 2: 794. 1891; Fischer 3: 1746. 1934; Bor, Grass. India 252. 1960. var. **quadrivalvis.** *Andropogon quadrivalvis* L. in Murr., Syst. Veg. ed. 13. 758. 1774. *Anthisteria ciliata* L. f., Suppl.113. 1781; FBI 7: 213. 1896.

An annual, slender grass; culms up to 1 m high, geniculately ascending and rooting from the lower nodes; nodes glabrous. Leaf sheaths 3-7 cm long, glabrous, ligule a narrow, ciliate membrane; blades 7-15 x 0.3-0.5 cm, lanceolate, flat, base cordate, margins scaberulous, apex acuminate, glabrous, midrib prominent. Spatheolate panicle up to 20 cm long, pedunculate racemes up to 5 cm; spatheole 4 mm, lanceolate. *Involucral spikelets:* 4 mm, oblong, tuberculate pilose, male, whorled;

lower glume oblong, tuberculate pilose on outer surface and margins; upper glume linear; lemma linear, hyaline. *Pedicelled spikelet:* 8 mm, linear-lanceolate, similar to above. *Sessile spikelet:* 7 mm, linear, bisexual; callus pungent; lower glume oblong-obtuse; upper glume lanceolate, hairy; lower lemma oblong, glabrous; upper lemma linear, awned, hispid; stamens 3; styles 2, slender, free, stigmas plumose. Caryopsis linear, grooved on the front.

Common on hill slopes. Fl. & Fr:August- December.

Samarajyamkonda–Gownivaipalli (ATP), *BR* & *ANS* 38478; Horsley hills (CTR), *PVP* 4668; Pamuleru, Rampa hills (EG), *TP & MSG* 12271; Diguvametta (PKM), *CAB* 8094 (MH); Kondapalle (KSN), *CAB* 4599 (MH); Thotamaamidi, near Devarapalli (EG), *NRR* 76819 (BSID); Karaka (VSKP), *CAB* 1702 (MH); Aditavalasa (VSKP), *GVS* 21623 (MH & CAL); Ramial dam site (Dhenkenal), *sine coll.*11108 (BSID); Rajgangpur (Sundargarh), *AC* 10254 (BSID); Betkote, Karlapat Wildlife Sanctuary (KHD), *AC* 2459 (BSID); Digi (GJM), *CAB* 1324 (MH).

INDIA: Drier regions of India.

WORLD: Myanmar, Tropical and South Africa, Indo-China.

Themeda saxicola Bor, Kew Bull. 1951: 452. 1952 & Grass. India 252. 1960; Saxena & Brahmam, Fl. Orissa 4: 2508. 1996.

Perennial, selender grass with a woody root stock; culms up to 35 cm, slender, terete, smooth, glabrous. Leaves mostly clustered at the base; lamina linear-lanceolate, up to 6 cm x 2.5-3 mm wide, strongly attenuate towards apex, glabrous beneath, pilose above with tubercle-based hairs, margin minutely scabrous; upper leaves with shorter or rudimentary, erect lamina; sheaths tight, keeled, glabrous, smooth or pilose between the veins; ligule short, 1 mm long, membranous with ciliate margin above. Raceme solitary, terminating the unbranched culms, with a spathe-like sheath. Pairs of involucral spikelets inserted at different levels, glabrous, spikelets inserted at different levels, glabrous, spikelets unequal in length, 1.1-1.5 cm long; lower glume 1.1–1.5 mm long, obliquely elliptic-acute, dorsally slightly convex, many-nerved, 2-keeled, keels narrowly winged, glabrous, smooth, purple or green; upper glume one-third shorter, hyaline, 3-nerved, smooth, glabrous, margin ciliate. Lower floret male; lemma hyaline, margin ciliate; palea equal in length but narrower; stamens 3; anthers long, callus 2 mm long, densely brown hairy; lower glume oblong, truncate, coriaceous but finally crustaceous, densely brown pilose, 5-nerved, dorsally scabrellous, margins scabrellus; upper glume oblong, truncate, membranous; lower glume empty; lemma triangular, hyaline, 4 mm long, palea 0; upper floret female; lemma strongly awned, base hyaline, 3 cm long, column of awn 2 cm long, brown, scabrous, twisted; styles 2, stigmas plumose. Pedicelled spikelets: pedicel 2.5 mm long, slender, compressed; lower glume 1 cm long, elliptic, acute, many-nerved, sides rounded, margins ciliate; upper glume *c.* 9 mm long, 3-nerved, dorsally rounded, margins ciliate; lower lemma hyaline; palea 0; upper lemma hyaline, short; stamens 3; anthers 3-5-4 mm long.

Koraput, in shallow pockets of soil overlying granitic-gneiss (Bor, 1960).

INDIA: Odisha. Endemic.

Themeda tremula (Nees ex Steud.) Hack, in DC. Monogr. Phan. 6: 667. 1889; Fischer 3:1746. 1934; Bor, Grass. India 254. 1960. *Anthistiria tremula* Nees ex Steud., Syn. Pl. Glumac, 1: 401. 1855; FBI 7: 214. 1896.

A perennial, tufted grass. Culms up to 1 m high, ascending from a creeping rootstock, slender, leafy. Leaf sheaths slightly compressed; ligule membranous; blades 10-40 x 3-6 cm, base broad, margins scabrid, apex acuminate, glabrous. Panicle up to 60 cm long, racemes in fascicles, 8 cm long, distant, enveloped by spathes. *Involucral spikelets:* 4, 8 mm; lower glume linear-lanceolate, tuberculate based hairs on outside, many-nerved; upper glume oblong-lanceolate, 1-nerved, ciliate above the middle, margins incurved; lower lemma narrowly linear, hyaline, 1-nerved; male. *Pedicelled spikelet:* similar to involucral spikelet. *Sessile spikelet:* 2, 4 mm, linear- oblong, obtuse, bisexual; callus small, bearded; lower glume deeply channelled outside, truncate, scabrid, nerveless, apex ciliate awned.

Occasional in northern Eastern Ghats. Fl. & Fr.: November- January.

Chinnaudasakarru (EG), *VMS* 29 & 52 (CAL); Ebul, North of Gudem (VSKP), *VNS* 5759 (CAL).

INDIA: Peninsular India.

WORLD: Sri Lanka.

Themeda triandra Forssk., Fl. Aegypt.-Arab. 123 & 178. 1775; Fischer 3: 1746. 1934; Bor, Grass. India 254. 1960; Matthew, Fl. Tamilnadu Carnatic 3: 1908. 1983. *Anthistiria imberbis* Retz., Observ. Bot. 3: 11. 1783; FBI 7: 211. 1896. *Themeda imberbis* (Retz.) T. Cooke, Fl. Pres. Bombay 2: 993. 1908.

A perennial, tufted grass; culms up to 2 m high, robust, solid; nodes glabrous. Leaf sheaths 4-12 cm long, compressed, glabrous; ligule a ciliate membrane; blades 8-25 x 0.3-0.5 cm, linear, flat, base rounded, margin scaberulous, apex acuminate, scabrid on both surfaces, midrib prominent. Spatheolate panicle up to 30 cm long, lax, bearing clusters of racemes up to 3 cm, distant; spatheolates 1.5 cm, lanceolate, pilose at base. *Involucral spikelet:* 4, 4 mm, oblong, tubercle-based hairs at apex, whorled, all arise at same level, male; lower glume oblong, ridge, keel winged, ciliate, coriaceous; upper glume linear, margins incurved; lemma linear. *Pedicelled spikelet:* 8 mm, linear-lanceolate, similar to above. *Sessile spikelet:* 6.5 mm, linear; callus densely pilose with yellow hairs; lower glume oblong, crustaceous, apex scabrid outside; upper glume oblong, obtuse, keeled; lower lemma oblong, acute, hyaline, ciliate at apex; upper lemma stipiform, awned, hispid, ciliate at apex; stamens 3; stigmas plumose. Caryopsis linear, grooved on the front.

Occasional grass in hilly areas slopes and in plains near the streams. Fl. & Fr.: August-December. Tam.: *Erighai thattupul.*

Batrepalli (ATP), *KRKS* & *BSS* 40337; Ahobilam (KNL), *BR* & *BSS* 29314; Horsley hills (CTR), *MSG & AA* 12218, *PVP* 4668; Talakona RF (CTR), *PVP* 2320; Pinjarikonda (EG), *TP & MSG* 12287; Ramanapenta (KNL), *JLE* 32624 (CAL & MH); Palamaneru (CTR), *KCJ* 438 (MH); Isukagundalu, GBM WLS (PKM), *J.Swamy* & *L.Rasingam* 7591 (BSID); Loya, Kondapalle (KSN), *PV* 5876 (MH); On the way to Papi hills (WG),

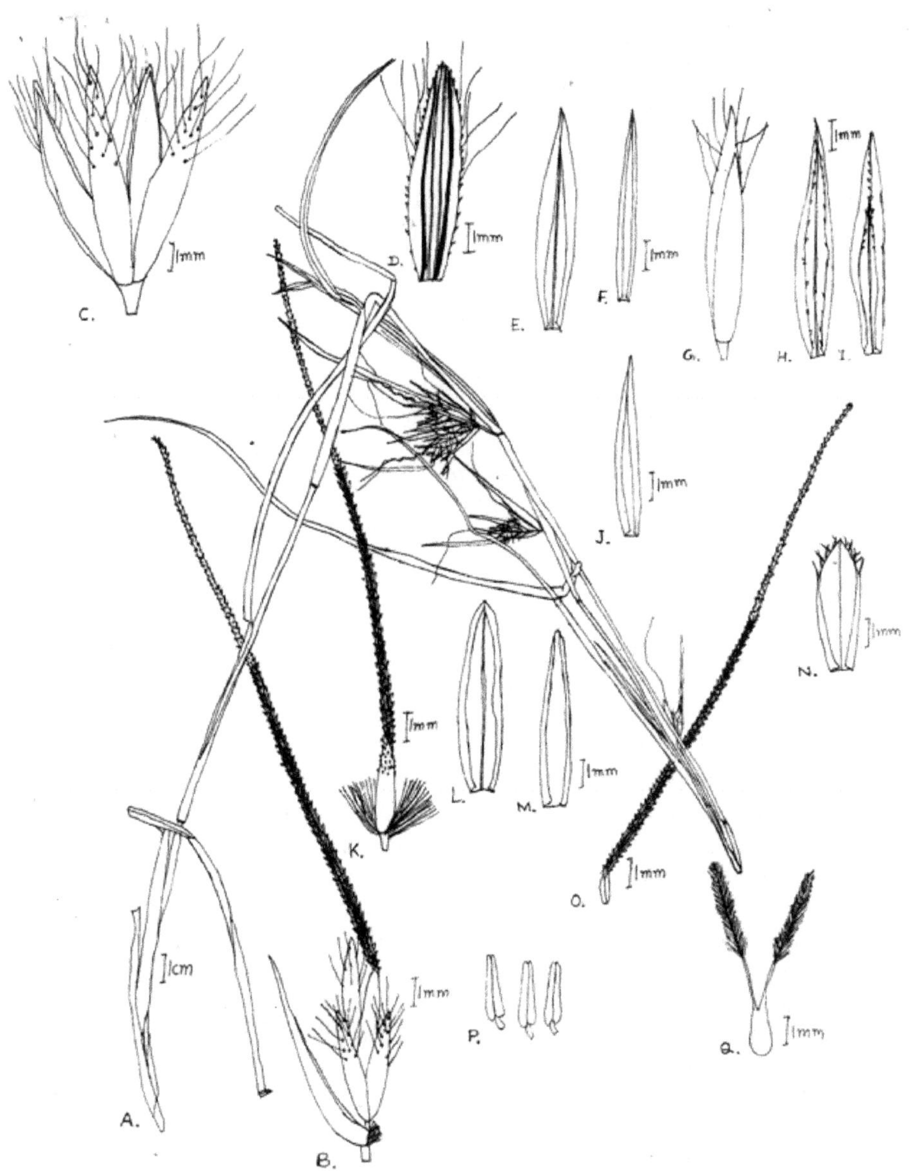

Figure 107. **Themeda triandra** Forssk.

A. Twig, B. Spikelets, C. Involucral spikelets, D. Lower glume, E. Upper glume, F. Lemma, G. Pedicelled spikelet, H. Lower glume, I. Upper glume, J. Lemma, K. Sessile spikelet, L. Lower glume, M. Upper glume, N. Lower lemma, O. Upper lemma, P. Stamens, Q. Pistil.

DCSR 452 (CAL); Bison hills (EG), *CAB* 5066 (MH); Rayapalli (EG), *VNS* 17 (CAL); Ebul RF, North of Gudem (VSKP), *VNS* 558 (CAL); Araku valley (VSKP), *NPBK* 589 (CAL); Mallelatheertham (MBNR), *SRS* 108857, 111412 (BSID); Sirumalai hills (DGL), *SKS* 344 (SGH); Kaveri peak, Yercaud (SLM), *DBD* 31253 (MH); Kalpadai (SA), *S.India Flora* 17862 (MH); Melpat (SA), *CAB* 1063 (MH); Baghmunda, Satkosia Tiger Reserve (AGL), *KCM* 7003 (BSID); Belghar, Kotgarh Wildlife Sanctuary (KDM), *AC* 2416 (BSID); Dimili, Kotgarh Wildlife Sanctuary (KDM), *AC* 2418, 2419 (BSID); Jenabil range, Similipal B.R. (MBJ), *AC* 2312, 2344 (BSID); Purnakote, Satkosia Wildlife Sanctuary (AGL), *AC* 2379 (BSID); Kabrapali (SBP), *AC* 10237 (BSID); Dimli, Kotgarh Wildlife sanctuary (KDM), *AC* 2427 (BSID).

INDIA: Hotterr parts of India.

WORLD: Warm and Tropical regions of Old World.

Themeda villosa (Poir.) A. Camus in Lecomte, Fl. Indo–Chine 7: 364.1922; Bor, Grass. India 254. 1960; Moulik, Grass. Bamb. India 1: 234. 1997; B.R.P.Rao *et al.*, J. Econ. Taxon. Bot. 36: 383. 2012. *Anthistiria villosa* Poir. in Lam., Encycl. Suppl. 1: 396. 1814. *Themeda gigantea* (Car.) Hack. subsp. *villosa* (Poir.) Hack. in DC. Monogr. Phan. 6: 675.1889; FBI 7: 217. 1896.

Perennials; culms erect or ascending, tufted, to 3 m high, smooth, glabrous, polished, slightly compressed. Leaf sheaths 5-12 cm, smooth and glabrous, keeled, auricled at the mouth; ligule 2 mm, membranous; leaf blades 30-70 x 1-1.5 cm, linear-lanceolate, acuminate, tapering to a narrow base, usually glabrous, margins scaberulous, midrib white. Inflorescence a single raceme embraced by a boat-shaped spatheole, these solitary racemes gathered into a narrow compound panicle. Racemes with 2 large involucral homogamous pairs at the base, 1 or 2 sessile spikelets and pedicelled spikelets. Peduncle terete, 8-15 mm, villous at the tip. Involucral spikelets: four, homogamous, male, inserted in pairs at different levels, each spikelet 10-15 x 0.7 mm, lanceolate, acuminate, puberulous; lower glume 10-13 x 1.7 mm, lanceolate, many-nerved, 2-keeled, one narrowly winged, margins inflexed, keels scabrid, puberulous; upper glume 8-10 x 1 mm, lanceolate, acute, hyaline, 3-nerved, ciliate on the margins, nerves and back scaberulous. Florets 2; lemmas 6-8 x 1mm, lanceolate, acute, ciliate on the margins; paleas oblong-lanceolate, 5-7 x 1 mm, hyaline, ciliate; stamens 3, anthers 4.5 mm. *Sessile spikelets*: oblong-lanceolate, 5-8 x 2.5 mm, bisexual, hairy, callus densely hispid with short brown hairs; lower glume 5-8 x 2.3 mm, oblong-lanceolate, tip truncate, margins inflexed, keels rounded, densely covered on the back and sides with appressed short hairs, 7-nerved; upper glume 5-7 x 1.5 mm, narrowly lanceolate, 3-nerved, tip obtuse, densely hairy on the dorsal surface. Florets 2; lemmas 6 x 1.5 mm, oblong-lanceolate, hyaline, ciliate; paleas minute; stamens 3, anthers 3mm; ovary ovate-oblong, 1.5mm. *Pedicelled spikelet:* linear-lanceolate, male; pedicels 2 mm, ciliate hispid on angles; lower glume 12-15 x 1.7 mm, narrowly lanceolate, acuminate, scabrid on the nerves, sparsely hairy on the dorsal surface, 14-nerved; upper glume 10-13 x 1.5 mm, lanceolate, acuminate, translucent, scabrid on the margins, 7-nerved, almost aristate. Florets 2; lemmas 5-7 x 1.5 mm, lanceolate, acute, hyaline, ciliate on the margins; paleas lanceolate, 5-6 x 1 mm, hyaline; stamens 3, anthers 3 mm.

Rare, along water courses and on exposed slopes in Visakhapatnam district. Fl. & Fr.: June–December.

G. Madugula (VSKP), *AMR* & *BR* 24600.

INDIA: NE. India, West Bengal, Odisha, Andhra Pradesh.

WORLD: SE. Asia.

THYSANOLAENA Nees

Thysanolaena latifolia (Roxb. ex Hornem.) Honda, J. Fac. Sci. Univ. Tokyo Sect. 3, Bot. 3: 312. 1930. *Melica latifolia* Roxb. ex Hornem., Hort. Bot. Hafn. Suppl. 117. 1819. *Thysanolaena maxima* (Roxb.) Kuntze, Revis. Gen. Pl. 2: 784, 1891; Fischer 3: 1797. 1934; Bor, Grass. India 650. 1960; Matthew, Fl. Tamilnadu Carnatic 2:1909. 1983; Saxena & Brahmam, Fl. Orissa 4: 25122. 1996. *Agrostis maxima* Roxb., Fl. Ind. 1: 319 1820. *Thysanolaena agrostis* Nees, Edinb. New Philos. J. 18: 180. 1835; FBI 7: 61. 1896.

A perennial, tufted grass; culms up to 3 m high, woody, polished or shining; nodes glabrous. Leaf sheaths glabrous, ciliate near the mouth; ligule small, ciliate; leaf blades 20-50 x 2-4 cm, lanceolate, base cordate and semi-amplexicaul, margins scaberulous, apex acuminate, glabrous. Panicle up to 60 cm long, large, effuse, drooping; branches capillary; spikelets 2 mm, ovoid-lanceolate, paired, pedicelled, 2- or 3-flowered, lower sterile, median bisexual, upper male or sterile; glumes subequal, ovate, hyaline, 1-nerved; lemma lanceolate, membranous, 1-nerved, epaleated, empty; upper lemma ovate-lanceolate, ciliate with long white hairs; stamens 2; ovary ovoid, biconvex, styles 2, free, slender, stigmas plumose. Caryopsis ellipsoid, reddish-brown

Fairly common on shady slopes in the forests, especially on damp steep banks along the ravines and streams. Fl. & Fr.: March- August. Tel.: *Konda cheepuru.*

Maredumilli (EG), *TP & MSG* 12255; Near Roygudem (WG), *DN* 85504 (BSID); Marripakala (EG), *GVS* 27422 (CAL); Yarlagadda (EG), *VNS* 155 (CAL); Chinturu, Maredumilli (EG), *MM* 100708 (MH); Sunkarimetta (VSKP), *NPBK* 697 (CAL); Way to Chintapalli (VSKP), *JLE* 37131 (MH); Lambasingi (VSKP), *S.India Flora* 11796 (MH); Chinagora RF (VSKP), *KCJ* 17201 (MH); Punyagiri hill (VJN), *GVS* 19428 (CAL); way to Odrubangi, G.W.Puram (VJN), *NRR* 83626 (BSID); Boragharo (GJM), *VNS* 5935 (MH); Digi (GJM), *CAB* 1325 (MH); Mahendragiri (GJM), *JSG* 13906 (MH).

INDIA: Wamer regions in low hills.

WORLD: SE. Asia.

TRACHYS Persoon

1. Clusters of spiketlets 2 or 3 at each rachis internodes .... **T.muricata**
1. Clusters of spikelets single at each rachis internodes .... **T. deccanensis**

Trachys deccanensis M. Anil Kumar & B.R.P. Rao, Rheedea 27(2): 79. 2017.

Stoloniferous annual or short-lived perennial. Culms to 80 cm long; nodes hairy, rooting from lower nodes. Leaves elliptic to linear-lanceolate, 4–11 x *c.*

0.8 cm, obtuse to rounded, margins undulate, apex sharply acute to acuminate, flat, glabrous; ligules 1–2 mm long, hyaline, membranous; leaf sheath laterally compressed, sharply keeled in lower leaves, 3–10 cm long, enclose up to half of internode length, glabrous, margins thin-hyaline, mouth with few tuberculate hairs or glabrous. Racemes spike-like, three in digitate, 6–9 cm long, spikelets in clusters, arranged alternately on the under surface of rachis without any vestigial structures, clusters of spikelets arranged apart 4.5–6 mm distance, pubescent at the divergence zone; joints triquetrous, 4.5–6 x 1.2–2.2 mm, straight to curved, margins winged, glabrous, green, many-veined; peduncles 6–12 cm long, terete, bearded near apex. Raceme joints bulged, 2–3 mm across, comprising one cluster of spikelets; cluster size 6.5–7 x 2–4 mm, each cluster with three bisexual spikelets which are similar in size, dissimilar in texture; entire cluster disarticulating at maturity. Spikelets narrowly lanceolate, 5.3–5.6 x 2–3 mm, base truncate, apex acute-acuminate, coriaceous, yellow-brown; pedicels stout, 1–1.5 mm, greenish yellow. Glumes shorter than spikelets; lower glume reduced, scale-like, 0.2–0.5 x 0.2–0.3 mm, base truncate, apex obtuse, subcoriaceous, veinless; upper glume lanceolate, 2–3.7 x 0.5–0.8 mm, base truncate, apex obtuse, chartaceous, glabrous or sparsely ciliate at margins, green, prominently 3-veined; spikelets with 2 florets; lower barren with 2 lodicules; upper bisexual; lower sterile lemma lanceolate, 3.5–5.5 x 2–3 mm, base truncate, apex obtuse, coriaceous, yellow-brown, inconspicuously 9–11-veined, margins thick, in-rolled; lower palea oblong, often reduced, 0.3–0.5 x 0.2–0.3 mm; upper lemma lanceolate, 3-3.2 x 1–2 mm, base truncate, apex obtuse, hyaline, 3-veined, margins in-rolled, envelop the margins of palea; upper palea lanceolate, similar to its lemma in shape, 3–3.1 x 1–1.5 mm, thin-chartaceous; lodicules 2; stamens 3; anthers 1–1.2 mm long, yellow; ovary oblong, *c.* 0.7 x 0.3 mm; style *c.* 2 mm long; stigma plumose, 1.5–2 mm long, light brown. Caryopsis not seen. Pedicelled spikelets similar to sessile spikelets in size and sex, but are hairy, hairs from dorsal side of lower sterile lemmas, usually in tow prominent lines in the middle.

Rare, found in rainfed cultivated fields associated with *Trachys muricata, Alloteropsis cimicina* and *Digitaria* spp. Fl. & Fr.: June–December.

Cultivated fields of Kadivella village near Yemmiganur (KNL), *BRR, K.V. Subbaiah, M. Anil Kumar* 52099 (SKU).

INDIA: Andhra Pradesh (Endemic).

Trachys muricata (L.) Pers. ex Trin., Mem. Acad. St. Petersbourg Ser. 6. Sci. Maths. 3: 189. 1834; Fischer 3:1813. 1934; Bor, Grass. India 369. 1960; Matthew, Fl. Tamilnadu Carnatic 3: 1910. 1983. *Cenchrus muricatus* L., Mant. 2: 302. 1771. *Trachys mucronata* Pers., Syn. Pl. 1:85. 1805; FBI 7: 96. 1896.

An annual, diffuse grass; culms up to 60 cm, weak, ascending or prostrate, rooting from lower nodes; nodes villous. Leaf sheaths 3-5 cm, loose, villous; ligule a thin membrane; blades 4-10 x 0.5-1.5 cm, lanceolate, flat, base rounded, apex acute, villous on both surfaces. Spiciform raceme, paired, up to 7 cm long; rachis angular, clusters of spikelets sunk in cavity of rachis; spikelets intermixed with small scale like empty glumes; spikelets 6 mm, globose; glumes unequal, lanceolate; lower coriaceous; upper hyaline; lower lemma broadly ovate, coriaceous, many-nerved,

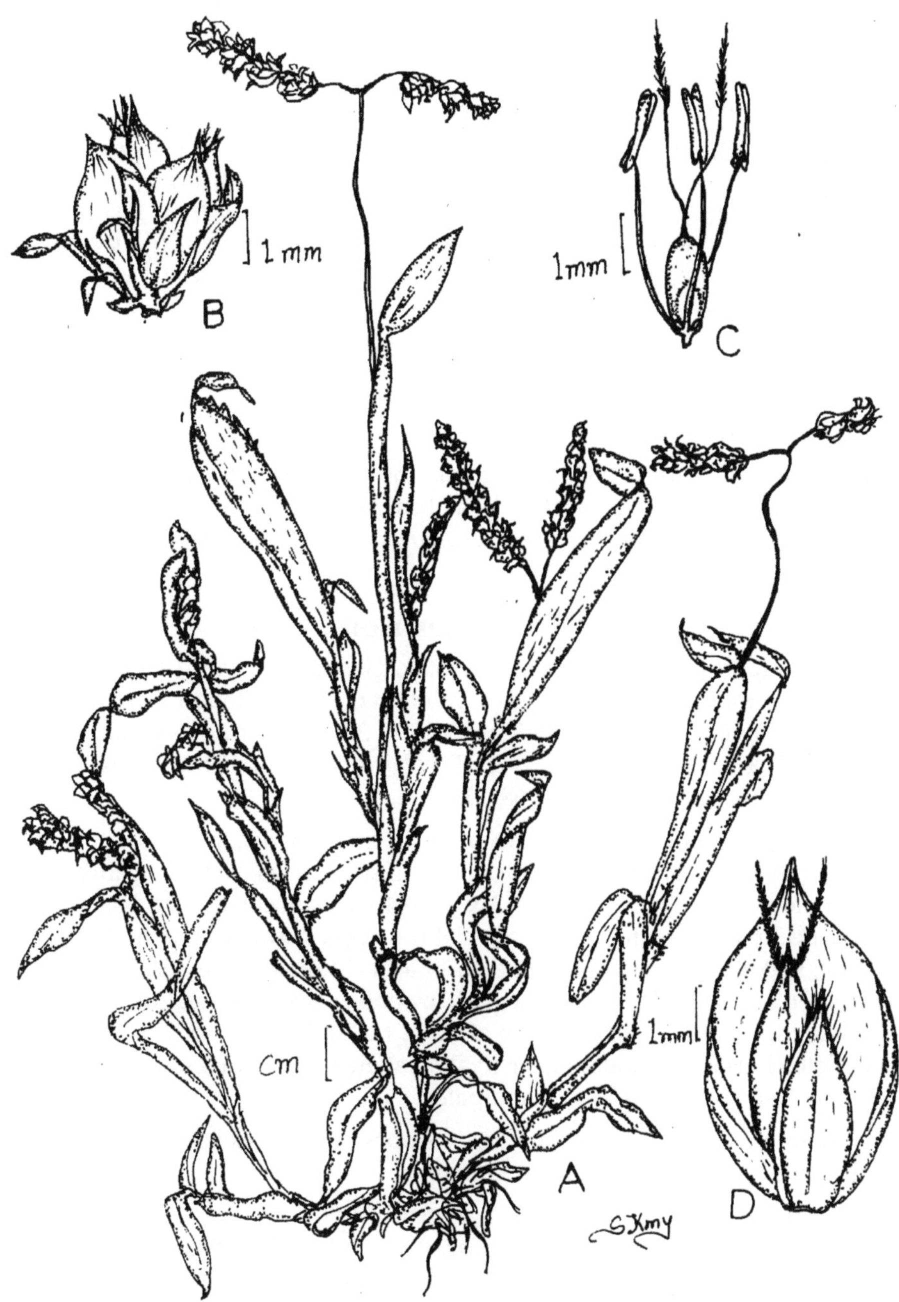

Figure 108. **Trachys muricata** (L.) Pers. ex Trin.

A. Habit, B. Spikelet, C. Pistil & Stamens, D. Flower.

cuspidately acute; palea lanceolate, hyaline, nerveless, empty; upper lemma ovate lanceolate, chartaceous, 3-nerved; palea lanceolate, nerveless, bisexual; stamens 3; stigmas plumose. Caryopsis broadly ovate.

Common weed in waste lands and in cultivated fields of all districts. Fl. & Fr.: August-November. Tam.: *Vennaithirattipul.*

Kalasamudram RF (ATP), *NY* 650; South Dhone RF (KNL), *RVR* 1599; Thummalapalle (KDP), *KRKS* & *KRS* 36978; V.P South (GNT), *VRK* 5794; Mekalavanikunta (CTR), *GVS* 46841 (MH); Palamaneru (CTR), *KCJ* 466 (MH); Somasila (NLR), *P.Venu* 113611 (BSID); Ethamukala, Ongole (PKM), *KR* & *RG* 88851 (BSID); Vandalur (CPT), *S.India Flora* 11547 (MH); Tippukadu RF (NA), *KRM* 17688 (MH); Devanpatnam (SA), *VNS* 5107 (MH); Hogainakkal RF (SLM), *EV* 21906 (MH); Tikarpada section, Satkosia Tiger Reserve (AGL), *KCM* 6119 (BSID).

INDIA: South India.

WORLD: Myanmar.

Note: Persoon (Syn. Pl. 1:85. 1805) made the combination *Trachys mucronata* by citing *Cenchrus mucronatus.* L. as basionym. There is no Linnean species C. *mucronatus* but, there is a *C. muricatus* L. Trinius (1834) corrected this mistake.

TRAGUS Heller *nom. cons.*

Tragus mongolorum Ohwi, Acta Phytotax. Geobot. 10: 268. 1941. *T. roxburghii* Panigr., Kew Bull. 29: 495. 1974; Matthew, Fl. Tamilnadu Carnatic 3: 1910. 1983; Pullaiah, Fl. Andhra Pradesh 5: 2296. 2018. *T. racemosus* Hook., f., Fl. Brit. India 7: 97. 1896 non Scop. 1777. *T. biflorus auct.* non Schult. 1824; Fischer 3: 1813. 1934; Bor, Grass. India 682. 1960.

A perennial, tufted grass; culms up to 25 cm high, decumbent or geniculately ascending from the base; nodes glabrous. Leaf sheaths 2-4 cm long, glabrous; ligule membranous ciliate rim; blades 1-3 x 0.2-0.4 cm, lanceolate, flat, base rounded, margin pectinately ciliate, apex acute, glabrous. Spiciform racemes up to 4 cm long, solitary; spikelets 3 mm, lanceolate, 2, in a deciduous subsessile clusters facing each other, 1-flowered; glumes unequal; lower 0 or scarious, hyaline; upper glume lanceolate, prickled, prickles terete at base, hooked at the apex; lemma oblong-lanceolate, chartaceous; palea linear, hyaline; stamens 3; stigmas plumose. Caryopsis ellipsoid, free within the lemma and palea.

Common in waste plains and lower slopes of Rayalaseema districts and rare in Coastal districts. Fl. & Fr.: July-September

Samathagram (ATP), *NY* 198; Batrepalli (ATP), *BR* & *P. Priyadarsini* 37070; North Dhone RF (KNL), *PVP* 2695; Gani RF (KNL), *RVR* 1525; Orampad (KDP), *PVP* 3482; Thummalapalli (KDP), *KRKS* & *SSR* 39024; Kondapuram (NLR), *PMR* 19551; Anupu (GNT), *VRK* 3508; Gooty (ATP), *JSG* 16390 (MH); Agasthyeswarakona, Lingapuram (KDP), *J.Swamy* & *S.Nagaraju* 14716 (BSID); Near Inakurthi (NLR), *JLW s.n.* (CAL); Devarapalli, Maredumilli (EG), *MM* 105064 (BSID); Hogainakkal (SLM), *EV* 20527 (MH).

Figure 109. **Tragus mongolorum** Ohwi.
(Syn.: *Tragus roxburghii* Panigr.)

A. Twig, B. Spikelet, C. Single spikelet, D. Upper glume, E. Lemma, F. Palea, G. Pistil and stamens, H. Caryopsis.

INDIA: Dry plains of India.

WORLD: SE. Asia and E. Africa.

TRIPOGON Roemer & Schultes

1. Lemmas 4-lobed .... **T. bromoides**
1. Lemmas 2-lobed:
    2. Median awn as long as or longer than the lemma:
        3. Spikelets 1-4 cm long; paleas broadly winged:
            4. Culm less than 30 cm tall; spikelets many-flowered .... **T. wightii**
            4. Culm more than 70 cm tall; spikelets 4-12-flowered .... **T. trifidus**
        3. Spikelets not more than 1 cm long; paleas narrowly winged:
            5. Plant up to 75 cm tall; leaf blades 20-30 cm long; inflorescence 15-20 cm long, upper glume unlobed unawned .... **T. tirumalae**
            5. Plant up to 100 cm tall; leaf blades 30-70 cm long; inflorescence 30-35 cm long, upper glume unlobed, unawned .... **T. uma-ganeshii**
    2. Median awn shorter than the lemma or absent:
        6. Leaf blades filiform, convolute:
            7. Leaves glabrous; spikelets with 8-20-flowered .... **T. jacquemontii**
            7. Leaves covered with white hairs; spikelets with 4 or 5-flowered .... **T. mahendragiriensis**
        6. Leaf blades linear to linear-lanceolate .... **T. purpurascens**

Tripogon bromoides Roem. & Schult., Syst. Veg. 2: 600. 1817; FBI 7:287. 1896; Fischer 3: 1834. 1934; Bor, Grass. India 521. 1960; Matthew, Fl. Tamilnadu Carnatic 3: 1911. 1983; Moulik, Grass. Bamb. India 2: 622. 1997.

Perennials, culms erect, tufted, to 30 cm high, nodes glabrous. Leaf sheaths glabrous, striate, bearded at mouth; ligules very short, membranous, ciliate; leaf blades 5-25 x 0.2-0.4 cm, linear, base cordate, margins scabrid, apex acuminate, hairy. Spike terminal, solitary, up to 20 cm long, secund; rachis flattened, margins scabrid; spikelets sessile, 6-13-flowered; lower glume oblong-lanceolate, chartaceous, 1-nerved, shortly awned, keel scabrid; upper glume lanceolate, ovate-oblong, 2-lobed, shortly awned, 1-nerved, keel glabrous; lemmas ovate, chartaceous, 3-nerved, 3-awned; paleas elliptic-oblong, keeled, 2-nerved, hyaline; stamens 3; ovary elongate, styles 2, slender, free, stigmas feathery. Caryopsis oblong, orange coloured.

Common on hill slopes and forest plains in Srikakulam, Vizianagaram and Rayalaseema districts. Fl. & Fr.:August-February.

Kekathi RF (ATP), *TP* & *NY* 786; Penukonda (ATP), *JNP* & *AL* 27348; Rachakuntapalli (KDP), *KRKS* 41812; Horsley hills (CTR), *PVP* 3500, *MSG* &*AA* 12222; Pachipenta (VJN), *AMR* 24543; Kadapa, *JSG* 21325 (DD); 4 km from Anantagiri (VSKP), *GVS* 44297 (MH); Araku valleuy (VSKP), DDSR 21885 (MH); Irlapenta (MBNR), *BR* & *BSS* 32289 (BSID); Mahendragiri (GJM), *VNS* 6063 (MH).

INDIA: Peninsular India.

WORLD: Sri Lanka, Nepal.

Tripogon jacquemontii Stapf, Kew Bull. 85.1892; FBI 7: 286. 1896; Fischer 3: 1833. 1934; Bor, Grass. India 522.1960; Moulik, Grass. Bamb. India 2: 622.1997.

Perennials, culms erect, tufted, to 90 cm high; nodes glabrous. Leaf sheaths glabrous; ligules fimbriate; leaf blades 5-20 cm long, linear, base narrow, apex acuminate, glabrous. Spikelets oblong, 8-20-flowered; lower glume ovate-lanceolate, 1-nerved, lobed on one side, sub coriaceous; upper glume ovate-lanceolate, 1-nerved, chartaceous; lemma elliptic-obovate, awned; palea lanceolate, hyaline, 2-keeled, winged; stamens 3. Caryopsis oblong.

Rare in barren lands and forest plains. Endemic to Peninsular India. Fl. & Fr.: September–December.

Horsley hills (CTR), *MSG* & *AA* 1222; Mallelatheertham (MBNR), *BR* & *BSS* 32217.

INDIA: Western India from Gujarat to the Deccan. Endemic.

Note: Although, Mitra (1958) mentioned the taxon occurrence in Assam, but has not been recorded by Shukla (1996). Further, Ahmedullah & Nayar (1986) and Nayar (1996) considered the taxon endemic to Peninsular India.

Tripogon mahendragiriensis Chorghe *et al.*, Nordic. J. Bot. 33: 655. 2015

Tufted perennial; culms 15–20 cm tall, erect; nodes glabrous. Leaf sheaths 3–5 cm long, glabrous; ligules 0.5 mm long, hairy; leaf blades 4–10 x 0.05 cm, linear, filiform, sparsely covered with white hairs. Inflorescences 5–10 cm long, spiciform; rachis flat, glabrous; peduncles 7–12 cm long. Spikelets 4.5–5.2 x 1.0 mm, linear-lanceolate, laterally compressed, 4- or 5-flowered with reduced florets near apex; lower glumes 2.00–2.25 x 0.50 mm, linear-lanceolate, shallowly lobed on one side; upper glumes 3.5–3.7 x 0.5 mm, linear-lanceolate; lemmas 2.8–3.1 x 0.7–0.9 mm, lanceolate, bearded at base, 3-nerved, keeled, unlobed, 3-awned, with median awn 1.8–2.2 mm long, shorter than the lemma, arising from the median nerve, and the other 2 awns arising from the lateral nerves, 0.5 mm long; palea 2.6–2.8 x 0.6 mm, oblanceolate, 2-keeled, with keels ciliate and hyaline; stamens 2; anthers 1.3 mm long; styles 2; stigma plumose; ovary 0.4 mm long. Caryopsis oblanceolate, 1.5 mm long.

Tripogon mahendragiriensis is currently only known from the Mahendragiri Hills in the Gajapati District of Odisha State. It was found growing on rocks above 1400 m a.s.l., associated with *Arundinella setosa, Garnotia tenella* and *Jansenella griffithiana*.

INDIA: Odisha. Endemic.

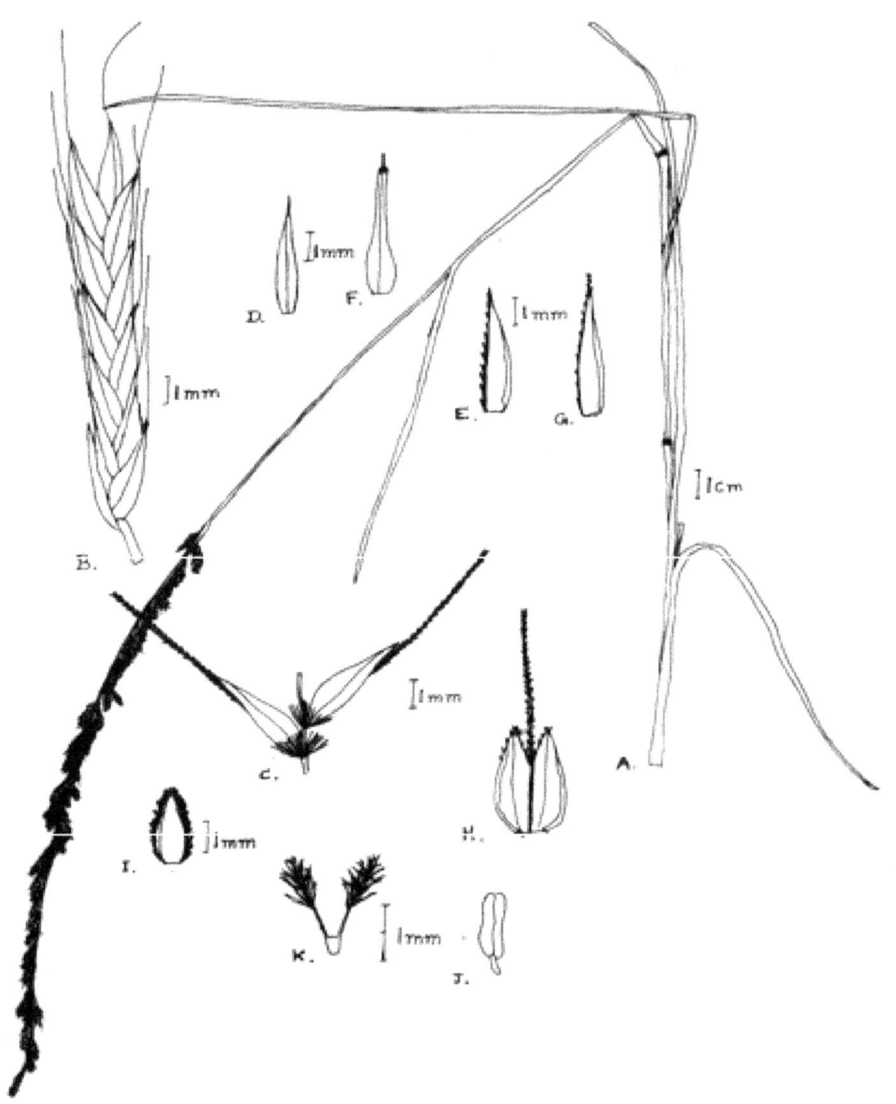

Figure 110. **Tripogon jacquemontii** Stapf

A. Twig, B. Spikelet, C. Open spikelet, D & E. Lower glume front and side views, F & G. Upper glume front and side views, H. Lemma, I. Palea, J. Stamen, K. Pistil.

Tripogon purpurascens Duthie, Ann., Roy. Bot. Gard. Cal. 9:75. 1901; Bor, Grass. India 522. 1960; Moulik, Grass. Bamb. India 2:623. 1997; B.R.P. Rao *et al.*, J. Econ. Taxon. 33: 435. 2009. *T. abyssinicus* Hook. f., Fl. Brit. India. 7: 287. 1896 in part non Nees (1854).

Perennials, culms erect, tufted, slender, to 15 cm high; nodes glabrous. Leaf sheaths glabrous; ligules ciliate, membranous; leaf blades filiform, convolute, tip obtuse with purple colour, densely hairy near the base. Inflorescence terminal, contracted; spikelets 2-3-flowered; lower glume small, triangular, 1-nerved, slightly awned; upper glume lanceolate, 1-nerved, membranous, central thickened, coriaceous, margin hyaline, ciliate upper portion; lemmas 3, 3-nerved, cleft, median nerve slightly awned in the cleft, margin hyaline; paleas 3, 2-keeled, tip 2-lobed except third palea; stamens 2-3. Caryopsis linear.

Occasional in dry and moist rocky habitats of forests. Fl. & Fr.: July-November.

Kalasamudram (ATP), *KRKS* 39175, *BSS & KRKS* 40458; Batrepalli (ATP), *KRKS* 39573; Devaragattu (KNL), *BR* & *SS* 21221; Pachipenta (VJN), *AMR* 24543; Kuntimaddi (ATP), *AL* & *BSS* 27357; Kalasamudram (ATP), *KVS* & *BSS* 29702; Amrabad (MBNR), *BSS* & *SKB* 32201; Appapur (MBNR), *BSS* & *SKB* 32220.

INDIA: NW to C Himalayas, Andhra Pradesh.

WORLD: Saudi Arabia, Yemen, Afghanistan, Pakistan, Bhutan, China.

Tripogon tirumalae Chorge *et al.*, Phytotaxa 131: 17. 2013.

Tufted perennials, culms 60-75 cm high, erect; nodes glabrous, 1 mm long. Leaf blades 20-30 x 0.1-0.2 cm, linear-lanceolate, filiform, glabrous; ligules 1 mm long, truncate, membranous, ciliate at apex; leaf sheaths 6-8 cm long, glabrous. Inflorescence 15-20 cm long, 0.5-0.7 mm wide, spiciform; rachis flat, glabrous; peduncle 15-20 cm long; spikelets 7-7.5 x 1.8-2 mm, linear-lanceolate, laterally compressed, 9-10-flowered with reduced florets near apex; lower glumes 2.5-3 x 1 mm, linear-lanceolate, shallowly lobed on one side; upper glumes 3.5-4 x 1 mm, linear-lanceolate; lemmas 2.8-3.1 x 1.2 mm, oblanceolate, bearded at base, 3-nerved, 1-keeled, 2-lobed, 1-awned, the awn 4-4.2 mm long, longer than the lemma, arising from sinus; paleas 2 x 0.5 mm, shorter than the lemma, oblanceolate, 2-keeled, the keels ciliate, hyaline; stamens 3, anthers 1 mm long; styles 2, stigmas feathery; ovary 0.1 mm long. Caryopsis not seen.

Grows on rocks at elevations between 800-900 m in Seshachalam Biosphere Reserve and Penchalakona of Veligonda hill ranges. Fl. & Fr.: September.

Tirumala hills, near Kumaradhara Pasupudara Dam (CTR), *L.Rasingam, M.S.Rao* & *AC* 2914 (holotype CAL, isotype BSID); Penchalakona hills (NLR), *L. Rasingam, M.S.Rao* & *AC* 2938 (BSID).

INDIA: Andhra Pradesh, Endemic.

Tripogon trifidus Munro ex Stapf, Kew Bull. 1892: 85.1892; FBI 7: 286.1896; Bor, Grass. 524.1960; Moulik, Grass. Bamb. India 2: 622.1997; B.R.P. Rao *et al.*, J. Econ. Taxon. Bot. 36: 383. 2012.

Tufted perennials; culms to 70 cm high. Leaf sheath 5-10 cm long, glabrous; ligule a ciliate rim, 0.2 cm; leaves linear-lanceolate, filiform, 5-30 x 0.2-0.4 cm, acute to attenuate at apex, glabrous. Raceme spiciform, erect or curved towards apex, 10-20 cm long; rhachis flat to triquetrous, glabrous; spikelets appressed, solitary, oblong-linear or lanceolate, 7-9 x 1.2-1.5 cm; glumes persistent; lower glume linear-lanceolate, 3.5 x 0.5 mm, asymmetrical, apex acute with a lobe below middle, mucronate, chartaceous, 1-nerved; upper glume linear-lanceolate, 4 x 5-1 mm, acute and mucronate at apex, chartaceous, 1-nerved. Florets 4-12; lemmas ovate-lanceolate, 2.2-2.5 x 0.8-1 mm, 2-lobed at apex, chartaceous, 3-nerved, awned from sinus, awn 2.5-4 mm long; lobes unawned, acute-acuminate; paleas linear-oblong, 2.5-0.3 mm, obtuse, membranous, 2-nerved, 2-keeled, keels scabrellate, narrowly winged; stamens 3; anthers *c.* 1mm long, yellowish.

Occasional in Lankamala hills among rocks. Fl. & Fr.: September-March

Lankamalla (KDP), *BSS & SKB* 30226.

INDIA: N.E.India, Andhra Pradesh.

WORLD: Bhutan, Nepal, China, Myanmar, Thailand, Vietnam.

Tripogon uma-ganeshii B.R.P.Rao & M.Anil Kumar, Indian J. Forestry 41: 97. 2018.

Tufted perennials, with dense basal leaves; culms up to 100 cm tall, erect; nodes glabrous, 1 mm long. Leaf blades 30-70 x 0.2-0.4 cm, linear, basal flat or margins slightly involute; cauline leaves infolded at times, glabrous, ligule a narrow ciliate membrane; leaf sheaths 8-12 cm long, glabrous. Inflorescence 30-35 cm long, 0.5-0.8 mm wide, spiciform; rachis flat, triquetrous, glabrous; peduncle 25-35 cm long. Spikelets leaden grey coloured, densely crowded, overlapping on primary axis, 1-1.6 x *c.* 0.2 cm, terete, with closely arranged florets, spreading and fragile at maturity, 13-20-flowered. Glumes 2, sub-equal, glabrous, membranous, easily fragile when dry; lower glumes 4-5(6) x *c.* 1 mm, linear-lanceolate, apex acuminate to shortly mucronate, upper glume 5-6.5 x *c.* 1mm, linear-lanceolate, apex bi-lobed, shortly mucronate, mucro from sinus. Lemmas similar in shape, but not in size, the size of lemmas gradually decreasing towards apex, lower fertile lemmas 4-4.5 x 1.2-1.5 mm, oblanceolate, bearded at base, hairs as long as internodes, 3-nerved, 2-lobed, 3-awned, the principal lemma awn 4.5-5.5 mm long, equal or slightly longer than lemma, arising from sinus, scabrid; lateral awns from the apex of lobes, shortly exserted, scabrid. Palea 2-2.5 x 0.6-0.8 mm, shorter than lemma, elliptic-lanceolate, 2-keeled; keels narrowly winged, wings ciliate, hyaline; stamens 3; anthers 1 mm long; styles 2; stigmas feathery; ovary *c.* 0.3 mm long. Caryopsis cylindrical, slightly bulged near base, 1.5 mm long, olive-yellowish brown in colour.

Rare on the rocky hill slopes associated with *Eragrostiella bifaria* Bor, *Heteropogon contortus* Roem. & Schult. and *Tripogon wightii* Hook.f. Fl. & Fr.: August–September.

Horsley hills (CTR), *BR & M. AnilKumar* 51980 (holotype SKU; isotype BSID).

INDIA: Andhra Pradesh. Endemic.

Tripogon wightii Hook. f., Fl. Brit. India 7:286. 1896; Fishcer 3: 1833. 1934; Bor, Grass. India 524. 1960; Moulik, Grass. Bamb. India 2: 623. 1997.

Perennials, culms erect, tufted, to 30 cm high; nodes glabrous. Leaf sheaths glabrous; leaf blades filiform. Inflorescence spike; spikelets oblong, many-flowered; glumes linear-oblong, hyaline, 2-lobed, mucronate; lemma narrow, base bearded, 2-lobed, awned in the cleft; palea oblong, winged, scabrid; stamens 3. Caryopsis narrow.

Rare in rock crevices and hill slopes in Horsley hills. Endemic to Peninsular India. Fl. & Fr.: August–December.

Horsley Konda (CTR), *JSG* 20920 (CAL & DD).

INDIA: South India, Endemic.

Note: Till date, the cited specimen is the only record from Andhra Pradesh.

TRIPSACUM Linnaeus

Tripsacum laxum Nash, N. Amer. Fl 17: 81. 1909; Bor, Grass. India 268. 1960; Matthew, Fl. Tamilnadu Carnatic 3: 1912. 1983.

A perennial, tufted grass; culms up to 4 m high. Leaf sheaths 20 cm long; ligule membranous; blades 50-100 x 4-9 cm, broadly lanceolate. Spikes 3, monoecious, male terminal and axillary, 5 cm long, falling entire; female basal, 8 cm long; rachis jointed only at female region, fragile. *Male spikelet:* 1 cm, oblong, 2-nate; lower glume elliptic, coriaceous, keeled; upper glume lanceolate; lower lemma lanceolate; palea lanceolate, chartaceous; upper lemma and its palea similar as above; stamens 3. *Female spikelet:* solitary, 8 mm, alternate, inserted in the hollow rachis-intemode; lower glume triangular; upper glume convex; lemmas hyaline, paleated.

Rare in Araku valley, introduced.

Araku valley (VSKP), *GVS* 32886 (MH).

INDIA: Introduced for Fodder.

WORLD: Mexico, Mesoamerica.

TRITICUM Linnaeus

1. Rachis fragile at maturity; caryopsis entirely enclosed between the lemma and palea and not removed by threshing .... **T.dicoccum**
1. Rachis tough at maturity; caryopsis separating from the gaping lemma and palea .... **T. aestivum**

Triticum aestivum L., Sp. Pl. 85. 1753 *nom. cons;* Bor, Grass. India 670. 1960. *T. vulgare* Vill., Hist. Pl. Bauph. 2:153. 1789; FBI 7: 367. 1896.

An annual, cultivated, tufted grass; culms up to 1 m high. Leaf blades linear-lanceolate, flat, apex acuminate. Spike up to 20 cm long, erect, curved at maturity, compact; spikelets solitary, laterally compressed, 3-5-flowered; glumes obtuse, shortly awned; lemma muticous, or shortly 1-3-awned, bisexual. Caryopsis oblong, ventrally grooved, free within the lemma and palea.

Occasionally cultivated in some districts. Fl. & Fr.: November-March. Tel.: *Godhumalu*; Tam.: *Gothumai*.

INDIA: Widely cultivated in the plains of India.

WORLD: Widely cultivated in most of the Subtropical and Temperate countries.

Triticum dicoccum (Schrank) Schubler, Char. & Descr. Cerealium in Hort. Tuling. 28. 1818; Fischer 3: 1856. 1934; Bor, Grass. India 679. 1960; Matthew, Fl. Tamilnadu Carnatic 3: 1912. 1983.

An annual, cultivated, tufted grass; culms up to 50 cm high, leafy; nodes glabrous. Leaf sheaths 6-15 cm long, glabrous; ligule membranous; leaf blades 10-30 x 0.5-1 cm, base rounded, apex acuminate, glabrous. Spikes up to 15 cm long, erect, compact; spikelets 1.2 cm, laterally compressed, ellipsoid; rachis fragile at maturity; spikelets in two rows, few-flowered; glumes broadly ovate, persistent, coriaceous, often unequal, shortly awned; lower lemma coriaceous, keeled, shortly awned; upper lemma boat-shaped, geniculately awned; palea elliptic, 2-keeled, keels ciliate, the lower enclosing a bisexual floret; the upper male or empty; stamens 3. Caryopsis oblong, often hairy, free within the lemma and palea.

Occasional cultivated crop in all districts. Fl. & Fr: January-March. Tel.: *Godhumalu.*

INDIA: Cultivated.

WORLD: Cultivated in most of the countries, especially in Temperate and subtropical regions.

UROCHLOA P. Beauvois

1. Perennials .... **U. mosambicensis**
1. Annuals:
    2. Palea ovate; caryopsis ovate, compressed ....**U. panicoides**
    2. Palea lanceolate; caryopsis oblong, not compressed...................... **U. setigera**

Urochloa mosambicensis (Hack.) Dandy, J. Bot. 69: 54. 1931; Bor, Grass. India 372. 1960. *Panicum mosambicense* Hack. in Bot. Soc. Brot. 6: 140.1888. *Urochloa pullulans* Stapf in Prain, Fl. Trop. Africa 9: 590. 1920.

Perennials; culms erect, tufted, *c.* 1 m high, rhizomatous, basal buds silky or white tomentose, nodes bearded. Leaf sheaths 4-10 cm, softly tubercle based hairy; ligule ciliate, 0.5 cm; leaf blades 3-25 x 0.5-1 cm, linear-lanceolate, base cordate, apex acute, tubercle based hairs on nerves on both sides and on margins. Racemes 3-4, each 3.5-4.5 cm long, rachis triquetrous, scabrid, 2-3 conspicuous fringe of bristles at the base of the spikelet, bristles persistent; spikelet solitary, ovate-elliptic, 5 x 2 mm, pubescent; lower glume ovate, 3 x 1.5 mm, apex obtuse, slightly keeled, 3-nerved, margin scabrid, hairy on back side; upper glume ovate-lanceolate, 5 x 2 mm, cuspidate-acuminate, 2-keeled, 5-nerved, softly hairy on back side, lower lemma ovate-lanceolate, 3 x 2 mm, cuspidate-acuminate, 2-keeled, 3-nerved, softly hairy on back side; upper lemma ovate-orbicular, 2 x 1.5 mm, rugose, 2-keeled, nerves obscure, mucronate, mucro *c.* 1 mm, lower palea ovate-oblanceolate, 3.5 x 1.5 mm, hyaline, 2-nerved, 2-keeled, scabrid above; upper palea ovate-orbicular, obtuse, 2 x 1.3 mm, 2-keeled, nerves obscure, rugose; stamens 3, 2 x 0.5 mm; ovary 0.5 mm, style 1 mm, stigma 1 mm.

Occasional in open forests at 350 m in Nallamalais. Fl. & Fr.: November–January.

Chelama forest, Nallamalais (KNL), *BR, BSS* & *SKB* 32945.

INDIA: Introduced, naturalized.

WORLD: Native to Africa, Naturalized in Tropics nd Subtropics in USA, Sri Lanka, Myanmar, Thailand, Indonesia, Fiji.

Urochloa panicoides P.Beauv., Ess. Agrost. 53, 6. 11, t. 1. 1812; Fischer 3:1775. 1934; Bor, Grass. India 372. 1960; Matthew, Fl. Tamilnadu Carnatic 3: 1912. 1983. *Panicum javanicum* Poir. in Lam. Encycl. 4: 274. 1816; FBI 7: 35. 1896. *Urochloa panicoides* P.Beauv. var. *velutina* (Henr.) Bor, Grass. India 373. 1960.

An annual, stoloniferous grass; culms up to 70 cm high, geniculately ascending and rooting at lower nodes; nodes pubescent. Leaf sheaths 2-5 cm long, loose, glabrous, keeled; ligule ciliate rim; leaf blades 4-12 x 0.6-1.2 cm, lanceolate, base cordate, margin tuberculate based hairs, apex acute, glabrous. Spiciform racemes of 4-9, up to 5 cm long; rachis serrate, flat, scabrid; spikelets 4 mm, ovate-elliptic, solitary, dorsally compressed; glumes unequal; lower very small, ovate-orbicular, hairy, 1-nerved, hyaline; upper glume ovate, acute, membranous, 7-nerved, densely ciliate; lower lemma ovate-oblong, acute, 3- nerved, densely ciliate, palea ovate, hyaline, keeled, male or empty; upper lemma ovate-obovate, coriaceous, nerveless, arista at apex, paleated, its palea similar to lemma, bisexual; stamens 3; stigmas plumose. Caryopsis ovate, compressed.

Common on the bunds of cultivated fields and roadsides. Fl. & Fr: August-October.

Sanevaripalli (ATP), *KRKS* 40164; Kalasamudram (ATP), *KRKS* 39551; Batrepalli (ATP), *KRKS* & *BSS* 40066; North Dhone RF (KNL), *RVR* 1613; Allagadda (KNL), *TP & RVR* 2577; Peapuly (KNL), *PVP* 2663 & 3170; Guvvalacheruvu (KDP), *PVP* 3417; Horsley hills (CTR), *MSG & AA* 12213; Anupu (GNT), *VRK* 3507; Panapakkam (CTR), *S. India Flora* 15766 (MH); Yerrabalem (KDP), *SCA* 162 (DD & CAL); Hanumanthapalli (KSN), *CAB s.n.* (MH); Kondapalli (KSN), *CAB* 8139 (MH); Salur (VJN), *NPBK* 1001 (CAL).

INDIA: Throughout India.

WORLD: Tropical East Africa, extending to South Africa.

Urochloa setigera (Retz.) Stapf in Prain, Fl. Trop. Afr. 9: 598. 1920; Fischer 3: 1775. 1934. *Brachiaria setigera* (Retz.) C.E Hubb. in Hook. f., Ic. Pl. 34, 2. 1938; Bor, Grass. India 286. 1960; Pullaiah, Fl. Andhra Pradesh 5: 2112. 2018. *Panicum setigerum* Retz., Obs. 4: 15. 1786; FBI 7: 36. 1896.

An annual, leafy, spreading grass; culms up to 60 cm high; nodes bearded. Leaf sheath 2.5-4 cm long, glabrous, ciliate at margin; ligule membranous; blades 3.5-

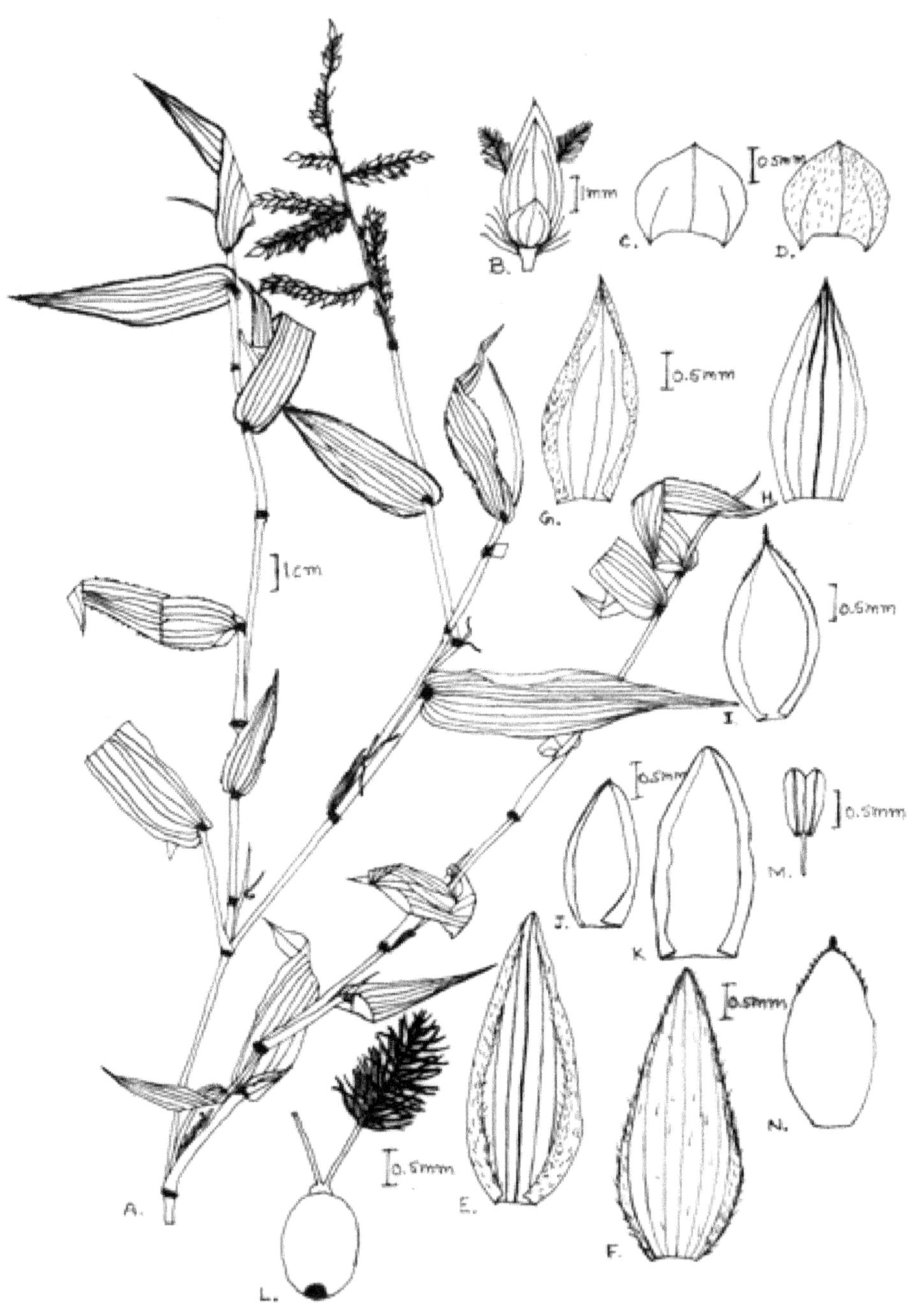

Figure 111. **Urochloa setigera** (Retz.) Stapf (Syn. *Brachiaria setigera* C.E.Hubb.).

A. Twig, B. Spikelet, C & D. Lower glume front and back views, G & H. Lower lemma front and back views, I. Upper lemma, J. Lower palea, K. Upper palea, L. Pistil, M. Stamen, N. Caryopsis.

8.5 x 0.8-2 cm, elliptic-lanceolate, base amplexicaul, margin basally ciliate, apically scaberulous, apex acuminate. Panicle of 3-10 racemes, 8 cm long; rachis trigonous, branches hairy; spikelets 0.4 cm, sparsely pubescent, acuminate, pedicelled, with long hairs; glumes dissimilar, slightly pubescent, lower ovate-sub orbicular, 3-nerved; upper elliptic-lanceolate, 7-nerved; lower lemma lanceolate, slightly pubescent, acute; palea lanceolate; upper lemma ovate, crustaceous, with mucronate and scaberulous tip; palea similar to upper lemma; stamens 3; stigmas plumose. Caryopsis oblong with mucronate tip.

Occasional in grasslands. Fl. & Fr.: September–January.

Tirumala hills (CTR), *RVR & GO* 2429, *MSG* 12137; Pamuleru (EG), *TP & MSG* 12269; Kondapalli (KSN), *CAB* 7993 (MH); Anantagiri (VSKP), *GVS* 19476 (CAL), Simhachalam hill (VSKP), *GVS* 19355 (MH); Pennagaram RF (SLM), *EV* 22401 (MH).

INDIA: Peninsular and North-East India.

WORLD: Nepal, China, Sri Lanka,Myanmar, Thailand.

VULPIA C.C.Gmelin

Vulpia myuros (L.) C.C.Gmel., Fl. Bad. 1: 8. 1806; Fischer 3: 1852. 1934; Bor, Grass. India 564. 1960; B.R.P.Rao *et al*., Indian J. Forestry 41: 113.2018. *Festuca myuros* L., Sp. Pl. 74. 1753.

Annuals, culms simple, erect, up to 20 cm high. Leaf sheaths glabrous or scabrous, ligules membranous; blades filiform-linear, inrolled, apex pungent acute. Inflorescence panicle, linear, branches appressed to the main axis; spikelets 1-5-flowered, awned; glumes linear-lanceolate, membranous, acute, acuminate or serrate at apex; lower glumes 1-nerved, 2-3.5 x *c*. 1 mm; upper glumes 3-nerved, 5-5.5 x 1-2 mm long; lemmas linear-oblong, chartaceous-coriaceous, involute, scabrid in upper half, 5-nerved, 1-awned, awn 3-5 mm long, antrorsely scabrid, straight. Paleas as long as or shorter than lemmas, linear-oblong, membranous-hyaline, dorsally strongly compressed, 2-nerved; nerves distinctly scabrid, 4.8-5.5 mm long.

Occasional in Macherla forest in Nallamalais. Fl. & Fr.: September–March.

Near Macherla (GNT), *BR, BSS* & *SKB* 34805 (SKU).

INDIA: Andhra Pradesh, Himachal Pradesh, Jammu & Kashmir, Karnataka, Tamil Nadu and Uttar Pradesh.

WORLD: Europe, Africa, Asia- Temporate.

ZEA Linnaeus

Zea mays L., Sp. Pl. 971. 1753; FBI 7:102. 1896; Fischer 3:1704. 1934; Bor, Grass. India 482. 1960; Matthew, Fl. Tamilnadu Carnatic 3: 1914. 1983.

An annual, cultivated, monoecious grass; culms up to 2 m high, robust; nodes bearded. Leaf sheaths 5-20 cm long, terete, densely pubescent along the margins; ligule a pubescent ring; blades 20-90 x 3-10 cm, lanceolate; flat, broad, base cordate, margin ciliate, apex acute, sparsely hairy on both surfaces. Panicle monoecious; male panicle up to 40 cm long, terminal, racemosely arranged; female panicle up to 2.5 cm long, solitary, axillary; spikelets 2-nate, 2-flowered.

Cultivated in all districts. Fl. &Fr.: July - October. Tel.: *Mokkajonna*; Tam.: *Makkacholam*.

INDIA: Cultivated throughout the hotter parts of India.

WORLD: Native of America, now cultivated in all warm countries.

ZENKERIA Trinius

Zenkeria elegans Trin., Linnaea 11: 150. t.3. 1837; FBI 7: 270. 1896; Fischer 3: 1805. 1934; Bor, Grass. India 482. 1960; Matthew, Fl. Tamilnadu Carnatic 3: 1914. 1983.

Erect grass; culms to 1 m, nodes sparsely pilose. Leaf sheath 6-20 cm; ligule fimbriate, membranous, to 0.5 mm; blades linear, elongate, 20-40 x 0.5-0.8 cm, convolute. Spikelets to 4 mm, 2-flowered, laterally compressed; lower glume ovate, to 1.5 mm, hyaline, 3-nerved, keeled, glabrous; upper glume lanceolate, to 2 mm, hyaline, 3-nerved, apiculate; lemmas similar, to 4 mm, membranous, 7-9-nerved, hairy below the middle; palea to 3.5 mm, 2-keeled, long ciliate; stamens 3.

Hills above 900 mm in Sirumalai hills, Kolli hills and Shervarayans of Tamil Nadu. Fl. & Fr.: October–January.

Sirumalai hills (DGL), *SKS* 692 (SGH); Kolli hills (NMK), *SKS* 1174 (SGH); Krishnagiri (SLM), *CAB* 14924 (MH).

INDIA: Peninsula.

GYMNOSPERMS

1. Plants palm like; leaves large, 1 to 2.5 m long, pinnate .... **Cycadaceae**
2. Plants dicotyledonous like; leaves small, to 21 cm long, simple, reticulately veined .... **Gnetaceae**

CYCADACEAE

CYCAS Linnaeus

1\. Leaf margins revolute; median leaflets less than 5 mm wide .... **C. beddomei**

1\. Leaf margins flat; median leaflets more than 6 mm wide:

  2. Lateral teeth of megasporophyll less than 3 mm long................. **C. circinalis**

  2. Lateral teeth of megasporophyll more than 3 mm long

    3. Ovules 2-5, ovule bearing lamina lanceolate........................... **C. sphaerica**

    3. Ovules 5-8 , ovule bearing lamina linear .... **C. seshachalamensis**

Cycas beddomei Dyer, Trans. Linn. Soc. Ser. II. Ii, 5: 85, t. 17. 1881. FBI 5: 658. 1888; Fischer 3: 1394. 1928.

Shrub, up to 1.5 m high; trunk cylindrical, dark. Leaves up to 1 m long; petioles *c*. 15 cm long, the upper one-third with few minute spines, base covered with tufted tomentum; leaflets 10-18 x 0.5 cm, linear, apex acuminate, margins strongly revolute. Male cone *c*. 20 x 8 cm, rusty brown, apical portion of microsporophyll

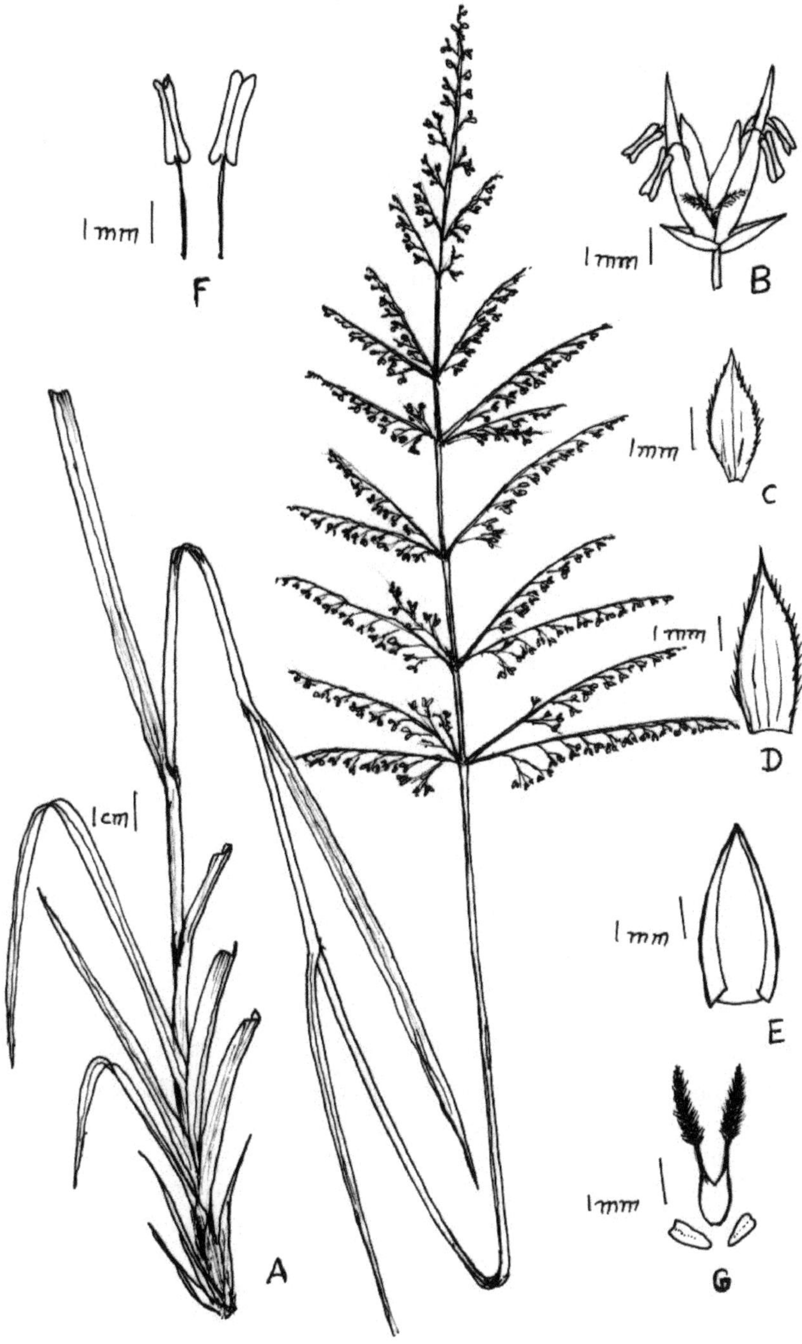

Figure 112. **Zenkeria elegans** Trin.

A. Habit, B. Spikelet, C. Upper glume, D. Lower glume, E. Lemma, F. Stamens, G. Lodicules with pistil.

of the upper half of the cone strongly deflexed, blade of megasporophyll usually narrow-lanceolate, long-acuminate; margins dentate-lobate, ovules 2 on either side. Seeds globose, *c.* 4 cm across, yellow.

Endemic to the Seshachalam range (Kadapa–Tirupati) of Southern Eastern Ghats. The plant grows in dry deciduous forest occurring on the exposed rocky slopes and in the denuded valleys of Tirumala hills between 300-900 m altitudes. It is found along the rock-strewn holy streams (known as '*theerthams*' in the local vernacular) of Tirumala hills. Tel.: *Paireetha, Perita.*

Status: Vulnerable in the wild. Causes for its decline are habitat destruction, forest clearing and excessive collection/ exploitation due to its professed medicinal properties (Nayar and Sastry, 1987-90).

Kadapa hills (KDP), *JSG* 21278 (MH); Talakona (CTR), *CSR* & *K.N.Reddy* 1734, *VSR* 3059 (KU).

INDIA: Andhra Pradesh, Endemic.

Cycas circinalis L., Sp. Pl. 1188. 1753; FBI 5: 656. 1888; Fischer 3: 1394. 1928. var. **circinalis**

A small evergreen palm-like tree, up to 2 m high, bark brown, smooth below, tessellated above by the small diamond-shaped scars of fallen leaves. Leaves 1 m long; petiole up to 25 cm long with short distant spines at right angles to the petiole, slightly deflexed, leaflets 10 -25 cm long, 5-7 cm wide, blunt or acute, not spinous-acuminate, margin flat, acumen of all the antheriferous scales upturned, blade of carpophyll ovate-lanceolate, long-acuminate, margins pectinate. Ovules 3-5 on either side; seeds ovoid-oblong, 7-12 cm long.

Southern hills of Chittoor district. Endemic to Peninsular India. Tel.: *Paireetha, Perita, Ranaguvva;* Tam.: *Kadavazhai, Chalampanai.*

Panapakam (CTR), *GVS* 31854 (MH), *VSR* 2162 (KU); Sirumalai hills (DGL), *SKS* 941 (SGH); Kolli hills (NMK), *SKS* 1037 (SGH).

INDIA: Peninsular India, Endemic.

Cycas seshachalamensis Chalapathi Rao *et al.*, Asian J. Conserv. Biol. 5: 55. 2016.

Subterranean spreading stem on which a cylindric trunk, base abruptly swollen, 2–4.5 m tall, *c.* 104 cm diam, branched or unbranched, bark light grey; bark thick with persistent leaf base. Leaves pinnate, deep green, glossy, 1.57–1.86 m long, leaflets 145–231, flat in section, not keeled, with brown tomentum shedding as the leaf expands, rachis consistently not terminated by a spine; petiole 26–37 cm long (15–30% of total leaf), glabrous, spinescent for 30–60% of length; spines short, straight, 2 mm long; basal leaflets not reduced to spines, linear, 14–18 cm x 6–8 mm; median leaflets strongly falcate, linear, 21–31.5 cm x 9–10 mm, decurrent for 8–10 mm, narrowed to 3–5 mm at base, 8–11 mm apart on rachis, margins slightly recurved; apex acute or spinescent; midrib strongly raised below and slightly raised above; apical leaflets paired, not reduced into spines, straight, linear, 12–16 x 0.6–0.7 cm. Cataphylls densely orange-brown pilose, pungent, narrowly triangular,

5.5–6.5 cm long, persistent. Pollen cones orange-brown or light brown, fusiform, 26–30.5 cm long, 9–12 cm wide, sessile; microsporophyll lamina firm, not thickened dorsiventrally, 4–5.5 cm long, 1.6–2 cm wide, fertile zone 2.5–3.5 cm long, sterile apex 0.6–1.1 cm long; apical spine prominent, gradually raised, spreading or slightly upturned, 1.1–1.3 cm long. Megasporophylls covered with orange-brown or grey tomentose, tomentum persistent, 15–20 cm long; ovules 8–12, glabrous; ovule bearing fertile lamina linear, 7–9 cm long; sterile lamina broadly ovate, shortly dentate with 18–30 lateral spines; lateral spines short, 2–4 mm long, 1–2 mm wide; apical spine distinct from lateral spines, 8–12 mm long, 2–3 mm wide at base. Seeds subglobose, 2–3 cm long, 2–2.5 cm wide; sarcotesta green when young, turns yellow at maturity, 3–4 mm thick; fibrous layer absent; sclerotesta smooth, spongy layer absent.

Rarely found in the foot hills of Seshachalam forests in Chittoor district of Andhra Pradesh. Generally grows in dry deciduous forest at an elevation range from 200–600 m.

Andhra Pradesh: Chittoor district, Seshachalam hills, Chamala Range, Bakarapet forest area, Mayurasarovaram, 520 m alt., 13° 38′ 92.1″ N, 79° 11′ 26.1″ E, 24 May 2012, *N.V. Sivaram Prasad* & *P. Madhu Babu* 401 (Holotype: CAL; Isotype: 401 BSID); Near Mayurasarovaram, 540 m alt., 27 Dec 2012, *N.V. Sivaram Prasad* 409 (paratype BSID; SVUTY); *N.V. Sivaram Prasad, P.V. Prasanna* & *M. Sankar Rao* 010309 (paratype BSID); 500 m alt., 09 Oct 2014 *K. Prasad* & *P.V. Prasanna* 006416 (paratype BSID); 510 m alt., 30 Nov 2014, *K. Prasad* 006447 (BSID).

Cycas sphaerica Roxb., Fl. Ind. 747. 1832; C.S.Reddy *et al*., J. Plant Sci. 2: 362. 2007.

Stems arborescent, up to 4.5 m high. Leaves bright green, semi-glossy, 150-270 cm long, tomentum shedding as leaf expands. Petiole 45-60 cm long, glabrous, spinescent for 90% length. Basal leaflets not gradually reducing to spines. Median leaflets simple, weakly discolorous, 20-30 cm long, 0.6-1.2 cm wide; section flat; margins flat; apex softly acuminate, not spinescent; midrib rised above, flat below. Cataphyll narrowly triangular, soft, persistent. Pollen cones narrowly ovoid, orange, 45 cm long, 10 cm diameter; microsporophyll lamina firm, not dorsiventrally thickened, 3.2–3.8 cm long, apical spine prominent, gradually raised, 1.7 cm long. Megasporophylls 15-25 cm long, brown-tomentose; ovules 2-5, glabrous; lamina lanceolate, 38-55 mm long, regularly dentate, with pungent lateral spines 0.5-1 cm long, apical spine distinct from lateral spines, 1.7-2.5 cm long. Seeds subglobose, 25 mm long; sarcotesta yellow, fibrous layer present, sclerotesta smooth; spongy endocarp absent.

Occasional in deciduous forests and woodlands of northern Eastern Ghats. Tel.: *Orguna chettu, Naasi chettu.*

Donubayi (SKLM), *CSR* 1954 (KU, RRL-B).

INDIA: Odisha, Andhra Pradesh. Endemic

Note: *Cycas sphaerica* was first described by Roxburgh in 1832. Two Roxburgh collections from the Calcutta gardens are now in the Natural History Museum, London (BM), representing *Cycas sphaerica* and *Cycas rumphii*, respectively. The latter is annotated '*Cycas circinalis*', and the former bears only the annotation '*Cycas planifolia* Solander MS. H.H. Haines also described this taxon in 1924 as a variety of closely related *Cycas circinalis*. No type was cited, but Haines stated 'Wild in the hill forests of the Mals of Puri, especially on the tops of ridges with heavy rainfall! Extending to Angul, in open forest, where it is less common! Fl. July–August'. Haines" practice was to add the mark "!" when he had seen the plant in the wild in that locality, and does not record the existence of a specimen. In Flora of Srikakulam it was mentioned under the name of *Cycas circinalis* (Rao and Sreeramulu, 1986).

GNETACEAE

GNETUM Linnaeus

1. Female inflorescence simple.. **G. ula**
1. Female inflorescence branched .... **G. montanum**

Gnetum montanum Markgraf, Bull. Jard. Bot. Buitenzorg. Ser. 3. 10: 466. 1930. *G. scandens auct. non* Roxb.; Hook.f., Fl. Brit. India 5: 642. 1888 *p.p.*

Large evergreen, dioecious climber; branches smooth, slender, swollen at nodes; bark peeling off in pieces. Leaves dark green, turning black on drying, oblong–ovate, secondary veins distant, tertiaries indistinctly reticulate. Male spike branched twice, branches aggregate at the nodes, fertile part of the spike 3 cm long; flowers up to 20. Female inflorescence branched twice or thrice; flowers 5-7 in each bract. Fruit ellipsoid, 1.5 x 1 cm, shining, smooth, red, fleshy.

Rare, occurs at high altitude (1100–1300 m) areas of semievergreen forests in Visakhapatnam district. Tel.: *Milagandi.*

Sambarikonda (VSKP), *CSR* & *VSR* 3164.

INDIA: Arunachal Pradesh, Assam, West Bengal, Bihar, Sikkim, Odisha, Andhra Pradesh.

WORLD: Nepal, Myanmar, Thailand, China, Indo-China.

Gnetum ula Brongn. in Duperr.,Voy. Monde 7: 121. 1829; Fischer 3: 1885. 1936; Matthew, Fl. Tamilnadu Carnatic 3: 1916. 1983. *G. funiculare* Sm. ex Wight, Icon. Pl. Ind. Orient. t. 1955. 1853 non Blume 1834. *G. scandens* auct. non Roxb.; FBI 5: 642. 1881 *p.p.*; Fischer 3: 1392. 1928.

Large dichotomously branched woody climber, branchlets terete, nodes dilated, bark thick, brown, rough. Leaves elliptic, elliptic-ovate, elliptic-oblong or ovate, 7.5-21 x 4.5-10 cm, entire, shortly acuminate, glabrous, midrib stout, base rounded or very obtuse; petiole 0.7-1.8 cm. Flowers monoecious (or dioecious). Male spike 2.5-5 cm long, peduncled, in 3-chotomous panicles; flowers 20-30 in a ring, embedded when young, exerted later, stamen single. Female inflorescence not branched; ovules 8 per ring, exerted, globose, papilla acuminate, fruiting spike up to 25 cm long. 'Fruit' orange when mature, ellipsoid, 2.5-4.5 x 1.5-2.5 cm.

Rare along valleys. Fl.: April–May, Fr.: October-November. Tel.: *Lolari.*

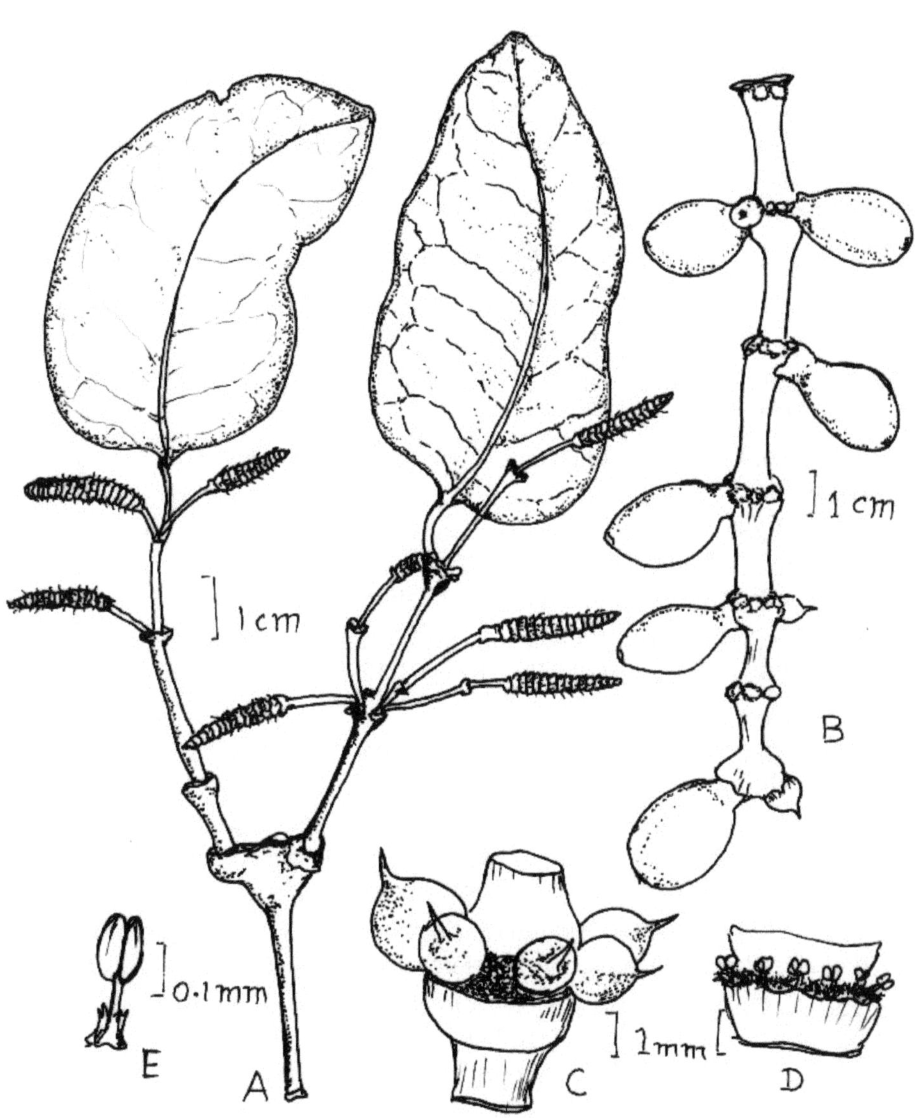

Figure 113. **Gnetum ula** Brong.

A. Male twig, B. Femlae cone, C. Female cupule with ovules, D. Male cupule, E. Stamen.

Maredumilli (EG), *CSR* &*VSR* 1511 (KU); Sirumalai hills (DGL), *SKS* 1132 (SGH).

INDIA: Peninsular India, Madhya Pradesh, Odisha.

WORLD: Myanmar

In addition to the above five wild species of Gymnosperms, following 14 cultivated species are often seen in gardens: *Araucaria columnaris* (Forst.) Hook.f. (Araucariaceae), *Cycas revoluta* Thunb., *Cycas rumphii* Miq. (Cycadaceae), *Callitris cupressiformis* Vent., *Callitris glauca* R.Br. ex Baker & Smith, *Cupressus sempervirens* L., *Cryptomeria japonica* D.Don, *Juniperus communis* L., *Thuja occidentalis* L., *Thuja orientalis* L. (Cupressaceae), *Pinus caribaea* Morelet, *Pinus excelsa* Wall. ex D. Don, *Pinus roxburghii* Sarg. and *Pinus rudis* Lindl. (Pinaceae).

REFERENCES

Benjamin, J. H. F. and G. V. S. Murthy. 2013. Flora of Sri Venkateswara National Park, Andhra Pradesh. Botanical Survey of India, Kolkata.

Bhat, K.G. and C.R. Nagendran. 1983. New Grasses for Karnataka. Indian J. Forestry 6 (4): 332.

Bor, N. L. 1960. The Grasses of Burma, Ceylon, India and Pakistan. Pergamon Press. Oxford.

Chalapathi Rao, P.V., Sivaram Prasad, N.V., Madhu Babu, P., Prasad, K. and Prasanna, P.V. 2016. *Cycas seshachalamensis* (Cycadaceae), a new cycad from Andhra Prdesh, India. Asian J. Conservation Biol. 5(1): 55-58.

Chandramohan, K. and P.V. Prasanna. 2017. A new variety of *Dimeria connivens* Hack. (Poaceae) from India. Bangladesh J. Plant Taxon. 24: 237-240.

Chandramoham, K., P.V. Prasanna and P.R. Reddy. 2016. Note on the distribution of *Lophopogon kingii* Hook.f. (Poaceae), an endemic grass from Eastern Ghats. Int. J. Adv. Res. Sci. Tech. 5(2): 604-605.

Chorghe, A., S. Dey, K. Prasad, K., P.V. Prasanna and Y.V. Rao. 2015. *Tripogon mahendragiriensis* sp. nov. (Poaceae) from the Eastern Ghats of Odisha (Orissa) State, India. Nordic. J. Bot. 33: 655-658.

Chorghe, A., K. Prasad, P.V. Prasanna and Y.V. Rao. 2016. A new species of *Themeda* (Poaceae: Panicoideae) from the Eastern Ghats, India. Phytotaxa 245(2): 183-186.

Chorghe, A., P.V. Prasanna and Y.V. Rao 2013. Additions to the flora of Odisha state. J. Econ. Taxon. Bot. 37(2): 320-323.

Chorghe, A., P.V. Prasanna and Y.V. Rao. 2015. Two new additions to the grasses of Odisha state. J. Econ. Taxon. Bot. 39(1): 81-84.

Chorghe, A., L. Rasingam, P. V. Prasanna and M. Sankara Rao. 2013. *Tripogon tirumalae* (Poaceae), a new species from the Seshachalam hills of Andhra Pradesh, Inda. Phytotaxa 131(1): 17-22.

Clayton, W.D. 1972. Studies in the Gramineae-XXIX. Kew Bull. 27 (3): 448.

Clayton, W. D and S. A. Renvoize. 1986. Genera *Graminum*–Grasses of the world. Kew Bulletin Additional series XIII, RBG, Kew; HMSO, London.

Clayton, W. D., M. Vorontsova, K. T. Harman and H. Williamson. 2017. Grass Base–The online World Grass Flora. Royal Botanic Gardens, Kew. www. kew. org/data/grassdb/www/imp05781.htm accessed on 17-2-2017.

Cook, C. D. K. 1996. Aquatic and wetland plants of India: A reference book and identification manual for the vascular plants found in permanent or seasonal fresh water in the subcontinent of India south of the Himalayas. Oxford University Press, New Delhi.

Deshpande, U. R. 1984. Poaceae: Andropogoneae. Flora of India, Fascicles. Fascicle 15. BSI, Calcutta.

Gad, H.S. and M.K. Janarthanam. 2007. Extended distribution of *Hemisorghum venustum* (Thw.) Clayton–A rare grass of Sri Lanka and India in Western Ghats. J. Econ. Taxon. Bot. 31: 344-347.

Gamble, J. S. and C. E. C. Fischer. 1915-1935. Flora of the Presidency of Madras. London. Rep. ed. 1957. Botanical Survey of India, Calcutta.

Gayathri, M. S. and T. Pullaiah. 1995. *Axonopus compressus* (Sw.) Beauv. and *Phalaris minor* Retz. (Poaceae)–new records for Andhra Pradesh. J. Bombay Nat. Hist. Soc. 92 (2): 291-293.

Gayathri, M. S. and T. Pullaiah. 1996. On the occurrence of *Dimeria kanjirapalliana* K. C. Jacob (Poaceae) in Andhra Pradesh. J. Bombay Nat. Hist. Soc. 93(1): 122-123.

Hooker, J. D. (ed.) 1872-1897. Flora of British India. (7 vols.) London. Hooker, J. D. (ed.) 1872-1897. Flora of British India. (7 vols.) London.

Kabeer, K.A.A. and V.J. Nair 2009. Flora of Tamil Nadu - Grasses. Botanical Survey of India, Kolkata.

Kalidass, C. 2015. *Eragrostis nairii* (Poaceae: Eragrostideae): A new species from Odisha, India.J. Econ. Taxon. Bot. 39(1): 126-127.

Kumar, M.A., P. Anjaneyulu and B.R.P. Rao. 2017. A new record of grass *Ottochloa* (Poaceae) to the Eastern Ghats, India. J. Threatened Taxa 9(11): 10974-10976.

Kumar, M.A. and B.R.P.Rao. 2018. *Lepturus* R.Br. (Poaceae: Chloridoideae: Leptureae): A new generic record to Andhra Pradesh, India. Indian J. Forestry 41(3): 307-310.

Kumar, M.A. and B.R.P.Rao. 2019. Three grasses (Poaceae), additions to the flora of Andhra Pradesh, India. J. Threatened Taxa. 11(12): 14606-14611.

Kumar, M.A., B. Sadasivaiah, P. Priyadarsini and B.R.P.Rao. 2017. *Apocopis mangalorensis* (Hochst. ex Steud.) Henrard and *Sporobolus ioclados* (Trin.) Nees, two new records for Andhra Pradesh, India. Indian J. Forestry 40(1): 79-81.

Kumar, M.A., S. Salamma, M. C. Naik and B. R. P. Rao. 2016. *Aristida stocksii* and *Dimeria bialata* (Poaceae): New distributional records to Eastern Ghats. Rheedea 26(2): 161-164.

Kumar, M.A., A. Sreenath, P.Anjaneyulu, S.Saravanan and B.R.P.Rao. 2018. *Themeda helferi* (Panicoideae: Poaceae): A new record for India. Rheedea 28(1): 57-61.

Lakshmaiah, A., M. V. Suresh Babu and B. R. P. Rao. 2012. Agrestals of irrigated fields of Rayalaseema region of Andhra Pradesh. In: Systematics, Biotechnology and Bioprospecting. pp. 735-747.

Matthew, K. M. 1983. The Flora of the Tamilnadu Carnatic. (3 vols.). Rapinat Herbarium, Tiruchirapalli.

Moulik, S. 1997. Grasses and Bamboos of India. 2 vols. Scientific Publishers, Jodhpur.

Murthy, U. R., and U. Sathyavathi. 1968. Two new species of *Iseilema* Anders. from India. J. Bombay Nat. Hist. Soc. 65: 664-669.

Naithani, H. B. 1993. Contributions to the Taxonomic studies of Indian Bamboos. Vol. II. Ph. D. thesis. HNB Garhwal University, Srinagar, Garhwal.

Naithani, H. B., P. V. C. Rao and B. V. P. Reddy. 2012a. *Schizostachyum polymorphum* (Munro) Majumdar- a bamboo new to South India for Andhra Pradesh. Indian Forester 138 (1): 96-97.

Naithani, H. B., P. V. C. Rao and B. V. P. Reddy. 2012b. *Bambusa tulda* Roxb., report of gregarious flowering from South India. Indian Forester 138(5): 484-485.

Naithani, H. B., P. V. C. Rao, B. V. P. Reddy and C. K. Rao. 2011. Note on the rediscovery of *Cephyalostachyum pergracile* in Andhra Pradesh, India. Indian Forester 137(4): 528-531.

Noltie, H. J. 2000. Flora of Bhutan. Vol. 3 part 2. The Grasses of Bhutan. Royal Botanic Garden, Edinburgh and Royal Government of Bhutan.

Pullaiah, T. 2018. Flora of Andhra Pradesh. Volume 5. Scientific Publishers, Jodhpur.

Raju, V.S. and V.S.Kumar. 2018. *Polytrias indica* (Poaceae: Andropogoneae): the name, species identity and its distribution in India. J. Threat. Taxa 10(7): 11969-11972.

Rao, B.R.P. and M. A. Kumar. 2018. *Tripogon uma-ganeshii* (Poaceae: Chloridoideae: Tripogoninae), a new species from India. Indian J. Forestry 41(1): 97-101.

Rao, B. R. P., M. V. S. Babu, B. Sadasivaih, S. K. Basha, and K. N. Ganeshaiah. 2009. Current threat status of *Cycas beddomei* Dyer, an endemic species of the Tirupati-Kadapa Hills, Andhra Pradesh, India. Encephalartos 97: 21–25.

Rao, B. R. P., M. V. S. Babu and J. Donaldson. 2010. A reassessment of the conservation status of *Cycas beddomei* Dyer (Cycadaceae), an endemic of the Tirupati-Kadapa Hills, Andhra Pradesh, India, and comments on its CITES Status. Encephalartos 102: 19–24.

Rao, B. R. P., P. Priyadarsini and K. V. Subbaiah. 2011. *Andropogon lividus* (Poaceae): a new distributional record for Andhra Pradesh. J. Econ. Taxon. Bot. 35(3): 606-608.

Rao, B. R. P. and A. M. Reddy. 2010. Two little known grasses of India. Indian Forester 136: 1230-1233.

Rao, B. R. P., A. M. Reddy, P. Priyadarsini, B. Sadasivaiah and S. K. Basha. 2012. *Themeda villosa* (Poiret) A. Camus, *Tripogon trifidus* Munro ex Stapf (Poaceae) new distributional records for South India. J. Econ. Taxon. Bot. 36(2): 383-386.

Rao, B. R. P., A. M. Reddy, B. Sadasivaiah, K. V. Subbaiah and S. Sunitha. 2009. Two new distributional records of Poaceae for Peninsular India. J. Econ. Taxon. Bot. 33(2): 434- 436.

Rao, B. R. P., A. M. Reddy, S. K. Basha and B. Sadasivaiah. 2010. Two endemic grasses of India: Additions to Flora of Andhra Pradesh. J. Econ Taxon. Bot. 34(2): 394-397.

Rao, B. R. P., B. Sadasivaiah, P. Priyadarsini, A. M. Reddy, S. Sunitha, A. Lakshmaiah, K. V. Subbaiah, J. N. Prasad, S. K. Basha, M. V. Suresh Babu and P. V. Prasanna. 2011. A conspectus of the Tribe Eragrostideae (Poaceae) in Andhra Pradesh, India. In: T. Pullaiah (ed.) Biodiversity in India. Vol. 5. Daya Publishing House, New Delhi. P. p. 259-291.

Rao, B. R. P., K.V.Subbaiah and P. Priyadarsini. 2017. Grasses of Arid Zone of Peninsular India. Bishen Singh Mahendra Pal Singh, Dehera Dun.

Rao, B. R. P., K.V.Subbaiah, P. Priyadarsini and A. Narayanaswamy. 2012. *Chrysopogon hamiltonii* (Hook. f.) Haines and *Cymbopogon pendulus* (Steud.) Wats. (Poaceae), new distributional records for South India. Indian J. Forestry 35(4): 517-520.

Ravi, N., N. Mohanan, T. Shaju, M. S. Kiran Raj and R. Rajesh. 2001. Three more new species of *Ischamum* L. (Poaceae) from Kerala, India. Bot. Bull. Acad. Sin. **42**: 223-230.

Reddy, A. M. 2002. Grasses of Eastern Ghats of Andhra Pradesh. Ph. D., Thesis submitted to Sri Krishnadevaraya University, Anantapur.

Reddy, A. M. and B. R. P. Rao 2009. *Isachne albens* (Poaceae) - a new distributional record for peninsular India. J. Indian Bot. Soc. 88 (3&4): 158-159.

Reddy, A. M., B. Sadasivaiah, K. V. Subbaiah, M. V. Suresh Babu and B. R. P. Rao 2009. Three new distributional records of Poaceae for Andhra Pradesh, India. J. Econ. Taxon. Bot. 33(2): 411-413.

Reddy, A. M., S. Sunitha and B. R. P. Rao. 2002. *Brachiaria brizantha* (A. Rich) Stapf, a new report to Peninsular India. J. Econ. Taxon. Bot. 26 (1): 166-168.

Reddy, A. M., S. Sunitha and B. R. P. Rao. 2004.. *Cyrtococcum deccanense* Bor (Poaceae), a new record to Andhra Pradesh, India. J. Econ. Taxon. Bot. 28(1): 84-86.

Reddy, A. M., S. Sunitha and B. R. P. Rao. 2005. *Paspalum conjugatum* Berg. (Poaceae)-A new record to Andhra Pradesh. J. Bombay Nat. Hist. Soc. 102(2): 259-260.

Sadasivaiah, B., P. Priyadarsini and B. R. P. Rao. 2013. Three rare grasses (*Eragrostis ciliaris* var. *clarkei*, *E. tenella* var. *insularis* and *Urochloa panicoides* var. *velutina*) from Andhra Pradesh, India. J. Sci. Trans. Environ. Technov. 6(3): 166-167.

Sadasivaiah, B., K. V. Subbaiah, S. K. Basha, M. V. S. Babu and B. R. P. Rao. 2008. *Arundinella tuberculata* Munro ex Lisboa- A new distributional record for Eastern Ghats, India. Proc. of A. P. Akademi of Sciences12 (3): 289-291.

Sadasivaiah, B., P. Priyadarshini, S. Khadar Basha and B. R. P. Rao. 2010. Occurrence of *Urochloa mosambicensis* (Hack.) Dandy (Poaceae) in Andhra Pradesh, India. Indian Forester 136 (1): 402-405.

Sandhya Rani, S., K. Sri Rama Murthy, D. Muralidhara Rao and T. Pullaiah. 2004. On the occurrence of *Dimeria connivens* Hack. in Andhra Pradesh. J. Bombay Nat. Hist. Soc. 101 (3): 478-479.

Saxena,H.O. and M. Brahmam. 1995-96. The Flora of Orissa. Orissa Forest Development Corporation Ltd., Bhubaneswar.

Srivastava, S. K. and V. J. Nair. 2010. Genus *Ischaemum* L. (Poaceae) in India. *Nelumbo* 52: 63-92.

Stapf, O. 1917-1934. Gramineae. in D. Prain (ed.), Flora of Tropical Africa, vol. 9. L. Reeve and co., Ltd., Kent.

Subbaiah, K. V. 2008. Grass diversity of Arid Zone of Peninsular India. Ph. D. thesis, submitted to Sri Krishnadevaraya University, Anantapur.

Sur, P. R. 2001. A revision of the genus *Ischaemum* Linn. (Poaceae) in India. J. Econ. Taxon. Bot. 25: 407-438.

Vivek, C.P., G.V.S. Murthy and V.J. Nair, 2013. *Eragrostis collinensis* (Poaceae; Eragrostideae): a new species from the hills of Kerala and Tamil Nadu, India. Indian Journal of Forestry 36(3): 401-404.

While the book is in press the following grass taxa have been reported as additions to Eastern Ghats Flora.

Bothriochloa insculpta (A. Rich.) A. Camus, Ann. Soc. Linn. Lyon n. s., 76: 165. 1931; Bor, Grass. India 107. 1960; Moulik, Grass. Bamb. India 1: 266. 1997; Kumar & Rao, J.Threat. Taxa 11(12): 14607. 2019. *Andropogon insculptus* Hochst. ex A. Rich., Tent. Fl. Abyss. 2: 458. 1851. *Andropogon pertusus* var. *insculptus* (A. Rich.) Hack., Monogr. Phan.6: 482. 1889; Hook. f., Fl. Brit. India 7: 174. 1896. *Amphilophis insculpta* (Hochst.) Stapf, Fl. Trop. Afr. 9: 176. 1917; C.E.C. Fisch. Fl. Madras 3: 1732. 1934.

Perennials. Culms erect or rambling, up to 2.5m high; nodes hairy, basal nodes stilt rooted. Leaf sheaths glabrous, as long as or longer than nodes, 7–14 cm long, shortly ciliate at mouth; ligule membranous, shortly ciliate at apex; blades linear-lanceolate, glabrous, mid nerve prominent, 20–28 x 0.6–0.8 cm long. Inflorescence of racemes, racemes digitate or sub digitate, racemes rachis internodes with translucent canal. Spikelets binate; sessile bisexual; pedicelled male. Sessile spikelet: 2–flowered, oblong–lanceolate, 4.25–4.5 mm long, awned. Lower glumes oblong–lanceolate, membranous–thinly chartaceous, flat, glabrous on dorsal surface, with a pit, margin narrowly winged in upper half, wings ciliate, apex shortly 2-lobed, 9–11-nerved, nerved inconspicuous; upper glumes lanceolate, membranous, boat-shaped, glabrous, lower margins sparsely ciliate hairy, apex acuminate, 1-keeled, 3-nerved. Florets 2; lower barren; upper bisexual. Lower lemmas hyaline, nerveless, 2.8–3.2 mm long. Lower paleas minute or absent. Upper lemmas reduced to the base of awn, principal lemma awn from the apex, geniculate, 11–14 mm long over all; column twisted, scabrid on margins, 6–8 mm long; bristle 4–6 mm long. Paleas minute or absent, stamen 3, anthers 1–1.5 mm long, ovary ovate-oblong, stigmas 2, plumose. Caryopsis not seen. *Pedicelled spikelets*: oblong-lanceolate, chartaceous, male, unawned; pedicel of pedicelled spikelets 2–3 mm long with a translucent canal, hairy on margins, 0.75 length of sessile spikelet. Lower glumes oblong-lanceolate, cartilaginous, glabrous on dorsal surface, pitted, pits 3 (2–4), glandular, margins narrowly winged in upper half, ciliate on margins, 11-13-nerved, nerved conspicuous; upper glumes more or less akin to upper glumes of sessile spikelets, 4–4.2 x c.1 mm long. Lemmas hyaline, nerveless, 2.5–2.8 mm long, unawned. Stamens 3, anthers 1–1.5 mm long.

Usually grows at high altitudes (above1100m). Fl. & Fr.: November–March.

Horsley Hills (CTR), *BR & M. Anil Kumar* 51982 (SKU).

INDIA: Bihar, Maharashtra and peninsular India.

WORLD: Africa; western Indian ocean; Australasia; Europe; South America; temperate and tropical Asia.

Cyrtococcum patens (L.) A. Camus, Bull. Mus. Natl. Hist. Nat. 27: 118. 1921, var. **patens**; C.E.C. Fisch. Fl. Madras 3: 1786. 1934; Bor, Grass. India 292. 1960; Moulik, Grass. Bamb. Ind. 1. 86. 1997; Kumar & Rao, J.Threat. Taxa 11(12): 14608. 2019. *Panicum patens* L., Sp. Pl.: 58. 1753. *Cyrtococcum radicans* (Retz.)

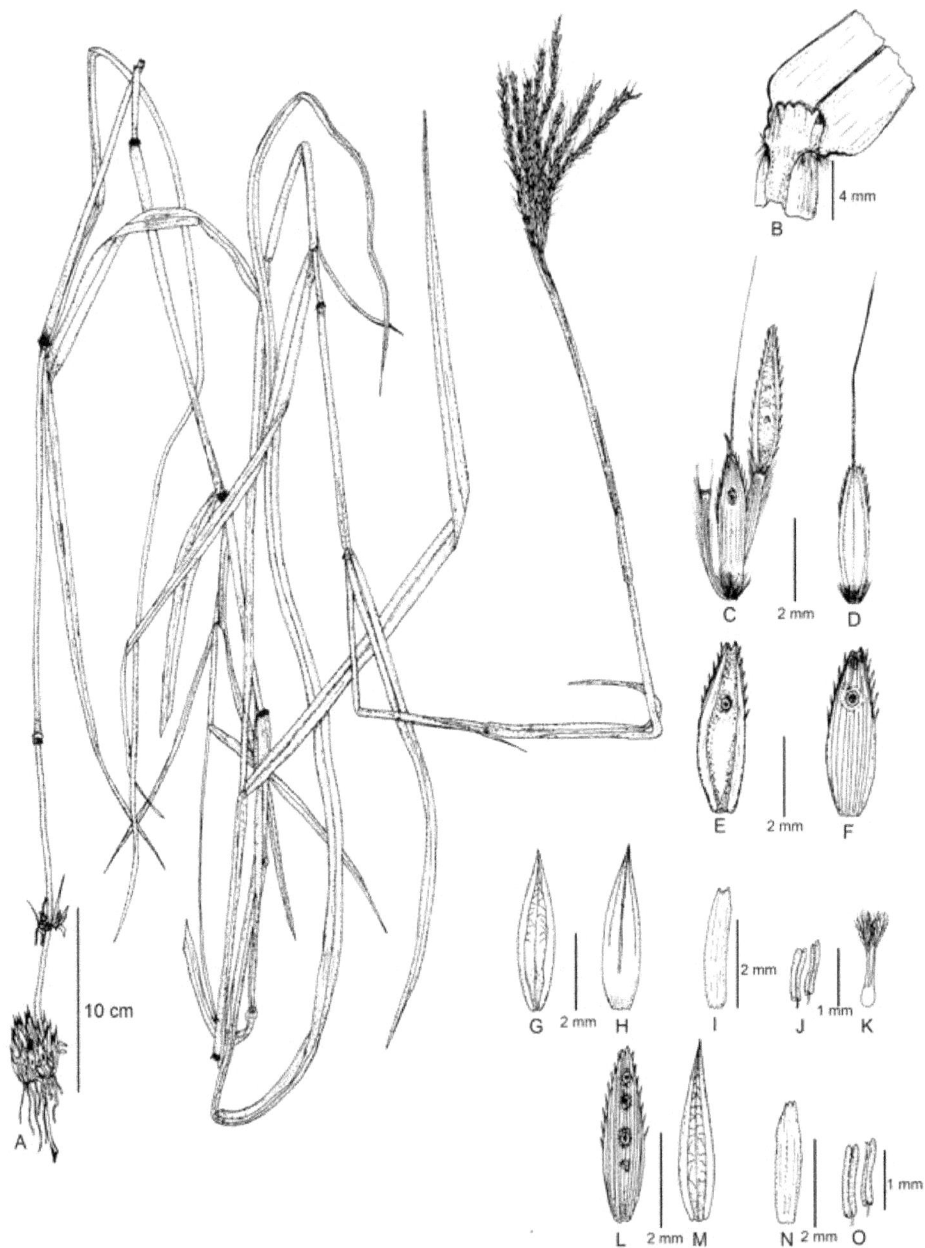

Figure 114. **Bothriochloa insculpta** (A. Rich.) A. Camus

A. Habit, B. Ligule, C. Spikelet pair, D. Sessile spikelet, E. Lower glume of sessile spikelet (ventral view), F. Lower glume of sessile spikelet (dorsal view), G. Upper glume of sessile spikelet (ventral view), H. Upper glume of sessile spikelet (dorsal view), I. Lower lemma of sessile spikelet, J. Stamen, K. Gynoecium, L. Lower glume of pedicelled spikelet (dorsal view), M. Upper glume of pedicelled spikelet (ventral view), N. Lower lemma of pedicelled spikelet, O. Stamen (After Kumar and Rao 2019).

Stapf., Hooker's Icon. Pl. 31: t. 3096. 1922; C.E.C. Fisch. in Fl. Madras 3: 1786. 1934. *Panicum radicans* Retz., Obsser. Bot. 4: 18. 1786. *Cyrtococcum muricatum* (Retz.) Bor, Grasses Burma, Ceylon, India & Pakistan: 291. 1960. *Panicum muricatum* Retz., Observ. Bot. 4: 18. 1786.

Annuals or perennials. Culms slender, erect, creeping, matt-forming, up to 40cm high. Leaf sheaths ciliate on one margin; ligules membranous 1–2 mm long; blades linear-lanceolate, dorsal surface ciliate with tubercle-based hairs, acuminate at apex, 2.5–12.5 x 0.5–1.2 cm long. Inflorescence of panicles, 5–10 cm long. Spikelets in pairs, one with short pedicel, another one with long pedicel, gibbose, 1.4–1.6 mm long. Lower glumes ovate, nearly as long as broad, margins much expanded or winged in the lower half, acute at apex, 3-nerved, c. 1 x 0.9 mm long; upper glumes helmet shaped, elliptic-oblong, membranous, tuberculate ciliate on surface, 3-nerved, c.1.5 x c. 0.5 mm long. Florets 2, lower sterile; upper bisexual. Lower lemmas similar to upper glumes, longer than fertile lemmas, tuberculate ciliate on surface, obtuse at apex, 3-nerved. Lower palea absent. Upper lemmas gibbose, crustaceous, obtuse or subcute, with an appendage at apex, scarcely 3-nerved, c.1.2 x 0.8 mm long. Paleas obtuse at apex, as long as its lemmas, coriaceous, 2-keeled, 2-nerved, 1.2 x c. 0.4 mm long; stamen 3; stigmas 2, plumose. Caryopsis not seen.

Found under the shades of trees in moist deciduous forests. Fl. & Fr.: July–May

Lambasingi Ghat (VSKP), *BR & M. Anil Kumar* 52962 (SKU),

INDIA: Andaman, Andhra Pradesh, Assam, Bihar, Karnataka, Kerala, Maharashtra, Manipur, Meghalaya, Odisha, Sikkim, Tamil Nadu, Tripura, Uttar Pradesh, and West Bengal.

WORLD: Sri Lanka, southeastern Asia.

Sacciolepis myosuroides (R.Br.) A. Camus in Fl. Indo–Chine 7: 460. 1922; C.E.C. Fisch. in Fl. Madras 3: 1786. 1934; Bor, Grasses Burma, Ceylon, India & Pakistan: 358. 1960; Moulik, Grass. Bamb. Ind. 1. 149. 1997; Kumar & Rao, J.Threat. Taxa 11(12): 14609. 2019. *Panicum myosuroides* R. Br., Prodr. Fl. Nov. Holl. 189. 1810; Hook. f., Fl. Brit. India. 7: 42. 1896.

Annuals. Culms erect, tufted or decumbent at base, up to 1.1m high, nodes glabrous. Leaf sheaths glabrous or scabrid, 5–8 cm long; ligules membranous, truncate; blades linear-lanceolate, glabrous or scabrid, base rounded, acuminate at apex, 10–20 x 4 cm long. Inflorescence of panicles, spiciform, usually dark purple when young, 3–20 cm long. Spikelets ovate-obovate to oblong, elliptic, obtuse at apex, 1.2–1.6 mm long. Lower glumes ovate, chartaceous, 5-nerved, 0.6–0.8 x c. 0.6–0.8 mm; upper glumes as long as lemmas, glabrous, 7–9-nerved, c.1.5 mm long. Lower lemmas akin to upper glumes, 5–7-nerved, 1.2–1.4 mm long. Lower paleas more or less enveloped, elliptic, hyaline, 2-nerved, c. 0.5 mm long. Upper lemmas elliptic, coriaceous, 3-nerved, nerves obscure, 0.8–1.1 mm long. Upper paleas as long as upper lemmas, elliptic, 2-nerved; stamen 3, anthers 0.4mm long; ovary 0.3mm long, elliptic; stigmas 2, plumose, 0.6mm long. Caryopsis ellipsoid, c. 0.5mm long.

Very common weed of cultivated fields, especially in paddy and similar swampy habitats. Fl. & Fr.: July–January.

Figure 115. **Cyrtococcum patens** (L.) A. Camus var. **patens**

A. Habit, B. Spikelets along with pedicels, C. Lower glume, D. Upper glume (dorsal view), E. Upper glume (side view), F. Lower lemma (dorsal view), F. Lower lemma (ventral view), G. Upper lemma (side view), H. Upper palea (After Kumar and Rao 2019).

Figure 116. **Sacciolepis myosuroides** (R.Br.) A. Camus

A. Habit, B. Spikelet, C. Lower glume, D. Upper glume, E. Lower lemma, F. Lower palea, G. Upper lemma, H. Upper palea, I. Stamens & gynoecium, J. Caryopsis (After Kumar and Rao, 2019).

Paderu fields (VSKP), *BR, M. Anil Kumar & P. Anjaneyulu*, 52840 (SKU).

INDIA: Andaman, Andhra Pradesh, Arunachal Pradesh, Assam, Bihar, Karnataka, Kerala, Madhya Pradesh, Maharashtra, Manipur, Meghalaya, Odisha, Sikkim, Tamil Nadu, Telangana, Uttar Pradesh, and West Bengal.

WORLD: Africa, Australasia, North & South America, Pacific, and temperate & tropical Asia.

Tripogon nallamalayanus Rasingam & J. Swamy, Phytotaxa 351: 296. 2018.

Tufted perennials. Culms 20–30 cm high, erect; nodes *c.* 1 mm long, glabrous. Leaf sheath terete, 6–11.5 x 0.12– 0.17 cm, glabrous; ligule 0.3–0.4 mm long, *c.* 1.3 mm wide, membranous, truncate, ciliate at apex; leaf blades 17–37 x 0.04–0.12 cm, linear, filiform, convolute, scabrid along margins, acuminate at apex, 5–8-veined, scabrid on lower surface, especially on veins beneath, glabrous on upper surface. Inflorescences 17–24 cm long, 0.4–0.8 mm wide, spiciform; rachis flat, glabrous. Spikelets 11–16.2 x 1.42–72 mm (excluding awns), narrowly elliptic, laterally compressed, 11–16-flowered. Lower glumes 3.11–3.65 x *c.* 0.65 mm, lanceolate, membranous, entire, acuminate at apex, glabrous, 1–veined, shallowly lobed on one side below the middle; upper glumes 4.38–4.88 x 0.58–0.69 mm, narrowly lanceolate, membranous, entire, acuminate at apex, 1–veined, glabrous. Lemmas 3.36–3.54 x 1.23–1.65 mm (excluding awn), membranous, ovate, glabrous at base, scabrid towards apex, 3–nerved, 1–keeled, 2–lobed, 1–awned, the awn 3.12–4.08 mm long, geniculate, arising from sinus, scabrid. Palea 2.44–2.48 x 0.83–0.93 mm, shorter than the lemma, obovate-oblanceolate, 2–keeled, the keels ciliate, hyaline, shortly bifid at apex. Lodicules 2, *c.* 0.31 x 0.12 mm, wedge shaped, glabrous. Stamens 3, *c.* 1.10 mm long; filaments *c.* 0.22 mm long; anthers *c.* 0.94 x 0.20 mm. Pistil *c.* 1.48 mm long; ovary *c.* 0.34 x 0.12 mm, glabrous; style 0.24–0.41 mm long; stigma feathery, 0.73–0.90 mm long; caryopsis not seen.

Rare in Nallamamalais. Fl. & Fr. August – November.

Umamaheshwaram (MBNR), *L. Rasingam & J. Swamy* 7396 (holotype CAL!, isotypes BSID!)

Additions to the Flora of Eastern Ghats

Flora of Eastern Ghats Volume 1 was published in 2002, Volume 2 in 2001, Volume 3 in 2010, Volume 4 in 2011 and Volume 5 in 2019. After publication of these volumes a few new species and several new records have been published. These are given in the following pages.

SNCH: Saraswathi Narayan College Herbarium, Madurai.

MCH: Madras Christian College Herbarium, Chennai.

ANNONACEAE

HUBERANTHA Chaowasku

Huberantha senjiana (R. Muralidharan, Naras. & N. Balach.) R. Muralidharan, Naras. & N. Balach., Phytotaxa 217(2): 200. 2015. *Hubera senjiana* R. Muralidharan, Naras. & N.Balach, Phytotaxa 205 (1): 129. 2015.

Trees, *c.* 5 m high, bark develops even on young branchlets, blackish–grey, apical bud covered by rusty brown tomentum. Leaves elliptic–oblong, 6.5–13.5 x 2.0–4.0 cm, base oblique or rounded, margin undulate, apex acute to acuminate; shiny above; midrib raised below, grooved above; pubescent on margins of midrib on the lower side of leaf; lateral veins *c.* 5 pairs; domatia absent; petiole *c.* 5 mm long; pubescent. Cymes axillary, 1–2–flowered, inflorescences on young and old, leafless branches, pendulous; flowers *c.* 3.5 cm across, pale yellowish green; pedicel slender, *c.* 7 mm long; puberulous; peduncle bract 1, ovate, *c.* 1 mm long; rusty pubescent on the outer side; sepals 3, ovate to sub orbicular, *c.* 3–4 mm long and recurved at margin, rounded at apex, hirsute outside, glabrous inside; petals 6; outer 3, elliptic-obovate, 1.5–2.0 x 0.7–1.0 cm, entire at margin, acute–obtuse at apex, puberulous on the outer side; inner 3, elliptic–oblong, 1.4–2.2 x 0.5–1.0 cm, entire at margin, obtuse at apex, yellow tinged at base; puberulous on the outer side; faintly nerved in the middle; stamens many, pale yellow; filaments *c.* 1 mm long; hirsute; connectives polygonal, flat at top, *c.* 1 mm across; anthers oblong, *c.* 1.2 x 1.0 mm; carpels *c.* 50, oblong, each *c.* 2 x 1 mm long, hirsute on ovary; ovule 1, basal, style indistinct; stigma capitate torus slightly raised, hairy. Fruitlets 20–30, densely packed, ellipsoid, 5 x 3 mm; pointed at apex; glabrous; stipe slender, *c.* 5 mm long; fruiting pedicel slender, *c.* 1 cm long; fruitlets red when ripe; seed 1, 2 x 1 mm, ellipsoid, brown; endosperm ruminations spiniform.

Restricted in Gingee hills of Villupuram district of Tamil Nadu. Fl. & Fr.: August–March. Tam.: *Kothukala.*

Pakkamalai RF., Gingee (VPM), *Muralidharan & Narasimhan* 7210A, (Holo! MH); *Muralidharan & Narasimhan* 7674 (MCH); *Balachandran* 12801 (MCH).

INDIA: Tamil Nadu (Endemic).

XYLOPIA Linnaeus

Xylopia parvifolia (Wight) Hook. f. & Thoms., Fl. Ind. 125. 1855; FBI 1: 84. 1872; Gamble, 1: 20. 1915.

Tree, 6-8 m tall, slightly buttressed; bark smooth, brown. Leaves oblong-elliptic/ lanceolate, 6-9 x 1.8-3 cm, glossy, margin entire, apex bluntly acuminate, acute at base; petioles purplish, 0.5-1 cm long. Inflorescence 2-4-flowered, axillary fascicle; peduncles short, up to 0.3 cm long. Flowers brownish yellow, scented; pedicels 0.5 cm long, glabrescent; bracts broadly ovate, 0.1-0.2 cm long, evanescent; sepals ovate-triangular, 0.2-0.3 x 0.1-0.3 cm, apex obtuse-apiculate, glabrous inside, pubescent outside; petals 6, linear, 1.5-2 cm long, base concave; stamens many, linear, 0.2 cm long; anthers concealed by connective; carpels 6-8, pubscent. Fruit a ring of obovoid monocarps, rugose, reticulate; seeds 6-10, ellipsoid, 0.6-1.2 x 0.5-0.8cm, dark or brown on drying.

Rare in Alagar hills, Karandamalai and Sirumalai hills of Tamil Nadu. Fl. & Fr.: February–May.

Along the streamlets of Kallanai valley, Karandamalai (DGL), *R. Kottaimuthu & V. Balasubramanian* 40025 (SNCH), *R. Kottaimuthu* & *R. Ponnuchamy* 40180 (SNCH); Nooparagangai, Alagarkoil R.F.(MDU), *Chandrabose 53415* (MH); Silambar valley, Alagar Hills (MDU), *K. Ravikumar 2168* (MH); On way to Periya aruvi, Alagar Hills (DGL), *R. Kottaimuthu, V. Meenakhsi Sundaram & K. Natarajan* 3018 (SNCH), *R. Kottaimuthu & M. Jeyakumar* 3180 (SNCH); On way to Puncholai, Sirumalai (DGL), *R. Kottaimuthu & V. Meenakshi Sundaram* 10019 (SNCH); *R. Kottaimuthu & K. Natarajan* 10057 (SNCH).

INDIA: Tamil Nadu (Endemic).

MENISPERMACEAE

COCCULUS DC.*nom. cons.*

Cocculus laurifolius DC., Syst. 1: 530. 1817; FBI 1: 101. 1872; Gamble 1: 29. 1915; Kottaimuthu *et al.*, Ethnobotanical Leaflets 12:300. 2008.

Trees up to 6 m tall; branches drooping; branchlets glabrous. Leaves simple, alternate, spiral; petiole 3-8 mm, slender; lamina 7-12 x 3-5.5 cm, elliptic or elliptic-oblanceolate, base acute or cuneate, apex acute or acuminate, margin entire, glabrous, shining above, chartaceous; basal nerves 3, strong, almost parallel to the margin beyond the middle of lamina; intercostae reticulate, slender, faint. Flowers unisexual, in axillary panicles; sepals 6, ovate, obtuse; petals 6, 2 x 1.5 mm, broadly obtuse, glabrous; stamens 6 in male flowers; free, anthers ovate, within curved margins; staminodes 12 in female flowers, seriate, outer staminodes ovate, tri lobed, with a small middle-lobe; inner smaller, cup shaped; ovaries 3, free, curved, 1-ovuled; stigma sessile, curved out. Fruit globose, 0.6-0.8 cm across, juicy, black when ripe.

Occasional in Sirumalai hills. Fr.: July.

Thumbi Sholai, Sirumalai (DGL), *R.Kottaimuthu & R.Ganesan* 5127 (ATREE).

IINDIA: Tamil Nadu.

WORLD: South-East Asia.

CAPPARACEAE

CAPPARIS Forssk.

Capparis kollimalayana M.B. Viswan., Kew Bull. 55 (1): 245. 2000.

Straggling shrub, irregularly and profusely branched, up to 3 m high. Indumentum rusty-pubescent when young, early glabrescent; hairs simple, both unicellular and multicellular. Twigs terete. Cataphylls absent. Stipular thorns always present, brown, recurved, mostly about 2 mm long and 1.5 mm broad at base. Petioles rusty-pubescent, 1.5 - 3.5 x 0.5 - 1.2 mm. Leaves alternate, often subopposite at ends of branches, 3.4 - 7.6 x 1 - 2.4 cm, ovate to elliptic or oblong, obliquely obtuse-truncate at base, with slightly undulate margin, apically abruptly subacute or subacuminate and retuse, subcoriaceous, glabrous on both surfaces, green and shining above, pale green or greenish brown beneath, midrib above sunken in the young leaves and slightly raised in mature leaves, raised beneath; lateral nerves 12- 16, prominent above, obscure beneath. Flowers zygomorphic, not fragrant, solitary and axillary or 1 - 8 borne in terminal or more rarely subterminal or axillary subumbels, *c.* 1.5 cm across; the subumbels sessile or pedunculate; peduncles 5 - 20 mm long; pedicels thin and slender, glabrous, 14- 22 mm long; sepals 4, deciduous, free at the base, not connate at apex, the outer pair *c.* 6 x 3.5 mm, oblong-obovate, concave, hood-shaped, truncate at base, margin entire and finely ciliate, obtuse at apex, glabrous within and without, green, the inner pair similar but *c.* 6 x 4.2 mm, suborbicular, obtusely rounded at apex, greenis white; petals 4, whitish to pale green, subsessile, subequal; dorsally orientated petals truncate and densely tomentose-sericeous within along midrib and sparsely to densely ciliate at margin, acute or subacute 2 on each side of the midrib, interconnected with each other, prominent; laterally oriented petals similar but larger, *c.* 8 x 4mm, suborbicular, irregularly undulate or distantly toothed at distal end; stamens 24-26, glabrous; filaments white, 9-13 mm long, anthers greyish white, sub-basifixed; gynophore *c.* 14.5 mm long, filiform, white; ovary green, ovoid, not beaked, glabrous, 1 mm thick; style absent; stigma obscure. Fruit stipitate, stipe *c.* 1 mm in diam., dry, pulp slimy, brown; seed globose, 6-8 mm dia, embedded in pulp.

Rare in Kolli hills. Fl. & Fr.: April - June.

Karadivangu, Kolli hills (SLM), *MBV* 17322 (MH - Holo!), 17240 (MH).

INDIA: Tamil Nadu (Endemic).

Capparis shevaroyensis Sundararagh., Kew Bull. 37: 72. 1982 & Fl. India 2: 290. 1993; Kottaimuthu *et al.*, Ethnobotanical Leaflets 12: 299. 2008. *C. parviflora* Hook.f. & Thomson in Hook. f., Fl. Brit. India1: 178. 1872. non Boiss. 1843; Gamble 1:46. 1915.

Straggling shrub, up to 3 m high; branchlets ribbed, pubescent when young, later glabrous. Stipular thorns minute or prominent (in old stems), sometimes absent. Leaves alternate, ovate-elliptic or oblong-lanceolate, 5-10.5 x 2-3.5 cm, base obtuse, margin entire, sometimes slightly undulate, apex acuminate with prominent cusp, sometimes obtusely acumiate and retuse, subcoriaceous, glabrous on both surfaces, paler beneath, midrib above sunken in young leaves and slightly raised in mature leaves; lateral nerves 8-14 pairs, impressed beneath; petioles 0.1-0.2 cm long. Flowers white in 5-10-flowered subterminal/axillary umbels; peduncles 0.5-3 cm long; pedicels 1-1.2 cm long, glabrous; sepals 4, ovate-oblong or obovate, 0.3-0.6 x 0.2-0.4 cm, concave; petals 4, white, oblong-elliptic, 0.5-0.7 x 0.2-0.4 cm, sericeous tomentose on both sides, margin ciliate; stamens 15-25, filaments curved, 0.6-1 cm long, anthers oblong, 0.1-0.2 cm long; gynophores 0.5-1.2 cm, purplish; ovary ovoid, 0.1-0.2 x 0.1 cm, style minute, bristle like, stigma obscure. Fruits globose, 1-1.2 cm in diameter; seed solitary, ovoid or globose, 0.6-0.8 cm in diameter.

Occasional in Sirumalai hills. Fl. & Fr.: July–August.

Nerkuthusholai, Sirumalai (DGL), *R.Kottaimuthu* & *R.Ganesan* 2179 (ATREE); Kannimaruthuodai (DGL), *R.Kottaimuthu* & *M.Vimala* 1077 (ATREE).

INDIA: Tamil Nadu (Endemic).

CLEOME Linnaeus

Cleome rutidosperma DC., Prodr. 1: 2411. 1824; Sundara Raghavan in Sharma & Balakrishnan, Fl. India 2: 313. 1993; Anjaneyulu & B.R.P.Rao, Indian J. Forestry 41(4), 373. 2018.

Annuals; stems erect or decumbent, weak, up to 50 cm high, glabrous, ribbed, with long, deciduous prickles. Leaves alternate, trifoliolate, petiolate; petiole 3-4 cm long; leaflets 0.5-3.5 x 0.5-1.5 cm, sessile to subsessile, rhomboid, elliptic-lanceolate to obovate, glabrous to glabrescent on adaxial surface, hairy on abaxial surface, base cuneate, margin entire, minutely crenulated, with pink line and a row of stiff hairs, apex obtuse-acuminate; stipules *c.* 0.5 mm long. Flowers axillary, solitary, pedicels 2.5-3.5 cm long, filiform; calyx 2-4 x 0.2 -0.3 mm, linear, subulate; corolla

0.6-10 x 1.0-2.2 mm, elliptic-oblanceolate, acute-apiculate at apex, showy pink to bluish-violet; stamens 6, 5-7mm long; anthers 1-2 mm long; gynophores 4-12 mm long, slightly curved. Capsules 4-8 x 0.2-0.4 cm, cylindrical, slightly compressed, glabrous to glabrescent, ribbed, beaked at apex; seeds many, suborbicular, reniform, elaisome white.

Rare in Palakonda area in Srikakulam district. Fl. & Fr.: Throughout the year.

Palakonda (SKLM), *BR* & *PA* 53169 (SKU).

INDIA: Andaman & Nicobar Islands, Andhra Pradesh, Assam, Bihar, Gujarat, Karnataka, Kerala, Maharashtra, Tamil Nadu, Uttar Pradesh and West Bengal.

WORLD: Bangladesh, Malaysia, Maldives, Myanmar, Philippines. Sri Lanka, Thailand, Tropical Africa and Tropical America.

VIOLACEAE

HYBANTHUS Jacq.

Hybanthus stellarioides (Domin) P.I. Forst., Muelleria 8(1): 18. 1993; Sadasivaiah & Pullaiah, Indian J. Trop. Biodiv. 20: 213. 2012. *H. enneaspermus* var. *stellarioides* Domin, Bibl. Bot. 89(4): 983. 1928. *H. enneaspermus* subsp. *stellarioides* (Domin) E.M. Benn., Nuytsia 1(3): 229. 1972; George, Fl. Australia 8: 103. 1982.

Annual herb, 15–30 cm high, stem erect, grooved, unbranched (rarely branched), hairy; hairs antrorse-divaricate. Leaves simple, alternate, clustered at apex, linear or linear to lanceolate, 1.5 x 0.1-0.4 cm, attenuate at base, entire with occasional marginal tooth and ciliate at margins, acute at apex, sparsely hairy; stipules linear, 1–2 mm long, densely hairy; hairs subsessile, gland-tipped. Flowers solitary, axillary; peduncle filiform, with scattered to sparse indumentum, *c.* 7 mm long; pedicels *c.* 2 mm long; a distinct joint exists between peduncle and pedicel; bracts triangular, *c.* 1 mm long, ciliate at margins; sepals 5, ovate-lanceolate, subequal, 2.5–4 mm long, acuminate, bent backwards at apex, keeled, hairy; petals 5, unequal; upper oblong, 3–4 mm long, pale yellow; lateral 2 falcate, 2–3 mm long, pale yellow; lower enlarged into a spathulate limb with a claw; limb broader than long, *c.* 6-10 mm, acute at middle, bright orange; claw *c.* 4 mm long; stamens 5, *c.* 3 mm long; filaments free; anterior 2 filaments with hairy appendages; anthers connate, 2 of them villous, others glabrous; connective extended vividly; pistil *c.* 4 mm long; ovary ovoid, glabrous; ovules 6–9; style suberect; stigma enlarged, flat. Capsules 3-angled, 4–6 mm long, 3-valved, with remnant petals; seeds 6–9, ellipsoid, *c.* 2 mm long, longitudinally ribbed, glabrous, pale yellow.

Rocky terrains on hill slopes along rock margins and in rocky crevices in shady areas in Anantapuram and Guntur districts and also in Nallamalai forests. Fl. & Fr.: July–September.

Nagarjunakonda (GNT), *BSS* 43843.

Hybanthus vatsavayii C.S. Reddy, J. Econ. Taxon. Bot. 25: 219. 2001.

Erect herb, branched from base, up to 20 cm high, stem pale yellow puberulous. Leaves linear to linear–lanceolate, 2-4 x 0.1-0.5 cm, base alternate, margin entire to distantly serrate, apex acute-apiculate. Flowers axillary, solitary, inconspicuous. Capsule 3-lobed, puberulous, seeds 5-8 per capsule, ovoid-ellipsoid

Rare in forest of Anantapur and Kadapa districts. Fl. & Fr.: August–December.

Idupulapaya (KDP), *KRKS* & *BSS* 40318.

INDIA: Andhra Preadesh (Endemic).

PITTOSPORACEAE

PITTOSPORUM Banks ex Solander

Pittosporum dasycaulon Miq., Anal. Bot. Ind. 3: 5 1852; FBI 1: 199. 1872; Cooke, Fl. Bombay 1: 58. 1901; Gamble 55. 1915; Nayar & Giri, Fasc. Fl. Ind 6: 6. 1980; Padal *et al.*, J.Econ. Taxon. Bot. 41: 70. 2017.

Small to medium-sized evergreen tree. Stem bark thin, brown, lenticellate; bitter and aromatic; young branchlets terete, lenticellate, minutely pubescent. Leaves simple, spiral, clustered at the end of branchlets, 3.8–13.5 x 1.5–4 cm, elliptic-lanceolate to oblanceolate, acute to acuminate at base, entire to slightly undulate, apex acute to acuminate, coriaceous, lateral veins 10–15 pairs, secondary and tertiary nerves form close reticulum, midrib distinct above but prominent beneath, glabrous; petiole puberulous, about 2–2.5 cm long, exstipulate. Inflorescence terminal to subterminal compound umbels, densely to sparsely brown tomentose. Flowers bisexual, whitish yellow; buds oval to oblong, light green; pedicels 4–15 mm long; calyx 5, subequal, elliptic-lanceolate to ovate, ciliate, apex rounded to acute, 3.5 x2 mm; corolla 5, light green, 9–11 x 2–2.5 mm, linear-oblong, obtuse to round at apex, entire; stamens 5, free, filaments glabrous, 7 mm long, oblong, sagittate, dorsifixed, 2 mm long; ovary superior, tomentose, unilocular, ovules on parietal placentation; style glabrous, 5 mm long; stigma capitate. Fruit capsule, globose, transversely ribbed, dehiscing with two valves, 8–10 mm long, pericarp woody with a rough surface. Seeds 5–8, ovoid, with red to brownish red resinous pulp.

Rare in the hill valleys of Gudem, Sapparla, Paderu and Gangaraju Madugula. Fl. & Fr.: January–August.

Gudem (VSKP), *S.B. Padal, K.V. Satish & J. Prakasa Rao* 22266 (AUV).

Vernacular names: Peeda Chettu (Andhra Pradesh), Sahyadri Pittosporum (Tamil Nadu), Boogri (Karnataka), and Gapasundi (Maharashtra).

INDIA: Endemic to the Western Ghats and Eastern Ghats; Maharashtra–Kolhapur, Satara, Sindhudurg; Goa–Kurdi Angod; Karnataka–Belgaum, Chikmagalur, Coorg, Dharwar, Hassan, Mysore, N. Kanara, Shimoga, S. Kanara; Kerala–Palakkad; Tamil Nadu–Nilgiri, Salem and northern Eastern Ghats.

Pittosporum neelgherrense Wight & Arn., Prodr. Fl. Ind. Orient. 154 1834; Gamble, Fl. Madras Pres. 1: 55. 1915; K.M. Matthew, Pocket Fl. Sirumalai hills 12. 2001.

Small trees to 8 m tall; bark brownish-black, lenticellate, with pungent aromatic smell. Leaves simple, alternate, estipulate; petiole 5-20 mm, slender, glabrous, grooved above; lamina 5-12 x 1.5-4 cm, narrowly elliptic, elliptic-obovate or elliptic-lanceolate, base attenuate or acute, apex acuminate, margin entire, glabrous, thinly coriaceous; lateral nerves 8-14 pairs, slender, faint, pinnate, midrib grooved above and slightly raised beneath, intercostae reticulate. Flowers bisexual, in slender simple racemes; pedicels 5 mm long, glabrous; sepals 5, 1.5 mm long, subulate, margin ciliate; petals yellow, 8 x 1.5 mm, oblong-lanceolate; ovary superior, densely tomentose with brown hairs, 1-celled; stigma 2-lobbed. Fruit a capsule, 2 valved, 1 cm across, yellow, woody, globular or slightly compressed; seeds 4-8, orange-red, smooth.

Rare in Sirumalai hills of Tamil Nadu. Fl. & Fr.: January–May.

Ooradi, Sirumalai hills (DGL), *SKS* 312 (SGH).

INDIA: Kerala and Tamil Nadu (Endemic).

CARYOPHYLLACEAE

DRYMARIA Schultes

Drymaria villosa Chamisso & Schlechtendal, Linnaea 5: 232. 1830; Sharma *et al.*, Biol. Mem. 2(1 & 2): 16. 1977; Sharma & Balakrishan, Fl. India 2: 533. 1993; Matthew, Suppl. III. Fl. Palni hills t. 760. 1998; Matthew, Fl. Palni hills, South India. Part- I, 78. 1999; Dhatchanomoorthy *et al.*, Int. J. Curr. Res. Biosci. Plant Biol. 9: 75. 2015.

Herbs; stems diffuse, slender, rarely rooting at lower nodes, pilose or villosus. Leaves opposite, ovate-reniform, 10 - 17 x 15 - 10 cm, obscurely nerved, minutely pilose or glabrous. Flowers dichotomous branching, cymes; peduncle 1.5 - 5.2 cm long; pedicel 2 cm long; sepals ovate to elliptic, 2.5 - 3 x 1 mm long, membranous at margin, acute at apex, hairy; corolla 2-lobed, white; oblong-linear, *c.* 3 mm long; stamens 5; filament shorter than the sepals, *c.* 2.5 mm long; ovary ovoid, 2 3 mm long; style 3, shortly connate; stigma sessile-subulate. Capsule ovoid subglobose, *c.* 3 mm long; seeds reniform, ovoid, tuberculate.

Along the stream margins; in-between rocky crevices of moist deciduous and semi-evergreen forests, at an elevation of 800m. Fl.: July-November; Fr.: March May.

Dindigul District, on the way of Puthur to Tholukadu, Sirumalai hills (DGL), *Dhatchanamoorthy*, AGCW BOT 119.

INDIA: West Bengal, Sikkim, Meghalaya, Karnataka and Tamil Nadu;

WORLD: Nepal, Bhutan, Malaysia, Africa, Mexico and Central and South America.

MALVACEAE

ABELMOSCHUS Medikus

Abelmoschus moschatus Medikus subsp. **tuberosus** (Span.) Borss., Blumea 14: 93. 1966; Chandramohan *et al.*, J. Econ. Taxon. Bot. 40: 35. 2016. *Hibiscus longifolius* var. *tuberosus* Span., Linnaea 15: 170. 1814. *Abelmoschus rugosus* Wall. ex Wight & Arn., Prodr. Fl. Ind. Orient. 53. 1834. *Hibiscus rugosus* (Wight & Arn.) Mast. in Hook. f., Fl. Brit. India 1: 342. 1875.

Tuberous herbs, 50 to 100 cm tall, with fusiform root stock, stem with spreading hairs, bristly at apex. Leaves extremely variable in shape and size, 3-14 x 2-12 cm, on proximal part of the stem, cordate, angular to 3-5-palmilobed, base sagittate-hastate, margins crenate to dentate, apex acute, upper surface hispid with simple strigose hairs and 3-4-armed stellate hairs on the lower surface, 5-7-nerved, lateral nerves clearly visible; petioles 1.5-15 cm long, solid, spreading hairy, caducous. Flowers solitary, axillary/terminal, pedicel up to 4 cm long, inarticulate, pubescent, accrescent in capsule, buds up to 2 cm long; epicalyx segments linear-filiform, 1.5 cm long, prominently setaceous; calyx spathaceous, *c.* 1.6 cm long, sericeous or densely puberulous; petals pale yellow turning pink, with purple centre; stamina column pale yellow, *c.* 1.5 cm long; ovary ovoid, style *c.* 1.5 cm long, distally 5-branched at apex, stigmas discoid, cushion like, dark purple. Capsule ovoid, 3-5 x 1.5-2.5 cm, acute or shortly beaked, dark brown, densely spreading hairy; seeds reniform, tomentose.

Rare, occurs along streams in dry deciduous forests. Fl. & Fr.: September-November.

INDIA: Andaman & Nicobar Islands, Andhra Pradesh, Kerala, Tamil Nadu. WORLD: Bangladesh, Cambodia, China,Indonesia, Laos, Malaysia, Papua New Guinea, Philippines, Sri Lanka, Taiwan, Thailand, Vietnam; Australasia.

ABUTILON Miller

Abutilon persicum var. **narasimhanii** Devanath., Rheedea 28: 92. 2018.

Shrubs, up to 5 m tall; branchlets softly stellate-tomentose mixed with glandular-trichomes, sub-glabrous when mature. Leaves alternate, broadly ovate to suborbicular near the base, ovate-lanceolate above, 7–24 x 6–22 cm, deeply cordate at base, irregularly crenate to shallowly-incised at margins, caudate-acuminate at apex, stellate-tomentose admixed with simple trichomes above, densely stellate-tomentose admixed with simple trichomes beneath; veins 5–9, lower two veins bending downwards, rest arching towards margins; venation palmately netted; petioles 10–26 cm long, lower leaves with longer petioles, upper leaves with shorter petioles, stellate-tomentose mixed with long simple trichomes, glabrous when mature. Inflorescence axillary, solitary cymes initially, becoming terminal panicles towards maturity due to reduction of upper leaves, up to 2.5 m long. Flowers 2.5–3 cm across, yellow; pedicels in flowers 3–4 cm long, *c.* 5 mm long above the point of articulation; fruiting pedicels 4–5 cm long; 5–10 mm long above the point of articulation; stellately pubescent admixed with glandular-trichomes; calyx campanulate, *c.* 1 cm across, stellate-tomentose admixed with glandular-trichomes outside, stellate with adpressed simple trichomes inside; tube 3–4 mm long, 5-lobed; lobes ovate-lanceolate, 6–10 x 4–5 mm, acute-acuminate at apex; fruiting calyx covering half or less than half of the schizocarp, persistent; petals 5, twisted, spreading, broadly obovate, *c.* 1.8 x 1 cm, cuneate at base, rounded-truncate at apex, glabrous inside, sparsely stellate with simple trichomes outside; stamens *c.* 50; staminal column *c.* 5 mm long, conical and glabrous at base, narrow and stellate-hirsute at apex; filaments 1–2.2 cm long, purple; anthers reniform, *c.* 3 x 2 mm, basifixed, tuberculate; ovary oblong-ovoid, *c.* 3 mm across, stellate-tomentose, (6–) 7 or 8 (–9) locular; styles (6–) 7 or 8 (–9), *c.* 2.5 cm long, free, shortly connate at base, greenish yellow; stigma (6–) 7 or 8 (–9), capitate. Schizocarps infundibuliform, 2–2.5 cm across, stellate-tomentose on sides, hirsute on ribs; mericarps (6–) 7 or 8 (–9), *c.* 20 x 5 mm, abruptly acute at apex; acumen 0.7–1 mm long; seeds 4 or 5 per mericarp, reniform, *c.* 3 x 2.3 mm, tuberculate, glabrous, brown.

Growing on the margins of deciduous forests and forest-cleared open lands at elevation ranging from 800 to 1100 m. Fl. & Fr.: December–May; peak during January–March.

Kolli Hills, Poogkulampatti (NMK), *Devanathan* 11330 (Holo MH; Iso, CALI); Kolli Hills (NMK), *K.M. Matthew, S.J. Britto & N. Rani* 28796 (RHT); Valayakaranchettipallam R.F. (NMK), *M.B. Viswanathan* 3312 (FRLH); Seekupaarai view point slopes (NMK), *Devanathan* 10403; Vilaram (Settur) (NMK), *Devanathan* 10478; Poogkulampatti (NMK), *Devanathan & D. Narasimhan* 11215 (MCH).

INDIA: Tamil Nadu.

WORLD: Myanmar.

STERCULIACEAE

ERIOLAENA DC.

Eriolaena stocksii Hook.f. & Thomson ex Mast. in Hook.f., Fl. Brit. India 1: 370. 1874; Malick in B.D.Sharma *et al.*, Fl. India 3: 419. 1993; Chandramohan *et al.*, Indian J., Forestry 40: 103. 2017.

Small tree; bark purplish; young branches stellately pubescent. Leaves petioled; petiole 5-8 cm long, terete, clothed with stellate hairs; leaf blade 8-12 x 6-11.5 cm, orbicular, base deeply cordate, margins irregularly crenate, apex shortly acuminate, 5-7-nerved from base, thinly dotted with stellate pubescence above, densely stellate-pubescent beneath; stipules linear-falcate, early deciduous. Flowers yellow, few in lax, axillary racemose cymes, 4-6 cm across, at the top of the long peduncle; peduncles up to 10 cm long, stellate pubescent; pedicels up to 1 cm long, more or less grooved, pubescent; involucral bracts multifid, segments many, linear or filiform, clothed with dense stellate pubescence; sepals lanceolate, up to 21 x 2.7 mm, cuspidate at apex, stellately pubescent outside, villous within, glandular near the base; petals obovate, cuneate; stamens many-seriate, on long, erect column; column up to 1.2 cm long, pubescent, ovary up to 5 mm long, 8-10-lobed, pubescent; style 2.5 mm long, erect, stigma 8-10-lobed. Capsules u to 5.4 x 2.4 cm, stellately pubescent, rough; seeds 2.6 x 0.8 cm, many, winged at the apex.

Rare along hills in deciduous forests in Odisha. Fl. & Fr.: May–Septmber.

Satkosia Wildlife Sanctuary, Tikarpada range (Orissa), *K.Chandramohan* 8303 (BSID).

INDIA: Bihar, Gujarat, Karnataka, Maharashtra, Orissa, Tamil Nadu. Endemic.

BALSAMINACEAE

IMPATIENS Linnaeus

Impatiens shevaroyensis Bhaskar, Tax. Monogrph Impatiens Western Ghats, S. India. 83. 2012.

Scapigerous perennial herbs, tuberous. Leaves radical, petioled to 14 cm long, lamina 3 - 7.5 x 2-5 cm, orbicular, deeply cordate or auricled at base, margin shallowly toothed, apex round, 3-8-nerved, sparsely hairy above, glabrous beneath. Flowers in racemose scapes, scape 18-22 cm long, many-flowered, deep pink, 2 cm across; bracts ovate, 3 mm long; pedicels 2.8 cm long, glabrous. Sepals inequilateral, 4-5-nerved, 3 x 2.5 mm, apex obtuse. Lip 6.5 mm long on top, with a long slender spur, spur s-shaped, up to 3 cm long; standard longer than broad, 0.8 x 0.5 mm, cupulate, deep pink out side and white inside, tip round; wings 3-lobed, the basal lobe half size of other two, usually overlapping or clasping on middle lobe, 1.5-2.1 cm along distal lobe, proximal lobe the smallest, 6 mm long, anticous lobe 6 x 3.5 mm, middle lobe broadest, 4-5 x 6 mm, base of all lobes white, dorsal auricle absent. Pollen white. Capsule ovate-oblong, 1 cm long when mature, red-tinged on a side, many seeded; seeds brown, with hairs all over the surface hairs filamentous, with closely packed single stranded helical banding.

Occur only in Shevaroy hills of Tamil Nadu. Fl. & Fr.: August - September.

Shevaroy hills (SLM), *Bhaskar* 871 (Holo! UASB) Bear's cave, Yercaud (SLM), *Bhaskar* 1975 (MGM).

INDIA: Tamil Nadu. Endemic.

Impatiens yercaudensis Bhaskar, in Tax. Monogrph Impatiens Western Ghats, S. India. 83. 2012.

Annual or perennial herbs, up to 60 cm tall; stems woody, stiff, terete or angled, erect, yellowish-red, often from a rooting prostrate or procumbent basal stem, whole plant densely hairy. Leaves opposite-decussate, aggregated at apex, stipulate, petiolate or sessile, leaf shape varies from rhomboidal to ovate to ovate-oblong or elliptic, 1.5 x 1 cm, acute or obtuse, rounded or subcordate, often pale reddish, prominently impressed venation above, scabrous hairy all over upper surface, hairy only veins beneath, serrate, serratures setaceous. Flowers axillary, solitary or fascicled, in opposite-decussate pattern as the leaves, 1.3-2.5 cm across pale pink with deep pinkish blotches at throat; pedicels longer than the leaves, 4.5 cm in flower, in fruits up to 5.5 cm long, deflexed in fruit, densely hairy; sepals linear-acute, hairy; lip boat-shaped, with prominent honey guides and yellow 'eye' at the centre inside, puberulous outside; spur slender, 2-3 cm long, slightly curved, distinctly hairy all along, often bifid, notched at the tip; standard orbicular, acuminate or horned, hairy above; wings 2-lobed, larger anterior lobe 1.5 x 2 cm, smaller basal lobe deltoid, 3-5 mm long, 3 mm wide, enclosed by hood-like standard, tip depressed and minutely mucronate, dorsal auricle short, round, 1 mm long, deep pink or dark purple in colour. Pollen light pink gradually during to creamish yellow. Capsule 1.2-2 cm long, ellipsoid, turgid in middle, fruiting pedicel hairy, fruit glabrous; seeds many, 16-18 per fruit, smooth, black, shiny.

Shevaroy hills of Tamil Nadu. Fl. & Fr.: November - January.

Bear's cave, Shevaroy hills, Yercaud (SLM), *Bhaskar* 434 (Holo, UASB); Bhaskar 872 (UASB); Shevaroy hills (SLM), *G.S. Perrottet*, K0006948877 (K).

INDIA: Tamil Nadu. Endemic.

SAPINDACAE

ALLOPHYLUS Linnaeus

Allophylus rheedei (Wight) Radlk. in Nat. Pflanzenfam. 3(5): 313. 1895; Gamble 1: 246. 1918; Chandramohan & Prasanna, Indian J. Forestry 40: 95. 2017.

Small tree, branches lenticellatee, yellowish brown. Leaves petioled, 3-foliolate, leaflets 6.5-12 x 3-6 cm, elliptic-obovate or obovate, base obstuse, margin obscurely dentate, apex acute to acuminate, ferruginous tomentose; lateral leaflets ovate and central one obovate, distinctly petiolulalte. Racemes longer than leaves, chiefly branched near the base. Flowers small, light brown, glabrous; sepals 4, unequal, elliptic; petals 4 small, hairy within; stamens 8, inserted inside the disc. Fruits globose, 5-7 mm in diam., orange-red when ripe, glabrous.

Occaasional along hills slopes in Odisha. Fl. & Fr.: July–November.

Satkosia Wildlife Sanctuary, Tikarpada range (Orissa), *K.Chandramohan* 7065 (BSID).

INDIA: Maharashtra, Tamil Nadu, Kerala and Odisha.

FABACEAE

CALOPOGONIUM Desv.

Calopogonium mucunoides Desv., Ann. Sci. Nat. (Paris) 9: 423. 1826; Sanjappa, Legumes Ind. 104. 1992; Kottaimuthu *et* al., Indian J. Forestry 39: 89. 2016.

Climbers, branches and branchlets coarsely brown-tomentose. Leaves alternate, 3-foliolate, leaflets elliptic, 3-7 x 2-4 cm, base rounded or oblique, apex obtuse-apiculate, densely pubescent on both surfaces, lateral nerves 4 pairs; petioles 5-8 cm long; stipels 3 mm long, subulate. Flowers pale blue, in axillary racemes; bracts and bracteoles small, caducous; pedicels to 3 mm long; calyx lobes unequal, 4-6 mm long, upper 2 connate, other 3 lobes linear to lanceolate; corolla papilionaceous, vexillum obovate, appendaged, with 2 inflexed auricles, keel petals shorter than wing petals; stamens diadelphous 9+1; ovary densely vllous, sessile, style slender, stigma capitate. Pods narrowly oblong, 2.5-3 cm long, brown-pubescent, compressed-convex, slightly constricted between the seeds; seeds 5-7.

Occasional in forests in Kadapa and East Godavari districts. Fl. & Fr.: February–March.

Kodur RF (KDP), *S.N.Chandrasekhara Iyyer* 93187 (MH); Maredumilli, near Tiger camp (EG), *MM* 101265 (MH).

INDIA: Kerala, Tamil Nadu, Andhra Pradesh, Karnataka, Maharashtra.

WORLD: Tropical Asia.

CENTROSEMA (DC.) Bentham

Centrosema pubescens Benth., Comm. Legum. Gen. 55. 1837; Sanjappa, Legumes Ind. 111. 1992; Kottaimuthu *et* al., Indian J. Forestry 39: 89. 2016.

Twining herbs, branchlets slender, terete, pubescent. Leaves pinnately 3-foliolate, leaflets ovate-oblong, 5-7.5 x 2.8-3.5 cm, base rounded, apex acuminate, sparsely pubescent on both surfaces, lateral nerves 5-6 pairs, distant, raised beneath; petiolules 3-5 mm; stipels linear-subulate, 1-2 mm long, persistent; petiole 3-10 cm long, stipules ovate. Flowers 3-6 together, in axillary racemes; peduncles 3-5 cm long; bracts large, ovate; bracteoles appressed to calyx; calyx campanulate, calyx lobes 5, subequal, 6 mm long, the upper two connate, shorter than the laterals; corolla papilionaceous, vexillum orbicular, 3-3.5 cm across, violet-pink, minutely hairy outside, wings falcate, 2.8-3 cm long, clawed, keel curved, acute; stamens diadelphous; ovary oblong, 3-4 mm long, pubescent, stye 1 mm long, curved, stigma fringed. Pods oblong, 8-12 cm long, sessile, glabrescent, compressed, with thick margins, apically beaked, 15-20-seeded.

Occasional in forests in East Godavari district. Fl. & Fr.: December–January

Chinthiru (EG), *MM* 100724 (MH).

INDIA: Kerala, Tamil Nadu, Andhra Pradesh.

WORLD: South and Central America, South East Asia, Java.

CROTALARIA Linnaeus

Crotalaria heyneana Grah. ex Wight & Arn., Prodr. 187-188. 1834; Baker in Hook.f., Fl. Br. India 2L 78. 1876; Gamble 1: 297. 1918; Sanjappa, Legumes India 121. 1992; Ansari, *Crotalaria* in India 265. 2008; Kottaimuthu, J. Econ. Taxon. Bot. 40: 156. 2017.

Erect, sparingly branched, glabrescent under shrubs, up to 1 m high. Leaves simple, subsessile, membranous, elliptic-ovate or elliptic-lanceolate, 7-15 x 4-6 cm, base acute or cuneate, apex acute or rarely acuminate, glabrous above, sparsely puberulous beneath, lateral veins conspicuous, stipules minute, semilunate, reflexed, caducous. Racemes terminal or rarely axillary; peduncles up to 4 cm long. Flowers 10-20, *c.* 2 cm across; pedicels 3-5 mm long, bracts minute, subulate, reflexed, bracteoles similar to bracts, placed at about the middle of the pedicel; calyx *c.* 3 cm long, pubescent, tube campanulate, lobes lanceeolate, acute, upper lip more deeply cleft; corolla papilionaceous, slightly exceeding calyx, vexillum suborbicular, *c.* 2 cm long, pale blue or white with purple streaks, at the base and with scattered hairs on the wing petals blue; stamens dimorphic, staminal sheath 6 mm long, glabrous, filaments 6 mm and 3 mm long alternately with ovoid (*c.* 1 mm long) and oblong anthers; ovary *c.* 7 mm long, pubescent, style 0.1 cm long, exceeding the anthers, hairy along the margin, stigma slightly expanded, hairy. Pods oblong, cbracts minute, subulate, reflexed,

Rare in Sirumalai hills. Fl. & Fr.: September–November.

Sirumalai hills, Nekuttu shoal, (DGL), *R.Kottimuthu* 30925, 30989 (SNCH).

INDIA: Karnataka, Kerala, Tamil Nadu.

DERRIS Loureiro

Derris mathewii Kottaim., Webbia 72: 97. 2017.

Woody climbers; branchlets dark brown, terete, lenticellate. Leaves compound, imparipinnate, 10-16 cm long, juvenile leaves red, mature leaves green, leaflets 7-9, coriaceous, oblong or obovate-oblong, 3-9 x 1.5- 4 cm, margin revolute, base obtuse or subcordate, apex acute to breviacuminate, glabrous above, glaucous below, lateral nervs 8-10 pairs; rachis 8-12 cm long, including 3-4.8 cm long petiole; petiolule 0.8-1 cm long. Inflorescence axillary, racemose panicles, 5-11 cm long; peduncles fasciculate, 2-4 cm long; pedicels verticillate, articulate, pink, twice the length of cayx, 0.8-1.4 cm long; bracts minute, ovate-deltoid, greenish, 1-2 mm long, deciduous, bracteoles minute, 1-1.5 mm long, appressed below the calyx, often caduceus. Flowers zygomorphic, 5-8 mm long, white, sweet-scented; calyx campanulate, reddish, 5-lobed, calyx teeth ovate-deltoid, 3 x 3 mm, apex acute-apiculate, glabrous; corolla white, papilionaceous; vexillum ovate or ovate-orbicular, 6 x 6 mm, glabrous, clawed, veins prominent, raised; two wing petals 6 x 3 mm, glabrous, clawed, two keel petals 5 x 3 mm, puberulent at apex, claws painted, 1 mm long; stamens 10, monadelphous but the vexillar filaments free at base, filaments white, glabrous, 5-6 mm long; staminal tube 3-4 mm long, glabrous, anthers yellow, ovoid-globose, 0.5 mm long; ovary oblong, glabrous, 5 mm long, two- to seven-ovuled, on marginal placenta, style 2-2.5 mm long, glabrous, incurved, stigma capitate. Pods ellipsoid, 3-6 x 1-2 cm, strap-shaped, winged on both sutures, glabrous, obtuse at both ends; seeds one or two, reniform, 1.5-2 x 0.5-0.7 cm, compressed, light brown.

Occasional in Alagar hills. Fl.: March-April; Fr.: May-June.

Alagar hills, Nupuragangai-Periaruvi valley (MDR), *R.Kottaimuthu* 2836 (holotype MH, isotype, SNCH); Karandamalai, on the way to Ayyanar hills, *Kottaimuthu* 52336 & 52689 (SNCH, paratype).

INDIA: Tamil Nadu. Endemic.

DESMODIUM Desv.

Desmodium zonatum Miq., Fl. Ned. Ind. 1: 250. 1855; Sanjappa, Legumes India 166. 1992; Chandramohan *et al.*, Rheedia 26: 26. 2016. *D. ormocorpoides* auct. non DC. 1825; FBI 2: 164. 1876; Gamble, 1: 345. 1918.

Undershrub, erect, 30-70 cm high; stems slightly angular; branches usually covered with fine hooked hairs intermixed with sparse appressed hairs when young, glabrescent at maturity. Leaves 1-foliolate, ovate or elliptic, 5-10.5 x 2.8-5 cm, base obtuse-rounded, margin entire, appressed hairy, apex acute to acuminate, chartaceous, sparsely hooked-hairy, especially on midvein or almost glabrescent above, densely appressed-hairy intermixed with sparse hooked hairs beneath; lateral veins 7-10 pairs, prominent; petioles 1-2.8 cm long, pulvinate for 2-3 mm long, spreading hairs intermixed with short hooked hairs; stipules 2, subulate, 6-9 x 2-2.5 mm, hairy at margins, acuminate at apex, 18-20-veined, persistent, stipels 2, subulate, 5-7 x 0.6-0.8 mm, 4 or 5-veined. Flowers in terminal, to 15 cm long raceme, rachis angular, with densely spreading hooked hairs intermixed with appressed hairs; pedicels 5-6 mm long, spreading hooked-hairy; bracts triangular, 3-7 x 0.8-1 mm, appressed hairy at margins, acuminate at apex, minutely hooked-hairy; calyx campanulate, *c.* 3 mm long, tube *c.* 15 mm long, strigose with minute hooked hairs outside, lobes 4, unequal, upper lobe deltoid, minutely 2-fid at apex, lateral lobes narrowly triangular, lower lobe triangular, longer than others; corolla papilionaceous, pale pink, standard obovate-orbicular, *c.* 5.5 x 3 mm, clawed, emarginated at apex, wings obovate, *c.* 5.2 x 1.2 mm, shortly clawd, round at apex, adnate to keel, keels oblanceolate, *c.* 6 x 2 mm, slightly auriculated at base, rounded at apex; stamens diadelphous, filaments to 6 mm long, anthers uniform; ovary *c.* 5 mm long, puberulent, style *c.* 1.2 mm long, incurved, glabrous, stigma capitate. Pod a lomentum, sessile, linear, to 10.5 cm long, densely hooked-hairy, 7- or 8-jointed, articles narrowly elliptic, *c.* 1.3 cm long, 5-6 times longer than wide; seeds transversely oblong, 1.5-2 x 4-5 mm.

Rare in Uppada Sacred Grove in Visakhapatnam district. Fl. & Fr.: August–November

Uppada (VSKP), *M.Talakeswara Naidu* & *M. Sankara Rao* 5043 (BSID), *M. Tarakeswara Naidu* 2975 (BSID).

INDIA: Andhra Pradesh, Kerala, Karnataka.

WORLD: China, Indonesia, Laos, Malaysia, Myanmar, Philippines, Pacific Islands, Sri Lanka, Thailand and Taiwan.

HYLODESMUM H. Ohasi & R.R. Mill

Hylodesmum repandum (Vahl) H.Ohashi & R.R.Mill., Edimburgh J. Bot. 57: 171. 2000; Candramohan *et al.*, J. Econ. Taxon. Bot. 38: 100. 2014. *Hedysarum repandum* Vahl, Symb. Bot, 2: 82. 1791. *Desmodium scalpe* DC., Prodr. 2: 334. 1825; FBI 2: 181. 1876.

Prostrate undershrubs, up to 80 cm tall, woody in the lower part, branchlets viscid pubescent. Stipules linear-lnceeolate, acute at apex, up to 1 cm long, pubescent outside. Stipels slender, 1-2 mm long. Leaves 3-foliolate, to 8 cm long, leaflets ovate or deltoid-rhomboid, upper leaflet bigger than lower ones, 3-5 x 2.5-4 cm, chartaceous, base sub deltoid, margins wavy, apex obtuse-acute, puberulous above, appressed pubescent below, petiole up to 5 cm long, petiolule of lower leaflets 2 mm long, petiolule of upper leaflet up to 1 cm long. Racemes up to 30 cm long, flowers 2-4 in cluster, peduncle to 5 cm, pedicels to 3 cm, bracts broadly ovate, 2 mm long, calyx pubescent, 3 mm long, calyx lobes divided to middle, unequal, persistent; corolla orange-brick red, standard petal orbicular, 8 x 1 mm, wings elliptic, 7 x 2 mm, keels 1-0.3 mm; stamens diadelphous (9+1), staminal sheath 7 mm long, ovary 5 mm long. Pod short stalked, articles 1-3, half-rhomboidal or crescent-shapted, to 2.5 cm long, hooked, glandular pubescent.

Rare in moist shady localities. Fl. & Fr.: January–March.

Rampa (EG), *Ramaswami* 1509 (CAL); Paderu (EG), *Chandra Mohan* & *M.R.Bhanja* 2781 (BSID).

INDIA: Karnataka, Kerala, Andhra Pradesh.

WORLD: Africa, Asia and Pacific Islands.

INDIGOFERA Linnaeus

Indigofera longiracemosa Boivin ex Baillon, Bull. Mems. Soc. Linn. Paris 1:399. 1883, Gamble 1:220.1918. Hemalatha *et al.*, Current Bot. 8: 1. 2017.

Subshrub. Leaves odd pinnate, leaflets 4 or 5 pairs, opposite, oblong, elliptic, glabrous above, pubescent below, base subacute, margin entire, apex obtuse, apiculate. Flowers pinkish in axillary racemes, racemes slender, up to 17 cm long, more than 12-flowered; calyx tube pubescent without; corolla pinkish. Pod oblong, thick, shortly mucronate, pubescent, about 6-seeded.

Rare in Ardhagiri hills in Chittoor district of Andhra Pradesh (Hemalatha *et al.*, 2017).

INDIA: Kerala, Tamil Nadu, Andhra Pradesh.

WORLD: Africa and Madagascar.

RHYNCHOSIA Lour.

Rhynchosia jacobii Chandrab. & B.V.Shetty, Bull. Bot. Surv. India 15: 139. 1973; Sanjappa, Legumes India 239. 1992; Swamy *et al.*, Indian J. Forestry 41: 4. 369. 2018.

Trailing herbs, stem, leaves, stipules, bracts, calyx and pods with dense, mixed short and long hairs; hairs up to 3 mm long. Stem woody at base; branching from base, slender, terete, faintly striate; internodes 0.5-7 cm long. Leaves alternate, pinnately trifoliolate;petioles 0.6-3.7 cm long; stipules 5.14-5.52 x 0.57-0.63 mm long, linear-lanceolate, glabrous on lower surface except at the tip, distinctly grooved in between the nerves; leaflets oblong or elliptic-oblong, subcoriaceous, 3-nerved from the basee, minutely yellowish gland-dotted beneath; lateral nerves 4-7 pairs from the midrib, nerves and nervules raised on both the surfaces; terminal leaflet 0.5-4.0 x 0.4-2.2 cm, base nearly rounded, margin entire, apex mucronate; lateral leaflets 0.6-3.7 x 0.4-1.9 cm, base unequal, margin entire, apex mucronate; petiolules up to 1.8 mm long; stipels absent. Flowers *c.* 6.93 x 2.48 mm, yellow, axillary, usually solitary, occasionally two on a slender rachis; rachis up to 2 cm long; pedicel *c.* 1.61 x 0.63 mm, bracts 3-4 mm long, ovate-lanceolate, glabrous on lower surface, distinctly grooved in between the nerves; calyx 4-5.71 x 2.47 mm; tube 2.43 x 2.16 mm; teeth 5, lanceolate, unequal, 2.76 -3.17 mm long; the lower one longer than the others; corolla yellow, papilionaceous; standard 6.2-6.8 x 3.3–4.34 mm, obovate, shortly clawed; the claw *c.* 1.13 x 0.28 mm with inflexed, *c.* 0.32 mm long auricles at the base, emarginated at apex; veins red; wings 5.99–7.28 x 1.40–2.26 mm, obliquely oblong–obovate, conspicuously clawed with *c.* 0.87 x 0.47 mm auricles; the claw 1.62–2.60 x 0.33 mm; keels 5.59–6.11 x 1.20–1.47 mm, obliquely obovate, distinctly clawed, with *c.* 0.83 x 0.46 mm auricle; the claw 131–1.41 mm long; stamens 5.26 x 0.78 mm long, 9+1; 9 filaments fused forming *c.*4.36 mm long sheath; 0.51–0.86 mm long, free at apex; anthers oblong-orbicular in outline, unequal, larger ones 0.46 -0.50 x 0.45–0.46 mm; smaller ones *c.* 0.47 x 0.29 mm.Pistil *c.* 5 mm long; ovary 1.95

x 0.83 mm, elliptic, subsessile, densely clothed with long, white silky hairs; ovules 2; style 3.02 x 0.09 mm, glabrous, white; stigma 0.03–0.04 mm in diam., capitate, brown. Pod 20 x 5.7 mm, slightly compressed, subfalcate, with short beak at apex; beak up to 5 mm long, 2-seeded, continuous within; seeds 3.58–4.27 x 4.41-4.81 mm, brown, subreniform, turgid, with an indistinct strophiole.

Rare in hills. Fl. & Fr.: July–December.

Janapalacheruvu section, Giddalur (PKM), *J. Swamy* 8836 (BSID).

INDIA: Western Ghats of Tamil Nadu and Eastern Ghats of Andhra Pradesh. Endemic.

Rhynchosia ravii K. Prasad & A. Narayana Swamy, Pytotaxa 175: 156. 2014.

Erect shrubs, *c.* 1.2 m high, with glandular tipped hairs, and with glands in the leaves; branches erect; mature branches woody, terete, tomentose or pubescent; young branches slender, terete, pale golden-yellow or greyish-white pubescent. Leaves 6–10 cm long, alternate, 3-foliolate; petiole *c.* 3.5 cm long, with pale golden-yellow hairs; stipule oblong-lanceolate or ovate-lanceolate, 3.5–5 x 1–1.5 mm, acuminate, with dense greyish-white hairs, caducous; leaflets dimorphic, with dense, short, fine, greyish-white hairs canescent on both sides, veins impressed above, strongly raised beneath; terminal leaflet ovate-oblong or ovate-elliptic, *c.* 5 x 2 cm, base obtuse or cuneate, margin entire, apex acute; petiolule 8–10 mm long; lateral leaflets obliquely oblong-ovate or oblong-lanceolate, *c.* 4 x 1.8 cm, base cuneate or obtuse, margin entire and slightly incurved, apex acute, petiolule *c.* 2 mm long; stipels 2 per leaf, linear or subulate, *c.* 1 x 0.4 mm, caducous. Flowers in axillary raceme, 1- 3 racemes per axil, slender, filiform, golden yellow pubescent; peduncle terete, 1 - 3.5 cm long; racemes 5–6 cm long, laxly 6-8-flowered; bracts ovate, *c.* 5 x 2.6 mm, shallowly or deeply 3-lobed at apex, pubescent; bracteoles absent; pedicels short, filiform, *c.* 1.2 mm long, sparsely golden-yellow hairy; calyx pale green with pale rose tinge, *c.* 1 cm long, campanulate, tube 2–4 mm long, with dense, short, fine, greyish-white hairs; lobes unequal, obliquely oblong-ovate or oblong-lanceolate, 6–8 x 4–4.5 mm, acute, 1-3-veined; corolla standard yellow, orbicular, 7–8 x 7–7.5 mm, rounded and emarginate at apex, with two inflexed auricles at the base; claw 1.6–2 mm long; wings yellow, oblong or narrowly oblong-obovate, 7.5–8.2 x 2–2.5 mm, rounded at apex, without auricles; claw *c.* 2 mm long; keels pale greenish-yellow or white, incurved, connate, falcately obovate, 7–7.5 x 2.6–3 mm, rounded at apex, with or without auricles; claw *c.* 2 mm long; stamens 10, white, diadelphous, 9 + 1, unequal; bundle 6–8 x 1.2–1.8 mm; individual filament *c.* 7.5 mm long; anthers yellow, dehiscent, oblong, *c.* 1 x 0.6 mm; ovary sessile, oblong-elliptic, *c.* 4 x 1.5 mm, with white hairs; ovule 2; style *c.* 6 mm long, glabrous; stigma pale yellow, capitate, *c.* 0.2 mm long. Pod compressed, longer than the calyx, septate between the seeds, oblong, *c.* 1.4 x 0.7 cm, incurved beak, tomentose; seeds 2, subreniform, *c.* 4 x 3 mm, strophiolate.

In red soils in grass dominated slopes of degraded dry deciduous open forests, at an elevation of 400–600m in Anantapuram and Kadapa districts. Fl. & Fr.: November–April.

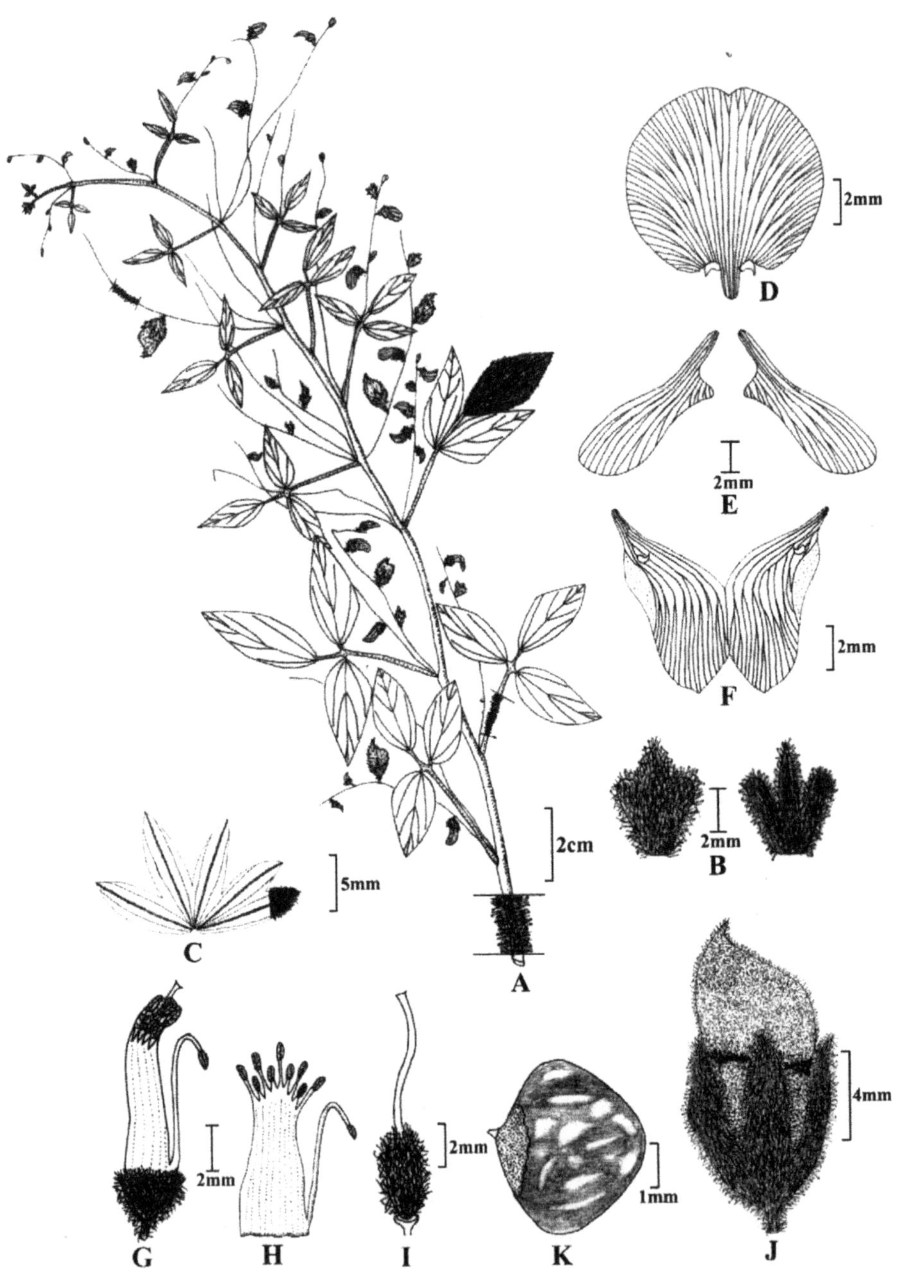

Figure 117. **Rhynchosia ravii** K. Prasad & A. Narayana Swamy.

A. Twig, B. Calyx opened out, C. Bracts, D. Standard petal, E. Wings, F. Keels, G. Androecium, H. Staminal column opened out, I. Gynoecium, J. Fruit, K. Seed (After Prasad and Narayana Swamy, 2014).

Talupula forest (ATP), *ANS* & *KP* 38496 (holotype BSID, isotypes SKU); Batrepalli RF (ATP), *BR & ANS* 36627 (BSID); Talupula RF (ATP), *ANS & KP* 44908; Mangapatnam (KDP), *BR &KP* 41602.

INDIA: Andhra Pradesh, Endemic.

Note: *Rhynchosia ravii* is closely allied to *Rhynchosia beddomei*, but differs in the following characters: plants with glandular hairs (versus glandular hairs absent); stipels present (versus stipels absent); racemes 5–6 cm long, laxlyflowered (versus racemes *c.* 2 cm long, densely flowered); bracts ovate, deeply or shallowly 3-lobed (versus bracts cordate or oblong-ovate, acute or obtuse); calyx tube 2–4 mm long (versus calyx tube less the 2 mm long); gynoecium as long as or longer than the staminal sheath (versus gynoecium shorter than the staminal sheath); ovary *c.* 4 mm long (versus ovary *c.* 1.5 mm long); style *c.* 6 mm long (versus style *c.* 1.5 mm long); and fruits oblong, two-seeded (versus suborbicular, one-seeded fruits)

VIGNA Savi

Vigna stipulacea (Lam.) Kuntze, Rev. Gen. 212. 1891. *Dolichos stipulaceus* Lam., Encycl. 2(1): 300. 1786.

Trailing herb, annual, to 2 m long. Root nodules cushion like, whitish brown, sub-globose. Stem twining, herbaceous, 4–6-angled, hairs sparsely covered, distinctly arranged, 0.2 mm long. Internode 0.5–5 cm long. Stipules 2, conspicuously large, opposite, ovate, 0.7–1.4 x 0.4–1 cm, glabrous, base deeply cordate, margin entire, tip acute, scarlet red color at base (when plant grows in shady places the color does not appear), papery appendages present below the stipules, ovate, transparent, 1.5 x 1 mm. Leaves trifoliate, 5–17.5 cm long (with petiole), petiole 3.3–12 cm, glabrous, round. Leaflets glabrescent above or distinctly hairy, pubescent below on veins; middle leaf let completely 3-lobed, 1.7–4.5 x 1.5–3.9 cm, rhomboid-ovate, middle lobe obovate, entire, tip obtuse, lateral lobes elliptic - oblong, entire, tip acute; lateral leaflets 2–3 lobed, 1.7–4.5 x 1.4–3.6 cm, rhomboid-ovate, middle lobe obovate, entire, tip obtuse, lateral lobe(s) elliptic-oblong, entire, tip acute. Petiolules 2 mm, densely hairy, light brown. Stipules 2, opposite, 2–2.5 x 1–1.5 mm, glabrous, ovate, base acute, margin entire, tip acuminate. Inflorescence axillary, pseudo raceme or terminal raceme 3–6-flowered, peduncle 10–30 cm, pubescent. Flowers golden yellow 1.1 x 0.5 cm; bracts 2, boat-shaped; primary bract elliptic-ovate, 1.8 x 1.8 mm, glabrous, peltate, base cordate, margin entire, tip acute; secondary bract elliptic-ovate, 2.1 x 1.2 mm, glabrous, peltate, base truncate, margin entire, tip acute; bracteoles 2, ovate-oblong, elliptic-oblong, 2.2 x 0.6 mm, hairy, base acute, margin entire, tip acuminate; sepals 4-lobed, connate, ovate, 2 x 5 mm, hairy, basally pinkish, tip rounded, calyx tube 2.2 mm long; standard petal golden yellow, 0.5 x 1.1 cm, glabrous, base cordate, margin entire, tip obcordate; wing petals yellow, orbicular, 0.5 x 0.6 cm, base divided, acute, pink tinge in lower margin, tip rounded; keel petals 2, incurved to the left, 8 x 2 mm, jointed at tip and forms tube or pocket, tube pink color, curved 180°, 1.2 mm long; stamens 8+1 or 9+1, nine stamens forms tube, 0.4–0.5 cm, anther filaments 2–3 mm long, one stamen with filament 8 mm long; ovary 2 mm long, flattened, style and stigma 4 mm long, stigma curved, densely

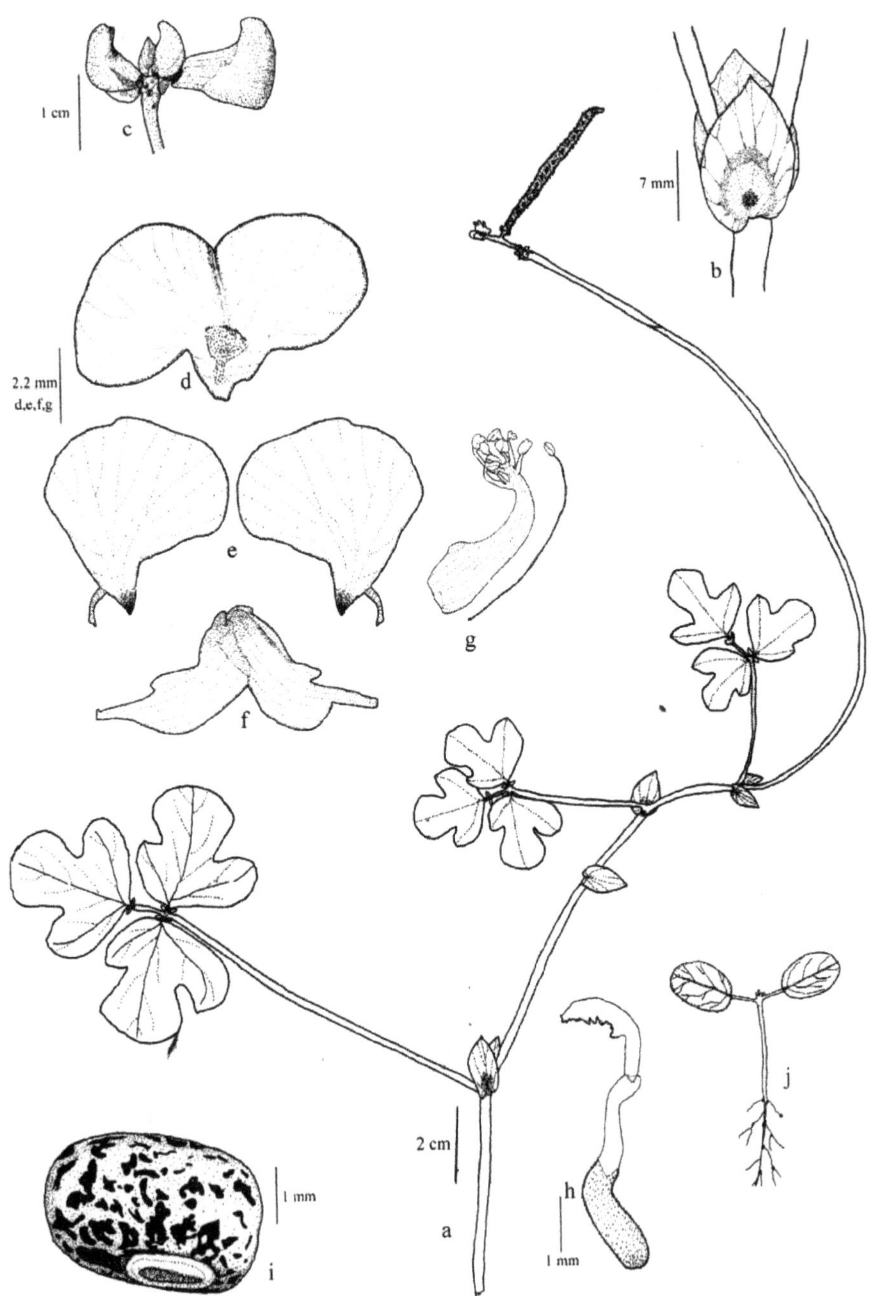

Figure 118. **Vigna stipulacea** (Lam.) Kuntze.

a. Habit, b. Stipule, c. Flower, d. Standard petal, e. Wing petals, f. Keel petals, g. Stamens, h. Pistil, i. Seed, j. Seedling.

hairy at lower side. Fruit sub cylindrical, linear 3.8–4.5 x 0.2–0.3 cm, hairy, hairs pink color, 1 mm long; seeds 15–23, 3 x 2 mm, ellipsoidal, green with black patches, fine reticulate, ends rounded; hilum oblong, protruding, 1.2–1.4 mm long, spongy like, white; aril cushion with little rim, moderately developed.

Common in moist places and near water bodies, sandy soil, grasses and paddy fields in Anantapuram district. It grows along with Cyperoids. Fl. & Fr.: June–September, sometimes May–July. Vern.: *Chitti pesalu.*

Chinnapalli (ATP), *KRKS & BSS* 37999; Kalasamudram (ATP), *KRKS* 39545.

MELASTOMATACEAE

CLIDEMIA D. Don

Clidemia hirta (L.) D. Don, Mem. Wern. Nat. Hist. Soc. 4: 309 309 1823. *Melastoma hirtum* L., Sp. Pl. 1: 390 390 1753.

Undershrubs, 1 - 1.5 m tall, stems branched, densely ciliate throughout. Leaves opposite and simple, sometimes unequal size; petiole, 1 to 2.5 cm long, densely hairy; lamina 7 - 15 x 4 - 8 cm, elliptical to broadly ovate to oblong, abruptly acuminate on top, rounded to subcordate at the base, loosely hairy on both sides, margin serrated and ciliated, main nerves 5 from the base, prominent on the beneath. Inflorescence axillary cyme, 5 to 12-flowered; pedicel 5 mm, hairy and equipped with two small papery linear bracts near the top. Receptacle forming a long cylindrical tube of 4 mm corresponding to the inferior ovary, surmounted by a membranous calyx slightly lobed and with 5 filiform appendages; corolla with 5 petals narrowly elliptical, about 9 mm long and 3.5 mm wide, white to pale purple; stamens 10, with short filament; ovary semi-inferior, with a simple style, cylindrical, glabrous, only fused to the base with the receptacle, with 5 multi-ovulated loculus. Fruit spherical to ellipsoid berry, about 7 mm long, pericarp fleshy, bluish purple when ripe, formed by the tube of the receptacle; seeds many.

Near the streams in high altitudes of Sirumalai hills. Fl. & Fr.: December–February.

Ooradi, Sirumalai hills (DGL), *SKS* 1245 (SGH).

INDIA: Peninsular India.

WORLD: Tropical Africa, Australia, invasive in all over tropics and S. Asia.

OSBECKIA Linnaeus

Osbeckia aspera (L.) Blume, Bijdr. Natuurk. Wetensch. 6: 220. 1831. var. **aspera**; FBI 2: 519. 1879; Gamble 1:492. 1919.

Undershrubs, 1-1.5 m high; branchlets adpressed pubescent. Leaves elliptic, 1.5-4 x 0.5-1.6 cm, coriaceous, velvety tomentose, 4-5-ribbed from base, apex acute-mucronate. Flowers pink, in 4-7-flowered terminal cymes; peduncles 0.5-1 cm long; hypanthium with prominent intersepalar emergences; petals 5, obovate, 1.5-1.8 x 1.8-2 cm; stamens 10, fiaments 0.5-0.6 cm long, incurved. Capsule urceolate, 1 cm across, many-seeded.

Occasional in Sirumalai hills.Fl. & Fr.: June–August.

Kannadiparai, Sirumalai (DGL), *R.Kottaimuthu* & *R. Ganesan* 2080 (ATREE).

INDIA: Peninsular India.

WORLD: Sri Lanka.

CRASSULACEAE

KALANCHOE Adanson

Kalanchoe bhidei T. Cooke, Fl. Bombay 1: 467. 1903; Gamble, Fl. Madras 1: 318. 1919; Gandhi in Saldanha & Nicolson, Fl. Hassan Dist. 209. 1976; Srinivasan in Nair & Henry, Fl. Tamil Nadu (Analysis) 1: 145. 1983; Sharma *et al.*, Fl. Karnataka 98. 1984; Godbole & Das Das in Singh *et al.*, Fl. Maharashtra 1: 840. 2000; N. B. Singh *et al.*, Crassulaceae of India 54, t. 2. 2011; Chorghe *et al.*, Nelumbo 59: 70. 2017.

Perennial, erect, succulent herbs, 70–100 cm high, stem terete, yellow-reddish in colour, glabrous. Leaves simple, opposite, ovate-oblong, 9 - 11 x 4–5 cm, obtuse at apex, attenuate at base, serrate along margins; petiole 1–2 cm long. Inflorescences terminal, cymose, paniculate, spreading, 18–22 x 13–17 cm. Flowers bisexual, 4-merous, 12–14 x 3–4 mm, whitish-pink in colour, pedicellate, pedicels 8–10 mm long; bracts linear-lanceolate, 2–3 mm long; sepals free, 3–4 x 1–2 mm, linear-lanceolate, acuminate at apex, broader at base, greenish-pink; corolla whitish-pink, 1.3–1.4 x 0.6–0.7 cm, tubular, 4-lobed at apex; lobes 4–4.5 x 1.8–2 mm, oblong, mucronate at apex; stamens 8, arranged in two rows, epipetalous; filaments 2–3 mm long; carpels 4, erect; style 3–4 mm long; nectar scales 4, linear, 2–2.5 mm long, present at the base of the carpels.

Rare along the rocky cliffs and exposed rocky areas of above 700m in deciduous forests. Fl. & Fr.: January-March.

Penchalakona (NLR), *L. Rasingam, M. Sankara Rao* & *Alok Chorghe* 2926 (BSID).

INDIA: Maharashtra, Karnataka, Tamil Nadu and Andhra Pradesh (Endemic).

ARALIACEAE

ARALIA Linnaeus

Aralia thomsonii Seem, ex C. B. Clarke, Fl. Brit. India 2: 723. 1879; Brandis, Ind. Trees 351. 1906; Kanjilal *et al.*, Fl. Assam 2: 345. 1938; Haridasan & R.R.Rao, Forest Fl. Meghalaya 1: 429. 1985; Wen *et al.*, Rheedea 12: 13. 2002; Chandramohan *et al.*, J. Econ. Taxon. Bot. 40: 159. 2016.

Prickly unbranched shrubs, 1.5-10 m high; bark brownish; young parts, inflorescence and leaves softly tomentose. Prickles short, strong and weaker on pinnae. Leaves 2-3-pinnate, large, up to 200 cm long, stipulate, rachis and its ramifications prickly; pinnae 5 pairs with the subsidiary pinnae, lower pinnae longer than above ones, leaflets chartaceous, ovate to lanceolate, 8-20 x 4-10 cm, apex acuminate, margins serrulate, base rounded or sub-cordate, densely covered with patent simple trichomes on adaxial surfaces, densely pubescent on abaxial surface;

lateral veins 8-10, pubescent; petiolules 0-13 mm long, pubescent. Inflorescence large panicle, terminal, compound umbel, densely pubescent or sparsely prickled pubescent, up to 90 cm long; bracts and bracteoles persistent; bracteoles up to 3.5 mm long, densely pubescent; pedicels up to 12 mm long, densely pubescent; umbels 8-25-flowered; sepals ovate to triangular, 0.5-0.8 x 0.6–0.7 mm; petals ovate to triangular, 1.8-2 x 1-1.2 mm; stamens 5; anthers white; filaments up to 4 mm long; ovary 5-locular; styles free. Fruits blackish or purple, ovoid to subglobose, 3.5-5 mm diam., ridged.

Rare in Maredumilli forest in East Godavai district. Fl. & Fr.: September-October.

Maredumilli, Near Tiger camp (EG), *M.R.Bhanja* & *C.S.Reddy* 7797 (BSID).

INDIA: Kerala, Tamil Nadu, Andhra Pradesh, Karnataka.

WORLD: Malaysia, Vietnam, Thailand, Myanmar.

APIACEAE

HYDROCOTYLE Linnaeus

Hydrocotyle kollimalayensis Karupp. *et al.*, Bangladesh J. Plant Taxon. 21(2): 167. 2014.

Perennial herbs. Stem weak, slender, filiform, silky setaceous, creeping, rooting at nodes. Leaves simple, pale green, sparsely hirsute above and dense below, membranous, base cordate, 1.5-2.0 cm wide, 1.0-1.5 cm long, angular, margin shallowly 5-lobed, lobes acute or obscurely triangular, middle lobe larger than the others, crenate, subequal. Petioles slender, 2-4 cm long, with white reflexed hirsute hairs above. Umbels simple, 6-12-flowered, globose; peduncle 1.0-2.5 cm long, glabrous, axillary, slender. Flowers sub-sessile or very shortly pedicillate; petals ovate, white, entire, minute, *c.* 0.2 mm long, triangular, acute; stamens 5, minute, slightly incurved in bud; filaments short, *c.* 0.1 mm long; anthers bilobed; ovary minute, ellipsoidal, greenish; style 1, persistent; stylopodium slightly elevated. Fruits ellipsoid, compressed, with obscure dorsal edges, glabrous, 3–ribbed, ribs obscure, subequal, *c.* 1.2 x 1 mm, glabrous above, minutely granulate at base, pale brown.

Rare in Kollimalai hills of Tamil Nadu. Fl. & Fr.: Throughout the year.

Solakkadu, Kolli hills (NMK), *SK* 24375 (Holo! MH).

INDIA: Tamil Nadu (Endemic).

RUBIACEAE

DIODELLA Small

Diodella teres (Walter) Small, Fl. Lancasteer Co. 271. 1913. *Diodia teres* Walt., Fl. Carol. 87. 1788; Soosairaj & Ananthi, J. Econ. Taxon. Bot. 38: 549. 2014

Annual, erect herbs, 15-20 cm high; stem terete, sparsely pubescent, hairs 1-3-cellular, 0.15 mm long. Leaves decussate, sessile, lanceolate or narrowly elliptic, 1.5-2.2 x 0.3-0.5 cm, base truncate or cuneate, margin fimbriate, apex acute, pubescent

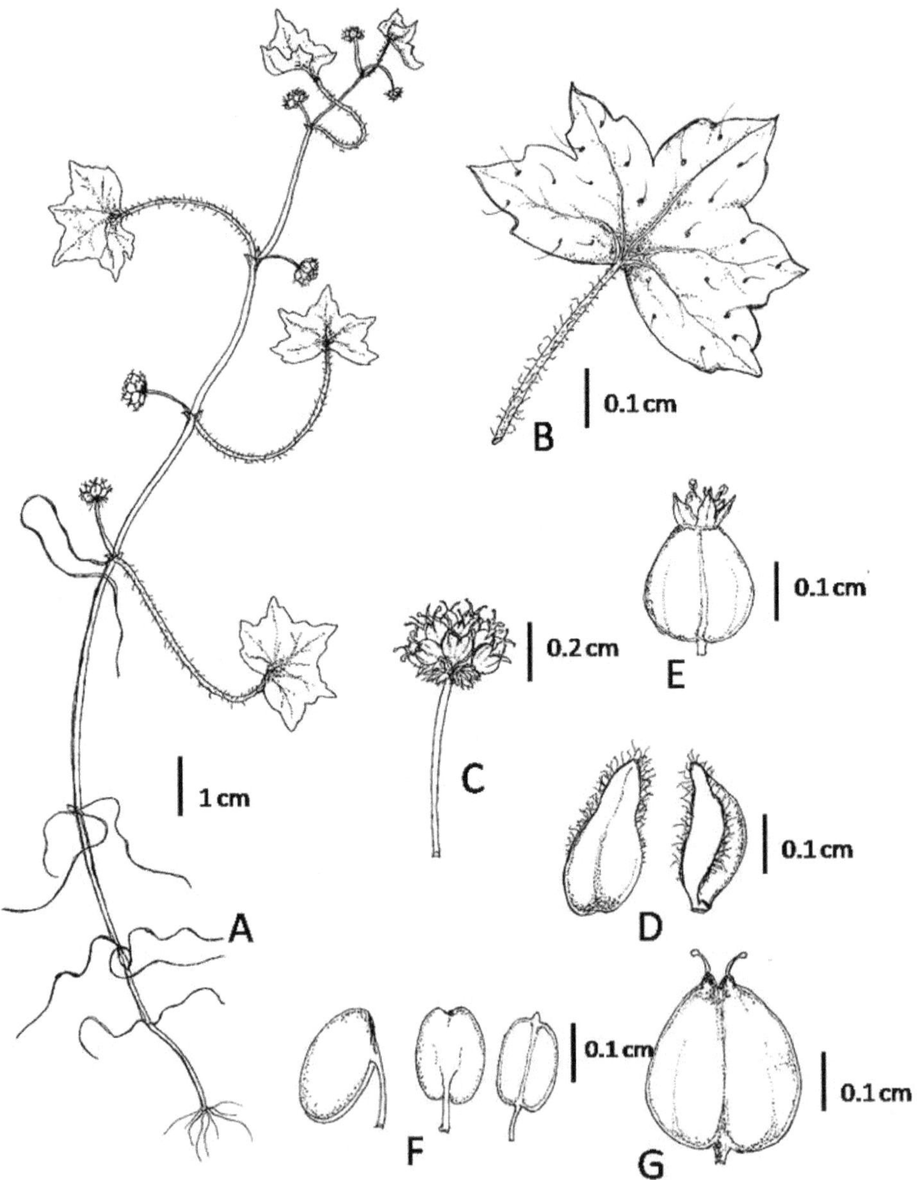

Figure 119. **Hydrocotyle kollimalayensis** Karupp. *et al.*

A. Habit, B. Leaf, C. Inflorescence, D. Bracts, E. Pistil with stamens, F. Stamens, G. Fruit.

on both surfaces, stipular sheath 2 mm long, purplish, adnate to leaf base, fimbriate, 6-8-lobed, bristle 2-6 mm long. Flowers 2-4 in axillary 3 mm long, subsessile cymes, pedicel 0.5 mm long; hypanthium 0.6 mm long; calyx 4-lobed, lobes lanceolate, 1.5 x 0.7 mm, fimbriate along the margin; corolla campanulate, 4.5 x 2 mm, purple, connate up to 2 mm, free lobes 1.5 mm, broadly ovate, apex mucronate; stamens 4, adnate to the throat of corolla, alternate to corolla lobes, introrse, filament 0.5 mm, anther 1.25 x 0.25 mm, basifixed; pistil 3 mm long, biicarpellary, style filiform, stigma biglobose, purple in colour. Capsule obcordate with four persistent calyx, 3 mm long, 3 mm diameter at apex, dehisces into two nutlets when mature; seeds circular, 2 x 2 mm, ventrally grooved.

In moist open places, loam soil along the streams at 1200 m altitude. Fl. & Fr.: November–January.

Pachamalai (TP), SJCBOT 1412, SJCBOT 1416, SJCBOT 1421 (St. Joseph College Herbarium, Tiruchirappalli).

INDIA: Kerala, Tamil Nadu.

WORLD: Tropical America and Africa.

PAVETTA Linnaeus

Pavetta travancorica Bremek., Repert. Spec. Nov. Regni Veg. 37: 81. 1934; Ahmed. & M.P.Nayar, End. Pl. Ind. Reg. 1: 164. 1987; Deb & Rout, J. Bombay Nat. Hist. Soc. 89(3): 349. 1992 & Bull. Bot. Surv. India 41: 161. 1999; Singh *et al.*, End. Vasc. Pl. India 236. 2015; Swamy *et al.*, Nelumbo 60 (2): 121.2018.

Shrubs up to 1.5 m high; stem stout, erect, branched; branches terete or sub-quadrangular, pubescent when young, becoming corky in age. Leaves petiolate, 5.5-15 x 1.9-5 cm, elliptic-lanceolate, apex acute-acuminate, margin entire, base attenuate, subcoriaceous, glabrous or puberulent on upper surface, hairy and with many bacterial leaf galls in some leaves and absent in others; leaf galls present on secondary and finer nerves; lateral nerves 10-13 pairs, alternate or sub-opposite, more prominent beneath; stipules interpetiolar, 4-7 x 3-5 mm, narrowly triangular, membranous, scarious, acute at apex, glabrous; petioles 1.2-3 cm, hairy. Inflorescence axillary, peduncled, trichotomously branched, loose corymbose cymes, 6-9 cm long; peduncle up to 5 cm long, slender, hairy; primary branches 3-5 cm long; secondary branches 0.5-1.5 cm long; tertiary braches 0.3-0.8 cm long; bracts traingular, connate at base, 4 x 2 mm, membranous, hairy on upper surface, pilose at base and glabrous towards apex on inner surface. Flowers white, 30-47 on slender pedicels; pedicels 2-7 mm long, hairy. Hypanthium 1 x 0.8 mm, sub-globose, densely hairy; calyx tube 0.5 x 1-1.5 mm, broader above, densely hairy outside; glabrous within; teeth 0.1–0.2 x 0.1 mm, dentate or triangular, acute at apex, pubescent without; glabrous within; corolla tube 7-16 x 0.5-0.8 mm, cylindrical, glabrous on upper surface, sparsely hairy on above 2/3 part on lower surface; lobes 4, oblong or oblong-obovate, 5-6 x 1.5-1.8 mm, obtuse at apex, glabrous; stamens 4, attached to the throat of the corolla tube, alternate to the petal lobes; filaments *c.* 0.8-1.2 mm long; anthers 4-5 x 0.4 mm, acute at apex; gynoecium 3-3.1 cm long; ovary 1 x 0.8 mm, glabrous; style *c.* 2.7 cm long, slender long exserted, glabrous, slightly pubescent at apex; stigma *c.* 3 mm, pubescentFruits not seen.

Rare along the waterfalls and shady habitats of the moist deciduous forest of Seshachalam Biosphere Reserve. Fl. & Fr.: February-June & March - November.

Seshachalam Biosphere Reserve, Papanasanam beat (CTR), *J. Swamy* 8842 (BSID).

INDIA: Karnataka, Tamil Nadu and Kerala and now from Andhra Pradesh.

PSYCHOTRIA Linnaeus

Psychotria gopalanii Samb., Rheedea 27(1): 42. 2017.

Shrub, *c.* 3 m high. Stems terete; branchlets lenticellate; internodes 1.5–2 cm long. Leaves simple, opposite-decussate, obovate-oblanceolate or elliptic-oblong, 1.5–11 x 0.8–6.5 cm, attenuate at base, entire at margins, acute-acuminate at apex, subcoraiceous, glabrous, dark green, shiny above, pale green beneath; lateral veins 8–12 pairs; midrib prominent, infossous above, raised beneath, intramarginally looped; domatia glabrous; stipules broadly ovate, *c.* 1.5 cm long, keeled, caducous, midrib rufous-hairy inside, base ligulate inside; ligules 4-lobed, densely rufous hairy, lobes acuminate at apex, persistent; petioles *c.* 1.5 cm long, canaliculate above, rounded beneath. Inflorescences terminal, paniculate racemose cymes, 8–13 cm long; peduncles *c.* 4.5 cm long, minutely grooved; flowering branches spirally arranged, stout, to 1 cm. Flowers 6–7 mm long; bracteoles linear, minute, caducous; pedicels 1–2 mm long; calyx tube 1–1.3 mm long; lobes 5, minute, glabrous; corolla tube 3–4 mm long, narrow, glabrous outside, with a rim of white villous-hairy above the middle inside; lobes 5, lanceolate, *c.* 2.5 mm long, thick, recurved at apex; stamens 5, exserted; filaments 1.5–2 mm long, erect, adnate to corolla throat; anthers oblong, 0.5–1 mm long, dorsifixed; ovary bilocular; ovule 1 in each locule; style *c.* 3 mm long, dilated towards apex, included; stigma bifid, ovate or tongue-shaped, smooth Fruits globose, *c.* 1 cm, 10-ribbed, apex with persistent calyx lobes; testa thin; pyrenes plano-convex; albumin ruminate.

Occassional in forests of Erode district in Tamil Nadu. F. & Fr.: March–September.

Anthiyur R.F. Erode District, Tamil Nadu, *Sambooranam* 686 (Holo! CAL, Iso, MH).

INDIA: Tamil Nadu (Endemic).

SPERMOCOCE Linnaeus

Spermacoce exilis (L.O. Williams) C.D.Adams ex W.C.Burger & C.M.Taylor, Fieldiana, Bot. N.s.33: 316. 1993; Chandramohan *et al.*, Indian J. Forestry 39(3): 295. 2016. *Borreria exilis* L.O.Williams, Phytologia 28: 227. 1974. *B. repens* DC., Prodr. 4: 542. 1830. *Spermacoce mauritiana* Gideon, Kew Bull. 37: 547. 1983.

Annual, branched, erect or decumbent herb, up to 40 cm high; branches trichotomous, 4-angled, pubescent. Leaves subsessile, opposite, decussate, ovate or oblong-elliptic, 0.7-2.5 x 0.6-1.2 cm, base obstuse, margin strongly revolute, apex acute, more or less hispid on the surface, much hispidulous on nerves beneath, mid nerve distinct, lateral nerve 3-5 pairs; stipules 2, 0.8-1.3 mm long, 5-6 mm broad, with 11-13 strigose bristles. Inflorescence axillary, in whorls or terminal heads,

1-1.3 cm across. Flowers many, white, up to 3.2 mm long; pedicels 0.5-0.6 mm long, glabrescent; bracts numerous, filiform, hyaline; calyx truncate, hyaline, strigose, along margins and sparsely strigose on apical parts, 4-lobed, lobes unequal, longer ones ovate to lanceolate, up to 2 x 0.9 mm long, acuminate to cuspidate at apex; shorter ones lanceolate, up to 1 mm long; corolla white, slightly exserted from calyx lobes, up to 2.2 mm long, tube 1.3 mm long, glabrous, 4-lobed, lobes ovate, 0.5-0.7 mm long, sparsely hairy at apex; stamens 4, partially exserted, inserted below the sinus of corolla lobes, filaments short; anthers oblong, 0.4 mm long, dorsifixed; ovary 1 mm long, style *c.* 2 mm long, stigma 2-lobed, lobes linear, 0.3 mm long, hairy. Capsules depressed globose, truncate at apex with persistent calyx, pale brown, crustaceous, sparsely strigose on upper portion; seeds 2, minute, brown.

Occasional on stream bunds. Fl. & Fr.: September–June.

Satkosia Wildlife Sanctuary, Tikarpada range, along river bank of Mahanadi, on the way to Bimki (Orissa), *K.Chandramohan* & *J.Swamy* 7631 (BSID).

INDIA: Karnataka, Maharashtra, Orissa, Tamil Nadu.

ASTERACEAE

ACMELLA Pers.

Acmella radicans var. **debilis** (Kunth) R.K. Jansen, Syst. Bot. Monogr. 8: 72. 1985; Chandramohan *et al.*, Indian J. Forestry 39: 295. 2016. *Spilanthes debilis* Kunth, Nov. Gen. Sp. 4(16): 165. 1820. *Acmella debilis* (Kunth) Cass., Dict. Sci. Nat. 24: 331. 1822. *Ceratophalus debilis* (Kunth) Kuntze, Revis. Gen.Pl. 1: 326. 1891.

Annual herb, up to 4 cm high; stem terete, minutely pubescent. Leaves simple, opposite, ovate to triangular, up to 7 x 4 cm, base obtuse, margins serrate, apex acute, 3-nerved from the base, sparsely pubescent on both the surfaces; petiole up to 1 cm long, densely hairy. Heads white, radiate, subglobose, 8-9 x 5-6 mm at maturity, borne on long, axillary or terminal peduncles; peduncles *c.* 5 cm long, receptacles columnar, glabrous, occasionally slightly hairy at apex. Involucral bracts 2-seriate, ciliate along margins, outer bracts 6, broadly ovate, 4-5 mm long, 4-5-nerved, inner 5, linear, 3-4 mm long, 3-nerved. Ray florets 4-6 below disc florets; corolla up to 3 mm long, white, corolla tube1-1.5 cm long, densely hairy, limb with three rounded lobes, lobes up to 2.2 mm long; style arms short, truncate. Disc florets many, bisexual; corolla funnel-shaped, 1.8-2 mm long, greenish white, distinctly 4-lobed, lobes sparsely papillate, base bulged; stamens 4, filaments up to 0.6 mm long, anther lobes linear, 0.5-0.7 mm long, sagittate with a terminal appendage, style 1.2-1.4 mm long, arms unequal, papillose. Achenes black, 1.6-2 x 0.6-1mm, dimorphic, of ray florets trigonous with corky margin, densely ciliate with straight tipped hairs, pappus three subequal awns; of disc florets flattened, ellipsoid, extensively ciliate with evident cork like margin; pappus of two unequal awns.

Occasional in moist habitats of perennial streams. Fl. & Fr.: September–May.

Satkosia Wildlife Sanctuary, Tikarpada range, along river bamk of Mahanandi (Orissa), *K.Chandramohan* & *J.Swamy* 7631 (BSID).

INDIA: Karnataka, Maharashtra, Orissa.

CALYPTOCARPUS Less.

Calyptocarpus vialis Less., Syst. Gen. Comp 221. 1832; Swetha & B.R.P.Rao, J. Biosci. Res. 4: 10. 2013. *Synedrella vialis* (Less.) A. Gray, Proc. Amer. Acad. 17: 217. 1882; Chowdhery in Hajra *et al.*, Fl. India 12: 415. 1995.

Perennial decumbent herb; stems prostrate, branched from the base, rooting at nodes, spreading, strigillose. Leaves opposite, ovate to 3 x 2 cm, basally 3-nerved, cuneate at the base, margins crenate-serrate, acute at apex, scabrid on both surfaces; petiole to 5mm. Heads axillary, solitary, sessile, heterogamous; involucral bracts four, in two rows, lanceolate, green, shortly hairy, ciliate along margins, outer one oblong- lanceolate, inner one linear. Ray florets 5-8, yellow, ligulate, unisexual-female, petal apex 2-3-lobed; disc florets 8-10, tubular, bisexual. Achenes dorsally compressed, oblanceolate, to 5 x 1.5 mm; achenes of disc florets narrower and covered with short spines; pappus usually of 2 divergent awns, shortly hairy.

Found occasional in clumps, in moist and shady localities in gardens, nurseries and households in Anantapur district and Alagar hills of Tamil Nadu. Fl. & Fr.: July - September.

S.K. University Campus (ATP), *BR* & *B.Swetha* 44549; Alagar hills (DGL), *Kottaimuthu* 22001 (SNCH).

INDIA: Andhra Pradesh, Tamil Nadu, Karnataka, Kerala.

WORLD: Native of North America and naturalized in Austalia and Thaiwan.

EMILIA Cass.

1. Weak herb; leaves radical, mostly cauline, sinuate-serrate; seeds red-brown..**E. sonchifolia**
1. Robust herb; leaves strictly cauline, dentate; seeds black .... **E. reddyi**

Emilia reddyi Satish & Prakasa Rao, J. Asia Pacific Biodiv. 10: 107. 2017.

Robust, annual herb, up to 85 cm high, stem erect, woody at the base, terete, 0.5-3 mm diameter, branched, densely pilose when young, sparsely pilose when mature, round, green gradually turning to deep purple, internodes up to 8 cm long. Leaves strictly cauline, alternate, simple, sessile, 1-5.5 x 0.5- 2 cm, variable in shape and size, the basal leaves spatulate, lyrate, middle leaves linear-oblong, apex subobtuse, upper leaves sagittate, apex acute, margins undulate, dentate, recurved, pigmented with deep purple color; glossy on adaxial side, leaf base auriculate, lobed, upper leaves somewhat clasping at the base, sparsely pilose on both sides with wavy hairs of unequal height. Peduncle up to 23 cm long, solitary or branched, with 1-4 homogamous heads. Capitulum base urceolate to globose, somewhat thickened, glossy. Involucre about three-quarters the corolla length; involucral bracts nine, lanceolate, 9-11 mm long, green, sometimes turning rosy brown, margins purple, sparsely pilose when young, glossy and glabrous when mature; heads up to 1.4 cm length, with up to 75 florets per each head, florets much exserted; receptacle somewhat pitted, width up to 4.5 mm; pappus finely barbellate, white, 9 mm long; corolla violet plum, tubular, 5.5-7.5 mm long, 0.1 mm wide, lobes linear to broadly oblong, 1-1.5 mm x 0.5 mm, apex obtuse; stamens 7 mm long, exserted about 3-4 mm from corolla, somewhat transparent, 0.1 mm long, apical appendage linear, <0.1 mm long, apex acute; pollen grains round, surface spiny, yellow; ovary 1.5 mm long, style 7.5 mm long, exserted from corolla up to 2 mm and forming a crown-like structure, pale purple, stigma lobes 0.1 mm long, capitate, pigmented with dark pink to violet color. Achene 2.5-2.75 mm long, angled, 5-ribbed, not deeply ribbed, densely white hairy on ridges; mature seeds black.

Occurs with in a patch of *Phoenix* savannah in Araku valley in Visakhapatnam district. Fl. & Fr.: September–February.

Galikonda hill top (VSKP), *K.V.Satish* & *J.Prakasa Rao* 21202 (CAL, MH).

INDIA: Andhra Pradesh. Endemic.

ASCLEPIADACEAE

BRACHYSTELMA R.Br.

Key to Species

1. Erect tuberous herbs, not branching;
    2. Leaves linear or linear-lanceolate; flowers terminal umbels, corolla reflexed, outer corona cupular:
        3. Stem glabrous at maturity:
            4. Herbs to 30 cm high; leaves to 2.8 cm long; flowers birght orange red:
                5. Corona with five subulate lobes, undulate or toothed; corolla lobes cage like globular ..... **B. ananthapuramense**
                5. Corona without any processes, shortly toothed; corolla lobe not cage-like:
                    6. Outer corona cupular, glabrous:
                        7. Corolla lobes narrowly lanceolate; cup-like outer corona raised above the corolla tube....... **B. saldanhae**
                        7. Corolla lobes linear; corona lobes not raised above corolla tube.. **B. glabrum**
                    6. Outer corona 5-lobed, densely ciliate .... **B. annamacharyae**
            4. Herbs to 80 cm high; leaves to 19 cm long; flowers pale wight with black spots .... **B. nallamalayana**
        3. Stem pubescent at maturity:
            8. Intrastaminal corona lobes deeply divided, erect; pollinarium with a winged corpusculum............................**B. vemanae**
            8. Intrastaminal corona lobes not divided, pendulous; pollinarium without winged corpusculum...........................**B. ciliatum**

2. Leaves filiform or linear; flowers axillary 3-flowerd cymes; corona 5-lobed yellow:

   9. Flowers solitary or in 2-3-flowered cymes corolla tube ash-coloured, corona pure yellow .... **B. pullaiahii**

   9. Flowers 3–5; corolla tube other than ash-coloured, corona greenish or pinkish-white .... **B. penchalakonense**

1. Climbing tuberous herbs; sometimes branching; flowers in axillary cymes:

   10. Flowers axillary or terminal umbels .... **B. beddomei**

   10. Flowers in axillary cymes:

   11. Corolla tube base quite flat, bulging limitedly with a constricted neck and in near urceolate shape, more than 1 cm long.......... **B. seshachalamense**

   11. Corolla tube campanulate, less than 5 mm long:

   12. Corona purplish; follicle divergent .... **B. volubile**

   12. Corona reddish; connate apically .... **B. nigidianum**

Note: *Brachystelma attenuatum* has not been included, as specimens or detailed description are not available.

Brachystelma ananthapuramense Prasad *et al.*, Kew Bull. 73: 16. 2018; Pullaiah *et al.*, Monogrph on *Brachystelma* and *Ceropegia* in India 11. 2019.

Dwarf, perennial herb, 2–5 cm tall. Tubers globose or subglobose, *c.* 3 x 2 cm, pale white, with few short roots. Stem short, erect, solitary, unbranched, pubescent or glabrous; internodes *c.* 3 mm, short. Leaves light greenish-brown, opposite-decussate, subsessile, oblong-lanceolate, elliptic-lanceolate or ovate-lanceolate, 2–5 x 0.5–1 cm, apex acute or acuminate, margins and midrib below with hairs or glabrous. Inflorescence extra-axillary, sessile, solitary flower either side of the node; bract small, greenish-pink, subulate, 0.2–0.4 mm long, acute, persistent, bracteoles absent. Flowers large, erect; pedicels short, greenish-pink, cylindrical, 3–4 mm long, less than 1 mm broad, pubescent; calyx gamosepalous, green or greenish- pink, glabrous; sepals ovate-lanceolate or linear-lanceolate or subulate, 1–1.5 x 0.6–0.8 mm, acute; corolla divided near to base, uniformly purple but darker inside; corolla tube much shorter than the lobes, *c.* 1 mm long; corolla lobes erect, connate at apex to form a globose cage, linear, 1.5–2.5 x 0.1 cm, acute, margins recurved, glabrous outside and puberulous and sparingly white pilose within; corona biseriate, 5-angled, uniformly dark purple and densely pubescent; interstaminal corona cup-shaped, 2–2.5 mm across, 5-lobed near to middle; lobes erect, triangular, thickened in middle, *c.* 1 mm long; staminal corona included in interstaminal corona cup, 5-lobed; lobes incurved, narrowly triangular, 0.6–0.8 mm long, incumbent and adpressed on the anther lobes. Pollinium yellow, subglobose or oblong-ovoid, 0.2–0.4 x 0.2 mm, with pellucid lateral margin, basally attached to corpusculum by short, tubular caudicles. Follicles solitary or paired, cylindrical, tapering towards apex, 6–8 cm long. Seeds comose, narrowly elliptic, 4–5 x 2–3 mm.

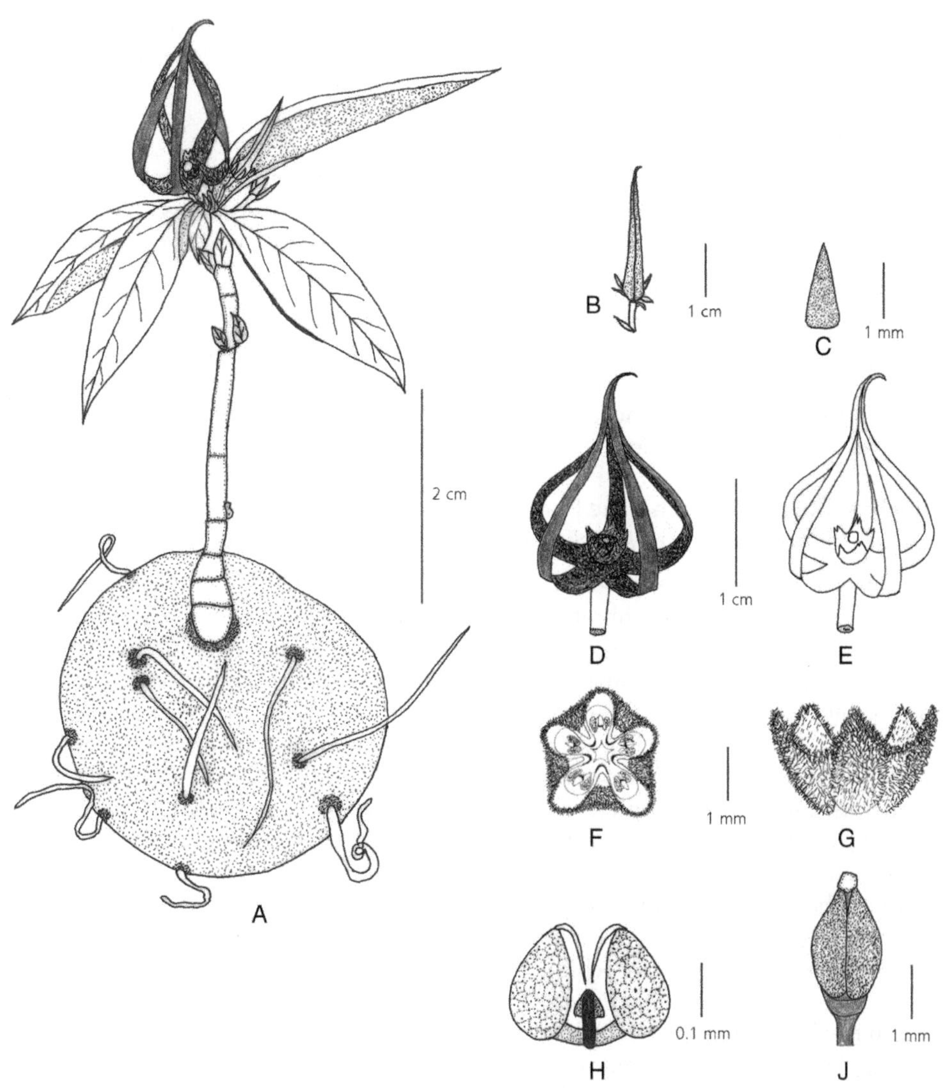

Figure 120. **Brachystelma ananthapuramense** Prasad *et al.*

A. Habit. B. Flower bud. C. Calyx lobe. D. Flower. E. Flower without hairs. F. Corona top view. G. Corona side view. H. Pollinia. J. Gynostegium (After Prasad *et al.*, 2018).

Found growing on black and red mixed soils in open dry deciduous forests at an altitude of 800 m a.s.l. The common associates include *Heteropogon contortus* (L.) P. Beauv. ex Roem. & Schult., *Tripogon bromoides* Roth, *Randia dumetorum* (Retz.) Lam., *Euphorbia fusiformis* Buch.-Ham. ex D. Don and *Drimia* spp. *Fl. & Fr.: May–July.*

Gorantla hills (ATP), 800 m, *A. N. Swamy* & *K. Prasad* 44922 (holotype CAL; isotype SKU); Gorantla hills (ATP), *A. N. Swamy* & *K. Prasad* 44965 (SKU), *B. R. P. Rao* & *A. N. Swamy* 45011 (SKU).

INDIA: Andhra Pradesh. Endemic.

Note: *B. ananthapuramense*, is morphologically similar to the *B. kolarense* complex but differs by its small habit, short internodes, large flowers with more than 1.5 cm long corolla lobes, globular cage and densely pubescent corona.

IUCN Category: The species has a restricted distribution with more than 100 individuals in the Gorantla hills of Ananthapuramu and further explorations in the adjacent hill tracts are necessary to ascertain its status. The current known habitat is prone to annual forest fires and grazing. Being edible, its tubers are harvested heavily by local people and also ploughed out by wild animals. It is currently categorised as Data Deficient (DD) (Prasad *et al.*, 2018).

Brachystelma annamacharyae Pradad *et al.*, Nordic J. Bot. 34: 360. 2015; Pullaiah *et al.*, Monogrph on *Brachystelma* and *Ceropegia* in India 14. 2019.

Dwarf, perennial herb, 4–12 cm tall. Tubers globose, *c.* 4 x 4 cm, dirty brown. Stem short, erect, solitary, unbranched, 1.5–2.0 mm thick, in juvenile stage pubescent becoming glabrous at maturity; internodes *c.* 1.5 cm long. Leaves light greenish-brown, simple, opposite-decussate, petiolate; petiole 1–2 mm long, glabrous; lamina lanceolate, ovate-lanceolate, ovate-oblong, 2.0–5.5 x 1.0–2.2 cm, acute, with puberulent margins; midrib puberulent and prominent below. Inflorescence extra-axillary, sessile, solitary; bract one, pinkish, linear, 0.6–0.8 x 0.2 mm, acute, glabrous, caducous. Flowers horizontally spreading with long pedicel; pedicel cylindrical, filiform, 3.0–3.5 cm long, pubescent; calyx gamosepalous, glabrous, greenish–yellow; sepals ovate, 1.0 x 0.5 mm, acute or obtuse, corolla yellowish, spotted with purple; corolla tube short, campanulate, pilose, *c.* 1.5 mm long; corolla lobes ovate, 2.0–2.5 x 1 mm, acute, reflexed, margins slightly recurved, densely pilose inside and glabrous outside; trichomes pinkish–white, 2–3 mm long; corona biseriate, staminal and interstaminal parts fused to a cup-shaped structure, *c.* 2 mm across, basally bright yellowish; interstaminal corona lobes erect, purplish, much taller than gynostegium, oblong, keeled, 0.8 x 0.4 mm, rounded at apex, with few white trichomes at apex margins; staminal corona basally fused to interstaminal corona ring, it's lobes triangular, *c.* 0.4 mm long. Pollinium subglobose, *c.* 0.20 x 0.18 mm, yellow, with pellucid lateral margin, basally attached to *c.* 0.1 mm long corpusculum by light-brown, short, tubular caudicles. Follicles solitary, purplish–yellow, cylindrical, tapering towards apex, 8–9 cm long; fruiting pedicels straight, spreading to erect; seeds comose, narrowly elliptic, 4–5 x 1.5–2.0 mm, light brown at centre.

Endemic to Seshachalam hills in Kadapa distict of Andhra Pradesh. Fl.: June–September; Fr.: December–January.

Sanipaya (KDP), *P.V. Prasanna, K. Prasad, M. Sankara Rao* and *T. Thulasaiah* 4568 (Holotype BSID, Isotype BSID).

Brachystelma ciliatum Arekal & T.M. Ramakrishna, Curr. Sci. 50: 145. 1981; Albers & Meve, Ill. Handb. Succ. Pl. (Asclepiadaceae) 2: 30. 2002; Pullaiah *et al.*, Monograph on *Brachystelma* and *Ceropegia* in India 22. 2019.

Perennial tuberous herb. Tuber brown, subglobose–globose, 4 cm across and dorsiventrally compressed, 5.6 x 5.6 x 1.5 cm. Stem erect, 30–35 cm high, 1.5–2 mm in dia., rarely branched, purplish, pubescent, buried part of stem 3–4 cm, white, glabrous. Leaf scales opposite, 3–4 pairs from the base, ovate, *c.* 3 x 2 mm, brown puberulent, sessile; leaves opposite in middle of the shoot, 4–5 pairs, linear, linear - lanceolate, 5–9 x 0.2–0.3 cm, nerves 9–12 pairs when young, semi succulent, brown, midrib pubescent beneath, base acute, margins entire, apex acuminate, pubescent, sessile with 2–3 ovate glands at base, green–pink; epidermal hairs unicellular. Inflorescence axillary, 2-4-flowered umbel, pendulous; pedicels filiform, 3.38–3.81 cm long, pubescent, stout and erect in fruits; bracts 4 at each node, each *c.*1 mm, ovate, pubescent; calyx 5-lobed, minute, *c.* 0.5 mm, connate at base, ovate, pubescent; corolla 5-lobed, connate at base, lobes oblong to ovate, *c.* 5 x 2 mm, glabrous, narrowing at apex with 2 mm long shiny cilia, greenish–yellow, dark pink at apex and middle; tube *c.* 1 mm long, lobes reflexed when mature; corona 2 whorled; outer whorl 5-lobed, *c.* 2.5 mm across, cupular, glabrous, dark pink dotted, 10-lobed at margin; inner whorl 5-lobed, alternate with stamens, ovate–oblong, extended to staminal column; staminal column 5-lobed, yellow, elliptic, winged; stigma light greenish-yellow, pentagonal, truncate, minute, sticky; pollinarium yellow, 60 µm across, erect; corpusculum brown. Follicles 1–2, linear-lanceolate, 7–8 x 0.2–0.3 cm, light reddish brown, glabrous, unequal, curved at apex when young, acuminate and stronger when mature. Seeds 8–12, brown, ovate, 7–8 x 2–25 mm, thin; coma white, 2–2.5 cm long.

Rare in Anantapuram district, Andhra Pradesh associated with *Ophioglossum gramineum, Heteropogon* sp., *Chrysopogon* sp., etc. in sandy moist places of dry deciduous forests. Fl. & Fr.: September - October.

Andhra Pradesh: Anantapuram Distr., Nigidi forest, *KRKS* 41870 A–D (SKU, MH & CAL)

INDIA: Andhra Pradesh, Karnataka (endemic).

Conservation status: As per IUCN *Red list* (IUCN 2001); it is Vulnerable VU, D 1+2. The extent of occurrence of the species is estimated to be less than 3 sq km that too, in a severely fragmented and degraded habitat while the number of individuals are less than a thousand. The tubers are favored food of several wild animals and this pressure poses a real threat to the species.

Note: Plants of *Brachystelma ciliatum* are growing among grasses and is very difficult to locate in vegetative and immature fruit stage. The plant can be recognized easily in the field in flower and in fruit particularly in dehiscence.

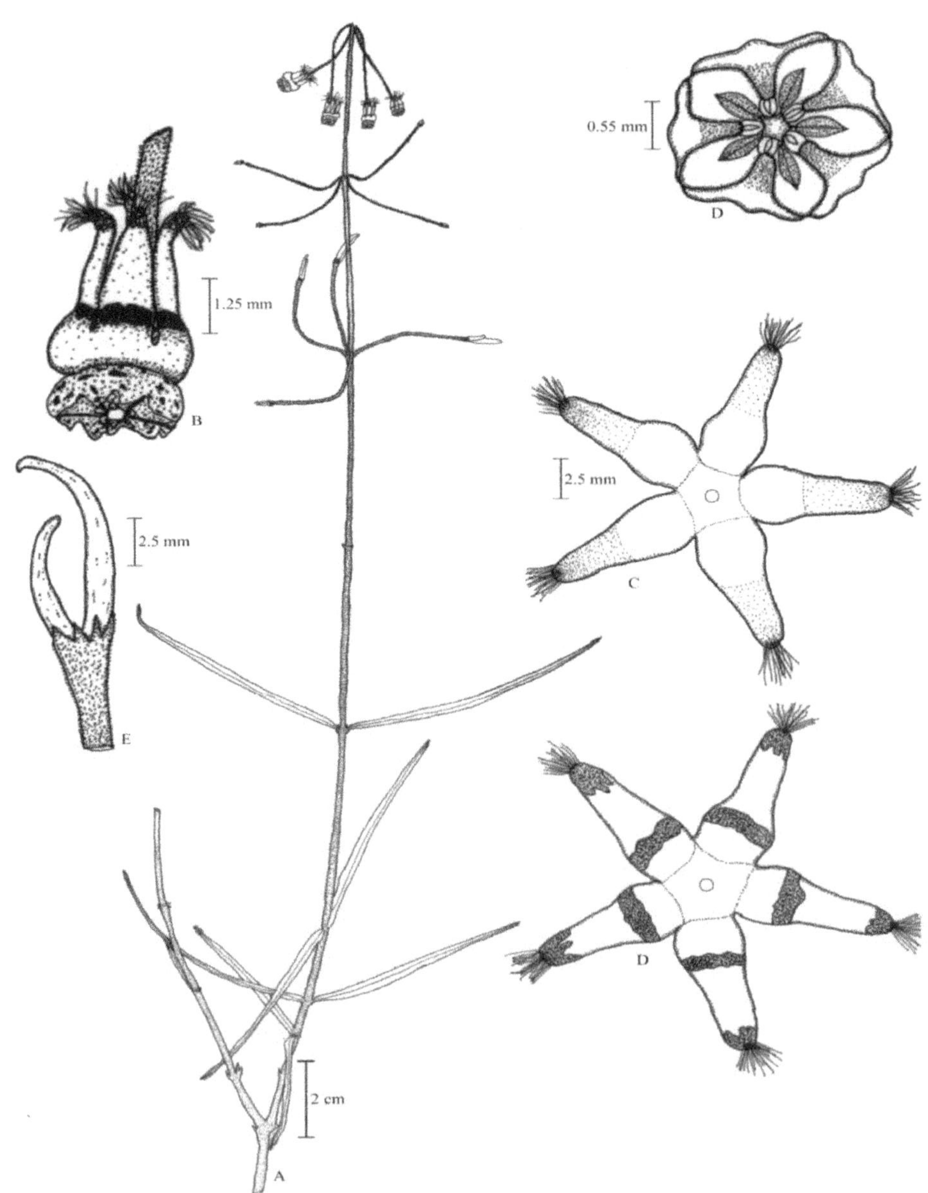

Figure 121. **Brachystelma ciliatum** Arekal & T.M. Ramakrishna.

A. Twig, B. Flower, C. Corolla dorsal view, D. Corolla ventral view, E. Corona, F. Young fruit.

Brachystelma nallamalayana Prasad & Rao, J. Threatened Taxa 5: 4904. 2013; Sadasivaiah *et al.*, Asian J. Plant Sci. Res. 6: 27. 2016; Pullaiah *et al.*, Monogrph on *Brachystelma* and *Ceropegia* in India 43. 2019.

Erect herbs, up to 80 cm high. Tubers in varied shapes, generally fusiform, to 6 x 3 cm diam., brownish. Stem solitary, terete, 1–2 mm in diam., unbranched, ridged, succulent, glabrous, glaucous green; internodes up to 10 cm long. Leaves opposite, decussate, fleshy, sessile; linear, up to 19 cm long and *c.* 2 mm diam., margin undulate, apex acute, glabrous, sometimes hairy on the mid vein. Inflorescence crowded at apical nodes, lateral, shortly peduncled, umbellate, 5–6-flowered, pendulous. Pedicels terete, filiform, up to 1.3 cm long, glabrous. Bracts and bracteoles persistent, linear, acute at apex, glabrous, greenish-pink; bracts 2 x 0.5 mm; bracteoles 1 x 0.2 mm; calyx 5-lobed, united at base; lobes linear, 3 x 0.5 mm, acute at apex, 3–5 veined, glabrous, pinkish-green; corolla 5-lobed; corolla tube up to 1mm long, shallow; corolla lobes erect, linear, 5–9 x 2 mm diam., acute but appears obtuse due to inturned apex, faintly 5-veined, margins involute with 1–2 mm long pink hairs throughout, more dense at apex; lobes basally pale white with black spots and yellow above without any blotches; corona *c.* 3 mm across, biseriate, glabrous; interstaminal corona obscurely 5-angled, cupular, forms a continuous ring around the gynostegium, brown; staminal corona 5-lobed, black; lobes appressed to the back of the anthers, exceeding and hooded on the style apex; pollinia waxy, yellow, globose, *c.* 200 ìm long, margins pellucid at apex, attached by light brown tubular caudicles to a red-coloured corpuscle; style apex pentangular. Follicles paired, linear, *c.* 6 x 0.3 cm, glabrous, glacous; seeds comose, 5-7 x 3-5 mm, ovate-oblong, 15–20 in each follicle, light brown at centre.

This species occurs amidst grasses on open hill slopes of dry deciduous forests at an altitude between 550–600 m, associated with *Cymbopogon coloratus* and *Heteropogon contortus* in the Nallamalai Hill ranges. Fl.: July–August.

Bheemudikolanu, Nallamalai (KNL), *BR* & *KP* 34667 A (Holotype), 34667 A-D (Isotype BSID); Way to Saleswaram (MBNR), *B. Sadasivaiah* & *B. Kalpana*, 2073; Way to Srisailam, Near Rasamola Baavi (MBNR), *B. Sadasivaiah* & *K. Prasad*, 2142 (Govt. Degree College, Wanaparthy).

INDIA: Andhra Pradesh, Telangna. Endemic.

Brachystelma nigidianum Raja Kullayisw. *et al.*, Kew Bull. 71: 1. 2016. *Brachystelma volubile* auct. non Hook. f.; Raja Kullayiswamy *et al.* Rheedea 22: 107. 2012; Pullaiah *et al.*, Monogrph on *Brachystelma* and *Ceropegia* in India 47. 2019.

Perennial climbing herb, to 1.5 m long. Tuber discoid-subglobose, 1 or 2 subterranean, brown, *c.* 2.8 x 3.5 cm; stem herbaceous, terete, glabrous, sometimes branched. Leaves linear-lanceolate, 5–16 x 0.2–0.5 cm, narrowed into 3–5 mm long petiole, base acute, margin entire, apex acuminate, scabrous above, hirsute beneath on raised midrib, margin ciliate, caducous, leaves reduced towards apex on stem, sessile *c.* 3 mm long, scabrous. Petiole scabrous, canaliculate with brown colleter (gland) at base. Inflorescence extra-axillary, 2- or 3-flowered; peduncles 5–6 mm long, glabrous; bracts lanceolate, glabrous, *c.* 4 mm long; bracteoles lanceolate, glabrous, *c.*

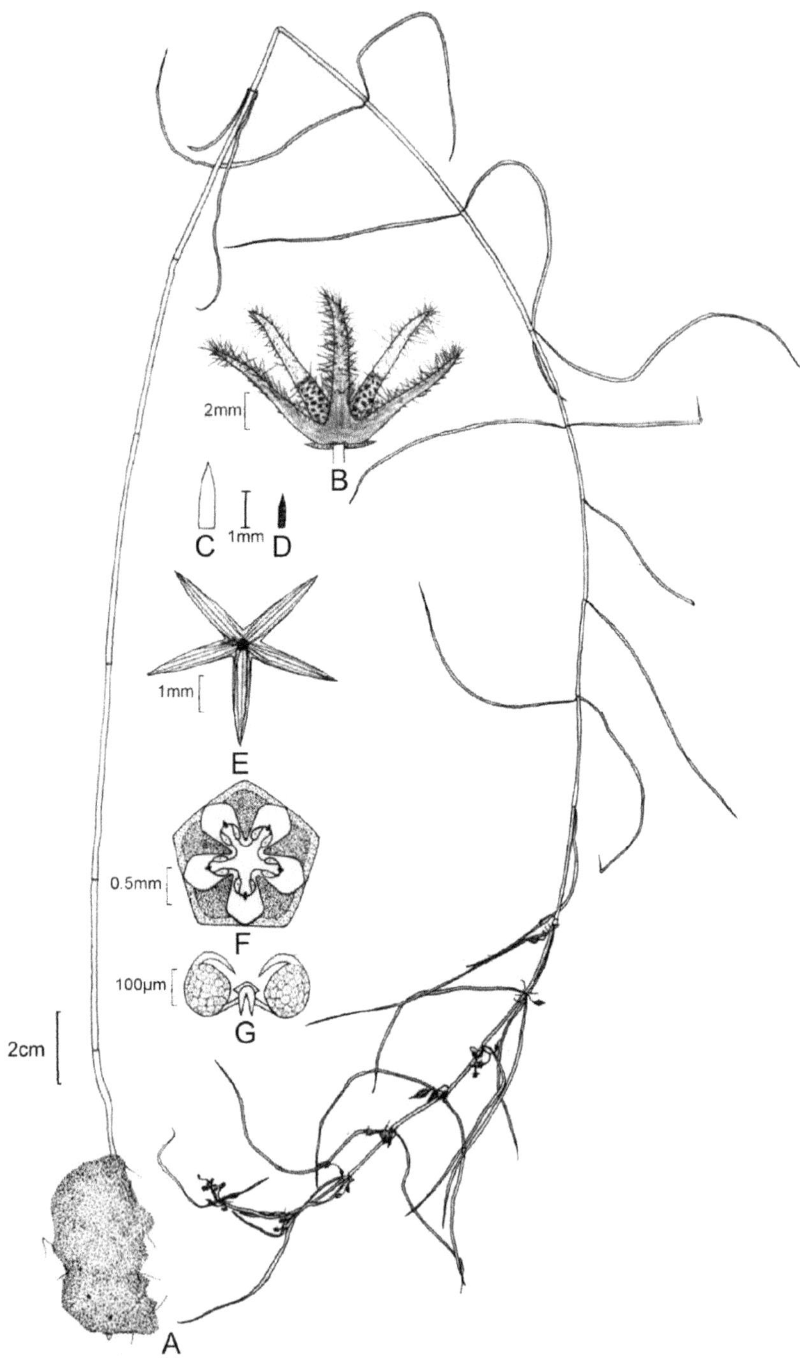

Figure 122. **Brachystelma nallamalayana** Prasad & Rao.

A. Habit, B. flower, C. Bract, D. Bracteole, E. Calyx, F. Corona top view, G. Pollinia (After Prasad and Rao, 2013).

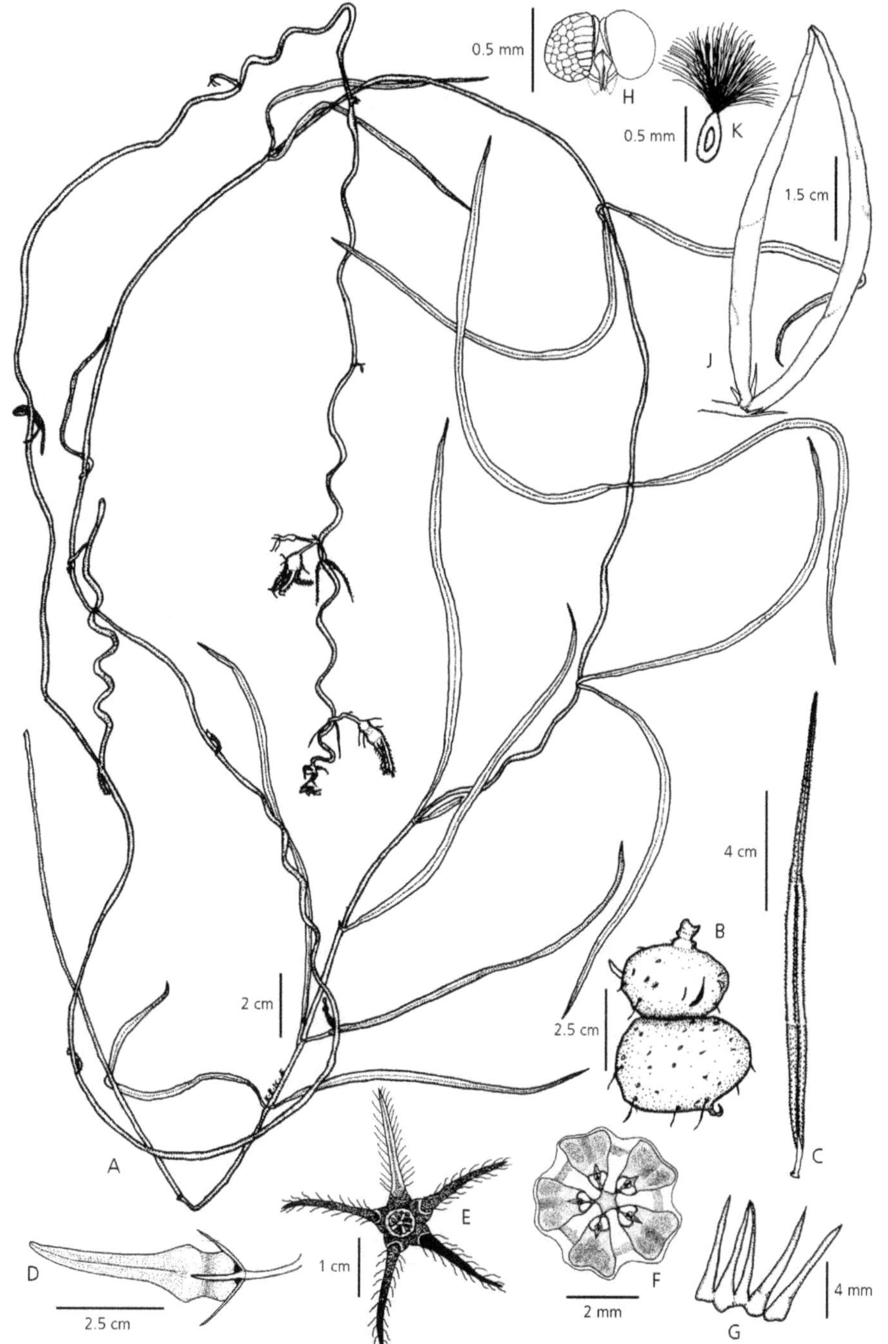

Figure 123. **Brachystelama nigidianum** Raja Kullayisw *et al.*

A. Habit, B. Tuber, C. Leaf, D. Flower bud, E. Open flower, F. Corona, G. Calyx, H. Pollinarium, J. Follicles, K. Seed. (After Kullayiswamy *et al.,* 2016).

3 mm long; pedicels *c.* 9 mm long, glabrous, purplish-green; calyx linear-lanceolate, subulate, 9–10 mm long, glabrous, purplish-green; corolla 3.3–3.5 cm long, *c.* 4.4 cm across (when open); corolla tube *c.* 9 mm long, urceolate or urn-shaped, base, middle and mouth of corolla tube *c.* 9 mm, *c.* 7 mm and *c.* 11 mm wide respectively, glabrous, greenish with brown spots without, and slightly ribbed, pubescent with fine hairs, dark brown with light yellow lines within; lobes deltoid at base (two times wider than long) tapering towards apex, *c.* 2.6 x 0.3 mm, cohering at apex when young and free when mature, acuminate at apex, glabrous, greenish outside, reddish maroon, densely pubescent within, scattered pilose along margin (hairs *c.* 2 mm long), margin slightly outcurved, yellow striations at base of lobes; corona *c.* 2 mm high, *c.* 4 mm across; outer corona cupular, reddish maroon, *c.* 2 mm across, pubescent, 5-lobed, each lobe slightly bifid; inner corona 5-lobed, dorsiventrally flattened, *c.* 1 mm long, basally fused with outer ring, lobes 5, free, oblong-ovate, yellow with maroon tinge at middle, apex obtuse, overlapping anthers with pollen chambers below; pollinaria *c.* 0.5 x 0.5 mm across; pollinia ovoid, *c.* 0.4 x 0.3 mm, yellow, basally attached by hallow hyaline caudicles to a brown corpuscle, pollinia with germination crests towards apex more than half as long as pollinium; ovary *c.* 2 mm long, style short. Follicles in pairs, slightly fusiform, 5.5–7.5 cm x 2.5–4 mm, tapering, connate at apex, slender, smooth, glabrous, pale green. Seeds flattened, ovate, *c.* 5 x 2.5 mm, light brown, outer margins thin, comose; coma 9–14 mm, white silky.

Rare in Nigidi forest in Anantapuram district of Andhra Pradesh. Fl. & Fr.: October - December.

Nigidi forest (ATP), *KRKS* 41853 (holotype SKU; isotypes BSID, CAL, K, MH).

INDIA: Andhra Pradesh. Endemic.

Brachystelma penchalakonense Rasingam *et al.*, Kew Bull. 68: 663. 2013; Pullaiah *et al.*, Monogrph on *Brachystelma* and *Ceropegia* in India 51. 2019.

Perennial herb, to 80 cm high. Tubers fusiform, brownish, *c.* 7 x 1.5 cm; stem erect, solitary, unbranched, greenish-yellow, terete at base, obtusely 4-angular towards apex, glabrous; internodes 3–9 cm long. Leaves sessile, opposite-decussate, acicular-filiform, 2–8 x 0.03–0.05 cm, margins reflexed, ciliate but otherwise glabrous; midrib prominent beneath, sunken above. Inflorescence slightly extra-axillary, sessile, 3-flowered. Buds ampulliform, *c.* 1 cm long, greenish-white with a red ring at base. Flowers pendant, *c.* 3.5 cm in diam.; bracts and bracteoles linear, acuminate, up to 2.5 mm long, glabrous. Pedicels filiform, *c.* 1.5 cm long, glabrous; calyx yellowish, linear, acuminate, 3.5 x 1 mm, single nerved, glabrous throughout; corolla very showy, rotate, *c.* 3.5 cm diam.; corolla tube convex with radial compressions, *c.* 3 mm long, glabrous, dark green-coloured, extended up to the lobes; lobes *c.* 1.7 cm long, 3 mm wide at base, deltoid bases gradually tapering to 0.5 mm into linear-lanceolate tails, spreading, margins reflexed, densely covered with spreading white hairs, *c.* 2 mm long; corona biseriate, *c.* 3 mm across, glabrous; outer corona cupular, wavy, ± 5 angled, *c.* 0.5 mm high, greenish; inner corona basally fused to outer corona ring, free corona lobes ovate, 0.8 mm long, subacute at apex, pinkish-white, overlapping the anther lobes; pollinia subglobose, 0.3 mm long, yellow, basally attached by light-

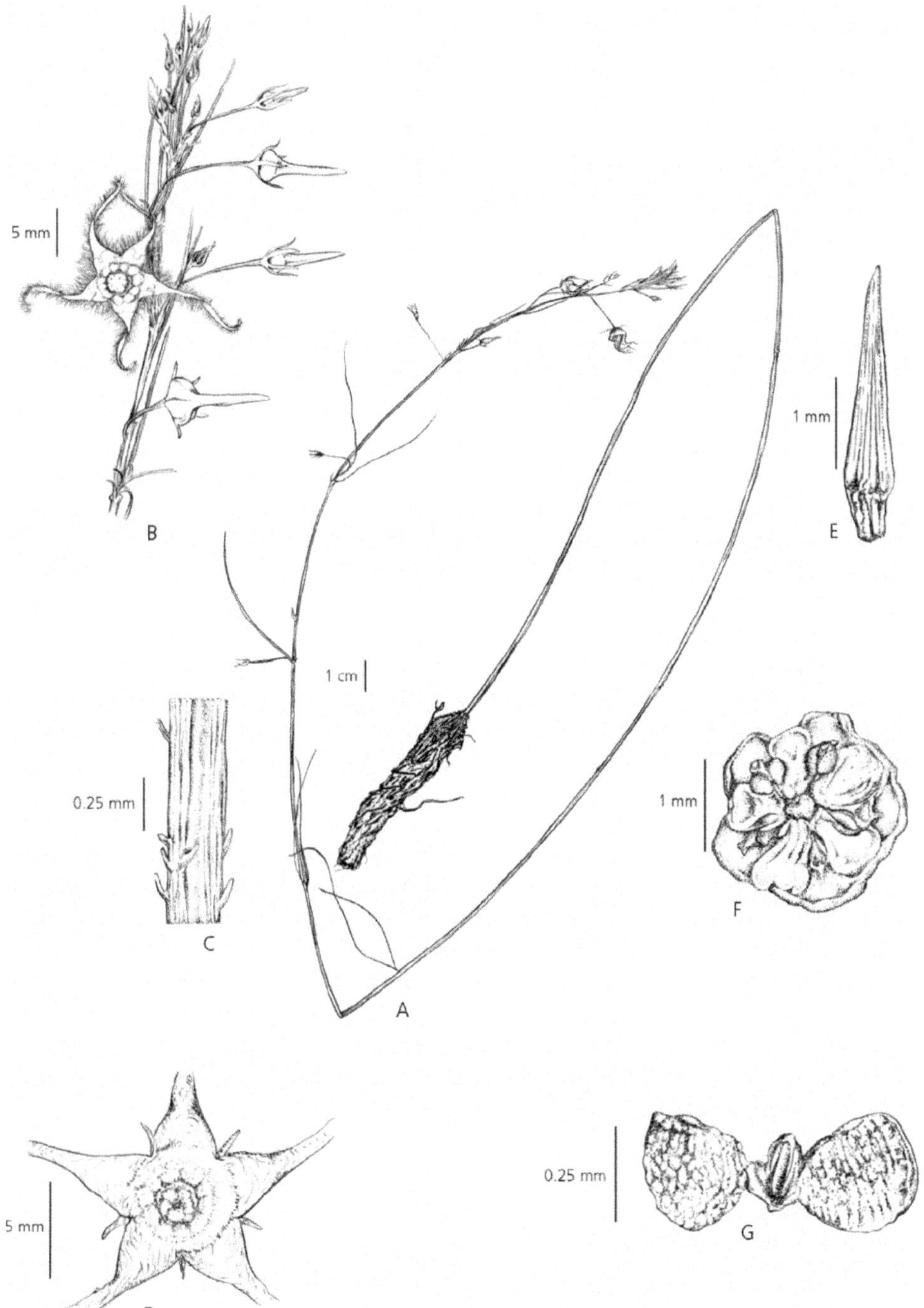

Figure 124. **Brachystelma penchalakonense** Rasingam *et al.*

A Habit, B. Flowering twig, C. Leaf portion with ciliate margins, D. Corolla, E. Calyx, F. Corona top view, G. Pollinarium (After Rasingam *et al.,* 2013).

brown, tubular caudicles to a brown-coloured corpuscle. Fruits were not available and hence details were not given by authors.

Rare in Penchalakona in Nellore district. Growing in crevices on rocky ground in open dry deciduous forests at an altitude of 645 m a.s.l. Common associates include *Chrysopogon verticillatus, Arundinella setosa, Boucerosia umbellata, Actinopteris radiata* and *Drimia* spp. Fl.: September

Penchalakona (NLR), *L.Rasingam, M. Sankara Rao* and *Alok Chorghe* 2923 (Holotype CAL; isotype BSID).

INDIA: Andhra Pradesh (endemic).

Brachystelma pullaiahii Rao *et al.*, Taiwania 56: 223. 2011; Pullaiah *et al.*, Monogrph on *Brachystelma* and *Ceropegia* in India 51. 2019.

Perennial herb, up to 1.6 m high. Tubers to 8 x 4 cm, fusiform or globose, brownish. Stem erect, solitary, terete, unbranched-sparsely branched, filled with watery sap, glabrous, green white glaucous downwards; nodes purple-tinged; internodes to 15 cm long. Leaves to 12 x 0.2 cm, linear or filiform, opposite, decussate, fleshy, sessile, margin entire, apex acute, glabrous. Inflorescence axillary, sessile or shortly pedicellate, usually with erect, solitary flower or 2-flowered, very rarely 3-flowered. Flowers with 1.5 cm long pedicels, terete, glabrous; bracts and bracteoles linear, acute, glabrous, pinkish-green; bracts to 2 x 0.4 mm; bracteoles 1 or 2 per flower, to 1 x 0.2 mm; calyx 5-lobed; lobes to 5 x 1 mm, linear, acute or acuminate, 5-lobed, rotate, star-shaped, ash-grey up to basal attenuated portions of lobes, striations white, appearing like a ring surrounding the corona; corolla tube shallow; corolla lobes to 3.6 x 0.2 cm, 5-veined, linear portion pale yellow, acute, occasionally obtuse due to inturned apex, margins folded back along the middle, prominently hairy throughout, occasionally puberulous or glabrous near base; corolla lobes hairs to 3 mm long, white; corona ca 3 mm across, biseriate, glabrous, yellow; inter staminal corona 5-lobed, cupular, forming a continuous ring around the gynostegium; staminal corona 5-lobed, lobes adpressed to back of the anther, incurved, linear, acute, as long as or slightly exceeding and hooded on the style apex; pollinia *c.* 0.2 mm long, yellow, waxy, globose-ellipsoid, margin pellucid at apex; attached by light-brown tubular caudicles to a red-coloured corpuscle; stigma pentangular. Follicles *c.* 10 cm long, paired, erect divergent, dehiscent; seeds 18-25, to 8 x 4 mm, with central dark embryonic region; outer margins hyaline, rounded at base, comose; coma silky white, *c.* 2.5 cm long.

Found in dry deciduous forests and grows on black and red soils mixed with slate rock on open hill slopes among grasses at an altitude range of 400-600 m in Nallamalais in Andhra Pradesh. Fl. & Fr.: June-December. Vern.: Tel.: *Nakshatramokka, Nemathigadda.*

Sikharam hills, Nallamalais (KNL), *B.R.P.Rao* & *Prasad* 34664 (SKU holo, BSID, MH, CAL iso).

INDIA: Andhra Pradesh. Endemic.

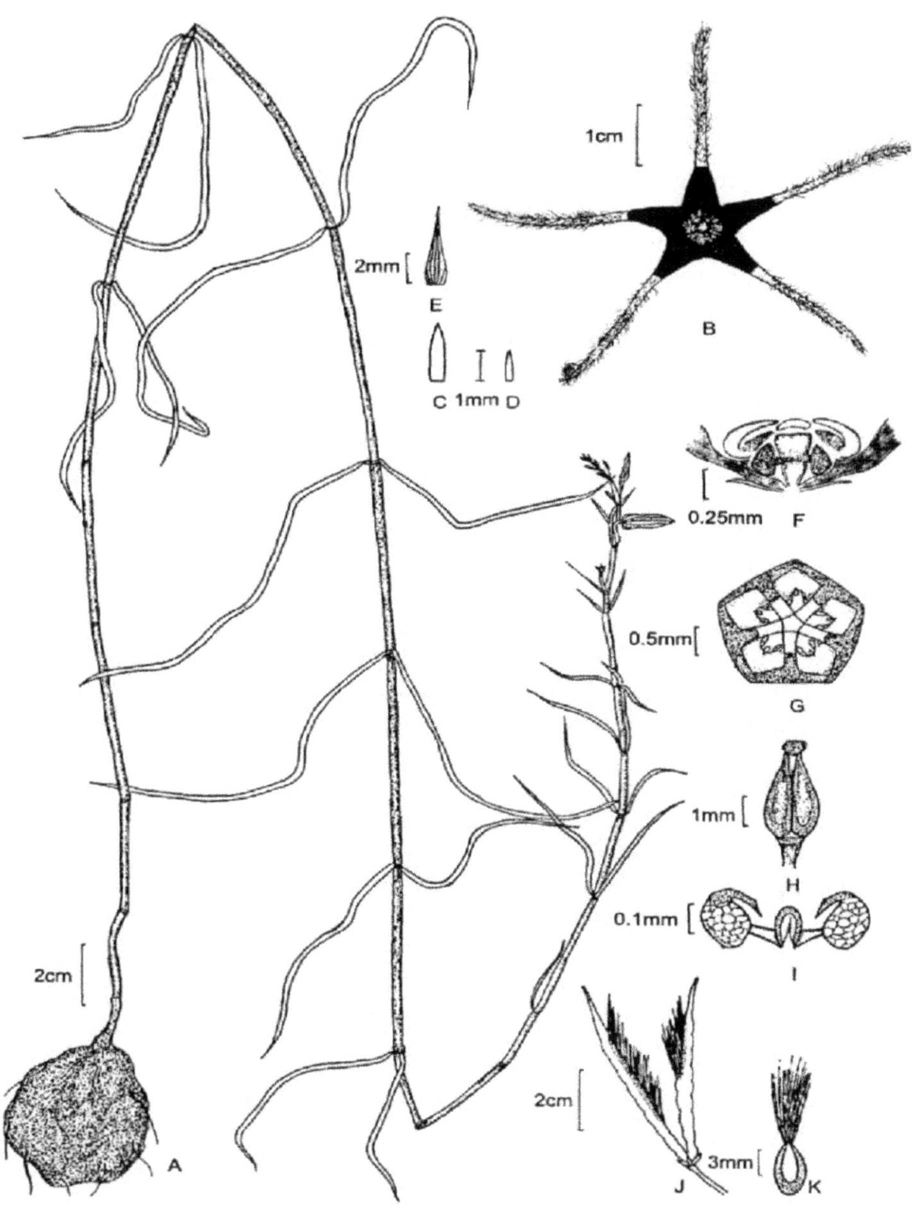

Figure 125. **Brachystelma pullaiahii** Rao *et al.*

A. Habit, B. Flower, C. Bract, D. Bracteole, E. Sepal, F. Flower L.s., G. Corona top view, H. Gynostegium, I. Pollinia, J. Fruit, K. Seed (After Rao *et al.*, 2011).

Note: *Brachystelma pullaiahii* Rao *et al.* is similar to *B. maculatum* Hook. f. but differs in its long stems, sessile leaves, solitary flowers, ash-coloured corolla base, shallow corolla tube, very long and non-keeled, pale yellow corolla lobes with white hairs, corona, biseriate and yellow.

Brachystelma saldanhae S.J. Britto & P.V. Bruyns, Haseltonia, 22: 52. 2016; Pullaiah *et al.*, Monogrph on *Brachystelma* and *Ceropegia* in India 57. 2019.

Perennial erect geophytic herb 5–10 (20) cm tall with 1–several stems arising from underground flattened-globose to spherical tuber 1.5–2.5 x 2–4 cm. Stem annual, unbranched or branching (sometimes densely) just above tuber, erect, slender, mainly purplish brown, glabrescent. Leaves subsessile, opposite, narrowly linear to linear-lanceolate, 1–4 (5) x 0.2–0.4 cm, green soon becoming brownish to purplish green, glandular-puberulous, narrowing to base, margins minutely ciliolate, acute-apiculate. Inflorescence in terminal and extra-axillary cymes alongside upper nodes, with 2–8 flowers often opening in quick succession in groups; peduncle ± absent; pedicel slender, spreading and often descending towards tip, 4–6 mm long, purplish-brown, glandular-puberulous; bracts linear, ± 1 mm long; sepals linear-lanceolate, 1–1.5 mm long, purplish green, glandular-puberulous; corolla ± rotate, 8–10 mm broad; outside greenish yellow becoming purplish brown towards base, glabrous; inside bright yellow from tips to middle of lobes and below this purple-brown or with irregular purple-brown blotches on yellow, glabrous; tube very shallowly plate-like (< 0.5 mm long); lobes convex inside, lanceolate-deltate, 3–4 x 1.5–2 mm, obtuse; corona 2-seriate; outer (inter staminal) series fused into broadly cup-like structure slightly narrowing towards mouth and much exceeding anthers, ± 1 x 1.5–2 mm, above with purple-brown blotches on inner and outer surface, below bright yellow, margin obscurely lobed, inner surface with downward-pointing stiff white hairs in tufts alternating with anthers; inner (staminal) series of five short bright yellow linear-obtuse lobules arising below middle of cupular structure and fused to it behind, pressed to but not exceeding anthers. Follicles to 6 cm long, puberulous, greenish to purplish brown, tapering to apices; seeds 8–10 per follicle, to 6 mm long, each with coma to 1 cm long.

Brachystelma saldanhae is known from exposed granitic domes at around 1600 m in the Servarayan (Shevaroy) Hills near Salem. Here it occurs in small clusters of individuals in patches of shallow (30–80 mm deep), seasonally very wet soil with small sedges and other minute herbs.

Servarayan hills, 6 km from Yercaud (SLM), *Britto* sub RHT 73892 (holotype RHT, isotype MH); Near Yercaud (SLM), *Bruyns* 5952 (E, K).

IUCN Red list Category: The type collection was made in July 2002, but it has been known in the area for many years and P.V.Bruyns was shown plants nearby in 1993 by A.A. Ansari of the Botanical Survey of India. At the same spot in August 2009 it was found that recent mining for bauxite has resulted in large piles of overburden being dumped close to the area where *B. saldanhae* occurs, partly covering and greatly reducing its habitat. It is rated Vulnerable (V, IUCN 2010).

INDIA: Tamil Nadu. Endemic.

Brachystelma seshachalamense Prasad & Prasanna, Bangladesh J. Plant Taxon. 23: 53. 2016; Pullaiah *et al.*, Monogrph on *Brachystelma* and *Ceropegia* in India 59. 2019.

Twining herbs, *c.* 2.5 m high. Tubers depressed-globose, 2.0–2.5 x 3.0–3.5 cm, dirty brownish without, white when cut open. Stem unbranched, slender, cylindric, glabrous. Leaves opposite-decussate, petiolate; petiole 2–5 mm long, puberulous; lamina linear or linear-lanceolate, 4–7 x 0.2–0.6 cm, acute at apex, cuneate at base, margin hairy; midrib prominent below, puberulous, glabrous above. Inflorescence extra-axillary, reduced cymes, usually solitary, rarely 2-flowered; peduncle pinkish-green, terete, 2.6–3.0 mm long, pubescent. Flowers drooping; floral bract linear, *c.* 1 mm long, acute, glabrous; bracteoles 2, linear, less than 1 mm long, acute, glabrous. Pedicels terete, 6–8 mm long, pubescent; calyx pinkish, glabrous; lobes linear, 2.0–2.4 x 0.8–1.0 mm, acute; *corolla* spreading; corolla tube campanulate, 4.5–5.0 mm long, a ring of hairs around the corona and pubescent inside, green with pink tinge outside, yellow between the lobes inside; corolla lobes greenish-yellow outside and dark pink inside, broad at base, narrowed progressively towards tip, 1.4–1.8 cm long, acute at apex, pubescent inside, densely white hairy along the margins; hairs 2–3 mm long; corona yellow, biseriate, *c.* 4 mm in diameter, staminal and interstaminal parts fused to a shallow cup shaped structure; interstaminal corona cupular, 5-angled, *c.* 0.6 mm long, shallowly bilobed with few hairs; staminal corona lobes 5, incumbent, overlapping the anther lobes, lemon yellow, ovate-oblong, 1.5 mm long, glabrous; pollinia yellow, obovate-oblong, *c.* 0.2 x 0.1 mm, with pellucid margin and basally attached light-brown, tubular caudicles to a brown corpusculum; carpels 2, *c.* 1 mm long, narrow, free; style absent; stigma head pentangular, discoid.

Rare in red or black soils of grass dominated dry deciduous open forests, at an of 400–600 m altitude in Seshachalam hills of Kadapa district in Andhra Pradesh (endemic). Fl.: December–January.

Sanipaya RF (KDP), *K.Prasad* 6454 (Holotype CAL; Isotype BSID).

INDIA: Andhra Pradesh. Endemic.

Brachystelma vemanae A. Madhusudhana Reddy, M.V. Suresh Babu & K. Prasad, Nordic J. Bot. 36(10): 1. 2018; Pullaiah *et al.*, Monogrph on *Brachystelma* and *Ceropegia* in India 65. 2019.

Geophytic herb, 5–20(–30) cm tall, white glaucous all over. Tubers globose or subglobose, with few short roots, 1.5–2.5 x 1.5–2.0 cm. Stem erect, solitary, slender, unbranched, pubescent, with watery sap; internodes *c.* 3.5 cm long. Leaves opposite, sessile or subsessile, linear-lanceolate to oblong, acute at apex, margins and midrib hairy, lamina glabrous or rarely hairy, 3–6 x 0.4–0.7 cm. Inflorescence on leafless terminal nodes; bract and bracteoles small, green, *c.* 0.5 mm long, subulate, acute. Flowers solitary, pendulous; pedicels dark pink, slender, cylindrical, 9–12 mm long, densely pubescent; calyx green or greenish-pink, pubescent adaxially; sepals 5, narrowly triangular, ovate-lanceolate or linear-lanceolate, acute, 1.4–2.0 x 0.6–1.0 mm; corolla divided almost to base; tube much shorter than the lobes, light brown or creamish with dark pink ornamentation, 1.2–2.0 mm long, densely pilose; lobes first patent, then completely reflexed, their basal portion light brown or creamish with dark rose ornamentation, later becoming uniformly brownish-pink, lanceolate or lanceolate- oblong, 7–10 x 1.2–1.8 mm, acute, with margins recurved and almost tubular, completely pilose adaxially and glabrous abaxially; hairs white, *c.* 2 mm long; corona biseriate, blackish-purple, 2.0–2.5 mm across, interstaminal and staminal corona basally fused to form a cup-shaped structure, densely pilose; interstaminal corona lobes deeply divided (bi-lobed), slightly longer or shorter than the staminal corona lobes, ascending-erect, triangular at base then narrower towards apex, sparsely pilose, 1.1–1.4 mm long, 0.4 mm broad at base; staminal corona 5-lobed; lobes incurved and connivent at apex over gynostegium, linear or linear-oblong, sparsely pilose, 1.1–1.4 mm long, 0.5–1.0 mm broad; pollinium yellow, subglobose or subglobose-ovoid, *c.* 325 x 300 ìm, inner pellucid margined, basally attached to corpusculum by short, hyaline, tubular caudicles; corpusculum *c.* 280 ìm long, sagittate, with distinct wing along each side. Follicles not seen.

Found growing on sandy and red mixed soils in margins of open dry deciduous forests at an elevation of 250–300 m a.s.l. Fl.: April–September.

Veligonda hills, Seetharamapuram (NLR), *A. Madhusudhana Reddy* 5071 (holotype: CAL, isotype: BSID, Yogi Vemana Univ.).

INDIA: Andhra Pradesh. Endemic.

CONVOLVULACEAE

IPOMOEA Linnaeus

Ipomoea clarkei Hook.f., Fl. Brit. India 4: 734 1885; Cooke, Fl. Bombay 2: 245.1908; P. Lakshminaras. & Sharma, Fl. Nasik Dist. 327. 1991; Almeida, Fl. Maharashtra 3B: 323. 2001; Venkanna & Das, Fl. Maharashtra (Dicot.) 2: 460. 2001; Rasingam *et al.*, J. Threat. Taxa 10: 11827. 2018.

Annual slender twining herb, stem slender, hairy, *c.* 1.2 mm in diam. Leaves 2.5–6 x 1.2–4 cm, ovate, deeply cordate with rounded 0.8–1 cm lobes at base, entire along margins, long acuminate at apex; acumen *c.* 1cm long, bulbous based hairy on both surfaces, 5-nerved at base; petioles up to 3cm long, slender, sparsely hairy. Inflorescence axillary solitary, 1-flowered cyme. Peduncle slender, 7–12 mm long, sparsely hairy; bract *c.* 2 mm long, subulate; bracteole 1-2, *c.* 1mm long; pedicel 8-15 mm long, slender, hairy, slightly thickned towards the apex in fruiting stage. Flowers yellow, 3–4 cm long; calyx 7–8 mm long, 5 lobed; lobes slightly connate at base, unequal, ovate-lanceolate, entire along margin, acute-acuminate at apex, bulbous based hairy on upper surface, glabrous on inner surface; corolla 3–4 cm long, infundibuliform, tube 2.5–3.2 cm long, limb *c.* 1.5 cm in diam., glabrous; stamens 5, unequal,1.2–1.5 cm long, epipetalous; filaments 0.9–1.2 cm long, slightly dilated and hairy at base, glabrous and narrow towards apex; anthers ovate *c.* 3 x 0.5 mm long; gynoecium up to 2.2cm long, inserted; ovary ovate, *c.* 1.8 x 1.1 mm long; style *c.* 2cm long; stigma globose, 0.5mm in diam. Capsule globose or sub orbicular, *c.* 1 x 0.9 cm, beaked, glabrous, 4-seeded; seeds *c.* 5 x 4 mm long, oblong, puberulous, dark brown.

Rare in Srisailam in mixed dry deciduous forests and open rocky grasslands in association with *Asparagus racemosus* Willd., *Heteropogon contortus* (L.) P.Beauv. ex Roem. & Schult. and *Pentanema indicum* (L.) Ling. Fl. & Fr.: September–December.

Neelganga, Srisailam (KNL), *L. Rasingam* & *M. Sankara Rao* 8005 (BSID).

INDIA: Maharashtra, Telangana, Andhra Pradesh. Endemic.

SOLANACEAE

LYCIANTHUS (Dunal) Hassler

Lycianthus neesiana (Wall. ex Nees) D'Arcy & Z.Y.Zhang, Novon 2(2): 127. 1992; Murugan *et al.*, NeBio 7: 127. 2016. *Solanum neesianum* Wall ex Nees, Trans. Linn. Soc. London 17(1): 42. 1837. *S. subtruncatum* Wall. ex Dunal in DC., Prodr. 13(1): 180. 1852; FBI 4: 231. 1883.

Shrubs or shuhsrubs, 1-1.5 m high, branchlets slender, simple or dichotomously branched, glabrescent. Leaves simple, 2 in each node, unequal, large leaf elliptic to lanceolate or ovate-lnceolate, 5-10 x 3-5 cm, apex acute to acuminate, margin entire, base cuneate or attenuate; petiole 2-6 mm long, slender, smaller leaf elliptic or ovate, 3-6 x 3-4 cm, apex acute, margin entire, base obtuse; petiole 2-6 mm long, slender. Berries globose, 5-10 mm across, with persistent calyx, *c.* 5 mm diam., 6-10-toothed, subulate, unequal, fruiting pedicels 1-2 cm long, linear; seeds 1-1.5 mm diam., deltoid or discoid, pale brownish yellow, compressed.

Occasional in Northern Eastern Ghats. Fr.: December.

Upper Barakamara–Jenabi, Similipal Biospheree Reserve (MBJ), *A.R.K.Sastry* & *G.P.Singh* 12411 (CAL).

INDIA: Arunachal Pradesh, Assam, Manipur, Meghalaya, Nagaland, Orissa, Sikkim, Tripura.

WORLD: Bangladesh, Bhutan, China, Indonesia, Thailand.

PHYSALIS Linnaeus

Physalis angulata L., Sp. Pl., 183. 1753; Deb, J. Econ. Taxon. Bot. 1: 43. 1980; Chitra in A.N.Henry *et al.*, Fl. Tamil Nadu 2:114. 1987; Murugan *et al.*, Indian J. For. 41(4): 395. 2018.

Annual herbs, branchlets erect, angular. Leaves simple, alternate, ovate-elliptic, 3-6 x 0.5-4 cm, base oblique, acute-decurrent into petiole, margin entire or irregularly dentate, apex acute, lateral nerves 7 pairs; petiole to 4 cm long. Flowers solitary, axillary; pedicels to 8 mm long; calyx copular, pubescent, 5-lobed; lobes triangular, entire and ciliate, acute, 1.5-2 mm long; corolla rotate, faintly pubescent outside, yellow, with brown blotches at throat, faintly 5-toothed, teeth ciliate; stamens 5, unequal; filaments 3-5 mm long, hairy at base, anthers oblong or ovoid; style *c.* 4 mm long, stigma discoid or capitate, green. Berries globose, *c.* 1 mm across, glabrous, red when ripe; fruiting calyx ovoid, 10-angled, minute-pubecent outside, green, purple-nerved; seeds reniform,pitted, yellowish brown.

Naturalized in Sirumalai hills. Fl. & Fr.: July–August.

Way to Kutladampatti falls, Sirumalai hills (DGL), *R.Kottaimuthu* 1950, 1962 (SNCH).

INDIA: Andaman & Nicobar Islands, Andhra Pradesh, Bihar, Gujarat, Haryana, Kerala, Maharashtra, Odisha, Punjab, Rajasthan, Sikkim, Tamil Nadu, Uttar Pradesh & West Bengal.

WORLD: Native to Tropical America.

Physalis lagascea Roem. & Schult., Syst. 4: 679. 1819; Murugan *et al.*, Indiam J. Forestry 41: 393. 2018. *P. micrantha* Link, Enum. Hort. Berol. 1: 181. 1821.

Annual herbs, up to 50 cm high; branchlets diffuse, angular, minutely pubescent. Leaves ovate, oblong-ovate, base oblique-cordate, margin entire, sinuate or irregularly dentate, apex acute to cuinate, 4-6.5 x 2-4 cm; petioles up to 5 cm long, minutely pubescent. Flowers solitary, terminal, nodding; pedicels up to 8 mm long, shortly pubescent; calyx campanulate; corolla campanulate to rotate, *c.* 5 mm across, pale yellow to white; lobes indistinct; stamens 5, filaments *c.* 2 mm long, flat, hairy, anthers oblong, yellow, 1-1.5 mm long; ovary subglobose or ovoid, 1.5-2 mm long, glabrous; style *c.* 3 mm long, stigma simple, green. Berris globose to subglobose, fruiting calyx inflated, 1.5-2.5 cm long, obscurely 10-angled, glabrous; seeds discoid, pale yellow.

Rare in Eastern Ghats of Odisha and Sirumalai hills in Tamil Nadu. Fl. & Fr.: May–September.

G. Udayagiri Forest range (KDM), *Kalidass & Murugan* 18021 (*Holotype*: CAL, *Isotypes*: MH, RPRC); Paburia (KDM), *Kalidass & Murugan* 18024 (RPRC); Kalinga Ghat (KDM), *Kalidass & Murugan* 18027 (RPRC); Sikulipadara (GJM), *Kalidass & Murugan* 18045 (RPRC); Parlakhemundi, Jajpur (GJP), *Kalidass & Murugan* 18067 (RPRC); Taptapani (GJM), *Kalidass & Murugan* 18087 (RPRC); Way to Kutladampatti falls, Sirumalai hills (DGL), *R.Kottaimuthu* 1935; Base of Sirumalai (DGL), *R.Kottaimuthu* 1990 (SNCH).

INDIA: Andaman & Nicobar Islands, Andhra Pradesh, Kerala, Madhya Pradessh, Maharashtra, Mizoraam, Odisha, Punjab, Rajasthan, Tamil Nadu, Uttar Pradesh and West Bengal.

WORLD: Sri Lanka.

SOLANUM Linnaeus

Solanum americanum Mill. var. **odishense** Kalidass & P. Murugan, Bangladesh J. Plant Taxon. 23(1): 83. 2016; Murugan *et al.*, Indian J. Forestry 41: 393. 2018.

Annual, erect herbs, up to 60 cm high; stem angular with prickly, young branches glabrescent. Leaves simple, alternate, ovate or ovate-lanceolate, 4-8 x 2.5-4 cm, base cuneate or obtuse, apex obtuse or acute, upper surface pubescent, lower surface pale, margin entire to sinuate; lateral nerves 408 pairs; petioles 1.5-3 cm long, flat or terete. Inflorescence extra-axillary, umbellate or subumbellate cymes, 5-9-flowered; flowers creamy white; calyx 5-lobed, lobes deltoid, *c.* 0.5 mm long, green, pubescent without, reflexeed away from mature berries, persistent; corolla

5-lobed, 2-3 mm across; lobes ovate, *c.* 2.5 mm long, glabrous, apex obtuse or acute, margin ciliate, reflexed; stamens 5, filaments *c.* 0.5 mm long, slender, glabrous, anthers oblong, *c.* 1 mm long, yellow, basifixed; ovary globose, less than 1 mm long, glabrous, style *c.* 1.2 mm long, filiform, densely villous in the middle, curved at apex, stigma capitate, green. Berries globose, 5-8 mm in diam., (immature) green, glabrous.

Rare in Sirumalai hills. Fl. & Fr.: May–Septeember.

Sirumali hill (DGL), *Kalidass* & *Murugan* 1711 (RPRC).

INDIA: Tamil Nadu. Endemic.

Solanum diphyllum L. Sp. Pl. 184. 1753; Paul & Biswas, Bull. Bot. Surv. India 37: 137. 1995; Kumari, Rheedea 23: 50. 2013; Murugan *et al.*, J. Econ. Taxon. Bot. 39: 527. 2015.

Undershrub, 1.5 m high, erect, unarmed, perennial, branches slender, greenish brown, doted, glabrous, internodes five to six cm long. Leaves simple and at each node one pair of leaves is present (one large and another small), obovate or elliptic, 3.5-6 x 2-3.5 cm, base attenuate or cuneate and decurrent on petiole, margin entire, apex acute or obtuse, glabrous; petiole to 0.5 mm long. Inflorescence a racemose fascicle, *c.* 10-flowered, peduncles to 7 mm long; calyx copular, 1.5-2 x 2-2.5 mm, minutesly pubescent, 5-lobed, lobes broadly triangular, acute, *c.* 0.5 x 0.8 mm, hirsute; corolla stellate, 5-8 mm across, creamy white, 5-lobed, lobes equal, elliptic or linear-ovate, acute, 4-5 x 2.5-3 mm, incurved, hyaline, tube 1-2 mm long; stamens 5, equal, epipetalous, introrse, filaments *c.* 1mm long, anthers oblong, initially dehiscing poricidally, later longitudinally; ovary globose, slightly constricted in middle, 7-12 mm diam., glabrous, yellow or reddish yellow; seeds reniform or discoid, *c.* 3 x 2 mm, thickened at margins, yellow or ovate.

Occasional in moist places. Fl. & Fr.: May–September. Vern.: Oriya: *Banner.*

Kalahandi, Goudthala, *Murugan* & *Kalidass* 18015 (RPRC); Khurda, Nayapalli, *Murugan* & *Kalidass* 18044 (RPRC).

INDIA: Karnataka, Maharashtra, West Bengal, Tamil Nadu, Odisha, Uttar Pradesh.

WORLD: Native to Mexico and Central America.

Solanum pseudocapsicum L., Sp. Pl. 184. 1753; Chithra in A.N.Henry *et al.*, Fl. Tamil Nadu 2: 116. 1987; Murugan *et al.*, Indian J. Forestry 41: 394. 2018.

Perennial, unarmed, erect, large herbs to undershrubs, up to 120 cm high; branchlets terete, young branches glabrous, sometime pubescent. Leaves simple, alternate, lanceolate or elliptic-lanceolate, 3-10 x 1.5–3 cm, base attenuate or decurrent, margin entire to repand, apex acute, upper surface glabrous, lateral nerves 4-8 pairs. Inflorescence extra-axillary or leaf-opposed, simple cyme, mostly solitary, sometimes 2 or 4-flowered; flowers regular, pentamerouss; peduncle 4-8 mm long, slender, glabrous; pedicels 8-10 mmlong; calyx 5-lobed, united at base, lobes lanceolate, 7 mm long, reflexed away from mature berries, persistent; corolla 5-lobed, creamy white, rotate, 1.5 cm across, lobes ovate or lanceolate, *c.* 6 mm long, apex obtuse; stamens 5, filaments 0.5 mm long, glabrous, anther oblong, 3.5 mm long, yellow, bsifixed, dehiscing apically; ovary globose, *c.* 1.5 mm long, glabrous, style slender, 4.5 mmlong, glabrous, stigma capitate, green. Berries globose, *c.* 1.5 cm across, bright orange red, glabrous, glossy; seeds 10-20, orbicular or concave, *c.* 2.5 mm long, brownish-white, compressed.

Occasional in Sirumalai hills (Murugn *et al.*, 2018) Fl. & Fr.: July–September.

Puthur, Sirumalai hills (DGL), *Kalidass & Murugan* 1705 (RPRC); Puthur to Jailani estate, Sirumalai hills (DGL), *R.Kottaimuthu* 2030 (SNCH).

INDIA: Kerala, Tamil Nadu.

WORLD: South West Asia, Australia and naturalized elsewhere.

Solanum villosum Mill., Gard. Disct. Ed. 8. No. 2. 1768; P.Murugan *et al.*, India J. Forestry 40: 73. 2017.

Annual, erect herbs, 40-100 cm high; branchlets many, terete, ascending, glabrescent. Leaves simple, alternate, ovate or ovate-lanceolate, 3.5-8 x 2.5-3.5 cm, apex acute to acuminate, margin entire or sinuate to coarsely dentate, base cuneate to decurrent, sparsely pubescent with pale beneath; venation pinnate, lateral nerves 4-8 pairs, petiole 1-2 cm long, flat, glabrous. Inflorescence extra-axillary or axillary, umbellate or subumbellate, 4-8-flowered; peduncle *c.* 1 cm long, erect, glabrous; pedicels 1-2 cm long, glabrous. Flowers regular, pentamerous; calyx 1.5-2 mm in diam., cup-shaped, 5-lobed, base united, lobes deltoid, 0.5 mm long, green, glabrescent, reflexed away from mature berries, persistent; corolla 3-6 mm in diam., 5-lobed, lobes ovate, 2.5 mm long, glabrous, with prominent purple line in the middle adaxially, apex obtuse or acute, reflexed; stamens 5, filaments *c.* 0.5 mm long, slender, glabrous, anthers oblong, *c.* 1.5 mm long, yellow, basifixed, dehiscing apically; ovary globose, less than 1.5 mm long, glabrous; style *c.* 3 mm long, filiform, densely villous in the middle; stigma capitate, green. Berries globose, 6-10 mm in diam., orange or orange to reddish when ripe, glabrous; seeds discoid, 1.5 mm in diam., brownish yellow, compressed.

Occasional along roads, water bodies in association with *Solanum americanum* Mill., *Argemone mexicana* L., *Datura metel* L., *Solanum erianthum* D.Don etc. Fl. & Fr.: September–December.

Khandapada RF (Nayagarh distr.), *Kalidass & Murugan* (RPRC); Chadeibas RF (Nayagarh distr.), *Kalidass & Murugaan* 10739 (RPRC); Dihagaon (Nayagarh distr.), *Kalidass & Murugan* 10747 (RPRC).

INDIA: Tamil Nadu, Kerala, Odisha, Maharashtra.

WORLD: Europe, western Asia, northern Africa, North America, Australia.

GESNERIACEAE

AESCHYANTHUS Jack

Aeschynanthus perrottetii A.DC. in DC., Prodr. 9: 261. 1845; FBI 4: 339. 1884; Gamble, 1: 985. 1924; Karuppusamy & Ravichandran, Kong. Res. J. 3(1): 30. 2016. *A. perrottetii* A. DC. var. *planiculmis* Clarke in Hook.f., Fl. Brit. India 4: 340. 1884. *A. planiculmis* (Clarke) Gamble, Fl. Pres. Madras 2: 985. 1924.

Epiphytic subshrubs; stems terete, glabrous, scandent. Leaves opposite, to 8 x 1.5 cm, elliptic, succulent, acute, pale below; petiole 5 mm long. Flowers axillary, solitary or rarely paired; pedicels 1 cm long; sepals 8 x 1 mm, free, lanceolate; corolla pink, 4.5 cm long, tubular, curved, widening above, oblique at mouth, lobes 5, subequal, ovate, obtuse, glandular-ciliate; stamens 4, all perfect; filaments 2.5 cm long, free, inserted at the middle of the corolla tube; upper stamens with smaller anthers, confluent in pairs; ovary long-stipitate; linear, 1-celled, ovules many; style slender, stigma terminal. Capsule to 15 cm long, linear; seeds linear, many.

Rare in Kolli hills. Fl. & Fr.: October–March.

Kulivalvavu shola, Kolli hills (NKL), *SKS & Ravichandran* 1242 (SGH).

INDIA: Peninsular India. Endemic.

CORDIACEAE

CORDIA Linnaeus

Cordia ramanujamii N. Balachandran & K. Rajendiran, Taiwania 61(2): 74. 2016.

Small tree, to 2.5 m tall. Bark greyish-green, smooth. Branchlets slender, lenticellate, tender parts stellate tomentose. Petiole 0.5–1.8 cm long, canaliculate, pubescent when young, glabrous at maturity. Leaves elliptic-oblanceolate, 1.8–7.8 x 0.9–4 cm, acute-acuminate at apex and shortly thick apiculate, base oblique, obtuse, lower margin entire, upper margin distantly serrulate, 3-nerved at base, lateral nerves 2–4 pairs, glabrous above, scattered pilose hairy below, especially near the base below. Inflorescence a cyme, 5–7 x 4.5–5.6 cm. Flowers bisexual, 1.7 x 0.8 cm; calyx tube 8 - 10 mm long, tube leathery, rusty tomentose half above the tube and dense near lobes, lobes 6 (4+2), unequal, 1 mm long, ciliate, densely brown-silky villous within; corolla 1.2–2 cm long, white, tube glabrous within and base; throat hyaline villous, lobes 6–7, 4 mm long; stamens 6–7, unequal, 3(4) shorter and 3 longer, alternate each other; anthers oblong, 1.5 mm, sagitate at base, filaments hairy at base, attached near throat; ovary glabrous, ellipsoid-oblong, 1.5–2 mm long; style 5–6 mm long, stigma 4, filiform, 8–10 mm long, slightly thickened at apex, glabrous. Fruit pale yellow, 1.7 cm across; seed 1, 8–10 mm across.

Rare in Gingee hills of Tamil Nadu. Fl. & Fr.: July–December.

Gingee (SA), *N. Balachandran* 12812 (holotype MH; isotype AURO, HIFP).

INDIA: Tamil Nadu. Endemic.

SCROPHULARIACEAE

LINDERNIA Lehm.

Lindernia molluginoides (Benth.) Wettst., Nat. Pflanzenfam. 4(3b): 80. 1891; Indira & Venkataraju, J.Indian Bot. Soc. 77: 135, 1998; Chandramohan *et al.*, Indian J. Forestry 40: 293-294. 2017. *Vandellia molluginoides* Benth., Scroph. India 35: 1835 & in DC., Prodr. 10: 413. 1846; Hook.f., Fl. Brit. India 4: 279. 1884.

Annual herb, up to 15 cmlong, high, erect or decumbent, rooting at nodes, glabrous but sparsely pubescent near the leaf axils. Leaves opposite, obscurely petiolate, ovate to oblong or oblong-lanceolate, 6-14 x 3-7 mm, apex obtuse or subacute, margin obscurely crenate, scabrid, base cuneate, glabrous on both surfaces except sparsely hairy veins. Flowers 3-10 in terminal, umbellate fascicles or rarely solitary and axillary; pedicels slender, 5-10 mm long, often deflexed in fruiting, glbrescent, bracts small, lanceolate, up to 1.5 mm long; calyx tubular, 3 mm long in flower, 4-5 mm long in fruit, 5-nerved, shallowly 5-lobed, sparsely adpressed hairy; corolla bluish-purple, 5-7 mm long, 2-lipped upper lip broadly elliptic, shallowly bilobed, lower lip with three rounded lobes; stamens 4, all fertile, posterior anthers each with connective of the lower locule produced into a short projection, filaments of anterior stamens each with a short, clavate spur, arising near the bas. Capsule oblong-ellipsoid, 3-4 x 1.5-2 mm, apiculate at apex, more or less shorter than persistent calyx; seeds oblong-ellpsoid, 0.3 x 0.2 mm, pale brown, rugulose.

Occasional in open moist grasslands in association with *Lindernia ciliata* (Colsm.) Pennell, *Eriocaulon truncatum* Buch.-Ham. ex Mart. and *Mollugo pentaphylla* L. Fl. & Fr.: September–November.

Kuanaria dam, Banigoccha forest range, Mahanadi Wildlife division, *Chandramohan* 8332 (BSID).

INDIA: Karnataka, Maharashtra, Andhra Pradesh, Odisha.

WORLD: Nnative to South East Asia, Myanmar.

LENTIBULARIACEAE

UTRICULARIA Linnaeus

Utricularia australis R. Br., Prodr. Fl. Nov. Holl. 430.1810; Saxena, Indian Forester 96: 249. 1970; Gandhi in Saldanha & Nicolson, Fl. Hassan 563. 1976; Taylor in Steenis, Fl. Males. I. 8; 299. 1977 & Kew Bull. Add. Ser. 14: 598. f. 184. 1989; Janarthanam & Henry, Bladderworts of India 30. 1992; Kullayiswamy *et al.*, J. Indian Bot. Soc. 92: 101. 2013. *U. flexuosa* sensu C.B.Clarke in Hook. f., Fl. Brit. India *p.p.* quoad Syn. *U. australis* R. Br., non Vahl 1804.

A small suspended aquatic herb with submerged free-floating branch, up to 30 cm in length, 0.5 mm thick, filiform, terete. Leaves absent, foliar organs equivalent of leaves, 1.5–2.5 cm long, primary segment divided into two equal segments, each divided into numerous alternate segments up to 10, pinnately arranged, corn segments capillary and spinulose-toothed utricles. These segments are thread like and have small bladders or traps, 1–1.5 mm, obliquely obovoid on nodes and foliar

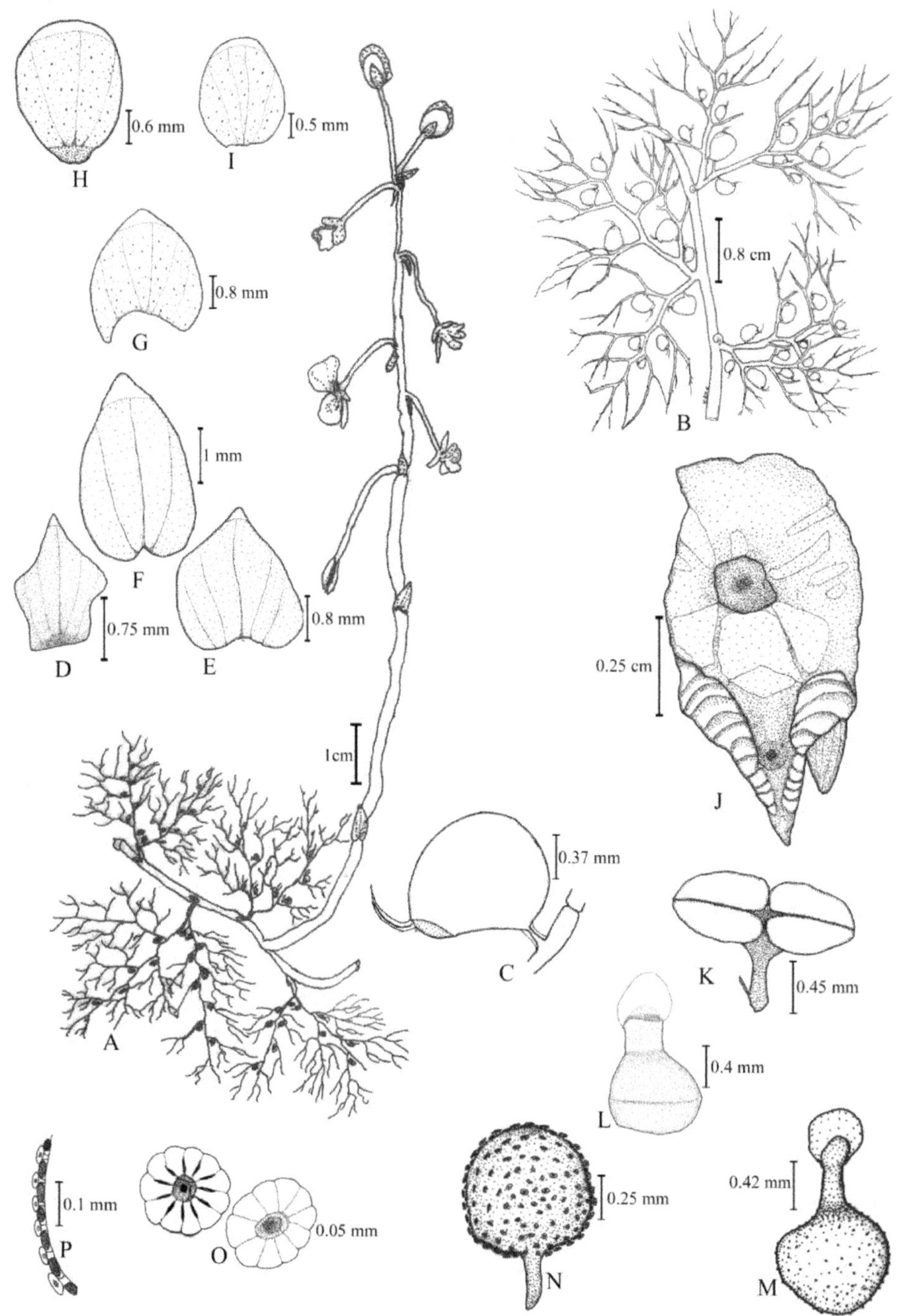

Figure 126. **Utricularia australis** R.Br.

A. Habit, B. Branching, C. Bladder, D, E & F. Scales on peduncle, G. Bract, H. & I. Calyx, J. Corolla, K. Stamen, L. Ovary, M. Fruit, N. Placenta, O. Seeds, P. Section of placenta.

segments, stalked, mouth laterals oblique; appendages 2, often simple setae present on lips, the bladders are hinged by a flap, which forms a door or valve that only opens inwards. Roots absent; sometimes root like organs present at the node to 2 mm long. Raceme or flower stalk weakly erect, emergent above water, 10–15 cm long, flexuous, the flower usually ending in a flower head of 3–6-flowers at regular intervals; axis initially short, elongating in fruit; scales 3 on peduncle, similar to the bracts in upper half of peduncle, 1st one obovate, 1.5 x 1 mm, base truncate, apex sharply acute, 2nd one ovate, 2.5 x 2.5 mm, base cordate, apex blunt acute, 3rd one ovate–oblong, 3 x 2 mm, base cordate, apex acute, bracts 3–3.2 x 3–3.2 mm, basifixed, ovate to suborbicular, cordate at base, acute to obtuse at apex, slightly glandular without; bracteoles absent. Pedicels 2–2.5 cm long, filiform, glabrous, pendulous in fruit; calyx lobes 2, slightly unequal, ovate, to oblong; upper lobe 2–2.5 x 1.5–1.8 mm, retuse; lower 1.5–2 x 1.5–2 mm, obtuse at apex, sparsely glandular without; corolla yellow, with reddish brown lines and spots on the darker lower lip, *c.* 10 mm long; upper lip inserted in lower lip, upper lip *c.* 5 x 3.5 mm, ovate to oblong, truncate at apex; lower *c.* 6 x 12 mm, transversely elliptic, auricled and prominently gibbous at base, truncate at apex. Spur *c.* 5–6 mm long, conical, obtuse at apex, glandular within; stamens 2, 1–1.5 mm long; filaments dilated near anthers; anther thecae confluent, sterile 1.5–1.8 mm; ovary *c.* 1.2 mm long, globose, glandular; style short; stigma 2-lipped, unequal, lower lip deltoid, upper lip obovate, thin and transparent. Capsule, 1.7 x 1 mm, black, obovate, capsule opens by valves, free central placentation; seeds abortive, minute, to 0.05 mm, flattened, appendages absent, inserted between the papillae.

Rare in ponds and tanks in Anantapuram district. Fl. & Fr.: June–August. Kalasamudram (ATP), *KRKS* 41861 (SKU, MH & CAL).

INDIA: Tamil Nadu, Kerala, Karnataka, Maharashtra.

WORLD: Europe, Africa, Sri Lanka, Indochina, Malaysia, Philippines, Japan and Australia.

ACANTHACEAE

MONOTHECIUM Hochst.

Monothecium aristatum (Wall. ex Nees) Anders. in Thw., Enum. Pl. Zeylan. 234. 1860; FBI 4: 524. 1885; Gamble 2: 1068. 1924; Kottaimuthu & Kalidass, J. Econ. Taxon. Bot. 38: 98. 2014. *Justicia aristata* Wall. ex Nees in Wall. Pl. Asiat. Rar. 3: 115. 1832.

Decumbent herbs; branches slender, terete, puberulous, rooting at nodes. Leaves opposite-decussate, ovate or elliptic-oblong, 3-6 x 1.5-2.3 cm, apex acute-acuminate, base acute, pubescent on nerves beneath; petiole 1-2 cm. Spike lax, terminal or on upper axils, 2-4 cm long, bracts linear-lanceolate, setaceous, 5-10 mm long, aristate, glandular-hirtellous without; sepals 5, equal, 5-8 mm long, linear-setaceous; corolla 2-lipped, white, tube 5-8 mm long; upper lip of limb linear-oblong, emarginated at apex; lower lip 3-lobed, ovate-oblong, brown-spotted; stamens 2, filaments glabrous, anthers 1-celled, muticous; ovary ovoid, ovules 2 in each locule; stigmas 2-lobed.

Capsule oblong, 5-6 mm long, shortly stalked, beaked, glandular-setaceous; seeds 4, orbicular, 1 x 0.5 mm across.

On the banks of the riparian forests in Dindigul hills in Tamil Nadu. Fl. & Fr.: December–March.

Karandamalai (DGL), *Kottaimuthu* & *Kalidass* 15486 & 15497 (SNCH).

INDIA: Kerala, Tamil Nadu.

WORLD: Sri Lanka.

STROBILANTHES Blume

Strobilanthes ciliatus Nees in Wall., Pl. Asiat. Rare. 3: 85. 1832 & in DC., Prodr. 11: 183. 1847; FBI 4: 439. 1884; Gamble 2: 1039. 1924. *Nilgirianthus ciliatus* (Nees) Bremek., Verh., Ned. Akad. Wetensch., Afd. Naturk. Tweede Sect. 41: 270. 1944; Kottaimuthu *et al.*, 12: 301.2008.

Shrub, 1-1.5 m tall; stem terete, fimbriate at nodes. Leaves elliptic-lanceolate, 10-18 x 3-5 cm, apex acute, margin serrate, base acute, decurrent, prominently lineolate above, not so beneath; secondary veins 5-7 pairs, raised above, impressed beneath; petioles 1.5-3.8 cm long, sometimes obscure due to decurrent lamina. Inflorescence spike, axillary, 2-2.5 cm long, glabrous; peduncles 2.3-3 cm long, articulate, bracteate below middle; bracts orbicular, 0.5 cm long, obtuse at apex; bracteoles linear-lanceolate, 0.5 cm long, apiculate; calyx lobes 5, oblanceolate, 0.5 cm long, glabrous; corolla tubular-campanulate, white with purple blotches, lobes obovate, 0.3 cm long, predominantly veined; stamens 4, didynamous, exserted, filaments arising from top of staminal sheath, longer ones up to 1.5 cm long; anthers purple, o.2 cm long; ovary on a disc, 0.3 cm long, glabroius, style slender, up to 1.5 cm long.

Occasional in Karandamalai in Tamil Nadu. Fl.: March.

Elaniikudai aruvi, Karandamalai (DGL), *Kottaimuthu* 870 (ATREE); on way to Kallanai valley, Karandamalai (DGL), *Kottaimuthu* 899 (ATREE).

INDIA: Maharashtra, Kerala, Tamil Nadu. Endemic.

THUNBERGIA Retzins

Thunbergia coccinea Wall. ex D.Don, Prodr. Fl. Nepal 120. 1825; FBI 4: 394. 1885; Kottaimuthu *et al.*, J. Econ. Taxon. Bot. 38: 652. 2014.

Woody climbers, stems angular, subglabrous, pubescent at nodes, internodes twisted. Leaves opposite, decussate, broadly ovate or ovate-lanceolate, 7-14 x 3.3-10 cm, base rounded to acute, margin crenulated, apex acuminate, pubescent on both sides, palmately 3-5-veined, petioles 3-7 cm long, grooved. Racemes pendulous, axillary or terminal, 20-35 cm long. Flowers showy, light red, 1-3 cm across, pedicels 1-2 cm, pubescent; bracts lanceolate, abaxially pubescent, adaxially glabrous; bracteoles 2, oblong, spathaceous, ovate-lanceolate, cohering along margin in bud, enclosing the corolla tube, 2.2-3 x 1-1.5 cm; calyx 2 mm long, reduced to a minute rim; corolla red, tube basally cylindric for 5-6 mm, throat 12-16 mm long, lobes suborbicular, 5-7 mm in diam.; stamens 4, didynamous, attached at the throat of the tube, filaments 1.2-2 cm long, glabrous but with a tuft of hairs at base, anther oblong; ovary 3-4 x 2-3 mm, glabrous, with a fleshy disc below the ovary; style 1-3 cm long, glabrous, exserted, stigma 2-cleft. Capsule glabrous, 1-1.2 x 1.5-2 cm; seeds compressed, ovoid, verrucose.

Rare in forest margins and also in disturbed forest sites in Sirumalai hills of Tamil Nadu. Fl. & Fr.: September–May.

Near Agasthypuram, Sirumalai hills (DGL), *R.Kottaimuthu* 5027 & 5148 (SNCH).

INDIA: Maharashtra, Kerala, Karnataka, Tamil Nadu.

WORLD: America.

VERBENACEAE

PHYLA Loureiro

Phyla nodiflora (L.) Greene var. **minor** (Hook.) N.O'Leary & Mulgura, Ann. Missouri Bot. Gard. 98(4): 589. 2011. *Lippia nodiflora* var. *minor* Gillies & Hook., Bot. Misc. 1: 171. 1830.

Herbs, creeping, 30-50 cm long; stem prostrate, rooting at nodes, branches slender, 4-angular, procumbent or ascending, greenish purple, nodes purple, internodes 0.3-2 cm long, densely canescent. Leaves oblanceolate-spathulate, 1.3-2 x 0.3-0.5 cm, base narrowed, margin sharply serrate above the middle with 3-5 pairs of teeth, apex acute, densely canescent on both surfaces with a prominent midrib, lateral nerves obscure; petiole absent. Spikes axillary, ovoid to oblong, 3-5 together on 2-4 cm long peduncle, moderate to densely-flowered, with imbricate floral bracts; spikes subtended by foliaceous bracts. Flowers zygomorphic, perfect, sessile and globose; floral bracts variable, ovate to obovate-elliptic, 0.2-0.3 x 0.1-0.15 cm, apex acute, densely canescent, imbricating corolla tube; calyx persistent, membranous, ovate, slightly compressed especially on one side, distinctly bifid, sub glabrous, keeled, 1.2-1.3 mm long and easily separable; corolla hypocrateriform, 2-lipped, 5-lobed, rose purple, upper lip 1-lobed, lower lip 3-lobed, suborbicular, glabroius, tube straight, slightly exserted from calyx, purplish white, glabrous; stamens 4, didynamous, included, anther ovoid, yellow, 0.2 mm long; cells of anthers of the uppermost stamens not parallel; ovary subglobose, *c.* 0.5 mm long, cream-coloured, style stout, stigma subcapitate, oblique. Drupe oblong, compressed, 1-2 mm long.

Occasional near moist places and periphery of ponds in association with *Phyla nodiflora* var. *nodiflora* and *Lindernia ciliata*. Fl.& Fr.: February–May.

Kuanaria dam, Banigoccha forest range, Mahanadi Wildlfe division, Satkosia Tiger reserve, *K. Chandramohan* 8369 (BSID).

INDIA: Assam, Bihar, Mahrastra, Madhya Pradesh, Manipur, Odisha, Rajasthan, Tamil Nadu, Uttar Pradesh.

WORLD: America, Baluchistan, China.

PLANTAGINACEAE

VERONICA Linnaeus

Veronica undulata Wall. in Roxb., Fl. Ind. (Carey & Wallich ed.) 1: 147. 1820; Pennell, Acad. Nat. Sci. Philad. 5: 91. 1943; Chandraamohan & Prasanna, J. Econ. Taxon. Bot. 41. 120. 2017. *V. anagallis-aquatica* subsp. *undulata* (Wall.) Elenevsky, Byull. Moskovsk. Obshch. Prir. Otd. Biol. 82(1): 153. 1977.

Perennial with root stock or rarely annual herbs, up to 70 cm long; stems fistular, erect or decumbent, branched near the base, succulent; rhizome inclined; stems, inflorescence axis, pedicels, calyces and capsule sparsely glandular. Leaves sessile, amplexicaul, elliptic to ovate or ovate to lanceolate, 3–10 x 1–3 cm, glabrous on both surfaces, irregularly serrate along margins, lower leaves rarely clasping stem, sub-acute at apex. Flowers small, white, in axillary racemes; racemes up to 17 cm long and 1–1.5 cm wide, longer than leaves; bracts lanceolate, *c.* 2 mm long, sparsely glandular hairy along margins; pedicels 3–5 mm long, as long as bracts, longer in fruit, straight or slightly curved at a right angle with inflorescence axis, spreading in fruit; calyx short, *c.* 3 mm long, deeply 4-lobed, persistent; lobes ovate to lanceolate, sub-equal, not appressed on capsule in fruit, sub-acute at apex, sparsely glandular hairy; corolla white, rotate, deeply 4-lobed, up to 1 mm long with unequal width; tube short, *c.* 0.3 mm, glabrous; stamens 2, opposite each other, connate near base of corolla tube, shorter than corolla; filaments *c.* 1.2 mm long, glabrous; anthers oblong, 0.2–0.3 mm long; style 1–1.5 mm long, shorter than capsule in fruit; stigma capitate. Capsules sub-globose, 2-celled, as long as wide, 2–2.8 mm, slightly retuse, more or less compressed, sparsely glandular hairy at the apex; seeds many, brownish, shiny, ellipsoid, more or less compressed, 0.3–0.4 mm.

Found in a single locality in marshy area along Mahanadi River at 53m altitude. This species grows in association with *Polygonum* spp., *Rumex dentatus* L. and *Schoenoplectiella articulata* (L.) Lye. Fl. & Fr.: February–May.

Satkosia Tiger Reserve Kusanga forest range, Nayagarh, *K.Chandramohan* 8379 (BSID).

INDIA: Northern India, Maharashtra, West Bengal.

WORLD: Afghanistan, Japan, Korea, Laos, Nepal, Pakistan, Thailand, Vietnam.

AMARANTHACEAE

ACHYRANTHES Linnaeus

Achyranthes coynei Santapau, Kew Bull. 3: 488 1948 publ. 1949; Karthik *et al.*, J. Econ. Taxon. Bot. 41: 123.2017.t

Perennial sub-erect under shrub, to 1.5 m high. Stem terete, woody; branches quadrangular, younger ones pubescent, green or with purple patches, older ones glabrous or glabrescent, brown to purple; nodes swollen. Leaves deciduous; dimorphic, with lower ones being larger 15–22 x 9.6–10 cm, whereas upper leaves gradually decreases in size, elliptic or elliptic-lanceolate, apex acute or acuminate; sub-glabrous above, pubescent beneath; slightly wavy, petiole 1–3 cm long, channeled, pink above, and green beneath. Flowers in spikes, densely arranged on

21–40 cm long, thick rachis densely pubescent and whitish; buds erect, pointing towards tip of the inflorescence; rosy or purplish flowers spreading up to1.2 cm in diameter; persistent, linear and membranous bract with distinct midrib, smaller than the longest tepal, up to 5 mm long, with rosy tinge at the base; two bracteoles, with thin papery wing like structure at the base; tepals five in two whorls, up to 8 mm long, 1 mm broad, glabrous, ribbed, margins scarious; stamens five, in number with filaments fused at base; staminodes fimbriate, half the length of the filaments, alternatively placed with pink orpurple fertile anthers, fused together, pink or purple; antheriferous filaments and lobes pink when young; ovary truncate; style and stigma pink; style 0.4 cm long, filiform, persistent with stigma capitellate. Capsule *c.* 5 mm, enclosed within persistent hardened sepals; seed one, cylindrical, smooth and brown.

This taxon has been reported by Karthik *et al.* (2017) from Chitthoor district of Andhra Pradesh. Fl. & Fr.: September–March.

INDIA: Maharashtra, Karnataka, Andhra Pradesh. Endemic.

VISCACEAE

VISCUM Linnaeus

Viscum subracemosum Sanjai & Balakrishnan, Rheedea 10: 113. 2000; Kottaimuthu *et al.*, Ethnobotanical Leaflets 12: 302. 2008.

Monoecious undershrub, up to 1 m high, branchlets trichotomousy branched, longitudinally grooved. Leaves opposite, elliptic-oblanceolate or lanceolate, 2.5-6.3 x1.5-2.8 cm, base attenuate, margin wavy, apex rounded, triplinerved, lateral pair converge towards apex. Inflorescence 3-6-flowered subraceme, 3-6 cm long; peduncles 1-2 cm long. Flowers enclosed in cupule; bracts ovate, up to 0.2 cm long. Berry smooth, globose, 0.5 cm in diameter.

Occasional in Karandamalai in Tamil Nadu. Fl. & Fr.: January–April.

On way to Kallanai valley, Karandamalai (DGL), *R.Kottaimuthu* & *M.Vimala* 1878 (ATREE), On way to Kallanai valley, Karandamalai (DGL), *R. Kottaimuthu* & *J. Brabhu* 1958 (ATREE), on way to Meenunuthi aruvi (DGL), *R,Kottaimuthu* & *K. Natarajan* 333 (ATREE).

INDIA: Karnatala, Tamil Nadu. Endemic.

Note on the Occurrence of *Ensete* species and *Musa shankarii* in Eastern Ghats

Joe and Sabu (2019) have recently published a book on Revision of Indian Musaceae. They are of the opinion that *Ensete glaucum* does not occur in Eastern Ghats, and it is *Ensete superbum* that occurs in Eastern Ghats. They say "Rao and Kumari (1967, 2008) recorded the natural occurrence of *E.glaucum* from Andhra Pradesh in Eastern Ghats of India, but these occurrence are of *E. superbum*, which is commonly found in Peninsular India; this has been confirmed from specimens held at Madras herbarium".

Musa shankarii Subba Rao & Kumari was described as new species. But Joe and Sabu (2019) reduced this species as a synonym under *Musa balbisiana* (see Joe and Sabu 2019 page 122).

REFERENCES

Anjaneyulu, P. and B.R.P.Rao. 2018. *Cleome rutidosperma* DC. (Cleomaceae) and *Nicotiana plumbaginifolia* Viv. (Solanaceae): New distributional records for Andhra Pradesh, India. Indian J. Forestry 41(4): 373-375.

Bhaskar, V. 2012. Taxonomic monograph on *Impatiens* L. (Balsaminaceae) of Western Ghats, South India. The key Genus for endemism. Centre for Plant Taxonomic Studies, Banglaore, India.

Chandramohan, K., M.R. Bhanja and P.S.Annamma. 2014. Recollection of *Hylodesmum repandum* (Vahl) H.Ohashi & R.R.Mill (Fabaceae-Papilionoideae) after a gap of 100 years. J. Econ. Taxon. Bot. 38: 100-101.

Chandramohan, K., M.R.Bhanja and C.R.Reddy. 2016. *Aralia thomsonii* (Arliaceae) a new distributional record for Peninsular India. J. Econ. Taxon. Bot. 40: 159-160.

Chandramohan, K. and P.V. Prasanna. 2017. Two new distributional records for Odisha state. Indian J.Forestry 40(1): 95-96.

Chandramohan, K. and P.V.Prasanna. 2017a. *Phyla nodiflora* var. *minor* (Hook) N.O'Leary and Mulgura (Verbenaceae)–a new distributional record for India. Indian J. Forestry 40(4): 371-373.

Chandramohan, K. and P.V.Prasanna. 2017b. The distribution of *Veronica undulata* Wall. (Plantaginaceae) in India. J. Econ. Taxon. Bot. 41: 120-122.

Chandramohan, K., P.V. Prasanna, P.R.Reddy and J.Swamy. 2016. Note on the distribution of *Abelmoschus moschatus* subs. *tuberosus* (Malvaceae) in Eastern Ghats, India. J. Econ. Taxon. Bot. 40: 35-37.

Chandramohan, K., P.V.Prasanna, P.R.Reddy and S. Nagaraju. 2017. *Eriolaena stocksii* Hook.f. & Thomson ex Mast.–A little known endemic species from Odisha state, India. Indian J. Forestry 40: 103-105.

Chandramohan, K., P.V.Prasanna, M.Sankara Rao and E. Kiran. 2017. *Lindernia molluginoides* (Benth.) Wettst. (Linderniaceae)–A rare species from Eastern Ghats. Indian J.Forestry 40(3): 293-294.

Chandramohan, K., P.V. Prasanna, J. Swamy and T. Rahman. 2016. Two new distributional records for Odisha state. Indian J. Forestry 39(3): 295-297.

Chandramohan, K., J. Swamy and M. T. Naidu. 2016. Notes on taxonomy and distribution of *Desmodium zonatum* (Leguminosae: Papilionoideae) in Peninsular India. Rheedea 26(1): 26-28.

Devanathan, K. 2018. A new variety of *Abutilon persicum* (Malvaceae) from Southern Estern Gats, India. Rheedea 28(2): 92-96.

Dhatchanamoorthy, N., K.Sambandan, P. Raja and S. Soosairaj. 2015. *Drymaria villosa* Chamisso & Schlechtendal (Caryophyllaceae)–A new report to the Eastern Ghats of Tamil Nadu, India. Int. J. Curr. Res. Biosci. Plant Biol. 9: 74-76.

Joe, A. and M. Sabu. 2019. Revision of Indian Musaceae. Indian Association for Angiosperm Taxonomy, University of Calicut.

Karthick, R., I. Baig, K. Mosachristas and E. Kowsalya. 2017. New distributional record of the endemic species *Achyranthes coynei* (Amaranthaceae) from Chittoor, Andhra Pradesh, India. J. Econ. Taxon. Bot. 41: 123-125.

Karuppusamy, **S**. and V. Ravichandran, 2016. *Aeschynanthus perrottetii* DC. (Gesneriaceae)–a new record for Eastern Ghats. Kong. Res. J. 3(1): 30-31.

Kottaimuthu, R. 2017. *Crotalaria heyneana* (Fabaceae) in Eastern Ghats of India: A new distributionak record. J. Econ. Taxon. Bot. 156-158

Kottaimuthu, R., R. Ganesan, K. Natarajan, J.Brabhu and M.Vinmala. 2008. Additions to the Flora of Eastern Ghats, Tamil Nadu, India. Ethnobotanical Leaflets 12: 299-304.

Kottaimuthu, R., G.Gnanasekharan and N. Vasudevan. 2016. *Calopogonium* and *Centrosema:* Two new generic records for Andhra Pradesh, India. Indian J. Forestry 39(1): 89-90.

Kottaimuthu, R., C. Kalidass and A. Saravanan. 2014. *Thunbergia coccinea* Wall. ex D.Don (Acanthaceae): A new record for South India. J. Econ. Tax. Bot. 38: 652-654.

Kottatimuthu, R. and N. Vasudevan. 2017. A new species of *Derris* Lour. (Fabaceae: Papilionoideae) from Tamil Nadu, India. Webbia 72: 97-100.

Kullayiswamy, K. R., S. Sandhyarani, S. Karuppusamy and T. Pullaiah. 2012. The rediscovery of *Brachystelma volubile* (Apocynaceae- Asclepiadoideae). Rheedea 22(2): 107-110.

Kullayiswamy, R.K., S. Sandhya Rani and T. Pullaiah. 2013. *Brachystelma ciliatum* (Apocynaceae): A new record for Eastern Ghats of Andhra Pradesh, India. Nelumbo 55: 191–195.

Kullayiswamy, K. R., S. Sandhya Rani and T.Pullaiah. 2016. *Brachystelma nigidianum* (Apocynaceae: Asclepiadoideae), a new species from India. Kew Bull. 71 (4): 1-7. DOI 10. 1007/s12225-016-9668-z

Murugan, P., C. Kalidass and P.C. Panda. 2015. A new record of *Solanum diphyllum* L. (Solanaceae) from Eastern Ghats of India. J. Econ. Taxon.Bot. 39: 527-529.

Murugan, P., C.Kalidass and P.C.Panda. 2016. A new variety of *Solanum americanum* Mill. (Solanaceae) from Eastern Ghats, India. Bangladesh J. Plant Taxon. 23(1): 83-85.

Murugan, P., C. Kalidass and P.C. Panda. 2017. *Solanum villosum* Mill. (Solanaceae): A new plant record for Eastern Ghats of India. Indian J. Forestry 40 (1): 73-75.

Murugan, P., R.Kottaimuthu, C. Kalidass and P.C. Panda. 2016. *Lycianthus* (Dunal) Hassl. (Solanaceae): A new generic record for Odisha, India. NeBio 7(4): 127-128.

Murugan, P., R. Kottaimuthu, C.Kalidass and P.C.Panda, P.C. 2018. Some additions to the Solanaceae of Sirumalai Hills, Southern Eastern Ghats, India. Indian J. Forestry 41(4); 391-395.

Padal, S.B., K.V.Satish, J.Prakasa Rao and K. Satyavathi. 2017. Extended distribution of medicinall tree species *Pittosporum dasycaulon* (Pittosporaceae) to northern Eastern Ghats, Andhra Pradesh, India. J. Econ. Taxon. Bot. 41: 70-73.

Prasad, K. and P.V.Prasanna. 2016b. *Brachystelma seshachalamensis* (Apocynaceae), a new species from Andhra Pradesh. Bangladesh J. Plant Taxon. **23(1)**: 53-57.

Prasad, K., P.V. Prasanna, U. Meve, M. Rao and T. Tulasaiah. 2015. *Brachystelma annamacharyae* (Apocynaceae), a new species from Seshachalam hils of Andhra Pradesh, India. Nordic J. Bot. 34: 360. Doi: 10. 1111/njb. 00887

Prasad, K. and B.R.P.Rao. 2013. *Brachystelma nallamalayanum sp. nov.* (Apocynaceae: Asclepiadoideae–Ceropegieae) from India. J. Threatened Taxa **5**(14): 4904–4906.

Pullaiah, T., Karuppusamy, S. and Srirama Murthy, K. 2019. Monograph on *Brachystelma* and *Ceropegia* in India. CRC Press, U.K.

Rao, B.R.P., K. Prasad, B. Sadasivaiah, K.S. Basha, M.V. Suresh Babu and P.V. Prasanna. 2011. A new species of *Brachystelma* R. Br. (Apocynaceae: Asclepiadoideae–Ceropegieae) from India. Taiwania **56**(3): 223–226.

Rasingam, L., A. Chorghe, U. Meve, M.S. Rao and P.V.Prasanna. 2013. *Brachystelma penchalakonense* (Apocynaceae: Asclepiadoideae), a new species from Andhra Pradesh, India. Kew Bull. **68**: 663–667.

Rasingam, L., J.Swamy and M.Sankara Rao. 2018. Clarke's Morning Glory *Ipomoea clarkei* Hook.f. (Convolvulaceae): addition to the flora of Eastern Ghats. J. Threat Taxa 10(6): 11827-11829.

Rasingam, L., J. Swamy and M.Sankara Rao, M. 2018. *Crotalaria nallamalayana* (Fabaceae: Crotalarieae): a new species from Telangana, India. Phytotaxa 345(2): 159-164. http://dx.doi.org/10.11646/phytotaxa.345.2.7

Sadasivaiah, B., B. Kalpana, A. Narasimha, M. Sharath Goud and N. Babu Rao. 2016. *Brachystelma nallamalayana* K. Prasad & B.R.P.Rao: A new distributional record for Telangana, India. Asian J. Plant Sci.Res. 6(2): 27-29.

Soosairaj, S. and P. Ananthi. 2014. *Diodia teres* Walt. (Rubiaceae): A new record to the Flora of Tamil Nadu. J. Econ.Taxon. Bot. 38: 549-551.

Sujana, K.A., P.A.Dhole and M. Mishra. 2018. New distributional record of *Toxocarpus longistigma* (Apocynaceae) from Bonda hill, Odisha, India. J. Econ. Taxon. Bot. 40: 165-167.

Swamy, J., L. Rasingam and R. Mane. 2018. *Rhynchosia jacobi* Chandrab. & B.V.Shetty (Leguminosae): An addition to the flora of Eastern Ghats, India. Indian J. Forestry 41(4): 369-371.

Swetha, B. and B.R.P.Rao. 2013. *Calyptocarpus vialis* Less. (Asteraceae), a new distributional record from Andhra Pradesh. J. Biosci. Res. 4(1): 10-11.

Venu, P. and K. Prasad. 2015. The existential crisis in Indian *Brachystelmas* (Apocynaceae). Current Science 109(4): 680–682.

Key to the Families

Synopsis of Groups

Ia. Seeds enclosed in fruit wall; ovules enclosed within carpel; stigma present .... **ANGIOSPERMS**

IIa. Floral parts in 3 to 5-merous, sometimes exceeds in pistil; vascular bundles of the stem concentric ring, enclosing in central pith; bark separable; cotyledons two .... **Dicotyledons**

IIIa. Perianth biseriate, outer sepaloid, inner petaloid .... **Dichlamydeae**

IVa. Petals free; stamens free from the petals .... **Polypetalae**

Va. Ovary superior:

VIa. Stamens more than twice as many as petals **1-25**

VIb. Stamens as many, or twice as many, rarely fewer **26-78**

Vb. Ovary inferior or half inferior **79-91**

IVb. Petals connate, at least at the base .... **Gamopetalae**

VIIa. Stamens epipetalous **92-123**

VIIb. Stamens free from the corolla **124-128**

IIIb. Perianth uniseriate, rarely apetalous, either sepaloid or petaloid .... **Monochlamydeae (129-158)**

IIb. Floral parts in usually 3-merous; vascular bundles of the stem scattered, without a central pith; bark not separable; cotyledon single .... **Monocotyledons (159-196)**

Ib. Seeds not enclosed in fruit wall; ovules borne on surface of megasporophyll; stigma absent.............................. **GYMNOSPERMS**

VIa. Stamens more than twice as many as petals and polypetaloid dicots with superior ovary

1. Pistil one or more, free, apocarpous:
   2. Stamens inserted on a convex receptacle:
      3. Plants aquatic; leaves peltate; flowers more than 10 cm across:
         4. Leaves cordate at base; filaments dilated; carpels united; stigma radiating .... **Nymphaeaceae**
         4. Leaves rounded at base; filaments linear; carpels free; stigma peltate.. **Nelumbonaceae**
      3. Plants terrestrial; leaves not peltate; flowers less than 7 cm across:
         5. Perianth 1 or 2-seriate; plants herbaceous or woody climbers... **Ranunculaceae**
         5. Perianth 3 or more-seriate; plants woody trees or climbers:
            6. Erect trees; flowers bisexual; stamens numerous:
               7. Sepals and petals valvate; flowers not showy; fruit indehiscent monocarps or a syncarp................ **Annonaceae**
               7. Sepals and petals imbricate; flowers showy; fruit usually follicles:
                  8. Leaves with parallel nerves and sheathing petiole .... **Dilleniaceae**
                  8. Leaves not as above... **Magnoliaceae**
            6. Climbers; flowers unisexual; stamens 6 ....**Menispermaceae**
   2. Stamens on a flat or concave hypanthium.. **Rosaceae**
1. Pistil syncarpous, one or more-celled:
   9. Stamens variously connate at base:
      10. Anthers 1-celled; epicalyx usually present (except *Abutilon* and *Sida*); staminal column adnate to petals:
         11. Leaves simple or lobed; seeds without silky fibres; stamen monadelphous... **Malvaceae**
         11. Leaves digitate; seeds with silky fibers; stamens 5 or many-adelphous .... **Bombacaceae**

10. Anthers 2-celled; epicalyx absent; staminal column free from petals:

12. Styles usually 5, terminal of the ovary; stamens in many bundles; leaves hairy rarely glandular.................... **Sterculiaceae**

12. Styles 3-5, variously united; stamens in a bundle; leaves with transluscent or blackish hairy............................ **Clusiaceae**

9. Stamens usually free:

13. Leaves lobed or compound:

14. Flowers more than 10 cm across; seeds with cottony hairs.. **Cochlospermaceae**

14. Flowers less than 5 cm across; seeds without cottony hairs:

15. Leaves prominently gland dotted; fruit with a woody rind .... **Rutaceae**

15. Leaves not gland dotted; fruit without woody rind... **Rosaceae (p.p.)**

13. Leaves simple:

16. Fruits winged; anthers with appendaged connectives .... **Dipterocarpaceae**

16. Fruits not winged; anthers without appendaged connectives:

17. Ovary borne on gynophores:

18. Herbs; fruit a capsule... **Cleomaceae**

18. Shrubs or trees; fruit a berry.................................... **Capparaceae**

17. Ovary not borne on gynophores:

19. Anthers dehiscence poricidal:

20. Disc enlarged in fruit; petals entire, yellow; seeds erect.. **Ochnaceae**

20. Disc not enlarged in fruit; petals laciniate or entire, cream-coloured; seeds pendulous:

21. Plants with coloured sap; seeds arillate................ **Bixaceae**

21. Plants without coloured sap; seeds not arillate

22. Plants woody; petals laciniate; fruit berry... **Elaeocarpaceae**

22. Plant herbaceous to subshrubby; petals not laciniate; fruit usually capsule .... **Hypericaceae**

19. Anthers dehiscence longitudinal:

23. Plant armed; latex yellow; stamens many:

24. Plant herbaceous; fruits spiny capsule .... **Papaveraceae**

24. Plant woody; fruits smooth berry:

25. Petals absent; flowers usually pale-greenish-yellow .... **Flacourtiaceae**

25. Petals present; flowers bright yellow .... **Berberidaceae**

23. Plants unarmed; latex watery; stamens 10-15 .... **Tiliaceae (p.p.)**

VIb. Polypetaloid dicots with superior ovary, and stamens as many or twice as many, as petals, rarely fewer

26. Stamens twice as many as petals, or fewer:

27. Fruit legume/lomentum; placentation marginal .... **Fabaceae**

27. Fruit not a legume/lomentum; placentation other than marginal:

28. Plants stragglers to climbers or trailers:

29. Fruit samaroid; petals fimbriate .... **Malphighiaceae**

29. Fruit not samaroid; petals entire:

30. Hook-climbers; leaves simple; flowers yellow, stamens 10, connate .... **Linaceae**

30. Tendril-climbers; leaves compound; flowers white, stamens 8, free (*Cardiospermum*) .... **Sapindaceae (p.p.)**

28. Plants not stragglers, usually erect herbs, shrubs and trees:

31. Leaves simple:

32. Stamens connate by filaments into staminal tube:

33. Leaves gland-dotted; plants armed (*Citrus, Atalantia*) .... **Rutaceae**

33. Leaves not gland-dotted; plants unarmed:

34. Petals adnate to staminal tube; inner sepals petaloid; seeds with caruncle .... **Polygalaceae**

34. Petals free from staminal tube; inner sepals not petaloid; seeds without caruncle:

35. Stipules prominent; petals with a double-ligule within; plant glabrous .... **Erythroxylaceae**

35. Stipules obscure; petals without ligule; plants with stellate hairy .... **Sterculiaceae (p.p.)**

32. Stamens free:

36. Fruits prickly; leaves palmi-nerved; plants stellate hairy (*Triumfetta*) .... **Tiliaceae (p.p.)**

36. Fruits smooth; leaves penni-nerved; plants not stellate hairy:

37. Petals crumbles; hypanthium present; fruit capsule .... **Lythraceae**

37. Petals cucullate; hypanthium absent; fruit drupe... **Anacardiaceae (p.p.)**

31. Leaves compound:

38. Stamens less than twice as many as petals:

39. Ovary on an elongate gynophore .... **Capparaceae (p.p.)**

39. Ovary not on an elongate gynophore:

40. Disc intra-staminal; flowers actinomorphic:

41. Trees; ovules pendulous (*Spondias*)............ **Anacardiaceae**

41. Herbs or shrubs; ovules parietal:

42. Leaves scale-like; seeds plumose or winged; corolla not cruciform................. **Tamaricaceae**

42. Leaves well-developed; seeds not as above; corolla cruciform .... **Brassicaceae**

40. Disc extra-staminal; flowers zygomorphic:

43. Plant sap watery; carpels 3, 3-locular; fruits other than samariod (except *Dodonaea*) .... **Sapindaceae (p.p.)**

43. Plant sap milky; carpels 2, 2-locular; fruit samaroid .... **Aceraceae**

38. Stamens twice as many as petals:

44. Flowers bisexual; stamens connate:

45. Flowers 3-merous; ovary 3-celled, uni-ovuled (*Canarium*) .... **Burseraceae**

45. Flowers other than 3-merous; ovary more than 3-celled:

46. Petals contorted, loosely connivent................ **Oxalidaceae**

46. Petals not contorted, free:

47. Stamens monadelphous; inflorescence not leaf-opposed................................... **Meliaceae (p.p.)**

47. Stamens free; inflorescence leaf-opposed...... **Leeaceae**

44. Flowers polygamous; stamens free:

48. Fruit winged; stamens 10-12; leaflets oblique.. **Simaroubaceae**

48. Fruit not winged; stamens 8-10; leaflets not oblique:

49. Leaves bifoliolate; plants armed .... **Balanitaceae**

49. Leaves more than 2-foliolate; plants unarmed .... **Burseraceae (p.p.)**

26. Stamens as many as petals, or fewer:

50. Plant insectivorous, glandular hairy... **Droseraceae**

50. Plants not insectivorous, not hairy:

51. Flowers zygomorphic:

52. Trees; fertile stamens 2 or 5, on a cupular scale or on calyx disc:

53. Fertile stamens 2, on a cupular scale; fruit other than capsule; seeds not winged................................**Sabiaceae**

53. Fertile stamens 5; on a calyx disc; fruit trivalved capsule; seeds three winged..**Moringaceae**

52. Herbs; fertile stamens 5, not on a cupular scale:

54. Ovary 1-celled; ovules parietal; dorsal sepal not spurred:

55. Sepals 2, scale-like; stamens 4 or 6, opposite to petals; connective anthers not produced beyond the anther cells ....**Fumariaceae**

55. Sepals 5, well developed; stamens 5, free; connective anthers produced beyond the anther cell ....**Violaceae**

54. Ovary 5-celled; ovules axile; dorsal sepal spurred ....**Balsaminaceae**

51. Flowers actinomorphic:

56. Plants tendrillar climbers:

57. Ovary 1-celled, sessile or on a gynophores; ovules many on parietal placenta; corona present .... **Passifloraceae**

57. Ovary 2- or more-celled, embedded in the disc; ovules 1 or 2 per cell, erect or pendulous; corona absent:

58. Inflorescence usually leaf-opposed; often tendrillar; leaves simple or compound .... **Vitaceae**

58. Inflorescence not as above; not tendrillar; leaves simple:

59. Fruit indehiscent, winged or not; seeds exarillate... **Rhamnaceae**

59. Fruit a dehiscent capsule, not winged; seeds arillate or not:

60. Flowers with a well-developed disc around the ovary:

61. Flowers with distinct calyx and corolla; seeds prominently arillate .... **Celastraceae**

61. Flowers with membranous calyx and corolla; seeds without arillate .... **Siphonodontaceae**

60. Flowers without a disc .... **Aquifoliaceae**

56. Plants other than tendrillar climbers:

62. Plants aquatic:

63. Ovary 2-celled; style single; fruit spinous, indehiscent.. **Trapaceae**

63. Ovary 2-4-celled; style 2 or 4; fruit furrowing 2 or 4 cocci... **Haloragaceae**

62. Plants terrestrial:

64. Leaves compound; fruits thorny......................... **Zygophyllaceae**

64. Leaves simple; fruits not thorny or prickly:

65. Flowers polygamo-dioecious or unisexual:

66. Plants acrid or resinous; ovary 1-celled, ovule pendulous ....**Anacardiaceae (p.p.)**

66. Plants neither acrid nor resinous; ovary 2- or 3-celled, ovules erect:

67. Armed shrub; leaves decussate, entire; flowers unisexual.........................**Salvadoraceae**

67. Unarmed straggler; leaves alternate, crenulate; flowers polygamous .... **Celastraceae (p.p.)**

65. Flowers bisexual:

68. Ovary 1- or 2-celled:

69. Ovule(s) 1 or 2, erect or pendulous:

70. Trees or shrubs; stamens alternating with petals:

71. Carpels 3, 1 or 2 of them sterile; the fertile containing 2-apical ovules .... **Icacinaceae**

71. Carpels 3 or more, all fertile; each containing 1 or 2 apical ovules .... **Aquifoliaceae**

70. Stragglers; stamens opposite to petals .... **Opiliaceae**

69. Ovules 3 or more; usually erect:

72. Trees or shrubs:

73. Ovary borne on gynophore (*Cadaba*) .... **Capparaceae (p.p.)**

73. Ovary not on gynophore:

74. Capsule resinous; plant without latex; petals free ...... **Pittosporaceae**

74. Capsule not resinous; plant with milky latex; petals united .... **Caricaceae**

72. Herbs:

75. Flowers usually yellow; placentation parietal .... **Turneraceae**

75. Flowers never pure yellow; placentation basal or free central:

76. Petals present; seeds reniform, variously striate, rugulose or muricate:

77. Leaves fleshy; sepals 2; petals 4-5. .... **Portulacaceae**

77. Leaves not fleshy; sepals and petals 5 each .... **Caryophyllaceae**

76. Petals absent; seeds smooth, not reniform:

78. Fruit 3-5-valved capsule or of 2-5, one seeded cocci .... **Molluginaceae**

78. Fruit transversely dehiscing; petiole often dilated at base or forming pouch...................... **Aizoaceae**

68. Ovary 3- or more-celled .... **Olacaceae**

Vb. Polypetaloid dicots with inferior or half inferior ovary (enclosed in a hypanthium)

79. Stamens more than 10:

80. Flowers unisexual; styles more than one .... **Begoniaceae**

80. Flowers bisexual; style single:

81. Plants armed either spiny or branch endings:

82. Succulent habit; leaves absent or reduced............................... **Cactaceae**

82. Woody trees; leaves alternate... **Alangiaceae**

81. Plants unarmed:

83. Leaves decussate, gland dotted .... **Myrtaceae**

83. Leaves alternate, not gland dotted:

84. Locules of ovary superposed; flowers red or orange; fruit berry with hard woody rind .... **Punicaceae**

84. Locules of ovary 4; flowers white; fruit a fibrous drupe .... **Barringtoniaceae**

79. Stamens 10 or fewer:

85. Styles more than one:

86. Herbs or shrubs; leaves simple or decompound:

87. Aromatic herbs; inflorescence umbel; fruit cremocarp .... **Apiaceae**

87. Non-aromatic herbs or shrubs; inflorescence cymose; fruit other than cremocarp:

88. Plants succulent herbs; fruit capsular follicle............. **Crassulaceae**

88. Plants erect shrubs or tendrillar climbers; fruit pepo... **Cucurbitaceae**

86. Plants woody; leaves pinnately compound.............................. **Araliaceae**

85. Style single:

89. Plants herbaceous:

90. Capsule 2-celled, subglobose.. **Vahliaceae**

90. Capsule 4-5-celled, linear-cylindric or obcylindric .... **Onagraceae**

89. Plants woody:

91. Inflorescence fascicuate; anther dehiscence poricidal; berry globose.. **Melastomataceae**

91. Inflorescence spicate; anther dehiscence longitudinal; fruit 4-winged .... **Combretaceae**

VIIa. Gamopetalous dicots with epipetalous stamens

92. Slender insectivorous herbs with bulbous traps .... **Lentibulariaceae**

92. Plants otherwise:

93. Ovary inferior or half inferior:

94. Inflorescence a capitulum; calyx reduced in to pappus; fruit an achene .... **Asteraceae**

94. Inflorescence other than capitulum; calyx well developed; fruits other than achenes:

95. Leaves decussate, stipulate:

96. Stipules conspicuous, inter or intra petiolar..................... **Rubiaceae**

96. Stipules inconspicuous, lateral .... **Caprifoliaceae**

95. Leaves alternate, estipulate:

97. Stamens numerous; placenta pendulous:

98. Staminal filaments unequal, anthers open longitudinally .... **Symplocaceae**

98. Staminal filaments equal, anthers produced at apex cylindrical tubes with terminal openings ...... **Vacciniaceae**

97. Stamens 5 or 10; placenta parietal, free-central or axile .... **Myrsinaceae (p.p.)**

93. Ovary superior:

99. Plants herbaceous, scapose .... **Gesneriaceae**

99. Plants non-scapose, usually herbs or shrubs:

100. Leaves alternate:

101. Stamens more than corolla-lobes .... **Sapotaceae**

101. Stamens as many as, or less than corolla-lobes:

102. Stamens opposite to corolla-lobes; ovary single-celled:

103. Herbs or subshrubby; corolla salver-form; ovule single or many; basal or free central placenta:

104. Corolla persistent and papery in fruit; ovule single on a long stalk base from the ovary .... **Plumbaginaceae**

104. Corolla not persistent and papery in fruit; ovules many on free central placenta .... **Primulaceae**

103. Shrubs or trees; corolla campanulate; ovules many, free central........................... **Myrsinaceae (p.p.)**

102. Stamens alternate with corolla-lobes; ovary 2 or more-celled:

105. Ovary 3 or 4-celled:

106. Climbers; corolla plicate:

107. Terrestrial plants; leaves well developed; corolla usually infundibuliform................................... **Convolvulaceae**

107. Parasitic twiners; leaves scaly or absent; corolla urceolate........................ **Cuscutaceae**

106. Erect plants; corolla not plicate:

108. Style arising from the depression between the four lobes of the ovary; fruit of four nutlets or more rarely a 1-4-seeded drupe:

109. Plant herbaceous; style 1 or 2; fruit usually four nutlets .... **Boraginaceae**

109. Plant woody shrubs; style 4; fruit 4-seeded drupe .... **Cordiaceae**

108. Style terminal; fruit a capsule, usually many-seeded....................................... **Hydrophyllaceae**

105. Ovary 2-celled:

110. Plants viscid-pubescent; placenta not swollen; fruit horned .... **Pedaliaceae**

110. Plants not viscid-pubescent; placenta swollen; fruit not horned:

111. Fruit berry; corolla actinomorphic; stamens 5 or more .... **Solanaceae**

111. Fruit a capsule; corolla zygomorphic; stamens 4 or 2:

112. Leaves scale-like, never green; root parasites ....**Orobanchaceae**

112. Leaves green, expanded; free living plants .... **Scrophulariaceae**

100. Leaves opposite (alternate in *Plumeria*, *Thevetia*):

113. Latex milky; fruit a follicle (except *Carissa*, *Thevitia*); seeds comose; anthers connate or connivent around stigma; pistil 2, free or partly connate:

114. Anthers inseparable, connate with stigma; pollen masses collected in pollinarium:

115. Corolla lobes valvate; pollinarium clip-shaped, adnate to the stigmatic lobes ....**Asclepiadaceae**

115. Corolla lobes contorted, rarely valvate; pollinarium spoon-shaped, adnate to the base of corolla tube ....**Periplocaceae**

114. Anthers separable, connivent around stigma; pollen free ....**Apocynaceae**

113. Latex absent or watery; fruit other than follicle; seeds without coma; anthers free, away from stigma; pistil syncarpous:

116. Corolla actinomorphic; stamens equal; staminodes absent (except Gentianaceae):

117. Herbs; ovary 1-celled; ovules many:

118. Plant terrestrial; placentation parietal........ **Gentianaceae**

118. Plant aquatic or marshy .... **Menyanthaceae**

117. Shrubs or trees; ovary 2-celled; ovules 2-20:

119. Stamens 2 .... **Oleaceae**

119. Stamens as many as corolla lobes................. **Loganiaceae**

116. Corolla zygomorphic (except *Symphorema*, *Tectona*); stamens didynamous, usually with two staminodes:

120. Fruit a capsule; seeds more than 4; placenta axile or parietal:

121. Seeds winged; leaves pinnately compound; placenta Parietal......................... **Bignoniaceae**

121. Seeds not winged; leaves simple; placenta axile:

122. Upper lip of corolla crested; capsule dehiscing elastically; seeds with jointed funicle; placenta never swollen; stem nodose... **Acanthaceae**

122. Upper lip of corolla not crested; capsule dehiscing otherwise; seeds with inconspicuous funicle; placenta swollen; stem otherwise.......... **Scrophulariaceae (p.p.)**

120. Fruit a drupe, pyrene or nutlet; seeds 4; placenta basal:

123. Plants strongly aromatic; calyx distinctly 2-lipped; ovary 4-lobed; style gynobasic (except *Teucrium*); fruit of 4 nutlets .... **Lamiaceae**

123. Plants non or scarcely aromatic; calyx not so above; ovary entire; style terminal; fruit a drupe or pyrene................................... **Verbenaceae**

VIIb. Gamopetalous dicots with free stamens:

124. Herbs; ovary half-inferior; petals united only basally:

125. Stamens 5; flowers usually bisexual:

126. Flowers actinomorphic; corolla lobes equal; filaments not connate with style... **Campanulaceae**

126. Flowers zygomorphic; corolla lobes unequal; filaments connate with style forming a column:

127. Plants with milky latex; stylar cup or pollen collecting inducium absent .... **Lobeliaceae**

127. Plants without milky latex; stylar cup or pollen collecting inducium present... **Goodeniaceae**

125. Stamens 2; flowers unisexual, bisexual or polygamous........... **Stylidiaceae**

124. Shrubs or trees; ovary fully inferior or superior; petals united from above middle:

128. Trees dioecious; stamens 3-20 or more, staminodes present; ovary 2-16- locular.. **Ebenaceae**

128. Trees monoecious: stamens 8-10 or fewer, staminodes absent; ovary 8-10-locular .... **Vacciniaceae**

IIIb. Perianth uniseriate, rarely apetalous, either sepaloid or petaloid (Monochlamydeae)

129. Plants parasites:

130. Achlorophyllous, root parasites; anthers connate .... **Balanophoraceae**

130. Chlorophyllous, stem parasites; anthers free:

131. Perianth limb tubular, more than 2 cm long; anthers dehiscence longitudinal .... **Loranthaceae**

131. Perianth limb campanulate, less than 0.5 cm long; anthers dehisce other than longitudinal .... **Viscaceae**

129. Plants autophytes:

132. Perianth absent:

133. Plants submerged aquatic, parts undifferentiated; flowers unisexual on elongated stipe.................................. **Podostemaceae**

133. Plants terrestrial, well developed plant parts; flowers unisexual on fleshy spike or cone-like catkin:

134. Leaves well-developed; male and female spikes similar; perianth completely absent in both flowers............ **Piperaceae**

134. Leaves reduced into scales; male and female spikes dissimilar; perianth absent in female flower .... **Casuarinaceae**

132. Perianth present:

135. Twining herbs or shrubs:

136. Ovary inferior; ovules many; perianth limb irregular; stamens adnate to the stylar column .... **Aristolochiaceae**

136. Ovary superior; ovules 1-6; perianth limb regular; stamens free from the style:

137. Flowers bisexual; stamens epiphyllous; fruiting perianth without stinging hairs .... **Basellaceae**

137. Flowers unisexual; stamens free; fruiting perianth with stinging hairs .... **Euphorbiaceae (p.p)**

135. Erect herbs or shrubs:

138. Leaves compound (*Bischofia*) .... **Euphorbiaceae (p.p)**

138. Leaves simple:

139. Inflorescence a syconium (*Ficus*) .... **Moraceae (p.p.)**

139. Inflorescence other than syconium:

140. Leaves opposite or whorled:

141. Stamens epiphyllous .... **Santalaceae**

141. Stamens free from perianth:

142. Perianth gamophyllous (limb coralline); fruit an anthocarp .... **Nyctaginaceae**

142. Perianth free or only basally connate; fruit other than anthocarp .... **Lauraceae**

140. Leaves alternate:

143. Flowers bisexual:

144. Stipule ochraceous .... **Polygonaceae**

144. Stipule absent or inconspicuous:

145. Anther dehiscence valvular:

146. Leaves cordiform; staminodes absent; fruit winged samara ..... **Hernandiaceae**

146. Leaves other than cordiform; staminodes present; fruit berry .... **Lauraceae (p.p.)**

145. Anther dehiscence longitudinal:

147. Leaves pinnatified; perianth limb irregular .... **Proteaceae**

147. Leaves simple; perianth limb regular:

148. Perianth scarious; stamens basally connate into a stamina sheath .... **Amaranthaceae**

148. Perianth calycine; stamens free:

149. Peltate scales on tender parts; stamens epiphyllous; plant stragglers .... **Elaeagnaceae**

149. Peltate scales absent; stamens free; plant herbaceous.................. **Chenopodiaceae**

143. Flowers unisexual:

150. Fruits more than 2.5 cm wide; endosperm ruminate; anthers connate in a column................................ **Myristicaceae**

150. Fruits less than 1 cm wide; endosperm not ruminate; anthers free:

151. Plants aquatic; leaves much divided; stamens 5-20 .... **Ceratophyllaceae**

151. Plant terrestrial; leaves not divided (except *Cannabis*); stamens 2-30:

152. Stamens more than perianth lobes; staminodes persent:

153. Perianth tube elongate, lobes 5; stamens without glands:

154. Leaves pinnately nerved; perianth well-developed ....**Thymelaeaceae**

154. Leaves palmately nerved; perianth cup-like or reduced ....**Cannabaceae**

153. Perianth tube short, lobes 2+2 or 3+3, stamens with or without glands:

155. Flowers bisexual; anthers with flap-like openings ....**Lauraceae (p.p)**

155. Flowers unisexual; anthers longitudinally split open:

156. Flowers irregular; ovary with 3 carpels......................**Buxaceae**

156. Flowers regular; ovary with 2 carpels ....**Salicaceae**

152. Stamens as many as perianth lobes; staminodes absent:

157. Stamens inflexed in bud; anthers reversed:

158. Latex watery; leaves prominently 3-nerved; often with stinging hairs; ovule basal..**Urticaceae**

158. Latex milky; leaves with basal nerves prominent; without stinging hairs; ovule pendulous.......................**Moraceae**

157. Stamens erect in bud; anthers not reversed.....................................**Ulmaceae**

IIb. Floral parts in usually 3-merous; vascular bundles of the stem scattered, without a central pith; bark not separable; cotyledon single (Monocotyledons)

159. Plants aquatic, marshy or riparian:

160. Plant body thalloid:

161. Flowers bisexual; free floating hydrophytes .... **Lemnaceae**

161. Flowers unisexual; plant body inseparable with substratum ....**Hydrocharitaceae**

160. Plant body well-differentiated:

162. Plants submerged aquatics:

163. Flowers bisexual...**Ptamogetonaceae**

163. Flowers unisexual ....**Najadaceae**

162. Plants not submerged, if submerged flowers above the water level:

164. Leaves scaly .... **Burmanniaceae**

164. Leaves well-developed:

165. Trees or shrubs; leaves with spiny margin................**Pandanaceae**

165. Herbs or subshrubs; leaves without spiny margin:

166. Perianth absent:

167. Flowers in two cylindrical superimposed brownish spike..**Typhaceae**

167. Flowers on erect fleshy spike covered with leafy spathe..**Araceae**

166. Perianth present:

168. Perianth 6 segments, scarious.................................**Juncaceae**

168. Perianth not of scarious 6 segments:

169. Flowers uniseriate, 2-lobed..................**Aponogetonaceae**

169. Flowers biseriate, 6-lobed; if uniseriate, 3-lobed:

170. Flowers in umbels, involucrate with spathaceous bracts or without:

171. Flowers in whorls, racemes or panicles; perianth differentiated into sepals and petals .... **Alismataceae**

171. Flowers on umbels; perianth undifferentiated petaloid ....**Butomaceae**

170. Flowers otherwise:

172. Ovary 1-celled...**Xyridaceae**

172. Ovary 3-celled................................ **Pontederiaceae**

159. Plants terrestrial or epiphytic:

173. Climbers or twiners:

174. Tendrils present:

175. Leaves 3-9-ribbed; flowers in umbels; fruit berry .... **Smilacaceae**

175. Leaves 1-ribbed; flowers solitary or in racemes; fruit capsule

176. Perianth petaloid .... **Liliaceae**

176. Perianth membranous.. **Flagellariaceae**

174. Tendrils absent:

177. Leaves well-developed, more than 1.5 cm wide:

178. Flowers unisexual; capsule winged..........................**Disocoreaceae**

178. Flowers bisexual; capsule not winged........................**Stemonaceae**

177. Leaves reduced, stem cladode; flowers bisexual .... **Liliaceae** (*Asparagus*)

173. Neither climbers nor twiners:

179. Inflorescence capitate..**Eriocaulaceae**

179. Inflorescence not capitate:

180. Leaves pinnate or palmate; inflorescence a spadix; drupes with fibrous pericarp .... **Arecaceae**

180. Leaves simple; inflorescence other than spadix; drupes without fibrous pericarp:

181. Flowers prominently zygomorphic:

182. Median petal produced a lip (labellum); stamens and style united in a column; fertile stamen 1 ....**Orchidaceae**

182. Median petal otherwise; stamens and style separate; fertile stamens 1-3:

183. Leaves pulvinate; anther 1-celled; ovary with single fertile cell; ovule 1 .... **Marantaceae**

183. Leaves not pulvinate; anther 2-celled; ovary with 3-fertile cells; ovules many:

184. Leaves spirally arranged; calyx lobes free.

185. Pseudostem less than 5 cm across; perianth petaloid, large; fertile stamen 1.... **Cannaceae**

185. Pseudostem more than 15 cm across; perianth sepaloid; fertile stamens 3 ....**Musaceae**

184. Leaves distichous; calyx lobes sometimes united...**Zingiberaceae**

181. Flowers actinomorphic:

186. Perianth well-developed:

187. Leaves septate; perianth lobes glumaceous...**Juncaceae**

187. Leaves not septate; perianth lobes not glumaceous:

188. Outer whorl of perianth sepaloid and inner petaloid...................................... **Commelinaceae**

188. Both whorls petaloid:

189. Root stock rhizomes, corms or bulbs:

190. Flowers umbellate; stamens inserted at the throat of perianth tube:

191. Bracts dimorphic, outer one leafy, inner one filiform..................................**Taccaceae**

191. Bracts not as above..................**Amaryllidaceae**

190. Flowers not umbellate; stamens inserted at the base of perianth tube:

192. Ovary inferior; leaves plicate....**Hypoxidaceae**

192. Ovary superior; leaves not plicate:

193. Root stock a rhizome............**Anthericaceae**

193. Root stock a bulb or corm .... **Liliaceae**

189. Root stock absent; roots fibrous:

194. Anthers foveolate at the joint of the filament; leaves radical....................... **Bromeliaceae**

194. Anthers not foveolate as above .... **Agavaceae**

186. Perianth absent:

195. Flowers in spikeletes; fruit a caryopsis or nut, rarely a utricle or berry:

196. Culm triquetrous, nodes absent with closed sheaths; flowers subtended with solitary bract .... **Cyperaceae**

196. Culm terete, nodose; leaves with open sheath; flowers subtended with 2-bracts .... **Poaceae**

195. Flowers in spadix; fruit otherwise............................. **Araceae**

Floristic Analysis of Eastern Ghats

Eastern Ghats are broken hillocks and discontinuous ranges, thus the species rich zones alternate with plains in between. This is the reason that the area is with rich number of wild and cultivated plant species which prevails hot tropical climate throughout the year. A total number of 3,417 species of seed plants have been recorded in the area; among them Dicotyledons are dominant with 870 genera with 2433 species. Monocotyledons are 34 families with 304 genera and 978 species. Gymnosperms are belonging two families and 2 genera with 6 species in the wild. Total number of plant species, including Orchidaceae, in Flora of Eastern Ghats is given in Table 1.

Table 1. Total Number of Plant Families, Genera and Species of Eastern Ghats

Plant Groups	*Families*	*Genera*	*Species*
Dicotyledons	144	870	2433
Monocotyledons	34	304	978
Gymnosperms	2	2	6
Total	**180**	**1172**	**3417**

Of the 178 families of Angiosperms, about 14 families have more than 40 species (Table 2) and they contribute the 53.6 % of the total number of plant species. Among them Poaceae has the maximum number of species 365 with 112 genera followed by Leguminosae (346 species with 72 genera). Family Euphorbiaceae and Cyperaceae have more than 150 species. Asteraceae and Rubiaceae also represented more than

100 species in Eastern Ghats. Remaining dominant families are Lamiaceae (98 species), Asclepiadaceae (77 species), Convolvulaceae (69 species), Malvaceae (64 species), Commelinaceae (42 species), Amaranthaceae (40 species) and Moraceae (40 species).

Table 2. Dominant Plant Families with Rich Number of Species

Families	*Total Number of Genera*	*Total Number of Species*
Poaceae	112	365
Fabaceae	72	346
Euphorbiaceae	49	168
Cyperaceae	19	156
Acanthaceae	37	138
Asteraceae	63	120
Rubiaceae	41	106
Lamiaceae	29	98
Asclepiadaceae	26	77
Convolvulaceae	14	69
Malvaceae	14	64
Commelinaceae	10	42
Amaranthaceae	17	40
Moraceae	5	40

Among the dominant genera, *Crotalaria* and *Cyperus* are highly diversified in Eastern Ghats with 49 and 44 species respectively. Followed by *Euphorbia* (40 species), *Fimbristylis* (32 species), *Ficus* (28 species), *Leucas* (28 species), and *Indigofera* (26 species). Remaining genera are *Ipomoea* (24 species), *Phyllanthus* (22 species), *Cassia* (22 species), *Eragrosits* (21 species), *Argyreia* (20 species), *Eriocaulon* (19 species), *Solanum* (19 species), *Grewia, Desmodium, Acacia* (18 species each), *Hibiscus, Alysicarpus, Hedyotis* (17 species each), *Barleria, Commelina, Jasminum* and *Rhyncosia* (15 species each) (Table 3).

Table 3. Dominant Genera of Eastern Ghats

Genera	*Total Number of Species*
Crotalaria	49
Cyperus	44
Euphorbia	40
Fimbristylis	32
Ficus	28
Leucas	27
Indigofera	26
Ipomoea	24
Phyllanthus	22
Cassia	22

Genera	Total Number of Species
Argyreia	20
Eragrostis	21
Eriocaulon	19
Solanum	19
Grewia	18
Desmodium	18
Acacia	18
Hibiscus	17
Alysicarpus	17
Hedyotis	17
Barleria	15
Commelina	15
Jasminum	15
Rhyncosia	15

The following genera were reported by different authors as new reports to the Flora of Eastern Ghats: *Huberantha, Xylopia, Calopogonium, Centrosema, Clidemia, Diodella, Aeschynanthus* and *Monothecium*. Genus *Odisha* is exclusively endemic to Eastern Ghats.

Endemism

A total of 166 endemic taxa, under 117 genera belonging to 43 families are known to occur in Eastern Ghats (Singh *et al.*, 2015). But recently many more new species described from the Eastern Ghats range and endemic species has sharply risen to 188 species under 120 genera with 46 families. The endemic taxa include 135 dicots and 50 monocots with 3 Gymnosperms. Poaceae have the maximum representation of 27 endemics followed by Acanthaceae (18 species), Asclepiadaceae (19 species) and Orchidaceae (15 species). The species richness and endemicity of Eastern Ghats may be divided into four zones or ranges as follows. All these four zones of Eastern Ghats are serving the habitat for a number of narrow endemic plants.

1. Ganjam–Kalahandi–Pulbhani–Gajapathi ranges

Bulbophullum panigrahianaum, Dimeria mooneyi, D. mahendragiriensis, D. orissae, Eragrstis nairii, Eria meghasaniensis, Eusteralis griffithi, Ficus concinna, Habenaria panigrahiana, Homonoia intermedia, Liparis espeevijii, Liparis udaii, Millettia orissae, Odisha cleistantha, Oryza jeyporensis, Peucedanum nagpurense, Themeda odishae, T. saxicola, Tripogon mahendragiriensis, Uvaria eucinata and *Zeuxine mooneyi.*

2. Vishakhapatnam–Araku–Koraput ranges

Argyreia arakuensis, Argyreia srinivasanii, Coleus subbaraoi, Curcuma longa var. *vanaharidra, Cycas sphaerica, Eulalia quadrinervis* var. *wightii, Globba orixensis* var. *pubescens, Kalanchoe cherukondensis, Leucas mukerjiana, Memecylon jadhavii, Memecylon madgolense, Millettia orissae, Musa shankarii, Ophiorrhiza chandrasekharanii, Orthisiphon rubicundus* var. *hohenackeri, Pavetta minor, Pimpinella bracteata, Pogostemon rugosus, Rhynchosia hainesiana,* and *Strobilanthes jeyporensis.*

3. Tiurpathi–Cuddapah (Kadapa)–Nallamalai ranges

Albizia thompsoni, Argyreia choisyana, Boswellia ovalifoliolata, Alysicarpus mahabubnagarensis, Andrographis beddomei, A. glandulosa, A. nallamalayana, Bupleurum andhricum, Brachystelma ananathapuramense, B. annamacharyae, B. brevitubulatum, B. nallamalayana, B. nigidianum, B. penchalakonense, B. seshachalamense, B. vemanae, B. pullaiahii, Bulbostylis subspinescens, Caralluma stalagmifera var. *intermedia, Chrysopogon velutinus, Croton scabiosus, Crotalaria candicans, C. paniculata* var. *nagarjunakondensis, Cycas beddomei, C. seshachalamensis, Decaschistia cuddapahensis, Dicliptera beddomei, Dimorphocalyx kurnoolensis, Drimia raogibikei, Emilia reddyi, Euphorbia cattimandoo, E. deccanensis* var. *nallamalayana, E. kadapensis, E. seshachalamensis, E. senguptae, E. venkatarajui, Glochidion tirupathiense, G. talakonense, Hildegardia populifolia, Hybanthus vatsavayii, Ischaemum fischeri, Iseilema venkateswarlui, Ledebouria hyderabadensis, Meyenia hawtayneana, Pancratium bhramarambae, P. telanganense Parahyparrhenia bellariensis, Pentanema indicum* var. *sivarajanianum, Pimpinella tirupatiensis, Pterocarpus santalinus, Rhynchosia filipes, R. ravii, Shorea tumbaggaia, Syzygium alternifolium, Terminalia pallida, Themeda mooneyi, Trachys deccanensis, Tripogon tirumalae* and *T. uma-ganeshii.*

4. Shevaroy–Kollimalai Ranges

Brachystelma saldanhae, Capparis shevaroyensis, C. kollimalayana, Cordia ramanujamii, Derris mathewii, Dimeria kollimalayana, Eleocharis swamyii, Epaltes pygmaea, Gymnema kollimalayanum, Huberantha senjiana, Hydrocotyle kollimalayensis, Impatiens shevaroyensis, I. yercaudensis, Kleinia shevaroyensis, Memecylon bremeri, Monosis shevaroyensis, Pavetta madrassica, Psychotria gopalanii, and *Tropidia hegderaoi.*

RED Listed Plants

The Indian Red Data book (Nayar and Sastry, 1987-1990) enlisted about 23 plant taxa from Eastern Ghats namely *Achyranthes coynei, Bulbophyllum kaitiense, Capparis shevaroyensis, Cayratia pedata* var. *glabra, Ceropegia spiralis, Chrysoglossum maculatum, Commelina hirsuta, Crotalaria pedata, C. filipes, C. priestleyoides, Cycas beddomei, Decaschistia rufa, Hedyotis swertioides, Hydrocotyle conferta, Leucas mukerjiana, Oianthus deccanensis, Phyllanthus narayanaswamii, Pimpinella tirupatiensis, Rhynchosia velutina, Urginea congesta, U. polyphylla, Vanilla wightiana,* and *Wendlandia angustifolia.*

Index to Families, Genera and Species
(Italicized names indicate Synonyms)

Index to Vernacular Names

TELUGU

TAMIL

ORIYA

www.ingramcontent.com/pod-product-compliance
Ingram Content Group UK Ltd.
Pitfield, Milton Keynes, MK11 3LW, UK
UKHW021326190726
13845UKWH00003B/49